Index of Applications

FINITE MATHEMATICS

FINITE MATHEMATICS

SEVENTH EDITION

Stefan Waner
Hofstra University

Steven R. Costenoble
Hofstra University

CENGAGE
Learning·

Australia · Brazil · Mexico · Singapore · United Kingdom · United States

CENGAGE
Learning®

Finite Mathematics, **Seventh Edition**
Stefan Waner, Steven R. Costenoble

Product Director: Terry Boyle

Product Manager: Rita Lombard

Content Developer: Morgan Mendoza

Product Assistant: Abby DeVeuve

Content Digitization Lead: Justin Karr

Marketing Manager: Ana Albinson

Content Project Manager:
    Teresa L. Trego

Art Director: Vernon Boes

Manufacturing Planner: Rebecca Cross

Production Service: Martha Emry
    BookCraft

Photo Researcher: Lumina Datamatics

Text Researcher: Lumina Datamatics

Text Designer: Diane Beasley

Cover Designer: Irene Morris

Cover Image: Large photo © CERN;
    Small photo © CERN

Compositor: Graphic World, Inc.

For product information and technology assistance, contact us at
Cengage Learning Customer & Sales Support, 1-800-354-9706.
For permission to use material from this text or product,
submit all requests online at **www.cengage.com/permissions.**
Further permissions questions can be e-mailed to
permissionrequest@cengage.com.

Library of Congress Control Number: 2016952197

Student Edition:
ISBN: 978-1-337-28042-6

Loose-leaf Edition:
ISBN: 978-1-337-29126-2

Cengage Learning
20 Channel Center Street
Boston, MA 02210
USA

Cengage Learning is a leading provider of customized learning solutions with employees residing in nearly 40 different countries and sales in more than 125 countries around the world. Find your local representative at **www.cengage.com**.

Cengage Learning products are represented in Canada by Nelson Education, Ltd.

To learn more about Cengage Learning Solutions, visit
www.cengage.com.

Purchase any of our products at your local college store or at our preferred online store **www.cengagebrain.com**.

About the Cover

The cover shows the innermost part of the Large Hadron Collider detector for the ATLAS (a toroidal LHC apparatus) high-energy particle experiment at the CERN Large Hadron Collider in Geneva, Switzerland, before the final insertion of the central pixel detector. The ongoing ATLAS experiment is one of two experiments that led to the discovery of the Higgs boson at CERN in 2012. CERN is the European Organization for Nuclear Research, where physicists and engineers are ⌡ the fundamental structure of the universe.

Printed in the United States of America
Print Number: 03   Print Year: 2017

Brief Contents

Contents

Preface

Finite Mathematics, Seventh Edition, is intended for a one- or two-term course for students majoring in business, the social sciences, or the liberal arts. Like the earlier editions, the seventh edition of *Finite Mathematics* is designed to address the challenge of generating enthusiasm and mathematical sophistication in an audience that is often underprepared and lacks motivation for traditional mathematics courses. We meet this challenge by focusing on real-life applications that students can relate to, many on topics of current interest; by presenting mathematical concepts intuitively and thoroughly; and by employing a writing style that is informal, engaging, and occasionally even humorous.

The seventh edition goes farther than earlier editions in implementing support for a wide range of instructional paradigms. On the one hand, the abundant pedagogical content available both in print and online, including comprehensive teaching videos and online tutorials, now allows us to be able to offer complete customizable courses for approaches ranging from on-campus and hybrid classes to distance learning classes. In addition, our careful integration of optional support for multiple forms of technology throughout the text makes it adaptable in classes with no technology, classes in which a single form of technology is used exclusively, and classes that incorporate several technologies.

We fully support three forms of technology in this text: TI-83/84 Plus graphing calculators, spreadsheets, and powerful online utilities we have created for the book. In particular, our comprehensive support for spreadsheet technology, both in the text and online, is highly relevant for students who are studying business and economics, in which skill with spreadsheets may be vital to their future careers.

New To This Edition

Content

- **Chapter 0:** We have added an entire new section on logarithms in the Precalculus Review, up through solving for unknowns in the exponent. Students can refer to this section for review when studying techniques involving the use of logarithms in the mathematics of finance (Chapter 2).

- **Chapter 1:** In our revision of this important introductory chapter, we have downplayed the algebra sophistication somewhat so as not to present artificial barriers to the mastery of the important new concepts we discuss.

- **Chapter 2:** The Mathematics of Finance chapter has been significantly revised: In the sections on simple and compound interest, we state and use both the year-based formulas and the compounding period-based versions. In the compound interest section, we now emphasize the latter formulation, as this helps with the segue to annuities, in which the period-based approach is the standard formulation. T-bills and zero coupon bonds are a bit esoteric, so the material on T-bills and further discussion of bonds has been moved to the end of the section to a subsection marked as "Optional." The section on annuities has been substantially reorganized: First, we have standardized the definition of "annuities" and now use more transparent and standard terminology to distinguish accumulation and annuitization (or payout). More important, we have added discussion, examples, and exercises on life insurance and mortgage refinancing, including a formula for calculating principal outstanding. The exercise sets have been radically reorganized and expanded, with numerous real-data based applications that follow the new organization of the section text.

Current Topics in the Applications

- We have added and updated numerous real data exercises and examples based on topics that are either of intense current interest or of general interest to our students, including many on social networks, and the 2009–2016 economic recovery, while retaining those of important historical interest, such as the 2008 economic crisis and resulting stock market panic, and many others.

Exercises

- We have added many new conceptual Communication and Reasoning exercises, including many dealing with common student errors and misconceptions.

Online Visualization and Practice Examples

- We have created a variety of web-based interactive apps available both on **www.wanermath.com** and in the new MindTap course that accompanies this edition. Instructors can use these to demonstrate important concepts such as calculating future values and present values of annuities, graphing inequalities, and solving linear programming problems graphically.

- Many key examples in the text are mirrored by web-based randomizable practice examples, which allow students to test their mastery of the textbook examples and provide instructors with material for interactive presentation and class discussion.

Our Approach to Pedagogy

Real-World Orientation   The diversity, breadth, and abundance of examples and exercises included in this edition continue to distinguish our book from others. A large number of these examples and exercises are based on real, referenced data from business, economics, the life sciences, and the social sciences. Our updated examples and exercises in the seventh edition are even more attuned to themes that students can identify with and relate to, from the technology used in their phones and tablets to the social networks in which they participate and many of the corporations they will instantly recognize as important in their lives. Notable events, such as the 1990s dot-com boom, the 2005–2006 real estate bubble, the resulting 2008 economic crisis and stock market panic, and many more, are addressed in examples and exercises throughout the book.

Adapting real data for pedagogical use can be tricky; available data can be numerically complex, intimidating for students, or incomplete. We have modified and streamlined many of the real-world applications, rendering them as tractable as any "made-up" application. At the same time, we have been careful to strike a pedagogically sound balance between applications based on real data and more traditional "generic" applications. Thus, the density and selection of real data-based applications have been tailored to the pedagogical goals and appropriate difficulty level for each section.

Readability   We would like students to read this book. We would like students to *enjoy* reading this book. Therefore, we have written the book in a conversational, student-oriented style and have made frequent use of question-and-answer dialogues to encourage the development of the student's mathematical curiosity and intuition. We hope that this text will give the student insight into how a mathematician develops and thinks about mathematical ideas and their applications to real life.

Pedagogical Aids   We have included our favorite unique and creative approaches to solving the kinds of problems that normally cause difficulties for students and headaches for instructors. To name just a few, we discuss a rewording technique in Chapters 3 and 5 to show how to translate phrases such as "there are (at least/at most) three times as many X as Y" directly into equations or inequalities, a technique of row reduction in Chapter 3 and tableau manipulation in Chapter 5 based on integer matrices (matrices with fractions are converted to integral matrices in the first step), a zooming-in technique to make solution of traffic flow problems in Chapter 3 almost routine, and decision algorithms in Chapter 6 that make calculations of real-life scenarios involving permutations and combinations almost mechanical.

Rigor   Mathematical rigor need not be antithetical to the kind of applied focus and conceptual approach that are hallmarks of this book. We have worked hard to ensure that we are always mathematically honest without being unnecessarily formal. Sometimes we do this through the question-and-answer dialogues and sometimes through the "Before we go on . . ." discussions that follow examples, but always in a manner designed to provoke the interest of the student.

Five Elements of Mathematical Pedagogy to Address Different Learning Styles   The "Rule of Four" is a common theme in many texts. Implementing this approach, we discuss many of the central concepts **numerically**, **graphically**, and **algebraically** and clearly delineate these distinctions. The fourth element, **verbal communication** of mathematical concepts, is emphasized through our discussions on translating English sentences into mathematical statements and in our extensive Communication and Reasoning exercises at the end of each section. A fifth element, **interactivity**, is implemented through expanded use of question-and-answer dialogues but is seen most dramatically in the eBook in the MindTap course that accompanies this edition and at **www.wanermath.com** through our new practice and learning modules. These are small interactive apps that help a student visualize new concepts or practice examples similar to those in the text. In addition, the wanermath .com website offers interactive tutorials in the form of games, interactive chapter summaries and chapter review exercises, and online utilities that automate a variety of tasks, from graphing to regression and matrix algebra.

Understand

Examples

Examples are a cornerstone of our approach. Many of the scenarios that we use in application examples and exercises are revisited several times throughout the book. In this way, students will find themselves analyzing the same application from a variety of different perspectives, such as systems of linear equations versus linear programming. Reusing scenarios and important functions provides unifying threads and shows students the complex texture of real-life problems. Complete solutions are provided with every example.

EXAMPLE 1   Savings Accounts

In December 2015, **Radius Bank** was paying 1.10% interest on savings accounts with balances of \$2,500 or more. If the interest is paid as simple interest, find the future value of a \$2,500 deposit after 6 years. What is the total interest paid over the period?[1]

Solution We use the future value formula:

$$FV = PV(1 + rt)$$
$$= 2,500[1 + (0.011)(6)] = 2,500[1.066] = \$2,665.$$

The total interest paid is given by the simple interest formula:

$$INT = PVrt$$
$$= (2,500)(0.011)(6) = \$165.$$

Note To find the interest paid, we could also have computed

$$INT = FV - PV = 2,665 - 2,500 = \$165. \ ■$$

Quick Examples

Most definition boxes include quick, straightforward examples that a student can use to solidify each new concept.

Quick Example

1. The simple interest on a \$5,000 investment earning 8% per year for 4 years is

$$INT = PVrt$$
$$= (5,000)(0.08)(4) = \$1,600.$$

Question-and-Answer Dialogues

We frequently use informal question-and-answer dialogues that anticipate the kinds of questions that may occur to the student and also guide the student through the development of new concepts.

FAQs

Which Formula to Use

Q: *How do I know when to use the formulas based on annual interest, such as INT = PVrt, as opposed to those based on general interest periods, such as INT = PVin?*

A: You can use either, as convenient. See, for instance, Quick Example 3 and the "Before we go on" discussion after Example 2.

Before We Go On . . .

Most examples are followed by supplementary discussions, which may include a check on the answer, a discussion of the feasibility and significance of a solution, or an in-depth look at what the solution means.

➡ **Before we go on . . .** In Example 1, we could look at the future value as a function of time:

$$FV = 2,500(1 + 0.011t) = 2,500 + 27.5t.$$

Thus, the future value is growing linearly at a rate of \$27.50 per year ■

Lecture Videos

Developed with Principal Lecturer, Jay Abramson, at Arizona State University, these video clips are flexible in their use as lecture starters in class or as an independent resource for students to review concepts on their own. Blending an introduction to concepts with specific examples, the videos let students quickly see the big picture of key concepts they are learning in class. Selected clips involve students and simulate a classroom-type interaction that creates a sense of the familiar and demystifies key concepts they are learning in their course. Frequently asked questions appear periodically throughout the video segments to further enhance learning. All videos are closed captioned and available in the new MindTap and Enhanced WebAssign courses that accompany the text. The topics for the lecture videos were carefully selected to accompany the subject areas that are most frequently taught and target the concepts that students struggle with most.

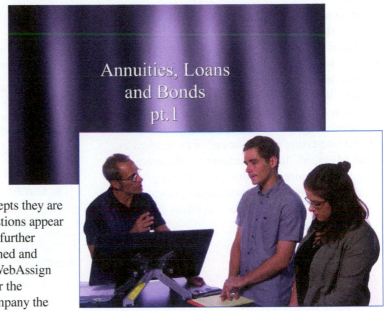

Online Visualization and Practice Examples

We have created a variety of web-based interactive apps that are available both on the wanermath.com website and in the new MindTap course accompanying this edition. Instructors can use these to demonstrate important concepts such as calculating future value and present value of annuities, graphing inequalities, and solving a linear programming problem graphically.

Many key examples in the text are mirrored by web-based randomizable practice examples that allow students to test their mastery of the textbook examples and provide instructors with material for interactive presentation and class discussion.

Linear programming graphically (three constraints)

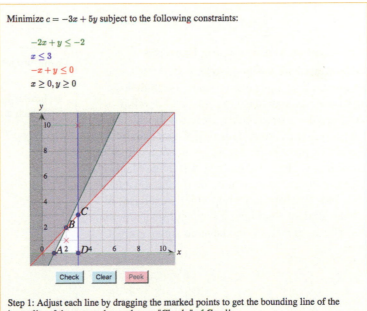

Minimize $c = -3x + 5y$ subject to the following constraints:

$-2x + y \leq -2$
$x \leq 3$
$-x + y \leq 0$
$x \geq 0, y \geq 0$

Step 1: Adjust each line by dragging the marked points to get the bounding line of the inequality of the same color, and press "Check." ✔ Good!

Step 2: Now identify the feasible region by clicking on a point in its interior and pressing "Check". ✔ Good! (Everything except the feasible region has been greyed out as in the textbook.)

Step 3: Now complete the table shown below.

Practice and Apply

Exercises

Our comprehensive collection of exercises provides a wealth of material that can be used to challenge students at almost every level of preparation and includes everything from straightforward drill exercises to interesting and challenging applications. The exercise sets have been carefully curated and ordered to move from straightforward basic exercises and exercises that are similar to examples in the text to more interesting and advanced ones, marked as "more advanced" for easy reference. There are also several much more difficult exercises, designated as "challenging." We have also included, in virtually every section of every chapter, exercises that are ideal for the use of technology.

Application Exercises

Exercises also include interesting applications based on real data to reinforce the applicability of math to real-life situations.

Communication and Reasoning Exercises

These exercises are designed to help students articulate mathematical concepts, broaden the student's grasp of the mathematical concepts, and develop modeling skills. They include exercises in which the student is asked to provide his or her own examples to illustrate a point or design an application with a given solution. They also include "fill in the blank" type exercises, exercises that invite discussion and debate, and—perhaps most important—exercises in which the student must identify and correct common errors. These exercises often have no single correct answer.

2.1   EXERCISES

▼ more advanced   ◆ challenging
T indicates exercises that should be solved using technology

In Exercises 1–10, compute the simple interest for the specified length of time and the future value at the end of that time. Round all answers to the nearest cent. [**HINT:** *See Quick Examples 1–5.*]

1. $2,000 is invested for 1 year at 6% per year.

2. $1,000 is invested for 10 years at 4% per year.

3. $4,000 is invested for 8 months at 0.5% per month.

Applications

In Exercises 17–36, compute the specified quantity. Round all answers to the nearest month, the nearest cent, or the nearest 0.001%, as appropriate.

17. *Simple Loans* You take out a 6-month, $5,000 loan at 8% annual simple interest. How much would you owe at the end of the 6 months? [**HINT:** See Example 2.]

18. *Simple Loans* You take out a 15-month, $10,000 loan at 1% monthly simple interest. How much would you owe at the end of the 15 months? [**HINT:** See Example 2.]

Communication and Reasoning Exercises

53. One or more of the following three graphs represents the future value of an investment earning simple interest. Which one(s)? Give the reason for your choice(s).

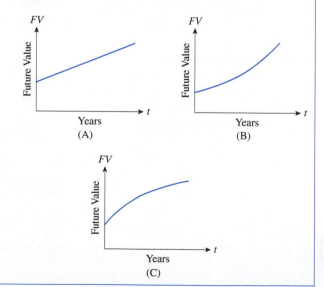

Review

At the end of every chapter is a comprehensive list of the key concepts that were covered in each section.

CHAPTER 2   REVIEW

KEY CONCEPTS

 www.WanerMath.com
Go to the Website to find a comprehensive and interactive Web-based summary of Chapter 2.

2.2 Compound Interest
Future value for compound interest:
$FV = PV(1 + i)^n$ [p. 145]
Present value for compound interest:

Payments to accumulate a future value:
$$PMT = FV\frac{i}{(1 + i)^n - 1} \text{ [p. 159]}$$
Annuitization: present value:

Review

Review exercises provide a great way to consolidate and check understanding and prepare for exams.

REVIEW EXERCISES

In Exercises 1–6, find the future value of the investment.

1. $6,000 for 5 years at 4.75% simple annual interest
2. $10,000 for 2.5 years at 5.25% simple annual interest

15. The monthly withdrawals possible over 5 years from an account earning 4.75% compounded monthly and starting with $6,000

Case Studies

Each chapter ends with a section entitled "Case Study," an extended application that uses and illustrates the central ideas of the chapter, focusing on the development of mathematical models appropriate to the topics. These applications are ideal for assignment as projects.

CASE STUDY

Adjustable Rate and Subprime Mortgages

The term **subprime mortgage** refers to mortgages given to home buyers with a heightened perceived risk of default, as when, for instance, the price of the home being purchased is higher than the borrower can reasonably afford. Such loans are typically **adjustable rate** loans, meaning that the lending rate varies through the duration of the loan.[*] Subprime adjustable rate loans typically start at artificially low "teaser rates" that the borrower can afford, but then increase significantly over the life of the mortgage. The U.S. real estate bubble of 2000–2005 led to a frenzy of subprime lending, the rationale being that a borrower having trouble meeting mortgage payments could either sell the property at a profit or refinance the loan, or the lending institution could earn a hefty profit by repossessing the property in the event of foreclosure.

Focus on Technology

Marginal Technology Notes

We give brief marginal technology notes to outline the use of graphing calculator, spreadsheet, and website technology in appropriate examples. When necessary, the reader is referred to more detailed discussion in the end-of-chapter Technology Guides.

End-of-Chapter Technology Guides

We continue to include detailed TI-83/84 Plus and Spreadsheet Guides at the end of each chapter. These Guides are referenced liberally in marginal technology notes at appropriate points in the chapter, so instructors and students can easily use this material or not, as they prefer.

Using Technology

TI-83/84 Plus
APPS 1:Finance, then
1:TVM Solver
N = 120, I% = 5, PV = −5000,
PMT = −100, P/Y = 12, C/Y = 12
With cursor on FV line,
ALPHA SOLVE
[More details in the Technology Guide.]

Spreadsheet
=FV(5%/12,10*12,−100,
−5000)
[More details in the Technology Guide.]

Website
www.WanerMath.com
→ Online Utilities
→ Time Value of Money Utility

TI-83/84 Plus   **Technology Guide**

Section 2.2

Example 1 (page 147) In December 2015, **Radius Bank** was paying 1.10% annual interest on savings accounts with balances of $2,500 and up. If the interest is compounded quarterly, find the future value of a $2,500 deposit after 6 years. What is the [...] paid over the time of the investment?

(such as the future value of your deposit, which the bank will give back to you) will be a positive number.

Solution

We could calculate the future value using

Spreadsheet   **Technology Guide**

Section 2.2

Example 1 (page 147) In December 2015, **Radius Bank** was paying 1.10% annual interest on savings accounts with balances of $2,500 and up. If the interest is compounded quarterly, find the future value of a

=FV(*i*, *n*, *PMT*, *PV*)

i = Interest per period   We use B2/B7 for the interest.
n = Number of periods   We use B3*B7 for the number of

Instructor Resources

MindTap: Through personalized paths of dynamic assignments and applications, MindTap is a digital learning solution and representation of your course that turns cookie cutter into cutting edge, apathy into engagement, and memorizers into higher-level thinkers.

The Right Content: With MindTap's carefully curated material, you get the precise content and groundbreaking tools you need for every course you teach. This course includes a dynamic Pre-Course Assessment that tests students on their prerequisite skills, an eBook, algorithmic assignments, and new lecture videos.

Personalization: Customize every element of your course—from rearranging the learning path to inserting videos and activities.

Improved Workflow: Save time when planning lessons with all of the trusted, most current content you need in one place in MindTap.

Tracking Students' Progress in Real Time: Promote positive outcomes by tracking students in real time and tailoring your course as needed based on the analytics.

Learn more at **www.cengage.com/mindtap**.

WebAssign: Exclusively from Cengage Learning, Enhanced WebAssign combines the exceptional mathematics content that you know and love with the most powerful online homework solution, WebAssign. Enhanced WebAssign engages students with immediate feedback, rich tutorial content, and eBooks, helping students to develop a deeper conceptual understanding of their subject matter. Quick Prep and Just In Time exercises provide opportunities for students to review prerequisite skills and content, both at the start of the course and at the beginning of each section. Flexible assignment options give instructors the ability to release assignments conditionally on the basis of students' prerequisite assignment scores. Visit us at **www.cengage.com/ewa** to learn more.

Cognero: Cengage Learning Testing Powered by Cognero is a flexible, online system that allows you to author, edit, and manage test bank content; create multiple test versions in an instant; and deliver tests from your LMS, your classroom, or wherever you choose.

Instructor Companion Site: This collection of book-specific lecture and class tools is available online at **www.cengage.com/login**. Access and download PowerPoint presentations, complete solutions manual, and more.

Student Resources

Student Solutions Manual (ISBN: 978-1-337-28047-1): Go beyond the answers—see what it takes to get there and improve your grade! This manual provides worked-out, step-by-step solutions to the odd-numbered problems in the text. You'll have the information you need to truly understand how the problems are solved.

MindTap: MindTap (assigned by the instructor) is a digital representation of your course that provides you with the tools you need to better manage your limited time, stay organized, and be successful. You can complete assignments whenever and wherever you are ready to learn, with course material specially customized for you by your instructor and streamlined in one proven, easy-to-use interface. With an array of study tools, you'll get a true understanding of course concepts, achieve better grades, and lay the groundwork for your future courses. Learn more at **www.cengage.com/mindtap**.

WebAssign: Enhanced WebAssign (assigned by the instructor) provides you with instant feedback on homework assignments. This online homework system is easy to use and includes helpful links to textbook sections, video examples, and problem-specific tutorials.

CengageBrain: Visit **www.cengagebrain.com** to access additional course materials and companion resources. At the cengagebrain.com home page, search for the ISBN of your title (from the back cover of your book) using the search box at the top of the page. This will take you to the product page where free companion resources can be found.

The Author Website

The authors' website, accessible through **www.wanermath.com**, has been evolving for close to two decades with growing recognition. Students, raised in an environment in which computers suffuse both work and play, can use their web browsers to engage with the material in an active way. The following features of the authors' website are fully integrated with the text and can be used as a personalized study resource:

- **Interactive Tutorials**   Highly interactive tutorials are included on major topics, with guided exercises that parallel the text and a great deal of help and feedback to assist the student.

- **Game Versions of Tutorials**   More challenging tutorials with randomized questions that work as games (complete with "health" scores, "health vials," and an assessment of one's performance at the end of the game) are offered alongside the traditional tutorials. These game tutorials, which mirror the traditional "more gentle" tutorials, randomize all the questions and do not give the student the answers but instead offer hints in exchange for "health points," so that just staying alive (not running out of health) can be quite challenging.

- **Learning and Practice Modules**   These interactive demos illustrate important concepts and randomizable "practice examples" that mirror many examples and quick examples in the text.

- **Detailed Chapter Summaries**   Comprehensive summaries with randomizable interactive elements review all the basic definitions and problem-solving techniques discussed in each chapter. These are a terrific pre-test study tool for students.

- **Downloadable Excel Tutorials**   Detailed Excel tutorials are available for almost every section of the book. These interactive tutorials expand on the examples given in the text.

- **Online Utilities**   Our collection of easy-to-use online utilities, referenced in the marginal notes of the textbook, allow students to solve many of the technology-based application exercises directly on the web. The utilities include a function grapher and evaluator that also does curve-fitting, regression tools, a time value of money calculator for annuities, a matrix algebra tool that also manipulates matrices with multinomial entries, a linear programming grapher that automatically solves two-dimensional linear programming problems graphically, and a powerful simplex method tool. These utilities require nothing more than a standard web browser.

- **Chapter True-False Quizzes**   Randomized quizzes that provide feedback for many incorrect answers based on the key concepts in each chapter assist the student in further mastery of the material.

- **Supplemental Topics**   We include complete interactive text and exercise sets for a selection of topics that are not ordinarily included in printed texts but are often requested by instructors.

- **Spanish**   A parallel Spanish version of almost the entire website is now deployed, allowing the user to switch languages on specific pages with a single mouse-click. In particular, all of the chapter summaries and most of the tutorials, game tutorials, and utilities are available in Spanish.

Acknowledgments

This project would not have been possible without the contributions and suggestions of numerous colleagues, students, and friends. We are particularly grateful to our colleagues at Hofstra and elsewhere who used and gave us useful feedback on previous editions and suggestions for this one, and to everyone at Cengage for their encouragement and guidance throughout the project. Specifically, we would like to thank Rita Lombard and Morgan Mendoza for their unflagging enthusiasm, Scott Barnett of Henry Ford Community College for his meticulous check of the mathematical accuracy, and Martha Emry and Teresa Trego for whipping the book into shape. Additionally, we would like to thank the creative force of Jay Abramson of Arizona State University for developing the new lecture videos that accompany our text, and Scott Barnett of Henry Ford Community College, Joe Rody of Arizona State University, Nada Al-Hanna of University of Texas at El Paso, and Kaat Higham of Bergen Community College for their thoughtful reviews and input into the scripts.

We would also like to thank Dario Menasce at CERN who helped us understand the fascinating new cover art, and the numerous reviewers and proofreaders who provided many helpful suggestions that have shaped the development of this book over time:

Christopher Brown, *California Lutheran University*

Melinda Camarillo, *El Paso Community College*

Nathan Carlson, *California Lutheran University*

Scott Fallstrom, *University of Oregon*

Irene Jai, *Raritan Valley Community College*

Latrice Laughlin, *University of Alaska Fairbanks*

Gabriel Mendoza, *El Paso Community College*

Charles Mundy-Castle, *Central New Mexico Community College*

Patrick Mutungi, *University of South Carolina*

Michael Price, *University of Oregon*

Christopher Quarles, *Everett Community College*

Leela Rakesh, *Central Michigan University*

Tom Rosenwinkel, *Concordia University Texas*

Bradley Stewart, *State University of New York at Oswego*

Larry Taylor, *North Dakota State University*

Daniel Wang, *Central Michigan University*

Stefan Waner
Steven R. Costenoble

FINITE MATHEMATICS

0

PRECALCULUS REVIEW

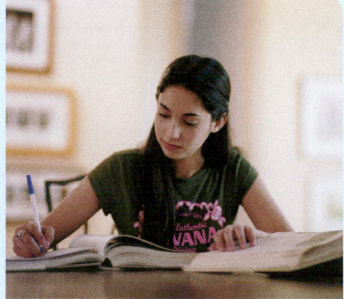

DreamPictures/Taxi/Getty Images

www.WanerMath.com

Introduction

In this chapter we review some topics from algebra that you need to know to get the most out of this book. This chapter can be used either as a refresher course or as a reference.

There is one crucial fact you must always keep in mind: The letters used in algebraic expressions stand for numbers. All the rules of algebra are just facts about the arithmetic of numbers. If you are not sure whether some algebraic manipulation you are about to do is legitimate, try it first with numbers. If it doesn't work with numbers, it doesn't work.

0.1   Real Numbers

The **real numbers** are the numbers that can be written in decimal notation, including those that require an infinite decimal expansion. The set of real numbers includes all integers, positive, negative, and zero; all fractions; and the irrational numbers, that is, those with decimal expansions that never repeat. Examples of irrational numbers are

$$\sqrt{2} = 1.414213562373\ldots$$

and

$$\pi = 3.141592653589\ldots$$

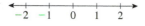

Figure 1

It is very useful to picture the real numbers as points on a line. As shown in Figure 1, larger numbers appear to the right, in the sense that if $a < b$, then the point corresponding to b is to the right of the one corresponding to a.

Intervals

Some subsets of the set of real numbers, called **intervals**, show up quite often, so we have a compact notation for them.

Interval Notation

Here is a list of types of intervals along with examples.

	Interval	Description	Picture	Example
Closed	$[a, b]$	Set of numbers x with $a \le x \le b$	a — b (includes end points)	$[0, 10]$
Open	(a, b)	Set of numbers x with $a < x < b$	a — b (excludes end points)	$(-1, 5)$
Half-Open	$(a, b]$	Set of numbers x with $a < x \le b$	a — b	$(-3, 1]$
	$[a, b)$	Set of numbers x with $a \le x < b$	a — b	$[0, 5)$

Infinite	$[a, +\infty)$	Set of numbers x with $a \le x$		$[10, +\infty)$
	$(a, +\infty)$	Set of numbers x with $a < x$		$(-3, +\infty)$
	$(-\infty, b]$	Set of numbers x with $x \le b$		$(-\infty, -3]$
	$(-\infty, b)$	Set of numbers x with $x < b$		$(-\infty, 10)$
	$(-\infty, +\infty)$	Set of all real numbers		$(-\infty, +\infty)$

Operations

There are five important operations on real numbers: addition, subtraction, multiplication, division, and exponentiation. "Exponentiation" means raising a real number to a power; for instance, $3^2 = 3 \cdot 3 = 9$; $2^3 = 2 \cdot 2 \cdot 2 = 8$.

A note on technology: Most graphing calculators and spreadsheets use an asterisk * for multiplication and a caret ^ for exponentiation. Thus, for instance, 3×5 is entered as `3*5`, $3x$ as `3*x`, and 3^2 as `3^2`.

When we write an expression involving two or more operations, such as

$$2 \cdot 3 + 4$$

or

$$\frac{2 \cdot 3^2 - 5}{4 - (-1)},$$

we need to agree on the order in which to do the operations. Does $2 \cdot 3 + 4$ mean $(2 \cdot 3) + 4 = 10$ or $2 \cdot (3 + 4) = 14$? We all agree to use the following rules for the order in which we do the operations.

Standard Order of Operations

Parentheses and Fraction Bars First, calculate the values of all expressions inside parentheses or brackets, working from the innermost parentheses out, before using them in other operations. In a fraction, calculate the numerator and denominator separately before doing the division.

Quick Examples

1. $6(2 + [3 - 5] - 4) = 6(2 + (-2) - 4) = 6(-4) = -24$
2. $\dfrac{(4 - 2)}{3(-2 + 1)} = \dfrac{2}{3(-1)} = \dfrac{2}{-3} = -\dfrac{2}{3}$
3. $3/(2 + 4) = \dfrac{3}{2 + 4} = \dfrac{3}{6} = \dfrac{1}{2}$
4. $(x + 4x)/(y + 3y) = (5x)/(4y)$

Exponents   Next, perform exponentiation.

> **Quick Examples**
>
> **5.** $2 + 4^2 = 2 + 16 = 18$ } Note the difference.
> **6.** $(2 + 4)^2 = 6^2 = 36$
>
> **7.** $2\left(\dfrac{3}{4-5}\right)^2 = 2\left(\dfrac{3}{-1}\right)^2 = 2(-3)^2 = 2 \times 9 = 18$
>
> **8.** $2(1 + 1/10)^2 = 2(1.1)^2 = 2 \times 1.21 = 2.42$

Multiplication and Division   Next, do all multiplications and divisions, from left to right.

> **Quick Examples**
>
> **9.** $2(3 - 5)/4 \cdot 2 = 2(-2)/4 \cdot 2$ Parentheses first
> $\qquad\qquad\qquad = -4/4 \cdot 2$ Leftmost product
> $\qquad\qquad\qquad = -1 \cdot 2 = -2$ Multiplications and divisions, left to right
>
> **10.** $2(1 + 1/10)^2 \times 2/10 = 2(1.1)^2 \times 2/10$ Parentheses first
> $\qquad\qquad\qquad\qquad = 2 \times 1.21 \times 2/10$ Exponent
> $\qquad\qquad\qquad\qquad = 4.84/10 = 0.484$ Multiplications and divisions, left to right
>
> **11.** $4\dfrac{2(4-2)}{3(-2 \cdot 5)} = 4\dfrac{2(2)}{3(-10)} = 4\dfrac{4}{-30} = \dfrac{16}{-30} = -\dfrac{8}{15}$

Addition and Subtraction   Last, do all additions and subtractions, from left to right.

> **Quick Examples**
>
> **12.** $2(3 - 5)^2 + 6 - 1 = 2(-2)^2 + 6 - 1 = 2(4) + 6 - 1$
> $\qquad\qquad\qquad\qquad\quad = 8 + 6 - 1 = 13$
>
> **13.** $\left(\dfrac{1}{2}\right)^2 - (-1)^2 + 4 = \dfrac{1}{4} - 1 + 4 = -\dfrac{3}{4} + 4 = \dfrac{13}{4}$
>
> **14.** $3/2 + 4 = 1.5 + 4 = 5.5$ } Note the difference.
> **15.** $3/(2 + 4) = 3/6 = 1/2 = 0.5$
>
> **16.** $4/2^2 + (4/2)^2 = 4/2^2 + 2^2 = 4/4 + 4 = 1 + 4 = 5$
>
> **17.** $-2\text{^}4 = (-1)2\text{^}4 = -16$ A negative sign before an expression means multiplication by -1.[1]

---

[1] Spreadsheets and some programming languages interpret `-2^4` (wrongly!) as `(-2)^4=16`. So when working with spreadsheets, write `-2^4` as `(-1)*2^4` to avoid this issue.

T indicates material discussing the use of technologies such as graphing calculators, spreadsheets, and web utilities.

T Entering Formulas

Any good calculator or spreadsheet will respect the standard order of operations. However, we must be careful with division and exponentiation and use parentheses as necessary. The following table gives some examples of simple mathematical expressions and their equivalents in the functional format used in most graphing calculators, spreadsheets, and computer programs.

Mathematical Expression	Formula	Comments
$\dfrac{2}{3-x}$	`2/(3-x)`	Note the use of parentheses instead of the fraction bar. If we omit the parentheses, we get the expression shown next.
$\dfrac{2}{3}-x$	`2/3-x`	The calculator follows the usual order of operations.
$\dfrac{2}{3\times 5}$	`2/(3*5)`	Putting the denominator in parentheses ensures that the multiplication is carried out first. The asterisk is usually used for multiplication in graphing calculators and computers.
$\dfrac{2}{x}\times 5$	`(2/x)*5`	Putting the fraction in parentheses ensures that it is calculated first. Some calculators will interpret `2/3*5` as $\dfrac{2}{3\times 5}$ but `2/3(5)` as $\dfrac{2}{3}\times 5.$
$\dfrac{2-3}{4+5}$	`(2-3)/(4+5)`	Note once again the use of parentheses in place of the fraction bar.
2^3	`2^3`	The caret ^ is commonly used to denote exponentiation.
2^{3-x}	`2^(3-x)`	Be careful to use parentheses to tell the calculator where the exponent ends. Enclose the *entire exponent* in parentheses.
2^3-x	`2^3-x`	Without parentheses, the calculator will follow the usual order of operations: exponentiation and then subtraction.
3×2^{-4}	`3*2^(-4)`	On some calculators, the negation key is separate from the minus key.
$2^{-4\times 3}\times 5$	`2^(-4*3)*5`	Note once again how parentheses enclose the entire exponent.
$100\left(1+\dfrac{0.05}{12}\right)^{60}$	`100*(1+0.05/12)^60`	This is a typical calculation for compound interest.
$PV\left(1+\dfrac{r}{m}\right)^{mt}$	`PV*(1+r/m)^(m*t)`	This is the compound interest formula. *PV* is understood to be a single number (present value) and not the product of *P* and *V* (or else we would have used `P*V`).
$\dfrac{2^{3-2}\times 5}{y-x}$	`2^(3-2)*5/(y-x)` or `(2^(3-2)*5)/(y-x)`	Notice again the use of parentheses to hold the denominator together. We could also have enclosed the numerator in parentheses, although this is optional. (Why?)
$\dfrac{2^y+1}{2-4^{3x}}$	`(2^y+1)/(2-4^(3*x))`	Here, it is necessary to enclose both the numerator and the denominator in parentheses.
$2^y+\dfrac{1}{2}-4^{3x}$	`2^y+1/2-4^(3*x)`	This is the effect of leaving out the parentheses around the numerator and denominator in the previous expression.

Accuracy and Rounding

When we use a calculator or computer, the results of our calculations are often given to far more decimal places than are useful. For example, suppose we are told that a square has an area of 2.0 square feet and we are asked how long its sides are. Each side is the square root of the area, which the calculator tells us is

$$\sqrt{2} \approx 1.414213562.$$

However, the measurement of 2.0 square feet is probably accurate to only two digits, so our estimate of the lengths of the sides can be no more accurate than that. Therefore, we round the answer to two digits:

Length of one side $\approx$ 1.4 feet.

The digits that follow 1.4 are meaningless. The following guide makes these ideas more precise.

Significant Digits, Decimal Places, and Rounding

The number of **significant digits** in a decimal representation of a number is the number of digits that are not leading zeros after the decimal point (as in .0005) or trailing zeros before the decimal point (as in 5,400,000). We say that a value is **accurate to n significant digits** if only the first n significant digits are meaningful.

When to Round

After doing a computation in which all the quantities are accurate to no more than n significant digits, round the final result to n significant digits.

Quick Examples

18. 0.00067 has two significant digits. The 000 before 67 are leading zeros.

19. 0.000670 has three significant digits. The 0 after 67 is significant.

20. 5,400,000 has two or more significant digits. We can't say how many of the zeros are trailing.[2]

21. 5,400,001 has seven significant digits. The string of zeros is not trailing.

22. Rounding 63,918 to three significant digits gives 63,900.

23. Rounding 63,958 to three significant digits gives 64,000.

24. $\pi = 3.141592653\ldots$ $\frac{22}{7} = 3.142857142\ldots$ Therefore, $\frac{22}{7}$ is an approximation of π that is accurate to only three significant digits: 3.14.

25. $4.02(1 + 0.02)^{1.4} \approx 4.13$ We rounded to three significant digits.

---

[2] If we obtained 5,400,000 by rounding 5,401,011, then it has three significant digits because the zero after the 4 is significant. On the other hand, if we obtained it by rounding 5,411,234, then it has only two significant digits. The use of scientific notation avoids this ambiguity: 5.40×10^6 (or 5.40E6 on a calculator or computer) is accurate to three digits, and 5.4×10^6 is accurate to two digits.

One more point, though: If, in a long calculation, you round the intermediate results, your final answer may be even less accurate than you think. As a general rule,

When calculating, don't round intermediate results. Rather, use the most accurate results obtainable, or have your calculator or computer store them for you.

When you are done with the calculation, *then* round your answer to the appropriate number of digits of accuracy.

0.1 EXERCISES

Calculate each expression in Exercises 1–24, giving the answer as a whole number or a fraction in lowest terms.

1. $2(4 + (-1))(2 \cdot -4)$

2. $3 + ([4 - 2] \cdot 9)$

3. `20/(3*4)-1`

4. `2-(3*4)/10`

5. $\dfrac{3 + ([3 + (-5)])}{3 - 2 \times 2}$

6. $\dfrac{12 - (1 - 4)}{2(5 - 1) \cdot 2 - 1}$

7. `(2-5*(-1))/1-2*(-1)`

8. `2-5*(-1)/(1-2*(-1))`

9. $2 \cdot (-1)^2/2$

10. $2 + 4 \cdot 3^2$

11. $2 \cdot 4^2 + 1$

12. $1 - 3 \cdot (-2)^2 \times 2$

13. `3^2+2^2+1`

14. `2^(2^2-2)`

15. $\dfrac{3 - 2(-3)^2}{-6(4 - 1)^2}$

16. $\dfrac{1 - 2(1 - 4)^2}{2(5 - 1)^2 \cdot 2}$

17. `10*(1+1/10)^3`

18. `121/(1+1/10)^2`

19. $3\left(\dfrac{-2 \cdot 3^2}{-(4 - 1)^2}\right)$

20. $-\left(\dfrac{8(1 - 4)^2}{-9(5 - 1)^2}\right)$

21. $3\left(1 - \left(-\dfrac{1}{2}\right)^2\right)^2 + 1$

22. $3\left(\dfrac{1}{9} - \left(\dfrac{2}{3}\right)^2\right)^2 + 1$

23. `(1/2)^2-1/2^2`

24. `2/(1^2)-(2/1)^2`

Convert each expression in Exercises 25–50 into its technology formula equivalent as in the table in the text.

25. $3 \times (2 - 5)$

26. $4 + \dfrac{5}{9}$

27. $\dfrac{3}{2 - 5}$

28. $\dfrac{4 - 1}{3}$

29. $\dfrac{3 - 1}{8 + 6}$

30. $3 + \dfrac{3}{2 - 9}$

31. $3 - \dfrac{4 + 7}{8}$

32. $\dfrac{4 \times 2}{\left(\frac{2}{3}\right)}$

33. $\dfrac{2}{3 + x} - xy^2$

34. $3 + \dfrac{3 + x}{xy}$

35. $3.1x^3 - 4x^{-2} - \dfrac{60}{x^2 - 1}$

36. $2.1x^{-3} - x^{-1} + \dfrac{x^2 - 3}{2}$

37. $\dfrac{\left(\frac{2}{3}\right)}{5}$

38. $\dfrac{2}{\left(\frac{3}{5}\right)}$

39. $3^{4-5} \times 6$

40. $\dfrac{2}{3 + 5^{7-9}}$

41. $3\left(1 + \dfrac{4}{100}\right)^{-3}$

42. $3\left(\dfrac{1 + 4}{100}\right)^{-3}$

43. $3^{2x-1} + 4^x - 1$

44. $2^{x^2} - (2^{2x})^2$

45. 2^{2x^2-x+1}

46. $2^{2x^2-x} + 1$

47. $\dfrac{4e^{-2x}}{2 - 3e^{-2x}}$

48. $\dfrac{e^{2x} + e^{-2x}}{e^{2x} - e^{-2x}}$

49. $3\left(1 - \left(-\dfrac{1}{2}\right)^2\right)^2 + 1$

50. $3\left(\dfrac{1}{9} - \left(\dfrac{2}{3}\right)^2\right)^2 + 1$

0.2 Exponents and Radicals

In Section 0.1 we discussed exponentiation, or "raising to a power"; for example, $2^3 = 2 \cdot 2 \cdot 2$. In this section we discuss the algebra of exponentials more fully. First, we look at *integer* exponents: cases in which the powers are positive or negative whole numbers.

Integer Exponents

Positive Integer Exponents

If a is any real number and n is any positive integer, then by a^n we mean the quantity $a \cdot a \cdot \ldots \cdot a$ (n times); thus, $a^1 = a$, $a^2 = a \cdot a$, $a^5 = a \cdot a \cdot a \cdot a \cdot a$. In the expression a^n the number n is called the **exponent**, and the number a is called the **base**.

Quick Examples

$$3^2 = 9 \qquad\qquad 2^3 = 8$$
$$0^{34} = 0 \qquad\qquad (-1)^5 = -1$$
$$10^3 = 1{,}000 \qquad\qquad 10^5 = 100{,}000$$

Negative Integer Exponents

If a is any real number *other than zero* and n is any positive integer, then we define

$$a^{-n} = \frac{1}{a^n} = \frac{1}{a \cdot a \cdot \ldots \cdot a} \quad (n \text{ times}).$$

Quick Examples

$$2^{-3} = \frac{1}{2^3} = \frac{1}{8} \qquad\qquad 1^{-27} = \frac{1}{1^{27}} = 1$$

$$x^{-1} = \frac{1}{x^1} = \frac{1}{x} \qquad\qquad (-3)^{-2} = \frac{1}{(-3)^2} = \frac{1}{9}$$

$$y^7 y^{-2} = y^7 \frac{1}{y^2} = y^5 \qquad\qquad 0^{-2} \text{ is not defined}$$

Zero Exponent

If a is any real number other than zero, then we define

$$a^0 = 1.$$

Quick Examples

$$3^0 = 1 \qquad\qquad 1{,}000{,}000^0 = 1$$

$$0^0 \text{ is not defined}$$

When combining exponential expressions, we use the following identities.

Exponent Identity	**Quick Examples**
1. $a^m a^n = a^{m+n}$	$2^3 2^2 = 2^{3+2} = 2^5 = 32$
	$x^3 x^{-4} = x^{3-4} = x^{-1} = \dfrac{1}{x}$
	$\dfrac{x^3}{x^{-2}} = x^3 \dfrac{1}{x^{-2}} = x^3 x^2 = x^5$
2. $\dfrac{a^m}{a^n} = a^{m-n}$ if $a \neq 0$	$\dfrac{4^3}{4^2} = 4^{3-2} = 4^1 = 4$
	$\dfrac{x^3}{x^{-2}} = x^{3-(-2)} = x^5$
	$\dfrac{3^2}{3^4} = 3^{2-4} = 3^{-2} = \dfrac{1}{9}$
3. $(a^n)^m = a^{nm}$	$(3^2)^2 = 3^4 = 81$
	$(2^x)^2 = 2^{2x}$
4. $(ab)^n = a^n b^n$	$(4 \cdot 2)^2 = 4^2 2^2 = 64$
	$(-2y)^4 = (-2)^4 y^4 = 16y^4$
5. $\left(\dfrac{a}{b}\right)^n = \dfrac{a^n}{b^n}$ if $b \neq 0$	$\left(\dfrac{4}{3}\right)^2 = \dfrac{4^2}{3^2} = \dfrac{16}{9}$
	$\left(\dfrac{x}{-y}\right)^3 = \dfrac{x^3}{(-y)^3} = -\dfrac{x^3}{y^3}$

Caution

- In the first two identities, the bases of the expressions must be the same. For example, the first identity gives $3^2 3^4 = 3^6$ but does *not* apply to $3^2 4^2$.
- People sometimes invent their own identities, such as $a^m + a^n = a^{m+n}$, which is wrong! (Try it with $a = m = n = 1$.) If you wind up with something like $2^3 + 2^4$, you are stuck with it; there are no identities around to simplify it further. (You can factor out 2^3, but whether or not that is a simplification depends on what you are going to do with the expression next.)

EXAMPLE 1   **Combining the Identities**

$$\frac{(x^2)^3}{x^3} = \frac{x^6}{x^3} \qquad \text{By identity (3)}$$

$$= x^{6-3} \qquad \text{By identity (2)}$$

$$= x^3$$

$$\frac{(x^4 y)^3}{y} = \frac{(x^4)^3 y^3}{y} \qquad \text{By identity (4)}$$

$$= \frac{x^{12} y^3}{y} \qquad \text{By identity (3)}$$

$$= x^{12} y^{3-1} \qquad \text{By identity (2)}$$

$$= x^{12} y^2$$

EXAMPLE 2   **Eliminating Negative Exponents**

Simplify the following and express the answer using no negative exponents.

a. $\dfrac{x^4 y^{-3}}{x^5 y^2}$ **b.** $\left(\dfrac{x^{-1}}{x^2 y}\right)^5$

Solution

a. $\dfrac{x^4 y^{-3}}{x^5 y^2} = x^{4-5} y^{-3-2} = x^{-1} y^{-5} = \dfrac{1}{x y^5}$

b. $\left(\dfrac{x^{-1}}{x^2 y}\right)^5 = \dfrac{(x^{-1})^5}{(x^2 y)^5} = \dfrac{x^{-5}}{x^{10} y^5} = \dfrac{1}{x^{15} y^5}$

Radicals

If a is any nonnegative real number, then its **square root** is the nonnegative number whose square is a. For example, the square root of 16 is 4, because $4^2 = 16$. We write the square root of n as $\sqrt{n}$. (Roots are also referred to as **radicals**.) It is important to remember that $\sqrt{n}$ is never negative. Thus, for instance, $\sqrt{9}$ is 3 and not -3, even though $(-3)^2 = 9$. If we want to speak of the "negative square root" of 9, we write it as $-\sqrt{9} = -3$. If we want to write both square roots at once, we write $\pm\sqrt{9} = \pm 3$.

The **cube root** of a real number a is the number whose cube is a. The cube root of a is written as $\sqrt[3]{a}$ so that, for example, $\sqrt[3]{8} = 2$ (because $2^3 = 8$). Note that we can take the cube root of any number, positive, negative, or zero. For instance, the cube root of -8 is $\sqrt[3]{-8} = -2$ because $(-2)^3 = -8$. Unlike square roots, the cube root of a number may be negative. In fact, the cube root of a always has the same sign as a.

Higher roots are defined similarly. The **fourth root** of the *nonnegative* number a is defined as the nonnegative number whose fourth power is a and is written $\sqrt[4]{a}$. The **fifth root** of any number a is the number whose fifth power is a, and so on.

Note We cannot take an even-numbered root of a negative number, but we can take an odd-numbered root of any number. Even roots are always positive, whereas odd roots have the same sign as the number we start with. ∎

EXAMPLE 3   *n*th **Roots**

$\sqrt{4} = 2$ Because $2^2 = 4$

$\sqrt{16} = 4$ Because $4^2 = 16$

$\sqrt{1} = 1$ Because $1^2 = 1$

If $x \geq 0$, then $\sqrt{x^2} = x$.   Because $x^2 = x^2$

$\sqrt{2} \approx 1.414213562$ $\sqrt{2}$ is not a whole number.

$\sqrt{1+1} = \sqrt{2} \approx 1.414213562$ First add, then take the square root.[3]

$\sqrt{9+16} = \sqrt{25} = 5$ Contrast with $\sqrt{9} + \sqrt{16} = 3 + 4 = 7$.

$\dfrac{1}{\sqrt{2}} = \dfrac{\sqrt{2}}{2}$ Multiply top and bottom by $\sqrt{2}$.

------

[3] In general, $\sqrt{a+b}$ means the square root of the *quantity* $(a+b)$. The radical sign acts as a pair of parentheses or a fraction bar, telling us to evaluate what is inside before taking the root. (See the Caution on the next page.)

$$\sqrt[3]{27} = 3 \qquad \text{Because } 3^3 = 27$$
$$\sqrt[3]{-64} = -4 \qquad \text{Because } (-4)^3 = -64$$
$$\sqrt[4]{16} = 2 \qquad \text{Because } 2^4 = 16$$

$\sqrt[4]{-16}$ is not defined.   Even-numbered root of a negative number

$\sqrt[5]{-1} = -1$, since $(-1)^5 = -1$.   Odd-numbered root of a negative number

$\sqrt[n]{-1} = -1$ if n is any odd number.

Q: *In the example we saw that $\sqrt{x^2} = x$ if x is nonnegative. What happens if x is negative?*

A: If x is negative, then x^2 is positive, so $\sqrt{x^2}$ is still defined as the nonnegative number whose square is x^2. This number must be $|x|$, the **absolute value of x**, which is the nonnegative number with the same size as x. For instance, $|-3| = 3$, while $|3| = 3$, and $|0| = 0$. It follows that

$$\sqrt{x^2} = |x|$$

for every real number x, positive or negative. For instance,

$$\sqrt{(-3)^2} = \sqrt{9} = 3 = |-3|$$

and $\quad \sqrt{3^2} = \sqrt{9} = 3 = |3|.$

In general, we find that

$$\sqrt[n]{x^n} = x \text{ if } n \text{ is odd and } \sqrt[n]{x^n} = |x| \text{ if } n \text{ is even.}$$

We use the following identities to evaluate radicals of products and quotients.

Radicals of Products and Quotients

If a and b are any real numbers (nonnegative in the case of even-numbered roots), then

$$\sqrt[n]{ab} = \sqrt[n]{a}\,\sqrt[n]{b} \qquad \text{Radical of a product = Product of radicals}$$
$$\sqrt[n]{\frac{a}{b}} = \frac{\sqrt[n]{a}}{\sqrt[n]{b}} \qquad \text{if } b \neq 0. \qquad \text{Radical of a quotient = Quotient of radicals}$$

Notes

- The first rule is similar to the rule $(a \cdot b)^2 = a^2 b^2$ for the square of a product, and the second rule is similar to the rule $\left(\dfrac{a}{b}\right)^2 = \dfrac{a^2}{b^2}$ for the square of a quotient.

- *Caution* There is no corresponding identity for addition. In general,

$$\sqrt{a + b} \text{ is } not \text{ equal to } \sqrt{a} + \sqrt{b}.$$

(Consider $a = b = 1$, for example.) Equating these expressions is a common error, so be careful! ∎

Quick Examples

1. $\sqrt{9 \cdot 4} = \sqrt{9}\sqrt{4} = 3 \times 2 = 6$ Alternatively, $\sqrt{9 \cdot 4} = \sqrt{36} = 6$.

2. $\sqrt{\dfrac{9}{4}} = \dfrac{\sqrt{9}}{\sqrt{4}} = \dfrac{3}{2}$

3. $\dfrac{\sqrt{2}}{\sqrt{5}} = \dfrac{\sqrt{2}\sqrt{5}}{\sqrt{5}\sqrt{5}} = \dfrac{\sqrt{10}}{5}$

4. $\sqrt{4(3 + 13)} = \sqrt{4(16)} = \sqrt{4}\sqrt{16} = 2 \times 4 = 8$

5. $\sqrt[3]{-216} = \sqrt[3]{(-27)8} = \sqrt[3]{-27}\sqrt[3]{8} = (-3)2 = -6$

6. $\sqrt{x^3} = \sqrt{x^2 \cdot x} = \sqrt{x^2}\sqrt{x} = x\sqrt{x}$ if $x \geq 0$

7. $\sqrt{\dfrac{x^2 + y^2}{z^2}} = \dfrac{\sqrt{x^2 + y^2}}{\sqrt{z^2}} = \dfrac{\sqrt{x^2 + y^2}}{|z|}$ We can't simplify the numerator any further.

Rational Exponents

We already know what we mean by expressions such as x^4 and a^{-6}. The next step is to make sense of *rational* exponents: exponents of the form p/q with p and q integers as in $a^{1/2}$ and $3^{-2/3}$.

Q: *What should we mean by $a^{1/2}$?*

A: The overriding concern here is that all the exponent identities should remain true. In this case the identity to look at is the one that says that $(a^m)^n = a^{mn}$. This identity tells us that

$$(a^{1/2})^2 = a^1 = a.$$

That is, $a^{1/2}$, when squared, gives us a. But that must mean that $a^{1/2}$ is the *square root* of a, or

$$a^{1/2} = \sqrt{a}.$$

A similar argument tells us that if q is any positive whole number, then

$$a^{1/q} = \sqrt[q]{a}, \text{ the } q\text{th root of } a.$$

Notice that if a is negative, this makes sense only for q odd. To avoid this problem, we usually stick to positive a.

Q: *If p and q are integers (q positive), what should we mean by $a^{p/q}$?*

A: By the exponent identities, $a^{p/q}$ should equal both $(a^p)^{1/q}$ and $(a^{1/q})^p$. The first is the qth root of a^p, and the second is the pth power of $a^{1/q}$.

These arguments give us the following formulas for conversion between rational exponents and radicals.

Conversion Between Rational Exponents and Radicals

If a is any nonnegative number, then

$$a^{p/q} = \sqrt[q]{a^p} = (\sqrt[q]{a})^p.$$

    ↑   ↑   ↑
Using exponents   Using radicals

In particular,

$$a^{1/q} = \sqrt[q]{a}, \text{ the } q\text{th root of } a.$$

Notes

- If a is negative, all of this makes sense only if q is odd.
- All of the exponent identities continue to work when we allow rational exponents p/q. In other words, we are free to use all the exponent identities even though the exponents are not integers. ■

Quick Examples

8. $4^{3/2} = (\sqrt{4})^3 = 2^3 = 8$

9. $8^{2/3} = (\sqrt[3]{8})^2 = 2^2 = 4$

10. $9^{-3/2} = \dfrac{1}{9^{3/2}} = \dfrac{1}{(\sqrt{9})^3} = \dfrac{1}{3^3} = \dfrac{1}{27}$

11. $\dfrac{\sqrt{3}}{\sqrt[3]{3}} = \dfrac{3^{1/2}}{3^{1/3}} = 3^{1/2-1/3} = 3^{1/6} = \sqrt[6]{3}$

12. $2^2 2^{7/2} = 2^2 2^{3+1/2} = 2^2 2^3 2^{1/2} = 2^5 2^{1/2} = 2^5 \sqrt{2}$

EXAMPLE 4   Simplifying Algebraic Expressions

Simplify the following.

a. $\dfrac{(x^3)^{5/3}}{x^3}$ **b.** $\sqrt[4]{a^6}$ **c.** $\dfrac{(xy)^{-3}y^{-3/2}}{x^{-2}\sqrt{y}}$

Solution

a. $\dfrac{(x^3)^{5/3}}{x^3} = \dfrac{x^5}{x^3} = x^2$

b. $\sqrt[4]{a^6} = a^{6/4} = a^{3/2} = a \cdot a^{1/2} = a\sqrt{a}$

c. $\dfrac{(xy)^{-3}y^{-3/2}}{x^{-2}\sqrt{y}} = \dfrac{x^{-3}y^{-3}y^{-3/2}}{x^{-2}y^{1/2}} = \dfrac{1}{x^{-2+3}y^{1/2+3+3/2}} = \dfrac{1}{xy^5}$

Radical Form, Positive Exponent Form, and Power Form

In calculus we must often convert algebraic expressions involving powers of x, such as $\dfrac{3}{2x^2}$, into expressions in which x does not appear in the denominator, such as $\dfrac{3}{2}x^{-2}$. Also, we must often convert expressions with radicals, such as $\dfrac{1}{\sqrt{1+x^2}}$, into expressions with no radicals and all powers in the numerator, such as $(1 + x^2)^{-1/2}$. In these cases, we are converting from **positive exponent form** or **radical form** to **power form**.

Radical Form

An expression is in **radical form** if it is written with integer powers and roots only.

Quick Examples

13. $\dfrac{2}{5\sqrt[3]{x}} + \dfrac{2}{x}$ is in radical form.

14. $\dfrac{2x^{-1/3}}{5} + 2x^{-1}$ is not in radical form because $x^{-1/3}$ appears.

15. $\dfrac{1}{\sqrt{1 + x^2}}$ is in radical form, but $(1 + x^2)^{-1/2}$ is not.

Positive Exponent Form

An expression is in **positive exponent form** if it is written with positive exponents only.

Quick Examples

16. $\dfrac{2}{3x^2}$ is in positive exponent form.

17. $\dfrac{2x^{-1}}{3}$ is not in positive exponent form because the exponent of x is negative.

18. $\dfrac{x}{6} + \dfrac{6}{x}$ is in positive exponent form.

Power Form

An expression is in **power form** if there are no radicals and all powers of unknowns occur in the numerator. We write such expressions as sums or differences of terms of the form

$$\text{Constant} \times (\text{Expression with } x)^p. \qquad \text{As in } \dfrac{1}{3}x^{-3/2}$$

Quick Examples

19. $\dfrac{2}{3}x^4 - 3x^{-1/3}$ is in power form.

20. $\dfrac{x}{6} + \dfrac{6}{x}$ is not in power form because the second expression has x in the denominator.

21. $\sqrt[3]{x}$ is not in power form because it has a radical.

22. $(1 + x^2)^{-1/2}$ is in power form, but $\dfrac{1}{\sqrt{1 + x^2}}$ is not.

EXAMPLE 5   **Converting from One Form to Another**

Convert the following to positive exponent form:

a. $\dfrac{1}{2}x^{-2} + \dfrac{4}{3}x^{-5}$

b. $\dfrac{2}{\sqrt{x}} - \dfrac{2}{x^{-4}}$

Convert the following to radical form:

c. $\dfrac{1}{2}x^{-1/2} + \dfrac{4}{3}x^{-5/4}$

d. $\dfrac{(3 + x)^{-1/3}}{5}$

Convert the following to power form:

e. $\dfrac{3}{4x^2} - \dfrac{x}{6} + \dfrac{6}{x} + \dfrac{4}{3\sqrt{x}}$

f. $\dfrac{2}{(x + 1)^2} - \dfrac{3}{4\sqrt[5]{2x - 1}}$

Solution For parts (a) and (b) we eliminate negative exponents, as we did in Example 2:

a. $\dfrac{1}{2}x^{-2} + \dfrac{4}{3}x^{-5} = \dfrac{1}{2} \cdot \dfrac{1}{x^2} + \dfrac{4}{3} \cdot \dfrac{1}{x^5} = \dfrac{1}{2x^2} + \dfrac{4}{3x^5}$

b. $\dfrac{2}{\sqrt{x}} - \dfrac{2}{x^{-4}} = \dfrac{2}{\sqrt{x}} - 2x^4$

For parts (c) and (d) we rewrite all terms with fractional exponents as radicals:

c. $\dfrac{1}{2}x^{-1/2} + \dfrac{4}{3}x^{-5/4} = \dfrac{1}{2} \cdot \dfrac{1}{x^{1/2}} + \dfrac{4}{3} \cdot \dfrac{1}{x^{5/4}}$

$$= \dfrac{1}{2} \cdot \dfrac{1}{\sqrt{x}} + \dfrac{4}{3} \cdot \dfrac{1}{\sqrt[4]{x^5}} = \dfrac{1}{2\sqrt{x}} + \dfrac{4}{3\sqrt[4]{x^5}}$$

d. $\dfrac{(3 + x)^{-1/3}}{5} = \dfrac{1}{5(3 + x)^{1/3}} = \dfrac{1}{5\sqrt[3]{3 + x}}$

For parts (e) and (f) we eliminate any radicals and move all expressions involving x to the numerator:

e. $\dfrac{3}{4x^2} - \dfrac{x}{6} + \dfrac{6}{x} + \dfrac{4}{3\sqrt{x}} = \dfrac{3}{4}x^{-2} - \dfrac{1}{6}x + 6x^{-1} + \dfrac{4}{3x^{1/2}}$

$$= \dfrac{3}{4}x^{-2} - \dfrac{1}{6}x + 6x^{-1} + \dfrac{4}{3}x^{-1/2}$$

f. $\dfrac{2}{(x + 1)^2} - \dfrac{3}{4\sqrt[5]{2x - 1}} = 2(x + 1)^{-2} - \dfrac{3}{4(2x - 1)^{1/5}}$

$$= 2(x + 1)^{-2} - \dfrac{3}{4}(2x - 1)^{-1/5}$$

Solving Equations with Exponents

EXAMPLE 6   **Solving Equations**

Solve the following equations:

a. $x^3 + 8 = 0$

b. $x^2 - \dfrac{1}{2} = 0$

c. $x^{3/2} - 64 = 0$

Solution

a. Subtracting 8 from both sides gives $x^3 = -8$. Taking the cube root of both sides gives $x = -2$.

b. Adding $\frac{1}{2}$ to both sides gives $x^2 = \frac{1}{2}$. Thus, $x = \pm\sqrt{\frac{1}{2}} = \pm\frac{1}{\sqrt{2}}$.

c. Adding 64 to both sides gives $x^{3/2} = 64$. Taking the reciprocal (2/3) power of both sides gives

$$(x^{3/2})^{2/3} = 64^{2/3}$$
$$x^1 = (\sqrt[3]{64})^2 = 4^2 = 16$$

so $x = 16$.

0.2   EXERCISES

Evaluate the expressions in Exercises 1–16.

1. 3^3 **2.** $(-2)^3$ **3.** $-(2 \cdot 3)^2$ **4.** $(4 \cdot 2)^2$

5. $\left(\dfrac{-2}{3}\right)^2$ **6.** $\left(\dfrac{3}{2}\right)^3$ **7.** $(-2)^{-3}$ **8.** -2^{-3}

9. $\left(\dfrac{1}{4}\right)^{-2}$ **10.** $\left(\dfrac{-2}{3}\right)^{-2}$ **11.** $2 \cdot 3^0$ **12.** $3 \cdot (-2)^0$

13. $2^3 2^2$ **14.** $3^2 3$ **15.** $2^2 2^{-1} 4^2 4^{-4}$ **16.** $5^2 5^{-3} 5^2 5^{-2}$

Simplify each expression in Exercises 17–30, expressing your answer in positive exponent form.

17. $x^3 x^2$ **18.** $x^4 x^{-1}$ **19.** $-x^2 x^{-3} y$ **20.** $-xy^{-1}x^{-1}$

21. $\dfrac{x^3}{x^4}$ **22.** $\dfrac{y^5}{y^3}$ **23.** $\dfrac{x^2 y^2}{x^{-1} y}$ **24.** $\dfrac{x^{-1}y}{x^2 y^2}$

25. $\dfrac{(xy^{-1}z^3)^2}{x^2 yz^2}$ **26.** $\dfrac{x^2 yz^2}{(xyz^{-1})^{-1}}$ **27.** $\left(\dfrac{xy^{-2}z}{x^{-1}z}\right)^3$

28. $\left(\dfrac{x^2 y^{-1}z^0}{xyz}\right)^2$ **29.** $\left(\dfrac{x^{-1}y^{-2}z^2}{xy}\right)^{-2}$ **30.** $\left(\dfrac{xy^{-2}}{x^2 y^{-1}z}\right)^{-3}$

Convert the expressions in Exercises 31–36 to positive exponent form.

31. $3x^{-4}$ **32.** $\dfrac{1}{2}x^{-4}$ **33.** $\dfrac{3}{4}x^{-2/3}$

34. $\dfrac{4}{5}y^{-3/4}$ **35.** $1 - \dfrac{0.3}{x^{-2}} - \dfrac{6}{5}x^{-1}$ **36.** $\dfrac{1}{3x^{-4}} + \dfrac{0.1x^{-2}}{3}$

Evaluate the expressions in Exercises 37–56, rounding your answer to four significant digits where necessary.

37. $\sqrt{4}$ **38.** $\sqrt{5}$ **39.** $\sqrt{\dfrac{1}{4}}$

40. $\sqrt{\dfrac{1}{9}}$ **41.** $\sqrt{\dfrac{16}{9}}$ **42.** $\sqrt{\dfrac{9}{4}}$

43. $\dfrac{\sqrt{4}}{5}$ **44.** $\dfrac{6}{\sqrt{25}}$ **45.** $\sqrt{9} + \sqrt{16}$

46. $\sqrt{25} - \sqrt{16}$ **47.** $\sqrt{9 + 16}$ **48.** $\sqrt{25 - 16}$

49. $\sqrt[3]{8 - 27}$ **50.** $\sqrt[4]{81 - 16}$ **51.** $\sqrt[3]{27/8}$

52. $\sqrt[3]{8 \times 64}$ **53.** $\sqrt{(-2)^2}$ **54.** $\sqrt{(-1)^2}$

55. $\sqrt{\dfrac{1}{4}(1 + 15)}$ **56.** $\sqrt{\dfrac{1}{9}(3 + 33)}$

Simplify the expressions in Exercises 57–64, given that x, y, z, a, b, and c are positive real numbers.

57. $\sqrt{a^2 b^2}$ **58.** $\sqrt{\dfrac{a^2}{b^2}}$ **59.** $\sqrt{(x + 9)^2}$

60. $\left(\sqrt{x + 9}\right)^2$ **61.** $\sqrt[3]{x^3(a^3 + b^3)}$ **62.** $\sqrt[4]{\dfrac{x^4}{a^4 b^4}}$

63. $\sqrt{\dfrac{4xy^3}{x^2 y}}$ **64.** $\sqrt{\dfrac{4(x^2 + y^2)}{c^2}}$

Convert the expressions in Exercises 65–84 to power form.

65. $\sqrt{3}$ **66.** $\sqrt{8}$ **67.** $\sqrt{x^3}$

68. $\sqrt[3]{x^2}$ **69.** $\sqrt[3]{xy^2}$ **70.** $\sqrt{x^2 y}$

71. $\dfrac{x^2}{\sqrt{x}}$ **72.** $\dfrac{x}{\sqrt{x}}$ **73.** $\dfrac{3}{5x^2}$

74. $\dfrac{2}{5x^{-3}}$ **75.** $\dfrac{3x^{-1.2}}{2} - \dfrac{1}{3x^{2.1}}$ **76.** $\dfrac{2}{3x^{-1.2}} - \dfrac{x^{2.1}}{3}$

77. $\dfrac{2x}{3} - \dfrac{x^{0.1}}{2} + \dfrac{4}{3x^{1.1}}$ **78.** $\dfrac{4x^2}{3} + \dfrac{x^{3/2}}{6} - \dfrac{2}{3x^2}$

79. $\dfrac{3\sqrt{x}}{4} - \dfrac{5}{3\sqrt{x}} + \dfrac{4}{3x\sqrt{x}}$ **80.** $\dfrac{3}{5\sqrt{x}} - \dfrac{5\sqrt{x}}{8} + \dfrac{7}{2\sqrt[3]{x}}$

81. $\dfrac{3\sqrt[5]{x^2}}{4} - \dfrac{7}{2\sqrt{x^3}}$ **82.** $\dfrac{1}{8x\sqrt{x}} - \dfrac{2}{3\sqrt[5]{x^3}}$

83. $\dfrac{1}{(x^2 + 1)^3} - \dfrac{3}{4\sqrt[3]{(x^2 + 1)}}$ **84.** $\dfrac{2}{3(x^2 + 1)^{-3}} - \dfrac{3\sqrt[3]{(x^2 + 1)^7}}{4}$

Convert the expressions in Exercises 85–96 to radical form.

85. $2^{2/3}$ **86.** $3^{4/5}$ **87.** $x^{4/3}$ **88.** $y^{7/4}$

89. $(x^{1/2}y^{1/3})^{1/5}$ **90.** $x^{-1/3}y^{3/2}$ **91.** $-\dfrac{3}{2}x^{-1/4}$ **92.** $\dfrac{4}{5}x^{3/2}$

93. $0.2x^{-2/3} + \dfrac{3}{7x^{-1/2}}$ **94.** $\dfrac{3.1}{x^{-4/3}} - \dfrac{11}{7}x^{-1/7}$

95. $\dfrac{3}{4(1-x)^{5/2}}$ **96.** $\dfrac{9}{4(1-x)^{-7/3}}$

Simplify the expressions in Exercises 97–106.

97. $4^{-1/2}4^{7/2}$ **98.** $2^{1/a}/2^{2/a}$ **99.** $3^{2/3}3^{-1/6}$

100. $2^{1/3}2^{-1}2^{2/3}2^{-1/3}$ **101.** $\dfrac{x^{3/2}}{x^{5/2}}$ **102.** $\dfrac{y^{5/4}}{y^{3/4}}$

103. $\dfrac{x^{1/2}y^2}{x^{-1/2}y}$ **104.** $\dfrac{x^{-1/2}y}{x^2y^{3/2}}$

105. $\left(\dfrac{x}{y}\right)^{1/3}\left(\dfrac{y}{x}\right)^{2/3}$ **106.** $\left(\dfrac{x}{y}\right)^{-1/3}\left(\dfrac{y}{x}\right)^{1/3}$

Solve each equation in Exercises 107–120 for x, rounding your answer to four significant digits where necessary.

107. $x^2 - 16 = 0$ **108.** $x^2 - 1 = 0$

109. $x^2 - \dfrac{4}{9} = 0$ **110.** $x^2 - \dfrac{1}{10} = 0$

111. $x^2 - (1+2x)^2 = 0$ **112.** $x^2 - (2-3x)^2 = 0$

113. $x^5 + 32 = 0$ **114.** $x^4 - 81 = 0$

115. $x^{1/2} - 4 = 0$ **116.** $x^{1/3} - 2 = 0$

117. $1 - \dfrac{1}{x^2} = 0$ **118.** $\dfrac{2}{x^3} - \dfrac{6}{x^4} = 0$

119. $(x-4)^{-1/3} = 2$ **120.** $(x-4)^{2/3} + 1 = 5$

0.3 Multiplying and Factoring Algebraic Expressions

Multiplying Algebraic Expressions

Distributive Law

The **distributive law** for real numbers states that

$$a(b \pm c) = ab \pm ac$$
$$(a \pm b)c = ac \pm bc$$

for any real numbers a, b, and c.

Quick Examples

1. $2(x-3)$ is *not* equal to $2x - 3$ but is equal to $2x - 2(3) = 2x - 6$.
2. $x(x+1) = x^2 + x$
3. $2x(3x-4) = 6x^2 - 8x$
4. $(x-4)x^2 = x^3 - 4x^2$
5. $(x+2)(x+3) = (x+2)x + (x+2)3$
   $$= (x^2 + 2x) + (3x + 6) = x^2 + 5x + 6$$
6. $(x+2)(x-3) = (x+2)x - (x+2)3$
   $$= (x^2 + 2x) - (3x + 6) = x^2 - x - 6$$

There is a quicker way of expanding expressions like the last two, called the FOIL method (First, Outer, Inner, Last). Consider, for instance, the expression $(x+1)(x-2)$. The FOIL method says: Take the product of the first terms: $x \cdot x = x^2$, the product of the outer terms: $x \cdot (-2) = -2x$, the product of the inner terms: $1 \cdot x = x$, and the product of the last terms: $1 \cdot (-2) = -2$, and then add them all up, getting $x^2 - 2x + x - 2 = x^2 - x - 2$.

EXAMPLE 1 **FOIL**

a. $(x - 2)(2x + 5) = 2x^2 + 5x - 4x - 10 = 2x^2 + x - 10$

   First   Outer   Inner   Last

b. $(x^2 + 1)(x - 4) = x^3 - 4x^2 + x - 4$

c. $(a - b)(a + b) = a^2 + ab - ab - b^2 = a^2 - b^2$

d. $(a + b)^2 = (a + b)(a + b) = a^2 + ab + ab + b^2 = a^2 + 2ab + b^2$

e. $(a - b)^2 = (a - b)(a - b) = a^2 - ab - ab + b^2 = a^2 - 2ab + b^2$

The formulas in parts (c), (d), and (e) of Example 1 are particularly important and worth memorizing, so let's repeat them.

Special Formulas

$$(a - b)(a + b) = a^2 - b^2 \qquad \text{Difference of two squares}$$

$$(a + b)^2 = a^2 + 2ab + b^2 \qquad \text{Square of a sum}$$

$$(a - b)^2 = a^2 - 2ab + b^2 \qquad \text{Square of a difference}$$

Quick Examples

7. $(2 - x)(2 + x) = 4 - x^2$

8. $(1 + a)(1 - a) = 1 - a^2$

9. $(x + 3)^2 = x^2 + 6x + 9$

10. $(4 - x)^2 = 16 - 8x + x^2$

Here are some longer examples that require the distributive law.

EXAMPLE 2   **Multiplying Algebraic Expressions**

a. $(x + 1)(x^2 + 3x - 4) = (x + 1)x^2 + (x + 1)3x - (x + 1)4$

$$= (x^3 + x^2) + (3x^2 + 3x) - (4x + 4)$$

$$= x^3 + 4x^2 - x - 4$$

b. $\left(x^2 - \dfrac{1}{x} + 1\right)(2x + 5) = \left(x^2 - \dfrac{1}{x} + 1\right)2x + \left(x^2 - \dfrac{1}{x} + 1\right)5$

$$= (2x^3 - 2 + 2x) + \left(5x^2 - \dfrac{5}{x} + 5\right)$$

$$= 2x^3 + 5x^2 + 2x + 3 - \dfrac{5}{x}$$

c. $(x - y)(x - y)(x - y) = (x^2 - 2xy + y^2)(x - y)$

$$= (x^2 - 2xy + y^2)x - (x^2 - 2xy + y^2)y$$

$$= (x^3 - 2x^2y + xy^2) - (x^2y - 2xy^2 + y^3)$$

$$= x^3 - 3x^2y + 3xy^2 - y^3$$

Factoring Algebraic Expressions

We can think of factoring as applying the distributive law in reverse—for example,

$$2x^2 + x = x(2x + 1),$$

which can be checked by using the distributive law. Factoring is an art that you will learn with experience and the help of a few useful techniques.

Factoring Using a Common Factor

To use this technique, locate a **common factor**—a term that occurs as a factor in each of the expressions being added or subtracted (for example, x is a common factor in $2x^2 + x$, because it is a factor of both $2x^2$ and x). Once you have located a common factor, factor it out by applying the distributive law.

Quick Examples

11. $2x^3 - x^2 + x$ has x as a common factor, so
$$2x^3 - x^2 + x = x(2x^2 - x + 1).$$

12. $2x^2 + 4x$ has $2x$ as a common factor, so
$$2x^2 + 4x = 2x(x + 2).$$

13. $2x^2y + xy^2 - x^2y^2$ has xy as a common factor, so
$$2x^2y + xy^2 - x^2y^2 = xy(2x + y - xy).$$

14. $(x^2 + 1)(x + 2) - (x^2 + 1)(x + 3)$ has $x^2 + 1$ as a common factor, so
$$\begin{aligned}
(x^2 + 1)(x + 2) - (x^2 + 1)(x + 3) &= (x^2 + 1)[(x + 2) - (x + 3)] \\
&= (x^2 + 1)(x + 2 - x - 3) \\
&= (x^2 + 1)(-1) = -(x^2 + 1).
\end{aligned}$$

15. $12x(x^2 - 1)^5(x^3 + 1)^6 + 18x^2(x^2 - 1)^6(x^3 + 1)^5$ has $6x(x^2 - 1)^5(x^3 + 1)^5$ as a common factor, so
$$\begin{aligned}
12x(x^2 - 1)^5&(x^3 + 1)^6 + 18x^2(x^2 - 1)^6(x^3 + 1)^5 \\
&= 6x(x^2 - 1)^5(x^3 + 1)^5[2(x^3 + 1) + 3x(x^2 - 1)] \\
&= 6x(x^2 - 1)^5(x^3 + 1)^5(2x^3 + 2 + 3x^3 - 3x) \\
&= 6x(x^2 - 1)^5(x^3 + 1)^5(5x^3 - 3x + 2).
\end{aligned}$$

We would also like to be able to reverse calculations such as $(x + 2)(2x - 5) = 2x^2 - x - 10$. That is, starting with the expression $2x^2 - x - 10$, we would like to **factor** it to get the expression $(x + 2)(2x - 5)$. An expression of the form $ax^2 + bx + c$, where a, b, and c are real numbers, is called a **quadratic** expression in x. Thus, given a quadratic expression $ax^2 + bx + c$, we would like to write it in the form $(dx + e)(fx + g)$ for some real numbers d, e, f, and g. There are some quadratics, such as $x^2 + x + 1$, that cannot be factored in this form at all. Here, we consider only quadratics that do factor and do so in such a way that the numbers d, e, f, and g are integers (other cases are discussed in Section 0.5). The usual technique of factoring such quadratics is a trial and error approach.

Factoring Quadratics by Trial and Error

To factor the quadratic $ax^2 + bx + c$, factor ax^2 as $(a_1x)(a_2x)$ (with a_1 positive) and c as c_1c_2, and then check whether or not $ax^2 + bx + c = (a_1x \pm c_1)(a_2x \pm c_2)$. If not, try other factorizations of ax^2 and c.

Quick Examples

16. To factor $x^2 - 6x + 5$, first factor x^2 as $(x)(x)$, and 5 as $(5)(1)$:

$$(x + 5)(x + 1) = x^2 + 6x + 5 \qquad \text{No good}$$
$$(x - 5)(x - 1) = x^2 - 6x + 5. \qquad \text{Desired factorization}$$

17. To factor $x^2 - 4x - 12$, first factor x^2 as $(x)(x)$, and -12 as $(1)(-12)$, $(2)(-6)$, or $(3)(-4)$. Trying them one by one gives

$$(x + 1)(x - 12) = x^2 - 11x - 12 \qquad \text{No good}$$
$$(x - 1)(x + 12) = x^2 + 11x - 12 \qquad \text{No good}$$
$$(x + 2)(x - 6) = x^2 - 4x - 12. \qquad \text{Desired factorization}$$

18. To factor $4x^2 - 25$, we can follow the above procedure or recognize $4x^2 - 25$ as the difference of two squares:

$$4x^2 - 25 = (2x)^2 - 5^2 = (2x - 5)(2x + 5).$$

Note Not all quadratic expressions factor. In Section 0.5 we look at a test that tells us whether or not a given quadratic factors. ■

Here are examples that require either a little more work or a little more thought.

EXAMPLE 3 Factoring Quadratics

Factor the following: **a.** $4x^2 - 5x - 6$ **b.** $x^4 - 5x^2 + 6$

Solution

a. Possible factorizations of $4x^2$ are $(2x)(2x)$ or $(x)(4x)$. Possible factorizations of -6 are $(1)(-6)$ or $(2)(-3)$. We now systematically try out all the possibilities until we come up with the correct one:

$$(2x)(2x) \text{ and } (1)(-6): \quad (2x + 1)(2x - 6) = 4x^2 - 10x - 6 \qquad \text{No good}$$
$$(2x)(2x) \text{ and } (2)(-3): \quad (2x + 2)(2x - 3) = 4x^2 - 2x - 6 \qquad \text{No good}$$
$$(x)(4x) \text{ and } (1)(-6): \quad (x + 1)(4x - 6) = 4x^2 - 2x - 6 \qquad \text{No good}$$
$$(x)(4x) \text{ and } (2)(-3): \quad (x + 2)(4x - 3) = 4x^2 + 5x - 6 \qquad \text{Almost!}$$
$$\text{Change signs:} \quad (x - 2)(4x + 3) = 4x^2 - 5x - 6. \qquad \text{Correct}$$

b. The expression $x^4 - 5x^2 + 6$ is not a quadratic, you say? Correct. It's a quartic (a fourth degree expression). However, it looks rather like a quadratic. In fact, it is quadratic *in* x^2, meaning that it is

$$(x^2)^2 - 5(x^2) + 6 = y^2 - 5y + 6,$$

where $y = x^2$. The quadratic $y^2 - 5y + 6$ factors as

$$y^2 - 5y + 6 = (y - 3)(y - 2),$$

so

$$x^4 - 5x^2 + 6 = (x^2 - 3)(x^2 - 2).$$

This is a sometimes useful technique.

Our last example is here to remind you why we should want to factor polynomials in the first place. We shall return to this in Section 0.5.

EXAMPLE 4 **Solving a Quadratic Equation by Factoring**

Solve the equation $3x^2 + 4x - 4 = 0$.

Solution We first factor the left-hand side to get

$$(3x - 2)(x + 2) = 0.$$

Thus, the product of the two quantities $(3x - 2)$ and $(x + 2)$ is zero. Now, if a product of two numbers is zero, one of the two must be zero. In other words, either $3x - 2 = 0$, giving $x = \frac{2}{3}$, or $x + 2 = 0$, giving $x = -2$. Thus, there are two solutions: $x = \frac{2}{3}$ and $x = -2$.

0.3 EXERCISES

Expand each expression in Exercises 1–22.

1. $x(4x + 6)$

2. $(4y - 2)y$

3. $(2x - y)y$

4. $x(3x + y)$

5. $(x + 1)(x - 3)$

6. $(y + 3)(y + 4)$

7. $(2y + 3)(y + 5)$

8. $(2x - 2)(3x - 4)$

9. $(2x - 3)^2$

10. $(3x + 1)^2$

11. $\left(x + \dfrac{1}{x}\right)^2$

12. $\left(y - \dfrac{1}{y}\right)^2$

13. $(2x - 3)(2x + 3)$

14. $(4 + 2x)(4 - 2x)$

15. $\left(y - \dfrac{1}{y}\right)\left(y + \dfrac{1}{y}\right)$

16. $(x - x^2)(x + x^2)$

17. $(x^2 + x - 1)(2x + 4)$

18. $(3x + 1)(2x^2 - x + 1)$

19. $(x^2 - 2x + 1)^2$

20. $(x + y - xy)^2$

21. $(y^3 + 2y^2 + y)(y^2 + 2y - 1)$

22. $(x^3 - 2x^2 + 4)(3x^2 - x + 2)$

In Exercises 23–30, factor each expression and simplify as much as possible.

23. $(x + 1)(x + 2) + (x + 1)(x + 3)$

24. $(x + 1)(x + 2)^2 + (x + 1)^2(x + 2)$

25. $(x^2 + 1)^5(x + 3)^4 + (x^2 + 1)^6(x + 3)^3$

26. $10x(x^2 + 1)^4(x^3 + 1)^5 + 15x^2(x^2 + 1)^5(x^3 + 1)^4$

27. $(x^3 + 1)\sqrt{x + 1} - (x^3 + 1)^2\sqrt{x + 1}$

28. $(x^2 + 1)\sqrt{x + 1} - \sqrt{(x + 1)^3}$

29. $\sqrt{(x + 1)^3} + \sqrt{(x + 1)^5}$

30. $(x^2 + 1)\sqrt[3]{(x + 1)^4} - \sqrt[3]{(x + 1)^7}$

In Exercises 31–48, (a) factor the given expression, and (b) set the expression equal to zero and solve for the unknown (x in the odd-numbered exercises and y in the even-numbered exercises).

31. $2x + 3x^2$

32. $y^2 - 4y$

33. $6x^3 - 2x^2$

34. $3y^3 - 9y^2$

35. $x^2 - 8x + 7$

36. $y^2 + 6y + 8$

37. $x^2 + x - 12$

38. $y^2 + y - 6$

39. $2x^2 - 3x - 2$

40. $3y^2 - 8y - 3$

41. $6x^2 + 13x + 6$

42. $6y^2 + 17y + 12$

43. $12x^2 + x - 6$

44. $20y^2 + 7y - 3$

45. $x^2 + 4xy + 4y^2$

46. $4y^2 - 4xy + x^2$

47. $x^4 - 5x^2 + 4$

48. $y^4 + 2y^2 - 3$

0.4   Rational Expressions

Rational Expression

A **rational expression** is an algebraic expression of the form $\dfrac{P}{Q}$, where P and Q are simpler expressions (usually polynomials) and the denominator Q is not zero.

Quick Examples

1. $\dfrac{x^2 - 3x}{x}$ $P = x^2 - 3x, Q = x$

2. $\dfrac{x + \frac{1}{x} + 1}{2x^2 y + 1}$ $P = x + \dfrac{1}{x} + 1, Q = 2x^2 y + 1$

3. $3xy - x^2$ $P = 3xy - x^2, Q = 1$

Algebra of Rational Expressions

We manipulate rational expressions in the same way that we manipulate fractions, using the following rules:

Algebraic Rule	**Quick Examples**
Product: $\dfrac{P}{Q} \cdot \dfrac{R}{S} = \dfrac{PR}{QS}$	$\dfrac{x + 1}{x} \cdot \dfrac{x - 1}{2x + 1} = \dfrac{(x + 1)(x - 1)}{x(2x + 1)} = \dfrac{x^2 - 1}{2x^2 + x}$
Sum: $\dfrac{P}{Q} + \dfrac{R}{S} = \dfrac{PS + RQ}{QS}$	$\dfrac{2x - 1}{3x + 2} + \dfrac{1}{x} = \dfrac{(2x - 1)x + 1(3x + 2)}{x(3x + 2)}$ $= \dfrac{2x^2 + 2x + 2}{3x^2 + 2x}$
Difference: $\dfrac{P}{Q} - \dfrac{R}{S} = \dfrac{PS - RQ}{QS}$	$\dfrac{x}{3x + 2} - \dfrac{x - 4}{x} = \dfrac{x^2 - (x - 4)(3x + 2)}{x(3x + 2)}$ $= \dfrac{-2x^2 + 10x + 8}{3x^2 + 2x}$
Reciprocal: $\dfrac{1}{\left(\frac{P}{Q}\right)} = \dfrac{Q}{P}$	$\dfrac{1}{\left(\frac{2xy}{3x - 1}\right)} = \dfrac{3x - 1}{2xy}$
Quotient: $\dfrac{\left(\frac{P}{Q}\right)}{\left(\frac{R}{S}\right)} = \dfrac{P}{Q} \cdot \dfrac{S}{R} = \dfrac{PS}{QR}$	$\dfrac{\left(\frac{x}{x - 1}\right)}{\left(\frac{y - 1}{y}\right)} = \dfrac{xy}{(x - 1)(y - 1)} = \dfrac{xy}{xy - x - y + 1}$
Cancellation: $\dfrac{P\cancel{R}}{Q\cancel{R}} = \dfrac{P}{Q}$	$\dfrac{(x - 1)(xy + 4)}{(x^2 y - 8)(x - 1)} = \dfrac{xy + 4}{x^2 y - 8}$

Caution Cancellation of summands is *invalid*. For instance,

$$\frac{\cancel{x} + (2xy^2 - y)}{\cancel{x} + 4y} = \frac{(2xy^2 - y)}{4y} \qquad \text{✗ WRONG!} \qquad \text{Do } not \text{ cancel a summand.}$$

$$\frac{\cancel{x}(2xy^2 - y)}{4\cancel{x}y} = \frac{(2xy^2 - y)}{4y}. \qquad \text{✔ CORRECT} \quad \text{Do cancel a factor.}$$

Here are some examples that require several algebraic operations.

EXAMPLE 1 **Simplifying Rational Expressions**

a. $\dfrac{\left(\frac{1}{x+y} - \frac{1}{x}\right)}{y} = \dfrac{\left(\frac{x - (x+y)}{x(x+y)}\right)}{y} = \dfrac{\left(\frac{-y}{x(x+y)}\right)}{y} = \dfrac{-y}{xy(x+y)} = -\dfrac{1}{x(x+y)}$

b. $\dfrac{(x+1)(x+2)^2 - (x+1)^2(x+2)}{(x+2)^4} = \dfrac{(x+1)(x+2)[(x+2) - (x+1)]}{(x+2)^4}$

$= \dfrac{(x+1)(x+2)(x+2-x-1)}{(x+2)^4} = \dfrac{(x+1)(x+2)}{(x+2)^4} = \dfrac{x+1}{(x+2)^3}$

c. $\dfrac{2x\sqrt{x+1} - \frac{x^2}{\sqrt{x+1}}}{x+1} = \dfrac{\left(\frac{2x(\sqrt{x+1})^2 - x^2}{\sqrt{x+1}}\right)}{x+1} = \dfrac{2x(x+1) - x^2}{(x+1)\sqrt{x+1}}$

$= \dfrac{2x^2 + 2x - x^2}{(x+1)\sqrt{x+1}} = \dfrac{x^2 + 2x}{\sqrt{(x+1)^3}} = \dfrac{x(x+2)}{\sqrt{(x+1)^3}}$

0.4 EXERCISES

Rewrite each expression in Exercises 1–16 as a single rational expression, simplified as much as possible.

1. $\dfrac{x-4}{x+1} \cdot \dfrac{2x+1}{x-1}$

2. $\dfrac{2x-3}{x-2} \cdot \dfrac{x+3}{x+1}$

3. $\dfrac{x-4}{x+1} + \dfrac{2x+1}{x-1}$

4. $\dfrac{2x-3}{x-2} + \dfrac{x+3}{x+1}$

5. $\dfrac{x^2}{x+1} - \dfrac{x-1}{x+1}$

6. $\dfrac{x^2-1}{x-2} - \dfrac{1}{x-1}$

7. $\dfrac{1}{\left(\frac{x}{x-1}\right)} + x - 1$

8. $\dfrac{2}{\left(\frac{x-2}{x^2}\right)} - \dfrac{1}{x-2}$

9. $\dfrac{1}{x}\left[\dfrac{x-3}{xy} + \dfrac{1}{y}\right]$

10. $\dfrac{y^2}{x}\left[\dfrac{2x-3}{y} + \dfrac{x}{y}\right]$

11. $\dfrac{(x+1)^2(x+2)^3 - (x+1)^3(x+2)^2}{(x+2)^6}$

12. $\dfrac{6x(x^2+1)^2(x^3+2)^3 - 9x^2(x^2+1)^3(x^3+2)^2}{(x^3+2)^6}$

13. $\dfrac{(x^2-1)\sqrt{x^2+1} - \frac{x^4}{\sqrt{x^2+1}}}{x^2+1}$

14. $\dfrac{x\sqrt{x^3-1} - \frac{3x^4}{\sqrt{x^3-1}}}{x^3-1}$

15. $\dfrac{\frac{1}{(x+y)^2} - \frac{1}{x^2}}{y}$

16. $\dfrac{\frac{1}{(x+y)^3} - \frac{1}{x^3}}{y}$

0.5   Solving Polynomial Equations

Polynomial Equation

A **polynomial equation** in one unknown is an equation that can be written in the form

$$ax^n + bx^{n-1} + \cdots + rx + s = 0,$$

where $a, b, \ldots, r$, and s are constants.

We call the largest exponent of x appearing in a nonzero term of a polynomial equation the **degree** of that polynomial equation.

Quick Examples

1. $3x + 1 = 0$ has degree 1 because the largest power of x that occurs is $x = x^1$. Degree 1 equations are called **linear** equations.

2. $x^2 - x - 1 = 0$ has degree 2 because the largest power of x that occurs is x^2. Degree 2 equations are also called **quadratic equations**, or just **quadratics**.

3. $x^3 = 2x^2 + 1$ is a degree 3 polynomial equation (or **cubic** equation) in disguise. It can be rewritten as $x^3 - 2x^2 - 1 = 0$, which is in the standard form for a degree 3 equation.

4. $x^4 - x = 0$ has degree 4. It is called a **quartic**.

Now comes the question: How do we solve these equations for x? This question was asked by mathematicians as early as 1600 BCE. Let's look at these equations one degree at a time.

Solution of Linear Equations

By definition a linear equation can be written in the form

$$ax + b = 0. \qquad \text{\textit{a} and \textit{b} are fixed numbers with } a \neq 0.$$

Solving this is a nice mental exercise: Subtract b from both sides, and then divide by a, getting $x = -b/a$. Don't bother memorizing this formula; just go ahead and solve linear equations as they arise. If you feel you need practice, see the exercises at the end of the section.

Solution of Quadratic Equations

By definition a quadratic equation has the form

$$ax^2 + bx + c = 0. \qquad \text{\textit{a}, \textit{b}, and \textit{c} are fixed numbers and } a \neq 0.[4]$$

The solutions of this equation are also called the **roots** of $ax^2 + bx + c$. We're assuming that you saw quadratic equations somewhere in high school but may be a

---

[4] What happens if $a = 0$?

little hazy about the details of their solution. There are two ways of solving these equations—one works sometimes, and the other works every time.

Solving Quadratic Equations by Factoring (works sometimes)

If we can factor[5] the left-hand side of a quadratic equation $ax^2 + bx + c = 0$, we can solve the equation by setting each factor equal to zero.

Quick Examples

5. $x^2 + 7x + 10 = 0$

   $(x + 5)(x + 2) = 0$ Factor the left-hand side.

   $x + 5 = 0$ or $x + 2 = 0$ If a product is zero, one or both factors are zero.

   Solutions: $x = -5$ and $x = -2$

6. $2x^2 - 5x - 12 = 0$

   $(2x + 3)(x - 4) = 0$ Factor the left-hand side.

   $2x + 3 = 0$ or $x - 4 = 0$

   Solutions: $x = -\dfrac{3}{2}$ and $x = 4$

Test for Factoring

The quadratic $ax^2 + bx + c$, with a, b, and c being integers (whole numbers), factors into an expression of the form $(rx + s)(tx + u)$ with r, s, t, and u integers precisely when the quantity $b^2 - 4ac$ is a perfect square. (That is, it is the square of an integer.) If this happens, we say that the quadratic **factors over the integers**.

Quick Examples

7. $x^2 + x + 1$ has $a = 1$, $b = 1$, and $c = 1$, so $b^2 - 4ac = -3$, which is not a perfect square. Therefore, this quadratic does not factor over the integers.

8. $2x^2 - 5x - 12$ has $a = 2$, $b = -5$, and $c = -12$, so $b^2 - 4ac = 121$. Because $121 = 11^2$, this quadratic does factor over the integers. (We factored it above.)

Solving Quadratic Equations with the Quadratic Formula (works every time)

The solutions of the general quadratic $ax^2 + bx + c = 0$ $(a \neq 0)$ are given by

$$x = \frac{-b \pm \sqrt{b^2 - 4ac}}{2a}.$$

---

[5] See Section 0.3 for a review of how to factor quadratics.

We call the quantity $\Delta = b^2 - 4ac$ the **discriminant** of the quadratic (Δ is the Greek letter delta), and we have the following general rules:

- If Δ is positive, there are two distinct real solutions.

- If Δ is zero, there is only one real solution: $x = -\dfrac{b}{2a}$. (Why?)

- If Δ is negative, there are no real solutions.

Quick Examples

9. $2x^2 - 5x - 12 = 0$ has $a = 2$, $b = -5$, and $c = -12$.

$$x = \frac{-b \pm \sqrt{b^2 - 4ac}}{2a} = \frac{5 \pm \sqrt{25 + 96}}{4} = \frac{5 \pm \sqrt{121}}{4} = \frac{5 \pm 11}{4}$$

$$= \frac{16}{4} \text{ or } -\frac{6}{4} = 4 \text{ or } -\frac{3}{2} \qquad \text{Δ is positive in this example.}$$

10. $4x^2 = 12x - 9$ can be rewritten as $4x^2 - 12x + 9 = 0$, which has $a = 4$, $b = -12$, and $c = 9$.

$$x = \frac{-b \pm \sqrt{b^2 - 4ac}}{2a} = \frac{12 \pm \sqrt{144 - 144}}{8} = \frac{12 \pm 0}{8} = \frac{12}{8} = \frac{3}{2}$$

$\text{$\Delta$ is zero in this example.}$

11. $x^2 + 2x - 1 = 0$ has $a = 1$, $b = 2$, and $c = -1$.

$$x = \frac{-b \pm \sqrt{b^2 - 4ac}}{2a} = \frac{-2 \pm \sqrt{8}}{2} = \frac{-2 \pm 2\sqrt{2}}{2} = -1 \pm \sqrt{2}$$

The two solutions are $x = -1 + \sqrt{2} = 0.414\ldots$ and
$x = -1 - \sqrt{2} = -2.414\ldots$. $\text{$\Delta$ is positive in this example.}$

12. $x^2 + x + 1 = 0$ has $a = 1$, $b = 1$, and $c = 1$. Because $\Delta = -3$ is negative, there are no real solutions. $\text{$\Delta$ is negative in this example.}$

Q: *This is all very useful, but where does the quadratic formula come from?*

A: To see where it comes from, we will solve a general quadratic equation using brute force. Start with the general quadratic equation,

$$ax^2 + bx + c = 0.$$

First, divide out the nonzero number a to get

$$x^2 + \frac{bx}{a} + \frac{c}{a} = 0.$$

Now we **complete the square**: Add and subtract the quantity $\dfrac{b^2}{4a^2}$ to get

$$x^2 + \frac{bx}{a} + \frac{b^2}{4a^2} - \frac{b^2}{4a^2} + \frac{c}{a} = 0.$$

We do this to get the first three terms to factor as a perfect square:

$$\left(x + \frac{b}{2a}\right)^2 - \frac{b^2}{4a^2} + \frac{c}{a} = 0.$$

(Check this by multiplying out.) Adding $\dfrac{b^2}{4a^2} - \dfrac{c}{a}$ to both sides gives

$$\left(x + \frac{b}{2a}\right)^2 = \frac{b^2}{4a^2} - \frac{c}{a} = \frac{b^2 - 4ac}{4a^2}.$$

Taking square roots gives

$$x + \frac{b}{2a} = \frac{\pm\sqrt{b^2 - 4ac}}{2a}.$$

Finally, adding $-\dfrac{b}{2a}$ to both sides yields the result

$$x = -\frac{b}{2a} + \frac{\pm\sqrt{b^2 - 4ac}}{2a}$$

or

$$x = \frac{-b \pm \sqrt{b^2 - 4ac}}{2a}.$$

Solution of Cubic Equations

By definition, a cubic equation can be written in the form

$$ax^3 + bx^2 + cx + d = 0. \qquad \textcolor{teal}{a, b, c, \text{ and } d \text{ are fixed numbers, and } a \neq 0.}$$

Now we get into something of a bind. Although there is a perfectly respectable formula for the solutions, it is very complicated and involves the use of complex numbers rather heavily.[6] So we discuss instead a much simpler method that *sometimes* works nicely. Here is the method in a nutshell.

Solving Cubics by Finding One Factor

Start with a given cubic equation $ax^3 + bx^2 + cx + d = 0$.

Step 1 By trial and error, find one solution $x = s$. If a, b, c, and d are integers, the only possible *rational* solutions[7] are those of the form $s = \pm$ (factor of d)/(factor of a).

Step 2 It will now be possible to factor the cubic as

$$ax^3 + bx^2 + cx + d = (x - s)(ax^2 + ex + f) = 0.$$

To find $ax^2 + ex + f$, divide the cubic by $x - s$, using long division.[8]

---

[6] It was when this formula was discovered in the sixteenth century that complex numbers were first taken seriously. Although we would like to show you the formula, it is too large to fit in this footnote.

[7] There may be *irrational* solutions, however; for example, $x^3 - 2 = 0$ has the single solution $x = \sqrt[3]{2}$.

[8] Alternatively, use synthetic division, a shortcut that would take us too far afield to describe.

Step 3 The factored equation says that either $x - s = 0$ or $ax^2 + ex + f = 0$. We already know that s is a solution, and now we see that the other solutions are the roots of the quadratic. Note that this quadratic may or may not have any real solutions, as usual.

> ### Quick Example
>
> 13. To solve the cubic $x^3 - x^2 + x - 1 = 0$, we first find a single solution. Here, $a = 1$ and $d = -1$. Because the only factors of ± 1 are ± 1, the only possible rational solutions are $x = \pm 1$. By substitution we see that $x = 1$ is a solution. Thus, $(x - 1)$ is a factor. Dividing by $(x - 1)$ yields the quotient $(x^2 + 1)$. Thus,
>
> $$x^3 - x^2 + x - 1 = (x - 1)(x^2 + 1) = 0,$$
>
> so either $x - 1 = 0$ or $x^2 + 1 = 0$.
>
> Because the discriminant of the quadratic $x^2 + 1$ is negative, we don't get any real solutions from $x^2 + 1 = 0$, so the only real solution is $x = 1$.

Possible Outcomes When Solving a Cubic Equation

If you consider all the cases, there are three possible outcomes when solving a cubic equation:

1. One real solution (as in Quick Example 13)
2. Two real solutions (try, for example, $x^3 + x^2 - x - 1 = 0$)
3. Three real solutions (see the next example)

EXAMPLE 1 Solving a Cubic

Solve the cubic $2x^3 - 3x^2 - 17x + 30 = 0$.

Solution First, we look for a single solution. Here, $a = 2$ and $d = 30$. The factors of a are ± 1 and ± 2, and the factors of d are ± 1, ± 2, ± 3, ± 5, ± 6, ± 10, ± 15, and ± 30. This gives us a large number of possible ratios: ± 1, ± 2, ± 3, ± 5, ± 6, ± 10, ± 15, ± 30, $\pm 1/2$, $\pm 3/2$, $\pm 5/2$, $\pm 15/2$. Undaunted, we first try $x = 1$ and $x = -1$, getting nowhere. So we move on to $x = 2$, and we hit the jackpot, because substituting $x = 2$ gives $16 - 12 - 34 + 30 = 0$. Thus, $(x - 2)$ is a factor. Dividing yields the quotient $2x^2 + x - 15$. Here is the calculation:

$$
\begin{array}{r}
2x^2 + x - 15 \\
x - 2 \overline{\smash{\big)}\ 2x^3 - 3x^2 - 17x + 30} \\
\underline{2x^3 - 4x^2} \\
x^2 - 17x \\
\underline{x^2 - 2x} \\
-15x + 30 \\
\underline{-15x + 30} \\
0.
\end{array}
$$

Thus,

$$2x^3 - 3x^2 - 17x + 30 = (x - 2)(2x^2 + x - 15) = 0.$$

Setting the factors equal to zero gives either $x - 2 = 0$ or $2x^2 + x - 15 = 0$. We could solve the quadratic using the quadratic formula, but luckily, we notice that it factors as

$$2x^2 + x - 15 = (x + 3)(2x - 5).$$

Thus, the solutions are $x = 2$, $x = -3$, and $x = 5/2$.

Solution of Higher Order Polynomial Equations

Logically speaking, our next step should be a discussion of quartics, then quintics (fifth degree equations), and so on forever. Well, we've got to stop somewhere, and cubics may be as good a place as any. On the other hand, since we've gotten this far, we ought to at least tell you what is known about higher order polynomials.

Quartics Just as in the case of cubics, there is a formula to find the solutions of quartics.[9]

Quintics and Beyond All good things must come to an end, we're afraid. It turns out that there is no "quintic formula." In other words, there is no single algebraic formula or collection of algebraic formulas that gives the solutions to all quintics. This question was settled by the Norwegian mathematician Niels Henrik Abel in 1824 after almost 300 years of controversy about this question. (In fact, several notable mathematicians had previously claimed to have devised formulas for solving the quintic, but these were all shot down by other mathematicians—this being one of the favorite pastimes of practitioners of our art.) The same negative answer applies to polynomial equations of degree 6 and higher. It's not that these equations don't have solutions; it's just that the solutions can't be found by using algebraic formulas.[10] However, there are certain special classes of polynomial equations that can be solved with algebraic methods. The way of identifying such equations was discovered around 1829 by the French mathematician Évariste Galois.[11]

----

[9] See, for example, *First Course in the Theory of Equations* by L. E. Dickson (New York: Wiley, 1922) or *Modern Algebra* by B. L. van der Waerden (New York: Frederick Ungar, 1953).

[10] What we mean by an "algebraic formula" is a formula in the coefficients using the operations of addition, subtraction, multiplication, division, and the taking of radicals. Mathematicians call the use of such formulas in solving polynomial equations "solution by radicals." If you were a math major, you would eventually go on to study this under the heading of Galois theory.

[11] Both Abel (1802–1829) and Galois (1811–1832) died young. Abel died of tuberculosis at the age of 26. Galois was killed in a duel at the age of 20.

0.5 EXERCISES

Solve the equations in Exercises 1–12 for x (mentally, if possible).

1. $x + 1 = 0$

2. $x - 3 = 1$

3. $-x + 5 = 0$

4. $2x + 4 = 1$

5. $4x - 5 = 8$

6. $\dfrac{3}{4}x + 1 = 0$

7. $7x + 55 = 98$

8. $3x + 1 = x$

9. $x + 1 = 2x + 2$

10. $x + 1 = 3x + 1$

11. $ax + b = c$ $(a \neq 0)$

12. $x - 1 = cx + d$ $(c \neq 1)$

By any method, determine all possible real solutions of each equation in Exercises 13–30. Check your answers by substitution.

13. $2x^2 + 7x - 4 = 0$

14. $x^2 + x + 1 = 0$

15. $x^2 - x + 1 = 0$

16. $2x^2 - 4x + 3 = 0$

17. $2x^2 - 5 = 0$

18. $3x^2 - 1 = 0$

19. $-x^2 - 2x - 1 = 0$

20. $2x^2 - x - 3 = 0$

21. $\frac{1}{2}x^2 - x - \frac{3}{2} = 0$

22. $-\frac{1}{2}x^2 - \frac{1}{2}x + 1 = 0$

23. $x^2 - x = 1$

24. $16x^2 = -24x - 9$

25. $x = 2 - \frac{1}{x}$

26. $x + 4 = \frac{1}{x - 2}$

27. $x^4 - 10x^2 + 9 = 0$

28. $x^4 - 2x^2 + 1 = 0$

29. $x^4 + x^2 - 1 = 0$

30. $x^3 + 2x^2 + x = 0$

Find all possible real solutions of each equation in Exercises 31–44.

31. $x^3 + 6x^2 + 11x + 6 = 0$

32. $x^3 - 6x^2 + 12x - 8 = 0$

33. $x^3 + 4x^2 + 4x + 3 = 0$

34. $y^3 + 64 = 0$

35. $x^3 - 1 = 0$

36. $x^3 - 27 = 0$

37. $y^3 + 3y^2 + 3y + 2 = 0$

38. $y^3 - 2y^2 - 2y - 3 = 0$

39. $x^3 - x^2 - 5x + 5 = 0$

40. $x^3 - x^2 - 3x + 3 = 0$

41. $2x^6 - x^4 - 2x^2 + 1 = 0$

42. $3x^6 - x^4 - 12x^2 + 4 = 0$

43. $(x^2 + 3x + 2)(x^2 - 5x + 6) = 0$

44. $(x^2 - 4x + 4)^2(x^2 + 6x + 5)^3 = 0$

0.6   Solving Miscellaneous Equations

Equations often arise in calculus that are not polynomial equations of low degree. Many of these complicated-looking equations can be solved easily if you remember the following, which we used in the previous section.

Solving an Equation of the Form $P \cdot Q = 0$

If a product is equal to 0, then at least one of the factors must be 0. That is, if $P \cdot Q = 0$, then either $P = 0$ or $Q = 0$.

Quick Examples

1. $x^5 - 4x^3 = 0$

$\quad x^3(x^2 - 4) = 0$ Factor the left-hand side.

$\quad$ Either $x^3 = 0$ or $x^2 - 4 = 0$ Either $P = 0$ or $Q = 0$.

$\quad x = 0, 2$ or -2.   Solve the individual equations.

2. $(x^2 - 1)(x + 2) + (x^2 - 1)(x + 4) = 0$

$\quad (x^2 - 1)[(x + 2) + (x + 4)] = 0$ Factor the left-hand side.

$\quad (x^2 - 1)(2x + 6) = 0$

$\quad$ Either $x^2 - 1 = 0$ or $2x + 6 = 0$ Either $P = 0$ or $Q = 0$.

$\quad x = -3, -1,$ or 1.   Solve the individual equations.

EXAMPLE 1   **Solving by Factoring**

Solve $12x(x^2 - 4)^5(x^2 + 2)^6 + 12x(x^2 - 4)^6(x^2 + 2)^5 = 0$.

Solution We start by factoring the left-hand side:

$$12x(x^2 - 4)^5(x^2 + 2)^6 + 12x(x^2 - 4)^6(x^2 + 2)^5$$
$$= 12x(x^2 - 4)^5(x^2 + 2)^5[(x^2 + 2) + (x^2 - 4)]$$
$$= 12x(x^2 - 4)^5(x^2 + 2)^5(2x^2 - 2)$$
$$= 24x(x^2 - 4)^5(x^2 + 2)^5(x^2 - 1).$$

Setting this equal to 0, we get

$$24x(x^2 - 4)^5(x^2 + 2)^5(x^2 - 1) = 0,$$

which means that at least one of the factors of this product must be zero. It certainly cannot be the 24, but it could be the x: $x = 0$ is one solution. It could also be that

$$(x^2 - 4)^5 = 0$$

or

$$x^2 - 4 = 0,$$

which has solutions $x = \pm 2$. Could it be that $(x^2 + 2)^5 = 0$? If so, then $x^2 + 2 = 0$, but this is impossible because $x^2 + 2 \geq 2$ no matter what x is. Finally, it could be that $x^2 - 1 = 0$, which has solutions $x = \pm 1$. This gives us five solutions to the original equation:

$$x = -2, -1, 0, 1, \text{ or } 2.$$

EXAMPLE 2   **Solving by Factoring**

Solve $(x^2 - 1)(x^2 - 4) = 10$.

Solution Watch out! You may be tempted to say that $x^2 - 1 = 10$ or $x^2 - 4 = 10$, but this does not follow. If two numbers multiply to give you 10, what must they be? There are lots of possibilities: 2 and 5, 1 and 10, and $-500,000$ and -0.00002 are just a few. The fact that the left-hand side is factored is nearly useless to us if we want to solve this equation. What we will have to do is multiply out, bring the 10 over to the left, and hope that we can factor what we get. Here goes:

$$x^4 - 5x^2 + 4 = 10$$
$$x^4 - 5x^2 - 6 = 0$$
$$(x^2 - 6)(x^2 + 1) = 0.$$

(Here, we used a sometimes useful trick that we mentioned in Section 0.3: We treated x^2 like x and x^4 like x^2, so factoring $x^4 - 5x^2 - 6$ is essentially the same as factoring $x^2 - 5x - 6$.) *Now* we are allowed to say that one of the factors must be 0: $x^2 - 6 = 0$ has solutions $x = \pm\sqrt{6} = \pm 2.449\ldots$, and $x^2 + 1 = 0$ has no real solutions. Therefore, we get exactly two solutions: $x = \pm\sqrt{6} = \pm 2.449\ldots$.

To solve equations involving rational expressions, the following rule is very useful.

Solving an Equation of the Form $P/Q = 0$

If $\dfrac{P}{Q} = 0$, then $P = 0$.

How else could a fraction equal 0? If that is not convincing, multiply both sides by Q (which cannot be 0 if the quotient is defined).

Quick Example

3. $\dfrac{(x + 1)(x + 2)^2 - (x + 1)^2(x + 2)}{(x + 2)^4} = 0$

$(x + 1)(x + 2)^2 - (x + 1)^2(x + 2) = 0$ If $\frac{P}{Q} = 0$, then $P = 0$.

$(x + 1)(x + 2)[(x + 2) - (x + 1)] = 0$ Factor.

$(x + 1)(x + 2)(1) = 0$

Either $x + 1 = 0$ or $x + 2 = 0$,

$x = -1$ or $x = -2$

$x = -1$ $x = -2$ does not make sense in the original equation: It makes the denominator 0. So it is not a solution, and $x = -1$ is the only solution.

EXAMPLE 3   **Solving a Rational Equation**

Solve $1 - \dfrac{1}{x^2} = 0$.

Solution Write 1 as $\frac{1}{1}$ so that we now have a difference of two rational expressions:

$$\frac{1}{1} - \frac{1}{x^2} = 0.$$

To combine these, we can put both over a common denominator of x^2, which gives

$$\frac{x^2 - 1}{x^2} = 0.$$

Now we can set the numerator, $x^2 - 1$, equal to zero. Thus,

$$x^2 - 1 = 0,$$

so

$$(x - 1)(x + 1) = 0,$$

giving $x = \pm 1$.

➡ **Before we go on . . .** This equation could also have been solved by writing

$$1 = \frac{1}{x^2}$$

and then multiplying both sides by x^2. ■

EXAMPLE 4   **Another Rational Equation**

Solve $\dfrac{2x-1}{x} + \dfrac{3}{x-2} = 0$.

Solution We *could* first perform the addition on the left and then set the top equal to 0, but here is another approach. Subtracting the second expression from both sides gives

$$\frac{2x-1}{x} = \frac{-3}{x-2}.$$

Cross-multiplying [multiplying both sides by both denominators—that is, by $x(x-2)$] now gives

$$(2x-1)(x-2) = -3x,$$

so

$$2x^2 - 5x + 2 = -3x.$$

Adding $3x$ to both sides gives the quadratic equation

$$2x^2 - 2x + 2 = 0.$$

The discriminant is $(-2)^2 - 4 \cdot 2 \cdot 2 = -8 < 0$, so we conclude that there is no real solution.

➡ **Before we go on . . .** Notice that when we said that $(2x-1)(x-2) = -3x$, we were *not* allowed to conclude that $2x-1 = -3x$ or $x-2 = -3x$. ∎

EXAMPLE 5   **A Rational Equation with Radicals**

Solve $\dfrac{\left(2x\sqrt{x+1} - \frac{x^2}{\sqrt{x+1}}\right)}{x+1} = 0$.

Solution Setting the top equal to 0 gives

$$2x\sqrt{x+1} - \frac{x^2}{\sqrt{x+1}} = 0.$$

This still involves fractions. To get rid of the fractions, we could put everything over a common denominator $(\sqrt{x+1})$ and then set the top equal to 0, or we could multiply the whole equation by that common denominator in the first place to clear fractions. If we do the second, we get

$$2x(x+1) - x^2 = 0$$
$$2x^2 + 2x - x^2 = 0$$
$$x^2 + 2x = 0.$$

Factoring, we have

$$x(x+2) = 0,$$

so either $x = 0$ or $x + 2 = 0$, giving us $x = 0$ or $x = -2$. Again, one of these is not really a solution. The problem is that $x = -2$ cannot be substituted into $\sqrt{x+1}$, because we would then have to take the square root of -1, and we are not allowing ourselves to do that. Therefore, $x = 0$ is the only solution.

0.6   EXERCISES

Solve the equations in Exercises 1–26.

1. $x^4 - 3x^3 = 0$

2. $x^6 - 9x^4 = 0$

3. $x^4 - 4x^2 = -4$

4. $x^4 - x^2 = 6$

5. $(x + 1)(x + 2) + (x + 1)(x + 3) = 0$

6. $(x + 1)(x + 2)^2 + (x + 1)^2(x + 2) = 0$

7. $(x^2 + 1)^5(x + 3)^4 + (x^2 + 1)^6(x + 3)^3 = 0$

8. $10x(x^2 + 1)^4(x^3 + 1)^5 - 10x^2(x^2 + 1)^5(x^3 + 1)^4 = 0$

9. $(x^3 + 1)\sqrt{x + 1} - (x^3 + 1)^2\sqrt{x + 1} = 0$

10. $(x^2 + 1)\sqrt{x + 1} - \sqrt{(x + 1)^3} = 0$

11. $\sqrt{(x + 1)^3} + \sqrt{(x + 1)^5} = 0$

12. $(x^2 + 1)\sqrt[3]{(x + 1)^4} - \sqrt[3]{(x + 1)^7} = 0$

13. $(x + 1)^2(2x + 3) - (x + 1)(2x + 3)^2 = 0$

14. $(x^2 - 1)^2(x + 2)^3 - (x^2 - 1)^3(x + 2)^2 = 0$

15. $\dfrac{(x + 1)^2(x + 2)^3 - (x + 1)^3(x + 2)^2}{(x + 2)^6} = 0$

16. $\dfrac{6x(x^2 + 1)^2(x^2 + 2)^4 - 8x(x^2 + 1)^3(x^2 + 2)^3}{(x^2 + 2)^8} = 0$

17. $\dfrac{2(x^2 - 1)\sqrt{x^2 + 1} - \dfrac{x^4}{\sqrt{x^2+1}}}{x^2 + 1} = 0$

18. $\dfrac{4x\sqrt{x^3 - 1} - \dfrac{3x^4}{\sqrt{x^3-1}}}{x^3 - 1} = 0$

19. $x - \dfrac{1}{x} = 0$

20. $1 - \dfrac{4}{x^2} = 0$

21. $\dfrac{1}{x} - \dfrac{9}{x^3} = 0$

22. $\dfrac{1}{x^2} - \dfrac{1}{x + 1} = 0$

23. $\dfrac{x - 4}{x + 1} - \dfrac{x}{x - 1} = 0$

24. $\dfrac{2x - 3}{x - 1} - \dfrac{2x + 3}{x + 1} = 0$

25. $\dfrac{x + 4}{x + 1} + \dfrac{x + 4}{3x} = 0$

26. $\dfrac{2x - 3}{x} - \dfrac{2x - 3}{x + 1} = 0$

0.7   The Coordinate Plane

Q: *Just what is the xy-plane?*

A: The *xy*-plane is an infinite flat surface with two perpendicular lines, usually labeled the **x-axis** and **y-axis**. These axes are calibrated as shown in Figure 2. (Notice also how the plane is divided into four **quadrants**.)

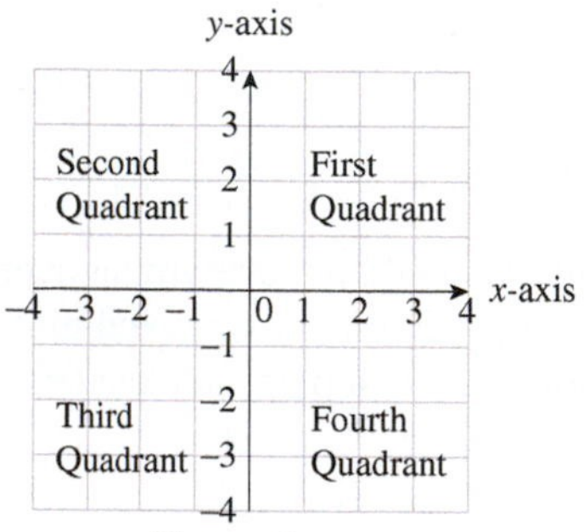

y-axis

Second Quadrant · First Quadrant · Third Quadrant · Fourth Quadrant

x-axis

The *xy*-plane

Figure 2

The *xy*-plane is nothing more than a very large—in fact, infinitely large—flat surface. The purpose of the axes is to allow us to locate specific positions, or **points**, on the plane, with the use of **coordinates**. (If Captain Picard wants to have himself beamed to a specific location, he must supply its coordinates, or he's in trouble.)

Q: *So how do we use coordinates to locate points?*

A: The rule is simple: Each point in the plane has two coordinates, an **x-coordinate** and a **y-coordinate**. These can be determined in two ways:

1. The *x*-coordinate measures a point's distance to the right or left of the *y*-axis. It is positive if the point is to the right of the axis, negative if it is to the left of the axis, and 0 if it is on the axis. The *y*-coordinate measures a point's distance above or below the *x*-axis. It is positive if the point is above the axis, negative if it is below the axis, and 0 if it is on the axis. Briefly, the *x*-coordinate tells us the *horizontal* position (distance left or right), and the *y*-coordinate tells us the *vertical* position (height).

2. Given a point P, we get its x-coordinate by drawing a vertical line from P and seeing where it intersects the x-axis. Similarly, we get the y-coordinate by extending a horizontal line from P and seeing where it intersects the y-axis.

This way of assigning coordinates to points in the plane is often called the system of **Cartesian** coordinates, in honor of the mathematician and philosopher René Descartes (1596–1650), who was the first to use them extensively.

Here are a few examples to help you review coordinates.

EXAMPLE 1 Coordinates of Points

a. Find the coordinates of the indicated points. (See Figure 3. The grid lines are placed at intervals of 1 unit.)

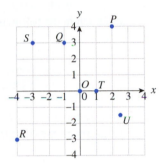

Figure 3

b. Locate the following points in the xy-plane:

$$A(2, 3), B(-4, 2), C(3, -2.5), D(0, -3), E(3.5, 0), F(-2.5, -1.5).$$

Solution

a. Taking the points in alphabetical order, we start with the origin O. This point has height zero and is also zero units to the right of the y-axis, so its coordinates are $(0, 0)$. Turning to P, dropping a vertical line gives $x = 2$ and extending a horizontal line gives $y = 4$. Thus, P has coordinates $(2, 4)$. For practice, determine the coordinates of the remaining points, and check your work against the list that follows:

$$Q(-1, 3), R(-4, -3), S(-3, 3), T(1, 0), U(2.5, -1.5).$$

b. To locate the given points, we start at the origin $(0, 0)$, and proceed as follows. (See Figure 4.)

To locate A, we move 2 units to the right and 3 up, as shown.

To locate B, we move -4 units to the right (that is, 4 to the *left*) and 2 up, as shown.

To locate C, we move 3 units right and 2.5 down.

We locate the remaining points in a similar way.

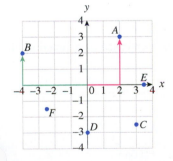

Figure 4

The Graph of an Equation

One of the more surprising developments of mathematics was the realization that equations, which are algebraic objects, can be represented by graphs, which are geometric objects. The kinds of equations that we have in mind are equations in x and y, such as

$$y = 4x - 1, \quad 2x^2 - y = 0, \quad y = 3x^2 + 1, \quad y = \sqrt{x - 1}.$$

The **graph** of an equation in the two variables x and y consists of all points (x, y) in the plane whose coordinates are solutions of the equation.

EXAMPLE 2 **Graph of an Equation**

Obtain the graph of the equation $y - x^2 = 0$.

Solution We can solve the equation for y to obtain $y = x^2$. Solutions can then be obtained by choosing values for x and then computing y by squaring the value of x, as shown in the following table:

x	-3	-2	-1	0	1	2	3
$y = x^2$	9	4	1	0	1	4	9

Plotting these points (x, y) gives the picture on the left side of Figure 5, suggesting the graph on the right in Figure 5.

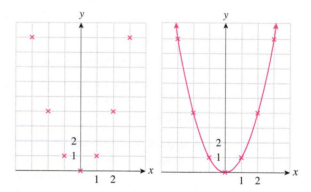

Figure 5

Distance

The distance between two points in the xy-plane can be expressed as a function of their coordinates, as follows.

Distance Formula

The distance between the points $P(x_1, y_1)$ and $Q(x_2, y_2)$ is

$$d = \sqrt{(x_2 - x_1)^2 + (y_2 - y_1)^2} = \sqrt{(\Delta x)^2 + (\Delta y)^2}.$$

Derivation

The distance d is shown in the figure below:

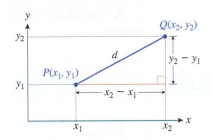

By the Pythagorean theorem applied to the right triangle shown, we get

$$d^2 = (x_2 - x_1)^2 + (y_2 - y_1)^2.$$

Taking square roots (d is a distance, so we take the positive square root), we get the distance formula. Notice that if we switch x_1 with x_2 or y_1 with y_2, we get the same result.

Quick Examples

1. The distance between the points $(3, -2)$ and $(-1, 1)$ is

$$d = \sqrt{(-1 - 3)^2 + (1 + 2)^2} = \sqrt{25} = 5.$$

2. The distance from (x, y) to the origin $(0, 0)$ is

$$d = \sqrt{(x - 0)^2 + (y - 0)^2} = \sqrt{x^2 + y^2}. \qquad \text{Distance to the origin}$$

The set of all points (x, y) whose distance from the origin $(0, 0)$ is a fixed quantity r is a circle centered at the origin with radius r. From the second Quick Example we get the following equation for the circle centered at the origin with radius r:

$$\sqrt{x^2 + y^2} = r. \qquad \text{Distance from the origin} = r$$

Squaring both sides gives the following equation.

Equation of the Circle of Radius r Centered at the Origin

$$x^2 + y^2 = r^2$$

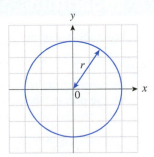

> **Quick Examples**
>
> **3.** The circle of radius 1 centered at the origin has equation $x^2 + y^2 = 1$.
> **4.** The circle of radius 2 centered at the origin has equation $x^2 + y^2 = 4$.

0.7 EXERCISES

1. Referring to the following figure, determine the coordinates of the indicated points as accurately as you can.

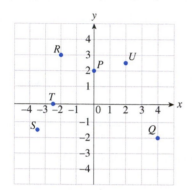

2. Referring to the following figure, determine the coordinates of the indicated points as accurately as you can.

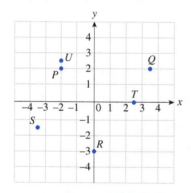

3. Graph the following points:

$P(4, 4), Q(4, -4), R(3, 0), S(4, 0.5), T(0.5, 2.5),$
$U(-2, 0), V(-4, 4)$

4. Graph the following points:

$P(4, -2), Q(2, -4), R(1, -3), S(-4, 2), T(2, -1),$
$U(-2, 0), V(-4, -4)$

Sketch the graphs of the equations in Exercises 5–12.

5. $x + y = 1$ **6.** $y - x = -1$

7. $2y - x^2 = 1$ **8.** $2y + \sqrt{x} = 1$

9. $xy = 4$ **10.** $x^2 y = -1$

11. $xy = x^2 + 1$ **12.** $xy = 2x^3 + 1$

In Exercises 13–16, find the distance between the given pairs of points.

13. $(1, -1)$ and $(2, -2)$ **14.** $(1, 0)$ and $(6, 1)$

15. $(a, 0)$ and $(0, b)$ **16.** (a, a) and (b, b)

17. Find the value of k such that $(1, k)$ is equidistant from $(0, 0)$ and $(2, 1)$.

18. Find the value of k such that (k, k) is equidistant from $(-1, 0)$ and $(0, 2)$.

19. Describe the set of points (x, y) such that $x^2 + y^2 = 9$.

20. Describe the set of points (x, y) such that $x^2 + y^2 = 0$.

0.8 Logarithms

From the equation

$$2^3 = 8$$

we can see that the power to which we need to raise 2 in order to get 8 is 3. We abbreviate the phrase "the power to which we need to raise 2 in order to get 8" as "$\log_2 8$." Thus, another way of writing the equation $2^3 = 8$ is

$$\log_2 8 = 3. \qquad \text{The power to which we need to raise 2 in order to get 8 is 3.}$$

This is read "The base 2 logarithm of 8 is 3" or "The log, base 2, of 8 is 3."

The general definition is as follows.

Base b Logarithm

The **base b logarithm of x**, $\log_b x$, is the power to which we need to raise b in order to get x. Symbolically,

$$\log_b x = y \qquad \text{means} \qquad b^y = x.$$

$\underset{\text{Logarithmic form}}{\log_b x = y} \qquad\qquad \underset{\text{Exponential form}}{b^y = x.}$

Quick Examples

The following table lists some exponential equations and their equivalent logarithmic forms:

Exponential Form	$10^3 = 1{,}000$	$4^2 = 16$	$5^1 = 5$	$7^0 = 1$	$4^{-2} = \dfrac{1}{16}$	$25^{1/2} = 5$
Logarithmic Form	$\log_{10} 1{,}000 = 3$	$\log_4 16 = 2$	$\log_5 5 = 1$	$\log_7 1 = 0$	$\log_4 \dfrac{1}{16} = -2$	$\log_{25} 5 = \dfrac{1}{2}$

1. $\log_2 8 = $ the power to which we need to raise 2 in order to get 8.

   Because $2^{\boxed{3}} = 8$, this power is 3, so $\log_2 8 = 3$.

2. $\log_2 16 = $ the power to which we need to raise 2 in order to get 16.

   Because $2^{\boxed{4}} = 16$, this power is 4, so $\log_2 16 = 4$.

3. $\log_2 2 = $ the power to which we need to raise 2 in order to get 2.

   Because $2^{\boxed{1}} = 2$, this power is 1, so $\log_2 2 = 1$.

4. $\log_3 9 = $ the power to which we need to raise 3 in order to get 9.

   Because $3^{\boxed{2}} = 9$, this power is 2, so $\log_3 9 = 2$.

5. $\log_3 27 = $ the power to which we need to raise 3 in order to get 27.

   Because $3^{\boxed{3}} = 27$, this power is 3, so $\log_3 27 = 3$.

6. $\log_3 3 = $ the power to which we need to raise 3 in order to get 3.

   Because $3^{\boxed{1}} = 3$, this power is 1, so $\log_3 3 = 1$.

7. $\log_{10} 10{,}000 = $ the power to which we need to raise 10 in order to get 10,000.

   Because $10^{\boxed{4}} = 10{,}000$, this power is 4, so $\log_{10} 10{,}000 = 4$.

8. $\log_{10} 1 = $ the power to which we need to raise 10 in order to get 1.

   Because $10^{\boxed{0}} = 1$, this power is 0, so $\log_{10} 1 = 0$.

9. $\log_2 1 = $ the power to which we need to raise 2 in order to get 1.

   Because $2^{\boxed{0}} = 1$, this power is 0, so $\log_2 1 = 0$.

10. $\log_3 1 = $ the power to which we need to raise 3 in order to get 1.

    Because $3^{\boxed{0}} = 1$, this power is 0, so $\log_3 1 = 0$.

11. $\log_2 \dfrac{1}{2}$ is the power to which we need to raise 2 in order to get $\dfrac{1}{2}$.

Because $2^{\boxed{-1}} = \dfrac{1}{2}$, this power is -1, so $\log_2 \dfrac{1}{2} = -1$.

12. $\log_2 \dfrac{1}{4}$ is the power to which we need to raise 2 in order to get $\dfrac{1}{4}$.

Because $2^{\boxed{-2}} = \dfrac{1}{4}$, this power is -2, so $\log_2 \dfrac{1}{4} = -2$.

13. $\log_3 \dfrac{1}{27}$ is the power to which we need to raise 3 in order to get $\dfrac{1}{27}$.

Because $3^{\boxed{-3}} = \dfrac{1}{27}$, this power is -3, so $\log_3 \dfrac{1}{27} = -3$.

Note The number $\log_b x$ is defined only if b and x are both positive and $b \neq 1$. For example, it is impossible to compute $\log_3(-9)$ (because there is no power of 3 that equals -9) or $\log_1 2$ (because there is no power of 1 that equals 2). ∎

The logarithm with base 10, $\log_{10}$, is called the **common logarithm** and is often written as "log" without the base.

Quick Examples

Logarithmic Form	**Exponential Form**
14. $\log 10{,}000 = \log_{10} 10{,}000 = 4$	$10^4 = 10{,}000$
15. $\log 10 = \log_{10} 10 = 1$	$10^1 = 10$
16. $\log \dfrac{1}{10{,}000} = -4$	$10^{-4} = \dfrac{1}{10{,}000}$

Algebraic Properties of Logarithms

The following are important properties of logarithms. The fourth property is suggested by some of the Quick Examples above.

Logarithm Identities

The following identities hold for all positive bases $a \neq 1$ and $b \neq 1$, all positive numbers x and y, and every real number r. These identities follow from the laws of exponents.

Identity	**Quick Examples**
1. $\log_b(xy) = \log_b x + \log_b y$	$\log_2 16 = \log_2 8 + \log_2 2$
	$\log_b 5 + \log_b 6 = \log_b 30$
2. $\log_b\left(\dfrac{x}{y}\right) = \log_b x - \log_b y$	$\log_2\left(\dfrac{5}{3}\right) = \log_2 5 - \log_2 3$
	$\log_b 5 - \log_b 6 = \log_b\left(\dfrac{5}{6}\right)$
3. $\log_b(x^r) = r \log_b x$	$\log_2(6^5) = 5 \log_2 6$
	$3 \log_b x = \log_b x^3$
4. $\log_b b = 1$ and $\log_b 1 = 0$	$\log_2 2 = 1; \log_3 3 = 1$
	$\log 10 = 1; \log_2 1 = 0; \log 1 = 0$
5. $\log_b\left(\dfrac{1}{x}\right) = -\log_b x$	$\log_2\left(\dfrac{1}{3}\right) = -\log_2 3$
6. $\log_b x = \dfrac{\log_a x}{\log_a b}$	$\log_2 5 = \dfrac{\log_{10} 5}{\log_{10} 2} = \dfrac{\log 5}{\log 2}$

EXAMPLE 1 **Using the Logarithm Identities**

Let $a = \log 2$, $b = \log 3$, and $c = \log 5$. Write the following in terms of a, b, and c:

a. $\log 6$ **b.** $\log 15$ **c.** $\log 30$ **d.** $\log 9$

e. $\log \dfrac{1}{9}$ **f.** $\log 1.5$ **g.** $\log 32$ **h.** $\log \dfrac{8}{81}$

Solution

a. We recognize 6 as the product of 2 and 3, so we use identity (1):

$$\log 6 = \log(2 \times 3) = \log 2 + \log 3 = a + b.$$

b. As in part (a), we recognize 15 as the product of 3 and 5, so we use again identity (1):

$$\log 15 = \log(3 \times 5) = \log 3 + \log 5 = b + c.$$

c. 30 can be written as $2 \times 15 = 2 \times 3 \times 5$, so we use identity (1) twice:

$$\log 30 = \log(2 \times 15) = \log 2 + \log 15 = \log 2 + (\log 3 + \log 5) = a + b + c.$$

More simply, we can simplify the above steps and write

$$\log 30 = \log(2 \times 3 \times 5) = \log 2 + \log 3 + \log 5 = a + b + c.$$

d. We can think of 9 either as a product 3×3 or as a power 3^2. Let us use the second interpretation, which would call for identity (3):

$$\log 9 = \log(3^2) = 2 \log 3 = 2b.$$

e. For a reciprocal we can use identity (5):

$$\log \frac{1}{9} = -\log 9 = -2b. \qquad \text{Using the answer to part (d)}$$

f. We recognize 1.5 as the ratio $\frac{3}{2}$, so we use identity (3):

$$\log 1.5 = \log \frac{3}{2} = \log 3 - \log 2 = b - a.$$

g. We recognize 32 as 2^5, so we use identity (3):

$$\log 32 = \log(2^5) = 5 \log 2 = 5a.$$

h. As $\frac{8}{81}$ is a ratio, we start by using identity (2):

$$
\begin{aligned}
\log \frac{8}{81} &= \log 8 - \log 81 && \text{By identity (2)} \\
&= \log(2^3) - \log(3^4) && \\
&= 3 \log(2) - 4 \log(3) && \text{By identity (5)} \\
&= 3a - 4b.
\end{aligned}
$$

Using Logarithms to Solve for Unknowns in the Exponent

One important use of logarithms is to solve equations in which the unknown is in the exponent.

EXAMPLE 2   **Solving for Unknowns in the Exponent**

Solve the following equations:

a. $4^{-x} = \dfrac{1}{64}$ **b.** $10(1.005)^{3x} = 200$ **c.** $4 \cdot 3^{4x-2} = 81$

Solution

a. As the base of the exponent is 4, we start by taking the base 4 logarithm of both sides. (We could also solve it by taking the logarithm to *any* base, as we will illustrate in part (b).)

$$\log_4(4^{-x}) = \log_4\left(\frac{1}{64}\right).$$

The left-hand side is

$$
\begin{aligned}
\log_4(4^{-x}) &= -x \log_4(4) && \text{By identity (3)} \\
&= -x \cdot 1 = -x. && \log_4(4) = 1 \text{ by identity (4)}
\end{aligned}
$$

while the right-hand side is

$$
\begin{aligned}
\log_4\left(\frac{1}{64}\right) &= -\log_4(64) && \text{By identity (5)} \\
&= -\log_4(4^3) = -3 \log_4(4) && \log_4(4) = 1 \text{ by identity (3)} \\
&= -3 \cdot 1 = -3. && \text{By identity (4)}
\end{aligned}
$$

Equating the left- and right-hand sides thus gives

$$-x = -3,$$

so

$$x = 3.$$

Alternatively, we can translate the given equation from exponent form into logarithmic form:

$$4^{-x} = \frac{1}{64} \qquad \text{Exponent form}$$

$$\log_4\left(\frac{1}{64}\right) = -x. \qquad \text{Logarithmic form}$$

Thus,

$$x = -\log_4\left(\frac{1}{64}\right) = -(-3) = 3. \qquad \text{We calculated this logarithm above.}$$

b. We solve this problem by taking the common logarithm of both sides (see the comment in the solution to part (a)). We *could* take the base 1.005 logarithm of both sides, but this would lead to a solution in terms of logarithms to that base. On the other hand, common logarithms are readily computable on a calculator (there is usually a button for it), so we take the common logarithm of both sides instead. But first, we simplify by dividing both sides by 10:

$$1.005^{3x} = \frac{200}{10} = 20 \qquad \text{Divide both sides by 10.}$$

$$\log 1.005^{3x} = \log 20 \qquad \text{Take the log of both sides.}$$

$$3x \log 1.005 = \log 20 \qquad \text{Apply identity (3) on the left.}$$

$$x = \frac{\log 20}{3 \log 1.005} \approx 200.21. \qquad \text{Solve for } x, \text{ and approximate with a calculator.}$$

Rather than using a calculator to approximate the answer by a decimal, we could have left it in the exact form $\dfrac{\log 20}{3 \log 1.005}$.

c. As in part (b), we solve this problem by taking the common logarithm of both sides; taking the base 3 logarithm will result in a solution in terms of $\log_3(20.25)$. (Try it!) But first, we simplify by dividing both sides by 4:

$$3^{4x-2} = \frac{81}{4} \qquad \text{Divide both sides by 4.}$$

$$\log 3^{4x-2} = \log\left(\frac{81}{4}\right) \qquad \text{Take the log of both sides.}$$

$$(4x - 2)\log 3 = \log 81 - \log 4 \qquad \text{Apply identity (3) on the left and identity (2) on the right.}$$

$$(4x - 2) = \frac{\log 81 - \log 4}{\log 3} \qquad \text{Divide.}$$

$$4x = \frac{\log 81 - \log 4}{\log 3} + 2 \qquad \text{Add 2 to both sides.}$$

$$x = \frac{1}{4}\left(\frac{\log 81 - \log 4}{\log 3} + 2\right) \approx 1.1845.$$

0.8   EXERCISES

In Exercises 1–4, complete the given tables.

1.

Exponential Form	$10^2 = 100$	$4^3 = 64$	$4^4 = 256$	$0.45^0 = 1$	$8^{1/2} = 2\sqrt{2}$	$4^{-3} = \dfrac{1}{64}$
Logarithmic Form						

2.

Exponential Form	$10^1 = 10$	$5^5 = 3{,}125$	$6^3 = 216$	$0.5^3 = 0.125$	$4^{1/4} = \sqrt{2}$	$3^{-4} = \dfrac{1}{81}$
Logarithmic Form						

3.

Exponential Form						
Logarithmic Form	$\log_{0.3} 0.09 = 2$	$\log_{1/2} 1 = 0$	$\log_{10} 0.001 = -3$	$\log_9 \dfrac{1}{81} = -2$	$\log_2 1{,}024 = 10$	$\log_{64} \dfrac{1}{4} = -\dfrac{1}{3}$

4.

Exponential Form						
Logarithmic Form	$\log_2 \dfrac{1}{2} = -1$	$\log_{10} 100{,}000 = 5$	$\log_{10} 0.00001 = -5$	$\log_2 2{,}048 = 11$	$\log_{16} \dfrac{1}{2} = -\dfrac{1}{4}$	$\log_{0.25} 0.0625 = 2$

In Exercises 5–16, evaluate the given quantity.

5. $\log_4 16$

6. $\log_5 125$

7. $\log_5 \dfrac{1}{25}$

8. $\log_4 \dfrac{1}{64}$

9. $\log 100{,}000$

10. $\log 1{,}000$

11. $\log_{16} 16$

12. $\log_{1/2} \dfrac{1}{2}$

13. $\log_4 \dfrac{1}{16}$

14. $\log_2 \dfrac{1}{8}$

15. $\log_2 \sqrt{2}$

16. $\log_4 \sqrt{2}$

In Exercises 17–28, use the logarithm identities to obtain the missing quantity.

17. $\log_b 3 + \log_b 4 = \log_b \boxed{}$

18. $\log_b 3 - \log_b 4 = \log_b \boxed{}$

19. $\log_b 2 - \log_b 5 - \log_b 4 = \log_b \boxed{}$

20. $\log_b 3 + \log_b 2 - \log_b 7 = \log_b \boxed{}$

21. $\log_b 3 - 3\log_b 2 = \log_b \boxed{}$

22. $3\log_b 2 + 2\log_b 3 = \log_b \boxed{}$

23. $4\log_b x + 5\log_b y = \log_b \boxed{}$

24. $3\log_b p - 2\log_b q = \log_b \boxed{}$

25. $2\log_b x + 3\log_b y - 4\log_b z = \log_b \boxed{}$

26. $4\log_b p - 3\log_b q - 2\log_b r = \log_b \boxed{}$

27. $x\log_b 2 - 2\log_b x = \log_b \boxed{}$

28. $p\log_b q + q\log_b p = \log_b \boxed{}$

Let $a = \log 2$, $b = \log 3$, and $c = \log 7$. In Exercises 29–46, use the logarithm identities to express the given quantity in terms of a, b, and c.

29. $\log 21$

30. $\log 14$

31. $\log 42$

32. $\log 28$

33. $\log\left(\dfrac{1}{7}\right)$

34. $\log\left(\dfrac{1}{3}\right)$

35. $\log\left(\dfrac{2}{3}\right)$

36. $\log\left(\dfrac{7}{2}\right)$

37. $\log\left(\dfrac{4}{7}\right)$

38. $\log\left(\dfrac{2}{9}\right)$

39. $\log 16$

40. $\log 81$

41. $\log 0.03$

42. $\log 7{,}000$

43. $\log 5$

44. $\log 25$

45. $\log \sqrt{7}$

46. $\log\left(\dfrac{2}{\sqrt{3}}\right)$

In Exercises 47–56, solve the given equation for the indicated variable.

47. $4 = 2^x$

48. $81 = 3^x$

49. $27 = 3^{2x-1}$

50. $4^{2-3x} = 256$

51. $5^{-x+1} = \dfrac{1}{125}$

52. $3^{x^2-12} = \dfrac{1}{27}$

53. $120 = 50(2^{3t})$
(Round the answer to four decimal places.)

54. $1{,}000 = 500(1.1^{2t})$
(Round the answer to four decimal places.)

55. $1{,}000 = 300(1.3^{4t-1})$
(Round the answer to four decimal places.)

56. $10{,}000 = 700(1.04^{3t+1})$
(Round the answer to four decimal places.)

1

FUNCTIONS AND APPLICATIONS

CASE STUDY

Modeling Spending on Mobile Advertising

You are the new director of *Impact Advertising Inc.*'s mobile division, which has enjoyed a steady 0.25% of the worldwide mobile advertising market. You have drawn up an ambitious proposal to expand your division in light of your anticipation that mobile advertising will continue to skyrocket. The VP for Financial Affairs feels that current projections (based on a linear model) do not warrant the level of expansion you propose.

How can you persuade the VP that those projections do not fit the data convincingly?

Es sarawuth/Shutterstock.com; (inset) Odua Images/Shutterstock.com

Introduction

To analyze recent trends in spending on mobile advertising and to make reasonable projections, we need a mathematical model of this spending. Where do we start? To apply mathematics to real-world situations like this, we need a good understanding of basic mathematical concepts. Perhaps the most fundamental of these concepts is that of a function: a relationship that shows how one quantity depends on another. Functions may be described numerically and, often, algebraically. They can also be described graphically—a viewpoint that is extremely useful.

The simplest functions—the ones with the simplest formulas and the simplest graphs—are linear functions. Because of their simplicity, they are also among the most useful functions and can often be used to model real-world situations, at least over short periods of time. In discussing linear functions, we will meet the concepts of slope and rate of change, which are the starting point of the mathematics of change.

In the last section of this chapter, we discuss *simple linear regression*: construction of linear functions that best fit given collections of data. Regression is used extensively in applied mathematics, statistics, and quantitative methods in business. The inclusion of regression utilities in computer spreadsheets like Excel makes this powerful mathematical tool readily available for anyone to use.

Precalculus Review

For this chapter, you should be familiar with real numbers and intervals (see Section 0.1) and exponents and radicals (see Section 0.2).

1.1 Functions from the Numerical, Algebraic, and Graphical Viewpoints

Introduction

The following table gives accumulated sales of **Nintendo** Wii game consoles from the end of December 2006 to the end of December 2014:[1]

Time t (years since Dec. 2006)	0	1	2	3	4	5	6	7	8
Accumulated Sales f (millions of consoles)	3	9	27	53	73	88	98	105	109

The table tells us that a total of around 3 million consoles had been sold by the end of December 2006, a total of 9 million had been sold by the end of the following year, and so on.

Let's write $f(0)$ for the accumulated sales (in millions of consoles) when $t = 0$, $f(1)$ for the accumulated sales when $t = 1$, and so on. (We read $f(0)$ as "f of 0.") Thus, $f(0) = 3$, $f(1) = 9$, $f(2) = 27, \ldots, f(8) = 109$. In general, we write $f(t)$ for the accumulated sales (in millions of consoles) at time t. We call f a **function** of the variable t, meaning that for each value of t from 0 through 8, f gives us a single corresponding number $f(t)$ that tells us how many consoles were sold up to time t.

---

[1] Accumulated sales of Wii and WiiU consoles since the Wii was first released in November 2006.
Sources: www.statista.com, Nintendo Inc.

Time
t

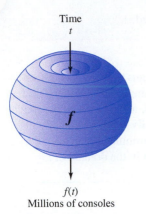

f(*t*)
Millions of consoles

Figure 1

✱ There is nothing special about the choice of the letter *t* as the argument of a function; we could have used any letter whatsoever to represent time. In general, it is customary to use *t* for time and *x* for (almost) anything else, as in *f*(*x*). Likewise, there is nothing special about our choice of the letter *f* for the function name except for tradition.

In general, we think of a function as a way of producing new objects from old ones. The functions we deal with in this text produce new numbers from old numbers. The numbers we have in mind are the real numbers, including not only positive and negative integers and fractions but also numbers like $\sqrt{2}$ or π. (See Chapter 0 for more on real numbers.) For this reason, the functions we use are called **real-valued functions of a real variable**. For example, the function *f* takes the time in years since December 2006 as input and returns the accumulated console sales at that time (Figure 1).

The letter *t* in *f*(*t*) is called the **argument** of the function *f*.✱ A function may be specified in several different ways. Here, we have specified the function *f* **numerically** by giving the values of the function for a number of values of the argument, as in the preceding table.

Q: *For which values of the argument t does it make sense to ask for f(t)? In other words, for which times t is f(t) defined?*

A: Because our function *f* refers to the accumulated sales over the period December 2006 ($t = 0$) through December 2014 ($t = 8$), *f*(*t*) should reasonably be defined when *t* is any number between 0 and 8, that is, when $0 \le t \le 8$. Using interval notation (see Chapter 0), we can say that *f*(*t*) is defined when *t* is in the interval $[0, 8]$.

The set of values of the argument for which a function is defined is called its **domain** and is a necessary part of the definition of the function. Notice that the preceding table gives the value of *f*(*t*) at only some of the infinitely many possible values in the domain $[0, 8]$. The domain of a function is not always specified explicitly; if no domain is specified for a function *f*, we take the domain to be the largest set of numbers *x* for which *f*(*x*) makes sense. This "largest possible domain" is sometimes called the **natural domain**.

The above Nintendo data can also be represented on a graph by plotting the given pairs of numbers $(t, f(t))$ in the *xy*-plane. (See Figure 2. We have connected successive points by line segments.) In general, the graph of a function *f* consists of all points $(x, f(x))$ in the plane with *x* in the domain of *f*.

In Figure 2 we specified the function *f* **graphically** by using a graph to display its values. Suppose now that we had only the graph without the table of data. We could use the graph to find approximate values of the function. For instance, to find *f*(2.5) from the graph, we do the following:

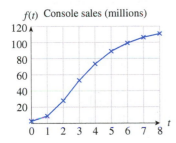

f(*t*) Console sales (millions)

Figure 2

1. Find the desired value of *t* at the bottom of the graph ($t = 2.5$ in this case).

2. Estimate the height (*y*-coordinate) of the corresponding point on the graph (around 40 in this case).

Thus, $f(2.5) \approx 40$ million consoles sold by the end of June 2009 ($t = 2.5$).†

In some cases we may be able to use an algebraic formula to calculate the values of a function, and we say that the function is specified **algebraically**. These are not the only ways in which a function can be specified; for instance, a function can sometimes be specified *verbally*, as in "Let *f*(*x*) be the number of miles my **Tesla** can go at 30 mph after being charged for *x* hours at the local station." Notice that any function can be represented graphically by plotting the points $(x, f(x))$ for a number of values of *x* in its domain.

† In a graphically defined function we can never know the *y*-coordinates of points exactly; no matter how accurately a graph is drawn, we can obtain only approximate values of the coordinates of points. That is why we have been using the word "estimate" rather than "calculate" and why we say $f(2.5) \approx 40$ rather than $f(2.5) = 40$.

Here is a summary of the terms we have just introduced.

Functions

A **real-valued function** f **of a real-valued variable** x assigns to each real number x in a specified set of numbers, called the **domain of** f, a single real number $f(x)$, read "f of x." The quantity x is called the **argument** of f, and $f(x)$ is called the **value of** f **at** x. A function is usually specified **numerically** using a table of values, **graphically** using a graph, or **algebraically** using a formula. The **graph of a function** consists of all points $(x, f(x))$ in the plane with x in the domain of f.

Quick Examples

Graph of c

Plotting the pairs $(t, c(t))$ gives the following graph:

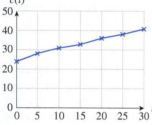

1. **A function specified numerically:** Take $c(t)$ to be the world emission of carbon dioxide in year t since 2000, represented by the following table:[2]

Year t (year since 2000)	CO_2 Emissions $c(t)$ (billions of metric tons)
0	24
5	28
10	31
15	33
20	36
25	38
30	41

The domain of c is $[0, 30]$, the argument is t, the number of years since 2000, and the values $c(t)$ give the world production of carbon dioxide in a given year. The graph of c is shown on the left. Some values of c are

$c(0) = 24$ 24 billion metric tons of CO_2 were produced in 2000.

$c(10) = 31$ 31 billion metric tons of CO_2 were produced in 2010.

$c(30) = 41.$ 41 billion metric tons of CO_2 were projected to be produced in 2030.

2. **A function specified graphically:** Take $q(p)$ to be the number of smartphones (in millions) sold worldwide in a year when the price is set at p dollars, as represented by the following graph:[3]

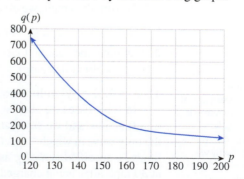

---

[2] Figures for 2015 and later are projections. Source: Energy Information Administration (EIA), www.eia.doe.gov.

[3] Based on prices and sales figures in www.businessweek.com and http://techcrunch.com.

The domain of q is $[120, 200]$, the argument is p, the price of a smartphone in dollars, and the values $q(p)$ give the number (in millions) of smartphones sold worldwide in a year at a given price.✱ Some values of q are

$$q(160) \approx 200$$ When the price is \$160, about 200 million smartphones are sold worldwide in a year.

$$q(130) \approx 550.$$ When the price is \$130, about 550 million smartphones are sold worldwide in a year.

3. **A function specified algebraically:** Let $f(x) = \dfrac{1}{x}$. The function f is specified algebraically. The argument is x, and the dependent variable is f. The natural domain of f consists of all real numbers except zero because $f(x)$ makes sense for all values of x other than $x = 0$. Some specific values of f are

$$f(2) = \frac{1}{2}, \quad f(3) = \frac{1}{3}, \quad f(-1) = \frac{1}{-1} = -1,$$

$f(0)$ is not defined because 0 is not in the domain of f.

4. **The graph of a function:** Let $f(x) = x^2$, with domain the set of all real numbers. To draw the graph of f, first choose some convenient values of x in the domain and compute the corresponding y-coordinates $f(x)$:

x	-3	-2	-1	0	1	2	3
$f(x) = x^2$	9	4	1	0	1	4	9

Plotting these points $(x, f(x))$ gives the picture on the left, suggesting the graph on the right.†

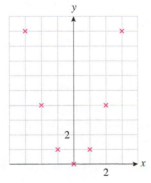
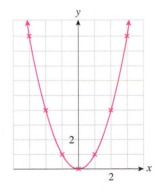

(This particular curve happens to be called a **parabola**, and its lowest point, at the origin, is called its **vertex**.)

EXAMPLE 1   iPod Sales

The total number of iPods sold by **Apple** in year x can be approximated by

$$f(x) = -2x^2 + 12x + 36 \text{ million iPods} \qquad (0 \le x \le 7),$$

where $x = 0$ represents 2005.[4]

a. What is the domain of f? Compute $f(0)$, $f(2)$, $f(4)$, and $f(6)$. What do these answers tell you about iPod sales? Is $f(-1)$ defined?

b. Compute $f(a)$, $f(-b)$, $f(a + h)$, and $f(a) + h$, assuming that the quantities a, $-b$, and $a + h$ are in the domain of f.

c. Sketch the graph of f. Does the shape of the curve suggest that iPod sales were accelerating or decelerating over the period 2005–2008?

Solution

a. The domain of f is the set of numbers x with $0 \le x \le 7$, that is, the interval $[0, 7]$. If we substitute 0 for x in the formula for $f(x)$, we get

$$f(0) = -2(0)^2 + 12(0) + 36 = 36. \qquad \text{Approximately 36 million iPods were sold in 2005.}$$

Similarly,

$$f(2) = -2(2)^2 + 12(2) + 36 = 52 \qquad \text{Approximately 52 million iPods were sold in 2007.}$$

$$f(4) = -2(4)^2 + 12(4) + 36 = 52 \qquad \text{Approximately 52 million iPods were sold in 2009.}$$

$$f(6) = -2(6)^2 + 12(6) + 36 = 36. \qquad \text{Approximately 36 million iPods were sold in 2011.}$$

As -1 is not in the domain of f, $f(-1)$ is not defined.

b. To find $f(a)$, we substitute a for x in the formula for $f(x)$ to get

$$f(a) = -2a^2 + 12a + 36. \qquad \text{Substitute } a \text{ for } x.$$

Similarly,

$$
\begin{aligned}
f(-b) &= -2(-b)^2 + 12(-b) + 36 & \text{Substitute } -b \text{ for } x. \\
&= -2b^2 - 12b + 36 & (-b)^2 = b^2 \\
f(a + h) &= -2(a + h)^2 + 12(a + h) + 36 & \text{Substitute } (a + h) \text{ for } x. \\
&= -2(a^2 + 2ah + h^2) + 12a + 12h + 36 & \text{Expand.} \\
&= -2a^2 - 4ah - 2h^2 + 12a + 12h + 36 \\
f(a) + h &= -2a^2 + 12a + 36 + h. & \text{Add } h \text{ to } f(a).
\end{aligned}
$$

Note how we placed parentheses around the quantities at which we evaluated the function. If we tried to do without any of these parentheses, we would likely get an error:

$$\text{Correct expression: } f(a + h) = -2(a + h)^2 + 12(a + h) + 36 \quad ✓$$
$$NOT -2a^2 + 12a + 36 + h \quad ✗$$

Also notice the distinction between $f(a + h)$ and $f(a) + h$: To find $f(a + h)$, we replace x by the quantity $(a + h)$; to find $f(a) + h$, we add h to $f(a)$.

---

[4] Sales are by calendar year. Source for data: Apple quarterly earnings reports at www.apple.com/investor.

c. To draw the graph of f, we plot points of the form $(x, f(x))$ for several values of x in the domain of f. Let us use the values at the integers from 0 to 7:

x	0	1	2	3	4	5	6	7
$f(x) = -2x^2 + 12x + 36$	36	46	52	54	52	46	36	22

Graphing these points gives the graph shown on the left in Figure 3, suggesting the curve shown on the right.

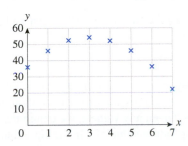 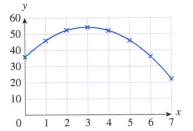

Figure 3

The period 2005–2008 is represented by the interval $[0, 3]$ on the x-axis, and the graph becomes less steep as we move from $x = 0$ to $x = 3$, suggesting that iPod sales were decelerating over the given period.

➡ **Before we go on . . .** The following table compares the value of f in Example 1 with the actual sales figures:

x	0	2	4	6
$f(x) = -2x^2 + 12x + 36$	36	52	52	36
Actual iPod Sales (millions)	32	53	52	39

The actual figures are stated here for only (some) integer values of x; for instance, $x = 4$ gives the sales in 2009. But what were, for instance, the sales in the year beginning in June 2009 ($x = 4.5$)? This is where our formula comes in handy: We can use the formula for f to **interpolate**—that is, to find sales at values of x between values that are stated:

$$f(4.5) = -2(4.5)^2 + 12(4.5) + 36 = 49.5 \text{ million iPods.}$$

We can also use the formula to **extrapolate**—that is, to predict sales at values of x *outside* the domain, say, for $x = 8$ (that is, sales in 2013):

$$f(8) = -2(8)^2 + 12(8) + 36 = 4 \text{ million iPods.}$$

As a general rule, extrapolation is far less reliable than interpolation. The extrapolated values depend heavily on the type of mathematical function used to model the data, so the same set of data can lead to vastly different extrapolated values. Further, the vagaries of the marketplace make the use of current data trends to predict the future problematic in the first place.

We call the algebraic function f an **algebraic model** of iPod sales because it uses an algebraic formula to model—or mathematically represent (approximately)—the annual sales. The particular kind of algebraic model we used is called a **quadratic model**. (See the end of this section for the names of some commonly used models.) ∎

Functions and Equations

To specify a function algebraically, we need to write down a defining equation, as in, say,

$$f(x) = 3x - 2.$$

If we replace the "$f(x)$" by "y" we get an equation with no explicit mention of any function:

$$y = 3x - 2. \qquad \text{An equation in two variables: } x \text{ and } y$$

Technically, $y = 3x - 2$ is an equation and not a function. However, an equation of this type, $y =$ expression in x, can be thought of as "specifying y as a function of x" as follows: Given any value x, we obtain the value of the function at x by calculating the corresponding value of y in the equation. So the value of the function at $x = 1$ is just

$$y = 3(1) - 2 = 1,$$

which is the same as $f(1)$ for our original function $f(x) = 3x - 2$.

Function Notation and Equation Notation

Instead of using the usual function notation to specify a function, as in, say,

$$f(x) = -2x^2 + 12x + 36, \qquad \text{Function notation}$$

we can specify it by an *equation with two variables* by replacing $f(x)$ by y or f or any letter we choose:

$$y = -2x^2 + 12x + 36 \qquad \text{Equation notation}$$

or we could choose

$$f = -2x^2 + 12x + 36.$$

In an equation of the form $y =$ expression in x, the variable x is called the **independent variable**, and y is called the **dependent variable** (because the value of y *depends on* a choice of the value for x). So we can think of the equation $y = -2x^2 + 12x + 36$ in two ways:

1. As specifying a function y of x, as in $y(x) = -2x^2 + 12x + 36$ (or, say, $f(x) = -2x^2 + 12x + 36$)

2. As simply an equation with independent variable x and dependent variable y

Note that when we think of a function as an equation in this way, the argument of the function becomes the independent variable, and the letter we choose for the left-hand side (sometimes the same letter as we use for the name of the function) becomes the dependent variable.

---

* We will discuss cost functions more fully in Section 1.2.

> ### Quick Example
>
> **5.** If the cost to manufacture x items is given by the "cost function"* C specified by
>
> $$C(x) = 40x + 2{,}000, \quad \text{Cost function}$$
>
> we could instead write
>
> $$C = 40x + 2{,}000 \quad \text{Cost equation}$$
>
> and still think of C, the cost, as a function of x.

Function notation and equation notation, sometimes using the same letter for the function name and the dependent variable, are often used interchangeably. It is important to be able to switch back and forth between function notation and equation notation, and we shall do so when it is convenient.

Piecewise-Defined Functions

Look again at the graph of the accumulated sales of **Nintendo** Wii game consoles in Figure 2. From year 0 through year 3, the sales seem to accelerate, but then, from year 3 to year 8, the sales seem to slow. In the following example, we model this behavior using a function with two different formulas: one that applies to the interval $[0, 3]$ and another for the interval $[3, 8]$. A function specified by two or more different formulas is called a **piecewise-defined function**.

EXAMPLE 2 A Piecewise-Defined Function: Nintendo Wii Sales

The accumulated sales of **Nintendo** Wii game consoles from December 2006 to December 2014 can be approximated by the following function of time t in years ($t = 0$ represents the end of December 2006):

$$f(t) = \begin{cases} 5t^2 + 2t + 2 & \text{if } 0 \le t \le 3 \\ -2t^2 + 33t - 28 & \text{if } 3 < t \le 8 \end{cases} \quad \text{million consoles.}$$

What were the accumulated sales of Nintendo Wii consoles at the end of December 2007, December 2009, and June 2010? Sketch the graph of f by plotting several points.

Solution We evaluate the given function at the corresponding values of t:

Dec. 2007 ($t = 1$): $f(1) = 5(1)^2 + 2(1) + 2 = 9$ Use the first formula because $0 \le t \le 3$.

Dec. 2009 ($t = 3$): $f(3) = 5(3)^2 + 2(3) + 2 = 53$ Use the first formula because $0 \le t \le 3$.

June 2010 ($t = 3.5$): $f(3.5) = -2(3.5)^2 + 33(3.5) - 28 = 63$ Use the second formula because $3 < t \le 9$.

Thus, Nintendo had sold around 9 million consoles by the end of December 2007, 53 million by the end of December 2009, and 63 million by the end of June 2010.

Using Technology

See the Technology Guides at the end of the chapter for detailed instructions on how to obtain the table of values and graph in Example 2 using a TI-83/84 Plus or Excel. Here is an outline:

TI-83/84 Plus
Table of values:
Y₁=(X≤3)*(5X^2+2X+2)
  +(X>3)*(-2X^2+33X-28)
2ND TABLE
Graph: WINDOW ; Xmin = 0,
Xmax = 8 ZOOM 0.
[More details in the Technology Guide.]

Spreadsheet
Table of values: Headings t and $f(t)$ in A1–B1; t-values 0, 1, . . . , 8 in A2–A10.
=(A2<=3)*(5*A2^2+2*A2+2)
+(A2>3)*(-2*A2^2+33*A2-28)
in B2; copy down through B10.
Graph: Highlight A1 through B10, and insert a scatter chart.
[More details in the Technology Guide.]

Website
www.WanerMath.com
Go to the Function evaluator and grapher under Online Utilities, and enter
(x≤3)*(5x^2+2x+2)
  +(x>3)*(-2x^2+33x-28)
for y1. To obtain a table of values, enter the x-values 0, 1, . . . , 8 in the Evaluator box, and press "Evaluate" at the top of the box.
Graph: Set Xmin = 0, Xmax = 8, and press "Plot Graphs".

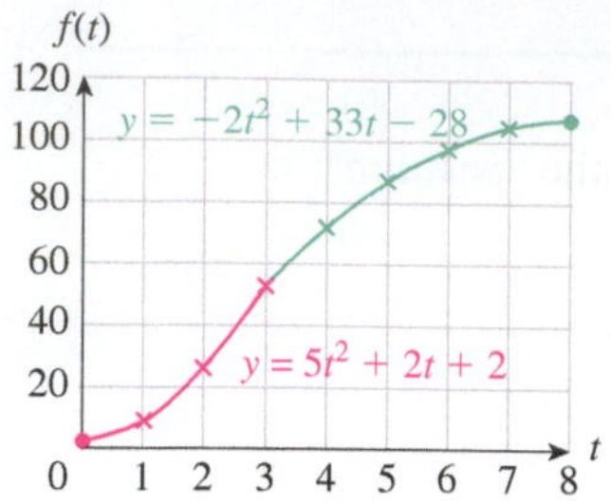

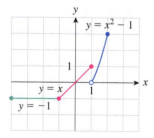

Figure 4

To sketch the graph of f, we use a table of values of $f(t)$ (two of which we have already calculated above), plot the points, and connect them to sketch the graph:

t	0	1	2	3	4	5	6	7	8
$f(t)$	2	9	26	53	72	87	98	105	108

First formula — Second formula

The graph (Figure 4) has the following features:

1. The first formula is used for $0 \le t \le 3$.
2. The second formula is used for $3 < t \le 8$.
3. The domain is $[0, 8]$, so the graph is cut off at $t = 0$ and $t = 8$.
4. The heavy solid dots at the ends indicate the endpoints of the domain.

EXAMPLE 3 **More Complicated Piecewise-Defined Functions**

Let f be the function specified by

$$f(x) = \begin{cases} -1 & \text{if } -4 \le x < -1 \\ x & \text{if } -1 \le x \le 1 \\ x^2 - 1 & \text{if } 1 < x \le 2. \end{cases}$$

Technology formula: `(X<-1)*(-1)+(-1≤X)*(X≤1)*X+(1<X)*(X^2-1)`

a. What is the domain of f? Find $f(-2)$, $f(-1)$, $f(0)$, $f(1)$, and $f(2)$.
b. Sketch the graph of f.

Solution

a. The domain of f is $[-4, 2]$, because $f(x)$ is specified only when $-4 \le x \le 2$.

$$f(-2) = -1 \qquad \text{We used the first formula because } -4 \le x < -1.$$
$$f(-1) = -1 \qquad \text{We used the second formula because } -1 \le x \le 1.$$
$$f(0) = 0 \qquad \text{We used the second formula because } -1 \le x \le 1.$$
$$f(1) = 1 \qquad \text{We used the second formula because } -1 \le x \le 1.$$
$$f(2) = 2^2 - 1 = 3 \qquad \text{We used the third formula because } 1 < x \le 2.$$

b. To sketch the graph by hand, we first sketch the three graphs $y = -1$, $y = x$, and $y = x^2 - 1$ and then use the appropriate portion of each (Figure 5). Note that solid dots indicate points on the graph, whereas open dots indicate points not on the graph. For example, when $x = 1$, the inequalities in the formula tell us that we are to use the middle formula (x) rather than the bottom one ($x^2 - 1$). Thus, $f(1) = 1$, not 0, so we place a solid dot at $(1, 1)$ and an open dot at $(1, 0)$.

Figure 5

Vertical Line Test

Every point in the graph of a function has the form $(x, f(x))$ for some x in the domain of f. Because f assigns a *single* value $f(x)$ to each value of x in the domain, it follows that, in the graph of f, there should be only one y corresponding to any such value of x—namely, $y = f(x)$. In other words, *the graph of a function cannot contain two or more points with the same x-coordinate—that is, two or more points on the same vertical line*. On the other hand, a vertical line at a value of x not in the domain will not contain any points in the graph. This gives us the following rule.

Vertical Line Test

For a graph to be the graph of a function, every vertical line must intersect the graph in *at most* one point.

Quick Examples

6. As illustrated below, only graph B passes the vertical line test, so only graph B is the graph of a function.

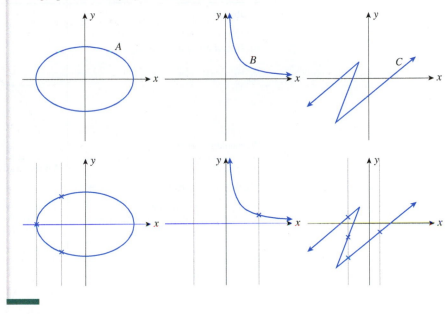

Common Functions

Table 1 lists some common types of functions that are often used to model real-world situations.

Table 1 A Compendium of Functions and Their Graphs

Type of Function	Examples	
Linear $f(x) = mx + b$ m, b constant Graphs of linear functions are straight lines. The quantity m is the **slope** of the line; the quantity b is the **y-intercept** of the line. (See Section 1.3.) Technology formulas:	$y = x$ x	$y = -2x + 2$ $-2*x+2$

Table 1 *(Continued)*

Type of Function	Examples	
Quadratic $f(x) = ax^2 + bx + c$ a, b, c constant $(a \neq 0)$ Graphs of quadratic functions are called **parabolas**. When a is positive, the parabola is **concave up** (example shown on the left). When a is negative, the parabola is **concave down** (example shown on the right).	$y = x^2$ 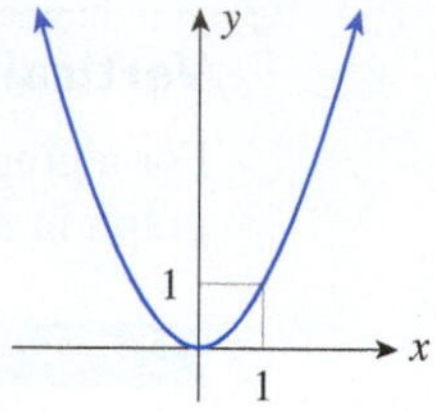	$y = -2x^2 + 2x + 4$ 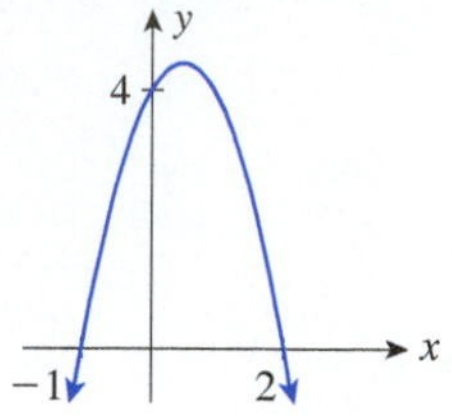
Technology formulas:	x^2	−2*x^2 + 2*x + 4
Cubic $f(x) = ax^3 + bx^2 + cx + d$ a, b, c, d constant $(a \neq 0)$	$y = x^3$ 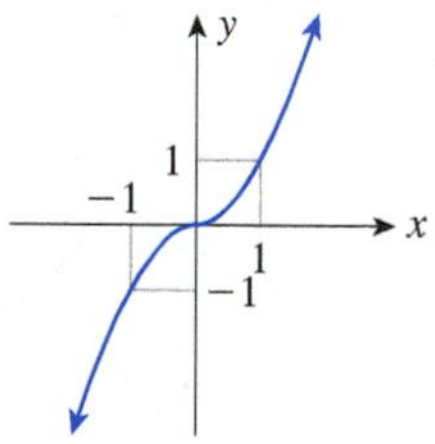	$y = -x^3 + 3x^2 + 1$ 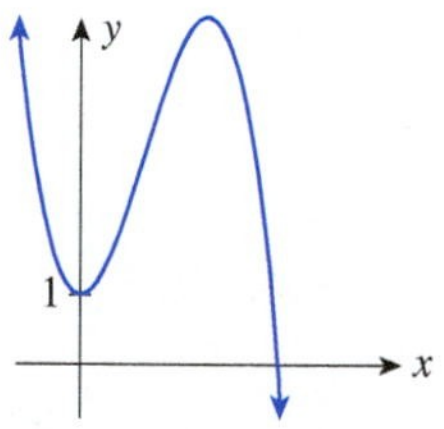
Technology formulas:	x^3	-x^3 + 3*x^2 + 1
Polynomial $f(x) = ax^n + bx^{n-1} + \cdots + rx + s$ $a, b, \ldots, r, s$ constant (includes all of the above functions)	All the above, and $f(x) = x^6 - 2x^5 - 2x^4 + 4x^2$ 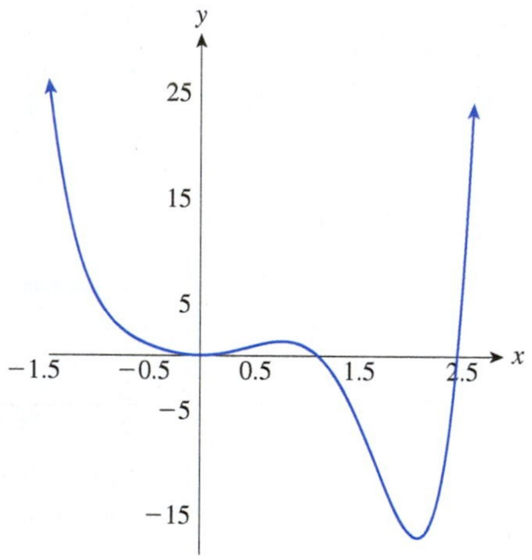	
Technology formula:	x^6-2x^5-2x^4+4x^2	
Exponential $f(x) = Ab^x$ A, b constant $(b > 0$ and $b \neq 1)$ The y-coordinate is multiplied by b every time x increases by 1.	$y = 2^x$ 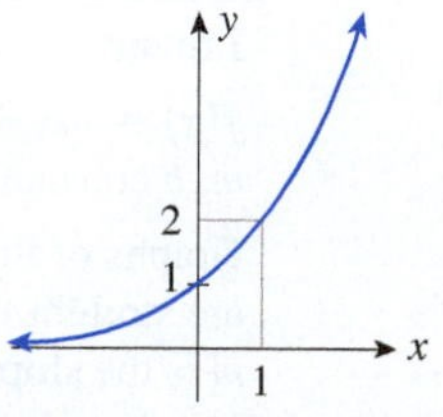 y is doubled every time x increases by 1.	$y = 4(0.5)^x$ 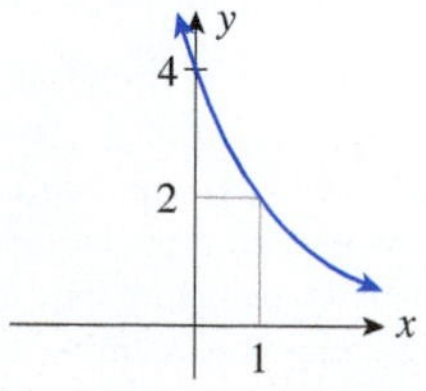 y is halved every time x increases by 1.
Technology formulas:	2^x	4*0.5^x

Table 1 (*Continued*)

Type of Function	Examples	
Logarithmic $f(x) = \log_b x + C$ b, C constant $(b > 0, b \neq 1)$	$y = \log_2 x$ $y = \log_{1/2} x$ x is doubled every time x is doubled every time y increases by 1. y decreases by 1.	
Technology formulas:	`ln(x)/ln(2)`	`ln(x)/ln(1/2)`
Rational $f(x) = \dfrac{P(x)}{Q(x)};$ $P(x)$ and $Q(x)$ polynomials The graph of $y = 1/x$ is a **hyperbola**. The domain excludes zero because $1/0$ is not defined.	$y = \dfrac{1}{x}$ 	$y = \dfrac{x}{x-1}$
Technology formulas:	`1/x`	`x/(x-1)`
Absolute Value For x positive or zero the graph of $y = \|x\|$ is the same as that of $y = x$. For x negative or zero it is the same as that of $y = -x$.	$y = \|x\|$ 	$y = \|2x + 2\|$
Technology formulas:	`abs(x)`	`abs(2*x+2)`
Square Root The domain of $y = \sqrt{x}$ must be restricted to the nonnegative numbers because the square root of a negative number is not real. Its graph is the top half of a horizontally oriented parabola.	$y = \sqrt{x}$ 	$y = \sqrt{4x - 2}$
Technology formulas:	`x^0.5` or $\sqrt{}$`(x)`	`(4*x-2)^0.5` or $\sqrt{}$`(4*x-2)`

Website
www.WanerMath.com
Follow the path
Online Text →
New Functions from Old:
Scaled and Shifted Functions
where you will find complete online interactive text, examples, and exercises on scaling and translating the graph of a function by changing the formula.

Functions and models other than linear ones are called **nonlinear**.

1.1 EXERCISES

▼ more advanced ◆ challenging

🅣 indicates exercises that should be solved using technology

In Exercises 1–4, evaluate each expression based on the follow-ing table. [**HINT:** See Quick Example 1.]

x	-3	-2	-1	0	1	2	3
$f(x)$	1	2	4	2	-1	-0.5	0.25

1. a. $f(0)$ **b.** $f(2)$ **2. a.** $f(-1)$ **b.** $f(1)$

3. a. $f(2) - f(-2)$ **b.** $f(-1)f(-2)$ **c.** $-2f(-1)$

4. a. $f(1) - f(-1)$ **b.** $f(1)f(-2)$ **c.** $3f(-2)$

In Exercises 5–8, use the graph of the function f to find approximations of the given values. [**HINT:** See Example 1.]

5.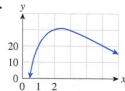

a. $f(1)$ **b.** $f(2)$
c. $f(3)$ **d.** $f(5)$
e. $f(3) - f(2)$ **f.** $f(3 - 2)$

6.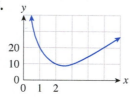

a. $f(1)$ **b.** $f(2)$
c. $f(3)$ **d.** $f(5)$
e. $f(3) - f(2)$ **f.** $f(3 - 2)$

7. a. $f(-1)$ **b.** $f(1)$ **c.** $f(3)$ **d.** $\dfrac{f(3) - f(1)}{3 - 1}$

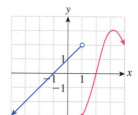

8. a. $f(-3)$ **b.** $f(-1)$ **c.** $f(1)$ **d.** $\dfrac{f(3) - f(1)}{3 - 1}$

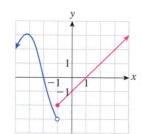

In Exercises 9–12, say whether or not $f(x)$ is defined for the given values of x. If it is defined, give its value. [**HINT:** See Quick Example 3.]

9. $f(x) = x - \dfrac{1}{x^2}$, with its natural domain

 a. $x = 4$ **b.** $x = 0$ **c.** $x = -1$

10. $f(x) = \dfrac{2}{x} - x^2$, with domain $[2, +\infty)$

 a. $x = 4$ **b.** $x = 0$ **c.** $x = 1$

11. $f(x) = \sqrt{x + 10}$, with domain $[-10, 0)$

 a. $x = 0$ **b.** $x = 9$ **c.** $x = -10$

12. $f(x) = \sqrt{9 - x^2}$, with domain $(-3, 3)$

 a. $x = 0$ **b.** $x = 3$ **c.** $x = -3$

13. Given $f(x) = 4x - 3$, find **a.** $f(-1)$ **b.** $f(0)$
 c. $f(1)$ **d.** $f(y)$ **e.** $f(a + b)$ [**HINT:** See Example 1.]

14. Given $f(x) = -3x + 4$, find
 a. $f(-1)$ **b.** $f(0)$ **c.** $f(1)$ **d.** $f(y)$ **e.** $f(a + b)$

15. Given $f(x) = x^2 + 2x + 3$, find
 a. $f(0)$ **b.** $f(1)$ **c.** $f(-1)$ **d.** $f(-3)$
 e. $f(a)$ **f.** $f(x + h)$ [**HINT:** See Example 1.]

16. Given $g(x) = 2x^2 - x + 1$, find
 a. $g(0)$ **b.** $g(-1)$ **c.** $g(r)$ **d.** $g(x + h)$

17. Given $g(s) = s^2 + \dfrac{1}{s}$, find

 a. $g(1)$ **b.** $g(-1)$ **c.** $g(4)$ **d.** $g(x)$ **e.** $g(s + h)$
 f. $g(s + h) - g(s)$

18. Given $h(r) = \dfrac{1}{r + 4}$, find

 a. $h(0)$ **b.** $h(-3)$ **c.** $h(-5)$ **d.** $h(x^2)$
 e. $h(x^2 + 1)$ **f.** $h(x^2) + 1$

In Exercises 19–24, graph the given functions. Give the tech-nology formula, and use technology to check your graph. We suggest that you become familiar with these graphs in addition to those in Table 1. [**HINT:** See Quick Example 4.]

19. $f(x) = -x^3$ (domain $(-\infty, +\infty)$)

20. $f(x) = x^3$ (domain $[0, +\infty)$)

21. $f(x) = x^4$ (domain $(-\infty, +\infty)$)

22. $f(x) = \sqrt[3]{x}$ (domain $(-\infty, +\infty)$)

23. $f(x) = \dfrac{1}{x^2}$ ($x \neq 0$)

24. $f(x) = x + \dfrac{1}{x}$ ($x \neq 0$)

In Exercises 25 and 26, match the functions to the graphs. (The gridlines are 1 unit apart.) Using technology to draw the graphs is suggested but not required.

25. a. $f(x) = x \quad (-1 \le x \le 1)$
  b. $f(x) = -x \quad (-1 \le x \le 1)$
  c. $f(x) = \sqrt{x} \quad (0 < x < 4)$
  d. $f(x) = x + \dfrac{1}{x} - 2 \quad (0 < x < 4)$
  e. $f(x) = |x| \quad (-1 \le x \le 1)$
  f. $f(x) = x - 1 \quad (-1 \le x \le 1)$

(A)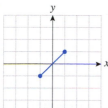

(B)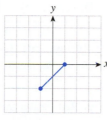

(C)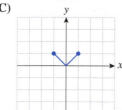

(D)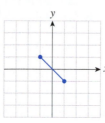

(E)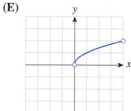

(F)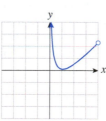

26. a. $f(x) = -x + 3 \quad (0 < x \le 3)$
  b. $f(x) = 2 - |x| \quad (-2 < x \le 2)$
  c. $f(x) = \sqrt{x + 2} \quad (-2 < x \le 2)$
  d. $f(x) = -x^2 + 2 \quad (-2 < x \le 2)$
  e. $f(x) = \dfrac{1}{x} - 1 \quad (0 < x \le 3)$
  f. $f(x) = x^2 - 1 \quad (-2 < x \le 2)$

(A)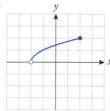

(B)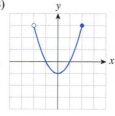

(C)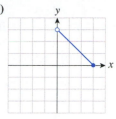

(D)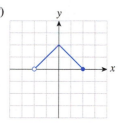

(E)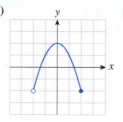

(F)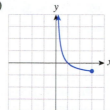

T *In Exercises 27–30, first give the technology formula for the given function, and then use technology to evaluate the function f for the given values of x (when f is defined there).*

27. T $f(x) = 0.1x^2 - 4x + 5; \; x = 0, 1, \dots, 10$

28. T $g(x) = 0.4x^2 - 6x - 0.1; \; x = -5, -4, \dots, 4, 5$

29. T $h(x) = \dfrac{x^2 - 1}{x^2 + 1}; \; x = 0.5, 1.5, 2.5, \dots, 10.5$ (Round all answers to four decimal places.)

30. T $r(x) = \dfrac{2x^2 + 1}{2x^2 - 1}; \; x = -1, 0, 1, \dots, 9$ (Round all answers to four decimal places.)

In Exercises 31–36, sketch the graph of the given function, evaluate the given expressions, and then use technology to duplicate the graphs. Give the technology formula.
[**HINT:** See Example 2.]

31. $f(x) = \begin{cases} x & \text{if } -4 \le x < 0 \\ 2 & \text{if } 0 \le x \le 4 \end{cases}$
  a. $f(-1)$ **b.** $f(0)$ **c.** $f(1)$

32. $f(x) = \begin{cases} -1 & \text{if } -4 \le x \le 0 \\ x & \text{if } 0 < x \le 4 \end{cases}$
  a. $f(-1)$ **b.** $f(0)$ **c.** $f(1)$

33. $f(x) = \begin{cases} x^2 & \text{if } -2 < x \le 0 \\ 1/x & \text{if } 0 < x \le 4 \end{cases}$
  a. $f(-1)$ **b.** $f(0)$ **c.** $f(1)$

34. $f(x) = \begin{cases} -x^2 & \text{if } -2 < x \le 0 \\ \sqrt{x} & \text{if } 0 < x < 4 \end{cases}$
  a. $f(-1)$ **b.** $f(0)$ **c.** $f(1)$

35. $f(x) = \begin{cases} x & \text{if } -1 < x \le 0 \\ x + 1 & \text{if } 0 < x \le 2 \\ x & \text{if } 2 < x \le 4 \end{cases}$
  a. $f(0)$ **b.** $f(1)$ **c.** $f(2)$ **d.** $f(3)$
  [**HINT:** See Example 3.]

36. $f(x) = \begin{cases} -x & \text{if } -1 < x < 0 \\ x - 2 & \text{if } 0 \le x \le 2 \\ -x & \text{if } 2 < x \le 4 \end{cases}$

   a. $f(0)$ **b.** $f(1)$ **c.** $f(2)$ **d.** $f(3)$

In Exercises 37–40, find and simplify (a) $f(x + h) - f(x)$
(b) $\dfrac{f(x + h) - f(x)}{h}$.

37. ▼ $f(x) = x^2$ **38.** ▼ $f(x) = 3x - 1$

39. ▼ $f(x) = 2 - x^2$ **40.** ▼ $f(x) = x^2 + x$

Applications

41. *Crude Oil Production: Mexico* The following table shows daily crude oil production by **Pemex**, Mexico's national oil company, for 2008–2014 ($t = 0$ represents 2008):[5]

Year t (year since 2008)	0	1	2	3	4	5	6
Crude Oil Production $p(t)$ (million barrels/day)	3.16	2.97	2.95	2.94	2.91	2.88	2.79

   a. Find $p(2)$, $p(3)$, and $p(6)$. Interpret your answers.

   b. Find $p(4) - p(2)$. Interpret your answer. [**HINT:** See Quick Example 1.]

42. *Offshore Crude Oil Production: Mexico* The following table shows daily offshore crude oil production by **Pemex**, Mexico's national oil company, for 2008–2014 ($t = 0$ represents 2008):[6]

Year t (year since 2008)	0	1	2	3	4	5	6
Offshore Crude Oil Production $s(t)$ (million barrels/day)	2.25	2.01	1.94	1.90	1.90	1.90	1.85

   a. Find $s(0)$, $s(2)$, and $s(4)$. Interpret your answers.

   b. Find $s(4) - s(0)$. Interpret your answer.

43. *Social Website Popularity: Twitter* The following table shows the popularity of **Twitter** among social media sites as rated by StatCounter.com (t is the number of years since the start of 2008):[7]

Year t (year since start of 2008)	1	2	4	5
Twitter Popularity $p(t)$ (%)	7	6	6	7

   a. Represent p graphically, and then use your graph to estimate $p(4.5)$. Interpret your answer.

   b. One of the following models fits the data almost exactly. Which model is it?

      (A) $p(t) = 0.4t^3 - 4t^2 + 12.5t + 15$

      (B) $p(t) = -0.33t^2 + 2t - 8.7$

      (C) $p(t) = -0.4t^3 + 4t^2 - 12.5t + 15$

      (D) $p(t) = 0.33t^2 - 2t + 8.7$

44. *Social Website Popularity: Delicious* The following table shows the popularity of **Delicious** among social media sites as rated by StatCounter.com (t is the number of years since the start of 2008):[8]

Year t (year since start of 2008)	1	2	3	4	5
Delicious Popularity $p(t)$ (%)	0.4	0.2	0.1	0.05	0.02

   a. Represent p graphically, and then use your graph to estimate $p(3.5)$. Interpret your answer.

   b. One of the following models fits the data exactly. Which model is it?

      (A) $p(t) = 0.8(2^{-t})$

      (B) $p(t) = 0.8(2^t)$

      (C) $p(t) = 0.02t^2 - 0.2t + 0.6$

      (D) $p(t) = -0.02t^2 + 0.2t - 0.6$

Housing Starts Exercises 45–48 refer to the following graph, which shows the number $f(t)$ of housing starts for single-family homes in the United States each year from 2000 through 2014 ($t = 0$ represents 2000, and $f(t)$ is in thousands of units):[9]

45. Estimate $f(7)$, $f(14)$, and $f(9.5)$. Interpret your answers.

46. Estimate $f(3)$, $f(6)$, and $f(8.5)$. Interpret your answers.

47. Estimate $f(7 - 3)$ and $f(7) - f(3)$. Interpret your answers.

48. Estimate $f(13 - 3)$ and $f(13) - f(3)$. Interpret your answers.

49. ▼ For which value or values of t is $f(t + 5) - f(t)$ greatest? Interpret your answer.

50. ▼ For which value or values of t is $f(t) - f(t - 1)$ least? Interpret your answer

[5] Source: www.pemex.com (March 2015).

[6] *Ibid.*

[7] Percentages are based on worldwide page views. Source: http://gs.statcounter.com.

[8] Figures are approximate. Source: *Ibid.*

[9] Data are approximate. Source: www.census.gov.

51. *Net Income: Casual Apparel* In the following graph, $n(t)$ is **Abercrombie & Fitch**'s approximate net income, in millions of dollars, for the year ending at time t (t is time in years since December 2004):[10]

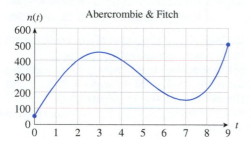

a. Estimate $n(2)$, $n(4)$, and $n(4.5)$ to the nearest 25. Interpret your answers.

b. At approximately which value of t in the interval $[3, 8]$ is $n(t)$ *increasing* most rapidly? Interpret your answer.

c. At approximately which value of t in the interval $[3, 8]$ is $n(t)$ *decreasing* most rapidly? Interpret your answer.

52. *Net Income: Casual Apparel* In the following graph, $n(t)$ is **Pacific Sunwear**'s approximate net income, in millions of dollars, for the year ending at time t (t is time in years since December 2004):[11]

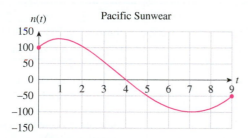

a. Estimate $n(0)$, $n(4)$, and $n(5.5)$ to the nearest 25. Interpret your answers.

b. At which of the following values of t is $n(t)$ *increasing* most rapidly: 1, 2, 4, 7, 8, or 9? Interpret your answer.

c. At which of the following values of t is $n(t)$ *decreasing* most rapidly: 1, 2, 4, 7, 8, or 9? Interpret your answer.

53. *Funding for NASA: 1958–1966* The percentage of the U.S. federal budget allocated to **NASA** from 1958 to 1966 can be approximated by

$$p(t) = \frac{4.5}{1.07^{(t-8)^2}} \text{ percentage points}$$

(t is time in years since 1958).[12] The following graph shows the data with the model:

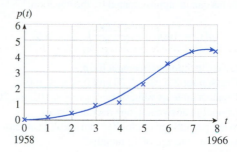

a. Find an appropriate domain of p. Is $t \geq 0$ an appropriate domain? Why or why not?

b. Compute $p(5)$ accurate to one decimal place. What does the answer say about the budget allocation to NASA?

c. At which of the following values of t is $p(t)$ increasing most rapidly: 0, 3, 5, 8? Interpret your answer.

54. *Funding for NASA: 1966–2015* The percentage of the U.S. federal budget allocated to **NASA** from 1966 to 2015 can be approximated by

$$p(t) = 0.03 + \frac{5}{t^{0.6}} \text{ percentage points} \quad (t \geq 1)$$

(t is time in years since 1965).[13] The following graph shows the data with the model:

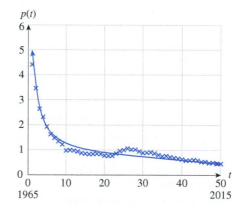

a. Find an appropriate domain of p. Is $[0, 50]$ an appropriate domain? Why or why not?

b. Compute $p(40)$ accurate to two decimal places. What does the answer say about the budget allocation to NASA?

c. If the model is extrapolated to larger and larger values of t, what does it suggest about long-term financing of NASA?

---

[10] "Net income" is an accounting term for profit (see Section 1.2). Model is the authors'. Source of data: www.wikinvest.com.
[11] *Ibid.*

[12] Model is the authors'. Source of data: U.S. Office of Management and Budget/www.wikipedia.org.
[13] *Ibid.*

55. **Ⓣ** *Acquisition of Language* The percentage $p(t)$ of children who can speak in at least single words by the age of t months can be approximated by the equation[14]

$$p(t) = 100\left(1 - \frac{12{,}200}{t^{4.48}}\right) \quad (t \geq 8.5).$$

a. Give a technology formula for p.
b. Graph p for $8.5 \leq t \leq 20$ and $0 \leq p \leq 100$.
c. Create a table of values of p for $t = 9, 10, \ldots, 20$ (rounding answers to one decimal place).
d. What percentage of children can speak in at least single words by the age of 12 months?
e. By what age are 90% or more children speaking in at least single words?

56. **Ⓣ** *Acquisition of Language* The percentage $p(t)$ of children who can speak in sentences of five or more words by the age of t months can be approximated by the equation[15]

$$p(t) = 100\left(1 - \frac{5.27 \times 10^{17}}{t^{12}}\right) \quad (t \geq 30).$$

a. Give a technology formula for p.
b. Graph p for $30 \leq t \leq 45$ and $0 \leq p \leq 100$.
c. Create a table of values of p for $t = 30, 31, \ldots, 40$ (rounding answers to one decimal place).
d. What percentage of children can speak in sentences of five or more words by the age of 36 months?
e. By what age are 75% or more children speaking in sentences of five or more words?

57. ▼ *Processor Speeds* The processor speed, in megahertz (MHz), of **Intel** processors during the period 1980–2010 could be approximated by the following function of time t in years since the start of 1980:[16]

$$v(t) = \begin{cases} 8(1.22)^t & \text{if } 0 \leq t < 16 \\ 400t - 6{,}200 & \text{if } 16 \leq t < 25 \\ 3{,}800 & \text{if } 25 \leq t \leq 30. \end{cases}$$

a. Evaluate $v(10)$, $v(16)$, and $v(28)$. Interpret the results.
b. Write down a technology formula for v.
c. **Ⓣ** Use technology to sketch the graph of v and to generate a table of values for $v(t)$ with $t = 0, 2, \ldots, 30$. (Round values to two significant digits.)
d. When, to the nearest year, did processor speeds reach 3 gigahertz (1 gigahertz = 1,000 megahertz), according to the model?

58. ▼ *Processor Speeds* The processor speed, in megahertz (MHz), of **Intel** processors during the period 1970–2000

could be approximated by the following function of time t in years since the start of 1970:[17]

$$v(t) = \begin{cases} 0.12t^2 + 0.04t + 0.2 & \text{if } 0 \leq t < 12 \\ 1.1(1.22)^t & \text{if } 12 \leq t < 26 \\ 400t - 10{,}200 & \text{if } 26 \leq t \leq 30. \end{cases}$$

a. Evaluate $v(2)$, $v(12)$, and $v(28)$. Interpret the results.
b. Write down a technology formula for v.
c. **Ⓣ** Use technology to sketch the graph of v and to generate a table of values for $v(t)$ with $t = 0, 2, \ldots, 30$. (Round values to two significant digits.)
d. When, to the nearest year, did processor speeds reach 500 MHz?

59. ▼ *Income Taxes* The U.S. federal income tax is a function of taxable income. Write $T(x)$ for the tax owed on a taxable income of x dollars. For tax year 2015 the function T for a single taxpayer was specified as follows:

If your taxable income was over ...	But not over ...	Your tax is ...	Of the amount over ...
$0	9,225	10%	$0
9,225	37,450	$922.50 + 15%	$9,225
37,450	90,750	5,156.25 + 25%	$37,450
90,750	189,300	18,481.25 + 28%	$90,750
189,300	411,500	46,075.25 + 33%	$189,300
411,500	413,200	119,401.25 + 35%	$411,500
413,200	—	119,996.25 + 39.6%	$413,200

a. Represent T as a piecewise-defined function of income x. [**HINT**: Each row of the table defines a formula with a condition.]
b. Use your function to compute the tax owed by a single taxpayer on a taxable income of $45,000.

60. ▼ *Income Taxes* Repeat Exercise 59 using the following information for tax year 2012:

If your taxable income was over ...	But not over ...	Your tax is ...	Of the amount over ...
$0	8,700	10%	$0
8,700	35,350	$870.00 + 15%	$8,700
35,350	85,650	4,867.50 + 25%	$35,350
85,650	178,650	17,442.50 + 28%	$85,650
178,650	388,350	43,482.50 + 33%	$178,650
388,350	—	112,683.50 + 35%	$388,350

---

[14] The model is the authors' and is based on data presented in the article *The Emergence of Intelligence* by William H. Calvin, *Scientific American*, October 1994, pp. 101–107.

[15] *Ibid.*

[16] Based on the fastest processors produced by Intel. Source for data: www.intel.com.

[17] *Ibid.*

Communication and Reasoning Exercises

61. Complete the following sentence: If the market price m of gold varies with time t, then the independent variable is ___, and the dependent variable is ___.

62. Complete the following sentence: If weekly profit P is specified as a function of selling price s, then the independent variable is ___, and the dependent variable is ___.

63. Complete the following: The function notation for the equation $y = 4x^2 - 2$ is ____.

64. Complete the following: The equation notation for $C(t) = -0.34t^2 + 0.1t$ is ____.

65. True or false? Every graphically specified function can also be specified numerically. Explain.

66. True or false? Every algebraically specified function can also be specified graphically. Explain.

67. True or false? Every numerically specified function with domain $[0, 10]$ can also be specified algebraically. Explain.

68. True or false? Every graphically specified function can also be specified algebraically. Explain.

69. ▼ True or false? Every function can be specified numerically. Explain.

70. ▼ Which supplies more information about a situation: a numerical model or an algebraic model?

71. ▼ Why is the following assertion false? "If $f(x) = x^2 - 1$, then $f(x + h) = x^2 + h - 1$."

72. ▼ Why is the following assertion false? "If $f(2) = 2$ and $f(4) = 4$, then $f(3) = 3$."

73. How do the graphs of two functions differ if they are specified by the same formula but have different domains?

74. How do the graphs of two functions f and g differ if $g(x) = f(x) + 10$? (Try an example.)

75. ▼ How do the graphs of two functions f and g differ if $g(x) = f(x - 5)$? (Try an example.)

76. ▼ How do the graphs of two functions f and g differ if $g(x) = f(-x)$? (Try an example.)

1.2 Functions and Models

The functions we used in Examples 1 and 2 in Section 1.1 are **mathematical models** of real-life situations because they model, or represent, situations in mathematical terms.

Mathematical Modeling

To mathematically model a situation means to represent it in mathematical terms. The particular representation used is called a **mathematical model** of the situation. Mathematical models do not always represent a situation perfectly or completely. Some (like Example 1 of Section 1.1) represent a situation only approximately; others represent only some aspects of the situation.

Quick Examples

1. The temperature is now 10°F and increasing by 20°F per hour.

   Model: $T(t) = 10 + 20t$ (t = time in hours, T = temperature)

2. I invest $1,000 at 5% interest compounded quarterly. Find the value of the investment after t years.

   Model: $A(t) = 1{,}000\left(1 + \dfrac{0.05}{4}\right)^{4t}$ (This is the compound interest formula we will study in Example 6.)

3. I am fencing a rectangular area whose perimeter is 100 feet. Find the area as a function of the width x.

Model: Take y to be the length, so the perimeter is

$$100 = x + y + x + y = 2(x + y).$$

This gives

$$x + y = 50.$$

Thus the length is $y = 50 - x$, and the area is

$$A = xy = x(50 - x).$$

4. You work 8 hours a day Monday through Friday and 5 hours on Saturday, and you have Sunday off. Model the number of hours you work as a function of the day of the week n, with $n = 1$ being Sunday.

Model: Take $f(n)$ to be the number of hours you work on the nth day of the week, so

$$f(n) = \begin{cases} 0 & \text{if } n = 1 \\ 8 & \text{if } 2 \le n \le 6 \\ 5 & \text{if } n = 7. \end{cases}$$

Note that the domain of f is $\{1, 2, 3, 4, 5, 6, 7\}$—a discrete set rather than a continuous interval of the real line.

5. The function

$$f(x) = -2x^2 + 12x + 36 \text{ million iPods sold } (x = \text{years since } 2005)$$

in Example 1 of Section 1.1 is a model of iPod sales.

6. The function

$$f(t) = \begin{cases} 5t^2 + 2t + 2 & \text{if } 0 \le t \le 3 \\ -2t^2 + 33t - 28 & \text{if } 3 < t \le 8 \end{cases} \text{ million consoles}$$

(t = years since the end of December 2006) in Example 2 of Section 1.1 is a model of accumulated Nintendo Wii game console sales.

Types of Models

Quick Examples 1–4 are **analytical models**, obtained by analyzing the situation being modeled. Quick Examples 5 and 6 are **curve-fitting models**, obtained by finding mathematical formulas that approximate observed data. All the models except for Quick Example 4 are **continuous models**, defined by functions whose domains are intervals of the real line. Quick Example 4 is a **discrete model**, as its domain is a discrete set, as was mentioned above. Discrete models are used extensively in probability and statistics.

Cost, Revenue, and Profit Models

EXAMPLE 1   Modeling Cost: Cost Function

As of March 2015, **Yellow Cab Chicago**'s rates amounted to $3.05 on entering the cab plus $1.80 for each mile.[18]

a. Find the cost C of an x-mile trip.

b. Use your answer to calculate the cost of a 40-mile trip.

c. What is the cost of the second mile? What is the cost of the tenth mile?

d. Graph C as a function of x.

Solution

a. We are being asked to find how the cost C depends on the length x of the trip, or to find C as a function of x. Here is the cost in a few cases:

Cost of a 1-mile trip: $C = 1.80(1) + 3.05 = 4.85$ 1 mile at $1.80 per mile plus $3.05

Cost of a 2-mile trip: $C = 1.80(2) + 3.05 = 6.65$ 2 miles at $1.80 per mile plus $3.05

Cost of a 3-mile trip: $C = 1.80(3) + 3.05 = 8.45$.   3 miles at $1.80 per mile plus $3.05

Do you see the pattern? The cost of an x-mile trip is given by the linear function*

$$C(x) = 1.80x + 3.05.$$

> * See the table of common functions at the end of the preceding section. Linear functions are discussed in detail in Section 1.3.

Notice that the cost function is a sum of two terms: the **variable cost** $1.80x$, which depends on x, and the **fixed cost** 3.05, which is independent of x:

$$\text{Cost} = \text{Variable cost} + \text{Fixed cost}.$$

The quantity 1.80 by itself is the incremental cost per mile; you might recognize it as the *slope* of the given linear function. In this context we call 1.80 the **marginal cost**. You might recognize the fixed cost 3.05 as the *C-intercept* of the given linear function.

b. We can use the formula for the cost function to calculate the cost of a 40-mile trip as

$$C(40) = 1.80(40) + 3.05 = \$75.05.$$

c. To calculate the cost of the second mile, we *could* proceed as follows:

Find the cost of a 1-mile trip: $C(1) = 1.80(1) + 3.05 = \$4.85$.

Find the cost of a 2-mile trip: $C(2) = 1.80(2) + 3.05 = \$6.65$.

Therefore, the cost of the second mile is $\$6.65 - \$4.85 = \$1.80$.

But notice that this is just the marginal cost. In fact, the marginal cost is the cost of each additional mile, so we could have done this more simply:

Cost of second mile $=$ Cost of tenth mile $=$ Marginal cost $= \$1.80$.

---

[18] According to their website at www.yellowcabchicago.com.

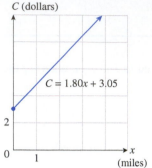

Figure 6

d. Figure 6 shows the graph of the cost function, which we can interpret as a *cost vs. miles* graph. The fixed cost is the starting height 3.05 on the left, while the marginal cost is the slope of the line: It rises 1.80 units per unit of *x*. (See Section 1.3 for a discussion of properties of straight lines.)

➡ **Before we go on . . .** The cost function in Example 1 is an example of an *analytical model:* We derived the form of the cost function from a knowledge of the cost per mile and the fixed cost.

As we discussed in Section 1.1, we can specify the cost function in Example 1 using equation notation:

$$C = 1.80x + 3.05. \quad \text{Equation notation}$$

Here, the independent variable is *x*, and the dependent variable is *C*. (This is the notation we have used in Figure 6. Remember that we will often switch between function and equation notation when it is convenient to do so.) ∎

Here is a summary of some terms we used in Example 1, along with an introduction to some new terms.

Cost, Revenue, and Profit Functions

A **cost function** specifies the cost *C* as a function of the number of items *x*. Thus, $C(x)$ is the cost of *x* items and has the form

$$\text{Cost} = \text{Variable cost} + \text{Fixed cost}$$

where the variable cost is a function of *x* and the fixed cost is a constant. A cost function of the form

$$C(x) = mx + b$$

is called a **linear cost function**; the variable cost is *mx*, and the fixed cost is *b*. The slope *m*, the **marginal cost**, measures the incremental cost per item.

The **revenue**, or **net sales**, resulting from one or more business transactions is the total income received. If $R(x)$ is the revenue from selling *x* items at a price of *m* each, then *R* is the linear function $R(x) = mx$, and the selling price *m* can also be called the **marginal revenue**.

The **profit**, or **net income**, on the other hand, is what remains of the revenue when costs are subtracted.* If the profit depends linearly on the number of items, the slope *m* is called the **marginal profit**. Profit, revenue, and cost are related by the following formula.

$$\text{Profit} = \text{Revenue} - \text{Cost}$$
$$P(x) = R(x) - C(x).^{†}$$

If the profit is negative, say −$500, we refer to a **loss** (of $500 in this case). To **break even** means to make neither a profit nor a loss. Thus, breakeven occurs when $P = 0$, or

$$R(x) = C(x). \quad \text{Breakeven}$$

The **break-even point** is the number of items *x* at which breakeven occurs.

---

* Note that taxes are also included in the costs. "Taxable profit" or "taxable net income" refers to profits before taxes are deducted.

† We say that the profit function *P* is the **difference** between the revenue and cost functions, and we express this fact as a formula about functions: $P = R − C$. (We will discuss this further when we talk about the algebra of functions at the end of this section.)

Quick Example

7. If the daily cost (including operating costs) of manufacturing x T-shirts is $C(x) = 8x + 100$ and the revenue obtained by selling x T-shirts is $R(x) = 10x$, then the daily profit resulting from the manufacture and sale of x T-shirts is

$$P(x) = R(x) - C(x) = 10x - (8x + 100) = 2x - 100.$$

Breakeven occurs when $P(x) = 0$, or $x = 50$.

EXAMPLE 2   Cost, Revenue, and Profit

The annual operating cost of *YSport Fitness* gym is estimated to be

$$C(x) = -2x^2 + 600x + 40,000 \text{ dollars} \qquad (0 \le x \le 150),$$

where x is the number of members. Annual revenue from membership averages \$800 per member. What is the variable cost? What is the fixed cost? What is the profit function? How many members must YSport have to make a profit? What will happen if it has fewer members? If it has more?

Solution The variable cost is the part of the cost function that depends on x:

$$\text{Variable cost} = -2x^2 + 600x.$$

The fixed cost is the constant term:

$$\text{Fixed cost} = 40,000.$$

The annual revenue YSport obtains from a single member is \$800. So if it has x members, it earns an annual revenue of

$$R(x) = 800x.$$

For the profit we use the formula

$$
\begin{aligned}
P(x) &= R(x) - C(x) && \text{Formula for profit}\\
&= 800x - (-2x^2 + 600x + 40,000) && \text{Substitute } R(x) \text{ and } C(x).\\
&= 2x^2 + 200x - 40,000.
\end{aligned}
$$

To make a profit, YSport needs to do better than break even, so let us find the breakeven point: the value of x such that $P(x) = 0$. All we have to do is set $P(x) = 0$ and solve for x:

$$
\begin{aligned}
2x^2 + 200x - 40,000 &= 0\\
2(x^2 + 100x - 20,000) &= 0\\
2(x + 200)(x - 100) &= 0 && \text{Factor the quadratic.}^*\\
x = -200 \quad &\text{or} \quad x = 100.
\end{aligned}
$$

* Had this quadratic not factored, we would have needed to use the quadratic formula. (See Section 0.5.)

We reject the negative solution (as the domain is $[0, 150]$) and conclude that $x = 100$ members. To make a profit, should YSport have more than 100 members or fewer than 100 members? To decide, take a look at Figure 7, which shows two graphs: On the left we see the graphs of revenue and cost, and on the right we see the graph of the profit function.

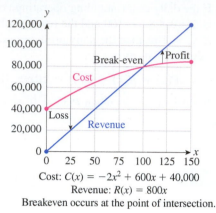

Cost: $C(x) = -2x^2 + 600x + 40{,}000$
Revenue: $R(x) = 800x$
Breakeven occurs at the point of intersection.

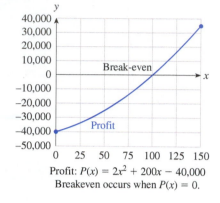

Profit: $P(x) = 2x^2 + 200x - 40{,}000$
Breakeven occurs when $P(x) = 0$.

Figure 7

Using Technology
Excel has a feature called "Goal Seek," which can be used to find the point of intersection of the cost and revenue graphs numerically rather than graphically. See the downloadable Excel tutorial for this section at the Website.

For values of x less than the break-even point of 100, $P(x)$ is negative, so the company will have a loss. For values of x greater than the break-even point, $P(x)$ is positive, so the company will make a profit. Thus, YSport needs at least 101 members to make a profit.

Demand and Supply Models

The demand for a commodity usually goes down as its price goes up. It is traditional to use the letter q for the (quantity of) demand as measured, for example, in sales. Consider the following example.

EXAMPLE 3   Demand: Private Schools

The demand, as meaured by total enrollment, for private schools in Michigan depends on the tuition cost and can be approximated by

$$q(p) = 32 + \frac{2{,}000}{p} \text{ thousand students enrolled} \qquad (200 \le p \le 2{,}200), \qquad \text{Demand function}$$

where p is the net tuition cost in dollars.[19] The graph of the demand function, shown in Figure 8, is called the associated **demand curve**.

What is the effect on demand if the tuition cost is increased from \$1,000 to \$2,000?

Solution The demand at tuition costs of \$1,000 and \$2,000 is

$$q(1{,}000) = 32 + \frac{2{,}000}{1{,}000} = 34 \text{ thousand students}$$

$$q(2{,}000) = 32 + \frac{2{,}000}{2{,}000} = 33 \text{ thousand students.}$$

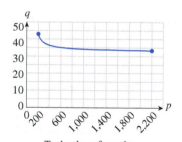

Technology formula:
`32+2000/x`

Figure 8

---

[19] The tuition cost is net cost: tuition minus tax credit. The model is based on projections of enrollment under a proposed tax credit in "The Universal Tuition Tax Credit: A proposal to Advance Personal Choice in Education," Patrick L. Anderson, Richard McLellan, J.D., Joseph P. Overton, J.D., Gary Wolfram, Ph.D., Mackinac Center for Public Policy, www.mackinac.org.

The change in demand is therefore

$$q(2{,}000) - q(1{,}000) = 33 - 34 = -1 \text{ thousand students,}$$

so demand decreases by around 1,000 students.

➡ **Before we go on . . .** As usual, we can represent the demand function in Example 3 as an equation:

$$q = 32 + \frac{2{,}000}{p}, \qquad \textcolor{blue}{\text{Demand equation}}$$

where the independent variable is the tuition cost p and the dependent variable is the demand (total enrollment) q. ∎

We have seen that a demand function gives the number of items consumers are willing to buy at a given price, and a higher price generally results in a lower demand. However, as the price rises, suppliers will be more inclined to produce these items (as opposed to spending their time and money on other products), so supply will generally rise. A **supply function** gives q, the number of items suppliers are willing to make available for sale,* as a function of p, the price per item.

* Although a bit confusing at first, it is traditional to use the same letter q for the quantity of supply and the quantity of demand, particularly when we want to compare them, as in the next example.

Demand, Supply, and Equilibrium Price

A **demand equation** or **demand function** expresses demand q (the number of items demanded) as a function of the unit price p (the price per item). A **supply equation** or **supply function** expresses supply q (the number of items a supplier is willing to make available) as a function of the unit price p (the price per item). It is usually the case that demand decreases and supply increases as the unit price increases.

Demand and supply are said to be in **equilibrium** when demand equals supply. The corresponding values of p and q are called the **equilibrium price** and **equilibrium demand**. To find the equilibrium price, determine the unit price p where the demand and supply curves cross (sometimes we can determine this value analytically by setting demand equal to supply and solving for p). To find the equilibrium demand, evaluate the demand (or supply) function at the equilibrium price.

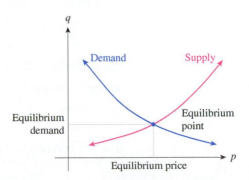

Quick Example

8. If the demand for your exclusive T-shirts is $q = -20p + 800$ shirts sold per day and the supply is $q = 10p - 100$ shirts per day, then the equilibrium point is obtained when demand = supply:

$$-20p + 800 = 10p - 100$$
$$30p = 900, \text{ giving } p = \$30.$$

The equilibrium price is therefore \$30, and the equilibrium demand is $q = -20(30) + 800 = 200$ shirts per day. What happens at prices other than the equilibrium price is discussed in Example 4.

Note   In economics it is customary to plot the independent variable (price) on the vertical axis and the dependent variable (demand or supply) on the horizontal axis, but in this book we follow the usual mathematical convention for all graphs and plot the independent variable on the horizontal axis. ∎

EXAMPLE 4   Demand, Supply, and Equilibrium Price

Continuing with Example 3, suppose that private school institutions are willing to create private schools to accommodate

$$q = 32 + 0.002p \text{ thousand students} \qquad (200 \le p \le 2{,}200) \qquad \text{\color{blue}Supply curve}$$

who pay a (net) tuition of p dollars.

a. What is the equilibrium tuition at private schools? Approximately how many students will be accommodated at that price?

b. What happens if the tuition is higher than the equilibrium tuition? What happens if it is lower?

c. Estimate the shortage or surplus of openings at private schools if the tuition is \$2,000.

Solution   The equilibrium point is obtained when demand = supply:

$$32 + \frac{2{,}000}{p} = 32 + 0.002p$$

$$32p + 2{,}000 = 32p + 0.002p^2 \qquad \text{\color{blue}Multiply both sides by } p.$$

$$0.002p^2 = 2{,}000 \qquad \text{\color{blue}Cancel the } 32p.$$

$$p^2 = \frac{2{,}000}{0.002} = 1{,}000{,}000$$

$$p = \sqrt{1{,}000{,}000} = 1{,}000.$$

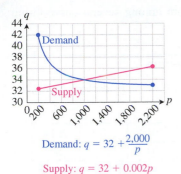

Demand: $q = 32 + \dfrac{2,000}{p}$

Supply: $q = 32 + 0.002p$

Figure 9

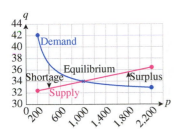

Figure 10

a. Figure 9 shows the graphs of demand $q = 32 + \dfrac{2,000}{p}$ and supply $q = 32 + 0.002p$. (See the margin note for a brief description of how to plot them.) The lines cross at $p = \$1,000$, so we conclude that demand = supply when $p = \$1,000$. This is the equilibrium tuition price. At that price we can calculate the demand or supply as

Demand: $q = 32 + 2,000/1,000 = 34$

Supply: $q = 32 + 0.002(1,000) = 34$,   Demand = Supply at equilibrium

or 34,000 students.

b. Take a look at Figure 10, which shows what happens if schools charge more or less than the equilibrium price. If tuition is, say, \$2,000, then the supply will be larger than demand, and there will be a surplus of available openings at private schools. Similarly, if tuition is less—say \$400—then the supply will be less than the demand, and there will be a shortage of available openings.

c. The discussion in part (b) shows that if tuition is set at \$2,000 there will be a surplus of available openings. To estimate that number, we calculate the projected demand and supply when $p = \$2,000$:

Demand: $q(2,000) = 32 + \dfrac{2,000}{2,000} = 33$ thousand seats

Supply: $q(2,000) = 32 + 0.002(2,000) = 36$ thousand seats

Surplus = Supply − Demand = 36 − 33 = 3 thousand seats.

So the models predict a surplus of 3,000 available seats.

➡ **Before we go on . . .** We saw in Example 4 that if tuition is less than the equilibrium price, there will be a shortage. If schools were to raise their tuition toward the equilibrium, they would create and fill more openings and increase revenue, since it is the supply equation—and not the demand equation—that determines what one can sell below the equilibrium price. On the other hand, if they were to charge more than the equilibrium price, they will be left with a possibly costly surplus of unused openings (and will want to lower tuition to reduce the surplus). Prices tend to move toward the equilibrium, so supply tends to equal demand. When supply equals demand, we say that the market **clears**. ∎

Modeling Change over Time

Things around us change with time. Thus, there are many quantities, such as your income or the temperature in Honolulu, that are natural to think of as functions of time. Example 1 (on iPod sales) and Example 2 (on Nintendo Wii sales) in Section 1.1 are models of change over time. Both of those models are curve-fitting models: We used algebraic functions to approximate observed data.

Note We usually use the independent variable t to denote time (in seconds, hours, days, years, etc.). If a quantity q changes with time, then we can regard q as a function of t. ∎

In the next example we are asked to select from among several curve-fitting models for given data.

EXAMPLE 5 ◨ Model Selection: Amazon Revenue

The following table shows some annual revenues, in billions of dollars, earned by **Amazon** from 2004 through 2014:[20]

Year	2004	2006	2008	2010	2012	2014
Revenue ($ billion)	7	11	19	34	61	89

Take t to be the number of years since 2004, and consider the following three models:

(1) $r(t) = 8t - 3$ Linear model

(2) $r(t) = 0.9t^2 - 0.6t + 7$ Quadratic model

(3) $r(t) = 7(1.3^t)$.   Exponential model

a. Which two models fit the data significantly better than the third?

b. If you extrapolate the models you selected in part (a) back to $t = -4$, what do you find? (Amazon's revenue in 2000 was $2.8 billion and increased in each subsequent year.)

Solution

a. The following table shows the original data together with the values, rounded to the nearest 0.1, for all three models:

t	0	2	4	6	8	10
Revenue ($ billion)	7	11	19	34	61	89
Linear: $r(t) = 8t - 3$ Technology: `8*x-3`	-3	13	29	45	61	77
Quadratic: $r(t) = 0.9t^2 - 0.6t + 7$ Technology: `0.9x^2-0.6x+7`	7	9.4	19	35.8	59.8	91
Exponential: $r(t) = 7(1.3^t)$ Technology: `7*1.3^x`	7	11.8	20	33.8	57.1	96.5

Notice that all three models give values that seem reasonably close to the actual sales values. However, the graphs show the quadratic and exponential functions to be a lot more accurate than the linear one (see Figure 11).

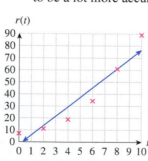
Linear: $r(t) = 8t - 3$

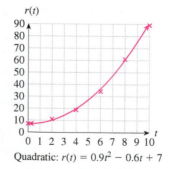
Quadratic: $r(t) = 0.9t^2 - 0.6t + 7$

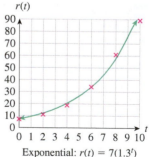
Exponential: $r(t) = 7(1.3^t)$

Figure 11

---

[20] Figures are rounded. Source: www.wikinvest.com.

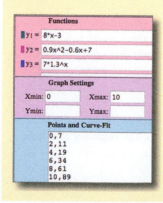

b. Although the quadratic and exponential models both appear to fit the data well, when we extrapolate, we find

$$\text{Quadratic model: } r(-4) = 0.9(-4)^2 - 0.6(-4) + 7 = 23.8$$

$$\text{Exponential model: } r(-4) = 7(1.3^{-4}) \approx 2.5.$$

Notice that the quadratic model predicts *more revenue* in 2000 than in 2004, whereas the exponential model predicts a value closer to the actual value of $2.8 billion. This discrepancy can be seen quite dramatically in Figure 12.

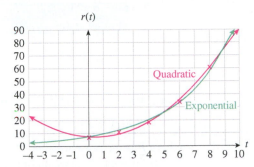

Figure 12

➡ **Before we go on . . .**

Q : *Does the result of Example 5(b) mean that the exponential model is more trustworthy than the quadratic model for projecting Amazon's revenue in years beyond 2013?*

A : The answer is a *qualified* "no:" No matter how well a mathematical curve fits a set of data over a specified period of time, there is nothing we can say with any certainty—either mathematically or statistically—as to whether that model applies to data beyond that period. On the other hand, the growth of many natural phenomena, from the revenues earned by a new company to population growth and the spread of an epidemic, tends to follow an exponential curve for a period of time,[*] so it is not surprising that extrapolating our exponential model backwards gave a good agreement with the real data. It is also quite reasonable to assume that our exponential model will also predict Amazon's revenue in the not-too-distant future, assuming that nothing out of the ordinary occurs. However, Amazon's revenue may begin to level off (as common sense predicts that it has to do eventually), experience a new spurt of growth due to the introduction of a popular new product, or take a nose-dive due to the emergence of a strong competitor.

[*] To model their behavior over long periods of time requires a *logistic* model.

Q : *So what's the use of mathematical models of time-dependent data in the first place?*

A : Curve-fitting models can be used to analyze the trends of data we already have. For instance, the slope in the linear model in Example 5 gives us information about the rate of increase in Amazon's revenue, the coefficient of t^2 in the quadratic model gives us information about its *acceleration,* and the exponent in the exponential model gives us information about its *percentage* rate of increase. All this information can guide us in shaping an *analytical* model of the data, and such models *can* give meaningful predictions of what will happen in the future if they are based on reasonable assumptions about ongoing trends and other external factors.

We now derive an analytical model of change over time based on the idea of **compound interest**. Suppose you invest $500 (the **present value**) in an investment account with an annual yield of 15% and the interest is reinvested at the end of every year (we say that the interest is **compounded** or **reinvested** once a year). Let t represent the number of years since you made the initial $500 investment. Each year, the investment is worth 115% of (or 1.15 times) its value the previous year. The **future value** A of your investment changes over time t, so we think of A as a function of t. The following table illustrates how we can calculate the future value for several values of t:

Year t	0	1	2	3
Future Value $A(t)$ ($)	500	575	661.25	760.44
A		$500(1.15)$	$500(1.15)^2$	$500(1.15)^3$

$\times 1.15 \qquad \times 1.15 \qquad \times 1.15$

Thus, $A(t) = 500(1.15)^t$. A traditional way to write this formula is

$$A(t) = P(1 + r)^t,$$

where P is the present value ($P = 500$) and r is the annual interest rate ($r = 0.15$).

If, instead of compounding the interest once a year, we compound it every three months (four times a year), we would earn one quarter of the interest ($r/4$ of the current investment) every three months. Because this would happen $4t$ times in t years, the formula for the future value becomes

$$A(t) = P\left(1 + \frac{r}{4}\right)^{4t}.$$

Compound Interest

If an amount (**present value**) P is invested for t years at an annual rate of r and if the interest is compounded (reinvested) m times per year, then the **future value** A is

$$A(t) = P\left(1 + \frac{r}{m}\right)^{mt}.$$

A special case is **interest compounded once a year**:

$$A(t) = P(1 + r)^t.$$

Quick Example

9. If $2,000 is invested for two and a half years in a mutual fund with an annual yield of 12.6% and the earnings are reinvested each month, then $P = 2,000$, $r = 0.126$, $m = 12$, and $t = 2.5$, which gives

$$A(2.5) = 2,000\left(1 + \frac{0.126}{12}\right)^{12 \times 2.5}$$

`2000*(1+0.126/12)^(12*2.5)`

$$= 2,000(1.0105)^{30} = \$2,736.02.$$

EXAMPLE 6   Compound Interest: Investments

Consider the scenario in Quick Example 9: You invest $2,000 in a mutual fund with an annual yield of 12.6%, and the interest is reinvested each month.

a. Find the associated exponential model.

b. Compute the value of your investment after 7, 8, and 9 years. During which year does the value of your investment reach $5,000?

Solution

a. Apply the formula

$$A(t) = P\left(1 + \frac{r}{m}\right)^{mt}$$

with $P = 2{,}000$, $r = 0.126$, and $m = 12$. We get

$$A(t) = 2{,}000\left(1 + \frac{0.126}{12}\right)^{12t}$$

$$= 2{,}000(1.0105)^{12t}. \qquad \text{2000*(1+0.126/12)^(12*t)}$$

This is the exponential model. (What would happen if we left out the last set of parentheses in the technology formula?)

b. To find the value of your investment after 7, 8, and 9 years, we calculate $A(t)$ for these values of t:

$$C(7) = 2{,}000\left(1 + \frac{0.126}{12}\right)^{12(7)} \approx \$4{,}809.29 \qquad \text{2000*(1+0.126/12)^(12*7)}$$

$$C(8) = 2{,}000\left(1 + \frac{0.126}{12}\right)^{12(8)} \approx \$5{,}451.51 \qquad \text{2000*(1+0.126/12)^(12*8)}$$

$$C(9) = 2{,}000\left(1 + \frac{0.126}{12}\right)^{12(9)} \approx \$6{,}179.49. \qquad \text{2000*(1+0.126/12)^(12*9)}$$

Because the balance first exceeds $5,000 at $t = 8$ (the end of year 8), your investment reaches $5,000 during year 8.

The compound interest examples we saw above are instances of **exponential growth**: a quantity whose magnitude is an increasing exponential function of time. The decay of unstable radioactive isotopes provides instances of **exponential decay**: a quantity whose magnitude is a *decreasing* exponential function of time. For example, carbon 14, an unstable isotope of carbon, decays exponentially to nitrogen. Because carbon 14 decay is extremely slow, it has important applications in the dating of fossils.

EXAMPLE 7   Exponential Decay: Carbon Dating

The amount of carbon 14 remaining in a sample that originally contained A grams is approximately

$$C(t) = A(0.999879)^t,$$

where t is time in years.

a. What percentage of the original amount remains after 1 year? After 2 years?

b. A fossilized plant unearthed in an archaeological dig contains 0.50 grams of carbon 14 and is known to be 50,000 years old. How much carbon 14 did the plant originally contain?

c. ⊤ Graph the function C for a sample originally containing 50 grams of carbon 14, and use your graph to estimate how long, to the nearest 1,000 years, it takes for half the original carbon 14 to decay.

Solution Notice that the given model is exponential as it has the form $f(t) = Ab^t$. (See the table at the end of Section 1.1.)

a. At the start of the first year, $t = 0$, so there are

$$C(0) = A(0.999879)^0 = A \text{ grams.}$$

At the end of the first year, $t = 1$, so there are

$$C(1) = A(0.999879)^1 = 0.999879\,A \text{ grams;}$$

that is, 99.9879% of the original amount remains. After the second year, the amount remaining is

$$C(2) = A(0.999879)^2 \approx 0.999758\,A \text{ grams,}$$

or about 99.9758% of the original sample.

b. We are given the following information: $C = 0.50$, $A =$ the unknown, and $t = 50{,}000$. Substituting gives

$$0.50 = A(0.999879)^{50{,}000}.$$

Solving for A gives

$$A = \frac{0.5}{0.999879^{50{,}000}} \approx 212 \text{ grams.}$$

Thus, the plant originally contained 212 grams of carbon 14.

c. For a sample originally containing 50 grams of carbon 14, $A = 50$, so $C(t) = 50(0.999879)^t$. Its graph is shown in Figure 13. We have also plotted the line $y = 25$ on the same graph. The graphs intersect at the point where the original sample has decayed to 25 grams: about $t = 6{,}000$ years.

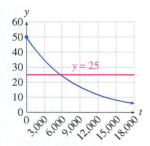

Technology formula:
50*0.999879^x

Figure 13

➡ **Before we go on . . .** The formula we used for A in Example 7(b) has the form

$$A(t) = \frac{C}{0.999879^t},$$

which gives the original amount of carbon 14 t years ago in terms of the amount C that is left now. A similar formula can be used in finance to find the present value, given the future value. ∎

Algebra of Functions

If you look back at some of the functions considered in this section, you will notice that we frequently constructed them by combining simpler or previously constructed functions. For instance:

Quick Example 3:   Area = Width × Length: $A(x) = x(50 - x)$

Example 1:   Cost = Variable cost + Fixed cost:
$$C(x) = 1.80x + 3.05$$

Quick Example 7:   Profit = Revenue − Cost:
$$P(x) = 10x - (8x + 100).$$

Let us look a little more deeply at each of the above examples.

Area Example: $A(x) = \text{Width} \times \text{Length} = x(50 - x)$:
Think of the width and length as separate functions of x:

$$\text{Width: } W(x) = x; \qquad \text{Length: } L(x) = 50 - x$$

so

$$A(x) = W(x)L(x). \qquad \text{\textcolor{blue}{Area = Width × Length}}$$

We say that the area function A is the **product of the functions** W and L, and we write

$$A = WL. \qquad \text{\textcolor{blue}{A is the product of the functions W and L.}}$$

To calculate $A(x)$, we multiply $W(x)$ by $L(x)$.

Cost Example: $C(x) = \text{Variable cost} + \text{Fixed cost} = 1.80x + 3.05$:
Think of the variable and fixed costs as separate functions of x:

$$\text{Variable cost: } V(x) = 1.80x; \qquad \text{Fixed cost: } F(x) = 3.05\text{\textcolor{red}{*}}$$

so

$$C(x) = V(x) + F(x). \qquad \text{\textcolor{blue}{Cost = Variable cost + Fixed cost}}$$

We say that the cost function C is the **sum of the functions V and F**, and we write

$$C = V + F. \qquad \text{\textcolor{blue}{C is the sum of the functions V and F.}}$$

To calculate $C(x)$, we add $V(x)$ to $F(x)$.

Profit Example: $P(x) = \text{Revenue} - \text{Cost} = 10x - (8x + 100)$:
Think of the revenue and cost as separate functions of x:

$$\text{Revenue: } R(x) = 10x; \qquad \text{Cost: } C(x) = 8x + 100$$

so

$$P(x) = R(x) - C(x). \qquad \text{\textcolor{blue}{Profit = Revenue − Cost}}$$

We say that the profit function P is the **difference between the functions R and C**, and we write

$$P = R - C. \qquad \text{\textcolor{blue}{P is the difference of the functions R and C.}}$$

To calculate $P(x)$, we subtract $C(x)$ from $R(x)$.

> **\textcolor{red}{*}** F is called a **constant function** as its value, 3.05, is the same for every value of x.

Algebra of Functions

If f and g are real-valued functions of the real variable x, then we define their **sum s, difference d, product p,** and **quotient q** as follows:

$$s = f + g \text{ is the function specified by } s(x) = f(x) + g(x).$$
$$d = f - g \text{ is the function specified by } d(x) = f(x) - g(x).$$
$$p = fg \text{ is the function specified by } p(x) = f(x)g(x).$$
$$q = \frac{f}{g} \text{ is the function specified by } q(x) = \frac{f(x)}{g(x)}.$$

Also, if f is as above and c is a constant (real number), then we define the associated **constant multiple m of f** by

$$m = cf \text{ is the function specified by } m(x) = cf(x).$$

Note on Domains

In order for any of the expressions $f(x) + g(x)$, $f(x) - g(x)$, $f(x)g(x)$, or $f(x)/g(x)$ to make sense, x must be simultaneously in the domains of both f and g. Further, for the quotient, the denominator $g(x)$ cannot be zero. Thus, we specify the domains of these functions as follows:

> **Domain of $f + g$, $f - g$, and fg:** All real numbers x simultaneously in the domains of f and g
>
> **Domain of f/g:** All real numbers x simultaneously in the domains of f and g such that $g(x) \neq 0$
>
> **Domain of cf:** Same as the domain of f

Quick Examples

10. If $f(x) = x^2 - 1$ and $g(x) = \sqrt{x}$ with domain $[0, +\infty)$, then the sum s of f and g has domain $[0, +\infty)$ and is specified by $s(x) = f(x) + g(x) = x^2 - 1 + \sqrt{x}$.

11. If $f(x) = x^2 - 1$ and $c = 3$, then the associated constant multiple m of f is specified by $m(x) = 3f(x) = 3(x^2 - 1)$.

12. If $c = -1$, then the associated constant multiple $(-1)f$ of f is often written as $-f$. Thus, if $f(x) = x^2 - 1$, then $(-f)(x) = (-1)(x^2 - 1) = -x^2 + 1$.

13. If there are $N = 1{,}000t$ Mars shuttle passengers in year t who pay a total cost of $C = 40{,}000 + 800t$ million dollars, then the cost per passenger is given by the quotient of the two functions:

$$\text{Cost per passenger} = q(t) = \frac{C(t)}{N(t)}$$

$$= \frac{40{,}000 + 800t}{1{,}000t} \text{ million dollars per passenger.}$$

The largest possible domain of C/N is $(0, +\infty)$, as the quotient is not defined if $t = 0$.

1.2 EXERCISES

▼ more advanced   ◆ challenging

T indicates exercises that should be solved using technology

Exercises 1–8 are based on the following functions:

$f(x) = x^2 + 1$ *with domain* $(-\infty, +\infty)$
$g(x) = x - 1$ *with domain* $(-\infty, +\infty)$
$h(x) = x + 4$ *with domain* $[10, +\infty)$
$u(x) = \sqrt{x + 10}$ *with domain* $[-10, 0)$
$v(x) = \sqrt{10 - x}$ *with domain* $[0, 10]$

In each exercise, (a) write a formula for the indicated function, (b) give its domain, and (c) specify its value at the given point a, if defined.

1. $s = f + g$; $a = -3$

2. $d = g - f$; $a = -1$

3. $p = gu$; $a = -6$

4. $p = hv$; $a = 1$

5. $q = \dfrac{v}{g}$; $a = 1$

6. $q = \dfrac{g}{v}$; $a = 1$

7. $m = 5f$; $a = 1$

8. $m = 3u$; $a = -1$

Applications

9. *Resources* You now have 200 music files on your hard drive, and this number is increasing by 10 music files each day. Find a mathematical model for this situation. [**HINT:** See Quick Example 1.]

10. *Resources* The amount of free space left on your hard drive is now 50 gigabytes (GB) and is decreasing by 5 GB per month. Find a mathematical model for this situation.

11. *Soccer* My rectangular soccer field site has a length equal to twice its width. Find its area in terms of its length x. [**HINT:** See Quick Example 3.]

12. *Cabbage* My rectangular cabbage patch has a total area of 100 square feet. Find its perimeter in terms of the width x.

13. *Vegetables* I want to fence in a square vegetable patch. The fencing for the east and west sides costs $4 per foot, and the fencing for the north and south sides costs only $2 per foot. Find the total cost of the fencing as a function of the length of a side x.

14. *Orchids* My square orchid garden abuts my house so that the house itself forms the northern boundary. The fencing for the southern boundary costs $4 per foot, and the fencing for the east and west sides costs $2 per foot. Find the total cost of the fencing as a function of the length of a side x.

15. *Study* You study math 4 hours a day on Sunday through Thursday and take Friday and Saturday off. Model the number of hours h you study math as a function of the day of the week n (with $n = 1$ being Sunday). [**HINT:** See Quick Example 6.]

16. *Recreation* You spend 5 hours per day on Saturdays and Sundays watching movies but only 2 hours per day during the week. Model the number of hours h you watch movies as a function of the day of the week n (with $n = 1$ being Sunday).

17. *Cost* A piano manufacturer has a daily fixed cost of $1,000 and a marginal cost of $1,500 per piano. Find the cost $C(x)$ of manufacturing x pianos in one day. Use your function to answer the following questions:
 a. On a given day, what is the cost of manufacturing three pianos?
 b. What is the cost of manufacturing the third piano that day?
 c. What is the cost of manufacturing the 11th piano that day?
 d. What is the variable cost? What is the fixed cost? What is the marginal cost?
 e. Graph C as a function of x. [**HINT:** See Example 1.]

18. *Cost* The cost of renting tuxes for the Choral Society's formal is $20 down plus $88 per tux. Express the cost C as a function of x, the number of tuxedos rented. Use your function to answer the following questions:
 a. What is the cost of renting two tuxes?
 b. What is the cost of the second tux?
 c. What is the cost of the 4,098th tux?
 d. What is the variable cost? What is the fixed cost? What is the marginal cost?
 e. Graph C as a function of x.

19. *Break-Even Analysis* Your college newspaper, *The Collegiate Investigator*, has fixed production costs of $70 per edition and marginal printing and distribution costs of 40¢ per copy. *The Collegiate Investigator* sells for 50¢ per copy.
 a. Write down the associated cost, revenue, and profit functions. [**HINT:** See Examples 1 and 2.]
 b. What profit (or loss) results from the sale of 500 copies of *The Collegiate Investigator*?
 c. How many copies should be sold to break even?

20. *Break-Even Analysis* The Audubon Society at *Enormous State University* (ESU) is planning its annual fund-raising "Eat-a-thon." The society will charge students 50¢ per serving of pasta. The only expenses the society will incur are the cost of the pasta, estimated at 15¢ per serving, and the $350 cost of renting the facility for the evening.
 a. Write down the associated cost, revenue, and profit functions.
 b. How many servings of pasta must the Audubon Society sell to break even?
 c. What profit (or loss) results from the sale of 1,500 servings of pasta?

21. *Break-Even Analysis Gymnast Clothing* manufactures expensive hockey jerseys for sale to college bookstores in runs of up to 200. Its cost (in dollars) for a run of x hockey jerseys is

$$C(x) = 2,000 + 10x + 0.2x^2 \quad (0 \le x \le 200).$$

Gymnast Clothing sells the jerseys at $100 each. Find the revenue and profit functions. How many jerseys should Gymnast Clothing manufacture to make a profit? [**HINT:** See Example 2.]

22. *Break-Even Analysis Gymnast Clothing* also manufactures expensive soccer cleats for sale to college bookstores in runs of up to 500. Its cost (in dollars) for a run of x pairs of cleats is

$$C(x) = 3,000 + 8x + 0.1x^2 \quad (0 \le x \le 500).$$

Gymnast Clothing sells the cleats at $120 per pair. Find the revenue and profit functions. How many pairs of cleats should Gymnast Clothing manufacture to make a profit?

23. *Break-Even Analysis: School Construction Costs* The cost, in millions of dollars, of building a two-story high school in New York State was estimated to be

$$C(x) = 1.7 + 0.12x - 0.0001x^2 \quad (20 \le x \le 400),$$

where x is the number of thousands of square feet.[21] Suppose that you are contemplating building a for-profit two-story high school and estimate that your total revenue will be $0.1 million per thousand square feet. What is the profit function? What size school should you build to break even?

---

[21] The model is the authors'. Source for data: *Project Labor Agreements and Public Construction Cost in New York State,* Paul Bachman and David Tuerck, Beacon Hill Institute at Suffolk University, April 2006, www.beaconhill.org.

24. *Break-Even Analysis: School Construction Costs* The cost, in millions of dollars, of building a three-story high school in New York State was estimated to be

$$C(x) = 1.7 + 0.14x - 0.0001x^2 \quad (20 \le x \le 400),$$

where x is the number of thousands of square feet.[22] Suppose that you are contemplating building a for-profit three-story high school and estimate that your total revenue will be $0.2 million per thousand square feet. What is the profit function? What size school should you build to break even?

25. ▼ *Profit Analysis: Aviation* The hourly operating cost of a **Boeing** 747-100, which seats up to 405 passengers, is estimated to be[23] $5,132. If an airline charges each passenger a fare of $100 per hour of flight, find the hourly profit P it earns operating a 747-100 as a function of the number of passengers x. (Be sure to specify the domain.) What is the least number of passengers it must carry to make a profit? [**HINT:** The cost function is constant (Variable cost = 0).]

26. ▼ *Profit Analysis: Aviation* The hourly operating cost of a **McDonnell Douglas** DC 10-10, which seats up to 295 passengers, is estimated to be[24] $3,885. If an airline charges each passenger a fare of $100 per hour of flight, find the hourly profit P it earns operating a DC 10-10 as a function of the number of passengers x. (Be sure to specify the domain.) What is the least number of passengers it must carry to make a profit? [**HINT:** The cost function is constant (Variable cost = 0).]

27. ▼ *Break-Even Analysis* *(based on a question from a CPA exam)* The *Oliver Company* plans to market a new product. Based on its market studies, Oliver estimates that it can sell up to 5,500 units in 2005. The selling price will be $2 per unit. Variable costs are estimated to be 40% of total revenue. Fixed costs are estimated to be $6,000 for 2005. How many units should the company sell to break even?

28. ▼ *Break-Even Analysis* *(based on a question from a CPA exam)* The *Metropolitan Company* sells its latest product at a unit price of $5. Variable costs are estimated to be 30% of the total revenue, while fixed costs amount to $7,000 per month. How many units should the company sell per month to break even, assuming that it can sell up to 5,000 units per month at the planned price?

29. ◆ *Break-Even Analysis* *(from a CPA exam)* Given the following notations, write a formula for the break-even sales level:

SP = Selling price per unit
FC = Total fixed cost
VC = Variable cost per unit.

30. ◆ *Break-Even Analysis* *(based on a question from a CPA exam)* Given the following notation, give a formula for the total fixed cost:

SP = Selling price per unit
VC = Variable cost per unit
BE = Break-even sales level in units.

31. ◆ *Break-Even Analysis: Organized Crime* The organized crime boss and perfume king Butch (Stinky) Rose has daily overheads (bribes to corrupt officials, motel photographers, wages for hit men, explosives, and so on) amounting to $20,000 per day. On the other hand, he has a substantial income from his counterfeit perfume racket: He buys imitation French perfume (Chanel № 22.5) at $20 per gram, pays an additional $30 per 100 grams for transportation, and sells the perfume via his street thugs for $600 per gram. Specify Stinky's profit function, $P(x)$, where x is the quantity (in grams) of perfume he buys and sells, and use your answer to calculate how much perfume should pass through his hands per day in order that he break even.

32. ◆ *Break-Even Analysis: Disorganized Crime* Butch (Stinky) Rose's counterfeit Chanel № 22.5 racket has run into difficulties: It seems that the *authentic* Chanel № 22.5 perfume is selling for less than his counterfeit perfume. However, he has managed to reduce his fixed costs to zero, and his overall costs are now $400 per gram plus $30 per gram transportation costs and commission. (The perfume's smell is easily detected by specially trained Chanel Hounds, and this necessitates elaborate packaging measures.) He therefore decides to sell the perfume for $420 per gram to undercut the competition. Specify Stinky's profit function, $P(x)$, where x is the quantity (in grams) of perfume he buys and sells, and use your answer to calculate how much perfume should pass through his hands per day in order that he break even. Interpret your answer.

33. *Demand: E-Readers* The demand for **Amazon**'s Kindle e-reader can be approximated by

$$q(p) = \frac{760}{p} - 1 \text{ million units per year} \quad (60 \le p \le 400),$$

where p is the price charged by Amazon.[25]
a. Graph the demand function.
b. What is the result on demand if the unit price is increased from $100 to $200? [**HINT:** See Example 3.]
c. According to the graph in part (a), if the price is $200 and successively increases in $10 increments, then the demand
   (A) increases at a greater and greater rate.
   (B) decreases at a greater and greater rate.
   (C) increases at a smaller and smaller rate.
   (D) decreases at a smaller and smaller rate.
   (E) increases at the same rate.
   (F) decreases at the same rate.

---

[22] See footnote for Exercise 23.

[23] In 1992. Source: Air Transportation Association of America.

[24] *Ibid.*

[25] Based on data from 2007 to 2013. Source: www.e-reader-info.com.

34. *Demand for Monorail Service: Mars* The demand for monorail service on the Utarek monorail, which links the three urbynes (or districts) of Utarek on Mars, can be approximated by

$$q(p) = 7.5 + \frac{30}{p} \text{ million rides per day} \quad (3 \le p \le 8),$$

where p is the cost per ride in zonars ($\overline{Z}$).[26]
   a. Graph the demand function.
   b. What is the result on demand if the cost per ride is decreased from $\overline{Z}5.00$ to $\overline{Z}3.00$?
   c. If the demand function is extrapolated, what does its graph suggest will be the effect of increasing the price to extremely large values?

35. ▼ *Demand: Smartphones* The worldwide demand for smartphones may be modeled by

$$q(p) = 0.17p^2 - 63p + 5{,}900 \text{ million units sold annually} \quad (100 \le p \le 200),$$

where p is the unit price in dollars.[27]
   a. Use the demand function to estimate, to the nearest million units, worldwide sales of smartphones if the price is $110.
   b. Extrapolate the demand function to estimate, to the nearest million units, worldwide sales of smartphones if the price is $90.
   c. Model the worldwide annual revenue from the sale of smartphones as a function of unit price, and use your model to estimate, to the nearest billion dollars, worldwide annual revenue when the price is set at $110. [**HINT:** Revenue = Price × Quantity = $p \cdot q(p)$.]
   d. Graph the function you obtained in part (c). According to the graph, does worldwide revenue increase or decrease as the price decreases past $110?

36. ▼ *Demand: Smartphones* (See Exercise 35.) Here is another model for worldwide demand for smartphones:

$$q(p) = 36{,}900(0.968^p) \text{ million units sold annually} \quad (100 \le p \le 200),$$

where p is the unit price in dollars.[28]
   a. Use the demand function to estimate, to the nearest million units, worldwide sales of smartphones if the price is $120.
   b. Extrapolate the demand function to estimate, to the nearest million units, worldwide sales of smartphones if the price is $210.

c. Model the worldwide annual revenue from the sale of smartphones as a function of unit price, and use your model to estimate, to the nearest billion dollars, worldwide annual revenue when the price is set at $120. [**HINT:** Revenue = Price × Quantity = $p \cdot q(p)$.]
   d. Graph the function you obtained in part (c). According to the graph, would increased worldwide revenue result from increasing or decreasing the price beyond $120?

37. *Equilibrium Price: Skateboards* The demand for your hand-made skateboards, in weekly sales, is

$$q = -3p + 700$$

if the selling price is p. You are prepared to supply $q = 2p - 500$ skateboards per week at the price p. At what price should you sell your skateboards so that there is neither a shortage nor a surplus? [**HINT:** See Quick Example 8.]

38. *Equilibrium Price: Skateboards* The demand for your factory-made skateboards, in weekly sales, is

$$q = -5p + 50$$

if the selling price is p. If you are selling them at that price, you can obtain $q = 3p - 30$ skateboards per week from the factory. At what price should you sell your skateboards so that there is neither a shortage nor a surplus?

39. *Equilibrium Price: Cell Phones* Worldwide quarterly sales of Nokia cell phones were approximately $q = -p + 156$ million phones when the wholesale price[29] was p.
   a. If Nokia was prepared to supply $q = 4p - 394$ million phones per quarter at a wholesale price of p, what would have been the equilibrium price?
   b. The actual wholesale price was $105 in the fourth quarter of 2004. Estimate the projected shortage or surplus at that price. [**HINT:** See Quick Example 8 and also Example 4.]

40. *Equilibrium Price: Cell Phones* Worldwide annual sales of all cell phones were approximately $-10p + 1{,}600$ million phones when the wholesale price[30] was p.
   a. If manufacturers were prepared to supply $q = 14p - 800$ million phones per year at a wholesale price of p, what would have been the equilibrium price?
   b. The actual wholesale price was projected to be $80 in the fourth quarter of 2008. Estimate the projected shortage or surplus at that price.

---

[26] The zonar ($\overline{Z}$) is the official currency in the city-state of Utarek, Mars (formerly www.Marsnext.com, a now extinct virtual society).

[27] The model is the authors' based on data available in 2013. Sources for data: www.businessweek.com, http://techcrunch.com, www.wikipedia.com, www.idc.com.

[28] *Ibid.*

---

[29] Source for data: Embedded.com/Company reports, December 2004.

[30] Wholesale price projections are the authors'. Source for sales prediction: I-Stat/NDR, December 2004.

41. *Demand: E-Readers* The demand for **Amazon**'s Kindle e-reader can be approximated by

$$q = \frac{760}{p} - 1 \text{ million units per year} \quad (60 \le p \le 400),$$

where p is the price charged by Amazon.[31] Assume that Amazon is prepared to supply

$$q = 0.019p - 1 \text{ million units per year} \quad (60 \le p \le 400)$$

at a price of $\$p$ per unit.
a. Calculate the equilibrium price and equilibrium demand.
b. 🖳 Graph the demand and supply functions to confirm your answer in part (a) graphically.
c. Estimate, to the nearest 0.1 million units, the surplus or shortage of Kindle e-readers if the price is set at $72.

42. *Equilibrium Price: Mars Monorail Service* The demand for monorail service on the Utarek monorail, which links the three urbynes (or districts) of Utarek on Mars, can be approximated by

$$q = 7.5 + \frac{30}{p} \text{ million rides per day} \quad (3 \le p \le 8),$$

where p is the fare the Utarek Monorail Cooperative charges in zonars. ($\overline{Z}$).[32] Assume that the cooperative is prepared to provide service for

$$q = 1.2p + 7.5 \text{ million rides per day} \quad (3 \le p \le 8)$$

at a fare of $\overline{Z}p$.
a. Calculate the equilibrium price and equilibrium demand.
b. 🖳 Graph the demand and supply functions to confirm your answer in part (a) graphically.
c. Estimate the shortage or surplus of monorail service at the December 2085 fare of $\overline{Z}6$ per ride.

43. ▼ *Toxic Waste Treatment* The cost of treating waste by removing PCPs goes up rapidly as the quantity of PCPs removed goes up. Here is a possible model:

$$C(q) = 2,000 + 100q^2,$$

where q is the reduction in toxicity (in pounds of PCPs removed per day) and $C(q)$ is the daily cost (in dollars) of this reduction.
a. Find the cost of removing 10 pounds of PCPs per day.
b. Government subsidies for toxic waste cleanup amount to

$$S(q) = 500q,$$

where q is as above and $S(q)$ is the daily dollar subsidy. The *net cost* function is given by $N = C - S$. Give a formula for $N(q)$, and interpret your answer.
c. Find $N(20)$, and interpret your answer.

44. ▼ *Dental Plans* A company pays for its employees' dental coverage at an annual cost C given by

$$C(q) = 1{,}000 + 100\sqrt{q},$$

where q is the number of employees covered and $C(q)$ is the annual cost in dollars.
a. If the company has 100 employees, find its annual outlay for dental coverage.
b. Assume that the government subsidizes coverage by an annual dollar amount of

$$S(q) = 200q.$$

The *net cost* function is given by $N = C - S$. Give a formula for $N(q)$, and interpret your answer.
c. Find $N(100)$, and interpret your answer.

45. *Spending on Corrections in the 1990s* The following table shows the annual spending by all states in the United States on corrections:[33]

Year t (year since 1990)	0	2	4	6	7
Spending ($ billion)	16	18	22	28	30

a. Which of the following functions best fits the given data? (Warning: None of them fits exactly, but one fits more closely than the others.) [**HINT:** See Example 5.]
  (A) $S(t) = -0.2t^2 + t + 16$
  (B) $S(t) = 0.2t^2 + t + 16$
  (C) $S(t) = t + 16$
b. Use your answer to part (a) to "predict" spending on corrections in 1998, assuming that the trend continued.

46. *Spending on Corrections in the 1990s* Repeat Exercise 45, this time choosing from the following functions:
  (A) $S(t) = 16 + 2t$
  (B) $S(t) = 16 + t + 0.5t^2$
  (C) $S(t) = 16 + t - 0.5t^2$

47. *Soccer Gear* The *East Coast College* soccer team is planning to buy new gear for its road trip to California. The cost per shirt depends on the number of shirts the team orders as shown in the following table:

Shirts Ordered x	5	25	40	100	125
Cost/Shirt $A(x)$ ($)	22.91	21.81	21.25	21.25	22.31

a. Which of the following functions best models the data?
  (A) $A(x) = 0.005x + 20.75$
  (B) $A(x) = 0.01x + 20 + \dfrac{25}{x}$

---

[31] Based on data from 2007 to 2013. Source: www.e-reader-info.com.
[32] The official currency of Utarek, Mars. (See the footnote to Exercise 34.)

[33] Data are rounded. Source: National Association of State Budget Officers/*New York Times*, February 28, 1999, p. A1.

(C) $A(x) = 0.0005x^2 - 0.07x + 23.25$
(D) $A(x) = 25.5(1.08)^{(x-5)}$

b. ▮ Graph the model you chose in part (a) for $10 \le x \le 100$. Use your graph to estimate the lowest cost per shirt and the number of shirts the team should order to obtain the lowest price per shirt.

48. Hockey Gear The *South Coast College* hockey team wants to purchase wool hats for its road trip to Alaska. The cost per hat depends on the number of hats the team orders as shown in the following table:

Hats Ordered x	5	25	40	100	125
Cost/Hat $A(x)$ ($)	25.50	23.50	24.63	30.25	32.70

a. Which of the following functions best models the data?
(A) $A(x) = 0.05x + 20.75$

(B) $A(x) = 0.1x + 20 + \dfrac{25}{x}$

(C) $A(x) = 0.0008x^2 - 0.07x + 23.25$
(D) $A(x) = 25.5(1.08)^{(x-5)}$

b. ▮ Graph the model you chose in part (a) with $5 \le x \le 30$. Use your graph to estimate the lowest cost per hat and the number of hats the team should order to obtain the lowest price per hat.

Cost: Hard Drive Storage *Exercises 49 and 50 are based on the following data showing how the approximate retail cost of a gigabyte of hard drive storage has fallen since 2000:*[34]

Year t (year since 2000)	Cost/Gigabyte $c(t)$ ($)
0	7.5
2	2.5
4	1.2
6	0.6
8	0.2
10	0.1
12	0.06

49. a. ▮ Graph each of the following models together with the data points above, and use your graph to decide which two of the models best fit the data: [**HINT:** See Example 5 and accompanying technology note.]
(A) $c(t) = 6.3(0.67)^t$
(B) $c(t) = 0.093t^2 - 1.6t + 6.7$
(C) $c(t) = 4.75 - 0.50t$

(D) $c(t) = \dfrac{12.8}{t^{1.7} + 1.7}$

b. Of the two models you chose in part (a), which predicts the lower price in 2020? What price does that model predict?

50. a. ▮ Graph each of the following models together with the data points above, and use your graph to decide which three of the models best fit the given data:

(A) $c(t) = \dfrac{15}{1 + 2^t}$

(B) $c(t) = (7.32)0.59^t + 0.10$
(C) $c(t) = 0.00085(t - 9.6)^4$
(D) $c(t) = 7.5 - 0.82t$

b. One of the three best-fit models in part (a) gives an unreasonable prediction for the price in 2020. Which is it, what price does it predict, and why is the prediction unreasonable?

51. Social Website Popularity: Pinterest The following table shows the popularity of **Pinterest** among social media sites as rated by StatCounter.com:[35]

Year t (year since start of 2008)	3	4	5
Pinterest Popularity $p(t)$ (%)	0.0	6.5	13.0

Which of the following kinds of models would best fit the given data? Explain your choice of model. (A, a, b, c, and m are constants.)
(A) $p(t) = mt + b$
(B) $p(t) = at^2 + bt + c$
(C) $p(t) = Ab^t$

52. Social Website Popularity: Twitter The following table shows the popularity of **Twitter** among social media sites as rated by StatCounter.com:[36]

Year t (year since start of 2008)	2	4	5
Twitter Popularity $p(t)$ (%)	6	6	7

Which of the following kinds of models would best fit the given data? Explain your choice of model. (A, a, b, c, and m are constants.)
(A) $p(t) = mt + b$
(B) $p(t) = a^2 + bt + c$
(C) $p(t) = Ab^t$

---

[34] The 2012 price is estimated. Source for data: "Cost of Hard Drive Storage Space," http://ns1758.ca/winch/winchest.html.

[35] Percentages are approximate and based on worldwide page views. Source: http://gs.statcounter.com.

[36] *Ibid.*

53. Demand for Gasoline The following table shows the demand for gasoline in the United States in terms of the price per gallon:[37]

Price p ($/gallon)	1.50	2.50	3	3.50
Demand q (gallons sold/person/day)	1.7	1.65	1.55	1.4

Which of the following kinds of models would best fit the given data? Explain your choice of model. (A, a, b, and c are constants.)
(A) $q = Ab^p$ $(b > 1)$
(B) $q = Ab^p$ $(b < 1)$
(C) $q = ap^2 + bp + c$ $(a > 0)$
(D) $q = ap^2 + bp + c$ $(a < 0)$

54. Demand for Mobile Data The following table shows the worldwide demand for mobile smartphone data in terms of its price per megabyte (MB):[38]

Price p ($/MB)	0.01	0.03	0.06	0.10	0.20	0.46
Demand q (MB/ smartphone user)	4,000	1,000	700	500	300	150

Which of the following kinds of models would best fit the given data? Explain your choice of model. (A, a, b, and c are constants.)
(A) $q = Ap^b$ $(b > 0)$
(B) $q = Ap^b$ $(b < 0)$
(C) $q = ap^2 + bp + c$ $(a > 0)$
(D) $q = ap^2 + bp + c$ $(a < 0)$

55. Investments In August 2013, **E*TRADE Financial** was offering only 0.05% interest on its online checking accounts, with interest reinvested monthly.[39] Find the associated exponential model for the value of a $5,000 deposit after t years. Assuming that this rate of return continued for 7 years, how much would a deposit of $5,000 in August 2013 be worth in August 2020? (Answer to the nearest $1.) [**HINT:** See Quick Example 8.]

56. Investments In August 2013, **Ally Bank** was offering 0.61% interest on its Online Savings Account, with interest reinvested daily.[40] Find the associated exponential model for

the value of a $4,000 deposit after t years. Assuming that this rate of return continued for 8 years, how much would a deposit of $4,000 in August 2013 be worth in August 2021? (Answer to the nearest $1.)

57. Investments Refer to Exercise 55. In August of which year will an investment of $5,000 made in August 2013 first exceed $5,050? [**HINT:** See Example 6.]

58. Investments Refer to Exercise 56. In August of which year will an investment of $4,000 made in August 2013 first exceed $4,400?

59. Carbon Dating A fossil originally contained 104 grams of carbon 14. Refer to the formula for $C(t)$ in Example 7 and estimate the amount of carbon 14 left in the sample after 10,000 years, 20,000 years, and 30,000 years. [**HINT:** See Example 7.]

60. Carbon Dating A fossil contains 4.06 grams of carbon 14. Refer to the formula for $A(t)$ at the end of Example 7, and estimate the amount of carbon 14 in the sample 10,000 years, 20,000 years, and 30,000 years ago.

61. Carbon Dating A fossil contains 4.06 grams of carbon 14. It is estimated that the fossil originally contained 46 grams of carbon 14. By calculating the amount left after 5,000 years, 10,000 years, ..., 35,000 years, estimate the age of the sample to the nearest 5,000 years. (Refer to the formula for $C(t)$ in Example 7.)

62. Carbon Dating A fossil contains 2.8 grams of carbon 14. It is estimated that the fossil originally contained 104 grams of carbon 14. By calculating the amount 5,000 years, 10,000 years, ..., 35,000 years ago, estimate the age of the sample to the nearest 5,000 years. (Refer to the formula for $C(t)$ at the end of Example 7.)

63. Radium Decay The amount of radium 226 remaining in a sample that originally contained A grams is approximately

$$C(t) = A(0.999567)^t$$

where t is time in years.
a. Find, to the nearest whole number, the percentage of radium 226 left in an originally pure sample after 1,000 years, 2,000 years, and 3,000 years.
b. Use a graph to estimate, to the nearest 100 years, when one half of a sample of 100 grams will have decayed.

64. Iodine Decay The amount of iodine 131 remaining in a sample that originally contained A grams is approximately

$$C(t) = A(0.9175)^t$$

where t is time in days.
a. Find, to the nearest whole number, the percentage of iodine 131 left in an originally pure sample after 2 days, 4 days, and 6 days.
b. Use a graph to estimate, to the nearest day, when one half of a sample of 100 grams will have decayed.

---

[37] Source: www.advisorperspectives.com.

[38] Data are approximate. Source: http://mobithinking.com.

[39] Source: https://us.etrade.com, August 2013.

[40] Interest rate based on annual percentage yield. Source: www.ally.com, August 2013.

Communication and Reasoning Exercises

65. If the population of the lunar station at Clavius is given by $P = 200 + 30t$, where t is time in years since the station was established, then the population is increasing by _____ per year.

66. My bank balance can be modeled by $B(t) = 5,000 - 200t$ dollars, where t is time in days since I opened the account. The balance on my account is _____ by \$200 per day.

67. Classify the following model as analytical or curve fitting, and give a reason for your choice: The price of gold was \$700 on Monday, \$710 on Tuesday, and \$700 on Wednesday. Therefore, the price can be modeled by $p(t) = -10t^2 + 20t + 700$ where t is the day since Monday.

68. Classify the following model as analytical or curve fitting, and give a reason for your choice: The width of a small animated square on my computer screen is currently 10 mm and is growing by 2 mm per second. Therefore, its area can be modeled by $a(t) = (10 + 2t)^2$ square mm where t is time in seconds.

69. Fill in the missing information for the following *analytical model* (answers may vary): _____. Therefore, the cost of downloading a movie can be modeled by $c(t) = 4 - 0.2t$, where t is time in months since January.

70. Repeat Exercise 69, but this time regard the given model as a *curve-fitting model*.

71. Fill in the blanks: In a linear cost function, the _____ cost is x times the _____ cost.

72. Complete the following sentence: In a linear cost function, the marginal cost is the _____.

73. ▼ We said in the discussion of demand and supply models that the demand for a commodity generally goes down as the price goes up. Assume that the demand for a certain commodity goes up as the price goes up. Is it still possible for there to be an equilibrium price? Explain with the aid of a demand and supply graph.

74. ▼ What would happen to the price of a certain commodity if the demand was always greater than the supply? Illustrate with a demand and supply graph.

75. You have a set of data points showing the sales of videos on your website versus time that are closely approximated by two different mathematical models. Give one criterion that would lead you to choose one over the other. (Answers may vary.)

76. Would it ever be reasonable to use a quadratic model $s(t) = at^2 + bt + c$ to predict long-term sales if a is negative? Explain.

77. If f and g are functions with $f(x) \geq g(x)$ for every x, what can you say about the values of the function $f - g$?

78. If f and g are functions with $f(x) > g(x) > 0$ for every x, what can you say about the values of the function $\frac{f}{g}$?

79. If f is measured in books and g is measured in people, what are the units of measurement of the function $\frac{f}{g}$?

80. If f and g are linear functions, then what can you say about $f - g$?

1.3 Linear Functions and Models

Linear functions are among the simplest functions and are perhaps the most useful of all mathematical functions.

Linear Function

A **linear function** is one that can be written in the form

Quick Example

$$f(x) = mx + b \qquad \text{Function form}$$

$$f(x) = 3x - 1$$

or

$$y = mx + b \qquad \text{Equation form}$$

$$y = 3x - 1$$

where m and b are fixed numbers. (The names m and b are traditional.*)

---

* Actually, c is sometimes used instead of b. As for m, there has even been some research into the question of its origin, but no one knows exactly why that particular letter is used.

Linear Functions from the Numerical and Graphical Point of View

The following table shows values of $y = 3x - 1$ ($m = 3$, $b = -1$) for some values of x:

x	-4	-3	-2	-1	0	1	2	3	4
y	-13	-10	-7	-4	-1	2	5	8	11

Notice that setting $x = 0$ gives $y = -1$, the value of b.

Numerically, b is the value of y when x = 0.

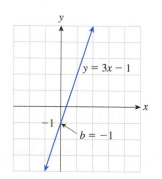

y-intercept = $b = -1$
Graphically, b is the y-intercept of the graph.

Figure 14

The graph of $f(x) = 3x - 1$ is shown in Figure 14. On the graph, the point $(0, b) = (0, -1)$ is the point where the graph crosses the y-axis, so we say that $b = -1$ is the **y-intercept** of the graph (Figure 14).

What about m? Looking once again at the table, notice that y increases by $m = 3$ units for every increase of 1 unit in x. This is caused by the term $3x$ in the formula: For every increase of 1 in x, we get an increase of $3 \times 1 = 3$ in y.

Numerically, y increases by m units for every 1-unit increase of x.

Likewise, for every increase of 2 in x, we get an increase of $3 \times 2 = 6$ in y. In general, if x increases by some amount, y will increase by three times that amount. We write

Change in $y = 3 \times$ Change in x.

The Change in a Quantity: Delta Notation

If a quantity q changes from q_1 to q_2, the **change in q** is just the difference:

Change in q = Second value − First value

$$= q_2 - q_1.$$

Mathematicians traditionally use Δ (delta, the Greek equivalent of the Roman letter D) to stand for change and write the change in q as Δq:

$$\Delta q = \text{Change in } q = q_2 - q_1.$$

Quick Examples

1. If x is changed from 1 to 3, we write

   $$\Delta x = \text{Second value} - \text{First value} = 3 - 1 = 2.$$

2. Looking at our linear equation $y = 3x - 1$, we see that, when x changes from 1 to 3, y changes from 2 to 8. So

   $$\Delta y = \text{Second value} - \text{First value} = 8 - 2 = 6.$$

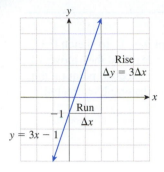

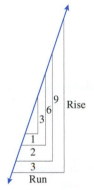

Slope = m = 3
Graphically, m is the slope of the graph.

Figure 15

Using delta notation, we can now write, for our linear equation $y = 3x - 1$,

$$\Delta y = 3\Delta x \qquad \text{Change in } y = 3 \times \text{Change in } x$$

or

$$\frac{\Delta y}{\Delta x} = 3.$$

Because the value of y increases by exactly 3 units for every increase of 1 unit in x, the graph is a straight line rising by 3 units for every 1 unit we go to the right. We say that we have a **rise** of 3 units for each **run** of 1 unit. Because the value of y changes by $\Delta y = 3\Delta x$ units for every change of Δx units in x, in general we have a rise of $\Delta y = 3\Delta x$ units for each run of Δx units (Figure 15). Thus, we have a rise of 6 for a run of 2, a rise of 9 for a run of 3, and so on. So $m = 3$ is a measure of the steepness of the line; we call m the **slope of the line**:

$$\text{Slope} = m = \frac{\Delta y}{\Delta x} = \frac{\text{Rise}}{\text{Run}}.$$

In general (replace the number 3 by a general number m), we can say the following.

The Roles of m and b in the Linear Function $f(x) = mx + b$

Role of m

Numerically   If $y = mx + b$, then y changes by m units for every 1-unit change in x. A change of Δx units in x results in a change of $\Delta y = m\Delta x$ units in y. Thus,

$$m = \frac{\Delta y}{\Delta x} = \frac{\text{Change in } y}{\text{Change in } x}.$$

Graphically m is the slope of the line $y = mx + b$:

$$m = \frac{\Delta y}{\Delta x} = \frac{\text{Rise}}{\text{Run}} = \text{Slope}.$$

For positive m the graph rises m units for every 1-unit move to the right, and rises $\Delta y = m\Delta x$ units for every Δx units moved to the right. For negative m the graph drops $|m|$ units for every 1-unit move to the right, and drops $|m|\Delta x$ units for every Δx units moved to the right.

Graph of $y = mx + b$

Positive m   *Negative m*

Role of b

Numerically   When $x = 0$, $y = b$.

Graphically b is the y-intercept of the line $y = mx + b$.

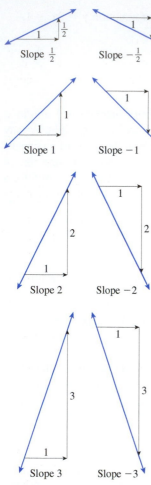

Figure 16

Quick Examples

3. $f(x) = 2x + 1$ has slope $m = 2$ and y-intercept $b = 1$. To sketch the graph, we start at the y-intercept $b = 1$ on the y-axis and then move 1 unit to the right and up $m = 2$ units to arrive at a second point on the graph. Now we connect the two points to obtain the graph on the left.

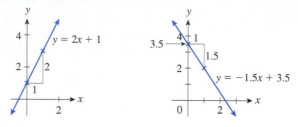

4. The line $y = -1.5x + 3.5$ has slope $m = -1.5$ and y-intercept $b = 3.5$. Because the slope is negative, the graph (above right) goes *down* 1.5 units for every 1 unit it moves to the right.

It helps to be able to picture what different slopes look like, as in Figure 16. Notice that the larger the absolute value of the slope, the steeper is the line.

EXAMPLE 1   Recognizing Linear Data Numerically and Graphically

Which of the following two tables gives the values of a linear function? What is the formula for that function?

x	0	2	4	6	8	10	12
$f(x)$	3	-1	-3	-6	-8	-13	-15

x	0	2	4	6	8	10	12
$g(x)$	3	-1	-5	-9	-13	-17	-21

Solution The function f cannot be linear. If it were, we would have $\Delta f = m \Delta x$ for some fixed number m. However, although the change in x between successive entries in the table is $\Delta x = 2$ each time, the change in f is not the same each time. Thus, the ratio $\Delta f/\Delta x$ is not the same for every successive pair of points.

On the other hand, the ratio $\Delta g/\Delta x$ is the same each time, namely,

$$\frac{\Delta g}{\Delta x} = \frac{-4}{2} = -2,$$

as we see in the following table:

Δx		$2-0=2$	$4-2=2$	$6-4=2$	$8-6=2$	$10-8=2$	$12-10=2$
x	0	2	4	6	8	10	12
$g(x)$	3	-1	-5	-9	-13	-17	-21
Δg		$-1-3$ $=-4$	$-5-(-1)$ $=-4$	$-9-(-5)$ $=-4$	$-13-(-9)$ $=-4$	$-17-(-13)$ $=-4$	$-21-(-17)$ $=-4$

Using Technology
See the Technology Guides at the end of the chapter for detailed instructions on how to obtain a table with the successive quotients $m = \Delta y/\Delta x$ for the functions f and g in Example 1 using a TI-83/84 Plus or Excel. These tables show at a glance that f is not linear. Here is an outline:

TI-83/84 Plus
STAT EDIT; Enter values of x and $f(x)$ in lists L_1 and L_2. Highlight the heading L_3 and then enter the following formula (including the quotes):
"ΔList$(L_2)/\Delta$List(L_1)"
[More details in the Technology Guide.]

Thus, g is linear with slope $m = -2$. By the table, $g(0) = 3$; hence, $b = 3$. Thus,

$$g(x) = -2x + 3. \qquad \text{Check that this formula gives the values in the table.}$$

If you graph the points in the tables defining f and g above, it becomes easy to see that g is linear and f is not; the points of g lie on a straight line (with slope -2), whereas the points of f do not lie on a straight line (Figure 17).

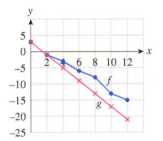

Figure 17

Finding a Linear Equation from Data

If we happen to know the slope and y-intercept of a line, writing down its equation is straightforward. For example, if we know that the slope is 3 and the y-intercept is -1, then the equation is $y = 3x - 1$. Sadly, the information we are given is seldom so convenient. For instance, we may know the slope and a point other than the y-intercept, two points on the line, or other information. We therefore need to know how to use the information we are given to obtain the slope and the intercept.

Computing the Slope

We can always determine the slope of a line if we are given two (or more) points on the line, because any two points—say, (x_1, y_1) and (x_2, y_2)—determine the line and hence its slope. To compute the slope when given two points, recall the formula

$$\text{Slope} = m = \frac{\text{Rise}}{\text{Run}} = \frac{\Delta y}{\Delta x}.$$

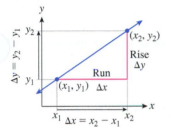

Figure 18

To find its slope, we need a run Δx and corresponding rise Δy. In Figure 18 we see that we can use $\Delta x = x_2 - x_1$, the change in the x-coordinate from the first point to the second, as our run, and $\Delta y = y_2 - y_1$, the change in the y-coordinate, as our rise. The resulting formula for computing the slope is given as follows.

Computing the Slope of a Line

We can compute the slope m of the line through the points (x_1, y_1) and (x_2, y_2) using

$$m = \frac{\Delta y}{\Delta x} = \frac{y_2 - y_1}{x_2 - x_1}.$$

Quick Examples

5. The slope of the line through $(x_1, y_1) = (1, 3)$ and $(x_2, y_2) = (5, 11)$ is

$$m = \frac{\Delta y}{\Delta x} = \frac{y_2 - y_1}{x_2 - x_1} = \frac{11 - 3}{5 - 1} = \frac{8}{4} = 2.$$

Notice that we can use the points in the reverse order: If we take $(x_1, y_1) = (5, 11)$ and $(x_2, y_2) = (1, 3)$, we obtain the same answer:

$$m = \frac{\Delta y}{\Delta x} = \frac{y_2 - y_1}{x_2 - x_1} = \frac{3 - 11}{1 - 5} = \frac{-8}{-4} = 2.$$

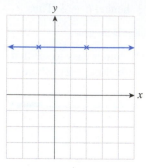

Figure 19

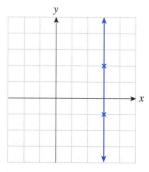

Vertical lines have undefined slope.

Figure 20

6. The slope of the line through $(x_1, y_1) = (1, 2)$ and $(x_2, y_2) = (2, 1)$ is

$$m = \frac{\Delta y}{\Delta x} = \frac{y_2 - y_1}{x_2 - x_1} = \frac{1 - 2}{2 - 1} = \frac{-1}{1} = -1.$$

7. The slope of the line through $(2, 3)$ and $(-1, 3)$ is

$$m = \frac{\Delta y}{\Delta x} = \frac{y_2 - y_1}{x_2 - x_1} = \frac{3 - 3}{-1 - 2} = \frac{0}{-3} = 0.$$

A line of slope 0 has zero rise, so it is a *horizontal* line, as shown in Figure 19.

8. The line through $(3, 2)$ and $(3, -1)$ has slope

$$m = \frac{\Delta y}{\Delta x} = \frac{y_2 - y_1}{x_2 - x_1} = \frac{-1 - 2}{3 - 3} = \frac{-3}{0},$$

which is undefined. The line passing through these points is *vertical*, as shown in Figure 20.

Computing the *y*-Intercept

Once we know the slope m of a line and also the coordinates of a point (x_1, y_1), then we can calculate its *y*-intercept b as follows: The equation of the line must be

$$y = mx + b,$$

where b is as yet unknown. To determine b, we use the fact that the line must pass through the point (x_1, y_1), and so (x_1, y_1) satisfies the equation $y = mx + b$. In other words,

$$y_1 = mx_1 + b.$$

Solving for b gives

$$b = y_1 - mx_1.$$

In summary, we have the following.

Computing the *y*-Intercept of a Line

The *y*-intercept of the line passing through (x_1, y_1) with slope m is

$$b = y_1 - mx_1.$$

Quick Example

9. The line through $(2, 3)$ with slope 4 has

$$b = y_1 - mx_1 = 3 - (4)(2) = -5.$$

Its equation is therefore

$$y = mx + b = 4x - 5.$$

EXAMPLE 2 Finding Linear Equations

Find equations for the following straight lines.

a. Through the points $(1, 2)$ and $(3, -1)$

b. Through $(2, -2)$ and parallel to the line $3x + 4y = 5$

c. Horizontal and through $(-9, 5)$

d. Vertical and through $(-9, 5)$

Solution

a. To write down the equation of the line, we need the slope m and the y-intercept b.

* *Slope* Because we are given two points on the line, we can use the slope formula:

$$m = \frac{y_2 - y_1}{x_2 - x_1} = \frac{-1 - 2}{3 - 1} = -\frac{3}{2}.$$

* *Intercept* We now have the slope of the line, $m = -3/2$, and also a point—we have two to choose from, so let us choose $(x_1, y_1) = (1, 2)$. We can now use the formula for the y-intercept:

$$b = y_1 - mx_1 = 2 - \left(-\frac{3}{2}\right)(1) = \frac{7}{2}.$$

Thus, the equation of the line is

$$y = -\frac{3}{2}x + \frac{7}{2}. \qquad y = mx + b$$

Using Technology

See the Technology Guides at the end of the chapter for detailed instructions on how to obtain the slope and intercept in Example 2(a) using a TI-83/84 Plus or a spreadsheet. Here is an outline:

TI-83/84 Plus
$\boxed{\text{STAT}}$ EDIT; Enter values of x and y in lists L_1 and L_2.
Slope: Home screen
$(L_2(2) - L_2(1)) / (L_1(2) - L_1(1)) \rightarrow M$
Intercept: Home screen
$L_2(1) - M * L_1(1)$
[More details in the Technology Guide.]

Spreadsheet
Enter headings x, y, m, b, in cells A1–D1 and the values (x, y) in cells A2–B3. Enter
$= (B3-B2) / (A3-A2)$
in cell C2 and
$=B2-C2*A2$
in cell D2.
[More details in the Technology Guide.]

b. Proceeding as before, we have the following.

* *Slope* We are not given two points on the line, but we are given a parallel line. We use the fact that *parallel lines have the same slope*. (Why?) We can find the slope of $3x + 4y = 5$ by solving for y and then looking at the coefficient of x:

$$y = -\frac{3}{4}x + \frac{5}{4}, \qquad \text{To find the slope, solve for } y.$$

so the slope is $-3/4$.

* *Intercept* We now have the slope of the line, $m = -3/4$ and also a point $(x_1, y_1) = (2, -2)$. We can now use the formula for the y-intercept:

$$b = y_1 - mx_1 = -2 - \left(-\frac{3}{4}\right)(2) = -\frac{1}{2}.$$

Thus, the equation of the line is

$$y = -\frac{3}{4}x - \frac{1}{2}. \qquad y = mx + b$$

c. We are given a point: $(-9, 5)$. Furthermore, we are told that the line is horizontal, which tells us that the slope is $m = 0$. Therefore, all that remains is the calculation of the y-intercept:

$$b = y_1 - mx_1 = 5 - (0)(-9) = 5,$$

so the equation of the line is

$$y = 5. \qquad y = mx + b$$

d. We are given a point: $(-9, 5)$. This time, we are told that the line is vertical, which means that the slope is undefined. Thus, we can't express the equation of the line in the form $y = mx + b$. (This formula makes sense only when the slope m of the line is defined.) What can we do? Well, here are some points on the desired line:

$$(-9, 1), (-9, 2), (-9, 3), \ldots$$

so $x = -9$ and $y = anything$. If we simply say that $x = -9$, then these points are all solutions, so the equation is $x = -9$.

Applications: Linear Models

Cost Functions

Using linear functions to describe or approximate relationships in the real world is called **linear modeling**. Recall from Section 1.2 that a **cost function** specifies the cost C as a function of the number of items x.

EXAMPLE 3   Linear Cost Function from Data

The manager of the *FrozenAir Refrigerator* factory notices that on Monday it cost the company a total of $25,000 to build 30 refrigerators and on Tuesday it cost $30,000 to build 40 refrigerators. Find a linear cost function based on this information. What is the daily fixed cost, and what is the marginal cost?

Solution We are seeking the cost C as a linear function of x, the number of refrigerators sold:

$$C = mx + b. \qquad \text{Linear cost function (equation form)}$$

We are told that $C = 25,000$ when $x = 30$, and this amounts to being told that $(30, 25,000)$ is a point on the graph of the cost function. Similarly, $(40, 30,000)$ is another point on the line (Figure 21).

We can use the two points on the line to construct the linear cost equation:

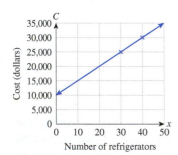

Figure 21

- **Slope** $\quad m = \dfrac{C_2 - C_1}{x_2 - x_1} = \dfrac{30{,}000 - 25{,}000}{40 - 30} = 500 \qquad$ *C plays the role of y.*

- **Intercept** $\quad b = C_1 - mx_1 = 25{,}000 - (500)(30) = 10{,}000.$ $\quad$ We used the point $(x_1, C_1) = (30, 25{,}000)$.

The linear cost function is therefore

$$C(x) = 500x + 10{,}000.$$

Because $m = 500$ and $b = 10{,}000$, the factory's fixed cost is $10,000 each day, and its marginal cost is $500 per refrigerator. (See the discussion of cost functions in Section 1.2.) These are illustrated in Figure 22.

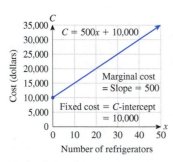

Figure 22

➡ **Before we go on . . .** Recall that, in general, the slope m measures the number of units of change in y per 1-unit change in x, so it is measured in units of y per unit of x:

Units of slope $=$ Units of y per unit of x.

Using Technology
To obtain the cost equation for Example 3 with technology, apply the Technology note for Example 2(a) to the given points (30, 25,000) and (40, 30,000) on the graph of the cost equation.

In Example 3, y is the cost C, measured in dollars, and x is the number of items, measured in refrigerators. Hence,

$$\text{Units of slope} = \text{Units of } y \text{ per unit of } x = \text{Dollars per refrigerator.}$$

The y-intercept b, being a value of y, is measured in the same units as y. In Example 3, b is measured in dollars. ■

Demand Functions

In Section 1.2 we saw that a **demand function** specifies the demand q as a function of the price p per item.

EXAMPLE 4   Linear Demand Function from Data

You run a small supermarket and must determine how much to charge for *Hot 'n' Spicy* brand baked beans. The following chart shows weekly sales figures (the demand) for Hot 'n' Spicy at two different prices:

Price ($/can)	0.50	0.75
Demand (cans sold/week)	400	350

a. Model these data with a linear demand function. (See Example 3 in Section 1.2.)

b. How do we interpret the slope and q-intercept of the demand function?

Solution

a. Recall that a demand equation—or demand function—expresses demand q (in this case, the number of cans of beans sold per week) as a function of the unit price p (in this case, dollars per can). We model the demand using the two points we are given: (0.50, 400) and (0.75, 350).

Using Technology
To obtain the demand equation for Example 4 with technology, apply the Technology note for Example 2(a) to the given points (0.50, 400) and (0.75, 350) on the graph of the demand equation.

$$\textit{Slope:} \quad m = \frac{q_2 - q_1}{p_2 - p_1} = \frac{350 - 400}{0.75 - 0.50} = \frac{-50}{0.25} = -200$$

$$\textit{Intercept:} \quad b = q_1 - mp_1 = 400 - (-200)(0.50) = 500$$

So the demand equation is

$$q = -200p + 500. \qquad {\color{blue} q = mp + b}$$

b. The key to interpreting the slope in a demand equation is to recall (see the "Before we go on" note at the end of Example 3) that we measure the slope in *units of y per unit of x*. Here, $m = -200$, and the units of m are units of q per unit of p, or the number of cans sold per \$1 change in the price. Because m is negative, we see that the number of cans sold decreases as the price increases. We conclude that the weekly sales will drop by 200 cans per \$1 increase in the price.

To interpret the q-intercept, recall that it gives the q-coordinate when $p = 0$. Hence, it is the number of cans the supermarket can "sell" every week if it were to give them away.*

* Does this seem realistic? Demand is not always unlimited if items are given away. For instance, campus newspapers are sometimes given away, yet piles of them are often left untaken. Also see the "Before we go on" discussion at the end of this example.

➡ **Before we go on . . .**

  : *Just how reliable is the linear model used in Example 4?*

A : The *actual* demand graph could in principle be obtained by tabulating demand figures for a large number of different prices. If the resulting points were plotted on the *pq*-plane, they would probably suggest a curve and not a straight line. However, if you looked at a small enough portion of any curve, you could closely *approximate* it by a straight line. In other words, *over a small range of values of p, a linear model is accurate.* Linear models of real-world situations are generally reliable only for small ranges of the variables. (This point will come up again in some of the exercises.)

■

Time Change Models

The next example illustrates modeling change over time t with a linear function of t.

EXAMPLE 5   Growth of Sales

Worldwide sales of tablet computers were expected to increase from around 130 million units in 2012 to around 350 million units in 2017.[41]

a. Use this information to model annual worldwide sales of tablet computers as a linear function of time t in years since 2012. What is the significance of the slope?

b. According to the model, in which year will tablet computer sales surpass 280 million units?

Solution

a. Since we are interested in worldwide sales s of tablet computers as a function of time, we take time t to be the independent variable (playing the role of x) and the annual sales s, in millions of units, to be the dependent variable (in the role of y). Notice that 2012 corresponds to $t = 0$ and 2017 corresponds to $t = 5$, so we are given the coordinates of two points on the graph of s as a function of t: $(0, 130)$ and $(5, 350)$. We model the sales using these two points:

$$m = \frac{s_2 - s_1}{t_2 - t_1} = \frac{350 - 130}{5 - 0} = \frac{220}{5} = 44$$
$$b = s_1 - mt_1 = 130 - (44)(0) = 130.$$

So

$$s = 44t + 130 \text{ million units.} \qquad s = mt + b$$

The slope m is measured in units of s per unit of t; that is, millions of tablet computers per year, and is thus the *rate of change* of annual tablet sales. To say that $m = 44$ is to say that annual sales are increasing at a rate of 44 million tablets per year.

---

[41] Source: ID Press Release, March 26, 2013, www.idc.com.

b. Our model of annual sales as a function of time is

$$s = 44t + 130 \text{ million units.}$$

Annual sales of tablet computers are 280 million when $s = 280$, or

$$280 = 44t + 130$$

Solving for t, we have

$$44t = 280 - 130 = 150$$

$$t = \frac{150}{44} \approx 3.4 \text{ years.}$$

Using Technology

To use technology to obtain s as a function of t in Example 5, apply the Technology note for Example 2(a) to the points (0, 130) and (5, 350) on its graph.

Thus, when $t = 3$ (2015), predicted sales are less than 280 million units, and when $t = 4$ (2016), predicted sales exceed 280 million units. Thus, sales will first surpass 280 million units in 2016. (Notice that, instead of rounding $t = 3.4$ to the nearest whole number, we needed to round upwards to $t = 4$.)

EXAMPLE 6 **Velocity**

You are driving down the Ohio Turnpike, watching the mileage markers to stay awake. Measuring time in hours after you see the 20-mile marker, you see the following markers each half hour:

Time (hours)	0	0.5	1	1.5	2
Marker (miles)	20	47	74	101	128

Find your location s as a function of t, the number of hours you have been driving. (The number s is also called your **position** or **displacement**.)

Solution If we plot the location s versus the time t, the five markers listed give us the graph in Figure 23. These points appear to lie along a straight line. We can verify this by calculating how far you traveled in each half hour. In the first half hour you traveled $47 - 20 = 27$ miles. In the second half hour you traveled $74 - 47 = 27$ miles also. In fact, you traveled exactly 27 miles each half hour. The points we plotted lie on a straight line that rises 27 units for every 0.5 units we go to the right, for a slope of $27/0.5 = 54$.

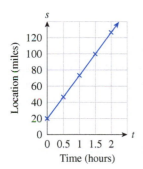

Figure 23

To get the equation of that line, notice that we have the s-intercept, which is the starting marker of 20. Thus, the equation of s as a function of time t is

$$s(t) = 54t + 20. \qquad \text{We used } s \text{ in place of } y \text{ and } t \text{ in place of } x.$$

Notice the significance of the slope: For every hour you travel, you drive a distance of 54 miles. In other words, you are traveling at a constant velocity of 54 mph. We have uncovered a very important principle:

In the graph of displacement versus time, velocity is given by the slope.

Using Technology

To use technology to obtain s as a function of t in Example 6, apply the Technology note for Example 2(a) to the points (0, 20) and (1, 74) on its graph.

Linear Change over Time

If a quantity q is a linear function of time t,

$$q = mt + b,$$

then the slope m measures the **rate of change** of q, and b is the quantity at time $t = 0$, the **initial quantity**. If q represents the position of a moving object, then the rate of change is also called the **velocity**.

Units of m and b

The units of measurement of m are units of q per unit of time. For instance, if q is income in dollars and t is time in years, then the rate of change m is measured in dollars per year.

The units of b are units of q. For instance, if q is income in dollars and t is time in years, then b is measured in dollars.

Quick Example

10. If the accumulated revenue from sales of your video game software is given by $R = 2{,}000t + 500$ dollars, where t is time in years from now, then you have earned $500 in revenue so far, and the accumulated revenue is increasing at a rate of $2,000 per year.

Examples 5 and 6 share the following common theme.

General Linear Models

If $y = mx + b$ is a linear model of changing quantities x and y, then the slope m is the rate at which y is increasing per unit increase in x, and the y-intercept b is the value of y that corresponds to $x = 0$.

Units of m and b

The slope m is measured in units of y per unit of x, and the intercept b is measured in units of y.

Quick Example

11. If the number n of spectators at a soccer game is related to the number g of goals your team has scored so far by the equation $n = 20g + 4$, then you can expect four spectators if no goals have been scored and 20 additional spectators per additional goal scored.

FAQs

What to Use as *x* and *y* and How to Interpret a Linear Model

Q : *In a problem where I must find a linear relationship between two quantities, which quantity do I use as x, and which do I use as y?*

A : The key is to decide which of the two quantities is the independent variable and which is the dependent variable. Then use the independent variable as *x* and the dependent variable as *y*. In other words, *y* depends on *x*.

Here are examples of phrases that convey this information, usually of the form *Find y [dependent variable] in terms of x [independent variable]*:

- Find the cost in terms of the number of items. $y = \text{Cost}, x = \text{Number of items}$
- How does color depend on wavelength? $y = \text{Color}, x = \text{Wavelength}$

If no information is conveyed about which variable is intended to be independent, then you can use whichever is convenient.

Q : *How do I interpret a general linear model $y = mx + b$?*

A : The key to interpreting a linear model is to remember the units we use to measure *m* and *b*:

The slope m is measured in units of y per unit of x. The intercept b is measured in units of y.

For instance, if $y = 4.3x + 8.1$ and you know that *x* is measured in feet and *y* in kilograms, then you can already say, "*y* is 8.1 kilograms when $x = 0$ feet and increases at a rate of 4.3 kilograms per foot" without knowing anything more about the situation.

1.3 EXERCISES

▼ more advanced   ◆ challenging
T indicates exercises that should be solved using technology

In Exercises 1–6, a table of values for a linear function is given. Fill in the missing value and calculate m in each case.

1.

x	−1	0	1
y	5	8	

2.

x	−1	0	1
y	−1	−3	

3.

x	2	3	5
y	−1	−2	

4.

x	2	4	5
y	−1	−2	

5.

x	−2	0	2
y	4		10

6.

x	0	3	6
y	−1		−5

In Exercises 7–10, first find $f(0)$, if not supplied, and then find the equation of the given linear function.

7.

x	−2	0	2	4
f(x)	−1	−2	−3	−4

8.

x	−6	−3	0	3
f(x)	1	2	3	4

9.

x	−4	−3	−2	−1
f(x)	−1	−2	−3	−4

10.

x	1	2	3	4
f(x)	4	6	8	10

In Exercises 11–14, decide which of the two given functions is linear, and find its equation. [**HINT:** See Example 1.]

11.

x	0	1	2	3	4
$f(x)$	6	10	14	18	22
$g(x)$	8	10	12	16	22

12.

x	−10	0	10	20	30
$f(x)$	−1.5	0	1.5	2.5	3.5
$g(x)$	−9	−4	1	6	11

13.

x	0	3	6	10	15
$f(x)$	0	3	5	7	9
$g(x)$	−1	5	11	19	29

14.

x	0	3	5	6	9
$f(x)$	2	6	9	12	15
$g(x)$	−1	8	14	17	26

In Exercises 15–24, find the slope of the given line if it is defined.

15. $y = -\dfrac{3}{2}x - 4$

16. $y = \dfrac{2x}{3} + 4$

17. $y = \dfrac{x + 1}{6}$

18. $y = -\dfrac{2x - 1}{3}$

19. $3x + 1 = 0$

20. $8x - 2y = 1$

21. $3y + 1 = 0$

22. $2x + 3 = 0$

23. $4x + 3y = 7$

24. $2y + 3 = 0$

In Exercises 25–38, graph the given equation.
[**HINT:** See Quick Examples 3 and 4.]

25. $y = 2x - 1$

26. $y = x - 3$

27. $y = -\frac{2}{3}x + 2$

28. $y = -\frac{1}{2}x + 3$

29. $y + \frac{1}{4}x = -4$

30. $y - \frac{1}{4}x = -2$

31. $7x - 2y = 7$

32. $2x - 3y = 1$

33. $3x = 8$

34. $2x = -7$

35. $6y = 9$

36. $3y = 4$

37. $2x = 3y$

38. $3x = -2y$

In Exercises 39–58, calculate the exact slope (rather than a decimal approximation) of the straight line through the given pair of points, if defined. Try to do as many as you can without writing anything down except the answer. [**HINT:** See Quick Example 5.]

39. $(0, 0)$ and $(1, 2)$

40. $(0, 0)$ and $(-1, 2)$

41. $(-1, -2)$ and $(0, 0)$

42. $(2, 1)$ and $(0, 0)$

43. $(4, 3)$ and $(5, 1)$

44. $(4, 3)$ and $(4, 1)$

45. $(1, -1)$ and $(1, -2)$

46. $(-2, 2)$ and $(-1, -1)$

47. $(2, 3.5)$ and $(4, 6.5)$

48. $(10, -3.5)$ and $(0, -1.5)$

49. $(300, 20.2)$ and $(400, 11.2)$

50. $(1, -20.2)$ and $(2, 3.2)$

51. $(0, 1)$ and $\left(-\frac{1}{2}, \frac{3}{4}\right)$

52. $\left(\frac{1}{2}, 1\right)$ and $\left(-\frac{1}{2}, \frac{3}{4}\right)$

53. (a, b) and (c, d) $(a \neq c)$

54. (a, b) and (c, b) $(a \neq c)$

55. (a, b) and (a, d) $(b \neq d)$

56. (a, b) and $(-a, -b)$ $(a \neq 0)$

57. $(-a, b)$ and $(a, -b)$ $(a \neq 0)$

58. (a, b) and (b, a) $(a \neq b)$

59. In the following figure, estimate the slopes of all line segments:

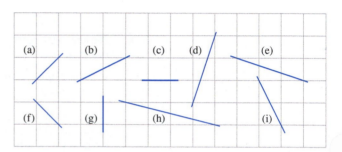

60. In the following figure, estimate the slopes of all line segments:

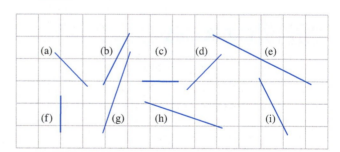

In Exercises 61–80, find a linear equation whose graph is the straight line with the given properties. [**HINT:** See Example 2.]

61. Through $(1, 3)$ with slope 3

62. Through $(2, 1)$ with slope 2

63. Through $\left(1, -\frac{3}{4}\right)$ with slope $\frac{1}{4}$

64. Through $\left(0, -\frac{1}{3}\right)$ with slope $\frac{1}{3}$

65. Through $(20, -3.5)$ and increasing at a rate of 10 units of y per unit of x

66. Through $(3.5, -10)$ and increasing at a rate of 1 unit of y per 2 units of x

67. Through $(2, -4)$ and $(1, 1)$

68. Through $(1, -4)$ and $(-1, -1)$

69. Through $(1, -0.75)$ and $(0.5, 0.75)$

70. Through $(0.5, -0.75)$ and $(1, -3.75)$

71. Through $(6, 6)$ and parallel to the line $x + y = 4$

72. Through $\left(\frac{1}{3}, -1\right)$ and parallel to the line $3x - 4y = 8$

73. Through $(0.5, 5)$ and parallel to the line $4x - 2y = 11$

74. Through $\left(\frac{1}{3}, 0\right)$ and parallel to the line $6x - 2y = 11$

75. ▼ Through $(0, 0)$ and (p, q) $(p \neq 0)$

76. ▼ Through (p, q) parallel to $y = rx + s$

77. ▼ Through (p, q) and (r, q) $(p \neq r)$

78. ▼ Through (p, q) and $(-p, -q)$ $(p \neq 0)$

79. ▼ Through $(-p, q)$ and $(p, -q)$ $(p \neq 0)$

80. ▼ Through (p, q) and (r, s) $(p \neq r)$

Applications

81. *Cost* The *RideEm Bicycles* factory can produce 100 bicycles in a day at a total cost of $10,500. It can produce 120 bicycles in a day at a total cost of $11,000. What are the company's daily fixed costs, and what is the marginal cost per bicycle? [**HINT:** See Example 3.]

82. *Cost* A soft-drink manufacturer can produce 1,000 cases of soda in a week at a total cost of $6,000 and 1,500 cases of soda at a total cost of $8,500. Find the manufacturer's weekly fixed costs and marginal cost per case of soda.

83. *Cost: iPhone 5 (16 GB)* If it costs **Apple** $2,070 to manufacture 10 iPhones per hour and $4,120 to manufacture 20 per hour at a particular plant,[42] obtain the corresponding linear cost function. What was the cost to manufacture each additional iPhone? Use the cost function to estimate the cost of manufacturing 40 iPhones in an hour.

84. *Cost: Kinects* If it costs **Microsoft** $1,230 to manufacture 8 Kinects per hour for the Xbox 360 and $2,430 to manufacture 16 per hour at a particular plant,[43] obtain the corresponding linear cost function. What was the cost to manufacture each additional Kinect? Use the cost function to estimate the cost of manufacturing 30 Kinects in an hour.

85. *Demand* Sales figures show that your company sold 1,960 pen sets each week when they were priced at $1.00 per pen set and 1,800 pen sets each week when they were priced at $5.00 per pen set. What is the linear demand function for your pen sets? [**HINT:** See Example 4.]

86. *Demand* A large department store is prepared to buy 3,950 of your tie-dye shower curtains per month for $5 each but only 3,700 per month for $10 each. What is the linear demand function for your tie-dye shower curtains?

87. *Demand for Smartphones* The following table shows worldwide sales of smartphones and their average selling prices in 2012 and 2013:[44]

Year	2012	2013
Selling Price ($)	385	335
Sales (millions)	720	1,010

a. Use the data to obtain a linear demand function for smartphones, and use your demand equation to predict sales if the price is lowered to $265.

b. Fill in the blanks: For every _____ increase in price, sales of smartphones decrease by ___ units.

88. *Demand for Smartphones* The following table shows worldwide sales of smartphones and their average selling prices in 2013 and 2017:[45]

Year	2013	2017
Selling Price ($)	335	265
Sales (millions)	1,010	1,710

a. Use the data to obtain a linear demand function for smartphones, and use your demand equation to predict sales if the price is raised to $385.

b. Fill in the blanks: For every _____ increase in price, sales of smartphones decrease by ___ units.

89. *Demand for Monorail Service: Las Vegas* In 2005 the Las Vegas monorail charged $3 per ride and had an average ridership of about 28,000 per day. In December 2005 the Las Vegas Monorail Company raised the fare to $5 per ride, and average ridership in 2006 plunged to around 19,000 per day.[46]

a. Use the given information to find a linear demand equation.

b. Give the units of measurement and interpretation of the slope.

c. What would have been the effect on ridership of raising the fare to $6 per ride?

90. *Demand for Monorail Service: Mars* The Utarek monorail, which links the three urbynes (or districts) of Utarek on Mars, charged Z̄5 per ride[47] and sold about 14 million rides per day. When the Utarek City Council lowered the fare to Z̄3 per ride, the number of rides increased to 18 million per day.

a. Use the given information to find a linear demand equation.

b. Give the units of measurement and interpretation of the slope.

c. What would have been the effect on ridership of raising the fare to Z̄10 per ride?

---

[42] Marginal costs are approximate, based on marginal cost data at www.isuppli.com and www.iphoneincanada.ca. Fixed costs are fictitious.

[43] Based on marginal cost data provided by a "highly-positioned, trusted source" (www.develop-online.net).

[44] Data are approximate. Source: IDC Worldwide Quarterly Mobile Phone Tracker, Nov. 26, 2013, www.zdnet.com.

[45] 2017 data are estimates. Source: *Ibid.*

[46] Source: *New York Times*, February 10, 2007, p. A9.

[47] The zonar (Z̄) is the official currency in the city-state of Utarek, Mars (formerly www.Marsnext.com, a now extinct virtual society).

91. *Pasta Imports in the 1990s* During the period 1990–2001, U.S. imports of pasta increased from 290 million pounds in 1990 ($t = 0$) by an average of 40 million pounds per year.[48]

a. Use this information to express y, the annual U.S. imports of pasta (in millions of pounds), as a linear function of t, the number of years since 1990.

b. Use your model to estimate U.S. pasta imports in 2005, assuming that the import trend continued.

92. *Mercury Imports in the 2210s* During the period 2210–2220, Martian imports of mercury (from the planet of that name) increased from 550 million kilograms in 2210 ($t = 0$) by an average of 60 million kilograms per year.

a. Use this information to express y, the annual Martian imports of mercury (in millions of kilograms), as a linear function of t, the number of years since 2210.

b. Use your model to estimate Martian mercury imports in 2230, assuming that the import trend continued.

93. *Net Income* The net income of **Amazon** decreased from $0.63 billion in 2011 to −$0.24 billion in 2014.[49]

a. Use this information to find a linear model for Amazon's net income N (in billions of dollars) as a function of time t in years since 2010.

b. Give the units of measurement and interpretation of the slope.

c. Use the model from part (a) to estimate the 2013 net income. (The actual 2013 net income was approximately $0.27 billion.)

94. *Operating Expenses* The operating expenses of **Amazon** increased from $3.6 billion in 2008 to $16.3 billion in 2012.[50]

a. Use this information to find a linear model for Amazon's operating expenses E (in billions of dollars) as a function of time t in years since 2010.

b. Give the units of measurement and interpretation of the slope.

c. Use the model from part (a) to estimate the 2011 operating expenses. (The actual 2011 operating expenses were $10.9 billion.)

95. *Velocity* The position of a model train, in feet along a railroad track, is given by

$$s(t) = 2.5t + 10$$

after t seconds.

a. How fast is the train moving?

b. Where is the train after 4 seconds?

c. When will the train be 25 feet along the track?

96. *Velocity* The height of a falling sheet of paper, in feet from the ground, is given by

$$s(t) = -1.8t + 9$$

after t seconds.

a. What is the velocity of the sheet of paper?

b. How high is the sheet of paper after 4 seconds?

c. When will the sheet of paper reach the ground?

97. ▼ *Fast Cars* A police car was traveling down Ocean Parkway in a high-speed chase from Jones Beach. The police car was at Jones Beach at exactly 10 pm ($t = 10$) and was at Oak Beach, 13 miles from Jones Beach, at exactly 10:06 pm.

a. How fast was the police car traveling? [**HINT:** See Example 6.]

b. How far was the police car from Jones Beach at time t?

98. ▼ *Fast Cars* The car that was being pursued by the police in Exercise 97 was at Jones Beach at exactly 9:54 pm ($t = 9.9$) and passed Oak Beach (13 miles from Jones Beach) at exactly 10:06 pm, where it was overtaken by the police.

a. How fast was the car traveling? [**HINT:** See Example 6.]

b. How far was the car from Jones Beach at time t?

99. *Textbook Sizes* The second edition of *Applied Calculus* by Waner and Costenoble was 585 pages long. By the time we got to the sixth edition, the book had grown to 755 pages.

a. Use this information to obtain the page length L as a linear function of the edition number n.

b. What are the units of measurement of the slope? What does the slope tell you about the length of *Applied Calculus*?

c. At this rate, by which edition will the book have grown to over 1,500 pages?

100. *Textbook Sizes* The second edition of *Finite Mathematics* by Waner and Costenoble was 603 pages long. By the time we got to the fifth edition, the book had grown to 690 pages.

a. Use this information to obtain the page length L as a linear function of the edition number n.

b. What are the units of measurement of the slope? What does the slope tell you about the length of *Finite Mathematics*?

c. At this rate, by which edition will the book have grown to over 1,000 pages?

101. *Fahrenheit and Celsius* In the Fahrenheit temperature scale, water freezes at 32°F and boils at 212°F. In the Celsius scale, water freezes at 0°C and boils at 100°C. Given that the Fahrenheit temperature F and the Celsius temperature C are related by a linear equation, find F in terms of C. Use your equation to find the Fahrenheit temperatures corresponding to 30°C, 22°C, −10°C, and −14°C, to the nearest degree.

102. *Fahrenheit and Celsius* Use the information about Celsius and Fahrenheit given in Exercise 101 to obtain a linear equation for C in terms of F, and use your equation to find the Celsius temperatures corresponding to 104°F, 77°F, 14°F, and −40°F, to the nearest degree.

---

[48] Data are rounded. Sources: Department of Commerce/*New York Times,* September 5, 1995, p. D4; International Trade Administration, March 31, 2002, www.ita.doc.gov.

[49] Recall that "net income" is another term for "profit." Data are approximate. Source: www.wikinvest.com.

[50] *Ibid.*

Airline Net Income *Exercises 103 and 104 are based on the following table, which compares the net incomes, in millions of dollars, of* **Southwest Airlines**, **JetBlue Airways**, *and* **Alaska Air Group**:[51]

Year	2010	2011	2012	2013	2014
Southwest Airlines	450	200	400	750	900
JetBlue Airways	100	90	130	170	400
Alaska Air Group	250	250	300	500	600

103. a. Use the 2012 and 2014 data to obtain the JetBlue net income *J* as a linear function of the Southwest Airlines net income *S*. (Use millions of dollars for all units, and round coefficients to two significant digits.)

b. How far off is your model in estimating the JetBlue net income based on the Southwest Airlines income in 2010?

c. What are the units of measurement of the slope? What does the slope of the linear function from part (a) suggest about the net incomes of these two airlines?

104. a. Use the 2010 and 2013 data to obtain the JetBlue net income *J* as a linear function of Alaska Air Group's net income *A*. (Use millions of dollars for all units, and round coefficients to two significant digits.)

b. In which of the remaining years does the model give the best prediction of JetBlue's net income?

c. What are the units of measurement of the slope? What does the slope of the linear function from part (a) suggest about the net incomes of these two airlines?

105. ▼ *Income* The well-known romance novelist Celestine A. Lafleur (a.k.a. Bertha Snodgrass) has decided to sell the screen rights to her latest book, *Henrietta's Heaving Heart*, to *Boxoffice Success Productions* for $50,000. In addition, the contract ensures Ms. Lafleur royalties of 5% of the net profits.[52] Express her income *I* as a function of the net profit *N*, and determine the net profit necessary to bring her an income of $100,000. What is her marginal income (share of each dollar of net profit)?

106. ▼ *Income* Because of the enormous success of the movie *Henrietta's Heaving Heart* based on a novel by Celestine A. Lafleur (see Exercise 105), *Boxoffice Success Productions* decides to film the sequel, *Henrietta, Oh Henrietta*. At this point, Bertha Snodgrass (whose novels now top the best-seller lists) feels that she is in a position to demand $100,000 for the screen rights and royalties of 8% of the net profits. Express her income *I* as a function of the net profit *N*, and determine the net profit necessary to bring

her an income of $1,000,000. What is her marginal income (share of each dollar of net profit)?

107. *Processor Speeds* The processor speed, in megahertz (MHz), of **Intel** processors during the period 1996–2010 could be approximated by the following function of time *t* in years since the start of 1990:[53]

$$v(t) = \begin{cases} 400t - 2,200 & \text{if } 6 \leq t < 15 \\ 3,800 & \text{if } 15 \leq t \leq 20. \end{cases}$$

How fast and in what direction was processor speed changing in 2000?

108. *Processor Speeds* The processor speed, in megahertz (MHz), of **Intel** processors during the period 1970–2000 could be approximated by the following function of time *t* in years since the start of 1970:[54]

$$v(t) = \begin{cases} 3t & \text{if } 0 \leq t < 20 \\ 174t - 3,420 & \text{if } 20 \leq t \leq 30. \end{cases}$$

How fast and in what direction was processor speed changing in 1995?

Superbowl Advertising *Exercises 109 and 110 are based on the following graph and data showing the increasing cost of a 30-second television ad during the Super Bowl.*[55]

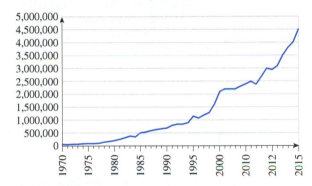

Year	1970	1980	1990	2000	2010
Cost ($1,000)	78	222	700	2,100	2,950

109. ▼ Take *t* to be the number of years since 1970 and *y* to be the cost, in thousands of dollars, of a Super Bowl ad.

a. Model the 1970 and 1990 data with a linear equation.

b. Model the 1990 and 2010 data with a linear equation.

c. Use the results of parts (a) and (b) to obtain a piecewise-linear model of the cost of a Super Bowl ad during 1970–2010.

---

[51] Net incomes from continuing operations. Data are rounded. Source for data: www.wikinvest.com.

[52] Percentages of net profit are commonly called "monkey points." Few movies ever make a net profit on paper, and anyone with any clout in the business gets a share of the *gross*, not the net.

---

[53] A rough model based on the fastest processors produced by Intel. Source: www.intel.com.

[54] *Ibid.*

[55] Source: Nielsen Media Research, http://superbowl-ads.com.

d. Use your model to estimate the cost in 2004. Is your answer in rough agreement with the graph? Explain any discrepancy.

110. ▼ Take t to be the number of years since 1980 and y to be the cost, in thousands of dollars, of a Super Bowl ad.
  a. Model the 1980 and 2000 data with a linear equation.
  b. Model the 2000 and 2010 data with a linear equation.
  c. Use the results of parts (a) and (b) to obtain a piecewise-linear model of the cost of a Super Bowl ad during 1980–2010.
  d. Use your model to estimate the cost in 1992. Is your answer in rough agreement with the graph? Explain any discrepancy.

111. ▼ *Employment in Mexico* The number of workers employed in manufacturing jobs in Mexico was 3 million in 1995, rose to 4.1 million in 2000, and then dropped to 3.5 million in 2004.[56] Model this number N as a piecewise-linear function of the time t in years since 1995, and use your model to estimate the number of manufacturing jobs in Mexico in 2002. (Take the units of N to be millions.)

112. ▼ *Mortgage Delinquencies* The percentage of borrowers in the highest risk category who were delinquent on their payments decreased from 9.7% in 2001 to 4.3% in 2004 and then shot up to 10.3% in 2007.[57] Model this percentage P as a piecewise-linear function of the time t in years since 2001, and use your model to estimate the percentage of delinquent borrowers in 2006.

Communication and Reasoning Exercises

113. How would you test a table of values of x and y to see whether it comes from a linear function?

114. You have ascertained that a table of values of x and y corresponds to a linear function. How do you find an equation for that linear function?

115. To what linear function of x does the linear equation $ax + by = c$ $(b \neq 0)$ correspond? Why did we specify $b \neq 0$?

116. Complete the following. The slope of the line with equation $y = mx + b$ is the number of units that ____ increases per unit increase in ____.

117. Complete the following. If, in a straight line, y is increasing three times as fast as x, then its ____ is ____.

118. Suppose that y is decreasing at a rate of 4 units per 3-unit increase of x. What can we say about the slope of the linear relationship between x and y? What can we say about the intercept?

119. If y and x are related by the linear expression $y = mx + b$, how will y change as x changes if m is positive? negative? zero?

120. Your friend April tells you that $y = f(x)$ has the property that, whenever x is changed by Δx, the corresponding change in y is $\Delta y = -\Delta x$. What can you tell her about f?

121. 🔳 Consider the following worksheet:

◇	A	B	C	D
1	x	y	m	b
2		1	2	=(B3-B2)/(A3-A2)	=B2-C2*A2
3		3	-1	Slope	Intercept

What is the effect on the slope of increasing the y-coordinate of the second point (the point whose coordinates are in row 3)? Explain.

122. 🔳 Referring to the worksheet in Exercise 121, what is the effect on the slope of increasing the x-coordinate of the second point (the point whose coordinates are in row 3)? Explain.

123. If y is measured in bootlags,[58] x is measured in zonars,[59] and $y = mx + b$, then m is measured in ____, and b is measured in ____.

124. If the slope in a linear relationship is measured in miles per dollar, then the independent variable is measured in ____, and the dependent variable is measured in ____.

125. If a quantity is changing linearly with time and it increases by 10 units in the first day, what can you say about its behavior in the third day?

126. The quantities Q and T are related by a linear equation of the form

$$Q = mT + b$$

Q is positive when $T = 0$ but decreases to a negative quantity when T is 10. What are the signs of m and b? Explain your answers.

127. ▼ The velocity of an object is given by $v = 0.1t + 20$ m/sec, where t is time in seconds. The object is
  (A) moving with fixed speed.
  (B) accelerating.
  (C) decelerating.
  (D) impossible to say from the given information.

128. ▼ The position of an object is given by $x = 0.2t - 4$, where t is time in seconds. The object is
  (A) moving with fixed speed.
  (B) accelerating.
  (C) decelerating.
  (D) impossible to say from the given information.

---

[56] Source: *New York Times*, Februrary 18, 2007, p. WK4.

[57] The 2007 figure was projected from data through October 2006. Source: *New York Times*, Februrary 18, 2007, p. BU9.

[58] An ancient Martian unit of length; one bootlag is the mean distance from a Martian's foreleg to its rearleg.

[59] The official currency of Utarek, Mars. (See the footnote to Exercise 90.)

129. If f and g are linear functions with slope m and n, respectively, then what can you say about $f + g$?

130. If f and g are linear functions, then is $\dfrac{f}{g}$ linear? Explain.

131. Give examples of nonlinear functions f and g whose product is linear.

132. Give examples of nonlinear functions f and g whose quotient is linear (on a suitable domain).

133. ▼ Suppose the cost function is $C(x) = mx + b$ (with m and b positive), the revenue function is $R(x) = kx$ ($k > m$), and the number of items is increased from the break-even quantity. Does this result in a loss or a profit, or is it impossible to say? Explain your answer.

134. ▼ You have been constructing a demand equation, and you obtained a (correct) expression of the form $p = mq + b$, whereas you would have preferred one of the form $q = mp + b$. Should you simply switch p and q in the answer, should you start again from scratch, using p in the role of x and q in the role of y, or should you solve your demand equation for q? Give reasons for your decision.

1.4 Linear Regression

Observed and Predicted Values

We have seen how to find a linear model given two data points: We find the equation of the line that passes through them. However, we often have more than two data points, and they will rarely all lie on a single straight line, but they may often come close to doing so. The problem is to find the line coming *closest* to passing through all of the points.

Suppose, for example, that we are conducting research for a company that is interested in expanding into Mexico. Of interest to us would be current and projected growth in that country's economy. The following table shows past and projected per capita gross domestic product (GDP)[60] of Mexico for 2000–2014:[61]

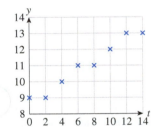

Figure 24(a)

Year t (year since 2000)	0	2	4	6	8	10	12	14
Per Capita GDP y ($1,000)	9	9	10	11	11	12	13	13

A plot of these data suggests a roughly linear growth of the GDP (Figure 24(a)). These points suggest a roughly linear relationship between t and y, although they clearly do not all lie on a single straight line. Figure 24(b) shows the points together with several lines, some fitting better than others. Can we precisely measure which lines fit better than others? For instance, which of the two lines labeled as "good" fits in Figure 24(b) models the data more accurately? We begin by considering, for each value of t, the difference between the actual GDP (the **observed value**) and the GDP predicted by a linear equation (the **predicted value**). The difference between the predicted value and the observed value is called the **residual**.

Figure 24(b)

$$\text{Residual} = \text{Observed value} - \text{Predicted value}$$

On the graph, the residuals measure the vertical distances between the (observed) data points and the line (Figure 25), and they tell us how far the linear model is from predicting the actual GDP.

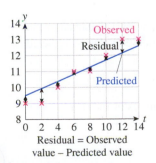

Residual = Observed value – Predicted value

Figure 25

---

[60] The GDP is a measure of the total market value of all goods and services produced within a country.

[61] Data are approximate and/or projected. Sources: CIA World Factbook, www.indexmundi.com, www.economist.com.

The more accurate our model, the smaller the residuals should be. We can combine all the residuals into a single measure of accuracy by adding their *squares*. (We square the residuals in part to make them all positive.[*]) The sum of the squares of the residuals is called the **sum-of-squares error**, **SSE**. Smaller values of SSE indicate more accurate models.

Here are some definitions and formulas for what we have been discussing.

[*] Why not add the absolute values of the residuals instead? Mathematically, using the squares rather than the absolute values results in a simpler and more elegant solution. Further, using the squares always results in a *single* best-fit line in cases where the *x*-coordinates are all different, whereas this is not the case if we use absolute values.

Observed and Predicted Values

Suppose we are given a collection of data points $(x_1, y_1), \ldots, (x_n, y_n)$. The n quantities $y_1, y_2, \ldots, y_n$ are called the **observed y-values**. If we model these data with a linear equation

$$\hat{y} = mx + b, \qquad \hat{y} \text{ stands for "estimated } y\text{" or "predicted } y\text{."}$$

then the y-values we get by substituting the given x-values into the equation are called the **predicted y-values**:

$$\hat{y}_1 = mx_1 + b \qquad \text{Substitute } x_1 \text{ for } x.$$
$$\hat{y}_2 = mx_2 + b \qquad \text{Substitute } x_2 \text{ for } x.$$
$$\vdots$$
$$\hat{y}_n = mx_n + b. \qquad \text{Substitute } x_n \text{ for } x.$$

Quick Example

1. Consider the three data points $(0, 2)$, $(2, 5)$, and $(3, 6)$. The observed y-values are $y_1 = 2$, $y_2 = 5$, and $y_3 = 6$. If we model these data with the equation $\hat{y} = x + 2.5$, then the predicted y-values are

$$\hat{y}_1 = x_1 + 2.5 = 0 + 2.5 = 2.5$$
$$\hat{y}_2 = x_2 + 2.5 = 2 + 2.5 = 4.5$$
$$\hat{y}_3 = x_3 + 2.5 = 3 + 2.5 = 5.5.$$

Residuals, Sum-of-Squares Error

Residuals and Sum-of-Squares Error (SSE)

If we model a collection of data $(x_1, y_1), \ldots, (x_n, y_n)$ with a linear equation $\hat{y} = mx + b$, then the **residuals** are the n quantities (Observed value – Predicted value):

$$(y_1 - \hat{y}_1), (y_2 - \hat{y}_2), \ldots, (y_n - \hat{y}_n).$$

The **sum-of-squares error (SSE)** is the sum of the squares of the residuals:

$$\text{SSE} = (y_1 - \hat{y}_1)^2 + (y_2 - \hat{y}_2)^2 + \cdots + (y_n - \hat{y}_n)^2.$$

Quick Example

2. For the data and linear approximation given in Quick Example 1, the residuals are

$$y_1 - \hat{y}_1 = 2 - 2.5 = -0.5$$
$$y_2 - \hat{y}_2 = 5 - 4.5 = 0.5$$
$$y_3 - \hat{y}_3 = 6 - 5.5 = 0.5,$$

and so SSE $= (-0.5)^2 + (0.5)^2 + (0.5)^2 = 0.75$.

EXAMPLE 1   **Computing SSE**

Using the data above on the GDP in Mexico, compute SSE for the linear models $y = 0.5t + 8$ and $y = 0.25t + 9$. Which model is the better fit?

Solution We begin by creating a table showing the values of t, the observed (given) values of y, and the values predicted by the first model:

Year t	Observed y	Predicted $\hat{y} = 0.5t + 8$
0	9	8
2	9	9
4	10	10
6	11	11
8	11	12
10	12	13
12	13	14
14	13	15

We now add two new columns for the residuals and their squares:

Year t	Observed y	Predicted $\hat{y} = 0.5t + 8$	Residual $y - \hat{y}$	Residual^2 $(y - \hat{y})^2$
0	9	8	$9 - 8 = 1$	$1^2 = 1$
2	9	9	$9 - 9 = 0$	$0^2 = 0$
4	10	10	$10 - 10 = 0$	$0^2 = 0$
6	11	11	$11 - 11 = 0$	$0^2 = 0$
8	11	12	$11 - 12 = -1$	$(-1)^2 = 1$
10	12	13	$12 - 13 = -1$	$(-1)^2 = 1$
12	13	14	$13 - 14 = -1$	$(-1)^2 = 1$
14	13	15	$13 - 15 = -2$	$(-2)^2 = 4$

Using Technology

See the Technology Guides at the end of the chapter for detailed instructions on how to obtain the tables and graphs in Example 1 using a TI-83/84 Plus or a spreadsheet. Here is an outline:

TI-83/84 Plus
STAT EDIT
Values of t in L_1 and y in L_2. Predicted y: Highlight L_3. Enter $0.5*L_1+8$
Squares of residuals: Highlight L_4. Enter the following (including the quotes):
"(L_2-L_3)^2"
SSE: Home screen sum($L4$)
Graph: $Y_1=0.5X+8$
$Y=$ screen: Turn on Plot 1 ZOOM (STAT) [More details in the Technology Guide.]

Spreadsheet
Headings t, y, y-hat, Residual^2, m, b, and SSE in A1–F1.
t-values in A2–A9, y-values in B2–B9; 0.25 for m and 9 for b in E2–F2
Predicted y: =$E$2*A2+$F$2 in C2 and copy down to C9.
Squares of residuals: =(B2-C2)^2 in D2 and copy down to D9.
SSE: =SUM(D2:D9) in G2
Graph: Highlight A1–C9. Insert a scatter chart.
[More details in the Technology Guide].

SSE, the sum of the squares of the residuals, is then the sum of the entries in the last column,

$$SSE = 8.$$

Repeating the process using the second model, $0.25t + 9$, yields the following table:

Year t	Observed y	Predicted $\hat{y} = 0.25t + 9$	Residual $y - \hat{y}$	Residual^2 $(y - \hat{y})^2$
0	9	9	$9 - 9 = 0$	$0^2 = 0$
2	9	9.5	$9 - 9.5 = -0.5$	$(-0.5)^2 = 0.25$
4	10	10	$10 - 10 = 0$	$0^2 = 0$
6	11	10.5	$11 - 10.5 = 0.5$	$0.5^2 = 0.25$
8	11	11	$11 - 11 = 0$	$0^2 = 0$
10	12	11.5	$12 - 11.5 = 0.5$	$0.5^2 = 0.25$
12	13	12	$13 - 12 = 1$	$1^2 = 1$
14	13	12.5	$13 - 12.5 = 0.5$	$0.5^2 = 0.25$

This time, SSE = 2, so the second model is a better fit. Figure 26 shows the data points and the two linear models in question.

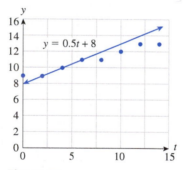 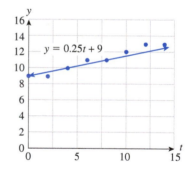

Figure 26

➡ **Before we go on . . .**

Q : *It seems clear from the figure that the second model in Example 1 gives a better fit. Why bother to compute SSE to tell me this?*

A : The difference between the two models we chose is so great that it is clear from the graphs which is the better fit. However, if we used a third model with $m = 0.25$ and $b = 9.1$, then its graph would be almost indistinguishable from that of the second but would be a slightly better fit as measured by SSE = 1.68.

■

The Regression Line

Among all possible lines, there ought to be one with the least possible value of SSE—that is, the greatest possible accuracy as a model. The line (and there is only one such line) that minimizes the sum of the squares of the residuals is called the **regression line**, the **least-squares line**, or the **best-fit line**.

To find the regression line, we need a way to find values of m and b that give the smallest possible value of SSE. As an example, let us take the second linear model in Example 1. We said in the "Before we go on" discussion that increasing b from 9 to 9.1 had the desirable effect of decreasing SSE from 2 to 1.68. We could then increase m to 0.26, further reducing SSE to 1.328. Imagine this as a kind of game: Alter the values of m and b alternately by small amounts until SSE is as small as you can make it. This works but is extremely tedious and time-consuming.

Fortunately, there is an algebraic way to find the regression line. Here is the calculation. To justify it rigorously requires calculus of several variables or linear algebra.

Regression Line

The **regression line** (**least squares line, best-fit line**) associated with the points $(x_1, y_1), (x_2, y_2), \ldots, (x_n, y_n)$ is the line that gives the minimum SSE. The regression line is

$$y = mx + b,$$

where m and b are computed as follows:

$$m = \frac{n(\sum xy) - (\sum x)(\sum y)}{n(\sum x^2) - (\sum x)^2}$$

$$b = \frac{\sum y - m(\sum x)}{n}$$

n = number of data points.

The quantities m and b are called the **regression coefficients**.

Here, "$\sum$" means "the sum of." Thus, for example,

$$\sum x = \text{Sum of the } x\text{-values} = x_1 + x_2 + \cdots + x_n$$
$$\sum xy = \text{Sum of products} = x_1 y_1 + x_2 y_2 + \cdots + x_n y_n$$
$$\sum x^2 = \text{Sum of squares of the } x\text{-values} = x_1{}^2 + x_2{}^2 + \cdots + x_n{}^2.$$

On the other hand,

$$(\sum x)^2 = \text{Square of } \sum x = \text{Square of the sum of the } x\text{-values.}$$

EXAMPLE 2   Per Capita Gross Domestic Product in Mexico

In Example 1 we considered the following data on the per capita gross domestic product (GDP) of Mexico:

Year x (year since 2000)	0	2	4	6	8	10	12	14
Per Capita GDP y ($1,000)	9	9	10	11	11	12	13	13

Find the best-fit linear model for these data, and use the model to predict the per capita GDP in Mexico in 2016.

Solution Let's organize our work in the form of a table, where the original data are entered in the first two columns and the bottom row contains the column sums:

x	y	xy	x^2
0	9	0	0
2	9	18	4
4	10	40	16
6	11	66	36
8	11	88	64
10	12	120	100
12	13	156	144
14	13	182	196
Σ (**Sum**)	56	88	670	560

Using Technology

See the Technology Guides at the end of the chapter for detailed instructions on how to obtain the regression line and graph in Example 2 using a TI-83/84 Plus or a spreadsheet. Here is an outline:

TI-83/84 Plus
STAT EDIT
Values of x in L_1 and y in L_2.
Regression equation: STAT CALC option #4: LinReg(ax+b)
Graph: Y= VARS 5 EQ 1, then ZOOM 9
[More details in the Technology Guide.]

Spreadsheet
x-values in A2–A9, y-values in B2–B9
Graph: Highlight A2–B9. Insert a scatter chart.
Regression line: Add a linear trendline. [More details in the Technology Guide.]

WM Website
www.WanerMath.com
The following two utilities will calculate and plot regression lines (link to either from Math Tools for Chapter 1):
Simple Regression Utility
Function Evaluator and Grapher

Because there are $n = 8$ data points, we get

$$m = \frac{n(\Sigma xy) - (\Sigma x)(\Sigma y)}{n(\Sigma x^2) - (\Sigma x)^2} = \frac{8(670) - (56)(88)}{8(560) - (56)^2} \approx 0.321$$

and

$$b = \frac{\Sigma y - m(\Sigma x)}{n} \approx \frac{88 - (0.321)(56)}{8} \approx 8.75.$$

So the regression line is

$$y = 0.321x + 8.75.$$

To predict the per capita GDP in Mexico in 2016, we substitute $x = 16$ and get $y \approx 14$, or $14,000 per capita.

Figure 27 shows the data points and the regression line (which has SSE ≈ 0.643, a lot lower than in Example 1).

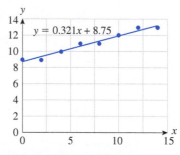

Figure 27

Coefficient of Correlation

If all the data points do not lie on one straight line, we would like to be able to measure how closely they can be approximated by a straight line. Recall that SSE measures the sum of the squares of the deviations from the regression line;

therefore, it constitutes a measurement of what is called goodness of fit. (For instance, if SSE = 0, then all the points lie on a straight line.) However, SSE depends on the units we use to measure y and also on the number of data points (the more data points we use, the larger SSE tends to be). Thus, while we can (and do) use SSE to compare the goodness of fit of two lines to the same data, we cannot use it to compare the goodness of fit of one line to one set of data with that of another to a different set of data.

To remove this dependency, statisticians have found a related quantity that can be used to compare the goodness of fit of lines to different sets of data. This quantity, called the **coefficient of correlation** or **correlation coefficient**, and usually denoted r, is between -1 and 1. The closer r is to -1 or 1, the better the fit. For an *exact* fit we would have $r = -1$ (for a line with negative slope) or $r = 1$ (for a line with positive slope). For a bad fit we would have r close to 0. Figure 28 shows several collections of data points with least-squares lines and the corresponding values of r.

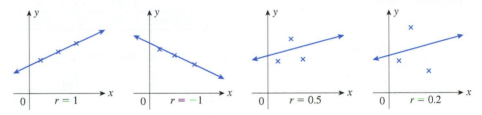

Figure 28

Correlation Coefficient

The coefficient of correlation of the n data points $(x_1, y_1), (x_2, y_2), \ldots, (x_n, y_n)$ is

$$r = \frac{n(\Sigma xy) - (\Sigma x)(\Sigma y)}{\sqrt{n(\Sigma x^2) - (\Sigma x)^2} \cdot \sqrt{n(\Sigma y^2) - (\Sigma y)^2}}.$$

It measures how closely the data points $(x_1, y_1), (x_2, y_2), \ldots, (x_n, y_n)$ fit the regression line. (The value r^2 is sometimes called the **coefficient of determination**.)

Interpretation

- If r is positive, the regression line has positive slope; if r is negative, the regression line has negative slope.

- If $r = 1$ or -1, then all the data points lie exactly on the regression line; if it is close to ± 1, then all the data points are close to the regression line.

- On the other hand, if r is not close to ± 1, then the data points are not close to the regression line, so the fit is not a good one. As a general rule of thumb, a value of $|r|$ less than around 0.8 indicates a poor fit of the data to the regression line.

EXAMPLE 3   **Computing the Coefficient of Correlation**

Find the correlation coefficient for the data in Example 2. Is the regression line a good fit?

Solution   The formula for r requires $\sum x$, $\sum x^2$, $\sum xy$, $\sum y$, and $\sum y^2$. We have all of these except for $\sum y^2$, which we find in a new column as shown:

x	y	xy	x^2	y^2
0	9	0	0	81
2	9	18	4	81
4	10	40	16	100
6	11	66	36	121
8	11	88	64	121
10	12	120	100	144
12	13	156	144	169
14	13	182	196	169
$\sum$ **(Sum)** 56	88	670	560	986

Substituting these values into the formula, we get

$$r = \frac{n(\sum xy) - (\sum x)(\sum y)}{\sqrt{n(\sum x^2) - (\sum x)^2} \cdot \sqrt{n(\sum y^2) - (\sum y)^2}}$$

$$= \frac{8(670) - (56)(88)}{\sqrt{8(560) - 56^2} \cdot \sqrt{8(986) - 88^2}}$$

$$\approx 0.982.$$

As r is close to 1, the fit is a fairly good one; that is, the original points lie nearly along a straight line, as can be confirmed from the graph in Example 2.

Using Technology

See the Technology Guides at the end of the chapter for detailed instructions on how to obtain the correlation coefficient in Example 3 using a TI-83/84 Plus or a spreadsheet. Here is an outline:

TI-83/84 Plus
[2ND] [CATALOG]
DiagnosticOn
Then [STAT] CALC option #4:
LinReg(ax+b) [More details in the Technology Guide.]

Spreadsheet
Add a trendline and select the option to "Display R-squared value on chart."
[More details and other alternatives in the Technology Guide.]

W Website
www.WanerMath.com
The following two utilities will show regression lines and also r^2 (link to either from Math Tools for Chapter 1):
   Simple Regression Utility
   Function Evaluator and Grapher

1.4   EXERCISES

▼ more advanced   ◆ challenging
T indicates exercises that should be solved using technology

In Exercises 1–4, compute the sum-of-squares error (SSE) by hand for the given set of data and linear model.
[**HINT:** See Example 1.]

1. $(1, 1), (2, 2), (3, 4)$; $y = x - 1$

2. $(0, 1), (1, 1), (2, 2)$; $y = x + 1$

3. $(0, -1), (1, 3), (4, 6), (5, 0)$; $y = -x + 2$

4. $(2, 4), (6, 8), (8, 12), (10, 0)$; $y = 2x - 8$

T *In Exercises 5–8, use technology to compute the sum-of-squares error (SSE) for the given set of data and linear models. Indicate which linear model gives the better fit.*

5. $(1, 1), (2, 2), (3, 4)$;   **a.** $y = 1.5x - 1$
                                  b. $y = 2x - 1.5$

6. $(0, 1), (1, 1), (2, 2)$;   **a.** $y = 0.4x + 1.1$
                                  b. $y = 0.5x + 0.9$

7. $(0, -1), (1, 3), (4, 6), (5, 0)$;   **a.** $y = 0.3x + 1.1$
                                            b. $y = 0.4x + 0.9$

8. $(2, 4), (6, 8), (8, 12), (10, 0)$;   **a.** $y = -0.1x + 7$
                                             b. $y = -0.2x + 6$

In Exercises 9–12, find the regression line associated with the given set of points. Graph the data and the best-fit line. (Round all coefficients to four decimal places.) [**HINT:** See Example 2.]

9. $(1, 1), (2, 2), (3, 4)$

10. $(0, 1), (1, 1), (2, 2)$

11. $(0, -1), (1, 3), (3, 6), (4, 1)$

12. $(2, 4), (4, 8), (8, 12), (10, 0)$

In Exercises 13 and 14, use correlation coefficients to determine which of the given sets of data is best fit by its associated regression line and which is fit worst. Is it a perfect fit for any of the data sets? [**HINT**: See Example 3.]

13. a. $(1, 3), (2, 4), (5, 6)$ **b.** $(0, -1), (2, 1), (3, 4)$
　　c. $(4, -3), (5, 5), (0, 0)$

14. a. $(1, 3), (-2, 9), (2, 1)$ **b.** $(0, 1), (1, 0), (2, 1)$
　　c. $(0, 0), (5, -5), (2, -2.1)$

Applications

15. *Mobile Broadband Subscriptions* The following table shows the number of mobile broadband subscribers worldwide (x is the number of years since 2010):[62]

Year x	0	2	4
Subscribers y (millions)	800	1,600	2,300

Complete the following table, and obtain the associated regression line. (Round coefficients to one decimal place.) [**HINT**: See Example 2.]

x	y	xy	x^2
0	800		
2	1,600		
4	2,300		
Σ (Sum)			

Use your regression equation to project the number in 2016.

16. *Fixed-Line Telephone Subscriptions* The following table shows the number of fixed-line telephone subscribers in the United Kingdom (x is the number of years since 2000):[63]

Year x	0	5	14
Subscribers y (millions)	35	34	33

Complete the following table, and obtain the associated regression line. (Round coefficients to one decimal place.) [**HINT**: See Example 2.]

x	y	xy	x^2
0	35		
5	34		
14	33		
Σ (Sum)			

Use your regression equation to project the number in 2015.

17. *Demand for Smartphones* The following table shows worldwide sales of smartphones and their average selling prices in 2012, 2013, and 2017:[64]

Year	2012	2013	2017
Selling Price p ($100)	4	3	2
Sales q (billions)	0.7	1	2

Find the regression line (round coefficients to one decimal place), and use it to estimate the demand (in millions of units sold) when the selling price was $350.

18. *Demand for Smartphones* The following table shows worldwide sales of smartphones and their average selling prices in 2010, 2012, and 2013:[65]

Year	2010	2012	2013
Selling Price p ($100)	5	4	3
Sales q (billions)	0.3	0.7	1

Find the regression line (round coefficients to one decimal place), and use it to estimate the demand (in millions of units sold) when the selling price was $450.

19. *Oil Recovery* The Texas Bureau of Economic Geology published a study on the economic impact of using carbon dioxide enhanced oil recovery (EOR) technology to extract additional oil from fields that have reached the end of their conventional economic life. The following table gives the approximate number of jobs for the citizens of Texas that would be created at various levels of recovery:[66]

Percent Recovery (%)	20	40	80	100
Jobs Created (millions)	3	6	9	15

Find the regression line, and use it to estimate the number of jobs that would be created at a recovery level of 50%.

20. *Oil Recovery* (Refer to Exercise 19.) The following table gives the approximate economic value associated with various levels of oil recovery in Texas:[67]

Percent Recovery (%)	10	40	50	80
Economic Value ($ billion)	200	900	1,000	2,000

Find the regression line, and use it to estimate the economic value associated with a recovery level of 70%.

---

[64] Data are approximate, and 2017 figures are the authors' estimates. Source: IDC Worldwide Quarterly Mobile Phone Tracker, Nov. 26, 2013, www.zdnet.com.

[65] *Ibid.*

[66] Source: "CO2–Enhanced Oil Recovery Resource Potential in Texas: Potential Positive Economic Impacts," Texas Bureau of Economic Geology, April 2004, www.rrc.state.tx.us/tepc/CO2-EOR_white_paper.pdf.

[67] *Ibid.*

---

[62] Data are rounded, and 2014 figure is estimated. Source: International Telecommunication Union, www.itu.int/ITU-D/ict/statistics.

[63] *Ibid.*

21. *Profit: Amazon* The following table shows **Amazon**'s approximate net sales (revenue) and net income (profit) in the period 2011–2014:[68]

Net Sales ($ billion)	50	60	70	80
Net Income ($ billion)	0.6	0.1	0.3	0.3

a. Use this information to find a linear regression model for Amazon's net income I (in millions of dollars) as a function of net sales S (in billions of dollars). Plot the data and regression line.

b. Give the units of measurement and interpretation of the slope.

c. What, according to the model, would Amazon need to earn in net sales for its net income to be $0.5 billion? (Round answer to the nearest billion dollars.)

d. Based on the graph, would you say that the linear model is reasonable? Why or why not?

22. *Operating Expenses: Amazon* The following table shows **Amazon**'s approximate net sales (revenue) and operating expenses in 2011–2014:[69]

Net Sales ($ billion)	50	60	70	80
Operating Expenses ($ billion)	11	16	23	25

a. Use this information to find a linear regression model for Amazon's operating expenses E (in billions of dollars) as a function of net sales S (in billions of dollars). Plot the data and regression line.

b. Give the units of measurement and interpretation of the slope.

c. What, according to the model, would Amazon need to earn in net sales for its operating expenses to be $5 billion? (Round answer to the nearest billion dollars.)

d. Based on the graph, would you say that the linear model is reasonable? Why or why not?

23. ▥ *Textbook Sizes* The following table shows the numbers of pages in previous editions of *Applied Calculus* by Waner and Costenoble:

Edition n	2	3	4	5	6
Number of Pages L	585	656	694	748	768

a. With the edition number as the independent variable, use technology to obtain a regression line and a plot of the points together with the regression line. (Round coefficients to two decimal places.)

b. Interpret the slope of the regression line.

24. ▥ *Textbook Sizes* Repeat Exercise 23 using the following corresponding table for *Finite Mathematics* by Waner and Costenoble:

Edition n	2	3	4	5	6
Number of Pages L	603	608	676	692	696

25. ▥ *Soybean Production: Cerrados* The following table shows soybean production, in millions of tons, in Brazil's Cerrados region as a function of the cultivated area, in millions of acres:[70]

Area (millions of acres)	25	30	32	40	52
Production (millions of tons)	15	25	30	40	60

a. Use technology to obtain the regression line and a plot of the points together with the regression line. (Round coefficients to two decimal places.)

b. Interpret the slope of the regression line.

26. ▥ *Soybean Production: United States* The following table shows soybean production, in millions of tons, in the United States as a function of the cultivated area, in millions of acres:[71]

Area (millions of acres)	30	42	69	59	74	74
Production (millions of tons)	20	33	55	57	83	88

a. Use technology to obtain the regression line and a plot of the points together with the regression line. (Round coefficients to two decimal places.)

b. Interpret the slope of the regression line.

27. ▥ *Airline Profits and the Price of Oil* A common perception is that airline profits are strongly correlated with the price of oil. Following are annual net incomes of **Continental Airlines** together with the approximate price of oil in the period 2005–2010:[72]

Year	2005	2006	2007	2008	2009	2010
Price of Oil ($/barrel)	56	63	67	92	54	71
Continental Net Income ($ million)	−70	370	430	−590	−280	150

---

[68] Figures are approximate, and 2014 figures are estimates. Source: www.wikinvest.com.

[69] *Ibid.*

[70] Source: Brazil Agriculture Ministry/*New York Times*, December 12, 2004, p. N32.

[71] Data are approximate. Source for data: L. David Roper, June 2010, *Crop Production in the World & the United States*, www.roperld.com/science/cropsworld&us.htm.

[72] Figures are rounded, and oil prices are inflation adjusted. Sources: www.wikinvest.com, www.inflationdata.com.

a. Use technology to obtain a regression line showing Continental's net income as a function of the price of oil and also the coefficient of correlation r.

b. What does the value of r suggest about the relationship of Continental's net income to the price of oil?

c. Support your answer to part (b) with a plot of the data and regression line.

28. ⊤ *Airline Profits and the Price of Oil* Repeat Exercise 27 using the following corresponding data for **American Airlines**:[73]

Year	2005	2006	2007	2008	2009	2010
Price of Oil ($/barrel)	56	63	67	92	54	71
American Net Income ($ million)	−850	250	450	−2,100	−1,450	−700

⊤ *Doctorates in Mexico* Exercises 29–32 are based on the following table showing the annual number of PhD graduates in Mexico in various fields:[74]

	Natural Sciences	Engineering	Social Sciences	Education
1990	70	10	60	30
1995	130	40	110	40
2000	330	130	280	130
2005	490	370	460	210
2010	590	550	830	520
2012	690	590	1,000	900

29. a. With $x =$ the number of natural science doctorates and $y =$ the number of engineering doctorates, use technology to obtain the regression equation and graph the associated points and regression line. (Round coefficients to three significant digits.)

b. What does the slope tell you about the relationship between the number of natural science doctorates and the number of engineering doctorates?

c. Use technology to obtain the coefficient of correlation r. Does the value of r suggest a strong correlation between x and y?

d. Does the graph suggest a roughly linear relationship between x and y? Why or why not?

30. a. With $x =$ the number of social science doctorates and $y =$ the number of education doctorates, use technology to obtain the regression equation, and graph the associated points and regression line. (Round coefficients to three significant digits.)

b. What does the slope tell you about the relationship between the number of social science doctorates and the number of education doctorates?

c. Use technology to obtain the coefficient of correlation r. Does the value of r suggest a strong correlation between x and y?

d. Does the graph suggest a roughly linear relationship between x and y? Why or why not?

31. ▼ **a.** Use technology to obtain the regression equation and the coefficient of correlation r for the number of natural science doctorates as a function of time t in years since 1990, and graph the associated points and regression line. (Round coefficients to three significant digits.)

b. What does the slope tell you about the number of natural science doctorates?

c. Judging from the graph, would you say that the number of natural science doctorates is increasing at a faster and faster rate, a slower and slower rate, or a more-or-less constant rate? Why?

d. If r had been equal to 1, could you have drawn the same conclusion as in part (c)? Explain.

32. ▼ **a.** Use technology to obtain the regression equation and the coefficient of correlation r for the number of social science doctorates as a function of time t in years since 1990, and graph the associated points and regression line. (Round coefficients to three significant digits.)

b. What does the slope tell you about the number of social science doctorates?

c. Judging from the graph, would you say that the number of social science doctorates is increasing at a faster and faster rate, a slower and slower rate, or a more-or-less constant rate? Why?

d. If r had been equal to 1, could you have drawn the same conclusion as in part (c)? Explain.

33. ▼ **a.** Do the results of Exercises 29 and 31 suggest that the number of engineering doctorates is increasing at a faster and faster rate, a slower and slower rate, or a more-or-less constant rate? Why?

b. If the value of r in Exercise 29 had been equal to 1, would your conclusion change? Explain.

c. If the value of r in Exercise 31 had been equal to 1, would your conclusion change? Explain.

34. ▼ **a.** Do the results of Exercises 30 and 32 suggest that the number of education doctorates is increasing at a faster and faster rate, a slower and slower rate, or at a more-or-less constant rate? Why?

b. If the value of r in Exercise 30 had been equal to 1, would your conclusion change? Explain.

c. If the value of r in Exercise 32 had been equal to 1, would your conclusion change? Explain.

---

[73] See footnote for Exercise 27.

[74] Education includes humanities other than social sciences. Figures are rounded, and 2012 data are estimated. Source: Instituto Nacional de Estadística y Geografía, www.inegi.org.mx.

35. [T] ▼ *New York City Housing Costs at the Turn of the Century* The following table shows the average price of a two-bedroom apartment in downtown New York City from 1994 to 2004 ($t = 0$ represents 1994):[75]

Year t	0	2	4	6	8	10
Price p ($ million)	0.38	0.40	0.60	0.95	1.20	1.60

   a. Use technology to obtain the linear regression line and correlation coefficient r, with all coefficients rounded to two decimal places, and plot the regression line and the given points.

   b. Does the graph suggest that a nonlinear relationship between t and p would be more appropriate than a linear one? Why or why not?

   c. Use technology to obtain the residuals. What can you say about the residuals in support of the claim in part (b)?

36. [T] ▼ *Fiber-Optic Connections at the Turn of the Century* The following table shows the number of fiber-optic cable connections to homes in the United States from 2000 to 2004 ($t = 0$ represents 2000):[76]

Year t	0	1	2	3	4
Connections c (thousands)	0	10	25	65	150

   a. Use technology to obtain the linear regression line and correlation coefficient r, with all coefficients rounded to two decimal places, and plot the regression line and the given points.

   b. Does the graph suggest that a nonlinear relationship between t and c would be more appropriate than a linear one? Why or why not?

---

[75] Data are rounded, and 2004 figure is an estimate. Source: Miller Samuel/*New York Times*, March 28, 2004, p. RE 11.

[76] Source: Render, Vanderslice & Associates/*New York Times*, October 11, 2004, p. C1.

   c. Use technology to obtain the residuals. What can you say about the residuals in support of the claim in part (b)?

Communication and Reasoning Exercises

37. Why is the regression line associated with the two points (a, b) and (c, d) the same as the line that passes through both? (Assume that $a \neq c$.)

38. What is the smallest possible sum-of-squares error if the given points happen to lie on a straight line? Why?

39. If the points $(x_1, y_1), (x_2, y_2), \ldots, (x_n, y_n)$ lie on a straight line, what can you say about the regression line associated with these points?

40. If all but one of the points $(x_1, y_1), (x_2, y_2), \ldots, (x_n, y_n)$ lie on a straight line, must the regression line pass through all but one of these points?

41. ▼ Verify that the regression line for the points $(0, 0)$, $(-a, a)$, and (a, a) has slope 0. What is the value of r? (Assume that $a \neq 0$.)

42. ▼ Verify that the regression line for the points $(0, a)$, $(0, -a)$, and $(a, 0)$ has slope 0. What is the value of r? (Assume that $a \neq 0$.)

43. ▼ Must the regression line pass through at least one of the data points? Illustrate your answer with an example.

44. ▼ Why must care be taken in using mathematical models to extrapolate?

45. ▼ Your friend Imogen tells you that if r for a collection of data points is more than 0.9, then the most appropriate relationship between the variables is a linear one. Explain why she is wrong by referring to one of the exercises.

46. ▼ Your other friend Mervyn tells you that if r for a collection of data points has an absolute value less than 0.8, then the most appropriate relationship between the variables is a quadratic one and not a linear one. Explain why *he* is wrong.

KEY CONCEPTS

 www.WanerMath.com
Go to the Website to find a comprehensive and interactive Web-based summary of Chapter 1.

1.1 Functions from the Numerical, Algebraic, and Graphical Viewpoints

Real-valued function f of a real-valued variable x, domain [p. 48]
Numerically specified function [p. 48]
Graphically specified function [p. 48]
Algebraically defined function [p. 48]
Graph of the function f [p. 48]
Function notation and equation notation [p. 52]
Independent and dependent variables [p. 52]
Piecewise-defined function [p. 53]
Vertical line test [p. 55]
Common types of algebraic functions and their graphs [p. 55]

1.2 Functions and Models

Mathematical model [p. 63]
Analytical model [p. 64]
Curve-fitting model [p. 64]
Cost, revenue, and profit; marginal cost, revenue, and profit; break-even point [p. 66]
Demand, supply, and equilibrium price [p. 69]
Selecting a model [p. 72]
Compound interest [p. 74]
Exponential growth and decay [p. 75]
Algebra of functions (sum, difference, product, quotient) [p. 77]

1.3 Linear Functions and Models

Linear function $f(x) = mx + b$ [p. 85]
Change in q: $\Delta q = q_2 - q_1$ [p. 86]
Slope of a line:
$$m = \frac{\Delta y}{\Delta x} = \frac{\text{Change in } y}{\text{Change in } x} \quad \text{[p. 87]}$$

Interpretations of m [p. 87]
Interpretation of b: y-intercept [p. 87]
Recognizing linear data [p. 88]
Computing the slope of a line [p. 89]
Slopes of horizontal and vertical lines [p. 90]
Computing the y-intercept [p. 90]
Linear modeling [p. 92]
Linear cost [p. 92]
Linear demand [p. 93]
Linear change over time; rate of change; velocity [p. 96]
General linear models [p. 96]

1.4 Linear Regression

Observed and predicted values [p. 104]
Residuals and sum-of-squares error (SSE) [p. 104]
Regression line (least-squares line, best-fit line) [p. 107]
Correlation coefficient; coefficient of determination [p. 109]

REVIEW EXERCISES

In Exercises 1–4, use the graph of the function f to find approximations of the given values.

1.

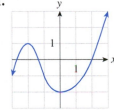

a. $f(-2)$ **b.** $f(0)$
c. $f(2)$ **d.** $f(2) - f(-2)$

2.

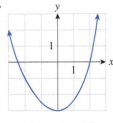

a. $f(-2)$ **b.** $f(0)$
c. $f(2)$ **d.** $f(2) - f(-2)$

3.

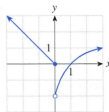

a. $f(-1)$ **b.** $f(0)$
c. $f(1)$ **d.** $f(1) - f(-1)$

4.

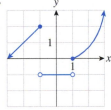

a. $f(-1)$ **b.** $f(0)$
c. $f(1)$ **d.** $f(1) - f(-1)$

In Exercises 5–8, graph the given function or equation.

5. $y = -2x + 5$

6. $2x - 3y = 12$

7. $y = \begin{cases} \frac{1}{2}x & \text{if } -1 \le x \le 1 \\ x - 1 & \text{if } 1 < x \le 3 \end{cases}$

8. $f(x) = 4x - x^2$ with domain $[0, 4]$

In Exercises 9–14, decide whether the specified values come from a linear, quadratic, exponential, or absolute value function.

9.

x	-2	0	1	2	4
$f(x)$	4	2	1	0	2

10.

x	-2	0	1	2	4
$g(x)$	-5	-3	-2	-1	1

11.

x	-2	0	1	2	4
$h(x)$	1.5	1	0.75	0.5	0

12.

x	-2	0	1	2	4
$k(x)$	0.25	1	2	4	16

13.

x	-2	0	1	2	4
$u(x)$	0	4	3	0	-12

14.

x	-2	0	1	2	4
$w(x)$	32	8	4	2	0.5

In Exercises 15–22, find the equation of the specified line.

15. Through $(3, 2)$ with slope -3

16. Through $(-2, 4)$ with slope -1

17. Through $(1, -3)$ and $(5, 2)$

18. Through $(-1, 2)$ and $(1, 0)$

19. Through $(1, 2)$ parallel to $x - 2y = 2$

20. Through $(-3, 1)$ parallel to $-2x - 4y = 5$

21. With slope 4 crossing $2x - 3y = 6$ at its x-intercept

22. With slope $1/2$ crossing $3x + y = 6$ at its x-intercept

In Exercises 23 and 24, determine which of the given lines better fits the given points.

23. $(-1, 1), (1, 2), (2, 0)$; $y = -\dfrac{x}{2} + 1$ or $y = -\dfrac{x}{4} + 1$

24. $(-2, -1), (-1, 1), (0, 1), (1, 2), (2, 4), (3, 3)$; $y = x + 1$ or $y = \dfrac{x}{2} + 1$

In Exercises 25 and 26, find the line that best fits the given points, and compute the correlation coefficient.

25. $(-1, 1), (1, 2), (2, 0)$

26. $(-2, -1), (-1, 1), (0, 1), (1, 2), (2, 4), (3, 3)$

Applications: OHaganBooks.com
[Try the game at www.OHaganBooks.com]

27. *Website Traffic* John Sean O'Hagan is CEO of the online bookstore OHaganBooks.com and notices that, since the establishment of the company website 6 years ago ($t = 0$), the number of visitors to the site has grown quite dramatically, as indicated by the following table:

Year t	0	1	2	3	4	5	6
Website Traffic $V(t)$ (visits/day)	100	300	1,000	3,300	10,500	33,600	107,400

  a. Graph the function V as a function of time t. Which of the following types of function seems to fit the curve best: linear, quadratic, or exponential?

  b. Compute the ratios $\dfrac{V(1)}{V(0)}, \dfrac{V(2)}{V(1)}, \ldots$, and $\dfrac{V(6)}{V(5)}$. What do you notice?

  c. Use the result of part (b) to predict website traffic next year (to the nearest 100).

28. *Publishing Costs* Marjory Maureen Duffin is CEO of publisher *Duffin House,* a major supplier of paperback titles to OHaganBooks.com. She notices that publishing costs over the past 5 years have varied considerably as indicated by

the following table, which shows the average cost to the company of publishing a paperback novel (t is time in years, and the current year is $t = 5$):

Year t	0	1	2	3	4	5
Cost $C(t)$	$5.42	$5.10	$5.00	$5.12	$5.40	$5.88

  a. Graph the function C as a function of time t. Which of the following types of function seems to fit the curve best: linear, quadratic, or exponential?

  b. Compute the differences $C(1) - C(0)$, $C(2) - C(1), \ldots$, and $C(5) - C(4)$, rounded to one decimal place. What do you notice?

  c. Use the result of part (b) to predict the cost of producing a paperback novel next year.

29. *Website Stability* John O'Hagan is considering upgrading the Web server equipment at OHaganBooks.com because of frequent crashes. The tech services manager has been monitoring the frequency of crashes as a function of website traffic (measured in thousands of visits per day) and has obtained the following model:

$$c(x) = \begin{cases} 0.03x + 2 & \text{if } 0 \le x \le 50 \\ 0.05x + 1 & \text{if } x > 50, \end{cases}$$

where $c(x)$ is the average number of crashes in a day in which there are x thousand visitors.

  a. On average, how many times will the website crash on a day when there are 10,000 visits? 50,000 visits? 100,000 visits?

  b. What does the coefficient 0.03 tell you about the website's stability?

  c. Last Friday, the website went down eight times. Estimate the number of visits that day.

30. *Book Sales* As OHaganBooks.com has grown in popularity, the sales manager has been monitoring book sales as a function of the website traffic (measured in thousands of visits per day) and has obtained the following model:

$$s(x) = \begin{cases} 1.55x & \text{if } 0 \le x \le 100 \\ 1.75x - 20 & \text{if } 100 < x \le 250, \end{cases}$$

where $s(x)$ is the average number of books sold in a day in which there are x thousand visitors.

  a. On average, how many books per day does the model predict that OHaganBooks.com will sell when it has 60,000 visits in a day? 100,000 visits in a day? 160,000 visits in a day?

  b. What does the coefficient 1.75 tell you about book sales?

  c. According to the model, approximately how many visitors per day will be needed to sell an average of 300 books per day?

31. *New Users* The number of registered users at OHaganBooks.com has increased substantially over the past few months.

The following table shows the number of new users registering each month for the past 6 months:

Month t	1	2	3	4	5	6
New Users (thousands)	12.5	37.5	62.5	72.0	74.5	75.0

a. Which of the following models best approximates the data?

(A) $n(t) = \dfrac{300}{4 + 100(5^{-t})}$

(B) $n(t) = 13.3t + 8.0$

(C) $n(t) = -2.3t^2 + 30.0t - 3.3$

(D) $n(t) = 7(3^{0.5t})$

b. What does each of the above models predict for the number of new users in the next few months: rising, falling, or leveling off?

32. Purchases OHaganBooks.com has been promoting a number of books published by *Duffin House*. The following table shows the number of books purchased each month from Duffin House for the past 5 months:

Month t	1	2	3	4	5
Purchases (books)	1,330	520	520	1,340	2,980

a. Which of the following models best approximates the data?

(A) $n(t) = \dfrac{3,000}{1 + 12(2^{-t})}$

(B) $n(t) = \dfrac{2,000}{4.2 - 0.7t}$

(C) $n(t) = 300(1.6^t)$

(D) $n(t) = 100(4.1t^2 - 20.4t + 29.5)$

b. What does each of the above models predict for the number of new users in the next few months: rising, falling, leveling off, or something else?

33. Internet Advertising Several months ago, John O'Hagan investigated the effect on the popularity of OHaganBooks.com of placing banner ads at well-known Internet portals. The following model was obtained from available data:

$$v(c) = -0.000005c^2 + 0.085c + 1,750 \text{ new visits per day,}$$

where c is the monthly expenditure on banner ads.

a. John O'Hagan is considering increasing expenditure on banner ads from the current level of $5,000 to $6,000 per month. What will be the resulting effect on website popularity?

b. According to the model, would the website popularity continue to grow at the same rate if he continued to raise expenditure on advertising $1,000 each month? Explain.

c. Does this model give a reasonable prediction of traffic at expenditures larger than $8,500 per month? Why or why not?

34. Production Costs Over at *Duffin House*, Marjory Duffin is trying to decide on the size of the print runs for the best-selling new fantasy novel *Larry Plotter and the Simplex Method*. The following model shows a calculation of the total cost to produce a million copies of the novel, based on an analysis of setup and storage costs:

$$c(n) = 0.0008n^2 - 72n + 2,000,000 \text{ dollars,}$$

where n is the print run size (the number of books printed in each run).

a. What would be the effect on cost if the run size was increased from 20,000 to 30,000?

b. Would increasing the run size in further steps of 10,000 result in the same changes in the total cost? Explain.

c. What approximate run size would you recommend that Marjory Duffin use for a minimum cost?

35. Internet Advertising When OHaganBooks.com actually went ahead and increased Internet advertising from $5,000 per month to $6,000 per month (see Exercise 33) it was noticed that the number of new visits increased from an estimated 2,050 per day to 2,100 per day. Use this information to construct a linear model giving the average number v of new visits per day as a function of the monthly advertising expenditure c.

a. What is the model?

b. Based on the model, how many new visits per day could be anticipated if OHaganBooks.com budgets $7,000 per month for Internet advertising?

c. The goal is to eventually increase the number of new visits to 2,500 per day. Based on the model, how much should be spent on Internet advertising to accomplish this?

36. Production Costs When *Duffin House* printed a million copies of *Larry Plotter and the Simplex Method* (see Exercise 34), it used print runs of 20,000, which cost the company $880,000. For the sequel, *Larry Plotter and the Simplex Method, Phase 2*, it used print runs of 40,000, which cost the company $550,000. Use this information to construct a linear model giving the production cost c as a function of the run size n.

a. What is the model?

b. Based on the model, what would print runs of 25,000 have cost the company?

c. Marjory Duffin has decided to budget $418,000 for production of the next book in the *Simplex Method* series. Based on the model, how large should the print runs be to accomplish this?

37. Recreation John O'Hagan has just returned from a sales convention in Puerto Vallarta, Mexico where, to win a bet he made with Marjory Duffin (who was also at the convention), he went bungee jumping at a nearby mountain retreat. The bungee cord he used had the property that a person weighing 70 kilograms would drop a total distance of 74.5 meters, while a 90 kg person would drop 93.5 meters. Express the distance d a jumper drops as a linear function of the jumper's weight w. John OHagan dropped 90 meters. What was his approximate weight?

38. Crickets The mountain retreat near Puerto Vallarta was so quiet at night that all one could hear was the chirping of the snowy tree crickets. These crickets behave in a rather interesting way: The rate at which they chirp depends linearly on the temperature. Early in the evening, John O'Hagan counted 140 chirps per minute and noticed that the temperature was 80°F. Later in the evening the temperature dropped to 75°F, and the chirping slowed down to 120 chirps per minute. Express the temperature T as a function of the rate of chirping r. The temperature that night dropped to a low of 65°F. At approximately what rate were the crickets chirping at that point?

39. Break-Even Analysis OHaganBooks.com has recently decided to start selling music albums online through a service it calls o'Tunes.[77] Users pay a fee to download an entire music album. Composer royalties and copyright fees cost an average of $5.50 per album, and the cost of operating and maintaining o'Tunes amounts to $500 per week. The company is currently charging customers $9.50 per album.
   a. What are the associated (weekly) cost, revenue, and profit functions?
   b. How many albums must be sold per week to make a profit?
   c. If the charge is lowered to $8.00 per album, how many albums must be sold per week to make a profit?

40. Break-Even Analysis OHaganBooks.com also generates revenue through its o'Books e-book service. Author royalties and copyright fees cost the company an average of $4 per novel, and the monthly cost of operating and maintaining the service amounts to $900 per month. The company is currently charging customers $5.50 per novel.
   a. What are the associated cost, revenue, and profit functions?
   b. How many novels must be sold per month to break even?
   c. If the charge is lowered to $5.00 per novel, how many books must be sold to break even?

41. Demand and Profit To generate a profit from its new o'Tunes service, OHaganBooks.com needs to know how the demand for music albums depends on the price it charges. During the first week of the service, it was charging $7 per album and sold 500. Raising the price to $9.50 had the effect of lowering demand to 300 albums per week.

---
[77] The (highly original) name was suggested to John O'Hagan by Marjory Duffin over cocktails one evening.

   a. Use the given data to construct a linear demand equation.
   b. Use the demand equation you constructed in part (a) to estimate the demand if the price were raised to $12 per album.
   c. Using the information on cost given in Exercise 39, determine which of the three prices ($7, $9.50, and $12) would result in the largest weekly profit, and the size of that profit.

42. Demand and Profit To generate a profit from its o'Books e-book service, OHaganBooks.com needs to know how the demand for novels depends on the price it charges. During the first month of the service, it was charging $10 per novel and sold 350. Lowering the price to $5.50 per novel had the effect of increasing demand to 620 novels per month.
   a. Use the given data to construct a linear demand equation.
   b. Use the demand equation you constructed in part (a) to estimate the demand if the price were raised to $15 per novel.
   c. Using the information on cost given in Exercise 40, determine which of the three prices ($5.50, $10, and $15) would result in the largest profit, and the size of that profit.

43. Demand OHaganBooks.com has tried selling music albums on o'Tunes at a variety of prices, with the following results:

Price	$8.00	$8.50	$10	$11.50
Demand (weekly sales)	440	380	250	180

   a. Use the given data to obtain a linear regression model of demand.
   b. Use the demand model you constructed in part (a) to estimate the demand if the company charged $10.50 per album. (Round the answer to the nearest album.)

44. Demand OHaganBooks.com has tried selling novels through o'Books at a variety of prices, with the following results:

Price	$5.50	$10	$11.50	$12
Demand (monthly sales)	620	350	350	300

   a. Use the given data to obtain a linear regression model of demand.
   b. Use the demand model you constructed in part (a) to estimate the demand if the company charged $8 per novel. (Round the answer to the nearest novel.)

Modeling Spending on Mobile Advertising

You are the new director of *Impact Advertising Inc.*'s mobile division, which has enjoyed a steady 0.25% of the worldwide mobile advertising market. You have drawn up an ambitious proposal to expand your division in light of your anticipation that mobile advertising will continue to skyrocket. However, upper management sees things differently and, judging by the following email, does not seem likely to approve the budget for your proposal.

TO: JCheddar@impact.com (J. R. Cheddar)
CC: CVODoylePres@impact.com (C. V. O'Doyle, CEO)
FROM: SGLombardoVP@impact.com (S. G. Lombardo, VP Financial Affairs)
SUBJECT: Your Expansion Proposal
DATE: May 30, 2017

Hi John:

Your proposal reflects exactly the kind of ambitious planning and optimism we like to see in our new upper management personnel. Your presentation last week was most impressive and obviously reflected a great deal of hard work and preparation.

I am in full agreement with you that mobile advertising is on the increase. Indeed, our Market Research department informs me that, based on a regression of the most recently available data, worldwide spending on mobile advertising will continue to grow at a rate of approximately $8.4 billion per year. This translates into $21 million in increased revenues per year for Impact, given our 0.25% market share. This rate of expansion is exactly what our planned 2018 budget anticipates. Your expansion proposal, on the other hand, is based on an increase in sales of almost five times what Market Research is projecting, even though your proposal provides no solid evidence to justify this projection.

At this stage, therefore, I am sorry to say that I am inclined not to approve the funding for your project, although I would be happy to discuss this further with you. I plan to present my final decision on the 2018 budget at next week's divisional meeting.

Regards, Sylvia

Refusing to admit defeat, you contact the Market Research department and request the details of their projections on mobile advertising. They fax you the following information:[78]

Year	2010	2011	2012	2013	2014	2015	2016	2017
Mobile Advertising Spending ($ billion)	2.34	4.02	8.80	15.82	24.91	35.55	47.16	59.67

Regression model: $y = 8.409x - 4.648$ (x = time in years since 2010)
Correlation coefficient: $r = 0.978$

Now you see where the VP got that $8.4 billion figure: The slope of the regression equation is close to 8.4, indicating a rate of increase of about $8.4 billion per year.

---

[78] Source for data through 2013: eMarketer. Figures from 2014 are forecasts by www.statista.com.

Mobile advertising spending
($ billion)

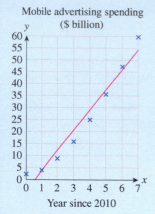

Figure 29

* Note that this r is not the linear correlation coefficient we defined in Section 1.4. What this r measures is how closely the quadratic regression model fits the data.

Mobile advertising spending
($ billion)

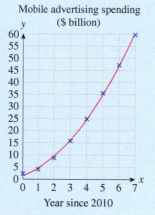

Figure 30

† The number of degrees of freedom in a regression model is 1 less than the number of coefficients. For a linear model, it is 1 (there are two coefficients: the slope m and the intercept b), and for a quadratic model, it is 2. For a detailed discussion, consult a text on regression analysis.

Also, the correlation coefficient is very high, an indication that the linear model fits the data well. In view of this strong evidence, it seems difficult to argue that revenues will increase by significantly more than the projected $8.4 billion per year.

To get a better picture of what's going on, you decide to graph the data together with the regression line in your spreadsheet. What you get is shown in Figure 29. You immediately notice that the data points in Figure 29 seem to suggest a curve, not a straight line. Then again, perhaps the suggestion of a curve is an illusion. Thus, you surmise that there are two possible interpretations of the data:

1. (Your first impression) As a function of time, mobile advertising revenue is non-linear, and is in fact accelerating (the rate of change is increasing), so a linear model is inappropriate.

2. (Devil's advocate) Mobile advertising revenue *is* a linear function of time; the fact that the points do not lie on the regression line is simply a consequence of random factors that do not reflect a long-term trend, such as world events, mergers and acquisitions, or short-term fluctuations in the economy or the stock market.

You suspect that the VP will probably opt for the second interpretation and discount the graphical evidence of accelerating growth by claiming that it is an illusion—a "statistical fluctuation." That is, of course, a possibility, but you wonder how likely it really is. For the sake of comparison you decide to try a regression based on the simplest nonlinear model you can think of—a quadratic function:

$$y = ax^2 + bx + c.$$

Your spreadsheet allows you to fit such a function with a click of the mouse. The result is the following:

$$y = 0.8842x^2 + 2.2193x + 1.5421 \qquad (x = \text{number of years since 2010})$$
$$r = 0.9995. \qquad \text{See Note.}*$$

Figure 30 shows the graph of the regression function together with the original data. Aha! The fit is visually far better, and the correlation coefficient is even higher! Further, the quadratic model predicts 2018 revenue as

$$y = 0.8842(8)^2 + 2.2193(8) + 1.5421 \approx \$75.9 \text{ billion},$$

which is about $16.2 billion above the 2017 spending figure in the table above. Given Impact Advertising's 0.25% market share, this translates into an increase in revenues of $40.5 million, which is about five times the increase predicted by the linear model!

You quickly draft an email to Sylvia Lombardo and are about to click "Send" when you decide, as a precaution, to check with a colleague who is knowledgeable in statistics. He tells you to be cautious: The value of r will always tend to increase if you pass from a linear model to a quadratic one because of the increase in degrees of freedom.† A good way to test whether a quadratic model is more appropriate than a linear one is to compute a statistic called the *p*-value associated with the coefficient of x^2. A low value of p indicates a high degree of confidence that the coefficient of x^2 cannot be zero (see below). Notice that if the coefficient of x^2 *is* zero, then you have a linear model.

You can, your colleague explains, obtain the *p*-value using your spreadsheet as follows. (The method we describe here works on all the popular spreadsheets, including Excel, Google Sheets, and OpenOffice Calc.)

First, set up the data in columns, with an extra column for the values of x^2:

	A	B	C
1	y	x	x^2
2	2.34	0	0
3	4.02	1	1
4	8.8	2	4
5	15.82	3	9
6	24.91	4	16
7	35.55	5	25
8	47.16	6	36
9	59.67	7	49

Then highlight a vacant 5×3 block (the block E1:G5, say), type the formula =LINEST(A2:A9,B2:C9,TRUE,TRUE), and press Cntl+Shift+Enter (not just Enter!). You will see a table of statistics like the following:

	E	F	G
1	=LINEST(A2:A9,B2:C9,TRUE,TRUE)		
2			
3			
4			
5			

Cntl+Shift+Enter →

	E	F	G
1	0.88422619	2.21934524	1.54208333
2	0.06156915	0.44823018	0.67164014
3	0.99897427	0.7980274	#N/A
4	2434.7854	5	#N/A
5	3101.17515	3.18423869	#N/A

(Notice the coefficients of the quadratic model in the first row.) The p-value is then obtained by the formula =TDIST(ABS(E1/E2),F4,2), which you can compute in any vacant cell. You should get $p \approx 0.0000295$.

Q: *What does p actually measure?*

A: *Roughly speaking, $1 - p \approx 0.9999705$ gives the degree of confidence you can have (99.99705%) in asserting that the coefficient of x^2 is not zero. (Technically, p is the probability—allowing for random fluctuation in the data—that if the coefficient of x^2 were in fact zero, the ratio E1/E2 could be as large as it is.)*

In short, you can go ahead and send your email with almost 100% confidence!

EXERCISES

Suppose you are given the following data for the spending on mobile advertising in a hypothetical country in which Impact Advertising also has a 0.25% share of the market.

Year	2010	2011	2012	2013	2014	2015	2016
Mobile Advertising Spending ($ billion)	0	0.3	1.5	2.6	3.4	4.3	5.0

1. Obtain a linear regression model and the correlation coefficient r. (Take t to be time in years since 2010.) According to the model, at what rate is spending on mobile advertising increasing in this country? How does this translate to annual revenues for Impact Advertising?

2. Use a spreadsheet or other technology to graph the data together with the best-fit line. Does the graph suggest a quadratic model (parabola)?

3. Test your impression in the preceding exercise by using technology to fit a quadratic function and graphing the resulting curve together with the data. Does the graph suggest that the quadratic model is appropriate?

4. Perform a regression analysis using the quadratic model, and find the associated p-value. What does it tell you about the appropriateness of a quadratic model?

Section 1.1

Example 1(a) and (c) (page 50)   The total number of iPods sold by **Apple** up to the end of year x can be approximated by $f(x) = -2x^2 + 12x + 36$ million iPods ($0 \le x \le 7$), where $x = 0$ represents 2005. Compute $f(0)$, $f(2)$, $f(4)$, and $f(6)$, and obtain the graph of f.

Solution

You can use the Y= screen to enter an algebraically defined function.

1. Enter the function in the Y= screen, as

$$Y_1 = -2X^2 + 12X + 36$$

or $$Y_1 = -2X^2 + 12X + 36$$

(See Chapter 0 for a discussion of technology formulas.)

2. To evaluate $f(0)$, for example, enter $Y_1(0)$ in the Home screen to evaluate the function Y_1 at 0. Alternatively, you can use the table feature: After entering the function under Y_1, press $\boxed{2ND}$ $\boxed{TBLSET}$, and set Indpnt to Ask. (You do this once and for all; it will permit you to specify values for x in the table screen.) Then press $\boxed{2ND}$ $\boxed{TABLE}$, and you will be able to evaluate the function at several values of x. Below (top) is a table showing the values requested:

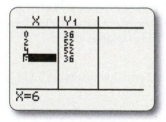

3. To obtain the graph above, press $\boxed{WINDOW}$, set Xmin = 0, Xmax = 7 (the range of x-values we are interested in), Ymin = 0, and Ymax = 60 (we estimated Ymin and Ymax from the corresponding set of y-values in the table), and press $\boxed{GRAPH}$ to obtain the curve. Alternatively, you can avoid having to estimate Ymin and Ymax by pressing ZoomFit

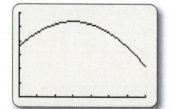

($\boxed{ZOOM}$ $\boxed{0}$), which automatically sets Ymin and Ymax to the smallest and greatest values of y, respectively, in the specified range for x.

Example 2 (page 53)   The accumulated sales of **Nintendo** Wii game consoles from December 2006 to December 2014 can be approximated by the following function of time t in years ($t = 0$ represents December 2006):

$$f(t) = \begin{cases} 5t^2 + 2t + 2 & \text{if } 0 \le t \le 3 \\ -2t^2 + 33t - 28 & \text{if } 3 < t \le 8 \end{cases}$$
$$\text{million consoles.}$$

Obtain a table showing the values $f(t)$ for $t = 0, \ldots, 8$, and obtain the graph of f.

Solution

You can enter a piecewise-defined function using the logical inequality operators $<$, $>$, $\le$, and $\ge$, which are found by pressing $\boxed{2ND}$ $\boxed{TEST}$:

1. Enter the function f in the Y=screen as

$Y_1 = (X \le 3) * (5X^2 + 2X + 2) + (X > 3) * (-2X^2 + 33X - 28)$

When x is less than or equal to 3, the logical expression $(X \le 3)$ evaluates to 1 because it is true, and the expression $(X > 3)$ evaluates to 0 because it is false. The value of the function is therefore given by the expression $(5X^2 + 2X + 2)$. When x is greater than 3, the expression $(X \le 3)$ evaluates to 0, while the expression $(X > 3)$ evaluates to 1, so the value of the function is given by the expression $(-2X^2 + 33X - 28)$.

2. As in Example 1, use the Table feature to compute several values of the function at once by pressing $\boxed{2ND}$ $\boxed{TABLE}$:

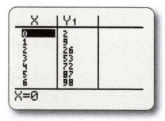

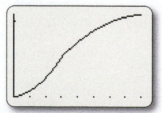

3. To obtain the graph above, we proceed as in Example 1: Press $\boxed{\text{WINDOW}}$, set Xmin = 0, Xmax = 8 (the range of x-values we are interested in), and press $\boxed{\text{ZOOM}}$ $\boxed{\text{10}}$ ("Zoomfit") to automatically compute Ymin and Ymax and plot the graph.

Section 1.2

Example 5(a) (page 72)   The following table shows some annual revenues, in billions of dollars, earned by **Amazon** from 2004 through 2014:

Year	2004	2006	2008	2010	2012	2014
Revenue ($ billion)	7	11	19	34	61	89

Take t to be the number of years since 2004, and consider the following three models:

(1) $r(t) = 8t - 3$		Linear model
(2) $r(t) = 0.9t^2 - 0.6t + 7$		Quadratic model
(3) $r(t) = 7(1.3^t).$		Exponential model

a. Which two models fit the data significantly better than the third?

b. If you extrapolate the models you selected in part (a) back to $t = -4$, what do you find? (Amazon's revenue in 2000 was $2.8 billion and subsequently increased each year.).

Solution

1. First enter the actual revenue data in the stat list editor ($\boxed{\text{STAT}}$ EDIT) with the values of t in L_1 and the values of $s(t)$ in L_2.

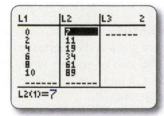

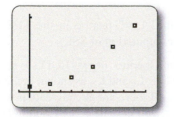

2. Now go to the Y= window, and turn `Plot1` on by selecting it and pressing $\boxed{\text{ENTER}}$. (You can also turn it on in the $\boxed{\text{2ND}}$ STAT PLOT screen.) Then press ZoomStat ($\boxed{\text{ZOOM}}$ $\boxed{9}$) to obtain a plot of the points (figure above).

3. To see any of the three curves plotted along with the points, enter its formula in the Y= screen (for instance, $Y_1=0.9X^2-0.6X+7$ for the second model), and press $\boxed{\text{GRAPH}}$ (figure on top below).

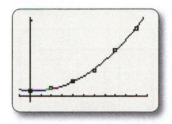

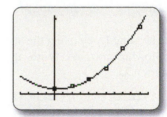

4. To see the extrapolation of the curve back to 2000, just change Xmin to -4 (in the $\boxed{\text{WINDOW}}$ screen), and press $\boxed{\text{GRAPH}}$ again (figure above).

5. Now change Y_1 to see similar graphs for the remaining curves.

6. When you are done, turn `Plot1` off again so that the points you entered do not show up in other graphs.

Section 1.3

Example 1 (page 88)   Which of the following two tables gives the values of a linear function? What is the formula for that function?

x	0	2	4	6	8	10	12
$f(x)$	3	-1	-3	-6	-8	-13	-15

x	0	2	4	6	8	10	12
$g(x)$	3	-1	-5	-9	-13	-17	-21

Solution

We can use the "List" feature in the TI-83/84 Plus to automatically compute the successive quotients $m = \Delta y/\Delta x$ for either f or g as follows:

1. Use the stat list editor ($\boxed{\text{STAT}}$ EDIT) to enter the values of x and $f(x)$ in the first two columns, called L_1 and L_2, as shown in the screenshot below. (If there are already data in a column you want to use, you can clear it by highlighting the column heading (e.g., L_1) using the arrow key, and pressing $\boxed{\text{CLEAR}}$ $\boxed{\text{ENTER}}$.)

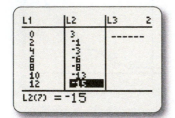

2. Highlight the heading L_3 by using the arrow keys, and enter the following formula (with the quotes, as explained below):

$$"\Delta\text{List}(L_2)/\Delta\text{List}(L_1)"$$

ΔList is found under $\boxed{\text{2ND}}$ $\boxed{\text{LIST}}$ OPS. L_1 is $\boxed{\text{2ND}}$ $\boxed{1}$

The "ΔList" function computes the differences between successive elements of a list, returning a list with one less element. The formula above then computes the quotients $\Delta y/\Delta x$ in the list L_3 as shown in the following screenshot. As you can see in the third column, $f(x)$ is not linear.

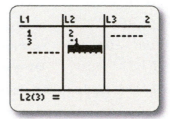

3. To redo the computation for $g(x)$, all you need to do is edit the values of L_2 in the stat list editor. By putting quotes around the formula we used for L_3, we told the calculator to remember the formula, so it automatically recalculates the values.

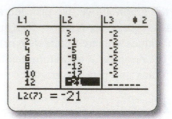

Example 2(a) (page 91) Find the equation of the line through the points $(1, 2)$ and $(3, -1)$.

Solution

1. Enter the coordinates of the given points in the stat list editor ($\boxed{\text{STAT}}$ EDIT) with the values of x in L_1 and the values of y in L_2.

2. To compute the slope, enter the following formula in the Home screen:

$$(L_2(2) - L_2(1))/(L_1(2) - L_1(1)) \to M$$

L_1 and L_2 are under $\boxed{\text{2ND}}$ $\boxed{\text{LIST}}$ and the arrow is $\boxed{\text{STO}}$

3. Then, to compute the y-intercept, enter

$$L_2(1) - M*L_1(1)$$

Section 1.4

Example 1 (page 105) Using the data on the per capita GDP in Mexico given at the beginning of Section 1.4, compute SSE, the sum-of-squares error, for the linear models $y = 0.5t + 8$ and $y = 0.25t + 9$, and graph the data with the given models.

Solution

We can use the "List" feature in the TI-83/84 Plus to automate the computation of SSE.

1. Use the stat list editor ($\boxed{\text{STAT}}$ EDIT) to enter the given data in the lists L_1 and L_2, as shown in the first screenshot below. (If there are already data in a column you want to use, you can clear it by highlighting the column heading (e.g., L_1) using the arrow key, and pressing $\boxed{\text{CLEAR}}$ $\boxed{\text{ENTER}}$.)

2. To compute the predicted values, highlight the heading L_3 using the arrow keys, and enter the following formula for the predicted values (figure on the top below):

$$0.5*L_1+8 \qquad L_1 \text{ is } \boxed{2\text{ND}}\boxed{1}$$

Pressing $\boxed{\text{ENTER}}$ again will fill column 3 with the predicted values (below bottom). Note that only seven of the eight data points can be seen on the screen at one time.

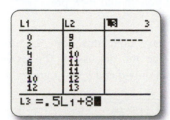

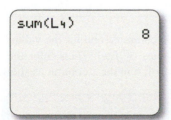

3. Highlight the heading L_4, and enter the following formula (including the quotes):

$$"(L_2-L_3)^2" \qquad \text{Squaring the residuals}$$

4. Pressing $\boxed{\text{ENTER}}$ will fill L_4 with the squares of the residuals. (Putting quotes around the formula will allow us to easily check the second model, as we shall see.)

5. To compute SSE, the sum of the entries in L_4, go to the Home screen, and enter sum(L_4). (See below; "sum" is under $\boxed{2\text{ND}}\boxed{\text{LIST}}\boxed{\text{MATH}}$.)

6. To check the second model, go back to the List screen, highlight the heading L_3, enter the formula for the second model, $0.25*L_1+9$, and press $\boxed{\text{ENTER}}$. Because

we put quotes around the formula for the residuals in L_4, the TI-83/84 Plus will remember the formula and automatically recalculate the values (below top). On the Home screen we can again calculate sum(L_4) to get SSE for the second model (below bottom).

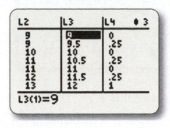

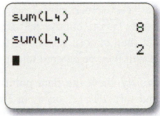

The second model gives a much smaller SSE, so it is the better fit.

7. You can also use the TI-83/84 Plus to plot both the original data points and the two lines (see below). Turn Plot1 on in the STAT PLOT window, obtained by pressing $\boxed{2\text{ND}}\boxed{\text{STAT PLOT}}$. To show the lines, enter them in the "Y=" screen as usual. To obtain a convenient window showing all the points and the lines, press $\boxed{\text{ZOOM}}$, and choose 9: ZoomStat.

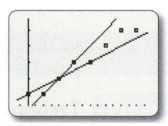

Example 2 (page 107) Use the data on the per capita GDP in Mexico to find the best-fit linear model.

Solution

1. Enter the data in the TI-83/84 Plus using the List feature, putting the x-coordinates in L_1 and the y-coordinates in L_2, just as in Example 1.

2. Press $\boxed{\text{STAT}}$, select CALC, and choose #4: LinReg(ax+b). Pressing $\boxed{\text{ENTER}}$ will cause the

equation of the regression line to be displayed in the Home screen:

So the regression line is $y \approx 0.321x + 8.75$.

3. To graph the regression line without having to enter it by hand in the "Y=" screen, press Y=, clear the contents of Y_1, press VARS, choose #5: Statistics, select EQ, and then choose #1:RegEQ. The regression equation will then be entered under Y_1.

4. To simultaneously show the data points, press 2ND STATPLOT, and turn Plot1 on as in Example 1. To obtain a convenient window showing all the points and the line (see below), press ZOOM and choose #9: ZoomStat.

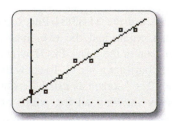

Example 3 (page 110)   Find the correlation coefficient for the data in Example 2.

Solution

To find the correlation coefficient using a TI-83/84 Plus, you need to tell the calculator to show you the coefficient at the same time that it shows you the regression line.

1. Press 2ND CATALOG, and select DiagnosticOn from the list. The command will be pasted to the Home screen, and you should then press ENTER to execute the command.

2. Once you have done this, the "LinReg(ax+b)" command (see the discussion for Example 2) will show you not only a and b, but r and r^2 as well:

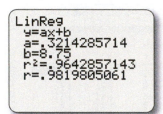

Spreadsheet   Technology Guide

Section 1.1

Example 1(a) and (c) (page 50)   The total number of iPods sold by **Apple** up to the end of year x can be approximated by $f(x) = -2x^2 + 12x + 36$ million iPods $(0 \le x \le 7)$, where $x = 0$ represents 2005. Compute $f(0)$, $f(2)$, $f(4)$, and $f(6)$, and obtain the graph of f.

Solution

To create a table of values of f using a spreadsheet, do the following:

1. Set up two columns: one for the values of x and one for the values of $f(x)$. Then enter the sequence of values 0, 2, 4, 6 in the x column as shown below.

	A	B
1	x	f(x)
2	0	
3	2	
4	4	
5	6	

2. Now we enter a formula for $f(x)$ in cell B2 (below). The technology formula is $-2x^2+12x+36$. To use this formula in a spreadsheet, we modify it slightly:

=-2*A2^2+12*A2+36   Spreadsheet version of tech formula

Notice that we have preceded the Excel formula by an equals sign ($=$) and replaced each occurrence of x by the name of the cell holding the value of x (cell A2 in this case).

	A	B	C
1	x	f(x)	
2		0 =-2*A2^2+12*A2+36	
3		2	
4		4	
5		6	

Note Instead of typing in the name of the cell "A2" each time, you can simply click on the cell A2, and "A2" will be automatically inserted. ■

3. Now highlight cell B2, and drag the **fill handle** (the little square at the lower right-hand corner of the selection) down until you reach row 5, as shown below on the top, to obtain the result shown on the bottom.

	A	B	C
1	x	f(x)	
2		0 =-2*A2^2+12*A2+36	
3		2	
4		4	
5		6	

	A	B
1	x	f(x)
2		0	36
3		2	52
4		4	52
5		6	36

4. To graph the data, highlight A1 through B5, and insert a scatter chart (the exact method of doing this depends on the specific version of the spreadsheet program). When choosing the style of the chart, choose a style that shows points connected by lines (if possible) to obtain a graph something like the following:

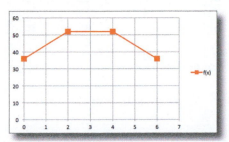

Example 2 (page 53) The accumulated sales of **Nintendo** Wii game consoles from December 2006 to December 2014 can be approximated by the following function of time t in years ($t = 0$ represents December 2006):

$$f(t) = \begin{cases} 5t^2 + 2t + 2 & \text{if } 0 \le t \le 3 \\ -2t^2 + 33t - 28 & \text{if } 3 < t \le 8 \end{cases}$$

million consoles.

Obtain a table showing the values $f(t)$ for $t = 0, \ldots, 8$, and also obtain the graph of f.

Solution

You can generate a table of values of $f(t)$ for $t = 0, 1, \ldots, 8$ as follows:

1. Set up two columns—one for the values of t and one for the values of $f(t)$—and enter the values $0, 1, \ldots, 8$ in the t column as shown below.

2. We must now enter the formula for f in cell B2. The following formula defines the function n in Excel:

= (A2<=3) * (5*A2^2+2*A2+2) + (A2>3) *
  (-2*A2^2+33*A2-28)

When x is less than or equal to 3, the logical expression (x<=3) evaluates to 1 because it is true, and the expression (x>3) evaluates to 0 because it is false. The value of the function is therefore given by the expression (5*x^2+2*x+2). When x is greater than 3, the expression (x<=3) evaluates to 0 while the expression (x>3) evaluates to 1, so the value of the function is given by the expression (-2*x^2+33*x-28). We therefore enter the formula

= (A2<=3) * (5*A2^2+2*A2+2) + (A2>3) *
  (-2*A2^2+33*A2-28)

in cell B2 and then copy down to cell B10 (first figure below) to obtain the result shown in the second figure:

	A	B	C	D	E	F
1	t	f(t)				
2		0 =(A2<=3)*(5*A2^2+2*A2+2)+(A2>3)*(-2*A2^2+33*A2-28)				
3		1				
4		2				
5		3				
6		4				
7		5				
8		6				
9		7				
10		8				

	A	B
1	t	f(t)
2	0	2
3	1	9
4	2	26
5	3	53
6	4	72
7	5	87
8	6	98
9	7	105
10	8	108

3. To graph the data, highlight A1 through B10, and insert a scatter chart as in Example 1 to obtain the result shown below:

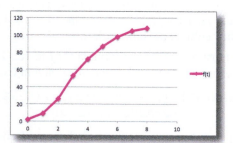

Section 1.2

Example 5(a) (page 72)   The following table shows some annual revenues, in billions of dollars, earned by **Amazon** from 2004 through 2014:

Year	2004	2006	2008	2010	2012	2014
Revenue ($ billion)	7	11	19	34	61	89

Take t to be the number of years since 2004, and consider the following three models:

   (1) $r(t) = 8t - 3$ Linear model

   (2) $r(t) = 0.9t^2 - 0.6t + 7$ Quadratic model

   (3) $r(t) = 7(1.3^t)$.   Exponential model

a. Which two models fit the data significantly better than the third?

b. If you extrapolate the models you selected in part (a) back to $t = -4$, what do you find? (Amazon's revenue in 2000 was $2.8 billion and subsequently increased each year.).

Solution

1. First create a scatter plot of the given data by tabulating the data as shown below, highlighting cells A1 through B7 and inserting a scatter chart:

	A	B
1	t	Revenue
2	0	7
3	2	11
4	4	19
5	6	34
6	8	61
7	10	89

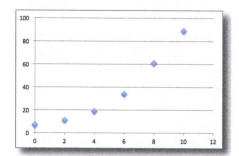

2. In column C, use the formula for the model you are interested in seeing; for example, model (2): `=0.9*A2^2-0.6*A2+7`.

	A	B	C	D
1	t	Revenue		
2	0	7	=0.9*A2^2-0.6*A2+7	
3	2	11		
4	4	19		
5	6	34		
6	8	61		
7	10	89		

	A	B	C
1	t	Revenue	
2	0	7	7
3	2	11	9.4
4	4	19	19
5	6	34	35.8
6	8	61	59.8
7	10	89	91

3. To adjust the graph to include the graph of the model you have added, you need to change the graph data from $A\$1:\$B\$7$ to $A\$1:\$C\$7$ to include column C. In Excel you can obtain this by right-clicking on the graph to select "Source Data". In OpenOffice, double-click on the graph, and then right-click it to choose "Data Ranges". In Excel you can also click once on the graph—the effect will be to outline the data you have graphed in columns A and B—and then use the fill handle at the bottom of column B to extend the selection to column C. The graph will now include markers showing the values of both the actual sales and the model you inserted in column C.

4. Right-click on any of the markers corresponding to column B in the graph (in OpenOffice you would first double-click on the graph), select "Format data series" to add lines connecting the points and remove the markers. The effect will be as shown below, with the model represented by a curve and the actual data points represented by dots (below):

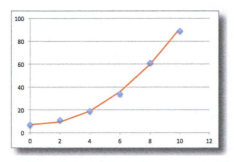

spreadsheet. If not, you will need to copy the formula in cell C6 up to C2. (Do not touch column B, as that contains the observed data starting at $t = 0$ only.) The result is shown below. Now change the graph data to $A\$1:\$C\$11$ by one of the techniques shown in Step 3 to obtain the graph shown below.

	A	B	C
1	t	Revenue	
2	-4		23.8
3	-3		16.9
4	-2		11.8
5	-1		8.5
6	0	7	7
7	2	11	9.4
8	4	19	19
9	6	34	35.8
10	8	61	59.8
11	10	89	91

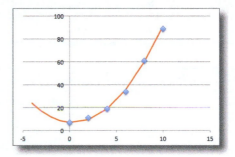

6. To see the plots for the remaining curves, change the formula in column C.

Section 1.3

Example 1 (page 88) Which of the following two tables gives the values of a linear function? What is the formula for that function?

x	0	2	4	6	8	10	12
$f(x)$	3	-1	-3	-6	-8	-13	-15

x	0	2	4	6	8	10	12
$g(x)$	3	-1	-5	-9	-13	-17	-21

Solution

1. The following worksheet shows how you can compute the successive quotients $m = \Delta y/\Delta x$ and hence check whether a given set of data shows a linear relationship, in which case all the quotients will be the same. (The

5. To see the extrapolation of the curve back to 2000, first move the data down by selecting A2 through C7, clicking in the middle of the lower edge of the selected region, and dragging the data down four cells. Then add the values -1, -2, -3, and -4 in column A as shown above. The values of $r(t)$ may automatically be computed in column C as you type, depending on the

	A	B	C
1	t	Revenue	
2	-4		
3	-3		
4	-2		
5	-1		
6	0	7	7
7	2	11	9.4
8	4	19	19
9	6	34	35.8
10	8	61	59.8
11	10	89	91

shading indicates that the formula is to be copied down only as far as cell C7. Why not cell C8?)

	A	B	C
1	x	f(x)	m
2	0	3	=(B3-B2)/(A3-A2)
3	2	-1	
4	4	-3	
5	6	-6	
6	8	-8	
7	10	-13	
8	12	-15	

2. Here are the results for both *f* and *g*:

	A	B	C
1	x	f(x)	m
2	0	3	-2
3	2	-1	-1
4	4	-3	-1.5
5	6	-6	-1
6	8	-8	-2.5
7	10	-13	-1
8	12	-15	

	A	B	C
1	x	g(x)	m
2	0	3	-2
3	2	-1	-2
4	4	-5	-2
5	6	-9	-2
6	8	-13	-2
7	10	-17	-2
8	12	-21	

Example 2(a) (page 91)   Find the equation of the line through the points $(1, 2)$ and $(3, -1)$.

Solution

1. Enter the *x*- and *y*-coordinates in columns A and B, as shown below.

	A	B
1	x	y
2	1	2
3	3	-1

	A	B	C	D
1	x	y	m	b
2	1	2	=(B3-B2)/(A3-A2)	=B2-C2*A2
3	3	-1	Slope	Intercept

2. Add the headings *m* and *b* in C1 and D1, and then the formulas for the slope and intercept in C2 and D2, as shown above. The result will be as shown below:

	A	B	C	D
1	x	y	m	b
2	1	2	-1.5	3.5
3	3	-1	Slope	Intercept

Section 1.4

Example 1 (page 105)   Using the data on the per capita GDP in Mexico given at the beginning of Section 1.4, compute SSE, the sum-of-squares error, for the linear models $y = 0.5t + 8$ and $y = 0.25t + 9$, and graph the data with the given models.

Solution

1. Begin by setting up your worksheet with the observed data in two columns, *t* and *y*, and the predicted data for the first model in the third.

	A	B	C	D	E	F
1	t	y (Observed)	y (Predicted)		m	b
2	0	9	=$E$2*A2+$F$2		0.5	8
3	2	9				
4	4	10				
5	6	11				
6	8	11				
7	10	12				
8	12	13				
9	14	13				

2. Notice that, instead of using the numerical equation for the first model in column C, we used absolute references to the cells containing the slope *m* and the intercept *b*. This way, we can switch from one linear model to the next by changing only *m* and *b* in cells E2 and F2. (We have deliberately left column D empty in anticipation of the next step.)

3. In column D we compute the squares of the residuals using the Excel formula = (B2-C2)^2.

	A	B	C	D	E	F
1	t	y (Observed)	y (Predicted)	Residual^2	m	b
2	0	9	8	=(B2-C2)^2	0.5	8
3	2	9	9			
4	4	10	10			
5	6	11	11			
6	8	11	12			
7	10	12	13			
8	12	13	14			
9	14	13	15			

4. We now compute SSE in cell F4 by summing the entries in column D.

	A	B	C	D	E	F
1	t	y (Observed)	y (Predicted)	Residual^2	m	b
2	0	9	8	1	0.5	8
3	2	9	9	0		
4	4	10	10	0	SSE:	=SUM(D2:D9)
5	6	11	11	0		
6	8	11	12	1		
7	10	12	13	1		
8	12	13	14	1		
9	14	13	15	4		

5. Here is the completed spreadsheet:

	A	B	C	D	E	F
1	t	y (Observed)	y (Predicted)	Residual^2	m	b
2	0	9	8	1	0.5	8
3	2	9	9	0		
4	4	10	10	0	SSE:	8
5	6	11	11	0		
6	8	11	12	1		
7	10	12	13	1		
8	12	13	14	1		
9	14	13	15	4		

6. Changing m to 0.25 and b to 9 gives the sum of squares error for the second model, SSE = 2.

	A	B	C	D	E	F
1	t	y (Observed)	y (Predicted)	Residual^2	m	b
2	0	9	9	0	0.25	9
3	2	9	9.5	0.25		
4	4	10	10	0	SSE:	2
5	6	11	10.5	0.25		
6	8	11	11	0		
7	10	12	11.5	0.25		
8	12	13	12	1		
9	14	13	12.5	0.25		

7. To plot both the original data points and each of the two lines, use a scatter plot to graph the data in columns A through C in each of the last two worksheets above.

$$y = 0.5t + 8 \qquad\qquad y = 0.25t + 9$$

Example 2 (page 107)   Use the data on the per capita GDP in Mexico to find the best-fit linear model.

Solution

Here are two spreadsheet shortcuts for linear regression; one is graphical, and one is based on a spreadsheet formula.

Using a Trendline

1. Start with the original data, and insert a scatter plot (below).

	A	B
1	t	y
2	0	9
3	2	9
4	4	10
5	6	11
6	8	11
7	10	12
8	12	13
9	14	13

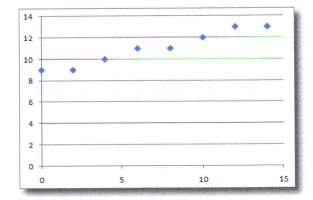

2. Insert a linear trendline, choosing the option to display the equation on the chart. The method for doing so varies from spreadsheet to spreadsheet.[79] In Excel you can right-click on one of the points in the graph and choose "Add Trendline". (In OpenOffice you would first double-click on the graph.) Then, under "Trendline Options", select "Display Equation on chart". The procedure for OpenOffice is almost identical, but you first need to double-click on the graph. The result is shown below.

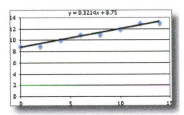

---

[79] At the time of this writing, Google Sheets has no trendline feature for its spreadsheet, so you would need to use the formula method.

Using a Formula

1. Enter your data as above, and select a block of unused cells two wide and one tall; for example, C2:D2. Then enter the formula

$$=LINEST(B2:B9,A2:A9)$$

as shown below. Then press Control+Shift+Enter. The result should appear as in the bottom figure below, with *m* and *b* appearing in cells C2 and D2 as shown:

◇	A	B	C	D
1	t	y	m	b
2	0	9	=LINEST(B2:B9,A2:A9)	
3	2	9		
4	4	10		
5	6	11		
6	8	11		
7	10	12		
8	12	13		
9	14	13		

◇	A	B	C	D
1	t	y	m	b
2	0	9	0.32143	8.75
3	2	9		
4	4	10		
5	6	11		
6	8	11		
7	10	12		
8	12	13		
9	14	13		

Example 3 (page 110) Find the correlation coefficient for the data in Example 2.

Solution

1. When you add a trendline to a chart, you can select the option "Display R-squared value on chart" to show the value of r^2 on the chart. (It is common to examine r^2, which takes on values between 0 and 1, instead of *r*.)

2. Alternatively, the LINEST function we used in Example 2 can be used to display quite a few statistics about a best-fit line, including r^2. Instead of selecting a block of cells two wide and one tall, as we did in Example 2, we select one that is two wide and *five* tall. We now enter the requisite LINEST formula with two additional arguments set to "TRUE" as shown, and press Control+Shift+Enter.

◇	A	B	C	D	E
1	t	y	m	b	
2	0	9	=LINEST(B2:B9,A2:A9,TRUE,TRUE)		
3	2	9			
4	4	10			
5	6	11			
6	8	11			
7	10	12			
8	12	13			
9	14	13			

◇	A	B	C	D
1	t	y	m	b
2	0	9	0.32143	8.75
3	2	9	0.02525	0.21129
4	4	10	0.96429	0.32733
5	6	11	162	6
6	8	11	17.3571	0.64286
7	10	12		
8	12	13		
9	14	13		

The values of *m* and *b* appear in cells C2 and D2 as before, and the value of r^2 appears in cell C4. (Among the other numbers shown is SSE in cell D6. For the meanings of the remaining numbers shown, do a web search for "LINEST"; you will see numerous articles, including many that explain all the terms. A good course in statistics wouldn't hurt either.)

2

THE MATHEMATICS OF FINANCE

CASE STUDY

Adjustable Rate and Subprime Mortgages

Mr. and Mrs. Wong have an appointment tomorrow with you, their investment counselor, to discuss their plan to purchase a $400,000 house in Orlando, Florida. Their combined annual income is $80,000 per year, and they estimate that it will increase by 4% annually over the foreseeable future. They are considering three different specialty 30-year mortgages:

Hybrid: The interest is fixed at a low introductory rate of 4% for 5 years.

Interest-Only: During the first 5 years, the rate is set at 4.2%, and no principal is paid.

Negative Amortization: During the first 5 years, the rate is set at 4.7% based on a principal of 60% of the purchase price of the home.

How would you advise them?

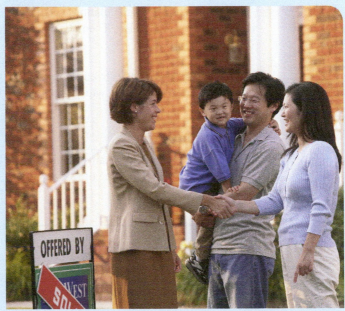

Ariel Skelley/Blend Images/Getty Images

Introduction

A knowledge of the mathematics of investments and loans is important not only for business majors but also for everyone who deals with money, which is all of us. This chapter is largely about *interest*: interest paid by an investment, interest paid on a loan, and variations on these.

We focus on three forms of investment: investments that pay simple interest, investments in which interest is compounded, and annuities. An investment that pays *simple interest* periodically gives interest directly to the investor, perhaps in the form of a monthly check. If, instead, the interest is reinvested, the interest is *compounded*, and the value of the account grows as the interest is added. An *annuity* is an investment earning compound interest into which periodic payments are made or from which periodic withdrawals are made. From the point of view of the lender, a loan is a kind of annuity.

We also look at bonds, the primary financial instrument used by companies and governments to raise money. Although bonds nominally pay simple interest, determining their worth, particularly in the secondary market, requires an annuity calculation.

2.1 Simple Interest

Calculating Interest

You deposit $1,000, called the **principal** or **present value**, into a savings account at *Gigantic Bank*. The bank pays you 5% interest, in the form of a check, each year. How much interest will you earn each year? Because the bank pays you 5% interest each year, your annual (or yearly) interest will be 5% of $1,000, or $1,000 \times 0.05 = \$50$.

Generalizing this calculation, call the present value PV and the interest rate (expressed as a decimal) r. Then INT, the annual interest paid to you, is given by*

$$INT = PVr.$$

If the investment continues to earn interest for t years, then the total interest accumulated is t times this amount, which gives us the following.

> ✱ Multiletter variables such as *PV* and *INT* used here may be unusual in a math textbook but are almost universally used in finance textbooks, calculators (such as the TI-83/84 Plus), and such places as study guides for the finance portion of the Society of Actuaries exams. Just watch out for expressions like *PVr*, which is the product of two things, *PV* and *r*, not three.

Simple Interest

The **simple interest** on an investment (or loan) of PV dollars at an annual interest rate of r for a time of t years is

$$INT = PVrt.$$

Quick Example

1. The simple interest on a $5,000 investment earning 8% per year for 4 years is

$$INT = PVrt$$
$$= (5,000)(0.08)(4) = \$1,600.$$

The above formula is based on *annual* interest, and we say that the interest **period** is 1 year. If the interest rate is given as, say, a monthly or weekly rate, we would need to adjust it to an annual rate in order to use the formula, and we would also need to adjust the length of time to be given in units of years. For example, to find the simple interest on a \$100 investment earning 0.5% per month for 7 months, we calculate the interest earned at a rate of $12 \times 0.5 = 6\%$ per year for $7/12$ years, which is

$$INT = 100(12 \times 0.005) \times \frac{7}{12} = 100(0.06) \times \frac{7}{12} = \$3.50.$$

However, in cases like this, it is often more convenient to use a formula based not on years but on arbitrary time periods. There was really nothing special about a year when we came up with the formula above, and we can substitute any time period in place of a year:

$$INT \quad = \quad PV \qquad\qquad r \qquad\qquad\qquad t$$

Interest = Present value $\times$ Interest rate per year $\times$ Number of years

or

Interest = Present value $\times$ Interest rate per period $\times$ Number of periods.

Now, the letter r is usually understood to mean interest rate *per year*. For a general, possibly different period, it is customary to instead use the letter i for interest rate *per period*. Similarly, t is usually understood to mean the number of *years*, whereas the letter n is used for the number of *periods*:

r = Interest rate per year, i = Interest rate per period,

t = Number of years, n = Number of periods.

So our simple interest formula becomes the following.

Simple Interest: General Interest Periods

The simple interest on an investment (or loan) of *PV* dollars at an interest rate of *i* per period for *n* periods is

$$INT = PVin.$$

Quick Examples

2. The simple interest on a \$100 investment earning 0.5% per month for 7 months is

$$INT = PVin$$
$$= (100)(0.005)(7) = \$3.50.$$

3. The simple interest on a 6-month \$5,000 loan at an interest rate of 1.5% per month is

$$INT = PVin$$
$$= (5,000)(0.015)(6) = \$450.$$

We can also use the formula based on annual interest:

$$r = \text{Annual rate} = 0.015 \times 12 = 0.18, t = \text{Time in years} = \frac{1}{2}$$

$$INT = PVrt = (5,000)(0.18)(1/2) = \$450.$$

Calculating Future Value

Let's return to your *Gigantic Bank* investment of $1,000 at 5% simple interest. How much money will you have after 2 years? To find the answer, we need to add the accumulated interest to the principal to get the **future value** (*FV*) of your deposit:

$$FV = PV + INT = \$1,000 + (1,000)(0.05)(2) = \$1,100.$$

In general, we can compute the future value as follows:

$$FV = PV + INT = PV + PVrt = PV(1 + rt).$$

Future Value for Simple Interest

The **future value** of an investment of *PV* dollars earning simple interest is given by

$$FV = PV(1 + rt) \qquad r = \text{Annual interest rate}, t = \text{Time in years}$$

or

$$FV = PV(1 + in). \qquad i = \text{Interest rate per period}, n = \text{Number of periods}$$

Quick Examples

4. The value, at the end of 4 years, of a $5,000 investment earning 8% simple interest per year is

$$FV = PV(1 + rt)$$
$$= 5,000[1 + (0.08)(4)] = \$6,600.$$

5. The total amount due after 2 years on a $10,000 loan charging 1% simple interest per month is

$$FV = PV(1 + in)$$
$$= 10,000[1 + (0.01)(24)] = \$12,400. \qquad \begin{aligned}&n = \text{Number of}\\&\text{periods (months)} = 24\end{aligned}$$

6. Writing the future value in Quick Example 4 as a function of time, we get

$$FV = 5,000(1 + 0.08t)$$
$$= 5,000 + 400t,$$

which is a linear function of time *t*. The intercept is *PV* = $5,000, and the slope is the annual interest, $400 per year.

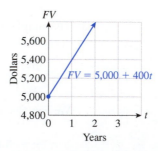

In general: *Simple interest growth is a linear function of the number of periods, with intercept = present value, and slope = interest per period.*

EXAMPLE 1 Savings Accounts

In December 2015, **Radius Bank** was paying 1.10% interest on savings accounts with balances of $2,500 or more. If the interest is paid as simple interest, find the future value of a $2,500 deposit after 6 years. What is the total interest paid over the period?[1]

Solution We use the future value formula:

$$FV = PV(1 + rt)$$
$$= 2,500[1 + (0.011)(6)] = 2,500[1.066] = \$2,665.$$

The total interest paid is given by the simple interest formula:

$$INT = PVrt$$
$$= (2,500)(0.011)(6) = \$165.$$

Note To find the interest paid, we could also have computed

$$INT = FV - PV = 2,665 - 2,500 = \$165. \quad \blacksquare$$

➡ **Before we go on . . .** In Example 1, we could look at the future value as a function of time:

$$FV = 2,500(1 + 0.011t) = 2,500 + 27.5t.$$

Thus, the future value is growing linearly at a rate of $27.50 per year ■

EXAMPLE 2 Bridge Loans

When moving to another location or "trading up," homeowners sometimes have to buy a new house before they sell their old house. One way to cover the costs of the new house until they get the proceeds from selling the old house is to take out a short-term *bridge loan*. Suppose a bank charges 12% simple annual interest on such a loan. How much will be owed at the maturation (the end) of a 3-month bridge loan of $90,000?

Solution As the length of the loan is given in months, we can use the formula

$$FV = PV(1 + in)$$

with $n = 3$. For the monthly interest rate we divide the annual rate by 12, so $i = 0.12/12 = 0.01$:

$$FV = 90,000[1 + (0.01)(3)]$$
$$= \$92,700.$$

➡ **Before we go on . . .** In Example 2 we could also have used the formula based on annual interest:

$$FV = PV(1 + rt)$$

with $r = 0.12$ and $t =$ time in years $= 3/12$:

$$FV = 90,000[1 + (0.12)(3/12)] = \$92,700.$$

---

[1] Source: Company website, www.radiusbank.com.

The astute reader might point out that dividing the interest rate by 12 to get the monthly rate is actually cheating; some months are longer than others, so the interest on some 3-month periods should be slightly more than on others. However, many financial institutions use the same simplifying assumption as we did: that the year is divided into 12 equal months.* ∎

* Another common assumption by financial institutions in calculating *daily* interest is that each month has 30 days, so a year has $12 \times 30 = 360$ days.

One of the primary ways companies and governments raise money is by selling **bonds**. At its most straightforward, a corporate bond promises to pay simple interest, usually twice a year, for a length of time until it **matures**, at which point it returns the original investment to the investor. (U.S. Treasury notes and bonds are similar.) Things get more complicated when the selling price is negotiable, as we will see later in this chapter.

EXAMPLE 3   **Corporate Bonds**

The *Megabucks Corporation* is issuing 10-year bonds paying an annual rate of 6.5%. If you buy $10,000 worth of bonds, how much interest will you earn every 6 months, and how much interest will you earn over the life of the bonds?

Solution Using the simple interest formula, every 6 months you will receive

$$INT = PVrt$$
$$= (10,000)(0.065)(0.5) = \$325.$$

Over the 10-year life of the bonds you will earn

$$INT = PVrt$$
$$= (10,000)(0.065)(10) = \$6,500$$

in interest. So at the end of 10 years, when your original investment is returned to you, your $10,000 will have turned into $16,500.

Calculating Present Value

We often want to turn an interest calculation around: Rather than starting with the present value and finding the future value, there are times when we know the future value and need to determine the present value. Solving the future value formula for *PV* gives us the following.

Present Value for Simple Interest

The present value of an investment with future value *FV* and earning simple interest is given by

$$PV = \frac{FV}{1 + rt}$$ $r =$ Annual interest rate, $t =$ Time in years

or

$$PV = \frac{FV}{1 + in}.$$ $i =$ Interest rate per period, $n =$ Number of periods

> **Quick Example**
>
> **7.** If an investment earns 5% simple interest and will be worth $1,000 in 4 years, then its present value (its initial value) is
>
> $$PV = \frac{FV}{1 + rt}$$
>
> $$= \frac{1,000}{1 + (0.05)(4)} = \$833.33.$$

Here is a typical example. U.S. Treasury bills (T-bills) are short-term investments (up to 1 year) that pay you a set amount after a period of time. What you pay to buy a T-bill depends on the interest rate.

EXAMPLE 4   Treasury Bills

A U.S. Treasury bill paying $10,000 after 6 months earns 3.67% simple annual interest. How much did it cost to buy?

Solution The future value of the T-bill is $10,000; the price we paid is its present value. We know that

$$FV = \$10,000$$
$$r = 0.0367$$

and

$$t = 0.5.$$

Substituting into the present value formula, we have

$$PV = \frac{10,000}{1 + (0.0367)(0.5)} = 9,819.81,$$

so we paid $9,819.81 for the T-bill.

➡ **Before we go on . . .** The simplest way to find the interest earned on the T-bill is by subtraction:

$$INT = FV - PV = 10,000 - 9,819.81 = \$180.19.$$

So after 6 months we received back $10,000, which is our original investment plus $180.19 in interest. ∎

Calculating the Interest Rate

Fees on loans can also be thought of as a form of interest, and sometimes it is eye-opening to know what rate of interest such fees represent.

EXAMPLE 5 Tax Refunds

You are expecting a tax refund of $800. Because it may take up to 6 weeks to get the refund, your tax preparation firm offers, for a fee of $40, to give you an "interest-free" loan of $800 to be paid back with the refund check. If we think of the fee as interest, what weekly interest rate is the firm actually charging? What is the corresponding annual interest rate?

Solution If we view the $40 as interest, then the future value of the loan (the value of the loan to the firm, or the total you will pay the firm) is $840. Thus, we have

$$FV = 840$$
$$PV = 800$$
$$n = 6, \qquad \text{Using weekly interest periods}$$

and we wish to find i, the weekly interest rate. Substituting, we get

$$FV = PV(1 + in)$$
$$840 = 800(1 + 6i) = 800 + 4{,}800i,$$

so

$$4{,}800i = 840 - 800 = 40$$
$$i = \frac{40}{4{,}800} \approx 0.00833,$$

or around 0.83% per week. The corresponding annual rate is 52 times that amount:

$$r = 52i \approx (52)(0.00833) \approx 0.43.$$

In other words, the firm is charging you 43% annual interest! Save your money, and wait 6 weeks for your refund.

More on Treasury Bills (Optional)

Here is some additional terminology on Treasury bills.

Treasury Bills (T-Bills): Maturity Value, Discount Rate, and Yield

The **maturity value** of a T-bill is the amount of money it will pay at the end of its life, that is, upon **maturity**.

Quick Example

8. A 1-year $10,000 T-bill has a maturity value of $10,000 and so will pay you $10,000 after 1 year.

* An exception occurred during the financial meltdown of 2008, when T-bills were heavily in demand as "safe haven" investments and were sometimes selling at—or even above—their maturity values.

The cost of a T-bill is generally less than its maturity value.* In other words, a T-bill will generally sell at a *discount*, and the **discount rate** is the *annualized* percentage of this discount; that is, the percentage is adjusted to give an annual percentage. (See Quick Examples 10 and 11.)

> **Quick Examples**
>
> 9. A 1-year $10,000 T-bill with a discount rate of 5% will sell for 5% less than its maturity value of $10,000, that is, for $9,500.
>
> 10. A 6-month $10,000 T-bill with a discount rate of 5% will sell at an actual discount of half of that—2.5% less than its maturity value, or $9,750—because 6 months is half of a year.
>
> 11. A 3-month $10,000 T-bill with a discount rate of 5% will sell at an actual discount of a fourth of that: 1.25% less than its maturity value, or $9,875.

The annual **yield** of a T-bill is the simple annual interest rate an investor earns when the T-bill matures, as calculated in the next example.

EXAMPLE 6 Treasury Bills

A T-bill paying $10,000 after 6 months sells at a discount rate of 3.6%. What does it sell for? What is the annual yield?

Solution The (annualized) discount rate is 3.6%; so for a 6-month bill, the actual discount will be half of that: $3.6\%/2 = 1.8\%$ below its maturity value. This makes the selling price

$$10,000 - (0.018)(10,000) = \$9,820. \qquad \text{Maturity value} - \text{Discount}$$

To find the annual yield, note that the present value of the investment is the price the investor pays, $9,820, and the future value is its maturity value, $10,000, six months later. So

$$PV = \$9,820, \quad FV = \$10,000, \quad t = 0.5,$$

and we wish to find the annual interest rate r. Substituting in the future value formula, we get

$$FV = PV(1 + rt)$$
$$10,000 = 9,820(1 + 0.5r),$$

so

$$1 + 0.5r = 10,000/9,820$$

and

$$r = (10,000/9,820 - 1)/0.5 \approx 0.0367.$$

Thus, the T-bill is paying 3.67% simple annual interest, so we say that its annual yield is 3.67%.

➡ **Before we go on ...** The T-bill in Example 6 is the same one as in Example 4 (with a bit of rounding). The yield and the discount rate are two different ways of telling what the investment pays. Exercise 62 asks you to find a formula for the yield in terms of the discount rate. ∎

FAQs

Which Formula to Use

Q: *How do I know when to use the formulas based on annual interest, such as INT = PVrt, as opposed to those based on general interest periods, such as INT = PVin?*

A: You can use either, as convenient. See, for instance, Quick Example 3 and the "Before we go on" discussion after Example 2.

2.1 EXERCISES

▼ more advanced ◆ challenging

T indicates exercises that should be solved using technology

In Exercises 1–10, compute the simple interest for the specified length of time and the future value at the end of that time. Round all answers to the nearest cent. [**HINT:** See Quick Examples 1–5.]

1. $2,000 is invested for 1 year at 6% per year.

2. $1,000 is invested for 10 years at 4% per year.

3. $4,000 is invested for 8 months at 0.5% per month.

4. $2,000 is invested for 12 weeks at 0.1% per week.

5. $20,200 is invested for 6 months at 5% per year.

6. $10,100 is invested for 3 months at 11% per year.

7. You borrow $10,000 for 10 months at 3% per year.

8. You borrow $6,000 for 5 months at 9% per year.

9. You borrow $12,000 for 10 months at 0.05% per month.

10. You borrow $8,000 for 5 weeks at 0.03% per week.

In Exercises 11–16, find the present value of the given investment. [**HINT:** See Quick Example 7.]

11. An investment earns 2% per year and is worth $10,000 after 5 years.

12. An investment earns 5% per year and is worth $20,000 after 2 years.

13. An investment earns 7% per year and is worth $1,000 after 6 months.

14. An investment earns 1% per month and is worth $5,000 after 3 months.

15. An investment earns 0.03% per week and is worth $15,000 after 15 weeks.

16. An investment earns 6% per year and is worth $30,000 after 20 months.

Applications

In Exercises 17–36, compute the specified quantity. Round all answers to the nearest month, the nearest cent, or the nearest 0.001%, as appropriate.

17. **Simple Loans** You take out a 6-month, $5,000 loan at 8% annual simple interest. How much would you owe at the end of the 6 months? [**HINT:** See Example 2.]

18. **Simple Loans** You take out a 15-month, $10,000 loan at 1% monthly simple interest. How much would you owe at the end of the 15 months? [**HINT:** See Example 2.]

19. **Investments** A corporate bond paying $1,000 after 12 weeks earns 0.02% simple interest per week. How much did it cost to buy? [**HINT:** See Example 4.]

20. **Simple Loans** Your total payment on a 4-year loan, which charged 9.5% annual simple interest, amounted to $30,360. How much did you originally borrow? [**HINT:** See Example 4.]

21. **Bonds** A 5-year bond costs $1,000 and will pay a total of $250 interest over its lifetime. What is its annual interest rate? [**HINT:** See Example 3.]

22. **Bonds** A 4-year bond costs $10,000 and will pay a total of $2,800 in interest over its lifetime. What is its annual interest rate? [**HINT:** See Example 3.]

Exercises 23–28 are based on the following table, which lists several corporate bonds issued during the second quarter of 2015:[2]

Company	AT&T	Bank of America	General Electric	Goldman Sachs	Verizon	Wells Fargo
Time to Maturity (years)	10	10	2	3	8	7
Annual Rate (%)	3.40	4.00	5.25	6.15	5.15	3.50

---

[2] Source: Financial Industry Regulatory Authority (www.finra.org).

23. If you spent $10,000 on AT&T bonds, how much interest would you earn every 6 months, and how much interest would you earn over the life of the bonds? [**HINT:** See Example 3.]

24. If you spent $5,000 on Bank of America bonds, how much interest would you earn every 6 months, and how much interest would you earn over the life of the bonds? [**HINT:** See Example 3.]

25. If the General Electric bonds you purchased had paid you a total of $8,840 at maturity, how much did you originally invest? (Round your answer to the nearest $1.)

26. If the Goldman Sachs bonds you purchased had paid you a total of $5,330.25 at maturity, how much did you originally invest? (Round your answer to the nearest $1.)

27. Which of Goldman Sachs and Wells Fargo would pay the most total interest on a $5,000 bond at maturity? How much interest would that be?

28. Which of Bank of America and Verizon would pay the most total interest on a $2,000 bond at maturity? How much interest would that be?

29. ▼ *Simple Loans* A $4,000 loan, taken now, with a simple interest rate of 8% per year, will require a total repayment of $4,640. When will the loan mature?

30. ▼ *Simple Loans* The simple interest on a $1,000 loan at 8% per year amounted to $640. When did the loan mature?

31. ▼ *Fees* You are expecting a tax refund of $1,000 in 4 weeks. A tax preparer offers you a $1,000 loan for a fee of $50 to be repaid by your refund check when it arrives in 4 weeks. Thinking of the fee as interest, what weekly simple interest rate would you be paying on this loan? What is the corresponding annual rate? [**HINT:** See Example 5.]

32. ▼ *Fees* You are expecting a tax refund of $1,500 in 3 weeks. A tax preparer offers you a $1,500 loan for a fee of $60 to be repaid by your refund check when it arrives in 3 weeks. Thinking of the fee as interest, what weekly simple interest rate would you be paying on this loan? What is the corresponding annual rate? [**HINT:** See Example 5.]

33. ▼ *Fees* You take out a 2-year, $5,000 loan at 9% simple annual interest. The lender charges you a $100 fee. Thinking of the fee as additional interest, what is the actual annual interest rate you will pay?

34. ▼ *Fees* You take out a 3-year, $7,000 loan at 8% simple annual interest. The lender charges you a $100 fee. Thinking of the fee as additional interest, what is the actual annual interest rate you will pay?

35. ▼ *Layaway Fees* Layaway plans allow you, for a fee, to pay for an item over a period of time and then receive the item when you finish paying for it. In November 2011, Senator Charles E. Schumer of New York warned that the holiday layaway programs recently reinstated by several popular retailers were, when the fees were taken into account,

charging interest at a rate significantly higher than the highest credit card rates.[3] Suppose that you bought a $69 item on November 15 on layaway, with the final payment due December 15, and that the retailer charged you a $5 service fee. Thinking of the fee as interest, what simple annual interest rate would you be paying for this layaway plan?

36. ▼ *Layaway Fees* Referring to Exercise 35, suppose that you bought a $99 item on October 15 on layaway, with the final payment due December 15, and that the retailer charged you a $10 service fee. Thinking of the fee as interest, what simple annual interest rate would you be paying for this layaway plan?

Stock Investments Exercises 37–42 are based on the following chart, which shows monthly figures for **Apple** stock in 2010:[4]

Marked are the following points on the chart:

Jan. 10	Feb. 10	Mar. 10	Apr. 10	May 10	June 10
211.98	195.46	218.95	235.97	235.86	255.96
July 10	**Aug. 10**	**Sep. 10**	**Oct. 10**	**Nov. 10**	**Dec. 10**
246.94	260.09	258.77	294.07	317.13	317.44

37. Calculate to the nearest 0.01% your monthly percentage return (on a simple interest basis) if you had bought Apple stock in June and sold in December.

38. Calculate to the nearest 0.01% your monthly percentage return (on a simple interest basis) if you had bought Apple stock in February and sold in June.

39. ▼ Suppose you bought Apple stock in April. If you later sold at one of the marked dates on the chart, which of those dates would have given you the largest monthly return (on a simple interest basis), and what would that return have been?

40. ▼ Suppose you bought Apple stock in May. If you later sold at one of the marked dates on the chart, which of those dates would have given you the largest monthly return (on a simple interest basis), and what would that return have been?

---

[3] Source: November 14, 2011 press release from the office of Senator Charles E. Schumer (http://schumer.senate.gov/Newsroom/ releases.cfm).
[4] Source: Yahoo! Finance (http://finance.yahoo.com).

41. ▼ Did Apple's stock undergo simple interest change in the period January through May? Give a reason for your answer.

42. ▼ If Apple's stock had undergone simple interest change from April to June at the same simple rate as from March to April, what would the price have been in June?

Population Exercises 43–48 are based on the following graph, which shows the population of San Diego County from 1950 to 2000:[5]

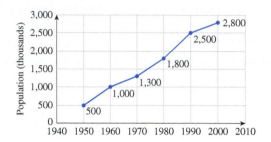

43. At what annual (simple interest) rate did the population of San Diego County increase from 1950 to 2000?

44. At what annual (simple interest) rate did the population of San Diego County increase from 1950 to 1990?

45. ▼ If you used your answer to Exercise 43 as the annual (simple interest) rate at which the population was growing since 1950, what would you predict the San Diego County population to be in 2010?

46. ▼ If you used your answer to Exercise 44 as the annual (simple interest) rate at which the population was growing since 1950, what would you predict the San Diego County population to be in 2010?

47. ▼ Use your answer to Exercise 43 to give a linear model for the population of San Diego County from 1950 to 2000. Draw the graph of your model over that period of time.

48. ▼ Use your answer to Exercise 44 to give a linear model for the population of San Diego County from 1950 to 2000. Draw the graph of your model over that period of time.

49. *Treasury Bills* In December 2008 (in the midst of the financial crisis) a $5,000 6-month T-bill was selling at a discount rate of only 0.25%.[6] What was its simple annual yield? [**HINT:** See Example 6.]

50. *Treasury Bills* In December 2008 (in the midst of the financial crisis) a $15,000 3-month T-bill was selling at a discount rate of only 0.06%.[7] What was its simple annual yield? [**HINT:** See Example 6.]

51. ▼ *Treasury Bills* At auction on August 18, 2005, 6-month T-bills were sold at a discount of 3.705%.[8] What was the simple annual yield? [**HINT:** See Example 6.]

52. ▼ *Treasury Bills* At auction on August 18, 2005, 3-month T-bills were sold at a discount of 3.470%.[9] What was the simple annual yield? [**HINT:** See Example 6.]

Communication and Reasoning Exercises

53. One or more of the following three graphs represents the future value of an investment earning simple interest. Which one(s)? Give the reason for your choice(s).

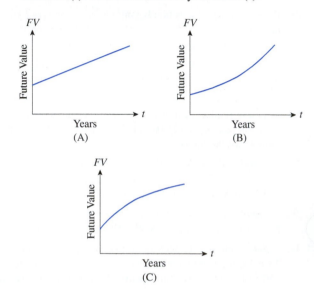

54. In the formula $FV = PV(1 + in)$, how is the slope of the graph of FV versus n related to the interest rate?

55. Given that $FV = 0.5n + 1,000$, for what monthly interest rate is this the equation of future value (in dollars) as a function of the number of months n?

56. Given that $FV = 5t + 400$, for what annual interest rate is this the equation of future value (in dollars) as a function of time t (in years)?

57. ▼ We said that the choice of which of the two formulas $FV = PV(1 + rt)$ and $FV = PV(1 + in)$ to use is a matter of convenience. Show that using the first formula to compute future value for a simple interest investment for n months at a monthly interest rate of i results in the second formula.

58. ▼ Refer to Exercise 57. Show that using the formula $FV = PV(1 + in)$ for interest periods of one month to compute future value for a simple interest investment for t years at an annual interest rate of r results in the formula $FV = PV(1 + rt)$.

[5] Source: Census Bureau/*New York Times*, April 2, 2002, p. F3.

[6] Discount rate on December 29, 2008. Source: U.S. Treasury (www.ustreas.gov/offices/domestic-finance/debt-management/interest-rate/daily_treas_bill_rates.shtml).

[7] *Ibid.*

[8] Source: The Bureau of the Public Debt's website (www.publicdebt.treas.gov).

[9] *Ibid.*

59. ▼ *Interpreting the News* You hear the following on your local radio station's business news: "The economy last year grew by 1%. This was the second year in a row in which the economy showed a 1% growth." Because the rate of growth was the same 2 years in a row, this represents simple interest growth—right? Explain your answer.

60. ▼ *Interpreting the News* You hear the following on your local radio station's business news: "The economy last year grew by 1%. This was the second year in a row in which the economy showed a 1% growth." This means that, in dollar terms, the economy grew more last year than the year before. Why?

61. ▼ Explain why simple interest is not the appropriate way to measure interest on a savings account that pays interest directly into your account.

62. ▼ Suppose that a 1-year T-bill sells at a discount rate d. Find a formula, in terms of d, for the simple annual interest rate r the bill will pay.

2.2 Compound Interest

Calculating Future Value

* Obviously not in the United States, where typical interest rates on savings accounts were around 0.08% at the time of this writing (December 2015). However, **Yes Bank** of India was offering 6% on its savings accounts at that time.

You deposit $10,000 into a savings account with a 6% annual interest rate.* At the end of each month the bank deposits, or **reinvests**, 1 month's worth of interest in your account. At the end of 6 months, how much money will you have accumulated? To calculate this amount, let's compute the amount you have at the end of each month.

The monthly interest is $i = 6\%/12 = 0.5\%$, so at the end of the first month the bank will pay you simple interest of 0.5% on your $10,000, which gives you a balance of

$$PV(1 + i) = 10{,}000(1 + 0.005) = \$10{,}050.00,$$

representing $50 of interest. At the end of the second month the bank will pay you another 0.5% interest, but this time it will be computed on the total in your account, which is $10,050.00. Thus, you will have a total of

$$10{,}050.00(1 + 0.005) = \$10{,}100.25.$$

If you were being paid simple interest on your original $10,000, you would have only $10,100 at the end of the second month. The extra 25¢ is the interest earned on the $50 interest added to your account at the end of the first month. Having interest earn interest is called **compounding** the interest; the length of time between the deposits of interest back into the account is called the **compounding period** (1 month in this case). We could continue like this until the end of the sixth month, but notice what we are doing: Each month we are multiplying the current balance by $1 + 0.005$. So at the end of 6 months you will have

$$10{,}000(1 + 0.005)^6 \approx \$10{,}303.78.$$

Compare this to the amount you would have earned if the bank had paid you simple interest:

$$PV(1 + in) = 10{,}000(1 + (0.005)(6)) = \$10{,}300.$$

The extra $3.78 is again the effect of compounding the interest.

> **Compound Interest: Future Value Formula**
>
> The future value of an investment of PV dollars earning compound interest at a rate of i per compounding period for n periods is
>
> $$FV = PV(1 + i)^n.$$

Quick Example

1. The future value of a $4,000 investment earning 2% interest per month for 8 months is

$$FV = PV(1 + i)^n = 4{,}000(1 + 0.02)^8 \approx \$4{,}686.64.$$

Q: *What about the corresponding formula for time measured in years rather than compounding periods?*

A: Suppose, for instance, that you invest PV dollars for t years in an account at an annual rate of r and the interest is compounded monthly. Then the interest per compounding period is $i = r/12$, and the number of compounding periods (months) is $n = 12 \times$ time in years $= 12t$, so

$$FV = PV(1 + i)^n = PV\left(1 + \frac{r}{12}\right)^{12t}.$$

We can generalize the above formula: If interest is compounded m times per year, we would get

$$FV = PV\left(1 + \frac{r}{m}\right)^{mt}.$$

Compound Interest: Future Value Formula Based on Years

The future value of an investment of PV dollars earning interest at an annual rate of r compounded m times per year for t years is

$$FV = PV\left(1 + \frac{r}{m}\right)^{mt}.$$

As in Section 2.1, the decision as to which of the two versions of the future value formula to use is a question of convenience. Practice using both, but note that, in Section 2.3, we will be using the simpler formula for general compounding periods exclusively.

Quick Examples

2. To find the future value after 5 years of a $10,000 investment earning 6% interest, with interest reinvested every month, we set $PV = 10{,}000$, $r = 0.06$, $m = 12$, and $t = 5$. Thus,

$$FV = PV\left(1 + \frac{r}{m}\right)^{mt} = 10{,}000\left(1 + \frac{0.06}{12}\right)^{60} \approx \$13{,}488.50.$$

3. Writing the future value in Quick Example 2 as a function of time, we get

$$FV = 10{,}000\left(1 + \frac{0.06}{12}\right)^{12t}$$
$$= 10{,}000(1.005)^{12t},$$

which is an **exponential** function of time t.

In general, *compound interest growth is an exponential function of time.*

Using Technology

All three technologies discussed in this book have built-in mathematics of finance capabilities. See the Technology Guides at the end of the chapter for details on using a TI-83/84 Plus or a spreadsheet to do the calculations in Example 1.

Website
www.WanerMath.com

→ Online Utilities
→ Time Value of Money Utility

This utility is similar to the TVM Solver on the TI-83/84 Plus. To compute the future value, enter the values shown, and press "Compute" next to FV.

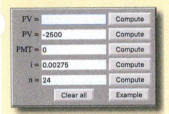

(For an explanation of the terms, see the Technology Guide for the TI-83/84 Plus.)

EXAMPLE 1   Savings Accounts

(Compare Example 1 in Section 2.1.) In December 2015, **Radius Bank** was paying 1.10% annual interest on savings accounts with balances of $2,500 and up. If the interest is compounded quarterly, find the future value of a $2,500 deposit after 6 years. What is the total interest paid over the time of the investment?[10]

Solution We can use either of the two formulas for future value, so let's use the first. The interest periods are quarters, so the interest per period is $i = 0.011/4 = 0.00275$, and the number of periods is $n = 4 \times 6 = 24$ quarters:

$$FV = PV(1 + i)^n$$
$$= 2,500(1 + 0.00275)^{24} \approx \$2,670.32.$$

The total interest paid is

$$INT = FV - PV = 2,670.32 - 2,500.00 = \$170.32,$$

$5.32 more than the $165.00 of simple interest in Example 1 in Section 2.1.

➡ **Before we go on . . .** If we use the formula based on years, we have $r = 0.011$, $t = 6$, and $m = 4$ (the number of times per year interest is compounded), so

$$FV = PV\left(1 + \frac{r}{m}\right)^{mt}$$
$$= 2,500\left(1 + \frac{0.011}{4}\right)^{4 \times 6}$$
$$= 2,500(1 + 0.00275)^{24} \approx \$2,670.32,$$

exactly the same calculation that we made using the simpler formula. ■

Example 1 illustrates the concept of the **time value of money**: A given amount of money received now will usually be worth a different amount to us than the same

---

[10] Source: Company website (www.radiusbank.com).

amount received some time in the future. In the example above, we can say that $2,500 received now is worth the same as $2,670.32 received 6 years from now, because if we receive $2,500 now, we can turn it into $2,670.32 by the end of 6 years.

Calculating Present Value

We often want to know, for some amount of money in the future, what is the equivalent value at present. As we did for simple interest, we can solve the future value formula for the present value and obtain the following formula.

> ### Present Value for Compound Interest
>
> The present value of an investment with future value FV is given by
>
> $$PV = \frac{FV}{(1 + i)^n} = FV(1 + i)^{-n}$$
>
> i = Interest rate per period
> n = Number of compounding periods
>
> or
>
> $$PV = \frac{FV}{(1 + \frac{r}{m})^{mt}}.$$
>
> r = Annual interest rate compounded m times per year
> t = Time in years
>
> #### Quick Example
>
> 4. To find the amount we need to invest in an investment earning 12% per year, compounded biannually (twice per year), so that we will have $1 million in 20 years, use $FV = \$1,000,000$, $i = 0.06$ per 6-month compounding period, and $n = 40$ compounding periods:
>
> $$PV = FV(1 + i)^{-n} = 1,000,000(1 + 0.06)^{-40} \approx \$97,222.19.$$
>
> Put another way, $1,000,000 twenty years from now is worth only $97,222.19 to us now if we have this 12% investment available.

In Section 2.1 we mentioned that a bond pays interest until it reaches maturity, at which point it pays you back an amount called its **maturity value** or **par value**.

EXAMPLE 2   Bonds

Megabucks Corporation is issuing 10-year corporate bonds paying no interest during their lifetime but promising to pay their maturity value at the end of the 10 years. (See the "Before we go on" discussion below.) How much would you pay for bonds with a maturity value of $10,000 if you wished to get a return of 6.5% compounded annually?*

Solution Think of the bond as if it were an account earning (compound) interest. We are asked to calculate the amount you will pay for the bond—the present value PV. We will use the formula based on general compounding periods. We have

$$FV = \$10,000$$
$$i = 0.065$$
$$n = 10.$$

*The return that investors look for depends on a number of factors, including risk (the chance that the company will go bankrupt and you will lose your investment); the higher the risk, the higher the return. U.S. Treasury bills are considered risk free because the federal government has never defaulted on its debts. On the other hand, so-called junk bonds are high-risk investments (below investment grade) and have correspondingly high yields.

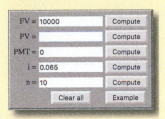

We can now use the present value formula:

$$PV = \frac{FV}{(1 + i)^n}$$

$$PV = \frac{10{,}000}{(1 + 0.065)^{10}} \approx \$5{,}327.26.$$

Thus, you should pay $5,327.26 to get a return of 6.5% annually.

➡ **Before we go on . . .** In Section 2.1 we said that a corporate bond usually pays simple interest over its lifetime as well as returning its maturity value at the end. The interest and the maturity value can actually be separated and sold and traded by themselves. A **zero coupon bond** is a form of corporate bond that pays no interest during its life but, like U.S. Treasury bills, promises to pay you the maturity value when it reaches maturity. Zero coupon bonds are often created by removing or *stripping* the interest coupons from an ordinary bond, and so are also known as **strips**. Zero coupon bonds sell for less than their maturity value, and the return on an investment is the difference between what the investor pays and the maturity value. Thus, in Example 2 an investor paying $5,327.26 for a 10-year $10,000 zero coupon bond would get a return of 6.5% on her investment. (The IRS refers to this kind of interest as **original issue discount (OID)** and taxes it as if it were interest actually paid to you each year.)

Particularly in financial applications you will see or hear the word "**discounted**" in place of "compounded" when discussing present value. Thus, the result of Example 2 might be phrased, "The present value of $10,000 to be received 10 years from now, with an interest rate of 6.5% discounted annually, is $5,327.26." ■

Inflation

Time value of money calculations are often done to take into account inflation, which behaves like compound interest. Suppose, for example, that inflation is running at 5% per year. Then prices will increase by 5% each year, so if PV represents the price now, the price 1 year from now will be 5% higher, or $PV(1 + 0.05)$. The price a year from then will be 5% higher still, or $PV(1 + 0.05)^2$. Thus, the effects of inflation are compounded, just as reinvested interest is.

EXAMPLE 3   **Inflation**

Inflation in East Avalon is 5% per year. *TruVision* television sets cost $200 today. How much will a comparable set cost 2 years from now?

Solution To find the price of a television set 2 years from now, we compute the future value of $200 at an inflation rate of 5% compounded yearly:

$$FV = 200(1 + 0.05)^2 = \$220.50.$$

EXAMPLE 4   **Constant Dollars**

Inflation in North Avalon is 6% per year. Which is really more expensive: a car costing $20,000 today or one costing $22,000 in 3 years?

Solution We cannot compare the two costs directly because inflation makes $1 today worth more (it buys more) than a dollar 3 years from now. We need the two prices expressed in comparable terms, so we convert to **constant dollars**. We take the car costing $22,000 three years from now and ask what it would cost in today's dollars. In other words, we convert the future value of $22,000 to its present value:

$$PV = FV(1 + i)^{-n}$$
$$= 22,000(1 + 0.06)^{-3}$$
$$\approx \$18,471.62.$$

Thus, the car costing $22,000 in 3 years actually costs less, after adjusting for inflation, than the one costing $20,000 now.

➡ **Before we go on . . .** In the presence of inflation the only way to compare prices at different times is to convert all prices to constant dollars. We pick some fixed time and compute future or present values as appropriate to determine what things would have cost at that time. ∎

Nominal and Effective Interest Rates

Different kinds of investments may have different compounding periods, so it is useful to have a way of comparing them.

EXAMPLE 5 **Comparing Investments**

You have just won $1 million in the lottery and are deciding what to do with it during the next year before you move to the South Pacific. *Bank Ten* offers 10% interest, compounded annually, while *Bank Nine* offers 9.8% compounded monthly. In which should you deposit your money?

Solution Let's calculate the future value of your $1 million after 1 year in each of the banks:

Bank Ten: $FV = 1(1 + 0.10)^1 = \$1.1$ million

Bank Nine: $FV = \left(1 + \dfrac{0.098}{12}\right)^{12} = \$1.1025$ million.

Bank Nine turns out to be better: It will pay you a total of $102,500 in interest over the year, whereas Bank Ten will pay only $100,000 in interest.

Another way of looking at the calculation in Example 5 is that Bank Nine gave you a total of 10.25% interest on your investment over the year. We call 10.25% the **effective interest rate** of the investment (also referred to as the **annual percentage yield**, or **APY**, in the banking industry); the stated 9.8% is called the **nominal** interest rate. In general, to best compare two different investments, it is wisest to compare their *effective*—rather than nominal—interest rates.

Notice that we got 10.25% by computing

$$\left(1 + \dfrac{0.098}{12}\right)^{12} = 1.1025$$

and then subtracting 1 to get 0.1025, or 10.25%. Generalizing, we get the following formula.

Nominal and Effective Interest Rate

The **nominal** annual interest rate of an investment paying compound interest is the original (quoted) annual rate. The **effective** annual interest rate is the actual percentage interest paid by the investment over 1 year and equals the annual rate that would result in the same interest if interest were compounded annually.

The effective interest rate r_{eff} of an investment paying a nominal interest rate of r_{nom} compounded m times per year is

$$r_{\text{eff}} = \left(1 + \frac{r_{\text{nom}}}{m}\right)^m - 1.$$

To compare rates of investments with different compounding periods, always compare the effective interest rates rather than the nominal rates.

Quick Example

5. To calculate the effective interest rate of an investment that pays 8% per year, with interest reinvested monthly, set $r_{\text{nom}} = 0.08$ and $m = 12$ to obtain

$$r_{\text{eff}} = \left(1 + \frac{0.08}{12}\right)^{12} - 1 \approx 0.0830, \text{ or } 8.30\%.$$

How Long to Invest

✳ See the Precalculus Review, Section 0.8.

Using Technology

See the Technology Guides at the end of the chapter for details on using TVM Solver on the TI-83/84 Plus or the built-in finance functions in spreadsheets to do the calculations in Example 6.

 Website
www.WanerMath.com

→ Online Utilities
→ Time Value of Money Utility

This utility is similar to the TVM Solver on the TI-83/84 Plus. To compute the time needed, enter the values shown, and press "Compute" next to n.

FV =	-6000	Compute
PV =	5000	Compute
PMT =	0	Compute
i =	0.005	Compute
n =		Compute
	Clear all	Example

EXAMPLE 6   **How Long to Invest**

You have $5,000 to invest at 6% annual interest compounded monthly. How long will it take for your investment to grow to $6,000?

Solution If we use the future value formula, we already have the values

$$FV = 6,000$$
$$PV = 5,000$$
$$i = \frac{0.06}{12} = 0.005$$
$$n = ?$$

Substituting, we get

$$6,000 = 5,000(1 + 0.005)^n = 5,000(1.005)^n.$$

Using logarithms, we can solve explicitly for n as follows:[*]

$$(1.005)^n = \frac{6,000}{5,000} = 1.2$$
$$\log(1.005)^n = \log 1.2$$
$$n \log(1.005) = \log 1.2$$
$$n = \frac{\log 1.2}{\log(1.005)} \approx 36.6 \text{ months},$$

which is $36.6/12 \approx 3$ years.

FAQs

What Formulas to Use, When to Use Compound Interest, and the Meaning of Present Value

Q: *How do I know when to use the formulas based on annual interest, such as*
$$PV\left(1 + \frac{r}{m}\right)^{mt}, \text{ as opposed to those based on general compounding periods, such}$$
as FV = PV(1 + i)^n?

A: You can use either, as convenient, as we have done in the examples. Know how to use both forms.

Q: *How do I distinguish a problem that calls for compound interest from one that calls for simple interest?*

A: Study the scenario to ascertain whether the interest is being withdrawn as it is earned or reinvested (deposited back into the account). If the interest is being withdrawn, the problem is calling for simple interest because the interest is not itself earning interest. If it is being reinvested, the problem is calling for compound interest.

Q: *How do I distinguish present value from future value in a problem?*

A: The present value always refers to the value of an investment before any interest is included (or, in the case of a depreciating investment, before any depreciation takes place). As an example, the future value of a bond is its maturity value. The value of $1 today in constant 2010 dollars is its present value (even though 2010 is in the past).

2.2 EXERCISES

▼ more advanced   ◆ challenging
T indicates exercises that should be solved using technology

In Exercises 1–10, calculate, to the nearest cent, the future value of an investment of $10,000 at the stated interest rate after the stated amount of time. [**HINT:** See Quick Examples 1 and 2.]

1. 0.2% per month, compounded monthly, after 15 months

2. 0.05% per week, compounded weekly, after 6 weeks

3. 0.2% per month, compounded monthly, after 10 years

4. 0.45% per month, compounded monthly, after 20 years

5. 3% per year, compounded annually, after 10 years

6. 4% per year, compounded annually, after 8 years

7. 2.5% per year, compounded quarterly (4 times per year), after 5 years

8. 1.5% per year, compounded weekly (52 times per year), after 5 years

9. 6.5% per year, compounded daily (assume 365 days per year), after 10 years

10. 11.2% per year, compounded monthly, after 12 years

In Exercises 11–16, calculate, to the nearest cent, the present value of an investment that will be worth $1,000 at the stated interest rate after the stated amount of time. [**HINT:** See Quick Example 4.]

11. 10 years, at 5% per year, compounded annually

12. 5 years, at 6% per year, compounded annually

13. 5 years, at 4.2% per year, compounded weekly (52 times per year)

14. 10 years, at 5.3% per year, compounded quarterly

15. 4 years, depreciating 5% each year

16. 5 years, depreciating 4% each year

In Exercises 17–22, find the effective annual interest rates of the given nominal annual interest rates. Round your answers to the nearest 0.01%. [**HINT:** See Quick Example 5.]

17. 5% compounded quarterly

18. 5% compounded monthly

19. 10% compounded monthly

20. 10% compounded daily (assume 365 days per year)

21. 10% compounded hourly (assume 365 days per year)

22. 10% compounded every minute (assume 365 days per year)

Applications

23. Savings You deposit $1,000 in an account at the *Lifelong Trust Savings and Loan* that pays 6% interest compounded quarterly. By how much will your deposit have grown after 4 years?

24. Investments You invest $10,000 in *Rapid Growth Funds*, which appreciate by 2% per year, with yields reinvested quarterly. By how much will your investment have grown after 5 years?

25. Depreciation: 2008 Financial Crisis During 2008 the S&P 500 index depreciated by 37.6%.[11] Assuming that this trend had continued, how much would a $3,000 investment in an S&P index fund have been worth after 3 years?

26. Depreciation: 2008 Financial Crisis During 2008 the NASDAQ Composite Index depreciated by 42%.[12] Assuming that this trend had continued, how much would a $5,000 investment in a NASDAQ Composite Index fund have been worth after 4 years?

27. Bonds You want to buy a 10-year bond with a maturity value of $5,000, and you wish to get a return of 5.5% annually. How much will you pay? [**HINT:** See Example 2.]

28. Bonds You want to buy a 15-year bond with a maturity value of $10,000, and you wish to get a return of 6.25% annually. How much will you pay? [**HINT:** See Example 2.]

29. ▼ Investments When I was considering what to do with my $10,000 lottery winnings, my broker suggested that I invest half of it in gold, the value of which was growing by 10% per year, and the other half in certificates of deposit (CDs), which were yielding 5% per year, compounded every 6 months. Assuming that these rates are sustained, how much will my investment be worth in 10 years?

30. ▼ Investments When I was considering what to do with the $10,000 proceeds from my sale of technology stock, my broker suggested that I invest half of it in municipal bonds, whose value was growing by 6% per year, and the other half in CDs, which were yielding 3% per year, compounded every 2 months. Assuming that these interest rates are sustained, how much will my investment be worth in 10 years?

31. ▼ Depreciation During a prolonged recession, property values on Long Island depreciated by 2% every 6 months. If my house cost $200,000 originally, how much was it worth 5 years later?

32. ▼ Depreciation Stocks in the health industry depreciated by 5.1% in the first 9 months of 1993.[13] Assuming that this trend had continued, how much would a $40,000 investment have been worth in 9 years? [**HINT:** Nine years corresponds to 12 nine-month periods.]

33. Present Value Determine the amount of money, to the nearest dollar, that you must invest at 6% per year, compounded annually, so that you will be a millionaire in 30 years.

34. Present Value Determine the amount of money, to the nearest dollar, that you must invest now at 7% per year, compounded annually, so that you will be a millionaire in 40 years.

35. Stocks Six years ago, I invested some money in *Dracubunny Toy Co.* stock, acting on the advice of a "friend." As things turned out, the value of the stock decreased by 5% every 4 months, and I discovered yesterday (to my horror) that my investment was worth only $297.91. How much did I originally invest?

36. Sales My recent marketing idea, the *Miracle Algae Growing Kit*, has been remarkably successful, with monthly sales growing by 6% every 6 months over the past 5 years. Assuming that I sold 100 kits the first month, how many kits did I sell in the first month of this year?

Exercises 37–42 are based on the following table, which lists several corporate bonds issued during the second quarter of 2015.[14] *Treat these as zero coupon bonds, as in Example 2.*

Company	AT&T	Bank of America	General Electric	Goldman Sachs	Verizon	Wells Fargo
Time to Maturity (years)	10	10	2	3	8	7
Annual Compound Interest Rate (%)	2.97	3.42	5.12	5.81	4.41	3.18

37. If you paid a total of $8,144.64 for General Electric bonds, what is their maturity value? (Round your answer to the nearest $1.)

38. If you paid a total of $8,441.50 for Goldman Sachs bonds, what is their maturity value? (Round your answer to the nearest $1.)

39. If you bought AT&T bonds with a maturity value of $10,000, how much did you originally pay? [**HINT:** See Example 2.]

40. If you bought Bank of America bonds with a maturity value of $5,000, how much did you originally pay? [**HINT:** See Example 2.]

41. ▼ What monthly simple interest is equivalent to the interest paid by Verizon corporate bonds over their lifetime?

42. ▼ What biannual simple interest is equivalent to the interest paid by Wells Fargo corporate bonds over their lifetime?

[11] Source: http://finance.google.com.
[12] *Ibid.*
[13] Source: *New York Times*, October 9, 1993, p. 37.

[14] Based on data in Financial Industry Regulatory Authority (www.finra.org).

43. ▼ *Retirement Planning* I want to be earning an annual salary of $100,000 when I retire in 15 years. I have been offered a job that guarantees an annual salary increase of 4% per year, and the starting salary is negotiable. What salary should I request in order to meet my goal?

44. ▼ *Retirement Planning* I want to be earning an annual salary of $80,000 when I retire in 10 years. I have been offered a job that guarantees an annual salary increase of 5% per year, and the starting salary is negotiable. What salary should I request in order to meet my goal?

45. ▼ *Inflation* Inflation has been running 2% per year. A car now costs $30,000. How much would it have cost 5 years ago? [**HINT:** See Example 3.]

46. ▼ *Inflation* (Compare Exercise 45.) Inflation has been running 1% every 6 months. A car now costs $30,000. How much would it have cost 5 years ago?

47. ▼ *Inflation* Housing prices have been rising 6% per year. A house now costs $200,000. What would it have cost 10 years ago?

48. ▼ *Inflation* (Compare Exercise 47.) Housing prices have been rising 0.5% each month. A house now costs $200,000. What would it have cost 10 years ago? [**HINT:** See Example 4.]

49. ▼ *Constant Dollars* Inflation is running 3% per year when you deposit $1,000 in an account earning interest of 5% per year compounded annually. In *constant dollars*, how much money will you have 2 years from now? [**HINT:** First calculate the value of your account in 2 years' time, and then find its present value based on the inflation rate.]

50. ▼ *Constant Dollars* Inflation is running 1% per month when you deposit $10,000 in an account earning 8% per year compounded monthly. In *constant dollars*, how much money will you have 2 years from now? [**HINT:** See Exercise 49.]

51. ▼ *Investments* You are offered two investments. One promises to earn 12% compounded annually. The other will earn 11.9% compounded monthly. Which is the better investment? [**HINT:** See Example 5.]

52. ▼ *Investments* You are offered three investments. The first promises to earn 15% compounded annually, the second will earn 14.5% compounded quarterly, and the third will earn 14% compounded monthly. Which is the best investment? [**HINT:** See Example 5.]

53. ▼ *History* Legend has it that a band of Lenape Indians known as the "Manhatta" sold Manhattan Island to the Dutch in 1626 for $24. In 2015 the total value of Manhattan real estate was estimated to be $362,524 million.[15]

Suppose that the Lenape had taken that $24 and invested it at 6.3% compounded annually (a relatively conservative investment goal). Could they then have bought back the island in 2015?

54. ▼ *History* Repeat Exercise 53, assuming that the Lenape had invested the $24 at 6.2% compounded annually.

Inflation *Exercises 55–62 are based on the following table, which shows the annual inflation rates in several Latin American countries in October 2015 (unless otherwise noted).[16] Assume that the rates shown continue indefinitely.*

Country	Argentina	Brazil	Bolivia	Nicaragua	Venezuela	Mexico	Uruguay
Currency	Peso	Real	Boliviano	Gold cordoba	Bolivar	Peso	Peso
Inflation Rate (%)	14.3	9.9	4.3	3.0	68.5	2.5	9.2

55. If an item in Brazil now costs 100 reals, what do you expect it to cost 5 years from now? (Answer to the nearest real.)

56. If an item in Argentina now costs 1,000 pesos, what do you expect it to cost 5 years from now? (Answer to the nearest peso.)

57. If an item in Bolivia will cost 1,000 bolivianos in 10 years, what does it cost now? (Answer to the nearest boliviano.)

58. If an item in Mexico will cost 20,000 pesos in 10 years, what does it cost now? (Answer to the nearest peso.)

59. ▼ You invest 1,000 bolivars in Venezuela at 8% annually, compounded twice a year. Find the value of your investment in 10 years, expressing the answer in constant bolivars. (Answer to the nearest bolivar.)

60. ▼ You invest 1,000 pesos in Uruguay at 8% annually, compounded twice a year. Find the value of your investment in 10 years, expressing the answer in constant pesos. (Answer to the nearest peso.)

61. ▼ Which is the better investment: an investment in Mexico yielding 5.3% per year, compounded annually, or an investment in Nicaragua yielding 6% per year, compounded every 6 months? Support your answer with figures that show the future value of an investment of 1 unit of currency in constant units.

62. ▼ Which is the better investment: an investment in Brazil yielding 10% per year, compounded annually, or an investment in Uruguay, yielding 9% per year, compounded every 6 months? Support your answer with figures that show the future value of an investment of 1 unit of currency in constant units.

---

[15] Source: Annual Report on the NYC Property Tax, Fiscal Year 2015, New York City Department of Finance, May 2015 (www.nyc.gov/finance).

[16] The Venezuela rate is as of December 2014. Source for data: www.tradingeconomics.com.

Stock Investments *Exercises 63–68 are based on the following chart, which shows monthly figures for* **Apple** *stock in 2010:*[17]

Marked are the following points on the chart:

Jan. 10	Feb. 10	Mar. 10	Apr. 10	May 10	June 10
211.98	195.46	218.95	235.97	235.86	255.96

July 10	Aug. 10	Sep. 10	Oct. 10	Nov. 10	Dec. 10
246.94	260.09	258.77	294.07	317.13	317.44

63. ▼ Calculate to the nearest 0.01% your annual percentage return (assuming annual compounding) if you had bought Apple stock in June and sold in December.

64. ▼ Calculate to the nearest 0.01% your annual percentage return (assuming annual compounding) if you had bought Apple stock in February and sold in June.

65. ▼ Suppose you bought Apple stock in April. If you later sold on one of the marked dates on the chart, which of those dates would have given you the largest annual return (assuming annual compounding), and what would that return have been?

66. ▼ Suppose you bought Apple stock in May. If you later sold on one of the marked dates on the chart, which of those dates would have given you the largest annual return (assuming annual compounding), and what would that return have been?

67. ▼ Did Apple's stock undergo compound interest change in the period January through May? Give a reason for your answer.

68. ▼ If Apple's stock had undergone compound interest change from April to June at the same monthly rate as from March to April, what would the price have been in June?

69. **T** ▼ *Competing Investments* I just purchased $5,000 worth of municipal bond funds that are expected to yield 5.4% per year, compounded every 6 months. My friend has just purchased $6,000 worth of CDs that will earn 4.8% per year, compounded every 6 months. Determine when, to the

nearest year, the value of my investment will be the same as hers and what this value will be. [**HINT:** This can be solved algebraically using logarithms; you can also graph the values of both investments or make a table of values.]

70. **T** ▼ *Investments* Determine when, to the nearest year, $3,000 invested at 5% per year, compounded daily, will be worth $10,000.

71. **T** ▼ *Epidemics* At the start of 1985 the incidence of AIDS was doubling every 6 months, and 40,000 cases had been reported in the United States. Assuming that this trend were to have continued, determine when, to the nearest tenth of a year, the number of cases would have reached 1 million.

72. **T** ▼ *Depreciation* My investment in *Genetic Splicing, Inc.*, is now worth $4,354 and is depreciating by 5% every 6 months. For some reason, I am reluctant to sell the stock and swallow my losses. Determine when, to the nearest year, my investment will drop below $50.

73. ◆ *Bonds* Once purchased, bonds can be sold in the secondary market. The value of a bond depends on the prevailing interest rates, which vary over time. Suppose that, in January 2020, you buy a 30-year zero coupon U.S. Treasury bond with a maturity value of $100,000 and a yield of 15% annually.
    a. How much do you pay for the bond?
    b. In January 2037, your bond has 13 years remaining until maturity. Rates on U.S. Treasury bonds of comparable length are now about 4.75%. If you sell your bond to an investor looking for a return of 4.75% annually, how much money do you receive?
    c. Using your answers to parts (a) and (b), what was the annual yield (assuming annual compounding) on your 17-year investment?

74. ◆ *Bonds* Suppose that, in January 2040, you buy a 30-year zero coupon U.S. Treasury bond with a maturity value of $100,000 and a yield of 5% annually.
    a. How much do you pay for the bond?
    b. Suppose that, 15 years later, interest rates have risen again, to 12%. If you sell your bond to an investor looking for a return of 12%, how much money will you receive?
    c. Using your answers to parts (a) and (b), what will be the annual yield (assuming annual compounding) on your 15-year investment?

Communication and Reasoning Exercises

75. Why is the graph of the future value of a compound interest investment as a function of time not a straight line (assuming a nonzero rate of interest)?

76. An investment that earns 10% (compound interest) every year is the same as an investment that earns 5% (compound interest) every 6 months—right?

[17] Source: Yahoo! Finance (http://finance.yahoo.com).

77. ▼ If a bacteria culture is currently 0.01 grams and increases in size by 10% each day, then its growth is linear—right?

78. ▼ At what point is the future value of a compound interest investment the same as the future value of a simple interest investment at the same annual rate of interest?

79. ▼ If two equal investments have the same effective interest rate and you graph the future value as a function of time for each of them, are the graphs necessarily the same? Explain your answer.

80. ▼ For what kind of compound interest investments is the effective rate the same as the nominal rate? Explain your answer.

81. ▼ For what kind of compound interest investments is the effective rate greater than the nominal rate? When is it smaller? Explain your answer.

82. ▼ If an investment appreciates by 10% per year for 5 years (compounded annually) and then depreciates by 10% per year (compounded annually) for 5 more years, will it have the same value as it had originally? Explain your answer.

83. ▼ You can choose between two investments that mature at different times in the future. If you knew the rate of inflation, how would you decide which is the better investment?

84. ▼ If you knew the various inflation rates for the years 2000–2011, how would you convert $100 in 2012 dollars to 2000 dollars?

85. Ⓣ ▼ On the same set of axes, graph the future value of a $100 investment earning 10% per year as a function of time over a 20-year period, compounded once a year, 10 times a year, 100 times a year, 1,000 times a year, and 10,000 times a year. What do you notice?

86. Ⓣ ▼ By graphing the future value of a $100 investment that is depreciating by 1% each year, convince yourself that, eventually, the future value will be less than $1.

2.3 Annuities, Loans, and Bonds

Annuities

Up to this point we have considered only situations in which a fixed amount of money earns interest over a period of time. However, investments like savings funds and pension funds also involve making periodic deposits into an account earning interest. For instance, a typical private sector pension fund works as follows: Every month while you work, you and your employer deposit a certain amount of money in an account. This money earns (compound) interest from the time it is deposited. When you retire, the account continues to earn interest, but you may then start withdrawing money at a rate calculated to reduce the account to zero after some number of years. Such an account is an example of an **annuity account**, an account earning interest into which you make periodic deposits or from which you make periodic withdrawals. The term **annuity** by itself refers to the sequence of deposits or withdrawals.*

*The term "annuity" is sometimes used to refer instead to the account or investment in question (see, for instance, Investopedia .com), so what we would call "an annuity account" or "annuity investment" might also just be called "an annuity."

Accumulation: Deposits and Future Value

The period during which you are contributing to your retirement account is the **accumulation phase**, during which the value of the account increases.

Suppose, for instance, that you make a payment of $100 at the end of every month into an account earning 3.6% interest per year, compounded monthly. This means that your investment is earning 3.6%/12 = 0.3% per month, so the interest per compounding period is $i = 0.003$. What will be the value of the investment at the end of 2 years (24 months)?

Think of the deposits separately. Each earns interest from the time it is deposited, and the total accumulated after 2 years is the sum of these deposits and the interest they earn. In other words, the accumulated value is the sum of the future values of the deposits, taking into account how long each deposit sits in the account. Figure 1 shows a timeline with the deposits and the contribution of each to the final value. For example, the very last deposit (at the end of month 24) has no time to earn interest, so it contributes only $100. The very first deposit, which earns interest for 23 months, by the future value formula for compound interest contributes $100(1 + 0.003)^{23}$ to the total. Adding together all of the future values gives us the total future value:

$$FV = 100 + 100(1 + 0.003) + 100(1 + 0.003)^2 + \cdots + 100(1 + 0.003)^{23}$$
$$= 100[1 + (1 + 0.003) + (1 + 0.003)^2 + \cdots + (1 + 0.003)^{23}].$$

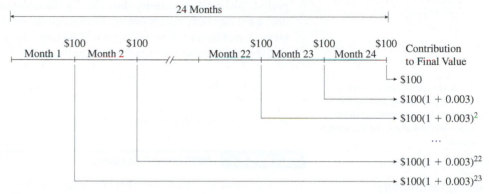

Figure 1

* It is called a **geometric series**.

† The quickest way to convince yourself that this formula is correct is to multiply out $(x - 1)(1 + x + x^2 + \cdots + x^{n-1})$ and see that you get $x^n - 1$. You should also try substituting some numbers. For example, $1 + 3 + 3^2 = 13 = (3^3 - 1)/(3 - 1)$.

Fortunately, this sort of sum is well-known (to mathematicians, anyway*), and there is a convenient formula for its value:†

$$1 + x + x^2 + \cdots + x^{n-1} = \frac{x^n - 1}{x - 1}.$$

In our case, with $x = 1 + 0.003$, this formula allows us to calculate the future value:

$$FV = 100\frac{(1 + 0.003)^{24} - 1}{(1 + 0.003) - 1} = 100\frac{(1.003)^{24} - 1}{0.003} \approx \$2{,}484.65$$

It is now easy to generalize this calculation.

Accumulation: Future Value

Suppose you make a payment of *PMT* at the end of each compounding period into an account with an interest rate of i per period. Then the future value of the account after n periods will be

$$FV = PMT\frac{(1 + i)^n - 1}{i}.$$

Quick Example

1. At the end of each month you deposit $50 into an account earning 2% annual interest compounded monthly. To find the future value after 5 years, we use $i = 0.02/12$ and $n = 12 \times 5 = 60$ compounding periods, so

$$FV = 50\frac{(1 + 0.02/12)^{60} - 1}{0.02/12} = \$3{,}152.37.‡$$

‡ Technology:
`50*((1+0.02/12)^60-1)/`
`(0.02/12)`

Notes

1. When a business or government accumulates money in an annuity for some future goal or obligation, the account is referred to as a **sinking fund**.

2. In general, when payments or withdrawals are made at the end of each compounding period, as we are discussing here, you have an **ordinary annuity**. If, instead, the payments or withdrawals are made at the beginning of each compounding period, you have an **annuity due**. During the accumulation phase for an annuity due, each payment occurs one period earlier, so there is one more period to earn interest, and hence the future value will be larger by a factor of $(1 + i)$. It follows that the future value formula for an annuity due is

$$FV = PMT(1 + i)\frac{(1 + i)^n - 1}{i}.* \qquad \text{\color{blue}Future value for an annuity due}$$

In this book we will concentrate on ordinary annuities. ∎

* Try to derive this formula from scratch yourself by making the appropriate adjustments to Figure 1 and adjusting the derivation above.

EXAMPLE 1 **Retirement Account**

Your retirement account has $5,000 in it and earns 5% interest per year compounded monthly. At the end of every month for the next 10 years you will deposit $100 into the account. How much money will there be in the account at the end of those 10 years?

Solution This is an annuity account with $PMT = \$100$, $i = 0.05/12$, and $n = 12 \times 10 = 120$. Ignoring for the moment the $5,000 already in the account, your payments have the following future value:

$$FV = PMT\frac{(1 + i)^n - 1}{i}$$

$$= 100\frac{(1 + 0.05/12)^{120} - 1}{0.05/12} \qquad \text{\color{blue}Technology:}$$
$$\text{\color{blue}100*((1+0.05/12)^120-1)/(0.05/12)}$$

$$\approx \$15,528.23.$$

What about the $5,000 that was already in the account? That sits there and earns interest, so we need to find its future value as well, using the compound interest formula:

$$FV = PV(1 + i)^n$$

$$= 5,000(1 + 0.05/12)^{120}$$

$$= \$8,235.05.$$

Hence, the total amount in the account at the end of 10 years will be

$$\$15,528.23 + 8,235.05 = \$23,763.28.$$

➡ **Before we go on . . .** We can combine the two calculations in Example 1 to obtain the following formula for the future value of an investment currently valued at PV into which deposits of PMT are made:

$$FV = PV(1 + i)^n + PMT\frac{(1 + i)^n - 1}{i}. \quad ∎$$

Sometimes we know what we want the future value of an annuity account to be and need to determine the payments necessary to achieve that goal. We can simply solve the future value formula for the payment, to get the following formula.

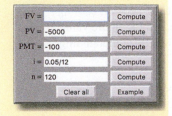

Payments to Accumulate a Future Value

Suppose you want to accumulate a total of *FV* after *n* interest periods in an account with an interest rate of *i* per period. Then the amount that must be paid into the account at the end of each period is

$$PMT = FV\frac{i}{(1 + i)^n - 1}.$$

Quick Example

2. To accumulate $100,000 after 17 years in a college education fund earning 4% annual interest compounded quarterly, the payments at the end of each quarter should be

$$PMT = 100,000\frac{0.01}{(1 + 0.01)^{68} - 1}\,\text{*}$$ $i = 0.04/4 = 0.01, n = 4 \times 17 = 68$

$$\approx \$1,033.89.$$

*Technology:
100000*0.01/(1.01^68-1)

Using Technology

To automate the computations in Example 2 using a graphing calculator or a spreadsheet, see the Technology Guides at the end of the chapter. Outline (for current age 20):

TI-83/84 Plus
APPS 1:Finance, then
1:TVM Solver
N = 59, I% = 3.6, PV = 0,
FV = 200000, P/Y = 12, C/Y = 12
With cursor on PMT line,
ALPHA SOLVE
[More details in the Technology Guide.]

Spreadsheet
=PMT(0.003,708,0,-200000)
[More details in the Technology Guide.]

 Website
www.WanerMath.com
→ Online Utilities
→ Time Value of Money Utility
Enter the values shown, and press "Compute" next to PMT.

FV = 200000	Compute
PV = 0	Compute
PMT =	Compute
i = 0.003	Compute
n = 12*59	Compute
Clear all	Example

EXAMPLE 2 Life Insurance

The average U.S. life expectancy is 79 years.[18] As an actuary in *Long Life Insurance,* you are trying to estimate the smallest monthly premiums the company can charge for a life insurance policy that can pay out $200,000 in the event of death at age 79. Assume that the payments are to be deposited at the end of each month in an annuity account earning 3.6% annual interest compounded monthly. Make a table showing the associated monthly premiums for clients currently of age 20, 30, 40, 50, 60, and 70 years.

Solution Think of the annuity account as an accumulating annuity with a future value of $200,000 and monthly interest rate $i = 0.036/12 = 0.003$. The number *n* of monthly payments depends on the age of the person and is given by

$$n = 12(79 - x),\quad 12 \times \text{Years expected to live} = 12(79 - \text{Current age})$$

where *x* is the current age. The monthy premiums that will result in a value of $200,000 at age 79 are therefore

$$PMT = FV\frac{i}{(1 + i)^n - 1} = 200,000\frac{0.003}{1.003^{12(79-x)} - 1}.$$

Substituting $x = 20, 30, 40, 50, 60$, and 70 results in the following values:

Current Age *x*	20	30	40	50	60	70
Number of Payment Periods $n = 12(79 - x)$	708	588	468	348	228	108
Monthly Premium ($) $200{,}000\dfrac{0.003}{1.003^n - 1}$	81.77	124.47	195.89	326.78	612.39	1,570.78

Technology: 200000*0.003/(1.003^(12(79-x))-1)

---

[18] Source: World Health Organization, 2015.

Annuitization: Withdrawals and Present Value

Let's go back to the scenario of a retirement account, but now assume that you have retired and begin to make monthly withdrawals from your account rather than making deposits. The period during which you are making the withdrawals is called the **annuitization phase**, or **payout phase**.

Suppose, for instance, it is the beginning of a particular month, and you currently have an amount PV in an account earning 3.6% interest per year, compounded monthly. Starting 1 month from now, you make monthly withdrawals of $100. What must PV be so that the account will be drawn down to $0 in exactly 2 years?

As before, we write $i = 0.036/12 = 0.003$, and we have $PMT = 100$. (We think of the withdrawals as periodic *payments* to you.) The first withdrawal of $100 will be made 1 month from now, so its present value is

$$\frac{PMT}{(1 + i)^n} = \frac{100}{1 + 0.003} = 100(1 + 0.003)^{-1} \approx \$99.70.$$

In other words, that much of the original PV goes toward funding the first withdrawal. The second withdrawal, 2 months from now, has a present value of

$$\frac{PMT}{(1 + i)^n} = \frac{100}{(1 + 0.003)^2} = 100(1 + 0.003)^{-2} \approx \$99.40.$$

That much of the original PV funds the second withdrawal. This continues for 2 years, at which point you make the last withdrawal, which has a present value of

$$\frac{PMT}{(1 + i)^n} = \frac{100}{(1 + 0.003)^{24}} = 100(1 + 0.003)^{-24} \approx \$93.06,$$

and that exhausts the account. Figure 2 shows a timeline with the withdrawals and the present value of each.

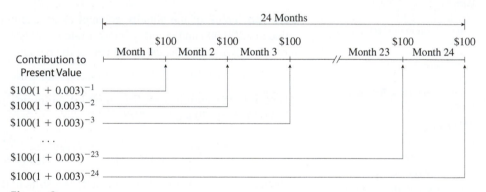

Figure 2

Because PV must be the sum of these present values, we get

$$PV = 100(1 + 0.003)^{-1} + 100(1 + 0.003)^{-2} + \cdots + 100(1 + 0.003)^{-24}$$
$$= 100[(1 + 0.003)^{-1} + (1 + 0.003)^{-2} + \cdots + (1 + 0.003)^{-24}].$$

We can again find a simpler formula for this sum:

$$x^{-1} + x^{-2} + \cdots + x^{-n} = \frac{1}{x^n}(x^{n-1} + x^{n-2} + \cdots + 1)$$

$$= \frac{1}{x^n} \cdot \frac{x^n - 1}{x - 1} = \frac{1 - x^{-n}}{x - 1}.$$

So in our case,

$$PV = 100\frac{1 - (1 + 0.003)^{-24}}{(1 + 0.003) - 1}$$

giving us

✱ Technology:
100*(1-1.003^(-24))/0.003

$$PV = 100\frac{1 - (1.003)^{-24}}{0.003} \approx \$2{,}312.29.✱$$

So you must have \$2,312.29 in the account initially so that, after you make withdrawals of \$100 per month for 2 years, your account will be exhausted at the end of that time. Generalizing, we get the following formula.

Annuitization: Present Value

Suppose you have an account earning interest at a rate of i per compounding period. If you receive a payment of PMT at the end of each compounding period, and the account is down to zero after n periods, then the starting balance of the account must be

$$PV = PMT\frac{1 - (1 + i)^{-n}}{i}.$$

Quick Example

3. At the end of each month you want to withdraw \$50 from an account earning 2% annual interest compounded monthly. If you want the account to last for 5 years (60 compounding periods), it must have the following amount to begin with:

$$PV = 50\frac{1 - (1 + 0.02/12)^{-60}}{0.02/12} = \$2{,}852.62.$$

Note Recall that what we are discussing are *ordinary annuities,* as withdrawals are being made at the end of each compounding period. If instead, the annuity is an *annuity due,* the withdrawals would be made at the beginning of each period, so there is one less period in which to earn interest. Thus, the present value must be larger by a factor of $(1 + i)$ to fund each payment, and the present value formula for an annuity due is

$$PV = PMT(1 + i)\frac{1 - (1 + i)^{-n}}{i}. \qquad \text{Present value for an annuity due} \quad ■$$

EXAMPLE 3   Trust Fund

You wish to establish a trust fund from which your niece can withdraw \$2,000 every 6 months for 15 years, at the end of which time she will receive the remaining money in the trust, which you would like to be \$10,000. The trust will be invested at 7% per year compounded every 6 months. How large should the trust be?

Solution We view this account as having two parts: one funding the semiannual payments and the other funding the \$10,000 lump sum at the end. The amount of

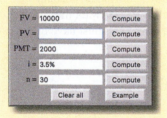
* Technology:
100000*0.01/(1-1.01^
(-16))

money necessary to fund the semiannual payments is the present value of an annuity, with $PMT = 2{,}000$, $i = 0.07/2 = 0.035$, and $n = 2 \times 15 = 30$. Substituting gives

$$PV = 2{,}000\frac{1 - (1 + 0.035)^{-30}}{0.035}$$

$$= \$36{,}784.09.$$

To fund the lump sum of $10,000 after 15 years, we need the present value of $10,000 under compound interest:

$$PV = 10{,}000(1 + 0.035)^{-30}$$

$$= \$3{,}562.78.$$

Thus the trust should start with $36,784.09 + 3,562.78 = $40,346.87.

Sometimes we know how much money we begin with and for how long we want to make withdrawals. We then want to determine the amount of money we can withdraw each period. For this we simply solve the present value formula for the payment.

Withdrawals to Annuitize a Present Value

Suppose that an account earns interest at a rate of i per compounding period. Suppose also that the account starts with a balance of PV. If you want to receive a payment of PMT at the end of each compounding period, and the account is down to $0 after n periods, then

$$PMT = PV\frac{i}{1 - (1 + i)^{-n}}.$$

Quick Example

4. The amount of money that should be withdrawn each quarter from a $100,000 college education fund earning 4% annual interest compounded quarterly, in order to draw down the account to zero at the end of 4 years, is

$$PMT = 100{,}000\frac{0.01}{1 - (1 + 0.01)^{-16}}*$$
$\qquad i = 0.04/4 = 0.01$, $n = 4 \times 4 = 16$

$$\approx \$6{,}794.46.$$

EXAMPLE 4   **Saving for Retirement**

Jane Q. Employee has just started her new job with *Big Conglomerate, Inc.*, and is already looking forward to retirement. BCI offers her as a pension plan an annuity account that is guaranteed to earn 6% annual interest compounded monthly. She plans to work for 40 years before retiring and would then like to be able to draw an income of $7,000 per month for 20 years. How much do she and BCI together have to deposit per month into the account to accomplish this?

Solution Here, we have the situation we described at the beginning of the section. This particular pension plan account begins with a 40-year accumulation phase followed by a 20-year annuitization phase. We know the desired payment out of the annuity account during the annuitization phase, so we work backward. The first thing we need to do is calculate the value in the account required to make the pension payments. We use the present value formula with $PMT = 7{,}000$, $i = 0.06/12 = 0.005$, and $n = 12 \times 20 = 240$:

$$PV = PMT\frac{1 - (1 + i)^{-n}}{i}$$

$$= 7{,}000\frac{1 - (1 + 0.005)^{-240}}{0.005} \approx \$977{,}065.40.$$

This is the total that must be accumulated during the 40-year accumulation phase. In other words, this is the *future* value, *FV*, of the annuity account during that phase. (Thus, the present value in the first step of our calculation is the future value in the second step.) To determine the payments necessary to accumulate this amount, we use the payment formula to accumulate a future value with $FV = 977{,}065.40$, $i = 0.005$, and $n = 12 \times 40 = 480$:

$$PMT = FV\frac{i}{(1 + i)^n - 1}$$

$$= 977{,}065.40\frac{0.005}{1.005^{480} - 1}$$

$$\approx \$490.62.$$

So if she and BCI collectively deposit $490.62 per month into her retirement account, she can retire with the income she desires.

Amortization: Loans and Mortgages

*The word "mortgage" comes from the French for "dead pledge."

In a typical installment loan, such as a car loan or a home mortgage,* we borrow an amount of money and then pay it back with interest by making fixed payments (usually every month) over some number of years. The process of paying off a loan is called **amortizing** the loan, meaning killing the debt owed. From the point of view of the lender, a loan is an annuitizing annuity: The lender "deposits" the amount of the loan with the borrower and makes periodic "withdrawals" (the loan payments) until the balance is zero. Thus, loan calculations are identical to annuity calculations.

EXAMPLE 5   Home Mortgages

Marc and Mira are buying a house and have taken out a 30-year, $90,000 mortgage at 8% interest per year. What will their monthly payments be?

Solution From the bank's point of view (see the comments before the example), the mortgage is a $90,000 annuity account from which it makes monthly withdrawals to draw the account down to zero after 30 years. Thus, the present value is $PV = \$90{,}000$, $i = 0.08/12$, and $n = 12 \times 30 = 360$. To find the mortgage payments, we use the formula for withdrawals:

$$PMT = 90{,}000\frac{0.08/12}{1 - (1 + 0.08/12)^{-360}} \approx \$660.39.$$

Month	Interest Payment	Payment on Principal	Outstanding Principal
0			$90,000.00
1	$600.00	$60.39	89,939.61
2	599.60	60.79	89,878.82
3	599.19	61.20	89,817.62
4	598.78	61.61	89,756.01
5	598.37	62.02	89,693.99
6	597.96	62.43	89,631.56
7	597.54	62.85	89,568.71
8	597.12	63.27	89,505.44
9	596.70	63.69	89,441.75
10	596.28	64.11	89,377.64
11	595.85	64.54	89,313.10
12	595.42	64.97	89,248.13
Total	**$7,172.81**	**$751.87**	

Using Technology

To automate the construction of the amortization schedule in Example 6 using a graphing calculator or a spreadsheet, see the Technology Guides at the end of the chapter.

EXAMPLE 6   Amortization Schedule

Continuing Example 5: Mortgage interest is tax deductible, so it is important to know how much of a year's mortgage payments represents interest. How much interest will Marc and Mira pay in the first year of their mortgage?

Solution Let us calculate how much of each month's payment is interest and how much goes to reducing the outstanding principal. At the end of the first month, Marc and Mira must pay 1 month's interest on $90,000, which is

$$\$90,000 \times \frac{0.08}{12} = \$600.$$

The remainder of their first monthly payment, $660.39 - 600 = \$60.39$, goes to reducing the principal. Thus, in the second month the outstanding principal is $90,000 - 60.39 = \$89,939.61$, and part of their second monthly payment will be for the interest on this amount, which is

$$\$89,939.61 \times \frac{0.08}{12} \approx \$599.60.$$

The remaining $660.39 - \$599.60 = \$60.79$ goes to further reduce the principal. If we continue this calculation for the 12 months of the first year, we get the beginning of the mortgage's **amortization schedule**, shown in the margin. As we can see from the totals at the bottom of the columns, Marc and Mira will pay a total of $7,172.81 in interest in the first year.

➡ **Before we go on . . .**

Q: Is there a formula to calculate the outstanding principal at the end of each month in Example 6?

A: We can use the present value formula. After k months there are $n - k$ payments of PMT remaining to make. The outstanding principal is the present value necessary to fund these payments. So

$$\text{Outstanding principal} = PV = PMT\frac{1 - (1 + i)^{-(n-k)}}{i}.$$

For instance, after 10 months there are $n - k = 360 - 10 = 350$ payments of $660.39 remaining, so the balance on the above mortgage is

$$660.39\frac{1 - (1 + 0.08/12)^{-350}}{0.08/12} = \$89,377.93.$$

Notice that it is 29¢ more than the value in the amortization schedule, which is due to the effects of rounding of the payment amount and the month-by-month rounding when we calculated the schedule in that example. Because banks will do all this rounding, the formula above is only an approximation of the actual outstanding principal (but a pretty good one).

EXAMPLE 7 Mortgage Refinancing

Continuing Example 5: Ten years into Marc and Mira's mortgage, interest rates have fallen considerably, and they are considering refinancing the total they still owe with a new 20-year mortgage at another bank at a rate of 5.5%. The bank where they have their current mortgage charges a prepayment penalty of 2% of the outstanding principal. What will their new monthly payments be if they go ahead with the refinance?

Solution Using the formula in the "Before we go on" discussion above, we calculate the outstanding principal after 10 years to be

$$\text{Outstanding principal} = PMT\frac{1 - (1 + i)^{-(n-k)}}{i}$$

$$= 660.39\frac{1 - (1 + 0.08/12)^{-(360-120)}}{0.08/12}$$

$$= \$78{,}952.46.$$

The 2% prepayment penalty boosts the amount they owe to

$$78{,}952.46 \times 1.02 = \$80{,}531.51.$$

Refinancing this amount at 5.5% interest for 20 years therefore results in monthly payments of

$$PMT = PV\frac{i}{1 - (1 + i)^{-n}}$$

$$= 80{,}531.51\frac{0.055/12}{1 - (1 + 0.055/12)^{-240}} = \$553.97,$$

so their monthly payments would decrease by $660.39 - 553.97 = \$106.42$.

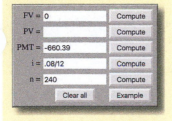

Bonds (Optional)

Suppose that a corporation offers a 10-year bond paying 6.5% with payments every 6 months. As we saw in Example 3 of Section 2.1, this means that if we pay $10,000 for bonds with a maturity value of $10,000, we will receive $6.5/2 = 3.25\%$ of $10,000, or $325, every 6 months for 10 years, at the end of which time the corporation will give us the original $10,000 back. But bonds are rarely sold at their maturity value. Rather, they are auctioned off and sold at a price the bond market determines they are worth.

For example, suppose that bond traders are looking for an investment that has a **rate of return** or **yield** of 7% rather than the stated 6.5% (sometimes called the **coupon interest rate** to distinguish it from the rate of return). How much would they be willing to pay for the bonds above with a maturity value of $10,000? Think of the bonds as an investment that will pay the owner $325 every 6 months for 10 years, and will pay an additional $10,000 on maturity at the end of the 10 years. We can treat the $325 payments as if they were an annuity and determine how much an investor would pay for such an annuity account if it earned 7% compounded semiannually. Separately, we determine the present value of an investment worth $10,000 ten years from now, if it earned 7% compounded semiannually. For the first calculation we use the annuity present value formula, with $i = 0.07/2$ and $n = 2 \times 10 = 20$:

$$PV = PMT\frac{1 - (1 + i)^{-n}}{i}$$

$$= 325\frac{1 - (1 + 0.07/2)^{-20}}{0.07/2}$$

$$= \$4{,}619.03.$$

For the second calculation we use the present value formula for compound interest:

$$PV = 10{,}000(1 + 0.07/2)^{-20}$$

$$= \$5{,}025.66.$$

Thus, an investor looking for a 7% return will be willing to pay $4,619.03 for the semiannual payments of $325 and $5,025.66 for the $10,000 payment at the end of 10 years, for a total of $4,619.03 + 5,025.66 = $9,644.69 for the $10,000 bond.

EXAMPLE 8   Bonds

Suppose that bond traders are looking for only a 6% yield on their investment. How much would they pay per $10,000 for the 10-year bonds above, which have a coupon interest rate of 6.5% and pay interest every 6 months?

Solution We redo the calculation with $r = 0.06$. For the annuity calculation we now get

$$PV = 325\frac{1 - (1 + 0.06/2)^{-20}}{0.06/2} = \$4{,}835.18.$$

For the compound interest calculation we get

$$PV = 10{,}000(1 + 0.06/2)^{-20} = \$5{,}536.76.$$

Thus, traders would be willing to pay a total of $4,835.18 + 5,536.76 = $10,371.94 for bonds with a maturity value of $10,000.

➡ **Before we go on ...** Notice how the selling price of the bonds behaves as the desired yield changes. As desired yield goes up, the price of the bonds goes down, and as desired yield goes down, the price of the bonds goes up. When the desired yield equals the coupon interest rate, the selling price will equal the maturity value. Therefore, when the yield is higher than the coupon interest rate, the price of the bond will be below its maturity value, and when the yield is lower than the coupon interest rate, the price will be above the maturity value.

As we've mentioned before, the desired yield depends on many factors, but it generally moves up and down with prevailing interest rates. And interest rates have historically gone up and down cyclically. The effect on the value of bonds can be quite dramatic (see Exercises 73 and 74 in Section 2.2). Because bonds can be sold again once bought, someone who buys bonds while interest rates are high and then resells them when interest rates decline can make a healthy profit. ∎

FAQs

Which Formula to Use

Q : *We have retirement accounts, trust funds, loans, bonds, and so on. How do we distinguish among them so that we can tell which formula(s) to use?*

A : In general, first decide whether the annuity account is growing (accumulating) or paying out (annuitizing). Investments into which payments are made are accumulating, while those from which money is being withdrawn are annuitizing. Here is a list of some of the annuity accounts we have discussed in this section:

- *Retirement Accounts* A retirement account is accumulating while payments are being made into the account (before retirement) and annuitizing while a pension is being withdrawn (after retirement).
- *Education Funds* These are similar to retirement accounts.
- *Trust Funds* A trust fund is annuitizing if periodic withdrawals are made.
- *Installment Loans* We think of an installment loan as an investment a bank makes in the lender. In this way, the lender's payments can be viewed as the bank's withdrawals, so a loan is an annuity account that is paying out, or annuitizing.
- *Bonds* A bond pays regular fixed amounts until it matures, at which time it pays its maturity value. We think of the bond as an annuitizing annuity account coupled with a compound interest investment funding the payment of the maturity value. We can then determine its present value based on the current market interest rate.

From a mathematical point of view, accumulating and annuitizing annuity accounts are really the same thing. See the Communication and Reasoning Exercises for more about this.

2.3   EXERCISES

▼ more advanced   ◆ challenging
T indicates exercises that should be solved using technology

In Exercises 1–6, find the amount accumulated in the given annuity account. (Assume end-of-period deposits and compounding at the same intervals as deposits.) [**HINT:** See Quick Example 1 and Example 1.]

1. $100 deposited monthly for 10 years at 5% per year

2. $150 deposited monthly for 20 years at 3% per year

3. $1,000 deposited quarterly for 20 years at 7% per year

4. $2,000 deposited quarterly for 10 years at 7% per year

5. ▼ $100 deposited monthly for 10 years at 5% per year in an account containing $5,000 at the start

6. ▼ $150 deposited monthly for 20 years at 3% per year in an account containing $10,000 at the start

In Exercises 7–12, find the periodic payments necessary to accumulate the given amount in an annuity account. (Assume end-of-period deposits and compounding at the same intervals as deposits.) [**HINT:** See Quick Example 2.]

7. $10,000 in a fund paying 5% per year, with monthly payments for 5 years

8. $20,000 in a fund paying 3% per year, with monthly payments for 10 years

9. $75,000 in a fund paying 6% per year, with quarterly payments for 20 years

10. $100,000 in a fund paying 7% per year, with quarterly payments for 20 years

11. ▼ $20,000 in a fund paying 5% per year, with monthly payments for 5 years, if the fund contains $10,000 at the start

12. ▼ $30,000 in a fund paying 3% per year, with monthly payments for 10 years, if the fund contains $10,000 at the start

In Exercises 13–18, find the present value of the annuity account necessary to fund the given withdrawals. (Assume end-of-period withdrawals and compounding at the same intervals as withdrawals.) [**HINT:** See Quick Example 3.]

13. $500 per month for 20 years, if the account earns 3% per year

14. $1,000 per month for 15 years, if the account earns 5% per year

15. $1,500 per quarter for 20 years, if the account earns 6% per year

16. $2,000 per quarter for 20 years, if the account earns 4% per year

17. ▼ $500 per month for 20 years, if the account earns 3% per year and if there is to be $10,000 left in the account at the end of the 20 years

18. ▼ $1,000 per month for 15 years, if the account earns 5% per year and if there is to be $20,000 left in the account at the end of the 15 years

In Exercises 19–24, find the periodic withdrawals for the given annuity account. (Assume end-of-period withdrawals and compounding at the same intervals as withdrawals.) [**HINT:** See Quick Example 4.]

19. $100,000 at 3%, paid out monthly for 20 years

20. $150,000 at 5%, paid out monthly for 15 years

21. $75,000 at 4%, paid out quarterly for 20 years

22. $200,000 at 6%, paid out quarterly for 15 years

23. ▼ $100,000 at 3%, paid out monthly for 20 years, leaving $10,000 in the account at the end of the 20 years

24. ▼ $150,000 at 5%, paid out monthly for 15 years, leaving $20,000 in the account at the end of the 15 years

In Exercises 25–32, determine the periodic payments on the given loan or mortgage. [**HINT:** See Example 5.]

25. $10,000 borrowed at 9% for 4 years, with monthly payments

26. $20,000 borrowed at 8% for 5 years, with monthly payments

27. $100,000 borrowed at 5% for 20 years, with quarterly payments

28. $1,000,000 borrowed at 4% for 10 years, with quarterly payments

29. A $100,000, 30-year, 4.3% mortgage with monthly payments

30. A $250,000, 15-year, 6.2% mortgage with monthly payments

31. A $1,000,000, 30-year, 5.4% mortgage with monthly payments

32. A $2,000,000, 15-year, 4.5% mortgage with monthly payments

In Exercises 33–38, determine the outstanding principal of the given mortgage. (Assume monthly interest payments and compounding periods.) [**HINT:** See Example 7.]

33. A $100,000, 30-year, 4.3% mortgage after 10 years

34. A $250,000, 15-year, 6.2% mortgage after 10 years

35. A $1,000,000, 30-year, 5.4% mortgage after 5 years

36. A $2,000,000, 15-year, 4.5% mortgage after 8 years

37. ▼ A $50,000, 200-year, 8.5% mortgage after 20 years. What do you notice? Explain.

38. ▼ A $100,000 200-year, 9.6% mortgage after 20 years. What do you notice? Explain.

In Exercises 39–44, determine the selling price, per $1,000 maturity value, of the given bond.[19] (Assume twice-yearly interest payments; do not round those payments to the nearest cent.) [**HINT:** See Example 8.]

39. ▼ 10-year, 4.875% bond, with a yield of 4.880%

40. ▼ 30-year, 5.375% bond, with a yield of 5.460%

41. ▼ 2-year, 3.625% bond, with a yield of 3.705%

42. ▼ 5-year, 4.375% bond, with a yield of 4.475%

43. ▼ 10-year, 5.5% bond, with a yield of 6.643%

44. ▼ 10-year, 6.25% bond, with a yield of 33.409%

Applications

Retirement Plans and Trusts *Exercises 45–54 are based on the following table, which shows the average returns for some of the largest mutual funds commonly found in retirement plans.[20] (Assume end-of-month deposits and withdrawals and monthly compounding, and assume that the quoted rate of return continues indefinitely.)*

Mutual Fund	Fidelity Growth Company	Vanguard 500 Index	PIMCO Total Return	Vanguard Total Bond Market Index
Rate of Return	14.83%	13.25%	2.77%	2.67%
Type	Stock fund	Stock fund	Bond fund	Bond fund

45. How much would be accumulated after 20 years in a retirement account invested entirely in the Fidelity stock fund with payments of $400 per month? [**HINT:** See Quick Example 1.]

---

[19] The first four are actual U.S. Treasury notes and bonds auctioned in 2001 and 2002. The last two are, respectively, a Spanish and a Greek government bond auctioned November 17, 2011, during the Eurozone crisis. Sources: The Bureau of the Public Debt's website (www.publicdebt.treas.gov) and *The Wall Street Journal*'s website (www.wsj.com).

[20] Quoted rates of return are net returns (return minus fund management expenses) based on 5-year returns as of September 30, 2015. Source: www.interest.com.

46. How much would be accumulated after 25 years in a retirement account invested entirely in the Vanguard stock fund with payments of $380 per month? [**HINT:** See Quick Example 1.]

47. For the past 15 years you have been depositing $500 per month in the PIMCO bond fund, and you have now transferred your current balance to the Vanguard bond fund, where you plan to continue depositing $500 per month until you retire in 20 years. How much can you anticipate having in the investment when you retire? [**HINT:** See Example 1.]

48. For the past 15 years you have been depositing $500 per month in the Vanguard bond fund, and you have now transferred your current balance to the PIMCO bond fund, where you plan to continue depositing $500 per month until you retire in 20 years. How much can you anticipate having in the investment when you retire? [**HINT:** See Example 1.]

49. How much should you pay each month into a retirement acount invested in the PIMCO bond fund if you wish to retire in 30 years with an investment valued at 1.5 million dollars? [**HINT:** See Quick Example 2.]

50. How much should you pay each month into a retirement account invested in the Vanguard bond fund if you wish to retire in 25 years with an investment valued at one million dollars? [**HINT:** See Quick Example 2.]

51. ▼ You will retire in 25 years and would like to accumulate a million dollars in your retirement account by that time using a mix of stock funds and bond funds, with 20% of your monthly payments being deposited into the Fidelity stock fund and the rest in the Vanguard bond fund. Assuming that the current interest rates continue until then, how much should you deposit in each fund each month? [**HINT:** Take x to be the payment deposited in the stock fund, and use Quick Example 1.]

52. ▼ Your spouse will retire in 25 years and would also like to accumulate a million dollars in his retirement account by that time, also using a mix of stock funds and bond funds but with 25% of his monthly payments being deposited into the Vanguard stock fund and the rest in the PIMCO bond fund. Assuming that the current interest rates continue until then, what is the total he should deposit each month? [**HINT:** Take x to be the payment deposited in the stock fund, and use Quick Example 1.]

53. You wish to establish a trust fund from which your heirs can withdraw $5,000 per month for 10 years, at the end of which time they will receive $30,000. How large should the trust be if invested in the Fidelity stock fund? [**HINT:** See Example 3.]

54. You wish to establish a trust fund from which your heirs can withdraw $2,000 per month for 20 years, at the end of which time they will receive $100,000. How large should the trust be if invested in the Vanguard stock fund? [**HINT:** See Example 3.]

55. *Pensions* Your pension plan is an annuity with a guaranteed return of 4% per year (compounded quarterly). You can afford to put $1,200 per quarter into the fund, and you will work for 40 years before retiring. After you retire, you will be paid a quarterly pension based on a 25-year payout. How much will you receive each quarter?

56. *Pensions* Jennifer's pension plan is an annuity with a guaranteed return of 5% per year (compounded monthly). She can afford to put $300 per month into the fund, and she will work for 45 years before retiring. If her pension is then paid out monthly based on a 20-year payout, how much will she receive per month?

57. *Pensions* Your pension plan is an annuity with a guaranteed return of 3% per year (compounded monthly). You would like to retire with a pension of $5,000 per month for 20 years. If you work 40 years before retiring, how much must you and your employer deposit each month into the fund? [**HINT:** See Example 5.]

58. *Pensions* Meg's pension plan is an annuity with a guaranteed return of 5% per year (compounded quarterly). She would like to retire with a pension of $12,000 per quarter for 25 years. If she works 45 years before retiring, how much money must she and her employer deposit each quarter? [**HINT:** See Example 5.]

Life Insurance *Exercises 59–64 are based on the following table, which shows the average life expectancies in several countries.*[21] *Assume that all premiums you calculate are based on end-of-month deposits in a fund yielding 4.8% annual interest compounded monthly to be paid out when a person reaches the life expectancy.* [**HINT:** See Example 2.]

Country	Japan	Canada	U.K.	U.S.	Mexico	China	India
Life Expectancy: Male	80	80	79	76	73	74	64
Life Expectancy: Female	87	84	83	81	79	77	68

59. Calculate the life insurance premium that a 30-year-old female in Japan would pay for a $1,000,000 policy.

60. Calculate the life insurance premium that a 30-year-old male in the United States would pay for a $1,000,000 policy.

61. Make a table showing the monthly premiums for males and females in China currently of age 30, 50, and 70 years for a policy that pays out $500,000.

62. Make a table showing the monthly premiums for males and females in Mexico currently of age 20, 40, and 60 years for a policy that pays out $800,000.

[21] Life expectancies at birth. Life expectancy is affected by factors including definition of stillbirth, healthcare quality, disease epidemics, and wars. Source: World Health Organization, 2015/www.wikipedia.org.

63. ▼ Joaquín Lopez purchased a $750,000 life insurance policy in Mexico when he began work at the age of 22 years. At age 30 he was transferred to Toronto, where his insurance company lowered the rate of his policy to reflect the greater life expectancy in Canada. How much lower were his monthly premiums in Canada? [**HINT:** See the formula in the "Before we go on" discussion after Example 1.]

64. ▼ April May purchased an $800,000 life insurance policy in the United Kingdom when she began work at the age of 25 years. At age 35 she was transferred to Calcutta, where her insurance company raised the rate of her policy to reflect the lower life expectancy in India. How much larger were her monthly premiums in India? [**HINT:** See the formula in the "Before we go on" discussion after Example 1.]

Exercises 65–76 are based on the following table, which shows annual rates for various types of loans in 2015.[22] *Assume monthly payments and compounding periods.* [**HINT:** See Examples 5 and 7.]

Loan Type	30-Year Mortgage	15-Year Mortgage	5-Year Car Loan	4-Year Car Loan	Credit Cards
October Rate (%)	3.93	3.14	4.30	4.24	13.10
November Rate (%)	4.09	3.31	4.31	4.26	13.10
December Rate (%)	4.09	3.34	4.34	4.29	13.10

65. **Mortgages** You were considering buying a home with a 30-year mortgage in November 2015 and could afford to make a down payment of $20,000 and up to $600 per month on mortgage payments. How much could you have afforded to pay for the home?

66. **Mortgages** You were considering buying a home with a 15-year mortgage in November 2015 and could afford to make a down payment of $50,000 and up to $900 per month on mortgage payments. How much could you have afforded to pay for the home?

67. **Mortgages** You were considering buying a $150,000 home with a 30-year mortgage in October 2015, but you suspected that the seller would agree to lower the price by $10,000 if you held out another 2 months. The real estate agent urged you to go ahead with the $150,000 purchase immediately, arguing that interest rates would likely move up if you waited. Which scenario would lead to the lower mortgage payments? By how much would they differ?

68. **Mortgages** Your friend was considering buying a $300,000 home with a 15-year mortgage in October 2015, but she suspected that the seller would agree to lower the price by

$3,000 if she held out another 2 months. The real estate agent urged her to go ahead with the $300,000 purchase immediately, arguing that interest rates would likely move up significantly if she waited. In retrospect, was the agent right? What would have been the resulting difference in payments?

69. **Mortgages** Fifteen years into your 30-year $150,000 mortgage begun in October 2015, you inherit your rich aunt's estate and decide to pay off the outstanding principal on your mortgage. What is that amount?

70. **Mortgages** Eight years into your 15-year $300,000 mortgage begun in October 2015, you inherit your rich uncle's estate and decide to pay off the outstanding principal on your mortgage. What is that amount?

71. **Refinancing** Ten years into your 30-year $250,000 mortgage begun in October 2015, you decide to refinance the mortgage using a 20-year loan at a rate of half the original rate. The bank charges a prepayment fee of 4% of the outstanding principal, which you finance by adding the amount of the fee to the principal of the new loan. How much will you save each month?

72. **Refinancing** Five years into your 15-year $500,000 mortgage begun in October 2015, you decide to refinance the mortgage using a 10-year loan at a rate of half the original rate. The bank charges a prepayment fee of 3% of the outstanding principal, which you finance by adding the amount of the fee to the principal of the new loan. How much will you save each month?

73. **Car Loans** You purchased a $35,000 car using a 5-year loan in October 2015. With the same monthly payments, how much could you have financed had you waited until November? until December?

74. **Car Loans** You purchased a $50,000 car using a 4-year loan in October 2015. With the same monthly payments, how much could you have financed had you waited until November? until December?

75. **Credit Card Payments** You currently owe $5,000 on your credit card, which charges interest at the October 2015 rate. What is the least you need to pay per month to pay off the card in 10 years?

76. **Credit Card Payments** You currently owe $8,000 on your credit card, which charges interest at the October 2015 rate. What is the least you need to pay per month to pay off the card in 1 year?

77. **Car Loans** While shopping for a car loan, you get the following offers: *Solid Savings & Loan* is willing to loan you $10,000 at 9% interest for 4 years. *Fifth Federal Bank & Trust* will loan you the $10,000 at 7% interest for 3 years. Both require monthly payments. You can afford to pay $250 per month. Which loan, if either, can you take? [**HINT:** See Example 7.]

---

[22] Car loan rates are for new cars. Source: www.bankrate.com.

78. _Business Loans_ You need to take out a loan of $20,000 to start up your T-shirt business. You have two possibilities: One bank is offering a 10% loan for 5 years, and another is offering a 9% loan for 4 years. Which will have the lower monthly payments? On which will you end up paying more interest total?

79. [T] ▼ **_Refinancing_** Your original mortgage was a $96,000, 30-year, 9.75% mortgage. After 4 years you refinance the remaining principal for 30 years at 6.875%. What was your original monthly payment? What is your new monthly payment? How much will you save in interest over the course of the loan by refinancing? [**HINT:** See Example 7. Total interest paid over a course of time = Sum of all payments during that time − Reduction in principal during that time.]

80. [T] ▼ **_Refinancing_** Kara and Michael take out a $120,000, 30-year, 10% mortgage. After 3 years they refinance the remaining principal with a 15-year, 6.5% loan. What were their original monthly payments? What is your new monthly payment? How much did they save in interest over the course of the loan by refinancing? [**HINT:** See Example 7. Total interest paid over a course of time = Sum of all payments during that time − Reduction in principal during that time.]

Note: In Exercises 81 and 82 we suggest the use of a spreadsheet to create the amortization table.

81. [T] ▼ **_Mortgages_** You take out a 15-year mortgage for $50,000, at 8%, to be paid off monthly. Construct an amortization table showing how much you will pay in interest each year and how much goes toward paying off the principal. [**HINT:** See Example 8.]

82. [T] ▼ **_Mortgages_** You take out a 30-year mortgage for $95,000 at 9.75%, to be paid off monthly. Construct an amortization table showing how much you will pay in interest each year for the first 15 years and how much goes toward paying off the principal. If you sell your house after 15 years, how much will you still owe on the mortgage according to the amortization table? [**HINT:** See Example 8.]

[T] _In Exercises 83–90, use a time value of money utility (a calculator, spreadsheet, or the Website). Such a utility can solve for any of the inputs, given values for the others._

83. ▼ **_Fees_** You take out a 2-year, $5,000 loan at 9% interest with monthly payments. The lender charges you a $100 fee that can be paid off, interest free, in equal monthly installments over the life of the loan. Thinking of the fee as additional interest, what is the actual annual interest rate you will pay?

84. ▼ **_Fees_** You take out a 3-year, $7,000 loan at 8% interest with monthly payments. The lender charges you a $100 fee that can be paid off, interest free, in equal monthly installments over the life of the loan. Thinking of the fee as additional interest, what is the actual annual interest rate you will pay?

85. ▼ **_Savings_** You wish to accumulate $100,000 through monthly payments of $500. If you can earn interest at an annual rate of 4% compounded monthly, how long (to the nearest year) will it take to accomplish your goal?

86. ▼ **_Retirement_** Alonzo plans to retire as soon as he has accumulated $250,000 through quarterly payments of $2,500. If Alonzo invests this money at 5.4% interest, compounded quarterly, when (to the nearest year) can he retire?

87. ▼ **_Loans_** You have a $2,000 credit card debt, and you plan to pay it off through monthly payments of $50. If you are being charged 15% interest per year, how long (to the nearest 0.5 years) will it take you to repay your debt?

88. ▼ **_Loans_** You owe $2,000 on your credit card, which charges you 15% interest. Determine, to the nearest 1¢, the minimum monthly payment that will allow you to eventually repay your debt.

89. ▼ **_Savings_** You are depositing $100 per month in an account that pays 4.5% interest per year (compounded monthly), while your friend Lucinda is depositing $75 per month in an account that earns 6.5% interest per year (compounded monthly). When, to the nearest year, will her balance exceed yours? [**HINT:** Graph the values of both investments.]

90. ▼ **_Car Leasing_** You can lease a $15,000 car for $300 per month. For how long (to the nearest year) should you lease the car so that your monthly payments are lower than if you were purchasing it with an 8%-per-year loan?

Communication and Reasoning Exercises

91. Your cousin Simon claims that you have wasted your time studying annuities: If you wish to retire on an income of $1,000 per month for 20 years, you need to save $1,000 per month for 20 years, period. Explain why he is wrong.

92. ▼ Your other cousin Cecilia claims that you will earn more interest by depositing $10,000 through smaller, more frequent payments than through larger less frequent payments. Is she correct? Give a reason for your answer.

93. ▼ You make equal payments of $200 per month into each of two accounts: one earning 5% and the other earning 6%. This is the same as investing $400 per month in a single investment earning 5.5%—right? (Justify your answer.)

94. ▼ You make equal payments of $400 per month into each of two accounts: one earning 5% and the other losing 5%. Thus, the two investments balance each other out, and no net interest is earned—right? (Justify your answer.)

95. ▼ A real estate broker tells you that doubling the period of a mortgage halves the monthly payments. Is he correct? Support your answer by means of an example.

96. ▼ Another real estate broker tells you that doubling the size of a mortgage doubles the monthly payments. Is he correct? Support your answer by means of an example.

97. ⊤ ▼ We have so far not taken advantage of the fact that time value of money utilities allow you to enter nonzero numbers for both the present value and the future value. Use this to determine the yield on a 5-year bond selling for $994.69 per $1,000 maturity value and paying 3.5% simple annual interest on the maturity value. Assume that interest payments are paid twice yearly, and compute the yield assuming twice-yearly compounding. [**HINT:** Think carefully about which values need to be entered as positive numbers and which as negative.]

98. ⊤ ▼ Repeat Exercise 97 for a 10-year, 3.375% bond selling for $991.20 per $1,000 maturity value.

99. ◆ Consider the formula for the future value of an accumulating annuity with given payments. Show algebraically that the present value of that future value is the same as the present value of the annuitizing annuity required to fund the same payments.

100. ◆ Give a nonalgebraic justification for the result from Exercise 99.

CHAPTER 2 REVIEW

KEY CONCEPTS

www.WanerMath.com
Go to the Website to find a comprehensive and interactive Web-based summary of Chapter 2.

2.1 Simple Interest

Simple interest: $INT = PVrt$ [p. 134]

General interest periods: $INT = PVin$ [p. 135]

Future value: $FV = PV(1 + rt)$,
$FV = PV(1 + in)$ [p. 136]

Present value: $PV = \dfrac{FV}{1 + rt}$,

$PV = \dfrac{FV}{1 + in}$ [p. 138]

Treasury bills [p. 140]

2.2 Compound Interest

Future value for compound interest:
$FV = PV(1 + i)^n$ [p. 145]

Present value for compound interest:

$PV = \dfrac{FV}{(1 + i)^n}$ [p. 148]

Inflation, constant dollars [p. 149]

Nominal and effective interest rates [p. 151]

How long to invest [p. 151]

2.3 Annuities, Loans, and Bonds

Annuities [p. 156]

Accumulation: future value

$FV = PMT\dfrac{(1 + i)^n - 1}{i}$ [p. 157]

Payments to accumulate a future value:

$PMT = FV\dfrac{i}{(1 + i)^n - 1}$ [p. 159]

Annuitization: present value:

$PV = PMT\dfrac{1 - (1 + i)^{-n}}{i}$ [p. 161]

Withdrawals to annuitize a present

value: $PMT = PV\dfrac{i}{1 - (1 + i)^{-n}}$

[p. 162]

Amortization: loans and mortgages [p. 163]

Mortgage refinancing [p. 165]

Bonds [p. 165]

REVIEW EXERCISES

In Exercises 1–6, find the future value of the investment.

1. $6,000 for 5 years at 4.75% simple annual interest

2. $10,000 for 2.5 years at 5.25% simple annual interest

3. $6,000 for 5 years at 4.75% compounded monthly

4. $10,000 for 2.5 years at 5.25% compounded semiannually

5. $100 deposited at the end of each month for 5 years at 4.75% interest compounded monthly

6. $2,000 deposited at the end of each half-year for 2.5 years at 5.25% interest compounded semiannually

In Exercises 7–12, find the present value of the investment.

7. Worth $6,000 after 5 years at 4.75% simple annual interest

8. Worth $10,000 after 2.5 years at 5.25% simple annual interest

9. Worth $6,000 after 5 years at 4.75% compounded monthly

10. Worth $10,000 after 2.5 years at 5.25% compounded semiannually

11. Funding $100 withdrawals at the end of each month for 5 years at 4.75% interest compounded monthly

12. Funding $2,000 withdrawals at the end of each half-year for 2.5 years at 5.25% interest compounded semiannually

In Exercises 13–18, find the amounts indicated.

13. The monthly deposits necessary to accumulate $12,000 after 5 years in an account earning 4.75% compounded monthly

14. The semiannual deposits necessary to accumulate $20,000 after 2.5 years in an account earning 5.25% compounded semiannually

15. The monthly withdrawals possible over 5 years from an account earning 4.75% compounded monthly and starting with $6,000

16. The semiannual withdrawals possible over 2.5 years from an account earning 5.25% compounded semiannually and starting with $10,000

17. The monthly payments necessary on a 5-year loan of $10,000 at 4.75%

18. The semiannual payments necessary on a 2.5-year loan of $15,000 at 5.25%

In Exercises 19–24, find the time requested, to the nearest 0.1 year.

19. The time it would take $6,000 to grow to $10,000 at 4.75% simple annual interest

20. The time it would take $10,000 to grow to $15,000 at 5.25% simple annual interest

21. The time it would take $6,000 to grow to $10,000 at 4.75% interest compounded monthly

22. The time it would take $10,000 to grow to $15,000 at 5.25% interest compounded semiannually

23. The time it would take to accumulate $10,000 by depositing $100 at the end of each month in an account earning 4.75% interest compounded monthly

24. The time it would take to accumulate $15,000 by depositing $2,000 at the end of each half-year in an account earning 5.25% compounded semiannually

25. How much would you pay for a $10,000, 5-year, 6% bond if you want a return of 7%? (Assume that the bond pays interest every 6 months.)

26. How much would you pay for a $10,000, 5-year, 6% bond if you want a return of 5%? (Assume that the bond pays interest every 6 months.)

27. **T** A $10,000, 7-year, 5% bond sells for $9,800. What return does it give you? (Assume that the bond pays interest every 6 months.)

28. **T** A $10,000, 7-year, 5% bond sells for $10,200. What return does it give you? (Assume that the bond pays interest every 6 months.)

Applications: OHaganBooks.com
[Try the game at www.OHaganBooks.com]

Stock Investments *Exercises 29–34 are based on the following table, which shows some values of ABCromD (ABCD) stock:*

Dec. 2002	Aug. 2004	Mar. 2005	May 2005	Aug. 2005	Dec. 2005
3.28	16.31	33.95	21.00	30.47	7.44
Jan. 2007	Mar. 2008	Oct. 2008	Nov. 2009	Feb. 2010	Aug. 2010
12.36	7.07	11.44	33.53	44.86	45.74

29. Marjory Duffin purchased ABCD stock in December 2002 and sold it in August 2010. Calculate her annual return on a simple interest basis to the nearest 0.01%.

30. John O'Hagan purchased ABCD stock in March 2005 and sold it in January 2007. Calculate his annual return on a simple interest basis to the nearest 0.01%.

31. Suppose Marjory Duffin had bought ABCD stock in January 2007. If she had later sold on one of the dates in the table, which of those dates would have given her the largest annual return on a simple interest basis, and what would that return have been?

32. Suppose John O'Hagan had purchased ABCD stock in August 2004. If he had later sold on one of the dates in the table, which of those dates would have given him the largest annual loss on a simple interest basis, and what would that loss have been?

33. Did ABCD stock undergo simple interest increase in the period December 2002 through March 2005? (Give a reason for your answer.)

34. If ABCD stock underwent simple interest increase from February 2010 through August 2010 and into 2011, what would the price have been in December 2011?

35. **Revenue** Total online revenues at OHaganBooks.com during 2009, its first year of operation, amounted to $150,000. After December 2009, revenues increased by a steady 20% each year. Track OHaganBooks.com's revenues for the subsequent 5 years, assuming that this rate of growth continued. During which year did the revenue surpass $300,000?

36. **Net Income** Unfortunately, the picture for net income was not so bright: The company lost $20,000 in the fourth quarter of 2009. However, the quarterly loss decreased at an annual rate of 15%. How much did the company lose during the third quarter of 2011?

37. **Stocks** To finance anticipated expansion, CEO John O'Hagan is considering making a public offering of OHaganBooks.com shares at $3.00 a share. O'Hagan is not sure how many shares to offer but would like the shares to reach a total market value of at least $500,000 six months after the initial offering. He estimates that the value of the stock will double in the first day of trading and then will appreciate at around 8% per month for the first 6 months. How many shares of stock should the company offer?

38. **Stocks** Unfortunately, renewed panic about the Monaco debt crisis caused the U.S. stock market to plunge on the very first day of OHaganBooks.com's initial public offering (IPO) of 600,000 shares at $3.00 per share, and the shares ended the trading day 60% lower. Subsequently, as the Monaco debt crisis worsened, the stock depreciated by 10% per week during the subsequent 5 weeks. What was the total market value of the stocks at the end of 5 weeks?

39. **Loans** OHaganBooks.com is seeking a $250,000 loan to finance its continuing losses. One of the best deals available is offered by *Industrial Bank,* which offers a 10-year 9.5% loan. What would the monthly payments be for this loan?

40. **Loans** (See Exercise 39.) *Expansion Loans* offers an 8-year 6.5% loan. What would the monthly payments be for the $250,000 loan from Expansion?

41. **Loans** (See Exercise 39.) OHaganBooks.com can afford to pay only $3,000 per month to service its debt. What, to the nearest dollar, is the largest amount the company can borrow from Industrial Bank?

42. **Loans** (See Exercise 40.) OHaganBooks.com can afford to pay only $3,000 per month to service its debt. What, to the nearest dollar, is the largest amount the company can borrow from Expansion Loans?

43. **T Loans** (See Exercise 39.) What interest rate would Industrial Bank have to offer in order to meet the company's loan requirements at a price (no more than $3,000 per month) it can afford?

44. **T Loans** (See Exercise 40.) What interest rate would Expansion Loans have to offer in order to meet the company's loan requirements at a price (no more than $3,000 per month) it can afford?

Retirement Planning *OHaganBooks.com has just introduced a retirement package for its employees. Under the annuity plan operated by* Sleepy Hollow, *the monthly contribution by the company on behalf of each employee is $800. Each employee can then supplement that amount through payroll deductions. The current rate of return of Sleepy Hollow's retirement fund is 7.3%. Use this information in Exercises 45–52.*

45. Jane Callahan, the website developer at OHaganBooks.com, plans to retire in 10 years. She contributes $1,000 per month to the plan (in addition to the company contribution

of $800). Currently, there is $50,000 in her retirement annuity. How much (to the nearest dollar) will it be worth when she retires?

46. Percy Egan, the assistant website developer at OHaganBooks .com, plans to retire in 8 years. He contributes $950 per month to the plan (in addition to the company contribution of $800). Currently, there is $60,000 in his retirement annuity. How much (to the nearest dollar) will it be worth when he retires?

47. When she retires, how much of Jane Callahan's retirement fund will have resulted from the company contribution? (See Exercise 45. The company did not contribute toward the $50,000 Callahan now has.)

48. When he retires, how much of Percy Egan's retirement fund will have resulted from the company contribution? (See Exercise 46. The company began contributing $800 per month to his retirement fund when he was hired 5 years ago. Assume that the rate of return from Sleepy Hollow has been unchanged.)

49. (See Exercise 45.) Jane Callahan actually wants to retire with $500,000. How much should she contribute each month to the annuity?

50. (See Exercise 46.) Percy Egan actually wants to retire with $600,000. How much should he contribute each month to the annuity?

51. (See Exercise 45.) On second thought, Callahan wants to be in a position to draw at least $5,000 per month for 30 years after her retirement. She feels that she can invest the proceeds of her retirement annuity at 8.7% per year in perpetuity. Given the information in Exercise 45, how much will she need to contribute to the plan, starting now?

52. (See Exercise 46.) On second thought, Egan wants to be in a position to draw at least $6,000 per month for 25 years after his retirement. He feels that he can invest the proceeds of his retirement annuity at 7.8% per year in perpetuity. Given the information in Exercise 46, how much will he need to contribute to the plan, starting now?

Actually, Jane Callahan is quite pleased with herself; 1 year ago she purchased a $50,000 government bond paying 7.2% per year (with interest paid every 6 months) and maturing in 10 years, and interest rates have come down since then.

53. The current interest rate on 10-year government bonds is 6.3%. If she were to auction the bond at the current interest rate, how much would she get?

54. If she holds onto the bond for 6 more months and the interest rate drops to 6%, how much will the bond be worth then?

55. **T** Jane suspects that interest rates will come down further during the next 6 months. If she hopes to auction the bond for $54,000 in 6 months' time, what will the interest rate need to be at that time?

56. **T** If, in 6 months' time, the bond is auctioned for only $52,000, what will the interest rate be at that time?

Adjustable Rate and Subprime Mortgages

The term **subprime mortgage** refers to mortgages given to home buyers with a heightened perceived risk of default, as when, for instance, the price of the home being purchased is higher than the borrower can reasonably afford. Such loans are typically **adjustable rate** loans, meaning that the lending rate varies through the duration of the loan.* Subprime adjustable rate loans typically start at artificially low "teaser rates" that the borrower can afford, but then increase significantly over the life of the mortgage. The U.S. real estate bubble of 2000–2005 led to a frenzy of subprime lending, the rationale being that a borrower having trouble meeting mortgage payments could either sell the property at a profit or refinance the loan, or the lending institution could earn a hefty profit by repossessing the property in the event of foreclosure.

Mr. and Mrs. Wong have an appointment tomorrow with you, their investment counselor, to discuss their plan to purchase a $400,000 house in Orlando, Florida. They have saved $20,000 for a down payment, so want to take out a $380,000 mortgage. Their combined annual income is $80,000 per year, which they estimate will increase by 4% annually over the foreseeable future, and they are considering three different specialty 30-year mortgages:

Ariel Skelley/Blend Images/Getty Images

* In an adjustable rate mortgage, the payments are recalculated each time the interest rate changes, based on the assumption that the new interest rate will be unchanged for the remaining life of the loan. We say that the loan is **re-amortized** at the new rate.

＊ The U.S. federal funds rate is the
rate banks charge each other for
loans and is often used to set
rates for other loans. Manipulating this rate is one way the U.S.
Federal Reserve regulates the
money supply.

Hybrid: The interest is fixed at a low introductory rate of 4% for 5 years and then adjusts annually to 5% over the U.S. federal funds rate.＊

Interest-Only: During the first 5 years the rate is set at 4.2%, and no principal is paid. After that time, the mortgage adjusts annually to 5% over the U.S. federal funds rate.

Negative Amortization: During the first 5 years the rate is set at 4.7% based on a principal of 60% of the purchase price of the home, with the result that the balance on the principal actually grows during this period. After that time, the mortgage adjusts annually to 5% over the U.S. federal funds rate.

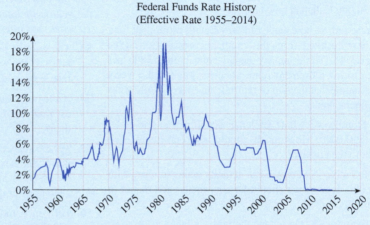

Federal Funds Rate History
(Effective Rate 1955–2014)

Figure 3

You decide that you should create an Excel worksheet that will compute the monthly payments for the three types of loan. Of course, you have no way of predicting what the U.S. federal funds rate will be over the next 30 years (see Figure 3 for historical values[23]), so you decide to include three scenarios for the federal funds rate in each case:

Scenario 1: Federal funds rate is 4.25% in year 6 and then increases by 0.25% per year.

Scenario 2: Federal funds rate is steady at 4% during the term of the loan.

Scenario 3: Federal funds rate is 10% in year 6 and then decreases by 0.25% per year.

Each worksheet will show month-by-month payments for the specific type of loan. Typically, to be affordable, payments should not exceed 28% of gross monthly income, so you will tabulate that quantity as well.

Hybrid Loan: You begin to create your worksheet by estimating 28% of the Wongs' monthly income, assuming a 4% increase each year. (The income is computed using the compound interest formula for annual compounding.)

[23] Source: Board of Governors of the Federal Reserve System (www.federalreserve.gov).

	A	B	C	D	E	F	G	H
1	Year	28% of Monthly Income		Interest Rate			Monthly Payment	
2			Scenario 1	Scenario 2	Scenario 3	Scenario 1	Scenario 2	Scenario 3
3	1	=0.28*80000*(1.04)^(A3-1)/12						
4	2							
5	3							
30	28							
31	29							
32	30							

The next sheet shows the result, as well as the formulas for computing the interest rate in each scenario.

	A	B	C	D	E	F	G	H
1	Year	28% of Monthly Income		Interest Rate			Monthly Payment	
2			Scenario 1	Scenario 2	Scenario 3	Scenario 1	Scenario 2	Scenario 3
3	1	$1,866.67	4	4	4			
4	2	$1,941.33	4	4	4			
5	3	$2,018.99	4	4	4			
6	4	$2,099.75	4	4	4			
7	5	$2,183.74	4	4	4			
8	6	$2,271.09	9.25	9	15			
9	7	$2,361.93	=C8+0.25	=D8	=E8-0.25			
10	8	$2,456.41						
30	28	$5,382.29						
31	29	$5,597.58						
32	30	$5,821.48						

To compute the monthly payment, you decide to use the built-in function PMT, which has the format

$$PMT(i, n, PV, [FV], [type]),$$

where i = interest per period, n = total number of periods of the loan, PV = present value, FV = future value (optional); the *type*, also optional, is 0 or omitted if payments are at the end of each period, and 1 if at the start of each period. The present value will be the outstanding principal owed on the home each time the rate is changed, so that too will need to be known. During the first 5 years we can use as the present value the original cost of the home, but each year thereafter, the loan is re-amortized at the new interest rate, so the outstanding principal will need to be computed. Although Excel has a built-in function that calculates payment on the principal, it returns only the payment for a single period (month), so without creating a month-by-month amortization table, it would be difficult to use this function to track the outstanding principal. On the other hand, the total outstanding principal at any point in time can be computed by using the future value formula *FV*. You decide to add three more columns to your Excel worksheet to show the principal outstanding at the start of each year. Here is the spreadsheet with the formulas for the payments and outstanding principal for the first 5 years.

	A	B	C	D	E	F	G	H	I	J	K
1	Year	28% of Monthly Income	Interest Rate			Monthly Payment			Balance on Principal		
2			Scenario 1	Scenario 2	Scenario 3	Scenario 1	Scenario 2	Scenario 3	Scenario 1	Scenario 2	Scenario 3
3	1	$1,866.67	4	4	4	=-PMT(C3/1200,360,I$3)			$380,000.00	$380,000.00	$380,000.00
4	2	$1,941.33	4	4	4				=FV(C3/1200,12*$A3,F3,-I$3)		
5	3	$2,018.99	4	4	4						
6	4	$2,099.75	4	4	4						
7	5	$2,183.74	4	4	4						
8	6	$2,271.09	9.25	9	15						
9	7	$2,361.93	9.5	9	14.75						
10	8	$2,456.41	9.75	9	14.5						

The two formulas will each be copied across to the adjacent two cells for the other scenarios. A few things to notice: The negative sign before *PMT* converts the negative quantity returned by *PMT* to a positive amount. The dollar sign in I$3 in the *PMT* formula fixes the present value for each year at the original cost of the home for the first 5 years, during which payments are computed as for a fixed rate loan. In the formula for the balance on principal at the start of each year, the number of periods is the total number of months up through the preceding year, and the present value is the same initial price of the home each year during the 5-year fixed rate period.

The next sheet shows the calculated results for the fixed rate period and the new formulas to be added for the adjustable rate period starting with the sixth year.

	C	D	E	F	G	H	I	J	K
1	Interest Rate			Monthly Payment			Balance on Principal		
2	Scenario 1	Scenario 2	Scenario 3	Scenario 1	Scenario 2	Scenario 3	Scenario 1	Scenario 2	Scenario 3
3	4	4	4	$1,814.18	$1,814.18	$1,814.18	$380,000.00	$380,000.00	$380,000.00
4	4	4	4	$1,814.18	$1,814.18	$1,814.18	$373,308.06	$373,308.06	$373,308.06
5	4	4	4	$1,814.18	$1,814.18	$1,814.18	$366,343.48	$366,343.48	$366,343.48
6	4	4	4	$1,814.18	$1,814.18	$1,814.18	$359,095.16	$359,095.16	$359,095.16
7	4	4	4	$1,814.18	$1,814.18	$1,814.18	$351,551.52	$351,551.52	$351,551.52
8	9.25	9	15	=-PMT(C8/1200,360-12*$A7,I8)			=FV(C7/1200,12,F7,-I7)		
9	9.5	9	14.75						
10	9.75	9	14.5						

Notice the changes: The loan is re-amortized each year starting with year 6, and the payment calculation needs to take into account the reduced, remaining lifetime of the loan each time. You now copy these formulas across for the remaining two scenarios and then copy all six formulas down the remaining rows to complete the calculation. Following is a portion of the complete worksheet showing, in red, those years during which the monthly payment will exceed 28% of the gross monthly income.

	A	B	C	D	E	F	G	H	I	J	K
1	Year	28% of Monthly Income		Interest Rate			Monthly Payment			Balance on Principal	
2			Scenario 1	Scenario 2	Scenario 3	Scenario 1	Scenario 2	Scenario 3	Scenario 1	Scenario 2	Scenario 3
3	1	$1,866.67	4	4	4	$1,814.18	$1,814.18	$1,814.18	$380,000.00	$380,000.00	$380,000.00
4	2	$1,941.33	4	4	4	$1,814.18	$1,814.18	$1,814.18	$373,308.06	$373,308.06	$373,308.06
5	3	$2,018.99	4	4	4	$1,814.18	$1,814.18	$1,814.18	$366,343.48	$366,343.48	$366,343.48
6	4	$2,099.75	4	4	4	$1,814.18	$1,814.18	$1,814.18	$359,095.16	$359,095.16	$359,095.16
7	5	$2,183.74	4	4	4	$1,814.18	$1,814.18	$1,814.18	$351,551.52	$351,551.52	$351,551.52
8	6	$2,271.09	9.25	9	15	$2,943.39	$2,884.32	$4,402.22	$343,700.55	$343,700.55	$343,700.55
9	7	$2,361.93	9.5	9	14.75	$3,001.60	$2,884.32	$4,336.46	$340,018.68	$339,866.12	$342,337.80
10	8	$2,456.41	9.75	9	14.5	$3,058.89	$2,884.32	$4,271.84	$336,135.05	$335,671.99	$340,686.40
11	9	$2,554.66	10	9	14.25	$3,115.17	$2,884.32	$4,208.48	$332,020.97	$331,084.43	$338,694.96
12	10	$2,656.85	10.25	9	14	$3,170.39	$2,884.32	$4,146.53	$327,643.98	$326,066.52	$336,305.10
13	11	$2,763.12	10.5	9	13.75	$3,224.44	$2,884.32	$4,086.13	$322,967.19	$320,577.89	$333,450.89
14	12	$2,873.65	10.75	9	13.5	$3,277.24	$2,884.32	$4,027.41	$317,948.51	$314,574.40	$330,058.36
15	13	$2,988.59	11	9	13.25	$3,328.70	$2,884.32	$3,970.53	$312,539.70	$308,007.73	$326,045.01
16	14	$3,108.14	11.25	9	13	$3,378.71	$2,884.32	$3,915.64	$306,685.30	$300,825.07	$321,319.42
17	15	$3,232.46	11.5	9	12.75	$3,427.16	$2,884.32	$3,862.89	$300,321.34	$292,968.62	$315,780.91
18	16	$3,361.76	11.75	9	12.5	$3,473.93	$2,884.32	$3,812.43	$293,373.72	$284,375.19	$309,319.29
19	17	$3,496.23	12	9	12.25	$3,518.89	$2,884.32	$3,764.40	$285,756.40	$274,975.63	$301,814.76
20	18	$3,636.08	12.25	9	12	$3,561.90	$2,884.32	$3,718.94	$277,369.15	$264,694.33	$293,137.86
21	19	$3,781.52	12.5	9	11.75	$3,602.81	$2,884.32	$3,676.20	$268,094.83	$253,448.57	$283,149.60
22	20	$3,932.79	12.75	9	11.5	$3,641.47	$2,884.32	$3,636.32	$257,796.15	$241,147.88	$271,701.73

In the third scenario the Wongs' payments would more than double at the start of the sixth year and would remain above what they can reasonably afford for 13 more years. Even if the federal funds rate were to remain at the low rate of 4%, the monthly payments would still jump to above what the Wongs can afford at the start of the sixth year.

Interest-Only Loan: Here, the only change in the worksheet constructed previously is the computation of the payments for the first 5 years; because the loan is interest-only during this period, the monthly payment is computed as simple interest at 4.2% on the $380,000 loan for a 30-year period:

$$INT = PVr = 380,000 \times .042/12 = \$1,330.00.$$

The formula you could use in the spreadsheet in cell F3 is =C3/1200*I$3, and you then copy this across and down the entire block of payments for the first 5 years. The rest of the spreadsheet (including the balance on principal) will adjust itself accordingly with the formulas you had for the hybrid loan. Below is a portion of the result, with a lot more red than in the hybrid loan case!

Year	28% of Monthly Income	Interest Rate			Monthly Payment			Balance on Principal		
		Scenario 1	Scenario 2	Scenario 3	Scenario 1	Scenario 2	Scenario 3	Scenario 1	Scenario 2	Scenario 3
1	$1,866.67	4.2	4.2	4.2	$1,330.00	$1,330.00	$1,330.00	$380,000.00	$380,000.00	$380,000.00
2	$1,941.33	4.2	4.2	4.2	$1,330.00	$1,330.00	$1,330.00	$380,000.00	$380,000.00	$380,000.00
3	$2,018.99	4.2	4.2	4.2	$1,330.00	$1,330.00	$1,330.00	$380,000.00	$380,000.00	$380,000.00
4	$2,099.75	4.2	4.2	4.2	$1,330.00	$1,330.00	$1,330.00	$380,000.00	$380,000.00	$380,000.00
5	$2,183.74	4.2	4.2	4.2	$1,330.00	$1,330.00	$1,330.00	$380,000.00	$380,000.00	$380,000.00
6	$2,271.09	9.25	9	15	$3,254.25	$3,188.95	$4,867.16	$380,000.00	$380,000.00	$380,000.00
7	$2,361.93	9.5	9	14.75	$3,318.45	$3,188.95	$4,794.45	$375,929.28	$375,760.60	$378,493.33
8	$2,456.41	9.75	9	14.5	$3,381.95	$3,188.95	$4,723.00	$371,635.48	$371,123.52	$376,667.51
9	$2,554.66	10	9	14.25	$3,444.18	$3,188.95	$4,652.96	$367,086.90	$366,051.44	$374,465.75
10	$2,656.85	10.25	9	14	$3,505.22	$3,188.95	$4,584.46	$362,247.64	$360,503.57	$371,823.49
11	$2,763.12	10.5	9	13.75	$3,564.98	$3,188.95	$4,517.68	$357,076.92	$354,435.28	$368,667.84
12	$2,873.65	10.75	9	13.5	$3,623.37	$3,188.95	$4,452.76	$351,528.19	$347,797.73	$364,917.01
13	$2,988.59	11	9	13.25	$3,680.26	$3,188.95	$4,389.88	$345,548.14	$340,537.54	$360,479.80
14	$3,108.14	11.25	9	13	$3,735.55	$3,188.95	$4,329.19	$339,075.44	$332,596.29	$355,255.12
15	$3,232.46	11.5	9	12.75	$3,789.12	$3,188.95	$4,270.87	$332,039.35	$323,910.09	$349,131.66
16	$3,361.76	11.75	9	12.5	$3,840.82	$3,188.95	$4,215.07	$324,357.97	$314,409.07	$341,987.61
17	$3,496.23	12	9	12.25	$3,890.53	$3,188.95	$4,161.97	$315,936.16	$304,016.79	$333,690.51
18	$3,636.08	12.25	9	12	$3,938.08	$3,188.95	$4,111.71	$306,663.10	$292,649.65	$324,097.21
19	$3,781.52	12.5	9	11.75	$3,983.32	$3,188.95	$4,064.46	$296,409.28	$280,216.18	$313,054.05
20	$3,932.79	12.75	9	11.5	$4,026.06	$3,188.95	$4,020.37	$285,022.93	$266,616.37	$300,397.13
21	$4,090.10	13	9	11.25	$4,066.11	$3,188.95	$3,979.57	$272,325.66	$251,740.81	$285,952.80
22	$4,253.70	13.25	9	11	$4,103.29	$3,188.95	$3,942.23	$258,107.21	$235,469.81	$269,538.33

In all three scenarios this type of mortgage is worse for the Wongs than the hybrid loan; in particular, their payments in Scenario 3 would jump to more than double what they can afford at the start of the sixth year.

Negative Amortization Loan: Again, the only change in the worksheet is the computation of the payments for the first 5 years. This time, the loan amortizes negatively during the initial 5-year period, so the payment formula in this period is adjusted to reflect this

$$=\text{-PMT}(C3/1200,360,I\$3*0.6)$$

Year	28% of Monthly Income	Interest Rate			Monthly Payment			Balance on Principal		
		Scenario 1	Scenario 2	Scenario 3	Scenario 1	Scenario 2	Scenario 3	Scenario 1	Scenario 2	Scenario 3
1	$1,866.67	4.7	4.7	4.7	$1,182.49	$1,182.49	$1,182.49	$380,000.00	$380,000.00	$380,000.00
2	$1,941.33	4.7	4.7	4.7	$1,182.49	$1,182.49	$1,182.49	$383,750.17	$383,750.17	$383,750.17
3	$2,018.99	4.7	4.7	4.7	$1,182.49	$1,182.49	$1,182.49	$387,680.45	$387,680.45	$387,680.45
4	$2,099.75	4.7	4.7	4.7	$1,182.49	$1,182.49	$1,182.49	$391,799.48	$391,799.48	$391,799.48
5	$2,183.74	4.7	4.7	4.7	$1,182.49	$1,182.49	$1,182.49	$396,116.32	$396,116.32	$396,116.32
6	$2,271.09	9.25	9	15	$3,431.01	$3,362.16	$5,131.53	$400,640.49	$400,640.49	$400,640.49
7	$2,361.93	9.5	9	14.75	$3,498.87	$3,362.16	$5,054.87	$396,348.66	$396,170.82	$399,051.98
8	$2,456.41	9.75	9	14.5	$3,565.64	$3,362.16	$4,979.54	$391,821.64	$391,281.87	$397,127.00
9	$2,554.66	10	9	14.25	$3,631.26	$3,362.16	$4,905.69	$387,025.99	$385,934.29	$394,805.64
10	$2,656.85	10.25	9	14	$3,695.62	$3,362.16	$4,833.48	$381,923.88	$380,085.08	$392,019.86
11	$2,763.12	10.5	9	13.75	$3,758.62	$3,362.16	$4,763.06	$376,472.30	$373,687.17	$388,692.80
12	$2,873.65	10.75	9	13.5	$3,820.18	$3,362.16	$4,694.62	$370,622.18	$366,689.09	$384,738.24
13	$2,988.59	11	9	13.25	$3,880.16	$3,362.16	$4,628.32	$364,317.31	$359,034.55	$380,060.01
14	$3,108.14	11.25	9	13	$3,938.46	$3,362.16	$4,564.34	$357,493.03	$350,661.95	$374,551.54
15	$3,232.46	11.5	9	12.75	$3,994.93	$3,362.16	$4,502.85	$350,074.77	$341,503.95	$368,095.48
16	$3,361.76	11.75	9	12.5	$4,049.45	$3,362.16	$4,444.02	$341,976.15	$331,486.66	$360,563.39
17	$3,496.23	12	9	12.25	$4,101.85	$3,362.16	$4,388.03	$333,096.99	$320,530.10	$351,815.60
18	$3,636.08	12.25	9	12	$4,151.99	$3,362.16	$4,335.05	$323,320.15	$308,545.52	$341,701.22
19	$3,781.52	12.5	9	11.75	$4,199.68	$3,362.16	$4,285.23	$312,509.37	$295,436.71	$330,058.23
20	$3,932.79	12.75	9	11.5	$4,244.74	$3,362.16	$4,238.74	$300,504.55	$281,098.20	$316,713.83
21	$4,090.10	13	9	11.25	$4,286.97	$3,362.16	$4,195.73	$287,117.60	$265,414.64	$301,484.92
22	$4,253.70	13.25	9	11	$4,326.17	$3,362.16	$4,156.36	$272,126.84	$248,259.85	$284,178.86
23	$4,423.85	13.5	9	10.75	$4,362.10	$3,362.16	$4,120.77	$255,270.30	$229,495.82	$264,594.34
24	$4,600.80	13.75	9	10.5	$4,394.53	$3,362.16	$4,089.09	$236,237.47	$208,971.60	$242,522.55

Clearly the Wongs should steer clear of this type of loan in order to be able to continue to afford making payments!

In short, it seems unlikely that the Wongs will be able to afford payments on any of the three mortgages in question, and you decide to advise them to either seek a less expensive home or wait until their income has appreciated to enable them to afford a home of this price.

EXERCISES

1. ▣ In the case of a hybrid loan, what would the federal funds rate have to be in Scenario 2 to ensure that the Wongs can afford to make all payments?

2. ▣ Repeat Exercise 1 in the case of an interest-only loan.

3. ▣ Repeat Exercise 1 in the case of a negative amortization loan.

4. ▣ What home price, to the nearest $5,000, could the Wongs afford if they took out a hybrid loan, regardless of the scenario? [**HINT:** Adjust the original value of the loan on your spreadsheet to obtain the desired result.]

5. ▣ What home price, to the nearest $5,000, could the Wongs afford if they took out a negative amortization loan, regardless of the scenario? [**HINT:** Adjust the original value of the loan on your spreadsheet to obtain the desired result.]

6. ▣ How long would the Wongs need to wait before they could afford to purchase a $400,000 home, assuming that their income continues to increase as above, they still have $20,000 for a down payment, and the mortgage offers remain the same?

Section 2.2

Example 1 (page 147) In December 2015, **Radius Bank** was paying 1.10% annual interest on savings accounts with balances of $2,500 and up. If the interest is compounded quarterly, find the future value of a $2,500 deposit after 6 years. What is the total interest paid over the time of the investment?

Solution

We could calculate the future value using the TI-83/84 Plus by entering

$$2500(1+0.011/4)^{\wedge}(4*6)$$

in the Home screen and pressing [ENTER]. However, the TI-83/84 Plus has this and other useful calculations built into its TVM (Time Value of Money) Solver.

1. Press [APPS], then choose item 1:Finance..., and then choose item 1:TVM Solver.... This brings up the TVM Solver window as shown on the left.

 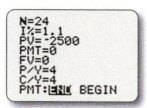

The screen on the right shows the values you should enter for this example. The various variables are

N	Number of compounding periods
I%	Annual interest rate, as percent, not decimal
PV	Negative of present value
PMT	Payment per period (0 in this section)
FV	Future value
P/Y	Payments per year
C/Y	Compounding periods per year
PMT:	Not used in this section.

Several things to notice:

- *I%* is the *annual* interest rate, corresponding to *r*, not *i*, in the compound interest formula.

- The present value, *PV*, is entered as a negative number. In general, when using the TVM Solver, any amount of money you give to someone else (such as the $2,500 you deposit in the bank) will be a negative number, whereas any amount of money someone gives to you

(such as the future value of your deposit, which the bank will give back to you) will be a positive number.

- *PMT* is not used in this example (it will be used in the next section) and should be 0.

- *FV* is the future value, which we shall compute in a moment; it doesn't matter what you enter now.

- *P/Y* and *C/Y* stand for payments per year and compounding periods per year, respectively. They should both be set to the number of compounding periods per year for compound interest problems. (Setting *P/Y* automatically sets *C/Y* to the same value.)

- *PMT*: *END* or *BEGIN* is not used in this example, and it doesn't matter which you select.

2. To compute the future value, use the up or down arrow to put the cursor on the *FV* line, then press [ALPHA] [SOLVE].

Example 2 (page 148) *Megabucks Corporation* is issuing 10-year bonds. How much would you pay for bonds with a maturity value of $10,000 if you wish to get a return of 6.5% compounded annually?

Solution

To compute the present value using a TI-83/84 Plus:

1. Enter the numbers shown below (left) in the TVM Solver window.

2. Put the cursor on the PV line, and press [ALPHA] [SOLVE].

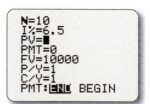 

Why is the present value given as negative?

Example 6 (page 151) You have $5,000 to invest at 6% annual interest compounded monthly. How long will it take for your investment to grow to $6,000?

Solution

1. Enter the numbers shown below (left) in the TVM Solver window.

2. Put the cursor on the N line, and press ALPHA SOLVE.

 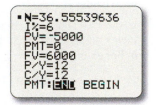

Recall that I% is the annual interest rate, corresponding to *r* in the formula, but N is the number of compounding periods, so number of months in this example. Thus, you will need to invest your money for about 36.6 months, or just over 3 years, before it grows to $6,000.

Section 2.3

Example 1 (page 158)   Your retirement account has $5,000 in it and earns 5% interest per year compounded monthly. At the end of every month for the next 10 years, you will deposit $100 into the account. How much money will there be in the account at the end of those 10 years?

Solution

We can use the TVM Solver in the TI-83/84 Plus to calculate future values like these:

1. The TVM Solver allows you to put the $5,000 already in the account as the present value of the account. Following the TI-83/84 Plus's usual convention, set PV to the *negative* of the present value because this is money you paid into the account.

2. Likewise, set PMT to −100, since you are paying $100 each month.

3. Set the number of payment and compounding periods to 12 per year.

4. Set the payments to be made at the end of each period.

5. With the cursor on the FV line, press ALPHA SOLVE to find the future value.

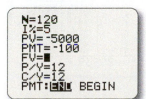 

Example 2 (page 159)   The average U.S. life expectancy is 79 years. As an actuary in *Long Life Insurance,* you are trying to estimate the smallest monthly premiums the company can charge for a life insurance policy that can pay out $200,000 in the event of death at age 79. Assume that the payments are to be deposited at the end of each month in an annuity account earning 3.6% annual interest compounded monthly. Make a table showing the associated monthly premiums for clients currently of age 20, 30, 40, 50, 60, and 70 years.

Solution

Let us do the calculation for the current age of 20, meaning $79 - 20 = 59$ years to age 79.

1. In the TVM Solver in the TI-83/84 Plus, enter the values shown below.

2. Solve for PMT.

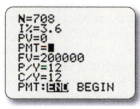 

Why is PMT negative?

Example 3 (page 161)   You wish to establish a trust fund from which your niece can withdraw $2,000 every 6 months for 15 years, at which time she will receive the remaining money in the trust, which you would like to be $10,000. The trust will be invested at 7% per year compounded every 6 months. How large should the trust be?

Solution

1. In the TVM Solver in the TI-83/84 Plus, enter the values shown below.

2. Solve for PV.

The payment and future value are positive because you (or your niece) will be receiving these amounts from the investment.

Note We have assumed that your niece receives the withdrawals at the end of each compounding period, so that the trust fund is an ordinary annuity. If, instead, she receives the payments at the beginning of each compounding period, it is an annuity due. You switch between the two types of annuity by changing PMT: END at the bottom to PMT: BEGIN.

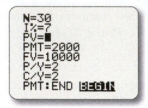 

As was mentioned in the text, the present value must be higher to fund payments at the beginning of each period, because the money in the account has less time to earn interest. ∎

Example 6 (page 164)   Marc and Mira are buying a house and have taken out a 30-year, $90,000 mortgage at 8% interest per year. Mortgage interest is tax deductible, so it is important to know how much of a year's mortgage payments represents interest. How much interest will Marc and Mira pay in the first year of their mortgage?

Solution

The TI-83/84 Plus has built-in functions to compute the values in an amortization schedule.

1. First, use the TVM Solver to find the monthly payment.

Three functions correspond to the last three columns of the amortization schedule given in the text: ΣInt, ΣPrn, and bal (found in the Finance menu accessed through APPS). They all require that the values of I%, PV, and PMT be entered or calculated ahead of time; calculating the payment in the TVM Solver in Step 1 accomplishes this.

2. Use ΣInt(*m*, *n*, 2) to compute the sum of the interest payments from payment *m* through payment *n*. For example,

$$\Sigma\text{Int}(1,12,2)$$

will return $-7,172.81$, the total paid in interest in the first year, which answers the question asked in this example. (The last argument, 2, tells the calculator to round all intermediate calculations to two decimal places—that is, the nearest cent—as would the mortgage lender.)

3. Use ΣPrn(*m*, *n*, 2) to compute the sum of the payments on principal from payment *m* through payment *n*. For example,

$$\Sigma\text{Prn}(1,12,2)$$

will return -751.87, the total paid on the principal in the first year.

4. Finally, bal(*n*, 2) finds the balance of the principal outstanding after *n* payments. For example,

$$\text{bal}(12,2)$$

will return the value 89,248.13, the balance remaining at the end of one year.

5. To construct an amortization schedule as in the text, make sure that FUNC is selected in the MODE window; then enter the functions in the Y= window as shown below.

6. Press 2ND TBLSET, and enter the values shown here.

7. Press 2ND TABLE to get the table shown here.

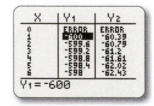

The column labeled X gives the month, the column labeled Y_1 gives the interest payment for each month, the column labeled Y_2 gives the payment on principal for each month, and the column labeled Y_3 (use the right arrow button to make it visible) gives the outstanding principal.

8. To see later months, use the down arrow. As you can see, some of the values will be rounded in the table, but by selecting a value (as the outstanding principal at the end of the first month is selected in the second screen), you can see its exact value at the bottom of the screen.

Spreadsheet   Technology Guide

Section 2.2

Example 1 (page 147) In December 2015, **Radius Bank** was paying 1.10% annual interest on savings accounts with balances of $2,500 and up. If the interest is compounded quarterly, find the future value of a $2,500 deposit after 6 years. What is the total interest paid over the time of the investment?

Solution

You can either compute compound interest directly or use financial functions built into your spreadsheet. The following worksheet has more than we need for this example, but will be useful for other examples in this and the next section.

	A	B	C	D
1		Entered	Calculated	
2	Rate	1.10%		
3	Years	6		
4	Payment	$0.00		
5	Present Value	$2,500.00		
6	Future Value		=FV(B2/B7,B3*B7,B4,B5)	
7	Periods per year	4		

For this example the payment amount in cell B4 should be 0. (We shall use it in the next section.)

1. Enter the other numbers as shown. As with other technologies, like the TVM Solver in the TI-83/84 Plus calculator, money that you pay to others (such as the $2,500 you deposit in the bank) should be entered as negative, whereas money that is paid to you is positive.

2. The formula entered in cell C6 uses the built-in FV function to calculate the future value based on the entries in column B. This formula has the following format:

$$=FV(i,n,PMT,PV)$$

i = Interest per period   We use B2/B7 for the interest.

n = Number of periods   We use B3*B7 for the number of periods.

PMT = Payment per period   The payment is 0 (cell B4).

PV = Present value.   The present value is in cell B5.

Instead of using the built-in FV function, we could use

$$=-B5*(1+B2/B7)^{\wedge}(B3*B7)$$

based on the future value formula for compound interest. After calculation the result will appear in cell C6.

	A	B	C	D
1		Entered	Calculated	
2	Rate	1.10%		
3	Years	6		
4	Payment	$0.00		
5	Present Value	$2,500.00		
6	Future Value		-$2,670.32	
7	Periods per year	4		

Note that we have formatted the cells B4:C6 as currency with two decimal places. If you change the values in column B, the future value in column C will be automatically recalculated.

Example 2 (page 148) *Megabucks Corporation* is issuing 10-year bonds. How much would you pay for bonds with a maturity value of $10,000 if you wished to get a return of 6.5% compounded annually?

Solution

You can compute present value in your spreadsheet using the PV worksheet function. The following worksheet is similar to the one in Example 1 except that we have entered a formula for computing the present value from the entered values.

	A	B	C	D
1		Entered	Calculated	
2	Rate	6.50%		
3	Years	10		
4	Payment	$0.00		
5	Present Value		=PV(B2/B7,B3*B7,B4,B6)	
6	Future Value	$10,000.00		
7	Periods per year	1		

The next worksheet shows the calculated value.

	A	B	C	D
1		Entered	Calculated	
2	Rate	6.50%		
3	Years	10		
4	Payment	$0.00		
5	Present Value		-$5,327.26	
6	Future Value	$10,000.00		
7	Periods per year	1		

Why is the present value negative?

Example 6 (page 151)   You have $5,000 to invest at 6% interest compounded monthly. How long will it take for your investment to grow to $6,000?

Solution

You can compute the requisite length of an investment in your spreadsheet using the NPER worksheet function. The following worksheets show the calculation.

	A	B	C	D
1		Entered	Calculated	
2	Rate	6.00%		
3	Years		=NPER(B2/B7,B4,B5,B6)/B7	
4	Payment	$0.00		
5	Present Value	-$5,000.00		
6	Future Value	$6,000.00		
7	Periods per year	12		

	A	B	C	D
1		Entered	Calculated	
2	Rate	6.00%		
3	Years		3.04628303	
4	Payment	$0.00		
5	Present Value	-$5,000.00		
6	Future Value	$6,000.00		
7	Periods per year	12		

The NPER function computes the number of compounding periods, months in this case, so we divide by B7, the number of periods per year, to calculate the number of years, which appears as 3.046. So you need to invest your money for just over 3 years for it to grow to $6,000.

Section 2.3

Example 1 (page 158)   Your retirement account has $5,000 in it and earns 5% interest per year compounded monthly. At the end of every month for the next 10 years you will deposit $100 into the account. How much money will there be in the account at the end of those 10 years?

Solution

We can use exactly the same worksheet that we used in Example 1 in Section 2.2. In fact, we included the "Payment" row in that worksheet just for this purpose.

	A	B	C	D
1		Entered	Calculated	
2	Rate	5.00%		
3	Years	10		
4	Payment	-$100.00		
5	Present Value	-$5,000.00		
6	Future Value		=FV(B2/B7,B3*B7,B4,B5)	
7	Periods per year	12		

	A	B	C	D
1		Entered	Calculated	
2	Rate	5.00%		
3	Years	10		
4	Payment	-$100.00		
5	Present Value	-$5,000.00		
6	Future Value		$23,763.28	
7	Periods per year	12		

Note that the FV function allows us to enter, as the last argument, the amount of money already in the account. Following the usual convention, we enter the present value and the payment as *negative*, because these are amounts you pay into the account.

Example 2 (page 159)   The average U.S. life expectancy is 79 years. As an actuary in *Long Life Insurance,* you are trying to estimate the smallest monthly premiums the company can charge for a life insurance policy that can pay out $200,000 in the event of death at age 79. Assume that the payments are to be deposited at the end of each month in an annuity account earning 3.6% annual interest compounded monthly. Make a table showing the associated monthly premiums for clients currently of age 20, 30, 40, 50, 60, and 70 years.

Solution

Let us do the calculation for the current age of 20, meaning $79 - 20 = 59$ years to age 79. We use the following worksheet, in which the PMT worksheet function is used to calculate the required payments.

	A	B	C	D
1		Entered	Calculated	
2	Rate	3.60%		
3	Years	59		
4	Payment		=PMT(B2/B7,B3*B7,B5,B6)	
5	Present Value	$0.00		
6	Future Value	$200,000.00		
7	Periods per year	12		

	A	B	C	D
1		Entered	Calculated	
2	Rate	3.60%		
3	Years	59		
4	Payment		-$81.77	
5	Present Value	$0.00		
6	Future Value	$200,000.00		
7	Periods per year	12		

	A	B	C	D
1		Entered	Calculated	
2	Rate	7.00%		
3	Years	15		
4	Payment	$2,000.00		
5	Present Value		=PV(B2/B7,B3*B7,B4,B6,1)	
6	Future Value	$10,000.00		
7	Periods per year	2		

	A	B	C	D
1		Entered	Calculated	
2	Rate	7.00%		
3	Years	15		
4	Payment	$2,000.00		
5	Present Value		-$41,634.32	
6	Future Value	$10,000.00		
7	Periods per year	2		

Why is the payment negative?

Example 3 (page 161)   You wish to establish a trust fund from which your niece can withdraw $2,000 every 6 months for 15 years, at which time she will receive the remaining money in the trust, which you would like to be $10,000. The trust will be invested at 7% per year compounded every 6 months. How large should the trust be?

Solution

You can use the same worksheet as in Example 2 in Section 2.2.

	A	B	C	D
1		Entered	Calculated	
2	Rate	7.00%		
3	Years	15		
4	Payment	$2,000.00		
5	Present Value		=PV(B2/B7,B3*B7,B4,B6)	
6	Future Value	$10,000.00		
7	Periods per year	2		

	A	B	C	D
1		Entered	Calculated	
2	Rate	7.00%		
3	Years	15		
4	Payment	$2,000.00		
5	Present Value		-$40,346.87	
6	Future Value	$10,000.00		
7	Periods per year	2		

The payment and future value are positive because you (or your niece) will be receiving these amounts from the investment.

Note We have assumed that your niece receives the withdrawals at the end of each compounding period, so the trust fund is an ordinary annuity. If, instead, she receives the payments at the beginning of each compounding period, it is an annuity due. You switch to an annuity due by adding an optional last argument of 1 to the PV function (and similarly for the other finance functions in spreadsheets).

As was mentioned in the text, the present value must be higher to fund payments at the beginning of each period, because the money in the account has less time to earn interest. ∎

Example 6 (page 164)   Marc and Mira are buying a house and have taken out a 30-year, $90,000 mortgage at 8% interest per year. Mortgage interest is tax deductible, so it is important to know how much of a year's mortgage payments represents interest. How much interest will Marc and Mira pay in the first year of their mortgage?

Solution

We construct an amortization schedule with which we can answer the question.

1. Begin with the worksheet below.

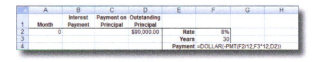

	A	B	C	D	E	F	G	H
1	Month	Interest Payment	Payment on Principal	Outstanding Principal				
2	0			$90,000.00	Rate	8%		
3					Years	30		
4					Payment	=DOLLAR(-PMT(F2/12,F3*12,D2))		

	A	B	C	D	E	F
1	Month	Interest Payment	Payment on Principal	Outstanding Principal		
2	0			$90,000.00	Rate	8%
3					Years	30
4					Payment	$660.39

Note the formula for the monthly payment:

=DOLLAR(-PMT(F2/12,F3*12,D2))

The function DOLLAR rounds the payment to the nearest cent, as the bank would.

2. Calculate the interest owed at the end of the first month using the formula

=DOLLAR(D2*F$2/12)

in cell B3.

3. The payment on the principal is the remaining part of the payment, so enter

$$=F\$4-B3$$

in cell C3.

4. Calculate the outstanding principal by subtracting the payment on the principal from the previous outstanding principal, by entering

$$=D2-C3$$

in cell D3.

5. Copy the formulas in cells B3, C3, and D3 into the cells below them to continue the table.

	A	B	C	D	E	F
1	Month	Interest Payment	Payment on Principal	Outstanding Principal		
2	0			$90,000.00	Rate	8%
3	1	$600.00	$60.39	$89,939.61	Years	30
4	2				Payment	$660.39
5	3					
6	4					
7	5					
8	6					
9	7					
10	8					
11	9					
12	10					
13	11					
14	12					

The result should be something like the following:

	A	B	C	D	E	F
1	Month	Interest Payment	Payment on Principal	Outstanding Principal		
2	0			$90,000.00	Rate	8%
3	1	$600.00	$60.39	$89,939.61	Years	30
4	2	$599.60	$60.79	$89,878.82	Payment	$660.39
5	3	$599.19	$61.20	$89,817.62		
6	4	$598.78	$61.61	$89,756.01		
7	5	$598.37	$62.02	$89,693.99		
8	6	$597.96	$62.43	$89,631.56		
9	7	$597.54	$62.85	$89,568.71		
10	8	$597.12	$63.27	$89,505.44		
11	9	$596.70	$63.69	$89,441.75		
12	10	$596.28	$64.11	$89,377.64		
13	11	$595.85	$64.54	$89,313.10		
14	12	$595.42	$64.97	$89,248.13		

6. Adding the calculated interest payments gives us the total interest paid in the first year: $7,172.81.

Note Spreadsheets have built-in functions that compute the interest payment (IPMT) or the payment on the principle (PPMT) in a given period. We could also have used the built-in future value function (FV) to calculate the outstanding principal each month. In fact, we used exactly this approach in Example 7. As we mentioned there, one thing to be aware of when using these functions is that, in a sense, they are too accurate. They do not take into account the fact that payments and interest are rounded to the nearest cent. Over time, this rounding causes the actual value of the outstanding principal to differ from what the FV function would tell us. In fact, because the actual payment is rounded slightly upward (to $660.39 from 660.38811...), the principal is reduced slightly faster than necessary, and a last payment of $660.39 would be $2.95 larger than needed to clear out the debt. The lender would reduce the last payment by $2.95 for this reason; Marc and Mira will pay only $657.44 for their final payment. This is common: The last payment on an installment loan is usually slightly larger or smaller than the others, to compensate for the rounding of the monthly payment amount. ■

3

SYSTEMS OF LINEAR EQUATIONS AND MATRICES

CASE STUDY

Hybrid Cars—Optimizing the Degree of Hybridization

You are involved in new model development at a major automobile company. The company is planning to introduce two new plug-in hybrid electric vehicles: the subcompact "Green Town Hopper" and the midsize "Electra Supreme," and your department must decide on the degree of hybridization (DOH) for each of these models that will result in the largest reduction in gasoline consumption. The data you have available show the gasoline savings for only three values of the DOH.

How do you estimate the optimal value?

Fedor Selivanov/Alamy Stock Photo

Introduction

In Chapter 1 we studied single functions and equations. In this chapter we seek solutions to **systems** of two or more equations. For example, suppose we need to *find two numbers whose sum is 3 and whose difference is 1*. In other words, we need to find two numbers x and y such that $x + y = 3$ and $x - y = 1$. The only solution turns out to be $x = 2$ and $y = 1$, a solution you might easily guess. But how do we know that this is the only solution, and how do we find solutions systematically? When we restrict ourselves to systems of *linear* equations, there is a very elegant method for determining the number of solutions and finding them all. Moreover, as we will see, many real-world applications give rise to just such systems of linear equations.

We begin in Section 3.1 with systems of two linear equations in two unknowns and some of their applications. In Section 3.2 we study a powerful matrix method, called *row reduction*, for solving systems of linear equations in any number of unknowns. In Section 3.3 we look at more applications.

Computers have been used for many years to solve the large systems of equations that arise in the real world. You probably already have access to devices that will do the row operations that are used in row reduction. Many graphing calculators can do them, as can spreadsheets and various special-purpose applications, including utilities available at the Website. Using such a device or program makes the calculations quicker and helps to avoid arithmetic mistakes. Then there are programs (and calculators) into which you simply feed the system of equations and out pop the solutions. We can think of what we do in this chapter as looking inside the "black box" of such a program. More important, we talk about how, starting from a real-world problem, to get the system of equations to solve in the first place. No computer will do this conversion for us yet.

3.1   Systems of Two Equations in Two Unknowns

Linear Equations and Solutions

Suppose you have \$3 in your pocket to spend on snacks and a drink. If x represents the amount you'll spend on snacks and y represents the amount you'll spend on a drink, you can say that $x + y = 3$. On the other hand, if for some reason you want to spend \$1 more on snacks than on your drink, you can also say that $x - y = 1$. These are simple examples of **linear equations in two unknowns**.

Linear Equations in Two Unknowns

A **linear equation in two unknowns** is an equation that can be written in the form

$$ax + by = c$$

with a, b, and c being real numbers. The number a is called the **coefficient of x**, and b is called the **coefficient of y**. A **solution** of an equation consists of a pair of numbers: a value for x and a value for y that satisfy the equation.

Quick Example

1. In the linear equation $3x - 2y = 12$, the coefficients are $a = 3$ and $b = -2$. The pair $(x, y) = (4, 0)$ is a solution, because $3(4) - 2(0) = 12$.

In fact, $(x, y) = (4, 0)$ is not the only solution to the linear equation $3x - 2y = 12$ in Quick Example 1. Linear equations have *infinitely many* solutions, as we now illustrate.

EXAMPLE 1 **Solutions of a Linear Equation**

Determine all solutions of $3x - 2y = 12$.

Solution If we rewrite the equation $3x - 2y = 12$ by solving for y, we get

$$y = \frac{3}{2}x - 6,$$

which expresses y as a linear function of x, as in Section 1.3. For every value of x that we choose, we can now get the corresponding value of y, giving a solution (x, y), as shown in the following table:

x	-2	0	2	4	6
$y = \dfrac{3}{2}x - 6$	-9	-6	-3	0	3

From the table we find the **particular solutions** $(-2, -9)$, $(0, -6)$, $(2, -3)$, $(4, 0)$, and $(6, 3)$. These are just five of the infinitely many solutions that are possible. Each of these solutions has the form $(x, y) = \left(x, \frac{3}{2}x - 6\right)$ for some arbitrary choice of x, so we call

$$(x, y) = \left(x, \frac{3}{2}x - 6\right); \quad x \text{ arbitrary} \qquad \text{\color{blue}{General solution parameterized by } } x$$

the **general solution**, as it incorporates all possible particular solutions: To get a particular solution from the general solution, just choose a value for x. When we write the general solution this way, we say that x is a **parameter**, and we have a solution **parameterized by** x.

Referring to Section 1.3, we see that these solutions are in fact the points on a straight line: the *graph* of $y = \frac{3}{2}x - 6$ (Figure 1). Geometrically, we have a "whole line of solutions," one for each point on the line.

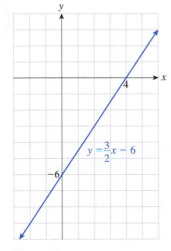

Figure 1

➡ **Before we go on ...** We could have solved the equation for x instead:

$$x = \frac{2}{3}y + 4$$

and obtained another form of the general solution, parameterized by y:

$$\left(\frac{2}{3}y + 4, y\right); \quad y \text{ arbitrary.} \qquad \text{\color{blue}{General solution parameterized by } } y \ ■$$

The following summarizes what we have just said.

Particular and General Solutions of a Linear Equation in Two Unknowns

A **particular solution** of the linear equation $ax + by = c$ is a specific pair of numbers (x, y) that satisfy the equation. If a and b are not both zero, the linear equation $ax + by = c$ has infinitely many such solutions; these solutions lie on a straight line, the **graph of the equation**.

The **general solution parameterized by** x of the linear equation $ax + by = c$ has the form $(x, f(x))$, where we solve the given equation for y as a function of x, $y = f(x)$ if possible.[*] Substituting any number for x then results in a particular solution.

Similarly, the **general solution parameterized by** y has the form $(g(y), y)$, where we solve the given equation for $x = g(y)$ if possible.[†] Substituting any number for y then results in a particular solution.

[*] That is, if $b \neq 0$, so that y appears in the equation. Otherwise, the general solution cannot be parameterized by x.

[†] That is, if $a \neq 0$, so that x appears in the equation. Otherwise, the general solution cannot be parameterized by y.

Quick Examples

2. $2x + y = 4$ has the two particular solutions $(0, 4)$ and $(1, 2)$, among infinitely many others. The graph of $2x + y = 4$ consists of all its solutions and is the line $y = -2x + 4$ with slope -2 and y-intercept 4.

General solution parameterized by x: Solve for y to obtain $y = -2x + 4$, giving the general solution $(x, -2x + 4)$; x arbitrary.

General solution parameterized by y: Solve for x to obtain $x = -\frac{1}{2}y + 2$, giving the general solution $\left(-\frac{1}{2}y + 2, y\right)$; y arbitrary.

3. $2y = 6$ has the two particular solutions $(0, 3)$ and $(1, 3)$. The graph of $2y = 6$ consists of all its solutions and is the horizontal line $y = 3$.

General solution parameterized by x: Solve for y to obtain $y = 3$, giving the general solution $(x, 3)$; x arbitrary.

General solution parameterized by y: As x does not occur in the equation, we cannot solve for x, so the general solution cannot be parameterized by y.

4. $3x = 5$ has the two particular solutions $\left(\frac{5}{3}, 0\right)$ and $\left(\frac{5}{3}, 1\right)$. The graph of $3x = 5$ consists of all its solutions and is the vertical line $x = \frac{5}{3}$.

General solution parameterized by x: As y does not occur in the equation, we cannot solve for y, so the general solution cannot be parameterized by x.

General solution parameterized by y: Solve for x to obtain $x = \frac{5}{3}$, giving the general solution $\left(\frac{5}{3}, y\right)$; y arbitrary.

Solutions to Systems of Linear Equations

In this section we are mainly concerned with pairs (x, y) that are solutions of two linear equations at the same time. For example, $(2, 1)$ is a solution of both of the equations $x + y = 3$ and $x - y = 1$, because substituting $x = 2$ and $y = 1$ into these equations gives $2 + 1 = 3$ (true) and $2 - 1 = 1$ (also true), respectively. So in the simple example at the beginning of this section you could spend \$2 on snacks and \$1 on a drink.

In the examples that follow, we see how to graphically and algebraically solve a system of two linear equations in two unknowns. Then we consider systems of more than two linear equations in two unknowns, and finally, we return to some more interesting applications.

EXAMPLE 2 **Solving a System Graphically and Algebraically**

Find all solutions (x, y) of the following system of two equations:

$$x + y = 3$$
$$x - y = 1.$$

Solution

Method 1: Graphical We already know that the solutions of a single linear equation are the points on its graph, which is a straight line. For a point to represent a solution of two linear equations, it must lie simultaneously on both of the corresponding lines. In other words, it must be a point where the two lines cross, or intersect. A look at Figure 2 should convince us that the lines cross only at the point $(2, 1)$, so this is the only possible solution.

Method 2: Algebraic: Solving by Elimination This will be our preferred approach. (See the "Before we go on" discussion at the end of this example.) To solve the system algebraically by elimination, we try to combine the equations in such a way as to eliminate one variable. In this case, notice that if we add the left-hand sides of the equations, the terms with y are eliminated. So we add the first equation to the second (that is, add the left-hand sides and add the right-hand sides*):

$$
\begin{array}{rcl}
x + y & = & 3 \\
\underline{x - y} & = & \underline{1} \\
2x + 0 & = & 4 \\
2x & = & 4 \\
x & = & 2.
\end{array}
$$

Now that we know that x has to be 2, we can substitute back into either equation to find y. Choosing the first equation (it doesn't matter which we choose), we have

$$2 + y = 3$$
$$y = 3 - 2 = 1.$$

We have found that the only possible solution is $x = 2$ and $y = 1$, or

$$(x, y) = (2, 1).$$

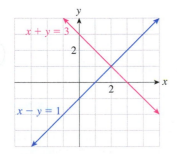

Figure 2

* We can add these equations because when we add equal amounts to both sides of an equation, the results are equal. That is, if $A = B$ and $C = D$, then $A + C = B + D$.

Method 3: Algebraic: Solving by Intersection Solution by intersection is based on the idea of the geometric approach: that the solution is the point of intersection of the graphs of two linear functions. To find this intersection point algebraically, we reason that the two functions must have the same value at the point of intersection, so we solve both equations for y,[*] set the resulting linear functions of x equal to each other, and solve for x:

[*] If y does not occur in an equation, it means that we can use that equation to obtain the value of x directly and then substitute this value in the other equation to obtain y.

$$y = -x + 3 \qquad \text{Solve } x + y = 3 \text{ for } y.$$
$$y = x - 1 \qquad \text{Solve } x - y = 1 \text{ for } y.$$
$$-x + 3 = x - 1 \qquad \text{Equate the resulting linear functions of } x.$$
$$-2x = -4$$
$$x = 2 \qquad \text{Solve for } x.$$

We can then substitute this value into either equation (or either linear function we found) to obtain y as in Method 2 and find the solution: $(x, y) = (2, 1)$.

Method 4: Algebraic: Solving by Substitution In the substitution method we solve one of the equations for one of the unknowns and then substitute the resulting expression in the other to solve for the other unknown:

$$y = -x + 3 \qquad \text{Solve } x + y = 3 \text{ for } y.$$
$$x - (-x + 3) = 1 \qquad \begin{array}{l}\text{Substitute the resulting expression in the}\\ \text{other equation, } x - y = 1.\end{array}$$
$$2x - 3 = 1 \quad \Rightarrow \quad 2x = 4$$
$$x = 2. \qquad \text{Solve for } x.$$

We can then substitute this value into either equation to obtain y as in the elimination method to obtain the solution: $(x, y) = (2, 1)$.

➡ **Before we go on . . .** As we said above, the method of elimination is our preferred algebraic method of solving systems of linear equations and is the algebraic method we will use in most of the examples that follow.

Q: *The intersection and substitution methods seem simpler and more direct. So why is the elimination method the "preferred" method?*

A: Two reasons: First, the elimination method extends more easily to systems with more equations and unknowns than the other two methods. It is the basis for the matrix method of solving systems—a method we will discuss in Section 3.2. So we shall use it exclusively for the rest of this section. Second, the system we considered in Example 2 is a simple one, so very little algebraic manipulation was required in the other methods. In general, the elimination method will, as we see below, allow us to completely avoid fractions and decimals even in systems presented with fractions and decimals, whereas the other methods may require manipulation of complicated expressions involving fractions and/or decimals.

The next example illustrates the drawbacks of the graphical method.

Using Technology

See the Technology Guides at the end of the chapter for details on the graphical solution of Example 1 using a TI-83/84 Plus or a spreadsheet. Here is an outline:

TI-83/84 Plus

$Y_1 = -X + 3$ $Y_2 = X - 1$
Graph: [WINDOW] ; Xmin = −4,
Xmax = 4; [ZOOM] [0]
Trace to estimate the point of intersection.
[More details in the Technology Guide.]

Spreadsheet

Headings x, $y1$, and $y2$ in A1–C1, x-values −4, 4 in A2, A3
= -A2+3 and =A2-1 in B1 and C1; copy down to B2–C2.
Graph the data in columns A–C with a line-segment scatter plot. To see a closer view, change the values in A2–A3.
[More details in the Technology Guide.]

W Website
www.WanerMath.com

→ Online Utilities

→ Function Evaluator and Grapher

Enter -x+3 for y_1 and x-1 for y_2.
Set Xmin = −4, Xmax = 4 and press "Plot Graphs".
Click on the graph, and use the trace arrows below to estimate the point of intersection.

EXAMPLE 3   Solving a System: Algebraically versus Graphically

Solve the following system:

$$\frac{x}{5} + \frac{y}{3} = 0$$

$$0.2x + 0.7y = 0.1.$$

Solution Fractions and decimals can significantly complicate any method used to solve the system, so our approach here—and throughout this chapter—will be to *get rid of fractions and decimals before doing anything:** To get rid of the fractions in the first equation, we can multiply both sides of the equation by a common multiple of the denominators such as 15; and to get rid of the decimals in the second equation, we can multiply both sides by 10:

$$(15)\left(\frac{x}{5} + \frac{y}{3}\right) = (15)(0)$$

$$(10)(0.2x + 0.7y) = (10)(0.1),$$

giving

$$3x + 5y = 0$$

$$2x + 7y = 1,$$

which looks a whole lot better! To solve, we try the graphical approach first.

Method 1: Graphical First, solve for y, obtaining $y = -\frac{3}{5}x$ and $y = -\frac{2}{7}x + \frac{1}{7}$. Graphing these equations, we get Figure 3. The lines appear to intersect slightly above and to the left of the origin. Redrawing with a finer scale (or zooming in using graphing technology), we can get the graph in Figure 4.

If we look carefully at Figure 4, we see that the graphs intersect near $(-0.45, 0.27)$. Is the point of intersection *exactly* $(-0.45, 0.27)$? (Substitute these values into the equations to find out.) In fact, it is extremely onerous to find the exact solution of this system graphically, but we now have a ballpark answer that we can use to help check the following algebraic solution.

Method 2: Algebraic: Solving by Elimination We first see that adding the equations is not going to eliminate either x or y. Notice, however, that if we multiply (both sides of) the first equation by 2 and the second by -3, the coefficients of x will become 6 and -6. *Then* if we add them, x will be eliminated. So we proceed as follows:

$$2(3x + 5y) = 2(0)$$

$$-3(2x + 7y) = -3(1)$$

gives

$$6x + 10y = 0$$

$$-6x - 21y = -3.$$

Adding these equations, we get

$$-11y = -3,$$

* If fractions and decimals are not necessary at each step of a calculation, why insist on carrying them through the entire calculation like dead weight, complicating all the calculations for no good reason?

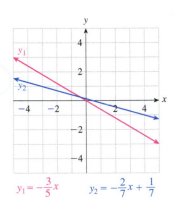

$$y_1 = -\frac{3}{5}x \qquad y_2 = -\frac{2}{7}x + \frac{1}{7}$$

Figure 3

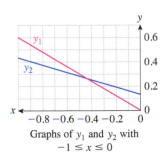

Graphs of y_1 and y_2 with $-1 \leq x \leq 0$

Figure 4

so

$$y = \frac{3}{11} = 0.\overline{27}.$$

Substituting $y = \frac{3}{11}$ in the first equation gives

$$3x + 5\left(\frac{3}{11}\right) = 0$$

$$3x = -\frac{15}{11}$$

$$x = -\frac{5}{11} = -0.\overline{45}.$$

The solution is $(x, y) = \left(-\frac{5}{11}, \frac{3}{11}\right) = (-0.\overline{45}, 0.\overline{27})$.

Notice that the algebraic method gives us the exact solution that we could not find with the graphical method. Still, we can check that the graph and our algebraic solution agree to the accuracy with which we can read the graph. To be absolutely sure that our answer is correct, we should check it:

$$3\left(-\frac{5}{11}\right) + 5\left(\frac{3}{11}\right) = -\frac{15}{11} + \frac{15}{11} = 0 \quad ✔$$

$$2\left(-\frac{5}{11}\right) + 7\left(\frac{3}{11}\right) = -\frac{10}{11} + \frac{21}{11} = 1. \quad ✔$$

Get in the habit of checking your answers.

➡ **Before we go on . . .**

Q: *In solving the system in Example 3, we multiplied (both sides of) the equations by numbers. How does that affect their graphs?*

A: Multiplying both sides of an equation by a nonzero number has no effect on its solutions, so the graph (which represents the set of all solutions) is unchanged.

Before doing some more examples, we summarize what we have said about solving systems of equations.

> ### Solving a System of Two Linear Equations in Two Unknowns
>
> Before starting, get rid of all decimals and fractions by multiplying (both sides of) each equation by a suitable nonzero number.
>
> ### Graphical Method of Solution
>
> Graph both equations on the same graph. (For example, solve each for y to find the slope and y-intercept.) A point of intersection gives the solution to the system. To find the point, you may need to adjust the range of x-values you use. To find the point accurately, you may need to use a smaller range (or zoom in if using technology).

Algebraic Methods of Solution

Solving by Elimination (Preferred): To eliminate x, multiply (both sides of) each equation by a nonzero number so that the coefficients of x are the same in absolute value but opposite in sign. Add the two equations to eliminate x; this gives an equation in y that we can solve to find its value. Substitute this value of y into one of the original equations to find the value of x (or we could first eliminate y to find x).

Solving by Intersection: Solve both equations for y, and equate the resulting functions of x to solve for x. Substitute this value of x into one of the original equations to find the value of y.

Solving by Substitution: Solve one equation for y, and then substitute the resulting expression in the other to solve for x. Substitute this value of x into one of the original equations to find the value of y.

Sometimes, something appears to go wrong with these methods. The following examples show what can happen.

EXAMPLE 4 **Inconsistent System**

Solve the system

$$x - 3y = 5$$
$$-2x + 6y = 8.$$

Solution To eliminate x, we multiply the first equation by 2 and then add:

$$2x - 6y = 10$$
$$-2x + 6y = 8.$$

Adding gives

$$0 = 18.$$

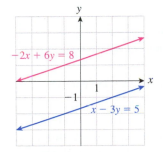

Figure 5

But this is absurd! This calculation shows that if we had two numbers x and y that satisfied both equations, it would be true that $0 = 18$. As 0 is *not* equal to 18, there can be no such numbers x and y. In other words, *the system has no solutions* and is called an **inconsistent system**.

In slope-intercept form, these lines are $y = \frac{1}{3}x - \frac{5}{3}$ and $y = \frac{1}{3}x + \frac{4}{3}$. Notice that they have the same slope but different y-intercepts. This means that they are parallel but different lines. Plotting them confirms this fact (Figure 5). Because they are parallel, they do not intersect. A solution must be a point of intersection, so we again conclude that there is no solution.

EXAMPLE 5 **Redundant System**

Solve the system

$$x + y = 2$$
$$2x + 2y = 4.$$

Solution Multiplying the first equation by -2 gives

$$-2x - 2y = -4$$
$$2x + 2y = 4.$$

Adding gives the not-very-enlightening result

$$0 = 0.$$

Now what has happened? Looking back at the original system, we note that the second equation is really the first equation in disguise. (It is the first equation multiplied by 2.) Put another way, if we solve both equations for y, we find that, in slope-intercept form, both equations become the same:

$$y = -x + 2,$$

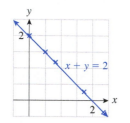

The line of solutions ($y = -x + 2$) and some particular solutions

Figure 6

so the two lines represented by the two equations are actually the same line (Figure 6). As these "two" lines intersect at every point, there is a solution for each point on the common line. In other words, we have infinitely many solutions, as in Example 1.

Algebraically, the second equation gives us the same information as the first, so we say that this is a **redundant** or **dependent system**. Put another way, we really have only one equation in two unknowns, so we refer to Example 1 to find the general solution by solving for y (which we did already above):

$$y = -x + 2.$$

Thus, the general solution is

$$(x, -x + 2); \quad x \text{ arbitrary.} \qquad \text{General solution parameterized by } x$$

As in Example 1, different choices of the parameter x lead to different particular solutions. For instance, choosing $x = 3$ gives the particular solution $(x, y) = (3, -1)$. Alternatively, we could solve for x to parameterize the general solution by y:

$$(-y + 2, y); \quad y \text{ arbitrary.} \qquad \text{General solution parameterized by } y$$

We summarize the three possible outcomes we have encountered.

> **Possible Outcomes for a System of Two Linear Equations in Two Unknowns**
>
> 1. **A single (or *unique*) solution:** This happens when the lines corresponding to the two equations are distinct and not parallel so they intersect at a single point. (See Example 2.)
>
> 2. **No solution:** This happens when the two lines are distinct and parallel. We say that the system is **inconsistent**. (See Example 4.)
>
> 3. **An infinite number of solutions:** This occurs when the two equations represent the same straight line, and we say that such a system is **redundant** or **dependent**. In this case we can represent the solutions by choosing one variable arbitrarily and solving for the other. (See Example 5.)
>
> In cases 1 and 3 we say that the system of equations is **consistent** because it has at least one solution.

You should think about straight lines and convince yourself that these are the only three possibilities.

Applications

EXAMPLE 6   **Resource Allocation**

Acme Baby Foods mixes two strengths of apple juice. One quart of Beginner's juice is made from 30 fluid ounces of water and 2 fluid ounces of apple juice concentrate. One quart of Advanced juice is made from 20 fluid ounces of water and 12 fluid ounces of concentrate. Every day Acme has available 30,000 fluid ounces of water and 3,600 fluid ounces of concentrate. If the company wants to use all the water and concentrate, how many quarts of each type of juice should it mix?

Solution   In all applications we follow the same general strategy:

1. ***Identify and label the unknowns.*** What are we asked to find? To answer this question, it is common to respond by saying, "The unknowns are Beginner's juice and Advanced juice." Quite frankly, this is a baffling statement. Just what is unknown about juice? We need to be more precise:

   *The unknowns are (1) the **number of quarts** of Beginner's juice and (2) the **number of quarts** of Advanced juice made each day.*

   So we label the unknowns as follows: Let

   x = number of quarts of Beginner's juice made each day

   y = number of quarts of Advanced juice made each day.

2. ***Use the information given to set up equations in the unknowns.*** This step is trickier, and the strategy varies from problem to problem. Here, the amount of juice the company can make is constrained by the fact that it has limited amounts of water and concentrate. This example shows a kind of application we will often see, and it is helpful in these problems to use a table to record the amounts of the resources used.

	Beginner's (x)	Advanced (y)	Available
Water (fluid ounces)	30	20	30,000
Concentrate (fluid ounces)	2	12	3,600

We can now set up an equation for each of the items listed in the left column of the table.

Water: We read across the first row. If Acme mixes x quarts of Beginner's juice, each quart using 30 fluid ounces of water, and y quarts of Advanced juice, each using 20 fluid ounces of water, it will use a total of $30x + 20y$ fluid ounces of water. But we are told that the total has to be 30,000 fluid ounces. Thus, $30x + 20y = 30,000$. This is our first equation.

Concentrate: We read across the second row. If Acme mixes x quarts of Beginner's juice, each using 2 fluid ounces of concentrate, and y quarts of Advanced juice, each using 12 fluid ounces of concentrate, it will use a total of $2x + 12y$ fluid ounces of concentrate. But we are told that the total has to be 3,600 fluid ounces. Thus, $2x + 12y = 3,600$.

Now we have two equations:

$$30x + 20y = 30,000$$
$$2x + 12y = 3,600.$$

To make the numbers easier to work with, let's divide (both sides of) the first equation by 10 and divide the second by 2:

$$3x + 2y = 3,000$$
$$x + 6y = 1,800.$$

We can now eliminate x by multiplying the second equation by -3 and adding:

$$3x + 2y = 3,000$$
$$\underline{-3x - 18y = -5,400}$$
$$-16y = -2,400.$$

So $y = 2,400/16 = 150$. Substituting this into the equation $x + 6y = 1,800$ gives $x + 900 = 1,800$, so $x = 900$. The solution is $(x, y) = (900, 150)$. In other words, the company should mix 900 quarts of Beginner's juice and 150 quarts of Advanced juice.

EXAMPLE 7   Blending

A medieval alchemist's love potion calls for a number of eyes of newt and toes of frog, the total being 20, but with twice as many newt eyes as frog toes. How many of each are required?

Solution As in Example 6, the first step is to identify and label the unknowns. Let

$$x = \text{number of newt eyes}$$
$$y = \text{number of frog toes.}$$

As for the second step—setting up the equations—a table is less appropriate here than in Example 6. Instead, we translate each phrase of the problem into an equation. The phrase "the total being 20" tells us that the total number of eyes and toes is 20. Thus,

$$x + y = 20.$$

The end of the first sentence gives us more information, but the phrase "twice as many newt eyes as frog toes" is a little tricky: Does it mean that $2x = y$ or that $x = 2y$? We can decide which by rewording the statement using the phrases "the *number of* newt eyes," which is x, and "the *number of* frog toes," which is y. Rephrased, the statement reads:

> The **number of** newt eyes is twice the **number of** frog toes.

(Notice how the word "twice" is forced into a different place.) With this rephrasing, we can translate directly into algebra:

$$x = 2y.$$

In standard form ($ax + by = c$), this equation reads

$$x - 2y = 0.$$

Thus, we have the two equations:

$$x + y = 20$$
$$x - 2y = 0.$$

SuperStock

To eliminate x, we multiply the second equation by -1 and then add:

$$x + y = 20$$
$$-x + 2y = 0.$$

We'll leave it to you to finish solving the system and find that $x = 13\frac{1}{3}$ and $y = 6\frac{2}{3}$.

So the recipe calls for exactly $13\frac{1}{3}$ eyes of newt and $6\frac{2}{3}$ toes of frog. The alchemist needs a very sharp scalpel and a very accurate balance (not to mention a very strong stomach).

We saw in Chapter 1 that the *equilibrium price* of an item (the price at which supply equals demand) and the *break-even point* (the number of items that must be sold to break even) can both be described as the intersection points of two graphs. If the graphs are straight lines, what we need to do to find the intersection is solve a system of two linear equations in two unknowns, as illustrated in the following problem.

EXAMPLE 8   Equilibrium Price

The demand for refrigerators in West Podunk is given by

$$q = -\frac{p}{10} + 100,$$

where q is the number of refrigerators that the citizens will buy each year if the refrigerators are priced at p dollars each. The supply of refrigerators is

$$q = \frac{p}{20} + 25,$$

where now q is the number of refrigerators the manufacturers will be willing to ship into town each year if the refrigerators are priced at p dollars each. Find the equilibrium price and the number of refrigerators that will be sold at that price.

Solution Figure 7 shows the demand and supply curves. The equilibrium price occurs at the point where these two lines cross, which is where demand equals supply. The graph suggests that the equilibrium price is $500, and zooming in confirms this.

To solve this system algebraically, the intersection method in Example 2 suggests itself, as both equations are already solved for q, so we just equate them:

$$-\frac{p}{10} + 100 = \frac{p}{20} + 25,$$

and solve for p:

$$-2p + 2{,}000 = p + 500 \qquad \text{\textcolor{blue}{Multiply by 20 to clear fractions.}}$$
$$-3p = -1{,}500$$
$$p = 500.$$

Substituting this value into the demand equation gives us the corresponding value of q:

$$q = -\frac{500}{10} + 100 = 50.$$

Thus, the equilibrium price is $500, and 50 refrigerators will be sold at this price.

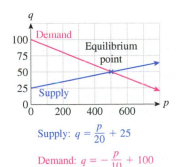

Supply: $q = \frac{p}{20} + 25$

Demand: $q = -\frac{p}{10} + 100$

Figure 7

➡ **Before we go on . . .** We could also have solved the system of equations in Example 8 by writing both equations in standard form with the unknowns on the left:

$$\frac{p}{10} + q = 100$$

$$-\frac{p}{20} + q = 25.$$

We could then have used elimination as in Example 3 (clearing fractions in the first step). ∎

FAQs

Setting Up the Equations

Q: *Looking through the example applications, I notice that in some, we can tabulate the information given and read off the equations (as in Example 6), whereas in others (such as Example 7), we have to reword each sentence to turn it into an equation. How do I know what approach to use?*

A: There is no hard-and-fast rule, and indeed some applications might call for a bit of each approach. However, it is generally not hard to see when it would be useful to tabulate values: Lists of the numbers of ingredients or components generally lend themselves to tabulation, whereas phrases such as "twice as many of these as those" generally require direct translation into equations (after rewording if necessary).

Q: *Help! Every time I write down the equation form of something like "there are twice as many apples as oranges" I get it backwards! What is going on?*

A: The most common reason for this kind of error is trying to write down an equation before rewording the information as we did in Example 7:

1. First, state what each unknown represents (using the phrase "the number of . . ."), as in

    "Let x be the number of apples, and let y be the number of oranges."

    (Don't just say "x = apples, y = oranges.")

2. Then reword all the given information *using the phrase "the number of,"* as in

    "The number of apples is twice the number of oranges."

3. Finally, translate the reworded information directly into symbols, using your statements of what the unknowns represent, as in "x is twice y," or

    $x = 2y.$

In the exercise set, you will find a group of exercises to practice this process.

3.1 EXERCISES

▼ more advanced ◆ challenging
T indicates exercises that should be solved using technology

In Exercises 1–6, find three different particular solutions of the given equation and also its general solution in two forms (if possible): parameterized by x and parameterized by y. [**HINT:** See Example 1 and Quick Examples 2–4.]

1. $2x - y = 1$ **2.** $x + 3y = 3$

3. $3x + 4y = 2$ **4.** $4x - 3y = 6$

5. $4x = -5$ **6.** $-3y = 2$

In Exercises 7–20, find all solutions of the given system of equations, and check your answer graphically. [**HINT:** See Examples 2–5.]

7. $x - y = 0$
$x + y = 4$

8. $x - y = 0$
$x + y = -6$

9. $x + y = 4$
$x - y = 2$

10. $2x + y = 2$
$-2x + y = 2$

11. $3x - 2y = 6$
$2x - 3y = -6$

12. $2x + 3y = 5$
$3x + 2y = 5$

[**HINT:** In Exercises 13–16, first eliminate all fractions and decimals; see Example 3.]

13. $0.5x + 0.1y = 0.7$
$0.2x - 0.2y = 0.6$

14. $-0.3x + 0.5y = 0.1$
$0.1x - 0.1y = 0.4$

15. $\dfrac{x}{3} - \dfrac{y}{2} = 1$

$\dfrac{x}{4} + y = -2$

16. $-\dfrac{2x}{3} + \dfrac{y}{2} = -\dfrac{1}{6}$

$\dfrac{x}{4} - y = -\dfrac{3}{4}$

17. $2x + 3y = 1$

$-x - \dfrac{3y}{2} = -\dfrac{1}{2}$

18. $2x - 3y = 1$

$6x - 9y = 3$

19. $2x + 3y = 2$

$-x - \dfrac{3y}{2} = -\dfrac{1}{2}$

20. $2x - 3y = 2$

$6x - 9y = 3$

T *In Exercises 21–30, use technology to obtain approximate solutions graphically. All solutions should be accurate to one decimal place. (Zoom in for improved accuracy.)*

21. $2x + 8y = 10$
$x + y = 5$

22. $2x - y = 3$
$x + 3y = 5$

23. $3.1x - 4.5y = 6$
$4.5x + 1.1y = 0$

24. $0.2x + 4.5y = 1$
$1.5x + 1.1y = 2$

25. $10.2x + 14y = 213$
$4.5x + 1.1y = 448$

26. $100x + 4.5y = 540$
$1.05x + 1.1y = 0$

27. ▼ Find the intersection of the line through $(0, 1)$ and $(4.2, 2)$ and the line through $(2.1, 3)$ and $(5.2, 0)$.

28. ▼ Find the intersection of the line through $(2.1, 3)$ and $(4, 2)$ and the line through $(3.2, 2)$ and $(5.1, 3)$.

29. ▼ Find the intersection of the line through $(0, 0)$ and $(5.5, 3)$ and the line through $(5, 0)$ and $(0, 6)$.

30. ▼ Find the intersection of the line through $(4.3, 0)$ and $(0, 5)$ and the line through $(2.1, 2.2)$ and $(5.2, 1)$.

In Exercises 31–38, translate the given statement into one or more linear equations in the form $ax + by = c$ using the indicated variable names. Do not try to solve the resulting equation(s). [**HINT:** See Example 7 and the end of section FAQ.]

31. There are twice as many soccer fans (x) as football fans (y).

32. The number of hockey players (x) is 90% of the number of lacrosse players (y).

33. The number of new clients (x) is 110% of the number of old clients (y).

34. There are half as many bondholders (x) as stockholders (y).

35. There are three times as many gas giants (x) as rocky planets (y) among a total of 12 gas giants and rocky planets in System X12.

36. Among the total of 25 planets in System L5, four times as many support some form of life (x) as do not (y).

37. Of your portfolio consisting of ordinary shares (x) and preferred shares (y), 20% of the total are preferred shares, and there are 15 more ordinary shares than preferred shares.

38. Of your customer base, consisting of paid-up customers (x) and customers who still owe money (y), 75% of your customer base is paid up, the result being that there are 14,000 more paid-up customers than customers who owe money.

Applications

39. *Resource Allocation* You manage an ice cream factory that makes two flavors: Creamy Vanilla and Continental Mocha. Into each quart of Creamy Vanilla go 2 eggs and 3 cups of cream. Into each quart of Continental Mocha go 1 egg and 3 cups of cream. You have in stock 500 eggs and 900 cups of cream. How many quarts of each flavor should you make in order to use up all the eggs and cream? [**HINT:** See Example 6.]

40. *Class Scheduling* *Enormous State University*'s Math Department offers two courses: Finite Math and Applied Calculus. Each section of Finite Math has 60 students, and each section of Applied Calculus has 50 students. The department will offer a total of 110 sections in a semester, and 6,000 students would like to take a math course. How many sections of each course should the department offer in order to fill all sections and accommodate all of the students? [**HINT:** See Example 6.]

41. *Nutrition* **Gerber Products**' Gerber Mixed Cereal for Baby contains, in each serving, 60 calories and 11 grams of carbohydrates. Gerber Mango Tropical Fruit Dessert contains, in each serving, 80 calories and 21 grams of carbohydrates.[1] If you want to provide your child with 200 calories and 43 grams of carbohydrates, how many servings of each should you use?

---

[1] Source: Nutrition information supplied with the products.

42. *Nutrition* Anthony Altino is mixing food for his young daughter and would like the meal to supply 1 gram of protein and 5 milligrams of iron. He is mixing together cereal, with 0.5 grams of protein and 1 milligram of iron per ounce, and fruit, with 0.2 grams of protein and 2 milligrams of iron per ounce. What mixture will provide the desired nutrition?

43. *Nutrition* One serving of **Campbell Soup Company**'s Campbell's Pork & Beans contains 5 grams of protein and 21 grams of carbohydrates.[2] A typical slice of white bread provides 2 grams of protein and 11 grams of carbohydrates per slice. The U.S. RDA (Recommended Daily Allowance) is 60 grams of protein each day.[3]
   a. I am planning a meal of beans on toast and wish to have it supply one-half of the RDA for protein and 139 grams of carbohydrates. How should I prepare my meal? (Fractions of servings are permitted.)
   b. Is it possible to have my meal supply the same amount of protein as in part (a) but only 100 grams of carbohydrates?

44. *Nutrition* One serving of **Campbell Soup Company**'s Campbell's Pork & Beans contains 5 grams of protein and 21 grams of carbohydrates.[4] A typical slice of "lite" rye bread contains 4 grams of protein and 12 grams of carbohydrates.
   a. I am planning a meal of beans on toast and wish to have it supply one-third of the U.S. RDA for protein (see Exercise 43) and 80 grams of carbohydrates. How should I prepare my meal? (Fractions of servings are permitted.)
   b. Is it possible to have my meal supply the same amount of protein as in part (a) but only 60 grams of carbohydrates?

Protein Supplements *Exercises 45–48 are based on the following data on three popular protein supplements. (Figures shown correspond to a single serving.)*[5]

	Protein (g)	Carbohydrates (g)	Sodium (mg)	Cost ($)
Designer Whey (**Next**)	18	2	80	0.50
Muscle Milk (**Cytosport**)	32	16	240	1.60
Pure Whey Protein Stack (**Champion**)	24	3	100	0.60

45. You are thinking of combining Designer Whey and Muscle Milk to obtain a 7-day supply that provides exactly 280 grams of protein and 56 grams of carbohydrates. How many servings of each supplement should you combine in order to meet your requirements? What will it cost?

46. You are thinking of combining Muscle Milk and Pure Whey Protein Stack to obtain a supply that provides exactly 640 grams of protein and 3,200 milligrams of sodium. How many servings of each supplement should you combine in order to meet your requirements? What will it cost?

47. ▼ You have a mixture of Designer Whey and Pure Whey Protein Stack that costs a total of $14 and supplies exactly 540 grams of protein. How many grams of carbohydrates does it supply?

48. ▼ You have a mixture of Designer Whey and Muscle Milk that costs a total of $14 and supplies exactly 104 grams of carbohydrates. How many grams of protein does it supply?

49. *Investments: Tech Stocks* In December 2014, **Twitter** (TWTR) stock decreased from $40 to $36 per share, and **Microsoft** (MSFT) stock decreased from $48 to $45 per share.[6] If you invested a total of $22,400 in these stocks at the beginning of the month and sold them for $20,700 at the end of the month, how many shares of each stock did you buy?

50. *Investments: Energy Stocks* In the 3-month period November 1, 2014, through January 31, 2015, **Hess Corp.** (HES) stock decreased from $80 to $64 per share, and **Exxon Mobil** (XOM) stock decreased from $96 to $80 per share.[7] If you invested a total of $21,600 in these stocks at the beginning of November and sold them for $17,600 3 months later, how many shares of each stock did you buy?

51. ▼ *Investments: Financial Stocks* During the first quarter of 2015, **Toronto Dominion Bank** (TD) stock cost $45 per share and was expected to yield 4% per year in dividends, while **CNA Financial Corp.** (CNA) stock cost $40 per share and was expected to yield 2.5% per year in dividends.[8] If you invested a total of $25,000 in these stocks and expected to earn $760 in dividends in a year, how many shares of each stock did you purchase?

52. ▼ *Investments: High Dividend Stocks* During the first quarter of 2015, **Plains All American Pipeline L.P.** (PAA) stock cost $50 per share and was expected to yield 5% per year in dividends, while **Total SA** (TOT) stock cost $50 per share and was expected to yield 6% per year in dividends.[9] If you invested a total of $45,000 in these stocks and expected to earn $2,400 in dividends in a year, how many shares of each stock did you purchase?

---

[2] According to the label information on a 16-ounce can.

[3] Recommended Daily Allowance for a person weighing 75 kilograms (165 pounds)

[4] According to the label information on a 16-ounce can.

[5] Source: Nutritional information supplied by the manufacturers (www.netrition.com). Cost per serving is approximate and varies.

[6] Approximate stock prices at or close to the dates cited. Source: http://finance.google.com.

[7] *Ibid.*

[8] Stock prices and yields are approximate. Source: http://finance.google.com.

[9] *Ibid.*

53. *Voting* An appropriations bill passed the U.S. House of Representatives with 49 more members voting in favor than against. If all 435 members of the House voted either for or against the bill, how many voted in favor and how many voted against?

54. *Voting* The U.S. Senate has 100 members. For a bill to pass with a supermajority, at least twice as many senators must vote in favor of the bill as vote against it. If each of the 100 senators votes either in favor of or against a bill, how many must vote in favor for it to pass with a supermajority?

55. *Intramural Sports* The best sports dorm on campus, Lombardi House, has won a total of 12 games this semester. Some of these games were soccer games, and the others were football games. According to the rules of the university, each win in a soccer game earns the winning house 2 points, whereas each win in a football game earns the house 4 points. If the total number of points Lombardi House earned was 38, how many of each type of game did it win?

56. *Law* Five years ago, *Enormous State University*'s campus publication, *The Campus Inquirer*, ran a total of 10 exposés dealing with alleged recruiting violations by the football team and with theft by the student treasurer of the film society. Each exposé dealing with recruiting violations resulted in a $4 million libel suit, and the treasurer of the film society sued the paper for $3 million as a result of each exposé about his alleged theft. Unfortunately for *The Campus Inquirer*, all the lawsuits were successful, and the paper wound up being ordered to pay $37 million in damages. (It closed down shortly thereafter.) How many of each type of exposé did the paper run?

57. *Purchasing* (*from the GMAT*) Elena purchased Brand X pens for $4.00 apiece and Brand Y pens for $2.80 apiece. If Elena purchased a total of 12 of these pens for $42.00, how many Brand X pens did she purchase?

58. *Purchasing* (*based on a question from the GMAT*) Earl is ordering supplies. Yellow paper costs $5.00 per ream, while white paper costs $6.50 per ream. He would like to order 100 reams total and has a budget of $560. How many reams of each color should he order?

59. *Equilibrium Price* The demand and supply functions for pet chias are $q = -60p + 150$ and $q = 80p - 60$, respectively, where p is the price in dollars. At what price should the chias be marked so that there is neither a surplus nor a shortage of chias? [**HINT:** See Example 8.]

60. *Equilibrium Price* The demand and supply functions for your college newspaper are $q = -10,000p + 2,000$ and $q = 4,000p + 600$, respectively, where p is the price in dollars. At what price should the newspapers be sold so that there is neither a surplus nor a shortage of papers? [**HINT:** See Example 8.]

61. *Supply and Demand* (*from the GRE Economics Test*) The demand curve for widgets is given by $D = 85 - 5P$, and the supply curve is given by $S = 25 + 5P$, where P is the

price of widgets. When the widget market is in equilibrium, what is the quantity of widgets bought and sold?

62. *Supply and Demand* (*from the GRE Economics Test*) In the market for soybeans the demand and supply functions are $Q_D = 100 - 10P$ and $Q_S = 20 + 5P$, where Q_D is quantity demanded, Q_S is quantity supplied, and P is price in dollars. If the government sets a price floor of $7, what will be the resulting surplus or shortage?

63. *Equilibrium Price* In June 2001 the retail price of a 25-kilogram bag of cornmeal was $8 in Zambia; by December the price had risen to $11. The result was that one retailer reported a drop in sales from 15 bags per day to 3 bags per day.[10] Assume that the retailer is prepared to sell 3 bags per day at $8 and 15 bags per day at $11. Find linear demand and supply equations, and then compute the retailer's equilibrium price.

64. *Equilibrium Price* At the start of December 2001 the retail price of a 25-kilogram bag of cornmeal was $10 in Zambia; by the end of the month the price had fallen to $6.[11] The result was that one retailer reported an increase in sales from 3 bags per day to 5 bags per day. Assume that the retailer is prepared to sell 18 bags per day at $8 and 12 bags per day at $6. Obtain linear demand and supply equations, and hence compute the retailer's equilibrium price.

65. *Pollution* Joe Slo, a college sophomore, neglected to wash his dirty laundry for 6 weeks. By the end of that time, his roommate had had enough and tossed Joe's dirty socks and T-shirts into the trash, counting a total of 44 items. (A pair of dirty socks counts as one item.) The roommate noticed that there were three times as many pairs of dirty socks as T-shirts. How many of each item did he throw out?

66. *Diet* The local sushi bar serves 1-ounce pieces of raw salmon (consisting of 50% protein) and $1\frac{1}{4}$-ounce pieces of raw tuna (40% protein). A customer's total intake of protein amounts to $1\frac{1}{2}$ ounces after consuming a total of three pieces. How many of each type did the customer consume? (Fractions of pieces are permitted.)

67. ▼ *Management* (*from the GMAT*) A manager has $6,000 budgeted for raises for four full-time and two part-time employees. Each of the full-time employees receives the same raise, which is twice the raise that each of the part-time employees receives. What is the amount of the raise that each full-time employee receives?

68. ▼ *Publishing* (*from the GMAT*) There were 36,000 hardback copies of a certain novel sold before the paperback version was issued. From the time the first paperback copy was sold until the last copy of the novel was sold, nine times as many paperback copies as hardback copies were

---

[10] The prices quoted are approximate. (Actual prices varied from retailer to retailer.) Source: *New York Times*, December 24, 2001, p. A4.

[11] *Ibid.*

sold. If a total of 441,000 copies of the novel were sold in all, how many paperback copies were sold?

Communication and Reasoning Exercises

69. A system of three equations in two unknowns corresponds to three lines in the plane. Describe how these lines might be positioned if the system has a unique solution.

70. A system of three equations in two unknowns corresponds to three lines in the plane. Describe several ways in which these lines might be positioned if the system has no solutions.

71. Both the supply and demand equations for a certain product have negative slope. Can there be an equilibrium price? Explain.

72. You are solving a system of equations with x representing the number of rocks and y representing the number of pebbles. The solution is $(200, -10)$. What do you conclude?

73. ▼ Referring to Exercise 39, but with different given data, suppose that the solution of the corresponding system of equations was 198.7 quarts of vanilla and 100.89 quarts of mocha. If your factory can produce only whole numbers of quarts, would you recommend rounding the answers to the nearest whole number? Explain.

74. ▼ Referring to Exercise 39, but using different data, suppose that the general solution of the corresponding system of equations was $(200 - y, y)$, where $x =$ number of quarts of vanilla and $y =$ number of quarts of mocha. Your factory can produce only whole numbers of quarts. There are infinitely many combinations of vanilla and mocha that satisfy the constraints—right? Explain.

75. ▼ Select one: Multiplying both sides of a linear equation by a nonzero constant results in a linear equation whose graph is
(A) parallel to   **(B)** the same as
(C) not always parallel to   **(D)** not the same as
the graph of the original equation.

76. ▼ Select one: If the addition or subtraction of two linear equations results in the equation $3 = 3$, then the graphs of those equations are
(A) equal.   **(B)** parallel.
(C) perpendicular.   **(D)** none of the above.

77. ▼ Select one: If the addition or subtraction of two linear equations results in the equation $0 = 3$, then the graphs of those equations are
(A) equal.   **(B)** parallel.
(C) perpendicular.   **(D)** not parallel.

78. ▼ Select one: If adding two linear equations gives $x = 3$ and subtracting them gives $y = 3$, then the graphs of those equations are
(A) equal.   **(B)** parallel.
(C) perpendicular.   **(D)** not parallel.

79. ▼ Invent an interesting application that leads to a system of two equations in two unknowns with a unique solution.

80. ▼ Invent an interesting application that leads to a system of two equations in two unknowns with no solution.

81. ◆ How likely do you think it is that a "random" system of two equations in two unknowns has a unique solution? Give some justification for your answer.

82. ◆ How likely do you think it is that a "random" system of three equations in two unknowns has a unique solution? Give some justification for your answer.

3.2   Using Matrices to Solve Systems of Equations

The Augmented Matrix of a System of Linear Equations

In this section we describe a systematic method for solving systems of equations that makes solving large systems of equations in any number of unknowns straightforward. Although this method may seem a little cumbersome at first, it will prove *immensely* useful in this and the next several chapters. First, we introduce some terminology.

Linear Equation

A linear equation in the n variables $x_1, x_2, \ldots, x_n$ has the form

$$a_1 x_1 + \cdots + a_n x_n = b \qquad (a_1, a_2, \ldots, a_n, b \text{ constants}).$$

The numbers $a_1, a_2, \ldots, a_n$ are called the **coefficients**, and the number b is called the **constant term**, or **right-hand side**.

Quick Examples

1. $3x - 5y = 0$

Linear equation in x and y
Coefficients: 3, -5; constant term: 0

2. $x + 2y - z = 6$

Linear equation in x, y, z
Coefficients: 1, 2, -1; constant term: 6

3. $30x_1 + 18x_2 + x_3 + x_4 = 19$

Linear equation in x_1, x_2, x_3, x_4
Coefficients: 30, 18, 1, 1; Constant term: 19

Note When the number of variables is small, we will almost always use $x, y, z, \ldots$ (as in Quick Examples 1 and 2) rather than $x_1, x_2, x_3, \ldots$ as the names of the variables. ■

Notice that a linear equation in any number of unknowns (for example, $2x - y = 3$) is entirely determined by its coefficients and its constant term. In other words, if we were simply given the row of numbers

$$[2 \quad -1 \quad 3],$$

we could easily reconstruct the original linear equation by multiplying the first number by x, multiplying the second by y, and inserting a plus sign and an equals sign, as follows:

$$2 \cdot x + (-1) \cdot y = 3$$

or $2x - y = 3.$

Similarly, the equation

$$-4x + 2y = 0$$

is represented by the row

$$[-4 \quad 2 \quad 0],$$

and the equation

$$-3y = \frac{1}{4}$$

is represented by the row

$$\left[0 \quad -3 \quad \frac{1}{4} \right].$$

As the last example shows, the first number is always the coefficient of x, and the second is the coefficient of y. If an x or a y is missing, we write a zero for its coefficient. We shall call such a row the **coefficient row** of an equation.

If we have a system of equations, for example, the system

$$2x - y = 3$$
$$-x + 2y = -4,$$

we can put the coefficient rows together like this:

$$\begin{bmatrix} 2 & -1 & 3 \\ -1 & 2 & -4 \end{bmatrix}.$$

We call this the **augmented matrix** of the system of equations. The term "augmented" means that we have included the right-hand sides 3 and -4. We will often drop the word "augmented" and simply refer to the matrix of the system. A **matrix** (plural: **matrices**) is nothing more than a rectangular array of numbers, as above.

Matrix, Augmented Matrix

A **matrix** is a rectangular array of numbers. The **augmented matrix** of a system of linear equations is the matrix whose rows are the coefficient rows of the equations.

Quick Example

4. The augmented matrix of the system

$$x + y = 3$$
$$x - y = 1$$

is $\begin{bmatrix} 1 & 1 & 3 \\ 1 & -1 & 1 \end{bmatrix}$.

We'll be studying matrices in more detail in Chapter 4.

Q: *What good are coefficient rows and matrices?*

A: Think about what we do when we multiply both sides of an equation by a number. For example, consider multiplying both sides of the equation $2x - y = 3$ by -2 to get $-4x + 2y = -6$. All we are really doing is multiplying the coefficients and the right-hand side by -2. This corresponds to *multiplying the row* $\begin{bmatrix} 2 & -1 & 3 \end{bmatrix}$ *by* -2, that is, multiplying every number in the row by -2. We shall see that any manipulation we want to do with equations can be done instead with rows. This fact leads to a method of solving equations that is systematic and generalizes easily to larger systems.

Here is the same operation in both the language of equations and the language of rows. (We refer to the equation here as *Equation* 1, or simply E_1 for short, and to the row as *Row* 1, or R_1.)

Equation	Row
E_1: $\quad 2x - y = 3$	$\begin{bmatrix} 2 & -1 & 3 \end{bmatrix}$ R_1
Multiply by -2: $\quad (-2)E_1$: $-4x + 2y = -6$	$\begin{bmatrix} -4 & 2 & -6 \end{bmatrix}$ $(-2)R_1$

Multiplying both sides of an equation by the number a corresponds to multiplying the coefficient row by a.

Now look at what we do when we add two equations.

Equation	Row
E_1: $2x - y = 3$	$\begin{bmatrix} 2 & -1 & 3 \end{bmatrix}$ R_1
E_2: $-x + 2y = -4$	$\begin{bmatrix} -1 & 2 & -4 \end{bmatrix}$ R_2
Add: $\quad E_1 + E_2$: $\quad x + y = -1$	$\begin{bmatrix} 1 & 1 & -1 \end{bmatrix}$ $R_1 + R_2$

All we are really doing is *adding the corresponding entries in the rows*, or *adding the rows*. In other words,

Adding two equations corresponds to adding their coefficient rows.

In short, the manipulations of equations that we saw in Section 3.1 can be done more easily with rows in a matrix because we don't have to carry x, y, and other unnecessary notation along with us; x and y can always be inserted at the end if desired.

The manipulations we are talking about are known as **row operations**. In particular, we use three **elementary row operations**.

Elementary Row Operations*

Type 1: Replacing R_i by aR_i (where $a \neq 0$)†
In words: multiplying or dividing a row by a nonzero number.

Type 2: Replacing R_i by $aR_i \pm bR_j$ (where $a \neq 0$)
In words: multiplying a row by a nonzero number and adding or subtracting a multiple of another row.

Type 3: Switching the order of the rows
This corresponds to switching the order in which we write the equations; occasionally, this will be convenient.

For Types 1 and 2 we write the instruction for the row operation *next to the row we wish to replace.* (See the Quick Examples below.)

Quick Examples

5. Type 1: $\begin{bmatrix} 1 & 3 & -4 \\ 0 & 4 & 2 \end{bmatrix} 3R_2 \rightarrow \begin{bmatrix} 1 & 3 & -4 \\ 0 & 12 & 6 \end{bmatrix}$ Replace R_2 by $3R_2$.

6. Type 2: $\begin{bmatrix} 1 & 3 & -4 \\ 0 & 4 & 2 \end{bmatrix} 4R_1 - 3R_2 \rightarrow \begin{bmatrix} 4 & 0 & -22 \\ 0 & 4 & 2 \end{bmatrix}$ Replace R_1 by $4R_1 - 3R_2$.

7. Type 3: $\begin{bmatrix} 1 & 3 & -4 \\ 0 & 4 & 2 \\ 1 & 2 & 3 \end{bmatrix} R_1 \leftrightarrow R_2 \rightarrow \begin{bmatrix} 0 & 4 & 2 \\ 1 & 3 & -4 \\ 1 & 2 & 3 \end{bmatrix}$ Switch R_1 and R_2.

One very important fact about the elementary row operations is that they do not change the solutions of the corresponding system of equations. In other words, the new system of equations that we get by applying any one of these operations will have exactly the same solutions as the original system: It is easy to see that numbers that make the original equations true will also make the new equations true, because each of the elementary row operations corresponds to a valid operation on the original equations. That any solution of the new system is a solution of the old system follows from the fact that these row operations are *invertible:* The effects of a row operation can be reversed by applying another row operation, called its **inverse**. Here are some examples of this invertibility. (Try them out in Quick Examples 5–7.)

Operation	Inverse Operation
Replace R_2 by $3R_2$.	Replace R_2 by $\frac{1}{3}R_2$.
Replace R_1 by $4R_1 - 3R_2$.	Replace R_1 by $\frac{1}{4}R_1 + \frac{3}{4}R_2$.
Switch R_1 and R_2.	Switch R_1 and R_2.

Our objective, then, is to use row operations to change the system we are given into one with exactly the same set of solutions in which it is easy to see what the solutions are.

---

*We are using the term "elementary row operations" a little more freely than most books do. Some mathematicians insist that $a = 1$ in an operation of Type 2, but the less restrictive version is very useful.

†Multiplying an equation or row by zero gives us the not very surprising result $0 = 0$. In fact, we lose any information that the equation provided, which usually means that the resulting system has more solutions than the original system.

Using Technology
See the Technology Guides at the end of the chapter to see how to do row operations using a TI-83/84 Plus or a spreadsheet.

Website
www.WanerMath.com
→ Online Utilities
→ Pivot and Gauss-Jordan Tool
Enter the matrix in columns $x1, x2, x3, \ldots$. To do a row operation, type the instruction(s) next to the row(s) you are changing as shown, and press "Do Row Ops" once.

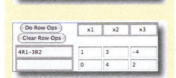

Solving Systems of Equations by Using Row Operations

Now we put row operations to work for us in solving systems of equations. Let's start with a complicated-looking system of equations:

$$-\frac{2x}{3} + \frac{y}{2} = -3$$

$$\frac{x}{4} - y = \frac{11}{4}.$$

We begin by writing the matrix of the system:

$$\begin{bmatrix} -\frac{2}{3} & \frac{1}{2} & -3 \\ \frac{1}{4} & -1 & \frac{11}{4} \end{bmatrix}.$$

Now what do we do with this matrix?

Step 1 *Clear the fractions and/or decimals (if any) using operations of Type 1.* To clear the fractions, we multiply the first row by 6 and the second row by 4. We record the operations by writing the symbolic form of an operation next to the row it will change, as follows:

$$\begin{bmatrix} -\frac{2}{3} & \frac{1}{2} & -3 \\ \frac{1}{4} & -1 & \frac{11}{4} \end{bmatrix} \begin{matrix} 6R_1 \\ 4R_2 \end{matrix}.$$

By this we mean that we will replace the first row by $6R_1$ and the second by $4R_2$. Doing these operations gives

$$\begin{bmatrix} -4 & 3 & -18 \\ 1 & -4 & 11 \end{bmatrix}.$$

Step 2 *Designate the first nonzero entry in the first row as the* **pivot.** In this case we designate the entry -4 in the first row as the "pivot" by putting a box around it:

$$\begin{bmatrix} \boxed{-4} & 3 & -18 \\ 1 & -4 & 11 \end{bmatrix}. \quad \leftarrow \text{Pivot row}$$

$\uparrow$
Pivot column

Q: *What is a "pivot"?*

A: A **pivot** is an entry in a matrix that is used to "clear a column." (See Step 3.) In this procedure we will always select the first nonzero entry of a row as our pivot. In Chapter 5, when we study the simplex method, we will select our pivots differently.

Step 3 *Use the pivot to clear its column using operations of Type 2.* By **clearing a column**, we mean changing the matrix so that the pivot is the only nonzero number in its column. The procedure of clearing a column using a designated pivot is also called **pivoting**.

$$\begin{bmatrix} \boxed{-4} & 3 & -18 \\ 0 & \# & \# \end{bmatrix} \quad \leftarrow \text{Desired row 2 (the "#"s stand for as yet unknown numbers)}$$

$\uparrow$
Cleared pivot column

We want to replace R_2 by a row of the form $aR_2 \pm bR_1$ to get a zero in column 1. Moreover—and this will be important when we discuss the simplex method in Chapter 5—*we are going to choose positive values for both a and b.** We need to choose a and b so that we get the desired cancellation. We can do this quite mechanically as follows:

* Thus, the only place a negative sign may appear is between aR_2 and bR_1 as indicated in the formula $aR_2 \pm bR_1$.

a. Write the name of the row you need to change on the left and that of the pivot row on the right:

$$R_2 \qquad R_1$$
$$\uparrow \qquad \uparrow$$
Row to change   Pivot row

b. Focus on the pivot column, $\begin{bmatrix} -4 \\ 1 \end{bmatrix}$. Multiply each row by the *absolute value* of the entry currently in the other. (We are not permitting a or b to be negative.)

$$4R_2 \qquad\qquad 1R_1$$
$$\uparrow \qquad\qquad \uparrow$$
From Row 1   From Row 2

The effect is to make the two entries in the pivot column numerically the same. Sometimes, you can accomplish this by using smaller values of a and b.

c. If the entries in the pivot column have opposite signs, insert a plus $(+)$. If they have the same sign, insert a minus $(-)$. Here, we get the instruction

$$4R_2 + 1R_1,$$

or simply $4R_2 + R_1$.

d. Write the operation next to the row you want to change, and then replace that row using the operation:

$$\begin{bmatrix} \boxed{-4} & 3 & -18 \\ 1 & -4 & 11 \end{bmatrix} \begin{matrix} \\ 4R_2 + 1R_1 \end{matrix} \rightarrow \begin{bmatrix} -4 & 3 & -18 \\ 0 & -13 & 26 \end{bmatrix}.$$

We have cleared the pivot column and completed Step 3.

† We are deviating somewhat from the traditional procedure here. It is traditionally recommended first to divide the pivot row by the pivot, turning the pivot into a 1. This allows us to use $a = 1$, but it usually results in fractions. The procedure we use here is easier for hand calculations and, we feel, mathematically more elegant because it illustrates what can be done with matrices whose entries are integers and also eliminates the possible need to work with fractions throughout the calculation. See the end of this section for an example done using the traditional procedure.

Note In general, the row operation you use should always have the following form†:

$$aR_c \qquad \pm \qquad bR_p$$
$$\uparrow \qquad\qquad \uparrow$$
Row to change   Pivot row

with a and b both positive. ■

The next step is one that can be performed at any time.

Simplification Step (Optional) *If, at any stage of the process, all the numbers in a row are multiples of an integer larger than* **1,** *divide by that integer*—a Type 1 operation.

This is an optional but extremely helpful step: It makes the numbers smaller and easier to work with. In our case, the entries in R_2 are divisible by 13, so we divide that row by 13. (Alternatively, we could divide by -13. Try it.)

$$\begin{bmatrix} -4 & 3 & -18 \\ 0 & -13 & 26 \end{bmatrix} \begin{matrix} \\ \frac{1}{13}R_2 \end{matrix} \rightarrow \begin{bmatrix} -4 & 3 & -18 \\ 0 & -1 & 2 \end{bmatrix}$$

Step 4 *Select the first nonzero number in the second row as the pivot, and clear its column.* Here we have combined two steps in one: selecting the new pivot and clearing the column (pivoting). The pivot is shown below, as well as the desired result when the column has been cleared:

$$\begin{bmatrix} -4 & 3 & -18 \\ 0 & \boxed{-1} & 2 \end{bmatrix} \rightarrow \begin{bmatrix} \# & 0 & \# \\ 0 & -1 & 2 \end{bmatrix}. \quad \leftarrow \text{Desired row}$$

↑   ↑
Pivot column   Cleared pivot column

We now wish to get a 0 in place of the 3 in the pivot column. Let's run once again through the mechanical steps to get the row operation that accomplishes this.

a. Write the name of the row you need to change on the left and that of the pivot row on the right:

R_1 R_2
↑   ↑
Row to change   Pivot row

b. Focus on the pivot column, $\begin{bmatrix} 3 \\ -1 \end{bmatrix}$. Multiply each row by the absolute value of the entry currently in the other:

$1R_1$ $3R_2$
↑   ↑
From Row 2   From Row 1

c. If the entries in the pivot column have opposite signs, insert a plus $(+)$. If they have the same sign, insert a minus $(-)$. Here, we get the instruction

$$1R_1 + 3R_2.$$

d. Write the operation next to the row you want to change, and then replace that row using the operation:

$$\begin{bmatrix} -4 & 3 & -18 \\ 0 & \boxed{-1} & 2 \end{bmatrix} \begin{matrix} R_1 + 3R_2 \\ \\ \end{matrix} \rightarrow \begin{bmatrix} -4 & 0 & -12 \\ 0 & -1 & 2 \end{bmatrix}.$$

Now we are essentially done, except for one last step.

Final Step *Using operations of Type 1, turn each pivot (the first nonzero entry in each row) into a 1.* We can accomplish this by dividing the first row by -4 and multiplying the second row by -1:

$$\begin{bmatrix} -4 & 0 & -12 \\ 0 & -1 & 2 \end{bmatrix} \begin{matrix} -\frac{1}{4}R_1 \\ -R_2 \end{matrix} \rightarrow \begin{bmatrix} 1 & 0 & 3 \\ 0 & 1 & -2 \end{bmatrix}.$$

The matrix now has the following nice form:

$$\begin{bmatrix} \boxed{1} & 0 & \# \\ 0 & \boxed{1} & \# \end{bmatrix}.$$

(This is the form we will always obtain with two equations in two unknowns when there is a unique solution.) This form is nice because, when we translate back into equations, we get

$$1x + 0y = 3$$
$$0x + 1y = -2.$$

In other words,

$$x = 3 \quad \text{and} \quad y = -2,$$

so we have found the solution, which we can also write as $(x, y) = (3, -2)$.

*Gauss-Jordan reduction is named after Carl Friedrich Gauss (1777–1855) and Wilhelm Jordan (1842–1899). Gauss was one of the great mathematicians, making fundamental contributions to number theory, analysis, probability, and statistics, as well as many fields of science. Gauss also made contributions to a method of solving systems of equations that has become known as Gaussian elimination, even though this method had been described by Isaac Newton in 1707 and, independently, had been known to the Chinese more than 2,000 years ago. (See *Mathematicians of Gaussian Elimination*, Notices of the American Mathematical Society, June/July 2011, for a history of Gaussian elimination.) The method we are showing you here, Gauss-Jordan reduction, is Jordan's variation on Gaussian elimination, first published in 1887.*

The procedure we've just demonstrated is called **Gauss-Jordan reduction*** or **row reduction**. It may seem too complicated a way to solve a system of two equations in two unknowns, and it is. However, for systems with more equations and more unknowns, it is very efficient.

In Example 1 below we use row reduction to solve a system of linear equations in *three* unknowns: x, y, and z. Just as for a system in two unknowns, a **solution** of a system in any number of unknowns consists of values for each of the variables that, when substituted, satisfy all of the equations in the system. Again, just as for a system in two unknowns, any system of linear equations in any number of unknowns has either no solution, exactly one solution, or infinitely many solutions. There are no other possibilities.

Solving a system in three unknowns graphically would require the graphing of planes (flat surfaces) in three dimensions. (The graph of a linear equation in three unknowns is a flat surface.) The use of row reduction makes three-dimensional graphing unnecessary.

EXAMPLE 1 **Solving a System by Gauss-Jordan Reduction**

Solve the system

$$\begin{aligned} x - y + 5z &= -6 \\ 3x + 3y - z &= 10 \\ x + 3y + 2z &= 5. \end{aligned}$$

Solution The augmented matrix for this system is

$$\begin{bmatrix} 1 & -1 & 5 & -6 \\ 3 & 3 & -1 & 10 \\ 1 & 3 & 2 & 5 \end{bmatrix}.$$

Note that the columns correspond to x, y, z, and the right-hand side, respectively. We begin by selecting the pivot in the first row and clearing its column. Remember that clearing the column means that we turn *all* other numbers in the column into zeros. Thus, to clear the column of the first pivot, we need to change two rows, setting up the row operations in exactly the same way as above:

$$\begin{bmatrix} \boxed{1} & -1 & 5 & -6 \\ 3 & 3 & -1 & 10 \\ 1 & 3 & 2 & 5 \end{bmatrix} \begin{matrix} \\ R_2 - 3R_1 \\ R_3 - R_1 \end{matrix} \rightarrow \begin{bmatrix} 1 & -1 & 5 & -6 \\ 0 & 6 & -16 & 28 \\ 0 & 4 & -3 & 11 \end{bmatrix}.$$

Notice that both row operations have the required form

$$aR_c \pm bR_1$$

↑   ↑
Row to change   Pivot row

with a and b both positive.

Now we use the optional simplification step to simplify R_2:

$$\begin{bmatrix} 1 & -1 & 5 & -6 \\ 0 & 6 & -16 & 28 \\ 0 & 4 & -3 & 11 \end{bmatrix} \begin{matrix} \\ \frac{1}{2}R_2 \\ \\ \end{matrix} \rightarrow \begin{bmatrix} 1 & -1 & 5 & -6 \\ 0 & 3 & -8 & 14 \\ 0 & 4 & -3 & 11 \end{bmatrix}.$$

Next, we select the pivot in the second row and clear its column:

$$\begin{bmatrix} 1 & -1 & 5 & -6 \\ 0 & \boxed{3} & -8 & 14 \\ 0 & 4 & -3 & 11 \end{bmatrix} \begin{matrix} 3R_1 + R_2 \\ \\ 3R_3 - 4R_2 \end{matrix} \rightarrow \begin{bmatrix} 3 & 0 & 7 & -4 \\ 0 & 3 & -8 & 14 \\ 0 & 0 & 23 & -23 \end{bmatrix}.$$

R_1 and R_3 are to be changed.
R_2 is the pivot row.

Using Technology
See the Technology Guides at the end of the chapter to see how to use a TI-83/84 Plus or a spreadsheet to solve this system of equations.

W **Website**
www.WanerMath.com
→ Online Utilities
→ Pivot and Gauss-Jordan Tool

Enter the augmented matrix in columns *x1*, *x2*, *x3*, ... At each step, type in the row operations exactly as written above next to the rows to which they apply. For example, for the first step, type:
R2 - 3R1 next to Row 2 and
R3 - R1 next to Row 3.
Press "Do Row Ops" once, and then press "Clear Row Ops" to prepare for the next step.
For the last step, type
(1/3) R1 next to Row 1 and
(1/3) R2 next to Row 2.
Press "Do Row Ops" once.
The utility also does other things, such as automatic pivoting and complete reduction in one step. Use these features to check your work.

We simplify R_3:

$$\begin{bmatrix} 3 & 0 & 7 & -4 \\ 0 & 3 & -8 & 14 \\ 0 & 0 & 23 & -23 \end{bmatrix} \begin{matrix} \\ \\ \frac{1}{23}R_3 \end{matrix} \rightarrow \begin{bmatrix} 3 & 0 & 7 & -4 \\ 0 & 3 & -8 & 14 \\ 0 & 0 & 1 & -1 \end{bmatrix}.$$

Now we select the pivot in the third row and clear its column:

$$\begin{bmatrix} 3 & 0 & 7 & -4 \\ 0 & 3 & -8 & 14 \\ 0 & 0 & \boxed{1} & -1 \end{bmatrix} \begin{matrix} R_1 - 7R_3 \\ R_2 + 8R_3 \\ \end{matrix} \rightarrow \begin{bmatrix} 3 & 0 & 0 & 3 \\ 0 & 3 & 0 & 6 \\ 0 & 0 & 1 & -1 \end{bmatrix}.$$

R_1 and R_2 are to be changed.
R_3 is the pivot row.

Finally, we turn all the pivots into 1s:

$$\begin{bmatrix} 3 & 0 & 0 & 3 \\ 0 & 3 & 0 & 6 \\ 0 & 0 & 1 & -1 \end{bmatrix} \begin{matrix} \frac{1}{3}R_1 \\ \frac{1}{3}R_2 \\ \end{matrix} \rightarrow \begin{bmatrix} 1 & 0 & 0 & 1 \\ 0 & 1 & 0 & 2 \\ 0 & 0 & 1 & -1 \end{bmatrix}.$$

The matrix is now reduced to a simple form, so we translate back into equations to obtain the solution:

$$x = 1, \quad y = 2, \quad z = -1, \quad \text{or} \quad (x, y, z) = (1, 2, -1).$$

Notice the form of the very last matrix in the example:

$$\begin{bmatrix} 1 & 0 & 0 & \# \\ 0 & 1 & 0 & \# \\ 0 & 0 & 1 & \# \end{bmatrix}.$$

The 1s are on the **(main) diagonal** of the matrix; the goal in Gauss-Jordan reduction is to reduce our matrix to this form. If we can do so, then we can easily read off the solution, as we saw in Example 1. However, as we will see in several examples in this section, it is not always possible to achieve this ideal state. After Example 6 we will give a form that is always possible to achieve.

EXAMPLE 2 **Solving a System by Gauss-Jordan Reduction**

Solve the system:

$$\begin{aligned} 2x + \ y + 3z &= 1 \\ 4x + 2y + 4z &= 4 \\ x + 2y + \ z &= 4. \end{aligned}$$

Solution

$$
\begin{bmatrix}
\boxed{2} & 1 & 3 & 1 \\
4 & 2 & 4 & 4 \\
1 & 2 & 1 & 4
\end{bmatrix}
\begin{array}{l} \\ R_2 - 2R_1 \to \\ 2R_3 - R_1 \end{array}
\begin{bmatrix}
2 & 1 & 3 & 1 \\
0 & 0 & -2 & 2 \\
0 & 3 & -1 & 7
\end{bmatrix}
$$

Now we have a slight problem: The number in the position where we would like to have a pivot—the second column of the second row—is a zero and thus cannot be a pivot. There are two ways out of this problem. One is to move on to the third column and pivot on the -2. Another is to switch the order of the second and third rows so that we can use the 3 as a pivot. We will do the latter:

$$
\begin{bmatrix}
2 & 1 & 3 & 1 \\
0 & 0 & -2 & 2 \\
0 & 3 & -1 & 7
\end{bmatrix}
R_2 \leftrightarrow R_3 \to
\begin{bmatrix}
2 & 1 & 3 & 1 \\
0 & \boxed{3} & -1 & 7 \\
0 & 0 & -2 & 2
\end{bmatrix}
\begin{array}{l} 3R_1 - R_2 \\ \\ \end{array}
$$

$$
\to
\begin{bmatrix}
6 & 0 & 10 & -4 \\
0 & 3 & -1 & 7 \\
0 & 0 & -2 & 2
\end{bmatrix}
\begin{array}{l} \frac{1}{2}R_1 \\ \\ -\frac{1}{2}R_3 \end{array}
\to
\begin{bmatrix}
3 & 0 & 5 & -2 \\
0 & 3 & -1 & 7 \\
0 & 0 & \boxed{1} & -1
\end{bmatrix}
\begin{array}{l} R_1 - 5R_3 \\ R_2 + R_3 \\ \end{array}
$$

$$
\to
\begin{bmatrix}
3 & 0 & 0 & 3 \\
0 & 3 & 0 & 6 \\
0 & 0 & 1 & -1
\end{bmatrix}
\begin{array}{l} \frac{1}{3}R_1 \\ \frac{1}{3}R_2 \to \\ \end{array}
\begin{bmatrix}
1 & 0 & 0 & 1 \\
0 & 1 & 0 & 2 \\
0 & 0 & 1 & -1
\end{bmatrix}.
$$

Thus, the solution is $(x, y, z) = (1, 2, -1)$, as you can check in the original system.

No Solutions and Infinitely Many Solutions

As in Section 3.1, a system of linear equations with three or more unknowns may have no solution or may have infinitely many solutions, as we see in the next two examples.

EXAMPLE 3   **Inconsistent System**

Solve the system:

$$
\begin{aligned}
x + y + z &= 1 \\
2x - y + z &= 0 \\
4x + y + 3z &= 3.
\end{aligned}
$$

Solution

$$
\begin{bmatrix}
\boxed{1} & 1 & 1 & 1 \\
2 & -1 & 1 & 0 \\
4 & 1 & 3 & 3
\end{bmatrix}
\begin{array}{l} \\ R_2 - 2R_1 \to \\ R_3 - 4R_1 \end{array}
\begin{bmatrix}
1 & 1 & 1 & 1 \\
0 & \boxed{-3} & -1 & -2 \\
0 & -3 & -1 & -1
\end{bmatrix}
\begin{array}{l} 3R_1 + R \\ \\ R_3 - R_2 \end{array}
$$

$$
\to
\begin{bmatrix}
3 & 0 & 2 & 1 \\
0 & -3 & -1 & -2 \\
0 & 0 & 0 & 1
\end{bmatrix}
$$

Stop. That last row translates into $0 = 1$, which is nonsense, so, as in Example 4 in Section 3.1, we can say that this system has no solution. We also say, as we did for systems with only two unknowns, that a system with no solution is **inconsistent**. A system with at least one solution is **consistent**.

➡ **Before we go on . . .**

Q : *How, exactly, does the nonsensical equation* $0 = 1$ *tell us that there is no solution of the system in Example 3?*

A : Here is an argument similar to that in Example 4 in Section 3.1: If there *were* three numbers x, y, and z satisfying the original system of equations, then manipulating the equations according to the instructions in the row operations above would lead us to conclude that $0 = 1$. Because 0 is *not* equal to 1, there can be no such numbers x, y, and z. ∎

EXAMPLE 4 **Infinitely Many Solutions**

Solve the system:

$$
\begin{aligned}
x + y + z &= 1 \\
\tfrac{1}{4}x - \tfrac{1}{2}y + \tfrac{3}{4}z &= 0 \\
x + 7y - 3z &= 3.
\end{aligned}
$$

Solution

$$
\begin{bmatrix}
1 & 1 & 1 & 1 \\
\tfrac{1}{4} & -\tfrac{1}{2} & \tfrac{3}{4} & 0 \\
1 & 7 & -3 & 3
\end{bmatrix}
\begin{matrix} \\ 4R_2 \\ \end{matrix} \rightarrow
\begin{bmatrix}
\boxed{1} & 1 & 1 & 1 \\
1 & -2 & 3 & 0 \\
1 & 7 & -3 & 3
\end{bmatrix}
\begin{matrix} \\ R_2 - R_1 \\ R_3 - R_1 \end{matrix}
$$

$$
\rightarrow
\begin{bmatrix}
1 & 1 & 1 & 1 \\
0 & -3 & 2 & -1 \\
0 & 6 & -4 & 2
\end{bmatrix}
\begin{matrix} \\ \\ \tfrac{1}{2}R_3 \end{matrix} \rightarrow
\begin{bmatrix}
1 & 1 & 1 & 1 \\
0 & \boxed{-3} & 2 & -1 \\
0 & 3 & -2 & 1
\end{bmatrix}
\begin{matrix} 3R_1 + R_2 \\ \\ R_3 + R_2 \end{matrix}
$$

$$
\rightarrow
\begin{bmatrix}
3 & 0 & 5 & 2 \\
0 & -3 & 2 & -1 \\
0 & 0 & 0 & 0
\end{bmatrix}
$$

There are no nonzero entries in the third row, so there can be no pivot in the third row. We skip to the final step and turn the pivots we did find into 1s:

$$
\begin{bmatrix}
3 & 0 & 5 & 2 \\
0 & -3 & 2 & -1 \\
0 & 0 & 0 & 0
\end{bmatrix}
\begin{matrix} \tfrac{1}{3}R_1 \\ -\tfrac{1}{3}R_2 \\ \end{matrix} \rightarrow
\begin{bmatrix}
1 & 0 & \tfrac{5}{3} & \tfrac{2}{3} \\
0 & 1 & -\tfrac{2}{3} & \tfrac{1}{3} \\
0 & 0 & 0 & 0
\end{bmatrix}.
$$

Now we translate back into equations and obtain

$$
\begin{aligned}
x \qquad + \tfrac{5}{3}z &= \tfrac{2}{3} \\
y - \tfrac{2}{3}z &= \tfrac{1}{3} \\
0 &= 0.
\end{aligned}
$$

But how does this help us find a solution? The last equation doesn't tell us anything useful, so we ignore it. The thing to notice about the other equations is that we can easily solve the first equation for x and the second for y, obtaining

$$
\begin{aligned}
x &= \tfrac{2}{3} - \tfrac{5}{3}z \\
y &= \tfrac{1}{3} + \tfrac{2}{3}z.
\end{aligned}
$$

This is the solution! We can choose z to be any number and get corresponding values for x and y from the formulas above. This gives us infinitely many different solutions. Thus, the general solution (see Example 5 in Section 3.1) is

$$x = \tfrac{2}{3} - \tfrac{5}{3}z$$
$$y = \tfrac{1}{3} + \tfrac{2}{3}z \qquad \text{General solution}$$

z is arbitrary.

We can also write the general solution as

$$\left(\tfrac{2}{3} - \tfrac{5}{3}z, \tfrac{1}{3} + \tfrac{2}{3}z, z\right); \quad z \text{ arbitrary.} \qquad \text{General solution}$$

This general solution has z as the parameter. Specific choices of values for the parameter z give particular solutions. For example, the choice $z = 6$ gives the particular solution

$$x = \tfrac{2}{3} - \tfrac{5}{3}(6) = -\tfrac{28}{3}$$
$$y = \tfrac{1}{3} + \tfrac{2}{3}(6) = \tfrac{13}{3} \qquad \text{Particular solution}$$
$$z = 6,$$

while the choice $z = 0$ gives the particular solution $(x, y, z) = \left(\tfrac{2}{3}, \tfrac{1}{3}, 0\right)$.

Note that, unlike the system given in Example 3, the system given in this example does have solutions and is thus *consistent*.

➡ **Before we go on ...** Why were there infinitely many solutions to Example 4? The reason is that the third equation was really a combination of the first and second equations to begin with, so we effectively had only two equations in three unknowns.* Choosing a specific value for z (say, $z = 6$) has the effect of supplying the "missing" equation. ∎

*In fact, you can check that the third equation, E_3, is equal to $3E_1 - 8E_2$. Thus, the third equation could have been left out because it conveys no more information than the first two. The process of row reduction always eliminates such a redundancy by creating a row of zeros.

Q : *How do we know when there are infinitely many solutions?*

A : When there are solutions (we have a consistent system, unlike the one in Example 3), and when the matrix we arrive at by row reduction has fewer pivots than there are unknowns. In Example 4 we had three unknowns but only two pivots.

Q : *How do we know which variables to use as parameters in a parameterized solution?*

A : The variables to use as parameters are those in the columns without pivots. In Example 4 there were pivots in the x and y columns but no pivot in the z column, and it was z that we used as a parameter.

EXAMPLE 5   Four Unknowns

Solve the system:

$$x + 3y + 2z - w = 6$$
$$2x + 6y + 6z + 3w = 16$$
$$x + 3y - 2z - 11w = -2$$
$$2x + 6y + 8z + 8w = 20.$$

Solution

$$
\begin{bmatrix}
\boxed{1} & 3 & 2 & -1 & 6 \\
2 & 6 & 6 & 3 & 16 \\
1 & 3 & -2 & -11 & -2 \\
2 & 6 & 8 & 8 & 20
\end{bmatrix}
\begin{matrix}
\\
R_2 - 2R_1 \\
R_3 - R_1 \\
R_4 - 2R_1
\end{matrix}
\rightarrow
\begin{bmatrix}
1 & 3 & 2 & -1 & 6 \\
0 & 0 & 2 & 5 & 4 \\
0 & 0 & -4 & -10 & -8 \\
0 & 0 & 4 & 10 & 8
\end{bmatrix}
$$

There is no pivot available in the second column, so we move on to the third column:

$$
\begin{bmatrix}
1 & 3 & 2 & -1 & 6 \\
0 & 0 & \boxed{2} & 5 & 4 \\
0 & 0 & -4 & -10 & -8 \\
0 & 0 & 4 & 10 & 8
\end{bmatrix}
\begin{matrix}
R_1 - R_2 \\
\\
R_3 + 2R_2 \\
R_4 - 2R_2
\end{matrix}
\rightarrow
\begin{bmatrix}
1 & 3 & 0 & -6 & 2 \\
0 & 0 & 2 & 5 & 4 \\
0 & 0 & 0 & 0 & 0 \\
0 & 0 & 0 & 0 & 0
\end{bmatrix}
\tfrac{1}{2}R_2
$$

$$
\rightarrow
\begin{bmatrix}
1 & 3 & 0 & -6 & 2 \\
0 & 0 & 1 & \frac{5}{2} & 2 \\
0 & 0 & 0 & 0 & 0 \\
0 & 0 & 0 & 0 & 0
\end{bmatrix}.
$$

Translating back into equations, we get

$$x + 3y - 6w = 2$$
$$z + \tfrac{5}{2}w = 2.$$

(We have not written down the equations corresponding to the last two rows, each of which is $0 = 0$.) There are no pivots in the y or w columns, so we use these two variables as parameters. We bring them over to the right-hand sides of the equations above and write the general solution as

$x = 2 - 3y + 6w$

y is arbitrary

$z = 2 - \tfrac{5}{2}w$

w is arbitrary

or

$$(x, y, z, w) = (2 - 3y + 6w, y, 2 - 5w/2, w); \quad y, w \text{ arbitrary.}$$

➡ **Before we go on ...** In Examples 4 and 5 you might have noticed an interesting phenomenon: If at any time in the process, two rows are equal or one is a multiple of the other, then one of those rows (eventually) becomes all zero. ■

Overdetermined Systems and Underdetermined Systems

Up to this point, we have always been given as many equations as there are unknowns. However, we shall see in Section 3.3 that some applications lead to systems in which the number of equations is not the same as the number of unknowns. Systems with more equations than unknowns are said to be **overdetermined**, while systems with fewer equations than unknowns are said to be **underdetermined**. As the following example illustrates, such systems can be handled in the same way as any other.

| EXAMPLE 6 | Number of Equations ≠ Number of Unknowns |

Solve the system:

$$x + y = 1$$
$$13x - 26y = -11$$
$$26x - 13y = 2.$$

Solution We proceed exactly as before and ignore the fact that there is one more equation than there are unknowns:

$$\begin{bmatrix} \boxed{1} & 1 & 1 \\ 13 & -26 & -11 \\ 26 & -13 & 2 \end{bmatrix} \begin{matrix} \\ R_2 - 13R_1 \\ R_3 - 26R_1 \end{matrix} \rightarrow \begin{bmatrix} 1 & 1 & 1 \\ 0 & -39 & -24 \\ 0 & -39 & -24 \end{bmatrix} \begin{matrix} \\ \frac{1}{3}R_2 \\ \frac{1}{3}R_3 \end{matrix}$$

$$\rightarrow \begin{bmatrix} 1 & 1 & 1 \\ 0 & \boxed{-13} & -8 \\ 0 & -13 & -8 \end{bmatrix} \begin{matrix} 13R_1 + R_2 \\ \\ R_3 - R_2 \end{matrix} \rightarrow \begin{bmatrix} 13 & 0 & 5 \\ 0 & -13 & -8 \\ 0 & 0 & 0 \end{bmatrix} \begin{matrix} \frac{1}{13}R_1 \\ -\frac{1}{13}R_2 \\ \end{matrix}$$

$$\rightarrow \begin{bmatrix} 1 & 0 & \frac{5}{13} \\ 0 & 1 & \frac{8}{13} \\ 0 & 0 & 0 \end{bmatrix}.$$

Thus, the solution is $(x, y) = \left(\frac{5}{13}, \frac{8}{13}\right)$.

If, instead of a row of zeros, we had obtained, say, $\begin{bmatrix} 0 & 0 & 6 \end{bmatrix}$ in the last row, we would immediately have concluded that the system was inconsistent.

The fact that we wound up with a row of zeros indicates that one of the equations was actually a combination of the other two; you can check that the third equation can be obtained by multiplying the first equation by 13 and adding the result to the second. Because the third equation therefore tells us nothing that we don't already know from the first two, we call the system of equations **redundant**, or **dependent.** (Compare Example 5 in Section 3.1.)

➡ **Before we go on ...** Example 5 above is another example of a redundant system; we could have started with the following smaller system of two equations in four unknowns:

$$x + 3y + 2z - w = 6$$
$$2x + 6y + 6z + 3w = 16$$

and obtained the same general solution as we did with the larger system. Verify this by solving the smaller system. ∎

Reduced Row Echelon Form

The preceding examples illustrated that we cannot always reduce a matrix to the form shown before Example 2, with pivots going all the way down the diagonal. What we *can* always do is reduce a matrix to the following form.

Reduced Row Echelon Form

A matrix is said to be in **reduced row echelon form** or to be **row-reduced** if it satisfies the following properties:

P1. The first nonzero entry in each row (called the **leading entry** of that row) is a 1.

P2. The columns of the leading entries are **clear** (i.e., they contain zeros in all positions other than that of the leading entry).

P3. The leading entry in each row is to the right of the leading entry in the row above, and any rows of zeros are at the bottom.

Quick Examples

8. $\begin{bmatrix} 1 & 0 & 0 & 2 \\ 0 & 1 & 0 & 4 \\ 0 & 0 & 1 & -3 \end{bmatrix}$, $\begin{bmatrix} 0 & 1 & -3 \\ 0 & 0 & 0 \end{bmatrix}$, and $\begin{bmatrix} 1 & 3 & 0 & -2 \\ 0 & 0 & 1 & 4 \\ 0 & 0 & 0 & 0 \end{bmatrix}$ are row-reduced.

9. $\begin{bmatrix} 1 & 1 & 0 & 2 \\ 0 & 1 & 0 & 4 \\ 0 & 0 & 1 & -3 \end{bmatrix}$ and $\begin{bmatrix} 0 & 1 & -3 \\ 0 & 0 & 1 \end{bmatrix}$ both violate P2 and so are not row-reduced.

(The column of the leading entry in the second row is not clear in either.)

10. $\begin{bmatrix} 0 & 0 & 1 & 4 \\ 1 & 0 & 0 & -2 \\ 0 & 0 & 0 & 0 \end{bmatrix}$ violates P3 and so is not row-reduced. (The leading entry of Row 2 is not to the right of the leading entry in Row 1.)

You should check in the examples we did that the final matrices were all in reduced row echelon form.

It is an interesting and useful fact, though not easy to prove, that any two people who start with the same matrix and row-reduce it will reach exactly the same row-reduced matrix, even if they use different row operations.

The Traditional Gauss-Jordan Method (Optional)

In the version of the Gauss-Jordan method we have presented, we eliminated fractions and decimals in the first step and then worked with integer matrices, partly to make hand computation easier and partly for mathematical elegance. However, complicated fractions and decimals present no difficulty when we use technology. The following example illustrates the more traditional approach to Gauss-Jordan reduction that is used in many of the computer programs that solve the huge systems of equations that arise in practice.*

* Actually, for reasons of efficiency and accuracy, the methods used in commercial programs are closer to the method presented above. To learn more, consult a text on numerical methods.

EXAMPLE 7   **Solving a System with the Traditional Gauss-Jordan Method**

Solve the following system using the traditional Gauss-Jordan method:

$$2x + y + 3z = 5$$
$$3x + 2y + 4z = 7$$
$$2x + y + 5z = 10.$$

Solution We make two changes in our method. First, there is no need to get rid of decimals (because computers and calculators can handle decimals as easily as they can integers). Second, after selecting a pivot, *divide the pivot row by the pivot value, turning the pivot into a* 1. It is easier to determine the row operations that will clear the pivot column if the pivot is a 1.

If we use technology to solve this system of equations, the sequence of matrices might look like this:

$$\begin{bmatrix} \boxed{2} & 1 & 3 & 5 \\ 3 & 2 & 4 & 7 \\ 2 & 1 & 5 & 10 \end{bmatrix} \begin{matrix} \frac{1}{2}R_1 \\ \\ \\ \end{matrix} \rightarrow \begin{bmatrix} \boxed{1} & 0.5 & 1.5 & 2.5 \\ 3 & 2 & 4 & 7 \\ 2 & 1 & 5 & 10 \end{bmatrix} \begin{matrix} \\ R_2 - 3R_1 \\ R_3 - 2R_1 \end{matrix}$$

$$\rightarrow \begin{bmatrix} 1 & 0.5 & 1.5 & 2.5 \\ 0 & \boxed{0.5} & -0.5 & -0.5 \\ 0 & 0 & 2 & 5 \end{bmatrix} 2R_2 \rightarrow \begin{bmatrix} 1 & 0.5 & 1.5 & 2.5 \\ 0 & \boxed{1} & -1 & -1 \\ 0 & 0 & 2 & 5 \end{bmatrix} \begin{matrix} R_1 - 0.5R_2 \\ \\ \\ \end{matrix}$$

$$\rightarrow \begin{bmatrix} 1 & 0 & 2 & 3 \\ 0 & 1 & -1 & -1 \\ 0 & 0 & \boxed{2} & 5 \end{bmatrix} \begin{matrix} \\ \\ \frac{1}{2}R_3 \end{matrix} \rightarrow \begin{bmatrix} 1 & 0 & 2 & 3 \\ 0 & 1 & -1 & -1 \\ 0 & 0 & \boxed{1} & 2.5 \end{bmatrix} \begin{matrix} R_1 - 2R_3 \\ R_2 + R_3 \\ \\ \end{matrix}$$

$$\rightarrow \begin{bmatrix} 1 & 0 & 0 & -2 \\ 0 & 1 & 0 & 1.5 \\ 0 & 0 & 1 & 2.5 \end{bmatrix}.$$

The solution is $(x, y, z) = (-2, 1.5, 2.5)$.

Q: *The solution to Example 7 looked quite easy. Why didn't we use the traditional method from the start like the other textbooks?*

A: It looked easy because we deliberately chose an example that leads to simple decimals. In all but the most contrived examples, the decimals or fractions involved get very complicated very quickly.

FAQs

Getting Unstuck, Going Around in Circles, and Knowing When to Stop

Q: *Help! I have been doing row operations on this matrix for half an hour. I have filled two pages, and I am getting nowhere. What do I do?*

A: Here is a way of keeping track of where you are at any stage of the process and also deciding what to do next.

Starting at the top row of your current matrix:

1. Scan along the row until you get to the leading entry: the first nonzero entry. If there is none—that is, the row is all zero—go to the next row.

2. Having located the leading entry, scan up and down its *column*. If its column is not clear (that is, it contains other nonzero entries), use your leading entry as a pivot to clear its column as in the examples in this section.

3. Now go to the next row, and start again at Step 1.

When you have scanned all the rows and find that all the columns of the leading entries are clear, all that remains to be done is to turn the leading entries into 1s (the "Final Step") and then possibly to reorder the rows so that the leading entries go from left to right as you read down the matrix and zero rows are at the bottom.

Q : *No good. I have been following these instructions, but every time I try to clear a column, I unclear a column I had already cleared. What is going on?*

A : Are you using *leading entries* as pivots? Also, are you *using the pivot* to clear its column? That is, are your row operations all of the following form?

$$aR_c \pm bR_p$$
$$\qquad \uparrow \qquad \uparrow$$
$$\text{Row to change} \quad \text{Pivot row}$$

The instruction next to the row you are changing should involve only that row and the pivot row, even though you might be tempted to use some other row instead.

Q : *Must I continue until I get a matrix that has 1s down the leading diagonal and 0s above and below?*

A : Not necessarily. You are completely done when your matrix is row-reduced: Each leading entry is a 1, the column of each leading entry is clear, and the leading entries go from left to right. You are done *pivoting* when the column of each leading entry is clear. After that, all that remains is to turn each pivot into a 1 (the "Final Step") and, if necessary, rearrange the rows.

3.2 EXERCISES

▼ more advanced ◆ challenging

T indicates exercises that should be solved using technology

In Exercises 1–42, use Gauss-Jordan row reduction to solve the given systems of equation. We suggest doing some by hand and others using technology. [**HINT:** See Examples 1–6.]

1. $x + y = 4$
$x - y = 2$

2. $2x + y = 2$
$-2x + y = 2$

3. $3x - 2y = 6$
$2x - 3y = -6$

4. $2x + 3y = 5$
$3x + 2y = 5$

5. $2x + 3y = 1$
$-x - \dfrac{3y}{2} = -\dfrac{1}{2}$

6. $2x - 3y = 1$
$6x - 9y = 3$

7. $2x + 3y = 2$
$-x - \dfrac{3y}{2} = -\dfrac{1}{2}$

8. $2x - 3y = 2$
$6x - 9y = 3$

9. $x + y = 1$
$3x - y = 0$
$x - 3y = -2$

10. $x + y = 1$
$3x - 2y = -1$
$5x - y = \dfrac{1}{5}$

11. $x + y = 0$
$3x - y = 1$
$x - y = -1$

12. $x + 2y = 1$
$3x - 2y = -2$
$5x - y = \dfrac{1}{5}$

13. $0.5x + 0.1y = 1.7$
$0.1x - 0.1y = 0.3$
$x + \quad y = \dfrac{11}{3}$

14. $-0.3x + 0.5y = 0.1$
$x - \quad y = 4$
$\dfrac{x}{17} + \dfrac{y}{17} = 1$

15. $-x + 2y - z = 0$
$-x - y + 2z = 0$
$2x \quad - z = 4$

16. $x + 2y \quad = 4$
$y - z = 0$
$x + 3y - 2z = 5$

17. $x + y + 6z = -1$
$\dfrac{1}{3}x - \dfrac{1}{3}y + \dfrac{2}{3}z = 1$
$\dfrac{1}{2}x \quad + z = 0$

18. $x - \dfrac{1}{2}y \quad = 0$
$\dfrac{1}{3}x + \dfrac{1}{3}y + \dfrac{1}{3}z = 2$
$\dfrac{1}{2}x \quad - \dfrac{1}{2}z = -1$

19. $-\dfrac{1}{2}x + y - \dfrac{1}{2}z = 0$
$-\dfrac{1}{2}x - \dfrac{1}{2}y + z = 0$
$x - \dfrac{1}{2}y - \dfrac{1}{2}z = 0$

20. $x - \dfrac{1}{2}y \quad = 0$
$\dfrac{1}{2}x \quad - \dfrac{1}{2}z = -1$
$3x - y - z = -2$

21. $x + y + 2z = -1$
$2x + 2y + 2z = 2$
$\dfrac{3}{5}x + \dfrac{3}{5}y + \dfrac{3}{5}z = \dfrac{2}{5}$

22. $x + y - z = -2$
$x - y - 7z = 0$
$\dfrac{2}{7}x \quad - \dfrac{8}{7}z = 14$

23. $-0.5x + 0.5y + 0.5z = 1.5$
$4.2x + 2.1y + 2.1z = 0$
$0.2x \qquad + 0.2z = 0$

24. $0.25x - 0.5y \qquad = 0$
$0.2x + 0.2y - 0.2z = -0.6$
$0.5x - 1.5y + \quad z = 0.5$

25. $2x - y + z = 4$ **26.** $3x - y - z = 0$
$3x - y + z = 5$ $x + y + z = 4$

27. $0.75x - 0.75y - \; z = 4$ **28.** $2x - \quad y + \quad z = 4$
$x - \quad y + 4z = 0$ $-x + 0.5y - 0.5z = 1.5$

29. ▼ $3x + y - z = 12$ **30.** ▼ $x + y - 3z = 21$
(Yes, one equation in three unknowns!)

31. ▼ $\quad x + \quad y + 2z = -1$
$2x + \quad 2y + 2z = 2$
$0.75x + 0.75y + \quad z = 0.25$
$-x \qquad - 2z = 21$

32. ▼ $\quad x + \quad y - \quad z = -2$
$x - \quad y - \quad 7z = 0$
$0.75x - 0.5y + 0.25z = 14$
$x + \quad y + \quad z = 4$

33. ▼ $x + \quad y + 5z \qquad = 1$
$y + 2z + \quad w = 1$
$x + 3y + 7z + 2w = 2$
$x + \quad y + 5z + \quad w = 1$

34. ▼ $x + \quad y \qquad + 4w = 1$
$2x - 2y - 3z + 3w = -1$
$4y + 6z + \quad w = 4$
$2x + 4y + 9z \qquad = 6$

35. ▼ $x + \quad y + 5z \qquad = 1$
$y + 2z + \quad w = 1$
$x + \quad y + 5z + \quad w = 1$
$x + 2y + 7z + 2w = 2$

36. ▼ $x + \quad y \qquad + 4w = 1$
$2x - 2y - 3z + 2w = -1$
$4y + 6z + \quad w = 4$
$3x + 3y + 3z + 7w = 4$

37. ▼ $x - 2y + \quad z - 4w = 1$
$x + 3y + 7z + 2w = 2$
$2x + \quad y + 8z - 2w = 3$

38. ▼ $x - 3y - 2z - \quad w = 1$
$x + 3y + \quad z + 2w = 2$
$2x \qquad - z + \quad w = 3$

39. ▼ $x + y + z + u + v = 15$
$y - z + u - v = -2$
$z + u + v = 12$
$u - v = -1$
$v = 5$

40. ▼ $x - y + z - u + v = 1$
$y + z + u + v = 2$
$z - u + v = 1$
$u + v = 1$
$v = 1$

41. ▼ $\quad x - y + z - u + \quad v = 0$
$y - z + u - \quad v = -2$
$x \qquad - 2v = -2$
$2x - y + z - u - 3v = -2$
$4x - y + z - u - 7v = -6$

42. ▼ $x + \quad y + \quad z + \quad u + \quad v = 15$
$y + \quad z + \quad u + \quad v = 3$
$x + 2y + 2z + 2u + 2v = 18$
$x - \quad y - \quad z - \quad u - \quad v = 9$
$x - 2y - 2z - 2u - 2v = 6$

T *In Exercises 43–46, use technology to solve the systems of equations. Express all solutions as fractions.*

43. $\quad x + 2y - \quad z + \quad w = 30$
$2x \qquad - z + 2w = 30$
$x + 3y + 3z - 4w = 2$
$2x - 9y \qquad + \quad w = 4$

44. $4x - 2y + \quad z + \quad w = 20$
$3y + 3z - 4w = 2$
$2x + 4y \qquad - \quad w = 4$
$x + 3y + 3z \qquad = 2$

45. $\quad x + 2y + 3z + 4w + 5t = 6$
$2x + 3y + 4z + 5w + \quad t = 5$
$3x + 4y + 5z + \quad w + 2t = 4$
$4x + 5y + \quad z + 2w + 3t = 3$
$5x + \quad y + 2z + 3w + 4t = 2$

46 $\quad x - 2y + 3z - 4w \qquad = 0$
$-2x + 3y - 4z \qquad + \quad t = 0$
$3x - 4y \qquad + \quad w - 2t = 0$
$-4x \qquad + \quad z - 2w + 3t = 0$
$y - 2z + 3w - 4t = 1$

⊤ *In Exercises 47–50, use technology to solve the system of equations. Express all solutions as decimals, rounded to one decimal place.*

47. $1.6x + 2.4y - 3.2z = 4.4$
$5.1x - 6.3y + 0.6z = -3.2$
$4.2x + 3.5y + 4.9z = 10.1$

48. $2.1x + 0.7y - 1.4z = -2.3$
$3.5x - 4.2y - 4.9z = 3.3$
$1.1x + 2.2y - 3.3z = -10.2$

49. $-0.2x + 0.3y + 0.4z - \quad t = 4.5$
$2.2x + 1.1y - 4.7z + \quad 2t = 8.3$
$9.2y \qquad\quad - 1.3t = 0$
$3.4x \qquad\quad + 0.5z - 3.4t = 0.1$

50. $1.2x - \quad 0.3y + 0.4z - \quad 2t = 4.5$
$1.9x \qquad\quad - 0.5z - 3.4t = 0.2$
$12.1y \qquad\quad - 1.3t = 0$
$3x + \quad 2y - 1.1z \qquad = 9$

Communication and Reasoning Exercises

51. What is meant by a pivot? What does pivoting do?

52. Give instructions to check whether or not a matrix is row-reduced.

53. You are row-reducing a matrix and have chosen a -6 as a pivot in Row 4. Directly above the pivot, in Row 1, is a 15. What row operation can you use to clear the 15?

54. You are row-reducing a matrix and have chosen a -4 as a pivot in Row 2. Directly below the pivot, in Row 4, is a -6. What row operation can you use to clear the -6?

55. In the matrix of a system of linear equations, suppose that two of the rows are equal. What can you say about the row-reduced form of the matrix?

56. In the matrix of a system of linear equations, suppose that one of the rows is a multiple of another. What can you say about the row-reduced form of the matrix?

57. ▼ Your friend Frans tells you that the system of linear equations you are solving cannot have a unique solution because the reduced matrix has a row of zeros. Comment on his claim.

58. ▼ Your other friend Hans tells you that because he is solving a consistent system of five linear equations in six unknowns, he will get infinitely many solutions. Comment on his claim.

59. ▼ If the reduced matrix of a consistent system of linear equations has five rows, three of which are zero, and five columns, how many parameters does the general solution contain?

60. ▼ If the reduced matrix of a consistent system of linear equations has five rows, two of which are zero, and seven columns, how many parameters does the general solution contain?

61. ▼ Suppose a system of equations has a unique solution. What must be true of the number of pivots in the reduced matrix of the system? Why?

62. ▼ Suppose a system has infinitely many solutions. What must be true of the number of pivots in the reduced matrix of the system? Why?

63. ▼ Give an example of a system of three linear equations with the general solution $x = 1$, $y = 1 + z$, z arbitrary. (Check your system by solving it.)

64. ▼ Give an example of a system of three linear equations with the general solution $x = y - 1$, y arbitrary, $z = y$. (Check your system by solving it.)

65. ◆ A system of linear equations is called **homogeneous** if the right-hand side of each equation in the system is 0. If a homogeneous system has a unique solution, what can you say about that solution? Why?

66. ◆ A certain homogeneous system of linear equations in three unknowns (see Exercise 65) has a solution $(1, 2, 0)$. Is this solution unique? Explain.

67. ◆ Can a homogeneous system (see Exercise 65) of linear equations be inconsistent? Explain.

68. ◆ Can a non-homogeneous system (see Exercise 65) of linear equations have the zero solution? Explain.

3.3   Applications of Systems of Linear Equations

In the examples and exercises of this section, we consider scenarios that lead to systems of linear equations in three or more unknowns. Some of these applications will strike you as a little idealized or even contrived in comparison with the kinds of problems you might encounter in the real world.[*] One reason is that we will not have tools to handle more realistic versions of these applications until we have studied linear programming in Chapter 5.

[*] See the discussion at the end of the first example below.

In each example that follows, we set up the problem as a linear system and then give the solution. The emphasis in this section is on modeling a scenario by a system of linear equations and then interpreting the solution that results, rather than on obtaining the solution. For practice, you should do the row reduction necessary to get the solution and also explore technologies such as the Pivot and Gauss-Jordan tool at the Website.

EXAMPLE 1   Blending

The *Arctic Juice Company* makes three juice blends: PineOrange, using 2 quarts of pineapple juice and 2 quarts of orange juice per gallon; PineKiwi, using 3 quarts of pineapple juice and 1 quart of kiwi juice per gallon; and OrangeKiwi, using 3 quarts of orange juice and 1 quart of kiwi juice per gallon. Each day the company has 800 quarts of pineapple juice, 650 quarts of orange juice, and 350 quarts of kiwi juice available. How many gallons of each blend should it make each day if it wants to use up all of the supplies?

Solution We take the same steps to understand the problem that we took in Section 3.1. The first step is to identify and label the unknowns. Looking at the question asked in the last sentence, we see that we should label the unknowns like this:

x = number of gallons of PineOrange made each day

y = number of gallons of PineKiwi made each day

z = number of gallons of OrangeKiwi made each day.

Next, we can organize the information we are given in a table:

	PineOrange (x)	PineKiwi (y)	OrangeKiwi (z)	Total Available
Pineapple Juice (quarts)	2	3	0	800
Orange Juice (quarts)	2	0	3	650
Kiwi Juice (quarts)	0	1	1	350

Notice how we have arranged the table: We have placed headings corresponding to the unknowns along the top, rather than down the side, and we have added a heading for the available totals. This gives us a table that is essentially the matrix of the system of linear equations we are looking for. (However, read the caution in the "Before we go on" discussion.)

Now we read across each row of the table. The fact that we want to use exactly the amount of each juice that is available leads to the following three equations:

$$
\begin{aligned}
2x + 3y &= 800 \\
2x + 3z &= 650 \\
y + z &= 350.
\end{aligned}
$$

The solution of this system is $(x, y, z) = (100, 200, 150)$, so Arctic Juice should make 100 gallons of PineOrange, 200 gallons of PineKiwi, and 150 gallons of OrangeKiwi each day.

➡ **Before we go on . . .** *Caution:* We do not recommend relying on the coincidence that the table we created to organize the information in Example 1 happened to be the matrix of the system; it is too easy to set up the table "sideways" and get the wrong matrix. You should always write down the system of equations *and be sure you understand each equation.* For example, the equation $2x + 3y = 800$ in Example 1 indicates that the number of quarts of pineapple juice that will be used $(2x + 3y)$ is equal to the amount available (800 quarts). By thinking of the reason for each equation, you can check that you have the correct system. If you have the wrong system of equations to begin with, solving it won't help you.

Q : *Just how realistic is the scenario in Example 1?*

A : This is a very unrealistic scenario, for several reasons:

1. Isn't it odd that we happened to end up with exactly the same number of equations as unknowns? Real scenarios are rarely so considerate. If there had been four equations, there would in all likelihood have been no solution at all. However, we need to understand these idealized problems before we can tackle the real world.
2. Even if a real-world scenario does give the same number of equations as unknowns, there is still no guarantee that there will be a unique solution consisting of positive values. What, for instance, would we have done in this example if x had turned out to be negative?
3. The requirement that we use exactly all the ingredients would be an unreasonable constraint in real life. When we discuss linear programming, we will be able to substitute the more reasonable constraint that you use no more than is available, and we will add the more reasonable objective that you maximize profit.

EXAMPLE 2   **Aircraft Purchases: Airbus and Boeing**

A new airline has recently purchased a fleet of Airbus A330-300s, Boeing 767-300ERs, and Boeing Dreamliner 787-9s to meet an estimated demand for 5,400 seats. The A330-300s seat 330 passengers and cost \$250 million each, the 767-300ERs seat 270 passengers and cost \$200 million each, while the 787-9s seat 240 passengers and cost \$250 million each.[12] The total cost of the fleet, which had twice as many 787-9s as 767s, was \$4,750 million. How many of each type of aircraft did the company purchase?

Solution   We label the unknowns as follows:

x = number of Airbus A330-300s

y = number of Boeing 767-300ERs

z = number of Boeing 787-9s.

We must now set up the equations. We can organize some (but not all) of the given information in a table:

	A330-300	767-300ER	787-9	Total
Capacity	330	270	240	5,400
Cost (\$ million)	250	200	250	4,750

---

[12] The prices are approximate 2015 list prices; actual selling prices are typically around 50% of list prices. Seating capacities depend on configuration and vary considerably from airline to airline. Sources for data: Company websites/Wikipedia.

Reading across, we get the equations expressing the facts that the airline needed to seat 5,400 passengers and that it spent \$4,750 million:

$$330x + 270y + 240z = 5{,}400$$
$$250x + 200y + 250z = 4{,}750.$$

There is an additional piece of information we have not yet used: The airline bought twice as many 787s as 767s. As we said in Section 3.1, it is easiest to translate a statement like this into an equation if we first reword it using the phrase "the number of." Thus, we say: "The number of 787s ordered was twice the number of 767s ordered," or

$$z = 2y$$
$$2y - z = 0.$$

We now have a system of three equations in three unknowns:

$$330x + 270y + 240z = 5{,}400$$
$$250x + 200y + 250z = 4{,}750$$
$$2y - z = 0.$$

Solving the system, we get the solution $(x, y, z) = (5, 5, 10)$. Thus, the airline ordered five A330-300s, five 767-300ERs, and ten 787-9s.

EXAMPLE 3 **Traffic Flow**

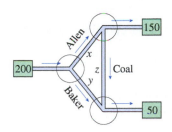

Figure 8

Traffic through downtown Urbanville flows through the one-way system shown in Figure 8. Traffic counting devices installed in the road (shown as boxes) count 200 cars entering town from the west each hour, 150 leaving town on the north each hour, and 50 leaving town on the south each hour.

a. From this information is it possible to determine how many cars drive along Allen, Baker, and Coal streets every hour?

b. What is the maximum possible traffic flow along Baker Street?

c. What is the minimum possible traffic along Allen Street?

d. What is the maximum possible traffic flow along Coal Street?

Solution

a. Our unknowns are

$x =$ number of cars per hour on Allen Street
$y =$ number of cars per hour on Baker Street
$z =$ number of cars per hour on Coal Street.

We now focus on the three intersections shown circled in Figure 9. Assuming that, at each of these intersections, cars do not fall into a pit or materialize out of thin air, then the number of cars entering each intersection has to equal the number exiting.

Intersection of Allen and Baker Streets: We consider *only what is inside the circle around this intersection,* as we see in the magnified view shown in Figure 10. In this zoomed-in view, we see that there are 200 cars entering and $x + y$ cars exiting:

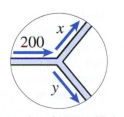

Intersection of Allen and Baker

Figure 10

Traffic in = Traffic out

$$200 = x + y.$$

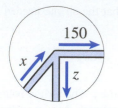

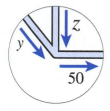

Intersection of Allen and Coal

Figure 11

Intersection of Baker and Coal

Figure 12

Intersection of Allen and Coal Streets (see Figure 11):

Traffic in = Traffic out

$$x = z + 150.$$

Intersection of Baker and Coal Streets (see Figure 12):

Traffic in = Traffic out

$$y + z = 50.$$

We now have the following system of equations:

$$
\begin{aligned}
x + y &= 200 \\
x \quad - z &= 150 \\
y + z &= 50.
\end{aligned}
$$

If we solve this system using the methods of the preceding section, we find that it has infinitely many solutions. The general solution is

$$x = z + 150$$
$$y = -z + 50$$
z is arbitrary.

Because we do not have a unique solution, it is *not* possible to determine how many cars drive along Allen, Baker, and Coal Streets every hour.

b. The traffic flow along Baker Street is measured by y. From the general solution,

$$y = -z + 50,$$

where z is arbitrary. How arbitrary is z? It makes no sense for any of the variables x, y, or z to be negative in this scenario, so $z \geq 0$. Therefore, the largest possible value y can have is

$$y = -0 + 50 = 50 \text{ cars per hour.}$$

c. The traffic flow along Allen Street is measured by x. From the general solution,

$$x = z + 150,$$

where $z \geq 0$, as we saw in part (b). Therefore, the smallest possible value x can have is

$$x = 0 + 150 = 150 \text{ cars per hour.}$$

d. The traffic flow along Coal Street is measured by z. Referring to the general solution, we see that z shows up in the expressions for both x and y:

$$x = z + 150$$
$$y = -z + 50.$$

In the first of these equations there is nothing preventing z from being as big as we like; the larger we make z, the larger x becomes. However, the second equation places a limit on how large z can be: If $z > 50$, then y is negative, which is impossible. Therefore, the largest value z can take is 50 cars per hour.

From the discussion above, we see that z is not completely arbitrary: We must have $z \geq 0$ and $z \leq 50$. Thus, z has to satisfy $0 \leq z \leq 50$ for us to get a realistic answer.

➡ **Before we go on . . .** Here are some questions to think about in Example 3: If you wanted to nail down x, y, and z to see where the cars are really going, how would you do it with only one more traffic counter? Would it make sense for z to be fractional? What if you interpreted x, y, and z as *average* numbers of cars per hour over a long period of time?

Traffic flow is only one kind of flow in which we might be interested. Water and electricity flows are others. In each case, to analyze the flow, we use the fact that the amount entering an intersection must equal the amount leaving it. ∎

EXAMPLE 4   **Transportation**

A car rental company has four locations in the city: Southwest, Northeast, Southeast, and Northwest. The Northwest location has 20 more cars than it needs, and the Northeast location has 15 more cars than it needs. The Southwest location needs 10 more cars than it has, and the Southeast location needs 25 more cars than it has. It costs $10 (in salary and gas) to have an employee drive a car from Northwest to Southwest. It costs $20 to drive a car from Northwest to Southeast. It costs $5 to drive a car from Northeast to Southwest, and it costs $10 to drive a car from Northeast to Southeast. If the company will spend a total of $475 rearranging its cars, how many cars will it drive from each of Northwest and Northeast to each of Southwest and Southeast?

Solution Figure 13 shows a diagram of this situation. Each arrow represents a route along which the rental company can drive cars. At each location is written the number of extra cars the location has (Northwest and Northeast) or the number it needs (Southwest and Southeast). Along each route is written the cost of driving a car along that route.

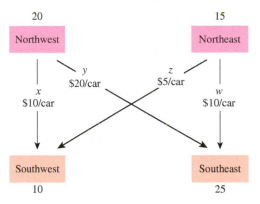

Figure 13

The unknowns are the number of cars the company will drive along each route, so we have the following four unknowns, as indicated in the figure:

x = number of cars driven from Northwest to Southwest

y = number of cars driven from Northwest to Southeast

z = number of cars driven from Northeast to Southwest

w = number of cars driven from Northeast to Southeast.

Consider the Northwest location. It has 20 more cars than it needs, so the total number of cars being driven out of Northwest should be 20. This gives us the equation

$$x + y = 20.$$

Similarly, the total number of cars being driven out of Northeast should be 15, so

$$z + w = 15.$$

Considering the number of cars needed at the Southwest and Southeast locations, we get the following two equations as well:

$$x + z = 10$$
$$y + w = 25.$$

There is one more equation that we should write down, the equation that says that the company will spend \$475:

$$10x + 20y + 5z + 10w = 475.$$

Thus, we have the following system of five equations in four unknowns:

$$
\begin{aligned}
x + y & = 20 \\
z + w & = 15 \\
x + z & = 10 \\
y + w & = 25 \\
10x + 20y + 5z + 10w & = 475.
\end{aligned}
$$

Solving this system, we find that $(x, y, z, w) = (5, 15, 5, 10)$. In words, the company will drive 5 cars from Northwest to Southwest, 15 from Northwest to Southeast, 5 from Northeast to Southwest, and 10 from Northeast to Southeast.

➡ **Before we go on . . .** A very reasonable question to ask in Example 4 is, Can the company rearrange its cars for less than \$475? Even better, what is the least possible cost? In general, a question asking for the optimal cost may require the techniques of linear programming, which we will discuss in Chapter 5. However, in this case we can approach the problem directly. If we remove the equation that says that the total cost is \$475 and solve the system consisting of the other four equations, we find that there are infinitely many solutions and that the general solution may be written as

$$
\begin{aligned}
x &= w - 5 \\
y &= 25 - w \\
z &= 15 - w
\end{aligned}
$$

w is arbitrary.

This allows us to write the total cost as a function of w:

$$
\begin{aligned}
\text{Cost} &= 10x + 20y + 5z + 10w \\
&= 10(w - 5) + 20(25 - w) + 5(15 - w) + 10w \\
&= 525 - 5w.
\end{aligned}
$$

So the larger we make w, the smaller the total cost will be. The largest we can make w is 15 (why?), and if we do so, we get $(x, y, z, w) = (10, 10, 0, 15)$ and a total cost of \$450. ■

3.3 EXERCISES

▼ more advanced   ◆ challenging

T indicates exercises that should be solved using technology

Exercises not marked with **T** *result in systems that can be solved by hand using the techniques of Section 3.2, though, of course, you can also use technology like the Pivot and Gauss-Jordan tool to solve them. What is important in all cases is setting up your system correctly and interpreting the solution.*

Applications

1. **Resource Allocation** You manage an ice cream factory that makes three flavors: Creamy Vanilla, Continental Mocha, and Succulent Strawberry. Into each batch of Creamy Vanilla go 2 eggs, 1 cup of milk, and 2 cups of cream. Into each batch of Continental Mocha go 1 egg, 1 cup of milk, and 2 cups of cream, while into each batch of Succulent Strawberry go 1 egg, 2 cups of milk, and 1 cup of cream. You have in stock 350 eggs, 350 cups of milk, and 400 cups of cream. How many batches of each flavor should you make in order to use up all of your ingredients? [**HINT:** See Example 1.]

2. **Resource Allocation** You own a hamburger franchise and are planning to shut down operations for the day, but you are left with 13 buns, 19 defrosted beef patties, and 15 opened cheese slices. Rather than throwing them out, you decide to use them to make burgers that you will sell at a discount. Plain burgers each require 1 beef patty and 1 bun; double cheeseburgers each require 2 beef patties, 1 bun, and 2 slices of cheese; while regular cheeseburgers each require 1 beef patty, 1 bun, and 1 slice of cheese. How many of each should you make? [**HINT:** See Example 1.]

3. **Resource Allocation** *Urban Community College* is planning to offer courses in Finite Math, Applied Calculus, and Computer Methods. Each section of Finite Math has 40 students and earns the college $40,000 in revenue. Each section of Applied Calculus has 40 students and earns the college $60,000, while each section of Computer Methods has 10 students and earns the college $20,000. Assuming that the college wishes to offer a total of six sections, accommodate 210 students, and bring in $260,000 in revenues, how many sections of each course should it offer? [**HINT:** See Example 2.]

4. **Resource Allocation** The *Enormous State University* History Department offers three courses—Ancient, Medieval, and Modern History—and the chairperson is trying to decide how many sections of each to offer this semester. The department is allowed to offer 45 sections total, there are 5,000 students who would like to take a course, and there are 60 professors to teach them. Sections of Ancient History have 100 students each, sections of Medieval History hold 50 students each, and sections of Modern History have 200 students each. Modern History sections are taught by a team of two professors, while Ancient and Medieval History need only one professor per section. How many sections of each course should the chair schedule in order to offer all the sections that are allowed, accommodate all of the students, and give one teaching assignment to each professor? [**HINT:** See Example 2.]

5. **Latin Music Sales (Digital)** In 2013, total revenues from digital sales of regional (Mexican/Tejano), pop/rock, and tropical (salsa/merengue/cumbia/bachata) Latin music in the United States amounted to $58 million. Regional music brought in four times as much as tropical music and pop/rock music brought in $10 million more than tropical music.[13] How much revenue was earned from digital sales in each of the three categories?

6. **Latin Music Sales (Digital)** In 2012, total revenues from digital sales of pop/rock, tropical (salsa/merengue/cumbia/bachata), and urban (reggaeton) Latin music in the United States amounted to $24 million. Pop/rock music brought in twice as much as the other two categories combined and $9 million more than tropical music.[14] How much revenue was earned from digital sales in each of the three categories?

7. **Purchasing Aircraft** In Example 2 we saw that Airbus A330-300s seat 330 passengers and cost $250 million each, Boeing 767-300ERs seat 270 passengers and cost $200 million each, while Boeing Dreamliner 787-9s seat 240 passengers and cost $250 million each. You are the purchasing manager of an airline company and have a spending goal of $4,300 million for the purchase of new aircraft to seat a total of 4,980 passengers. Your company has a policy of supporting U.S. industries, and you have been instructed to buy twice as many Boeings as Airbuses. Given the selection of three aircraft, how many of each should you order?

8. **Purchasing Aircraft** Refer to Exercise 7. René DuFleur has just been appointed the new CEO of your airline, and you have received instructions that the company policy is now to purchase as many Airbuses as Boeings. Further, the desired seating capacity has been revised downward to 2,910 passengers, and the cost target has been lowered to $2,400 million. Given the specifications of the three types of aircraft, how many of each should you order?

---

[13] Revenues are approximate. Source: Recording Industry Association of America (http://riaa.com).

[14] *Ibid.*

9. *Supply* A bagel store orders cream cheese from three suppliers: *Cheesy Cream Corp.* (CCC), *Super Smooth & Sons* (SSS), and *Bagel's Best Friend Co.* (BBF). One month, the total order of cheese came to 100 tons. (The store does do a booming trade.) The costs were $80, $50, and $65 per ton from the three suppliers, respectively, with total cost amounting to $5,990. Given that the store ordered the same amount from CCC and BBF, how many tons of cream cheese were ordered from each supplier?

10. *Supply* Refer to Exercise 9. The bagel store's outlay for cream cheese the following month was $2,310, when it purchased a total of 36 tons. Two more tons of cream cheese came from *Bagel's Best Friend Co.* than from *Super Smooth & Sons.* How many tons of cream cheese came from each supplier?

11. *Pest Control* Halmar the Great has boasted to his hordes of followers that many a notorious villain has fallen to his awesome sword: His total of 560 victims consists of evil sorcerers, trolls, and orcs. These he has slain with a total of 620 mighty thrusts of his sword, evil sorcerers and trolls each requiring two thrusts (to the chest) and orcs each requiring one thrust (to the neck). When asked about the number of trolls he has slain, he replies, "I, the mighty Halmar, despise trolls five times as much as I despise evil sorcerers. Accordingly, five times as many trolls as evil sorcerers have fallen to my sword!" How many of each type of villain has he slain?

12. *Manufacturing Perfume* The *Fancy French Perfume Company* recently had its secret formula divulged. It turned out that it was using, as the three ingredients, rose oil, oil of fermented prunes, and alcohol. Moreover, each 22-ounce econo-size bottle contained 4 more ounces of alcohol than oil of fermented prunes, while the amount of alcohol was equal to the combined volume of the other two ingredients. How much of each ingredient did the company use in an econo-size bottle?[15] [**HINT:** The answer is the brand-name of a famous eau de cologne.]

13. ▼ *Donations* The *Enormous State University Good Works Society* recently raised funds for three worthwhile causes: the Math Professors' Benevolent Fund (MPBF), the Society of Computer Nerds (SCN), and the NY Jets. Because the society's members are closet jocks, the society donated twice as much to the NY Jets as to the MPBF, and it donated equal amounts to the first two funds. (It is unable to distinguish between mathematicians and nerds.) Further, for every $1 it gave to the MPBF, it decided to keep $1 for itself; for every $1 it gave to the SCN, it kept $2; and for every $1 to the Jets, it also kept $2. The treasurer of the Society, Johnny Treasure, was required to itemize all donations for the Dean of Students but discovered to his consternation that he had lost the receipts! The only information

available to him was that the society's bank account had swelled by $4,200. How much did the society donate to each cause?

14. ▼ *Tenure* Professor Walt is up for tenure and wishes to submit a portfolio of written student evaluations as evidence of his good teaching. He begins by grouping all the evaluations into four categories: good reviews, bad reviews (a typical one being "GET RID OF WALT! THE MAN CAN'T TEACH!"), mediocre reviews (such as "I suppose he's OK, given the general quality of teaching at this college"), and reviews left blank. When he tallies up the piles, Walt gets a little worried: There are 280 more bad reviews than good ones and only half as many blank reviews as bad ones. The good reviews and blank reviews together total 170. On an impulse, he decides to even up the piles a little by removing 280 of the bad reviews, and this leaves him with a total of 400 reviews of all types. How many of each category of reviews were there originally?

▮*Airline Costs* *Exercises 15 and 16 are based on the following table, which shows the amount spent by four U.S. airlines to fly one available seat 1 mile in the second quarter of 2014.*[16] *Set up each system and then solve using technology.* [**HINT:** *See the technology note accompanying Example 1.*]

Airline	United Continental	American	JetBlue	Southwest
Cost (¢)	14.9	14.6	11.9	12.4

15. ▼ Suppose that, on a 3,000-mile New York to Los Angeles flight, United Continental, American, and Southwest flew a total of 210 empty seats, costing them a total of $89,760. If United Continental had three times as many empty seats as American, how many empty seats did each of these three airlines carry on its flight?

16. ▼ Suppose that, on a 2,000-mile Miami to Memphis flight, United Continental, JetBlue, and Southwest flew a total of 200 empty seats, costing them a total of $51,100. If JetBlue had twice as many empty seats as Southwest, how many empty seats did each of these three airlines carry on its flight?

Investing: Inverse Mutual Funds *Inverse mutual funds, sometimes referred to as "bear market" or "short" funds, seek to deliver the opposite of the performance of the index or category they track and can thus be used by traders to bet against the stock market. Exercises 17 and 18 are based on the following table, which shows the performance of three such funds as of February 27, 2015:*[17]

---

[15] Most perfumes consist of 10 to 20% perfume oils dissolved in alcohol. This may or may not be reflected in this company's formula.

[16] Costs are rounded to the nearest 0.1¢. Source: Company filings. The cost per available seat-mile (CASM) is a widely used operating statistic in the airline industry.

[17] Based on prices at the close of the stock market on February 27, 2015. YTD losses rounded to the nearest percentage point. Source: www.fidelity.com.

	Year-to-Date Loss (%)
SHPIX (Short Smallcap Profund)	4
RYURX (Rydex Inverse S&P 500)	3
RYCWX (Rydex Inverse Dow)	6

17. You invested a total of $9,000 in the three funds at the beginning of 2015, including equal amounts in RYURX and RYCWX. Your year-to-date loss from the first two funds amounted to $260. How much did you invest in each of the three funds?

18. You invested a total of $6,000 in the three funds at the beginning of 2015, including equal amounts in SHPIX and RYURX. Your total year-to-date loss amounted to $260. How much did you invest in each of the three funds?

Investing: Lesser-Known Stocks Exercises 19 and 20 are based on the following information about the stocks of **Whitestone REIT**, **HCC Insurance Holdings, Inc.,** *and* **SanDisk Corporation**:[18]

	Price ($)	Dividend Yield (%)
WSR (WSR Whitestone REIT)	16	7
HCC (HCC Insurance Holdings, Inc.)	56	2
SNDK (SanDisk Corporation)	80	2

19. ▼ You invested a total of $8,400 in shares of the three stocks at the given prices and expected to earn $248 in annual dividends. If you purchased a total of 200 shares, how many shares of each stock did you purchase?

20. ▼ You invested a total of $11,200 in shares of the three stocks at the given prices and expected to earn $304 in annual dividends. If you purchased a total of 250 shares, how many shares of each stock did you purchase?

21. **Ⓣ** *Internet Audience* At the end of 2003 the four companies with the largest number of home Internet users in the United States were **Microsoft**, **Time Warner**, **Yahoo**, and **Google**, with a combined audience of 284 million users.[19] Taking x to be the Microsoft audience in millions, y the Time Warner audience in millions, z the Yahoo audience in millions, and u the Google audience in millions, it was observed that

$$z - u = 3(x - y) + 6$$
$$x + y = 50 + z + u$$

and　　$x - y + z - u = 42.$

How large was the audience of each of the four companies in November 2003?

22. **Ⓣ** *Internet Audience* At the end of 2003 the four organizations ranking 5 through 8 in home Internet users in the United States were **eBay**, the U.S. government, **Amazon**, and **Lycos**, with a combined audience of 112 million users.[20] Taking x to be the eBay audience in millions, y the U.S. government audience in millions, z the Amazon audience in millions, and u the Lycos audience in millions, it was observed that

$$y - z = z - u$$
$$x - y = 3(y - u) + 5$$

and　　$x - y + z - u = 12.$

How large was the audience of each of the four organizations in November 2003?

23. ▼ *Market Share: Homeowners Insurance* Three market leaders in homeowners insurance in Missouri are **State Farm**, **American Family Insurance Group**, and **Allstate**. Based on data from 2007, two relationships between the Missouri homeowners insurance percentage market shares are found to be

$$x = 1 + y + z$$
$$z = 16 - 0.2w,$$

where x, y, z, and w are the percentages of the market held by State Farm, American Family, Allstate, and other companies, respectively.[21] Given that the four groups account for the entire market, obtain a third equation relating x, y, z, and w, and solve the associated system of three linear equations to show how the market shares of State Farm, American Family, and Allstate depend on the share held by other companies. Which of the three companies' market share is most affected by the share held by other companies?

24. ▼ *Market Share: Auto Insurance* Repeat Exercise 23 using the following relationships among the auto insurance percentage market shares:

$$x = -40 + 5y + z$$
$$z = 3 - 2y + w.$$

25. *Inventory Control* *Red Bookstore* wants to ship books from its warehouses in Brooklyn and Queens to its stores, one on Long Island and one in Manhattan. Its warehouse in Brooklyn has 1,000 books, and its warehouse in Queens has 2,000. Each store orders 1,500 books. It costs $5 to ship each book from Brooklyn to Long Island and $1 to ship each book from Brooklyn to Manhattan. It costs $4 to ship each book from

---

[18] Yields rounded to the nearest percentage point and stock prices at the close of the stock market on February 27, 2015, rounded to the nearest $1. Source: www.finance.yahoo.

[19] Source: Nielsen/NetRatings www.nielsen-netratings.com, January 1, 2004.

[20] *Ibid.*

[21] Source: Missouri Dept. of Insurance (www.insurance.mo.gov).

Queens to Long Island and $2 to ship each book from Queens to Manhattan.

a. If Red has a transportation budget of $9,000 and is willing to spend all of it, how many books should Red ship from each warehouse to each store in order to fill all the orders?

b. Is there a way of doing this for less money? [**HINT:** See Example 4.]

26. *Inventory Control* The *Tubular Ride Boogie Board Company* has manufacturing plants in Tucson, Arizona, and Toronto, Ontario. You have been given the job of coordinating distribution of the latest model, the Gladiator, to outlets in Honolulu and Venice Beach. The Tucson plant, when operating at full capacity, can manufacture 620 Gladiator boards per week, while the Toronto plant, beset by labor disputes, can produce only 410 boards per week. The outlet in Honolulu orders 500 Gladiator boards per week, while Venice Beach orders 530 boards per week. Transportation costs are as follows:

Tucson to Honolulu: $10 per board; Tucson to Venice Beach: $5 per board.

Toronto to Honolulu: $20 per board; Toronto to Venice Beach: $10 per board.

a. Assuming that you wish to fill all orders and ensure full-capacity production at both plants, is it possible to meet a total transportation budget of $10,200? If so, how many Gladiator boards are shipped from each manufacturing plant to each distribution outlet?

b. Is there a way of doing this for less money? [**HINT:** See Example 4.]

27. ▼ *Tourism in the 1990s* In the 1990s, significant numbers of tourists traveled from North America and Europe to Australia and South Africa. In 1998 a total of 1,390,000 of these tourists visited Australia, while 1,140,000 of them visited South Africa. Further, 630,000 of them came from North America, and 1,900,000 of them came from Europe.[22] (Assume that no single tourist visited both destinations or traveled from both North America and Europe.)

a. The given information is not sufficient to determine the number of tourists from each region to each destination. Why?

b. If you were given the additional information that a total of 2,530,000 tourists traveled from these two regions to these two destinations, would you now be able to determine the number of tourists from each region to each destination? If so, what are these numbers?

c. If you were given the additional information that the same number of people from Europe visited South Africa as visited Australia, would you now be able to determine the number of tourists from each region to each destination? If so, what are these numbers?

28. ▼ *Tourism in the 1990s* In the 1990s, significant numbers of tourists traveled from North America and Asia to Australia and South Africa. In 1998 a total of 2,230,000 of these tourists visited Australia, while 390,000 of them visited South Africa. Also, 630,000 of these tourists came from North America, and a total of 2,620,000 tourists traveled from these two regions to these two destinations.[23] (Assume that no single tourist visited both destinations or traveled from both North America and Asia.)

a. The given information is not sufficient to determine the number of tourists from each region to each destination. Why?

b. If you were given the additional information that a total of 1,990,000 tourists came from Asia, would you now be able to determine the number of tourists from each region to each destination? If so, what are these numbers?

c. If you were given the additional information that 200,000 tourists visited South Africa from Asia, would you now be able to determine the number of tourists from each region to each destination? If so, what are these numbers?

29. ▼ *Alcohol* The following table shows some data from a 2000 study on substance use among 10th graders in the United States and Europe:[24]

	Used Alcohol	Alcohol-Free	Totals
U.S.	x	y	14,000
Europe	z	w	95,000
Totals	63,550	45,450	

a. The table leads to a linear system of four equations in four unknowns. What is the system? Does it have a unique solution? What does this indicate about the given and the missing data?

b. [T] Given that the number of U.S. 10th graders who were alcohol-free was 50% more than the number who had used alcohol, find the missing data.

30. ▼ *Tobacco* The following table shows some data from the same study cited in Exercise 29:[25]

	Smoked Cigarettes	Cigarette-Free	Totals
U.S.	x	y	14,000
Europe	z	w	95,000
Totals		70,210	109,000

a. The table leads to a linear system of four equations in four unknowns. What is the system? Does it have a

---

[22] Figures are rounded to the nearest 10,000. Sources: South African Dept. of Environmental Affairs and Tourism; Australia Tourist Commission/*New York Times*, January 15, 2000, p. C1.

[23] *Ibid.*

[24] "Used Alcohol" indicates consumption of alcohol at least once in the past 30 days. Source: Council of Europe/University of Michigan. "Monitoring the Future"/*New York Times*, February 21, 2001, p. A10.

[25] "Smoked Cigarettes" indicates that at least one cigarette was smoked in the past 30 days. Source: *Ibid.*

unique solution? What does this indicate about the missing data?

b. **T** Given that 31,510 more European 10th graders smoked cigarettes than U.S. 10th graders, find the missing data.

31. *Traffic Flow* One-way traffic through *Enormous State University* is shown in the figure, where the numbers indicate daily counts of vehicles.

a. Is it possible to determine the daily flow of traffic along each of the three streets from the information given? If your answer is *yes*, what is the traffic flow along each street? If your answer is *no*, what additional information would suffice?

b. Is a flow of 60 vehicles per day along Southwest Lane consistent with the information given?

c. What is the minimum traffic flow possible along Northwest Lane consistent with the information given? [**HINT:** See Example 3.]

32. *Traffic Flow* The traffic through downtown East Podunk flows through the one-way system shown below.

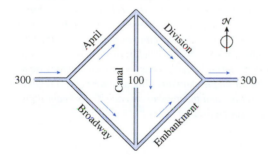

Traffic counters find that 300 vehicles enter town from the west each hour, and 300 leave town toward the east each hour. Also, 100 cars drive down Canal Street each hour.

a. Write down the general solution of the associated system of linear equations. Is it possible to determine the number of vehicles on each street per hour?

b. On which street could you put another traffic counter in order to determine the flow completely?

c. What is the minimum traffic flow along April Street consistent with the information given? [**HINT:** See Example 3.]

33. *Traffic Flow* The traffic through downtown Johannesburg follows the one-way system shown below, with traffic

movement recorded at incoming and outgoing streets (in cars per minute) as shown.

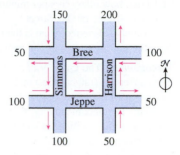

a. Set up and solve a system of equations to solve for the traffic flow along the middle sections of the four streets.

b. Is there sufficient information to calculate the traffic along the middle section of Jeppe Street? If so, what is it? If not, why not?

c. Given that 400 cars per minute flow down the middle section of Bree Street, how many cars per minute flow down the middle section of Simmons?

d. What is the minimum traffic flow along the middle section of Harrison Street?

e. Is there an upper limit to the possible traffic down Simmons Street consistent with the information given? Explain.

34. *Traffic Flow* Officials of the town of Hempstead were planning to make Hempstead Turnpike and the surrounding streets into one-way streets to prepare for the opening of Hofstra USA. In an experiment, they restricted traffic flow along the streets near Hofstra as shown in the diagram. The numbers show traffic flow per minute and the arrows indicate the direction of traffic.

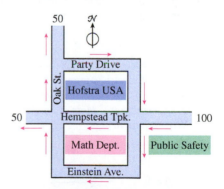

a. Set up and solve the associated traffic flow problem with the following unknowns:

x = Traffic per minute along the middle stretch of Oak St.

y = Traffic per minute along the middle stretch of Hempstead Tpk.

z = Traffic per minute along Einstein Ave.

u = Traffic per minute along Party Drive.

b. If 20 cars per minute drive along Party Drive, what is the traffic like along the middle stretch of Oak Street?

c. If 20 vehicles per minute drive along Einstein Ave. and 90 vehicles per minute drive down the middle stretch of Hempstead Tpk., how many cars per minute drive along the middle stretch of Oak Street?

d. If Einstein Avenue is deserted, what is the minimum traffic along the middle stretch of Hempstead Tpk.?

35. ▼ *Traffic Management* The Outer Village Town Council has decided to convert its (rather quiet) main street, Broadway, to a one-way street but is not sure of the direction of most of the traffic. The accompanying diagram illustrates the downtown area of Outer Village as well as the *net* traffic flow along the intersecting streets (in vehicles per day). (There are no one-way streets. A net traffic flow in a certain direction is defined as the traffic flow in that direction minus the flow in the opposite direction.)

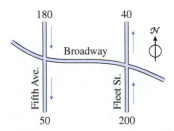

a. Is the given information sufficient to determine the net traffic flow along the three portions of Broadway shown? If your answer is *yes,* give the traffic flow along each stretch. If your answer is *no,* what additional information would suffice? [**HINT:** For the direction of net traffic flow, choose either east or west. If a corresponding value is negative, it indicates net flow in the opposite direction.]

b. Assuming that there is little traffic (fewer than 160 vehicles per day) east of Fleet Street, in what direction is the net flow of traffic along the remaining stretches of Broadway?

36. ▼ *Electric Current* Electric current measures (in **amperes**, or **amps**) the flow of electrons through wires. Like traffic flow, the current entering an intersection of wires must equal the current leaving it.[26] Here is an electrical circuit known as a **Wheatstone bridge.**

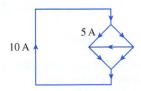

a. If the currents in two of the wires are 10 amps and 5 amps as shown, determine the currents in the unlabeled wires in terms of suitable parameters.

b. In which wire should you measure the current in order to know all of the currents exactly?

37. ▼ *Econometrics (from the GRE Economics Test)* This and the next exercise are based on the following simplified model of the determination of the money stock:

$$M = C + D$$
$$C = 0.2D$$
$$R = 0.1D$$
$$H = R + C,$$

where

$$M = \text{Money stock}$$
$$C = \text{Currency in circulation}$$
$$R = \text{Bank reserves}$$
$$D = \text{Deposits of the public}$$
$$H = \text{High-powered money}$$

If the money stock were \$120 billion, what would bank reserves have to be?

38. ▼ *Econometrics (from the GRE Economics Test)* With the model in Exercise 37, if *H* were equal to \$42 billion, what would *M* equal?

CAT Scans CAT (computerized axial tomographic) scans are used to map the exact location of interior features of the human body. CAT scan technology is based on the following principles: (1) Different components of the human body (water, gray matter, bone, etc.) absorb X-rays to different extents; and (2) to measure the X-ray absorption by a specific region of, say, the brain, it suffices to pass a number of line-shaped pencil beams of X-rays through the brain at different angles and measure the total absorption for each beam, which is the sum of the absorptions of the regions through which it passes. The accompanying diagram illustrates a simple example. (The number in each region shows its absorption, and the number on each X-ray beam shows the total absorption for that beam.)[27]

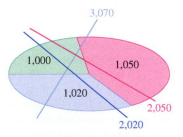

---

[26] This is known as **Kirchhoff's current law,** named after Gustav Robert Kirchhoff (1824–1887). Kirchhoff made important contributions to the fields of geometric optics, electromagnetic radiation, and electrical network theory.

[27] Based on a COMAP video, *Geometry: New Tools for New Technologies,* by J. Malkevitch, Video Applications Library, COMAP, 1992. The absorptions are actually calibrated on a logarithmic scale. In real applications the size of the regions is very small, and very large numbers of beams must be used.

In Exercises 39–44, use the table and the given X-ray absorption diagrams to identify the composition of each of the regions marked by a letter.

Type	Air	Water	Gray matter	Tumor	Blood	Bone
Absorption	0	1,000	1,020	1,030	1,050	2,000

39.

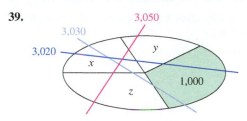

40.

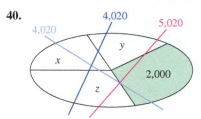

41.

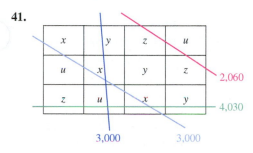

42.

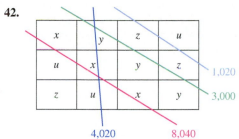

43. ▼ Identify the composition of site *x*.

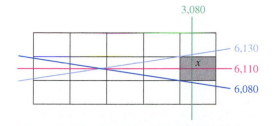

(The horizontal and slanted beams each pass through five regions.)

44. ▼ Identify the composition of site *x*.

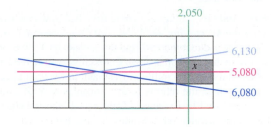

(The horizontal and slanted beams each pass through five regions.)

45. ◆ *Voting* In the 75th Congress (1937–1939) the U.S. House of Representatives had 333 Democrats, 89 Republicans, and 13 members of other parties. Suppose that a bill passed the House with 31 more votes in favor than against, with 10 times as many Democrats voting for the bill as Republicans, and with 36 more non-Democrats voting against the bill than for it. If every member voted either for the bill or against it, how many Democrats, how many Republicans, and how many members of other parties voted in favor of the bill?

46. ◆ *Voting* In the 75th Congress (1937–1939) there were in the Senate 75 Democrats, 17 Republicans, and 4 members of other parties. Suppose that a bill passed the Senate with 16 more votes in favor than against, with three times as many Democrats voting in favor as non-Democrats voting in favor, and with 32 more Democrats voting in favor than Republicans voting in favor. If every member voted either for the bill or against it, how many Democrats, how many Republicans, and how many members of other parties voted in favor of the bill?

47. ◆ *Investments* Things have not been going too well here at *Accurate Accounting, Inc.* since we hired Todd Smiley. He has a tendency to lose important documents, especially around April, when tax returns of our business clients are due. Today Smiley accidentally shredded *Colossal Conglomerate Corp.*'s investment records. We must therefore reconstruct them on the basis of the information he can gather. Todd recalls that the company earned an $8 million return on investments totaling $65 million last year. After a few frantic telephone calls to sources in Colossal, he learned that Colossal had made investments in four companies last year: X, Y, Z, and W. (For reasons of confidentiality we are withholding their names.) Investments in company X earned 15% last year, investments in Y depreciated by 20% last year, investments in Z neither appreciated nor depreciated last year, while investments in W earned 20% last year. Smiley was also told that Colossal invested twice as much in company X as in company Z, and three times as much in company W as in company Z. Does Smiley have sufficient information to piece together Colossal's investment portfolio before its tax return is due next week? If so, what does the investment portfolio look like?

48. ◆ *Investments* Things are going from bad to worse here at *Accurate Accounting, Inc.*! *Colossal Conglomerate Corp.*'s tax return is due tomorrow, and the accountant Todd Smiley seems to have no idea how Colossal earned a return of $8 million on a $65 million investment last year. It appears that, although the returns from companies X, Y, Z, and W were as listed in Exercise 47, the rest of the information there was wrong. What Smiley is now being told is that Colossal invested only in companies X, Y, and Z and that the investment in X amounted to $30 million. His sources in Colossal still maintain that twice as much was invested in company X as in company Z. What should Smiley do?

Communication and Reasoning Exercises

49. Are Exercises 1 and 2 realistic in their expectation of using up all the ingredients? What does your answer have to do with the solution(s) of the associated system of equations?

50. Suppose that you obtained a solution for Exercise 3 or 4 consisting of positive values that were not all whole numbers. What would such a solution signify about the situation in the exercise? Should you round these values to the nearest whole numbers?

In Exercises 51–56, x, y and z represent the weights of the three ingredients X, Y, and Z in a gasoline blend. Say which of the following is represented by a linear equation in x, y, and z, and give a form of the equation when it is.

51. The blend consists of 100 pounds of ingredient X.

52. The blend is free of ingredient X.

53. The blend contains 30% ingredient Y by weight.

54. The weight of ingredient X is the product of the weights of ingredients Y and Z.

55. There is at least 30% ingredient Y by weight.

56. There is twice as much ingredient X by weight as Y and Z combined.

57. Make up an entertaining word problem leading to the following system of equations:

$$
\begin{aligned}
10x + 20y + 10z &= 100 \\
5x + 15y &= 50 \\
x + y + z &= 10.
\end{aligned}
$$

58. Make up an entertaining word problem leading to the following system of equations:

$$
\begin{aligned}
10x + 20y &= 300 \\
10z + 20w &= 400 \\
20x + 10z &= 400 \\
10y + 20w &= 300.
\end{aligned}
$$

CHAPTER 3 REVIEW

KEY CONCEPTS

WM **www.WanerMath.com**
Go to the Website to find a
comprehensive and interactive
Web-based summary of Chapter 3.

3.1 Systems of Two Equations in Two Unknowns
Linear equation in two unknowns [p. 190]
Coefficient [p. 190]
Solution of an equation in two
unknowns [p. 190]
Particular and general (parameterized)
solutions of an equation in two
unknowns [p. 192]

Graphical method for solving a system
of two linear equations [p. 196]
Algebraic methods for solving a system
of two linear equations: elimination,
intersection, and substitution
[p. 197]
Possible outcomes for a system of two
linear equations [p. 198]
Consistent system [p. 198]
Redundant or dependent system [p. 198]

3.2 Using Matrices to Solve Systems of Equations
Linear equation (in any number of
unknowns) [p. 206]

Matrix [p. 208]
Augmented matrix of a system of linear
equations [p. 208]
Elementary row operations [p. 209]
Pivot [p. 210]
Clearing a column; pivoting [p. 210]
Gauss-Jordan or row reduction [p. 213]
Overdetermined and underdetermined
systems [p. 218]
Reduced row echelon form [p. 219]

3.3 Applications of Systems of Linear Equations
Resource allocation [p. 225]
(Traffic) flow [p. 227]
Transportation [p. 229]

REVIEW EXERCISES

In Exercises 1–6, graph the given equations and determine how many solutions the system has, if any.

1. $x + 2y = 4$
$2x - y = 1$

2. $0.2x - 0.1y = 0.3$
$0.2x + 0.2y = 0.4$

3. $\frac{1}{2}x - \frac{3}{4}y = 0$
$6x - 9y = 0$

4. $2x + 3y = 2$
$-x - \frac{3}{2}y = \frac{1}{2}$

5. $x + y = 1$
$2x + y = 0.3$
$3x + 2y = \frac{13}{10}$

6. $3x + 0.5y = 0.1$
$6x + y = 0.2$
$\frac{3x}{10} - 0.05y = 0.01$

In Exercises 7–18, solve the given system of linear equations.

7. $x + 2y = 4$
$2x - y = 1$

8. $0.2x - 0.1y = 0.3$
$0.2x + 0.2y = 0.4$

9. $\frac{1}{2}x - \frac{3}{4}y = 0$
$6x - 9y = 0$

10. $2x + 3y = 2$
$-x - \frac{3}{2}y = \frac{1}{2}$

11. $x + y = 1$
$2x + y = 0.3$
$3x + 2y = \frac{13}{10}$

12. $3x + 0.5y = 0.1$
$6x + y = 0.2$
$\frac{3x}{10} - 0.05y = 0.01$

13. $x + 2y \quad = -3$
$x \quad - z = 0$
$x + 3y - 2z = -2$

14. $x - y + z = 2$
$7x + y - z = 6$
$x - \frac{1}{2}y + \frac{1}{3}z = 1$
$x + y + z = 6$

15. $x - \frac{1}{2}y + z = 0$
$\frac{1}{2}x \quad - \frac{1}{2}z = -1$
$\frac{3}{2}x - \frac{1}{2}y + \frac{1}{2}z = -1$

16. $x + y - 2z = -1$
$-2x - 2y + 4z = 2$
$0.75x + 0.75y - 1.5z = -0.75$

17. $x = \frac{1}{2}y$
$\frac{1}{2}x = -\frac{1}{2}z + 2$
$z = -3x + y$

18. $x - y + z \quad = 1$
$y - z + w = 1$
$x \quad + z - w = 1$
$2x \quad + z \quad = 3$

Exercises 19–22 are based on the following equation relating the Fahrenheit and Celsius (or centigrade) temperature scales:

$$5F - 9C = 160,$$

where F is the Fahrenheit temperature of an object and C is its Celsius temperature.

19. What temperature should an object be if its Fahrenheit and Celsius temperatures are the same?

20. What temperature should an object be if its Celsius temperature is half its Fahrenheit temperature?

21. Is it possible for the Fahrenheit temperature of an object to be 1.8 times its Celsius temperature? Explain.

22. Is it possible for the Fahrenheit temperature of an object to be 30° more than 1.8 times its Celsius temperature? Explain.

239

In Exercises 23–28, let x, y, z, and w represent the population in millions of four cities A, B, C, and D, respectively. Express the given statement as an equation in x, y, z, and w. If the equation is linear, say so and express it in the standard form $ax + by + cz + dw = k.$

23. The total population of the four cities is 10 million people.

24. City A has three times as many people as cities B and C combined.

25. City D is actually a ghost town; there are no people living in it.

26. The population of City A is the sum of the squares of the populations of the other three cities.

27. City C has 30% more people than City B.

28. City C has 30% fewer people than City B.

Applications: OHaganBooks.com
[Try the game at www.OHaganBooks.com]

Purchasing You are the buyer for OHaganBooks.com and are considering increasing stocks of romance and horror novels at the new OHaganBooks.com warehouse in Texas. You have offers from two publishers: Duffin House *and* Higgins Press. *Duffin offers a package of 5 horror novels and 5 romance novels for $50, and Higgins offers a package of 5 horror and 11 romance novels for $150. Exercises 29–32 give different scenarios for your purchasing options in response to these offers.*

29. How many packages should you purchase from each publisher to get exactly 4,500 horror novels and 6,600 romance novels?

30. You want to spend a total of $50,000 on books and have promised to buy twice as many packages from Duffin as from Higgins. How many packages should you purchase from each publisher?

31. The accountant tells you that the company can actually afford to spend a total of $90,000 on romance and horror books. She also reminds you that you had signed an agreement to spend twice as much money for books from Duffin as from Higgins. How many packages should you purchase from each publisher?

32. Upon revising her records, the accountant now tells you that the company can afford to spend a total of only $60,000 on romance and horror books and that it is company policy to spend the same amount of money at both publishers. How many packages should you purchase from each publisher?

33. *Equilibrium* The demand for *Finite Math the OHagan Way* is given by $q = -1{,}000p + 140{,}000$ copies per year, where p is the price per book in dollars. The supply is given by $q = 2{,}000p + 20{,}000$ copies per year. Find the price at which supply and demand balance.

34. *Equilibrium* OHaganBooks.com CEO John O'Hagan announces to a stunned audience at the annual board meeting that he is considering expanding into the jumbo jet airline manufacturing business. The demand per year for jumbo jets is given by $q = -2p + 18$, where p is the price per jet in millions of dollars. The supply is given by $q = 3p + 3$. Find the price the envisioned O'Hagan jumbo jet division should charge to balance supply and demand.

35. *Feeding Schedules* Billy-Sean O'Hagan is John O'Hagan's son and a freshman in college. Billy's 36-gallon tropical fish tank contains three types of carnivorous creatures—baby sharks, piranhas, and squids—and he feeds them three types of delicacies: goldfish, angelfish, and butterfly fish. Each baby shark can consume 1 goldfish, 2 angelfish, and 2 butterfly fish per day; each piranha can consume 1 goldfish and 3 butterfly fish per day (the piranhas are rather large as a result of their diet); while each squid can consume 1 goldfish and 1 angelfish per day. After a trip to the local pet store, Billy-Sean was able to feed his creatures to capacity, and he noticed that 21 goldfish, 21 angelfish, and 35 butterfly fish were eaten. How many of each type of creature does he have?

36. *Resource Allocation* *Duffin House* is planning its annual Song Festival, when it will serve three kinds of delicacies: granola treats, nutty granola treats, and nuttiest granola treats. The following table shows the ingredients required (in ounces) for a single serving of each delicacy, as well as the total amount of each ingredient available:

	Granola	Nutty Granola	Nuttiest Granola	Total Available
Toasted Oats	1	1	5	1,500
Almonds	4	8	8	10,000
Raisins	2	4	8	4,000

The Song Festival planners at Duffin House would like to use up all the ingredients. Is this possible? If so, how many servings of each kind of delicacy can they make?

37. *Website Traffic* OHaganBooks.com has two principal competitors: *JungleBooks.com* and *FarmerBooks.com.* Combined website traffic at the three sites is estimated at 10,000 hits per day. Only 10% of the hits at OHaganBooks.com result in orders, whereas JungleBooks.com and FarmerBooks.com report that 20% of the hits at their sites result in book orders. Together, the three sites process 1,500 book orders per day. FarmerBooks.com appears to be the most successful of the three and gets as many book orders as the other two combined. What is the traffic (in hits per day) at each of the sites?

38. *Sales* As the buyer at OHaganBooks.com, you are planning to increase stocks of books about music, and have been monitoring worldwide sales. Last year, worldwide sales of books about rock, rap, and classical music amounted to $5.8 billion. Books on rock music brought in twice as much revenue as books on rap music and they brought in 900% the revenue of books on classical music. How much revenue was earned in each of the three categories of books?

39. Investing in Stocks Billy-Sean O'Hagan is the treasurer at his college fraternity, which recently earned $12,400 in its annual carwash fundraiser. Billy-Sean decided to invest all the proceeds in the purchase of three computer stocks: HAL, POM, and WELL.

	Price per Share ($)	Dividend Yield (%)
HAL	100	0.5
POM	20	1.50
WELL	25	0

If the investment was expected to earn $56 in annual dividends and he purchased a total of 200 shares, how many shares of each stock did he purchase?

40. Initial Public Offerings (IPOs) *Duffin House, Higgins Press,* and *Sickle Publications* all went public on the same day recently. John O'Hagan had the opportunity to participate in all three initial public offerings (partly because he and Marjory Duffin are good friends). He made a considerable profit when he sold all of the stock 2 days later on the open market. The following table shows the purchase price and percentage yield on the investment in each company:

	Purchase Price per Share ($)	Yield (%)
Duffin House (DHS)	8	20
Higgins Press (HPR)	10	15
Sickle Publications (SPUB)	15	15

He invested $20,000 in a total of 2,000 shares and made a $3,400 profit from the transactions. How many shares in each company did he purchase?

41. Degree Requirements During his lunch break, John O'Hagan decides to devote some time to assisting his son Billy-Sean, who is having a terrible time coming up with a college course schedule. One reason for this is the very complicated Bulletin of *Suburban State University.* It reads as follows:

> *All candidates for the degree of Bachelor of Science at SSU must take a total of 124 credits from the Sciences, Fine Arts, Liberal Arts, and Mathematics,[28] including an equal number of Science and Fine Arts credits, and*

---

[28] Strictly speaking, mathematics is not a science; it is the Queen of the Sciences, although we like to think of it as the Mother of all Sciences.

> *twice as many Mathematics credits as Science credits and Fine Arts credits combined, but with Liberal Arts credits exceeding Mathematics credits by exactly one-third of the number of Fine Arts credits.*

What are all the possible degree programs for Billy-Sean?

42. Degree Requirements Having finally decided on his degree program, Billy-Sean learns that the *Suburban State University* Senate (under pressure from the English Department) has revised the Bulletin to include a "Verbal Expression" component in place of the Fine Arts requirement in all programs (including the sciences):

> *All candidates for the degree of Bachelor of Science at SSU must take a total of 120 credits from the Liberal Arts, Sciences, Verbal Expression, and Mathematics, including an equal number of Science and Liberal Arts credits, and twice as many Verbal Expression credits as Science credits and Liberal Arts credits combined, but with Liberal Arts credits exceeding Mathematics credits by one quarter of the number of Verbal Expression Credits.*

What are now the possible degree programs for Billy-Sean?

43. Network Traffic All book orders received at the Order Department at OHaganBooks.com are transmitted through a small computer network to the Shipping Department. The following diagram shows the network (which uses two intermediate computers as routers), together with some of the average daily traffic measured in book orders:

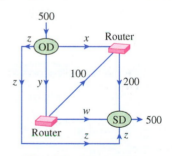

OD = Order department
SD = Shipping department

a. Set up a system of linear equations in which the unknowns give the average traffic along the paths labeled x, y, z, w, and find the general solution.
b. What is the minimum volume of traffic along y?
c. What is the maximum volume of traffic along w?
d. If there is no traffic along z, find the volume of traffic along all the paths.
e. If there is the same volume of traffic along y and z, what is the volume of traffic along w?

44. Business Retreats Marjory Duffin is planning a joint business retreat for *Duffin House* and OHaganBooks.com at Laguna Surf City, but she is concerned about traffic conditions. (She feels that too many cars tend to spoil the ambiance of a seaside retreat.) She managed to obtain the following map from the Laguna Surf City Engineering Department. (All the streets are one-way as indicated.) The counters show traffic every 5 minutes.

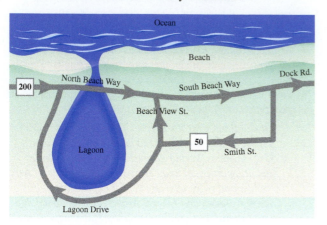

a. Set up and solve the associated system of linear equations. *Be sure to give the general solution.* (Take x = Traffic along North Beach Way, y = Traffic along South Beach Way, z = Traffic along Beach View St., u = Traffic along Lagoon Drive, and v = Traffic along Dock Road.)

b. Assuming that all roads are one-way in the directions shown, what, if any, is the maximum possible traffic along Lagoon Drive?

c. The Laguna Surf City Traffic Department is considering opening up Beach View Street to two-way traffic, but an environmentalist group is concerned that this will result in increased traffic on Lagoon Drive. What, if any, is the maximum possible traffic along Lagoon Drive assuming that Beach View Street is two-way and the traffic counter readings are as shown?

45. **Shipping** On the same day that the sales department at *Duffin House* received an order for 600 packages from the OHaganBooks.com Texas headquarters, it received an additional order for 200 packages from *FantasyBooks.com,* based in California. Duffin House has warehouses in New York and Illinois. The Illinois warehouse is closing down and must clear all 300 packages it has in stock. Shipping costs per package of books are as follows:

New York to Texas: $20   New York to California: $50

Illinois to Texas: $30   Illinois to California: $40

Is it possible to fill both orders and clear the Illinois warehouse at a cost of $22,000? If so, how many packages should be sent from each warehouse to each online bookstore?

46. **Transportation Scheduling** *Duffin House* is about to start a promotional blitz for its new book, *Physics for the Liberal Arts.* The company has 20 salespeople stationed in Chicago and 10 in Denver, and it would like to fly 15 of them to sales fairs at each of Los Angeles and New York. A round-trip plane flight from Chicago to LA costs $200; from Chicago to NY costs $150; from Denver to LA costs $400; and from Denver to NY costs $200. For tax reasons, Duffin House needs to budget exactly $6,500 for the total cost of the plane flights. How many salespeople should the company fly from each of Chicago and Denver to each of LA and NY?

Hybrid Cars—Optimizing the Degree of Hybridization

Fedor Selivanov/Alamy Stock Photo

You are involved in new model development at a major automobile company. The company is planning to introduce two new plug-in hybrid electric vehicles: the subcompact "Green Town Hopper" and the midsize "Electra Supreme," and your department must decide on the degree of hybridization (DOH) for each of these models that will result in the largest reduction in gasoline consumption. (The DOH of a vehicle is defined as the ratio of electric motor power to the total power; it typically ranges from 10% to 50%. For example, a model with a 20% DOH has an electric motor that delivers 20% of the total power of the vehicle.)

The tables below show the benefit for each of the two models, measured as the estimated reduction in annual gasoline consumption, as well as an estimate of retail cost increment, for various DOH percentages.[29] (The retail cost increment estimate

---

[29] The figures are approximate and based on data for two actual vehicles as presented in a 2006 paper entitled *Cost-Benefit Analysis of Plug-In Hybrid Electric Vehicle Technology* by A. Simpson. Source: National Renewable Energy Laboratory, U.S. Department of Energy (www.nrel.gov).

is given by the formula $5,000 + 50(DOH - 10)$ for the Green Town Hopper and $7,000 + 50(DOH - 10)$ for the Electra Supreme.)

Green Town Hopper

	10	20	50
DOH (%)	10	20	50
Reduction in Annual Consumption (gals)	180	230	200
Retail Cost Increment ($)	5,000	5,500	7,000

Electra Supreme

	10	20	50
DOH (%)	10	20	50
Reduction in Annual Consumption (gals)	220	270	260
Retail Cost Increment ($)	7,000	7,500	9,000

Notice that increasing the DOH toward 50% results in a decreased benefit. This is due in part to the need to increase the weight of the batteries while keeping the vehicle performance at a desirable level, thus necessitating a more powerful gasoline engine. The *optimum* DOH is the percentage that gives the largest reduction in gasoline consumption, and this is what you need to determine. Since the optimum DOH may not be 20%, you would like to create a mathematical model to compute the reduction R in gas consumption as a function of the DOH x. Your first inclination is to try linear equations—that is, an equation of the form

$$R = ax + b \qquad (a \text{ and } b \text{ constants}),$$

but you quickly discover that the data simply won't fit, no matter what the choice of the constants. The reason for this can be seen graphically by plotting R versus x (Figure 14). In neither case do the three points lie on a straight line. In fact, the data are not even *close* to being linear. Thus, you will need curves to model these data. After giving the matter further thought, you remember something your mathematics instructor once said: The simplest curve passing through any three points not all on the same line is a parabola. Since you are looking for a simple model of the data, you decide to try a parabola. A general parabola has the equation

$$R = ax^2 + bx + c,$$

where a, b, and c are constants. The problem now is: What are a, b, and c? You decide to try substituting the values of R and x for the Green Town Hopper into the general equation, and you get the following:

$$x = 10, R = 180 \qquad \text{gives} \quad 180 = 100a + 10b + c$$
$$x = 20, R = 230 \qquad \text{gives} \quad 230 = 400a + 20b + c$$
$$x = 50, R = 200 \qquad \text{gives} \quad 200 = 2,500a + 50b + c.$$

Now you notice that you have three linear equations in three unknowns! You solve the system:

$$a = -0.15, \quad b = 9.5, \quad c = 100.$$

Thus, your reduction equation for the Green Town Hopper becomes

$$R = -0.15x^2 + 9.5x + 100.$$

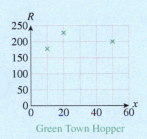

Green Town Hopper

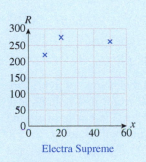

Electra Supreme

Figure 14

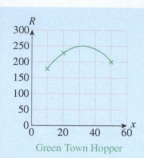

Green Town Hopper

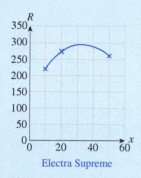

Electra Supreme

Figure 15

For the Electra Supreme you get

$$x = 10, R = 220: \quad 220 = 100a + 10b + c$$
$$x = 20, R = 270: \quad 270 = 400a + 20b + c$$
$$x = 50, R = 260: \quad 260 = 2{,}500a + 50b + c.$$

$$a = -0.1\overline{3}, \quad b = 9, \quad c = 143.\overline{3},$$

so

$$R = -0.1\overline{3}x^2 + 9x + 143.\overline{3}.$$

Figure 15 shows the parabolas superimposed on the data points. You can now estimate a value for the optimal DOH as the value of x that gives the largest benefit R. Recalling that the x-coordinate of the vertex of the parabola $y = ax^2 + bx + c$ is $x = -\frac{b}{2a}$, you obtain the following estimates:

$$\text{Green Town Hopper:} \quad \text{Optimal DOH} = -\frac{9.5}{2(-0.15)} \approx 31.67\%$$

$$\text{Electra Supreme:} \quad \text{Optimal DOH} = -\frac{9}{2(-0.1\overline{3})} \approx 33.75\%.$$

You can now use the formulas given earlier to estimate the resulting reductions in gasoline consumption and increases in cost:

Green Town Hopper:

Reduction in gasoline consumption = $R \approx -0.15(31.67)^2 + 9.5(31.67) + 100$
$$\approx 250.4 \text{ gallons per year}$$

Retail cost increment $\approx 5{,}000 + 50(31.67 - 10) \approx \$6{,}080.$

Electra Supreme:

Reduction in gasoline consumption = $R = -0.1\overline{3}(33.75)^2 + 9(33.75) + 143.\overline{3}$
$$\approx 295.2 \text{ gallons per year}$$

Retail cost increment $\approx 7{,}000 + 50(33.75 - 10) \approx \$8{,}190.$

You therefore submit the following estimates: The optimal degree of hybridization for the Green Town Hopper is about 31.67% and will result in a reduction in gasoline consumption of 250.4 gallons per year and a retail cost increment of around $6,080. The optimal degree of hybridization for the Electra Supreme is about 33.75% and will result in a reduction in gasoline consumption of 295.2 gallons per year and a retail cost increment of around $8,190.

EXERCISES

1. Repeat the computations above for the Earth Suburban using the following data:

Earth Suburban

DOH (%)	10	20	50
Reduction in Annual Consumption (gals)	240	330	300
Retail Cost Increment ($)	9,000	9,500	11,000

Retail cost increment $= 9,000 + 50(DOH - 10)$

2. [T] Repeat the analysis for the Green Town Hopper, but this time take x to be the cost increment, in thousands of dollars. (The curve of benefit versus cost is referred to as a *cost-benefit* curve.) What value of DOH corresponds to the optimal cost? What do you notice? Comment on the answer.

3. [T] Repeat the analysis for the Electra Supreme, but this time use the optimal values, DOH = 33.8, Reduction = 295.2 gallons per year in place of the 20% data. What do you notice?

4. Find the equation of the parabola that passes through the points $(1, 2)$, $(2, 9)$, and $(3, 19)$.

5. Is there a parabola that passes through the points $(1, 2)$, $(2, 9)$, and $(3, 16)$?

6. Is there a parabola that passes though the points $(1, 2)$, $(2, 9)$, $(3, 19)$, and $(-1, 2)$?

7. [T] You submit your recommendations to your manager, and she tells you, "Thank you very much, but we have additional data for the Green Town Hopper: A 30% DOH results in a saving of 255 gallons per year. Please resubmit your recommendations taking this into account by tomorrow." [**HINT:** You now have four data points on each graph, so try a general cubic instead: $R = ax^3 + bx^2 + cx + d$. Use a graph to estimate the optimal DOH.]

Section 3.1

Example 2 (page 193)   Find all solutions (x, y) of the following system of two equations:

$$x + y = 3$$
$$x - y = 1.$$

Solution

You can use a graphing calculator to draw the graphs of the two equations on the same set of axes and to check the solution. First, solve the equations for y, obtaining $y = -x + 3$ and $y = x - 1$. On the TI-83/84 Plus:

1. Set

   $Y_1 = -X+3$

   $Y_2 = X-1$

2. Decide on the range of x-values you want to use. As in Figure 1, let us choose the range $[-4, 4]$.[30]
3. In the WINDOW menu, set $Xmin = -4$ and $Xmax = 4$.
4. Press ZOOM, and select ZoomFit to set the y-range.

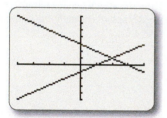

You can now zoom in for a more accurate view by choosing a smaller x-range that includes the point of intersection, such as $[1.5, 2.5]$, and using ZoomFit again. You can also use the trace feature to see the coordinates of points near the point of intersection.

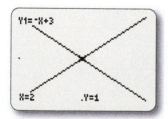

To check that $(2, 1)$ is the correct solution, use the table feature to compare the two values of y corresponding to $x = 2$:

1. Press 2ND TABLE.
2. Set $X = 2$, and compare the corresponding values of Y_1 and Y_2; they should each be 1.

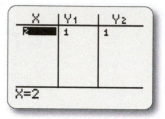

Q: *How accurate is the answer shown using the trace feature?*

A: That depends. We can increase the accuracy up to a point by zooming in on the point of intersection of the two graphs. But there is a limit to this: Most graphing calculators are capable of giving an answer correct to about 13 decimal places. This means, for instance, that, in the eyes of the TI-83/84 Plus, 2.000 000 000 000 1 is exactly the same as 2. (Subtracting them yields 0.) It follows that if you attempt to use a window so narrow that you need approximately 13 significant digits to distinguish the left and right edges, you will run into accuracy problems.

Section 3.2

Row Operations with a TI-83/84 Plus   Start by entering the matrix into [A] using MATRIX EDIT. You can then do row operations on [A] using the instructions found in the following table. (*row, *row+, and rowSwap are found in the MATRIX MATH menu.)

---

[30] How did we come up with this interval? Trial and error. You might need to try several intervals before finding one that gives a graph showing the point of intersection clearly.

Row Operation	TI-83/84 Plus Instruction (Matrix name is [A])
$R_i \rightarrow kR_i$	`*row(k,[A],i)→[A]`
Example: $R_2 \rightarrow 3R_2$	`*row(3,[A],2)→[A]`
$R_i \rightarrow R_i + kR_j$	`*row+(k,[A],j,i)→[A]`
Examples: $R_1 \rightarrow R_1 - 3R_2$	`*row+(-3,[A],2,1)→[A]`
$\qquad\qquad R_1 \rightarrow 4R_1 - 3R_2$	`*row(4,[A],1)→[A]` `*row+(-3,[A],2,1)→[A]`
Swap R_i and R_j	`rowSwap([A],i,j)→[A]`
Example: Swap R_1 and R_2	`rowSwap([A],1,2)→[A]`

Example 1 (page 213) Solve the system:

$$x - \quad y + 5z = -6$$
$$3x + 3y - \quad z = 10$$
$$x + 3y + 2z = 5.$$

Solution

1. Begin by entering the matrix into [A] using MATRIX EDIT. Only three columns can be seen at a time; you can see the rest of the matrix by scrolling left or right.

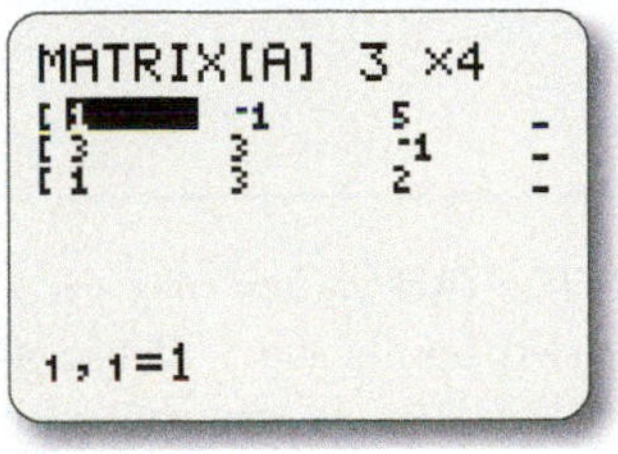

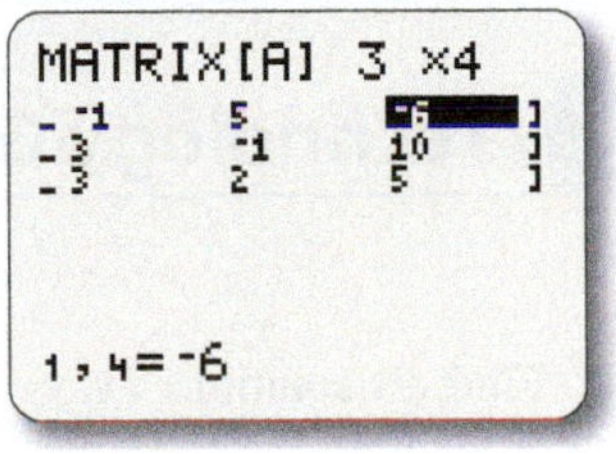

2. Now perform the operations given in Example 1.

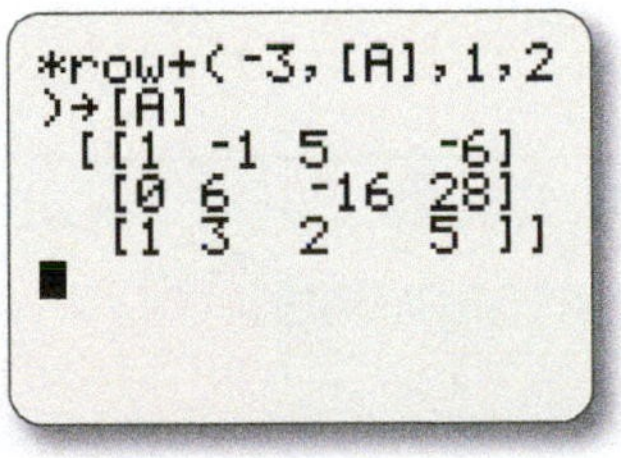

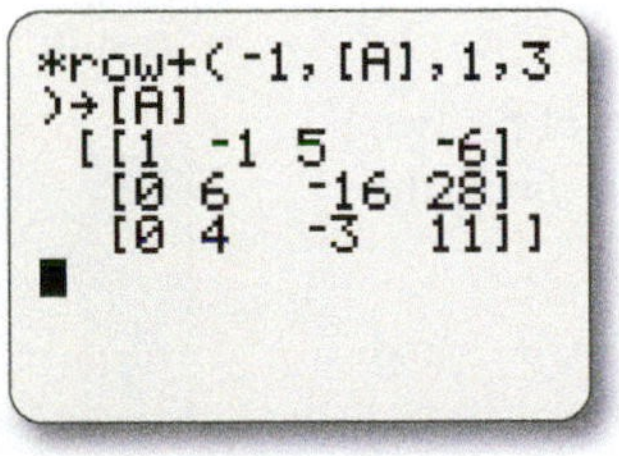

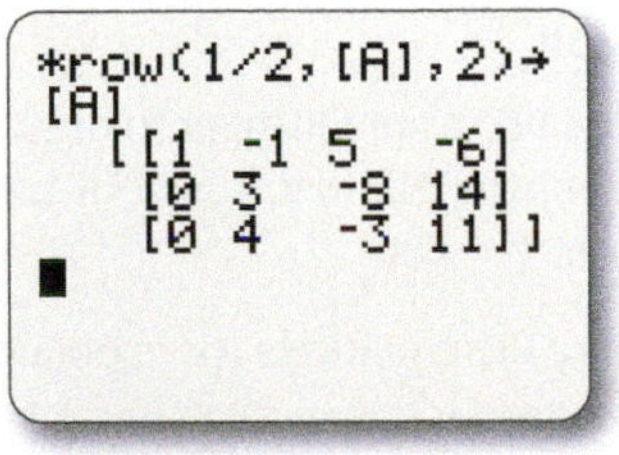

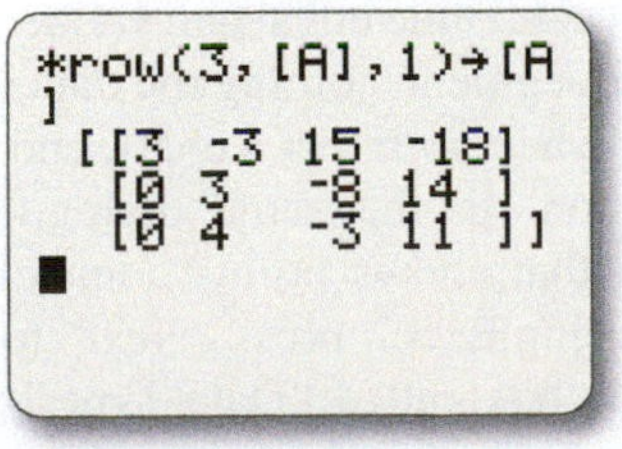

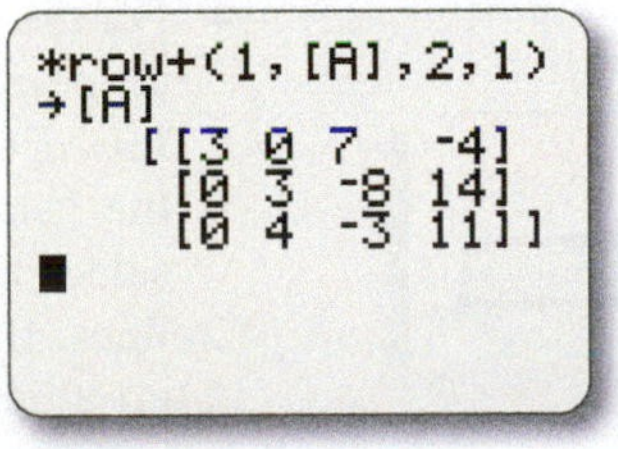

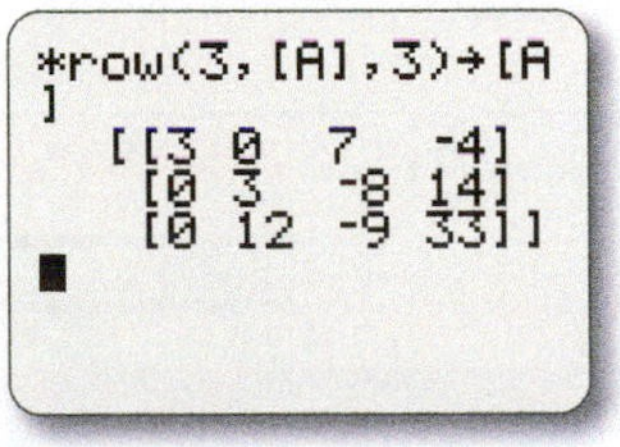

```
*row+(-4,[A],2,3
)→[A]
   [[3  0  7   -4 ]
    [0  3  -8  14 ]
    [0  0  23  -23]]
■
```

```
*row(1/23,[A],3)
→[A]
   [[3  0  7   -4]
    [0  3  -8  14]
    [0  0  1   -1]]
■
```

```
*row+(-7,[A],3,1
)→[A]
   [[3  0  0   3 ]
    [0  3  -8  14]
    [0  0  1   -1]]
■
```

```
*row+(8,[A],3,2)
→[A]
   [[3  0  0  3 ]
    [0  3  0  6 ]
    [0  0  1  -1]]
■
```

```
*row(1/3,[A],1)→
[A]
   [[1  0  0  1 ]
    [0  3  0  6 ]
    [0  0  1  -1]]
■
```

```
*row(1/3,[A],2)→
[A]
   [[1  0  0  1 ]
    [0  1  0  2 ]
    [0  0  1  -1]]
■
```

As in Example 1, we can now read the solution from the right-hand column: $x = 1$, $y = 2$, $z = -1$.

Note The TI-83/84 Plus has a function, `rref`, which gives the reduced row echelon form of a matrix in one step. (See the text for the definition of reduced row echelon form.) Internally, it uses a variation of Gauss-Jordan reduction to do this. ■

Spreadsheet   Technology Guide

Section 3.1

Example 2 (page 193) Find all solutions (x, y) of the following system of two equations:

$$x + y = 3$$
$$x - y = 1.$$

Solution

You can use a spreadsheet to draw the graphs of the two equations on the same set of axes and to check the solution.

1. Solve the equations for y, obtaining $y = -x + 3$ and $y = x - 1$.

2. To graph these lines, we can use the following simple worksheet:

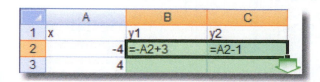

The two values of x give the x-coordinates of the two points we will use as endpoints of the lines. (We have—somewhat arbitrarily—chosen the range $[-4, 4]$ for x.) The formula for the first line, $y = -x + 3$, is in cell B2, and the formula for the second line, $y = x - 1$, is in cell C2.

3. Copy these two cells as shown to yield the following result:

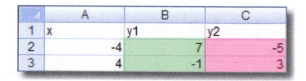

4. For the graph, select all nine cells, and create a scatter graph with line segments joining the data points. Instruct the spreadsheet to insert a chart, and select the "scatter" option. In the same dialogue box, select the option that shows points connected by lines. If you are using Excel, press "Next" to bring up a new dialogue box called "Data Type," where

you should make sure that the "Series in Columns" option is selected, telling the program that the *x*- and *y*-coordinates are arranged vertically, down columns. (Other spreadsheet programs have corresponding options you may need to set.) Your graph should appear as shown below.

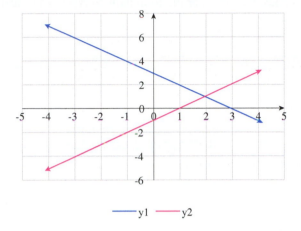

—— y1 —— y2

To zoom in:

1. First decide on a new *x*-range, say, $[1, 3]$.
2. Change the value in cell A2 to 1 and the value in cell A3 to 3, and the spreadsheet will automatically update the *y*-values and the graph.[31]

To check that $(2, 1)$ is the correct solution:

1. Enter the value 2 in cell A4 in your spreadsheet.
2. Copy the formulas in cells B3 and C3 down to row 4 to obtain the corresponding values of *y*.

	A	B	C
1	x	y1	y2
2	-4	7	-5
3	4	-1	3
4	2	1	1

Because the values of *y* agree, we have verified that $(2, 1)$ is a solution of both equations.

Section 3.2

Row Operations with a Spreadsheet   To a spreadsheet, a block of data with one or more rows or columns is an **array**, and a spreadsheet has built in the capability to handle arrays in much the same way that it handles single cells. Consider the following example:

$$\begin{bmatrix} 1 & 3 & -4 \\ 0 & 4 & 2 \end{bmatrix} 3R_2 \rightarrow \begin{bmatrix} 1 & 3 & -4 \\ 0 & 12 & 6 \end{bmatrix}.$$ Replace R_2 by $3R_2$.

1. Enter the original matrix in a convenient location, say, the cells A1 through C2.
2. To get the first row of the new matrix, which is simply a copy of the first row of the old matrix, decide where you want to place the new matrix, and highlight *the whole row of the new matrix,* say, A4:C4.
3. Enter the formula =A1:C1. (The easiest way to do this is to type "=" and then use the mouse to select cells A1 through C1, that is, the first row of the old matrix.)

Enter formula for new first row.   Press Control+Shift+Enter.

---

4. Press **Control+Shift+Enter** (instead of "Enter" alone), and the whole row will be copied.[32] Pressing Control+Shift+Enter tells the spreadsheet that your formula is an *array formula,* one that returns an array rather than a single number. Once entered, the spreadsheet will show an array formula enclosed in "curly braces." (Note that you must also use Control+Shift+Enter to delete any array you create: Select the block you wish to delete, and press Delete followed by Control+Shift+Enter.)

5. Similarly, to get the second row, select cells A5 through C5, where the new second row will go, enter the formula =3*A2:C2, and press Control+Shift+Enter. (Again, the easiest way to enter the formula is to type "=3*" and then select cells A2 through C2 using the mouse.)

Enter formula for new second row.

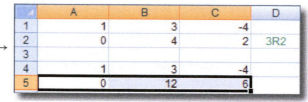

Press Control+Shift+Enter.

We can perform the following operation in a similar way:

$$\begin{bmatrix} 1 & 3 & -4 \\ 0 & 4 & 2 \end{bmatrix} \xrightarrow{4R_1 - 3R_2} \begin{bmatrix} 4 & 0 & -22 \\ 0 & 4 & 2 \end{bmatrix}.$$ Replace R_1 by $4R_1 - 3R_2$.

(To easily enter the formula for $4R_1 - 3R_2$, type "=4*", select the first row A1:C1 using the mouse, type "-3*", and then select the second row A2:C2 using the mouse.)

Example 1 (page 213)   Solve the system:

$$\begin{aligned} x - \; y + 5z &= -6 \\ 3x + 3y - \; z &= 10 \\ x + 3y + 2z &= 5. \end{aligned}$$

---

[32] Note that, on a Mac, Command-Enter and Command+Shift+Enter have the same effect as Control+Shift+Enter.

Solution

Here is the complete row reduction as it would appear in a spreadsheet.

	A	B	C	D	E
1	1	-1	5	-6	
2	3	3	-1	10	R2 - 3R1
3	1	3	2	5	R3 - R1
4					
5	=A1:D1				
6	=A2:D2-3*A1:D1				
7	=A3:D3-A1:D1				

	A	B	C	D	E
1	1	-1	5	-6	
2	3	3	-1	10	R2 - 3R1
3	1	3	2	5	R3 - R1
4					
5	1	-1	5	-6	
6	0	6	-16	28	(1/2)R2
7	0	4	-3	11	
8					
9	=A5:D5				
10	=(1/2)*A6:D6				
11	=A7:D7				

	A	B	C	D	E
8					
9	1	-1	5	-6	3R1 + R2
10	0	3	-8	14	
11	0	4	-3	11	3R3 - 4R2
12					
13	=3*A9:D9+A10:D10				
14	=A10*D10				
15	=3*A11:D11-4*A10:D10				

	A	B	C	D	E
12					
13	3	0	7	-4	
14	0	3	-8	14	
15	0	0	23	-23	(1/23)R3
16					
17	=A13:D13				
18	=A14:D14				
19	=(1/23)*A15:D15				

	A	B	C	D	E
16					
17	3	0	7	-4	R1 - 7R3
18	0	3	-8	14	R2 + 8R3
19	0	0	1	-1	
20					
21	=A17:D17-7*A19:D19				
22	=A18:D18+8*A19:D19				
23	=A19:D19				

	A	B	C	D	E
20					
21	3	0	0	3	(1/3)R1
22	0	3	0	6	(1/3)R2
23	0	0	1	-1	
24					
25	=(1/3)*A21:D21				
26	=(1/3)*A22:D22				
27	=A23:D23				

24				
25	1	0	0	1
26	0	1	0	2
27	0	0	1	-1

As in Example 1, we can now read the solution from the right-hand column: $x = 1$, $y = 2$, $z = -1$. What do you notice if you change the entries in cells D1, D2, and D3?

4

MATRIX ALGEBRA AND APPLICATIONS

CASE STUDY

Predicting Market Share

You are the sales director at *Selular,* a cellphone provider, and things are not looking good for your company: *iClone*, a recently launched competitor, is beginning to chip away at Selular's market share. Particularly disturbing are rumors of fierce brand loyalty by iClone customers, with several bloggers suggesting that iClone retains close to 100% of their customers. Worse, you will shortly be presenting a sales report to the board of directors, and the CEO has "suggested" that your report include 2-, 5-, and 10-year projections of Selular's market share given the recent impact on the market by iClone. The CEO also wants a worst-case scenario projecting what would happen if iClone customers were so loyal that none of them ever switch services. You have results from two market surveys, taken one quarter apart, of the major cellphone providers, which show the current market shares and the percentages of subscribers who switched from one service to another during the quarter.

How should you respond?

oneinchpunch/Shutterstock.com

Introduction

We used matrices in Chapter 3 simply to organize our work. It is time we examined them as interesting objects in their own right. There is much that we can do with matrices besides row operations: We can add, subtract, multiply, and even, in a sense, "divide" matrices. We use these operations to study game theory and input-output models in this chapter, and Markov chains in a later chapter.

Many calculators, spreadsheets, and other computer programs can do these matrix operations, which is a big help in doing calculations. However, we need to know how these operations are defined to see why they are useful and to understand which to use in any particular application.

4.1 Matrix Addition and Scalar Multiplication

Matrices

Let's start by formally defining what a matrix is and introducing some basic terms.

Matrix, Dimension, and Entries

An **$m \times n$ matrix** A is a rectangular array of real numbers with m rows and n columns. We refer to m and n as the **dimensions** of the matrix. The numbers that appear in the matrix are called its **entries**. We customarily use capital letters A, B, C, . . . for the names of matrices.

Quick Examples

1. $A = \begin{bmatrix} 2 & 0 & 1 \\ 33 & -22 & 0 \end{bmatrix}$ is a 2 × 3 matrix because it has two rows and three columns.

2. $B = \begin{bmatrix} 2 & 3 \\ 10 & 44 \\ -1 & 3 \\ 8 & 3 \end{bmatrix}$ is a 4 × 2 matrix because it has four rows and two columns.[*]

The entries of A are 2, 0, 1, 33, -22, and 0. The entries of B are the numbers 2, 3, 10, 44, -1, 3, 8, and 3.

---

[*] Remember that the number of rows is given first and the number of columns second. An easy way to remember this is to think of the acronym "RC" for "Row then Column."

Referring to the Entries of a Matrix

There is a systematic way of referring to particular entries in a matrix. If i and j are numbers, then the entry in the ith row and jth column of the matrix A is called the **ijth entry** of A. We usually write this entry as a_{ij} or A_{ij}. (If the matrix were called B, we would write its ijth entry as b_{ij} or B_{ij}.) Notice that this follows the "RC" convention: The row number is specified first, and the column number is specified second.

Quick Example

3. With $A = \begin{bmatrix} 2 & 0 & 1 \\ 33 & -22 & 0 \end{bmatrix}$,

$a_{13} = 1$ First row, third column

$a_{21} = 33$.   Second row, first column

According to the labeling convention, the entries of the matrix A above are

$$A = \begin{bmatrix} a_{11} & a_{12} & a_{13} \\ a_{21} & a_{22} & a_{23} \end{bmatrix}.$$

In general, the $m \times n$ matrix A has its entries labeled as follows:

$$A = \begin{bmatrix} a_{11} & a_{12} & a_{13} & \cdots & a_{1n} \\ a_{21} & a_{22} & a_{23} & \cdots & a_{2n} \\ \vdots & \vdots & \vdots & \ddots & \vdots \\ a_{m1} & a_{m2} & a_{m3} & \cdots & a_{mn} \end{bmatrix}.$$

We say that two matrices A and B are **equal** if they have the same dimensions and the corresponding entries are equal. Note that a 3×4 matrix can never equal a 3×5 matrix because they do not have the same dimensions.

EXAMPLE 1   **Matrix Equality**

Let $A = \begin{bmatrix} 7 & 9 & x \\ 0 & -1 & y+1 \end{bmatrix}$ and $B = \begin{bmatrix} 7 & 9 & 0 \\ 0 & -1 & 11 \end{bmatrix}$. Find the values of x and y such that $A = B$.

Solution For the two matrices to be equal, we must have corresponding entries equal, so

$$x \qquad = 0 \qquad\qquad a_{13} = b_{13}$$
$$y + 1 = 11 \quad \text{or} \quad y = 10. \qquad a_{23} = b_{23}$$

➡ **Before we go on . . .** Note in Example 1 that the matrix equation

$$\begin{bmatrix} 7 & 9 & x \\ 0 & -1 & y+1 \end{bmatrix} = \begin{bmatrix} 7 & 9 & 0 \\ 0 & -1 & 11 \end{bmatrix}$$

is really six equations in one: $7 = 7$, $9 = 9$, $x = 0$, $0 = 0$, $-1 = -1$, and $y + 1 = 11$. We used only the two that were interesting. ■

Row Matrix, Column Matrix, and Square Matrix

A matrix with a single row is called a **row matrix** or **row vector**. A matrix with a single column is called a **column matrix** or **column vector**. A matrix with the same number of rows as columns is called a **square matrix**.

Quick Examples

4. The 1×5 matrix $C = \begin{bmatrix} 3 & -4 & 0 & 1 & -11 \end{bmatrix}$ is a row matrix.

5. The 4×1 matrix $D = \begin{bmatrix} 2 \\ 10 \\ -1 \\ 8 \end{bmatrix}$ is a column matrix.

6. The 3×3 matrix $E = \begin{bmatrix} 1 & -2 & 0 \\ 0 & 1 & 4 \\ -4 & 32 & 1 \end{bmatrix}$ is a square matrix.

Matrix Addition and Subtraction

The first matrix operations we discuss are matrix addition and subtraction. The rules for these operations are simple.

Matrix Addition and Subtraction

Two matrices can be added (or subtracted) if and only if they have the same dimensions. To add (or subtract) two matrices of the same dimensions, we add (or subtract) the corresponding entries. More formally, if A and B are $m \times n$ matrices, then $A + B$ and $A - B$ are the $m \times n$ matrices whose entries are given by

$$(A + B)_{ij} = A_{ij} + B_{ij} \qquad \text{\textit{ij}th entry of the sum = Sum of the \textit{ij}th entries}$$
$$(A - B)_{ij} = A_{ij} - B_{ij}. \qquad \text{\textit{ij}th entry of the difference = Difference of the \textit{ij}th entries}$$

Visualizing Matrix Addition

$$\begin{bmatrix} 2 & -3 \\ 1 & 0 \end{bmatrix} + \begin{bmatrix} 1 & 1 \\ -2 & 1 \end{bmatrix} = \begin{bmatrix} 3 & -2 \\ -1 & 1 \end{bmatrix}$$

Quick Examples

7. $\begin{bmatrix} 2 & -3 \\ 1 & 0 \\ -1 & 3 \end{bmatrix} + \begin{bmatrix} 9 & -5 \\ 0 & 13 \\ -1 & 3 \end{bmatrix} = \begin{bmatrix} 11 & -8 \\ 1 & 13 \\ -2 & 6 \end{bmatrix}$ Corresponding entries added

8. $\begin{bmatrix} 2 & -3 \\ 1 & 0 \\ -1 & 3 \end{bmatrix} - \begin{bmatrix} 9 & -5 \\ 0 & 13 \\ -1 & 3 \end{bmatrix} = \begin{bmatrix} -7 & 2 \\ 1 & -13 \\ 0 & 0 \end{bmatrix}$ Corresponding entries subtracted

Using Technology

Technology can be used to enter, add, and subtract matrices. Here is an outline (see the Technology Guides at the end of the chapter for additional details on using a TI-83/84 Plus or a spreadsheet):

TI-83/84 Plus
Entering a matrix: MATRIX ; EDIT
Select a name, ENTER ; type in the entries.
Adding two matrices:
Home screen: [A] + [B]
(Use MATRIX ; NAMES to enter them on the Home screen.)
[More details in the Technology Guide.]

Spreadsheet
Entering a matrix: Type entries in a convenient block of cells.
Adding two matrices: Highlight block where you want the answer to appear.
Type "="; highlight first matrix; type "+"; highlight second matrix; press Control+Shift+Enter
[More details in the Technology Guide.]

WM Website
www.WanerMath.com
→ Online Utilities
→ Matrix Algebra Tool
Enter matrices as shown (use a single letter for the name).

```
      Enter your matrices here.
J = [20, 15
10, 12
8, 4]

F = [23, 12,
8, 12
4, 5]
```

To compute their difference, type F-J in the formula box, and press "Compute". (You can enter multiple formulas separated by commas in the formula box. For instance, F+J, F-J will compute both the sum and the difference. *Note:* The utility is case sensitive, so be consistent.)

EXAMPLE 2   Sales

The *A-Plus* auto parts store chain has two outlets, one in Vancouver and one in Quebec. Among other things, it sells wiper blades, windshield cleaning fluid, and floor mats. The monthly sales of these items at the two stores for 2 months are given in the following tables:

January Sales

	Vancouver	Quebec
Wiper Blades	20	15
Cleaning Fluid (bottles)	10	12
Floor Mats	8	4

February Sales

	Vancouver	Quebec
Wiper Blades	23	12
Cleaning Fluid (bottles)	8	12
Floor Mats	4	5

Use matrix arithmetic to calculate the change in sales of each product in each store from January to February.

Solution The tables suggest two matrices:

$$J = \begin{bmatrix} 20 & 15 \\ 10 & 12 \\ 8 & 4 \end{bmatrix} \quad \text{and} \quad F = \begin{bmatrix} 23 & 12 \\ 8 & 12 \\ 4 & 5 \end{bmatrix}.$$

To compute the change in sales of each product for both stores, we want to subtract corresponding entries in these two matrices. In other words, we want to compute the difference of the two matrices:

$$F - J = \begin{bmatrix} 23 & 12 \\ 8 & 12 \\ 4 & 5 \end{bmatrix} - \begin{bmatrix} 20 & 15 \\ 10 & 12 \\ 8 & 4 \end{bmatrix} = \begin{bmatrix} 3 & -3 \\ -2 & 0 \\ -4 & 1 \end{bmatrix}.$$

Thus, the change in sales of each product is the following:

	Vancouver	Quebec
Wiper Blades	3	−3
Cleaning Fluid (bottles)	−2	0
Floor Mats	−4	1

Scalar Multiplication

A matrix A can be added to itself because the expression $A + A$ is the sum of two matrices that have the same dimensions. When we compute $A + A$, we end up doubling every entry in A. So we can think of the expression $2A$ as telling us to *multiply every element in A by 2*.

In general, to multiply a matrix by a number, multiply every entry in the matrix by that number. For example,

$$6\begin{bmatrix} \frac{5}{2} & -3 \\ 1 & 0 \\ -1 & \frac{5}{6} \end{bmatrix} = \begin{bmatrix} 15 & -18 \\ 6 & 0 \\ -6 & 5 \end{bmatrix}.$$

It is traditional in talking about matrices to call individual numbers **scalars**. For this reason we call the operation of multiplying a matrix by a number **scalar multiplication**.

EXAMPLE 3   Sales

The revenue generated by sales in the Vancouver and Quebec branches of the *A-Plus* auto parts store (see Example 2) was as follows:

January Sales in Canadian Dollars

	Vancouver	Quebec
Wiper Blades	140.00	105.00
Cleaning Fluid	30.00	36.00
Floor Mats	96.00	48.00

If the Canadian dollar was worth $0.65 U.S. at the time, compute the revenue in U.S. dollars.

Solution We need to multiply each revenue figure by 0.65. Let A be the matrix of revenue figures in Canadian dollars:

$$A = \begin{bmatrix} 140.00 & 105.00 \\ 30.00 & 36.00 \\ 96.00 & 48.00 \end{bmatrix}.$$

The revenue figures in U.S. dollars are then given by the scalar multiple

$$0.65A = 0.65\begin{bmatrix} 140.00 & 105.00 \\ 30.00 & 36.00 \\ 96.00 & 48.00 \end{bmatrix} = \begin{bmatrix} 91.00 & 68.25 \\ 19.50 & 23.40 \\ 62.40 & 31.20 \end{bmatrix}.$$

In other words, in U.S. dollars, $91 worth of wiper blades was sold in Vancouver, $68.25 worth of wiper blades was sold in Quebec, and so on.

Formally, scalar multiplication is defined as follows.

Scalar Multiplication

If A is an $m \times n$ matrix and c is a real number, then cA is the $m \times n$ matrix obtained by multiplying all the entries of A by c. (We usually use lowercase letters $c, d, e, \ldots$ to denote scalars.) Thus, the ijth entry of cA is given by

$$(cA)_{ij} = c(A_{ij}).$$

In words, this rule is: To get the ijth entry of cA, multiply the ijth entry of A by c.

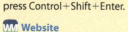

EXAMPLE 4 **Combining Operations**

Let $A = \begin{bmatrix} 2 & -1 & 0 \\ 3 & 5 & -3 \end{bmatrix}$, $B = \begin{bmatrix} 1 & 3 & -1 \\ 5 & -6 & 0 \end{bmatrix}$, and $C = \begin{bmatrix} x & y & w \\ z & t+1 & 3 \end{bmatrix}$.

Evaluate the following: $4A$, xB, and $A + 3C$.

Solution First, we find $4A$ by multiplying each entry of A by 4:

$$4A = 4\begin{bmatrix} 2 & -1 & 0 \\ 3 & 5 & -3 \end{bmatrix} = \begin{bmatrix} 8 & -4 & 0 \\ 12 & 20 & -12 \end{bmatrix}.$$

Similarly, we find xB by multiplying each entry of B by x:

$$xB = x\begin{bmatrix} 1 & 3 & -1 \\ 5 & -6 & 0 \end{bmatrix} = \begin{bmatrix} x & 3x & -x \\ 5x & -6x & 0 \end{bmatrix}.$$

We get $A + 3C$ in two steps as follows:

$$A + 3C = \begin{bmatrix} 2 & -1 & 0 \\ 3 & 5 & -3 \end{bmatrix} + 3\begin{bmatrix} x & y & w \\ z & t+1 & 3 \end{bmatrix}$$

$$= \begin{bmatrix} 2 & -1 & 0 \\ 3 & 5 & -3 \end{bmatrix} + \begin{bmatrix} 3x & 3y & 3w \\ 3z & 3t+3 & 9 \end{bmatrix}$$

$$= \begin{bmatrix} 2+3x & -1+3y & 3w \\ 3+3z & 3t+8 & 6 \end{bmatrix}.$$

Addition and scalar multiplication of matrices have nice properties, reminiscent of the properties of addition and multiplication of real numbers. Before we state them, we need to introduce some more notation.

If A is any matrix, then $-A$ is the matrix $(-1)A$. In other words, $-A$ is A multiplied by the scalar -1. This amounts to changing the signs of all the entries in A. For example,

$$-\begin{bmatrix} 4 & -2 & 0 \\ 6 & 10 & -6 \end{bmatrix} = \begin{bmatrix} -4 & 2 & 0 \\ -6 & -10 & 6 \end{bmatrix}.$$

For any two matrices A and B, $A - B$ is the same as $A + (-B)$. (Why?)

Also, a **zero matrix** is a matrix all of whose entries are zero. Thus, for example, the 2×3 zero matrix is

$$O = \begin{bmatrix} 0 & 0 & 0 \\ 0 & 0 & 0 \end{bmatrix}.$$

Now we state the most important properties of the operations that we have been talking about.

Properties of Matrix Addition and Scalar Multiplication

If A, B, and C are any $m \times n$ matrices and if O is the zero $m \times n$ matrix, then the following hold:

$$A + (B + C) = (A + B) + C \qquad \textit{Associative law}$$
$$A + B = B + A \qquad \textit{Commutative law}$$

$$
\begin{aligned}
A + O &= O + A = A & & \textit{Additive identity law} \\
A + (-A) &= O = (-A) + A & & \textit{Additive inverse law} \\
c(A + B) &= cA + cB & & \textit{Distributive law} \\
(c + d)A &= cA + dA & & \textit{Distributive law} \\
1A &= A & & \textit{Scalar unit} \\
0A &= O & & \textit{Scalar zero}
\end{aligned}
$$

These properties would be obvious if we were talking about addition and multiplication of *numbers*, but here we are talking about addition and multiplication of *matrices*. We are using "+" to mean something new: matrix addition. There is no reason why matrix addition has to obey *all* the properties of addition of numbers. It happens that it does obey many of them, which is why it is convenient to call it *addition* in the first place. This means that we can manipulate equations involving matrices in much the same way that we manipulate equations involving numbers. One word of caution: We haven't yet discussed how to multiply matrices, and it probably isn't what you think. It will turn out that multiplication of matrices does *not* obey all the same properties as multiplication of numbers.

Transposition

We mention one more operation on matrices.

Transposition

If A is an $m \times n$ matrix, then its **transpose** is the $n \times m$ matrix obtained by writing its columns as rows, so the ith column of the original matrix becomes the ith row of the transpose. We denote the transpose of the matrix A by A^T.

Visualizing Transposition

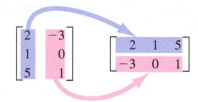

Quick Examples

9. Let $A = \begin{bmatrix} 2 & 0 & 1 & 0 \\ 33 & -22 & 0 & 5 \\ 1 & -1 & 2 & -2 \end{bmatrix}$. Then $A^T = \begin{bmatrix} 2 & 33 & 1 \\ 0 & -22 & -1 \\ 1 & 0 & 2 \\ 0 & 5 & -2 \end{bmatrix}$.

3×4 matrix $\qquad\qquad\qquad\qquad$ 4×3 matrix

10. $\begin{bmatrix} -1 & 1 & 2 \end{bmatrix}^T = \begin{bmatrix} -1 \\ 1 \\ 2 \end{bmatrix}$.

1×3 matrix $\quad$ 3×1 matrix

Properties of Transposition

If A and B are $m \times n$ matrices, then the following hold:

$$(A + B)^T = A^T + B^T$$
$$(cA)^T = c(A^T)$$
$$(A^T)^T = A.$$

To see why the laws of transposition are true, let us consider the first one: $(A + B)^T = A^T + B^T$. The left-hand side is the transpose of $A + B$ and so is obtained by first adding A and B and then writing the rows as columns. This is the same as first writing the rows of A and B individually as columns before adding, which gives the right-hand side. Similar arguments can be used to establish the other laws of transposition.

4.1 EXERCISES

▼ more advanced   ◆ challenging

⊤ indicates exercises that should be solved using technology

In Exercises 1–10, find the dimensions of the given matrix, and identify the given entry.

1. $A = \begin{bmatrix} 1 & 5 & 0 & \frac{1}{4} \end{bmatrix}; a_{13}$

2. $B = \begin{bmatrix} 44 & 55 \end{bmatrix}; b_{12}$

3. $C = \begin{bmatrix} \frac{5}{2} \\ 1 \\ -2 \\ 8 \end{bmatrix}; C_{11}$

4. $D = \begin{bmatrix} 15 & -18 \\ 6 & 0 \\ -6 & 5 \\ 48 & 18 \end{bmatrix}; d_{31}$

5. $E = \begin{bmatrix} e_{11} & e_{12} & e_{13} & \cdots & e_{1q} \\ e_{21} & e_{22} & e_{23} & \cdots & e_{2q} \\ \vdots & \vdots & \vdots & \ddots & \vdots \\ e_{p1} & e_{p2} & e_{p3} & \cdots & e_{pq} \end{bmatrix}; E_{22}$

6. $A = \begin{bmatrix} 2 & -1 & 0 \\ 3 & 5 & -3 \end{bmatrix}; A_{21}$

7. $B = \begin{bmatrix} 1 & 3 \\ 5 & -6 \end{bmatrix}; b_{12}$

8. $C = \begin{bmatrix} x & y & w & e \\ z & t+1 & 3 & 0 \end{bmatrix}; C_{23}$

9. $D = \begin{bmatrix} d_1 & d_2 & \cdots & d_n \end{bmatrix}; D_{1r}$ (any r with $1 \le r \le n$)

10. $E = \begin{bmatrix} d & d & d & d \end{bmatrix}; E_{1r}$ (any r with $1 \le r \le 4$)

11. Solve for x, y, z, and w. [**HINT:** See Example 1.]
$$\begin{bmatrix} x+y & y+z \\ z+w & w \end{bmatrix} = \begin{bmatrix} 3 & 5 \\ 7 & 4 \end{bmatrix}$$

12. Solve for x, y, z, and w. [**HINT:** See Example 1.]
$$\begin{bmatrix} x-y & x-z \\ y-w & w \end{bmatrix} = \begin{bmatrix} 0 & 0 \\ 0 & 6 \end{bmatrix}$$

In Exercises 13–20, evaluate the given expression. Take

$$A = \begin{bmatrix} 0 & -1 \\ 1 & 0 \\ -1 & 2 \end{bmatrix}, \quad B = \begin{bmatrix} 0.25 & -1 \\ 0 & 0.5 \\ -1 & 3 \end{bmatrix}, \quad and$$

$$C = \begin{bmatrix} 1 & -1 \\ 1 & 1 \\ -1 & -1 \end{bmatrix}.$$

[**HINT:** See Example 4 and Quick Examples 7 and 8.]

13. $A + B$

14. $A - C$

15. $A + B - C$

16. $12B$

17. $2A - C$

18. $2A + 0.5C$

19. $2A^T$

20. $A^T + 3C^T$

In Exercises 21–28, evaluate the given expression. Take

$$A = \begin{bmatrix} 1 & -1 & 0 \\ 0 & 2 & -1 \end{bmatrix}, \quad B = \begin{bmatrix} 3 & 0 & -1 \\ 5 & -1 & 1 \end{bmatrix}, \quad and$$

$$C = \begin{bmatrix} x & 1 & w \\ z & r & 4 \end{bmatrix}.$$

[**HINT:** See Example 4 and Quick Examples 7 and 8.]

21. $A + B$

22. $B - C$

23. $A - B + C$

24. $\frac{1}{2}B$

25. $2A - B$

26. $2A - 4C$

27. $3B^T$

28. $2A^T - C^T$

T *In Exercises 29–36, use technology to evaluate the given expression. Take*

$$A = \begin{bmatrix} 1.5 & -2.35 & 5.6 \\ 44.2 & 0 & 12.2 \end{bmatrix}, \quad B = \begin{bmatrix} 1.4 & 7.8 \\ 5.4 & 0 \\ 5.6 & 6.6 \end{bmatrix}, \quad and$$

$$C = \begin{bmatrix} 10 & 20 & 30 \\ -10 & -20 & -30 \end{bmatrix}.$$

29. $A - C$

30. $C - A$

31. $1.1B$

32. $-0.2B$

33. $A^T + 4.2B$

34. $(A + 2.3C)^T$

35. $(2.1A - 2.3C)^T$

36. $(A - C)^T - B$

Applications

37. *Sales* The following table shows the number of Mac computers, iPods, and iPhones sold by **Apple**, in millions of units, in 2012, as well as the year-over-year changes in 2013 and 2014:[1]

	Macs	iPhones	iPads
2012	17.0	135.8	65.7
Change in 2013	0.2	17.6	8.5
Change in 2014	2.4	39.2	−10.8

Use matrix algebra to find the sales of Macs, iPhones, and iPads in 2013 and 2014. [**HINT**: See Example 2.]

38. *Home Prices* The following table shows median home prices, in thousands of dollars, in four regions of the United States in May 2013, as well as the year-over-year changes in 2014 and 2015:[2]

	Northeast	Midwest	South	West
May 2013	270	160	183	276
Change in 2014	−13	6	0	18
Change in 2015	12	16	15	30

Use matrix algebra to find the median home price in each region in May 2014 and 2015. [**HINT**: See Example 2.]

39. *Inventory* The *Left Coast Bookstore* chain has two stores, one in San Francisco and one in Los Angeles. It stocks three kinds of book: hardcover, softcover, and plastic (for infants). At the beginning of January the central computer showed the following books in stock:

	Hard	Soft	Plastic
San Francisco	1,000	2,000	5,000
Los Angeles	1,000	5,000	2,000

Suppose its sales in January were as follows: 700 hardcover books, 1,300 softcover books, and 2,000 plastic books sold in San Francisco, and 400 hardcover, 300 softcover, and 500 plastic books sold in Los Angeles. Write these sales figures in the form of a matrix, and then show how matrix algebra can be used to compute the inventory remaining in each store at the end of January.

40. *Inventory* The *Left Coast Bookstore* chain discussed in Exercise 39 actually maintained the same sales figures for the first 6 months of the year. Each month, the chain restocked the stores from its warehouse by shipping 600 hardcover, 1,500 softcover, and 1,500 plastic books to San Francisco and 500 hardcover, 500 softcover, and 500 plastic books to Los Angeles.

a. Use matrix operations to determine the total sales over the 6 months, broken down by store and type of book.

b. Use matrix operations to determine the inventory in each store at the end of June.

41. *Profit* Annual revenues and production costs at *Luddington's Wellington Boots & Co.* are shown in the following spreadsheet.

	A	B	C	D
1	Revenue			
2		2004	2005	2006
3	Full Boots	$10,000	$9,000	$11,000
4	Half Boots	$8,000	$7,200	$8,800
5	Sandals	$4,000	$5,000	$6,000
6				
7	Production Costs			
8		2004	2005	2006
9	Full Boots	$2,000	$1,800	$2,200
10	Half Boots	$2,400	$1,440	$1,760
11	Sandals	$1,200	$1,500	$2,000

Use matrix algebra to compute the profits from each sector each year.

42. *Revenue* The following spreadsheet gives annual production costs and profits at *Gauss-Jordan Sneakers, Inc*:

	A	B	C	D
1	Production Costs			
2		2004	2005	2006
3	Gauss Grip	$1,800	$2,200	$2,400
4	Air Gauss	$1,400	$1,700	$1,200
5	Gauss Gel	$1,500	$2,000	$1,300
6				
7	Profit			
8		2004	2005	2006
9	Gauss Grip	$10,000	$14,000	$16,000
10	Air Gauss	$8,000	$12,000	$14,000
11	Gauss Gel	$9,000	$14,000	$12,000

Use matrix algebra to compute the revenues from each sector each year.

43. *Population Movement* In 2000 the U.S. population, broken down by regions, was 53.6 million in the Northeast, 64.4 million in the Midwest, 100.2 million in the South, and

[1] Source for data: Apple company reports (www.apple.com/investor).

[2] Source for data: National Association of Realtors (www.realtor.org).

63.2 million in the West.[3] In 2010 the population was 55.3 million in the Northeast, 66.9 million in the Midwest, 114.6 million in the South, and 71.9 million in the West. Set up the population figures for each year as a row vector, and then show how to use matrix operations to find the net increase or decrease of population in each region from 2000 to 2010. Assuming the same population growth from 2010 to 2020 as from 2000 to 2010, use matrix operations to predict the population in each region in 2020.

44. *Population Movement* In 1990 the U.S. population, broken down by regions, was 50.8 million in the Northeast, 59.7 million in the Midwest, 85.4 million in the South, and 52.8 million in the West.[4] Between 1990 and 2000 the population in the Northeast grew by 2.8 million, the population in the Midwest grew by 4.7 million, the population in the South grew by 14.8 million, and the population in the West grew by 10.4 million. Set up the population figures for 1990 and the growth figures for the decade as row vectors. Assuming the same population growth from 2000 to 2010 as from 1990 to 2000, use matrix operations to estimate the population in each region in 2010. Compare the predicted 2010 population in the Northeast with the actual population given in Exercise 43.

Foreclosure Crisis *Starting in 2010, on the heels of the 2007–2009 subprime mortgage crisis, the United States saw an epidemic of mortgage foreclosures, often initiated improperly by large financial institutions. Exercises 45–48 are based on the following table, which shows the numbers of foreclosures in three states in April through August of 2011.[5]*

	April	May	June	July	Aug.
California	55,900	51,900	54,100	56,200	59,400
Florida	19,600	19,200	23,800	22,400	23,600
Texas	8,800	9,100	9,300	10,600	10,100

45. Use matrix algebra to determine the total number of foreclosures in each of the given months.

46. Use matrix algebra to determine the total number of foreclosures in each of the given states during the entire period shown.

47. Use matrix algebra to determine in which month the difference between the number of foreclosures in California and in Florida was greatest.

48. Use matrix algebra to determine in which region the difference between the number of foreclosures in April and August was greatest.

[3] Source: U.S. Census Bureau (http://2010.census.gov/2010census/data/apportionment-pop-text.php).

[4] *Ibid.*

[5] Figures are rounded. Source: www.realtytrac.com.

49. ▼ *Inventory* Microbucks Computer Company makes two computers, the Pomegranate II and the Pomegranate Classic, at two different factories. The Pom II requires 2 processor chips, 16 memory chips, and 20 vacuum tubes, while the Pom Classic requires 1 processor chip, 4 memory chips, and 40 vacuum tubes. Microbucks has in stock at the beginning of the year 500 processor chips, 5,000 memory chips, and 10,000 vacuum tubes at the Pom II factory and 200 processor chips, 2,000 memory chips, and 20,000 vacuum tubes at the Pom Classic factory. It manufactures 50 Pom IIs and 50 Pom Classics each month.
   a. Find the company's inventory of parts after 2 months, using matrix operations.
   b. When (if ever) will the company run out of one of the parts?

50. ▼ *Inventory* Microbucks Computer Company, besides having the stock mentioned in Exercise 49, gets shipments of parts every month in the amounts of 100 processor chips, 1,000 memory chips, and 3,000 vacuum tubes at the Pom II factory and 50 processor chips, 1,000 memory chips, and 2,000 vacuum tubes at the Pom Classic factory.
   a. What will the company's inventory of parts be after 6 months?
   b. When (if ever) will the company run out of one of the parts?

51. ▼ *Tourism in the 1990s* The following table gives the number of people (in thousands) who visited Australia and South Africa in 1998:[6]

	To	Australia	South Africa
From	North America	440	190
	Europe	950	950
	Asia	1,790	200

It was predicted that in 2008, 20,000 fewer people from North America would visit Australia and 40,000 more would visit South Africa, 50,000 more people from Europe would visit each of Australia and South Africa, and 100,000 more people from Asia would visit South Africa, but there would be no change in the number of people from Asia visiting Australia.
   a. Represent the changes predicted in 2008 in the form of a matrix, and use matrix algebra to predict the number of visitors from the three regions to Australia and South Africa in 2008.
   b. Take A to be the 3×2 matrix whose entries are the 1998 tourism figures, and take B to be the 3×2 matrix whose entries are the 2008 tourism figures. Give a formula (in terms of A and B) that predicts the average of the numbers of visitors from the three regions to Australia and South Africa in 1998 and 2008. Compute its value.

[6] Figures are rounded to the nearest 10,000. Sources: South African Dept. of Environmental Affairs and Tourism; Australia Tourist Commission/*New York Times*, January 15, 2000, p. C1.

52. ▼ *Tourism in the 1990s* Referring to the 1998 tourism figures given in Exercise 51, assume that the following (fictitious) figures represent the corresponding numbers from 1988:

	To	Australia	South Africa
From North America		500	100
Europe		900	800
Asia		1,400	50

Take A to be the 3×2 matrix whose entries are the 1998 tourism figures, and take B to be the 3×2 matrix whose entries are the 1988 tourism figures.
   a. Compute the matrix $A - B$. What does this matrix represent?
   b. Assuming that the changes in tourism over 1988–1998 are repeated in 1998–2008, give a formula (in terms of A and B) that predicts the number of visitors from the three regions to Australia and South Africa in 2008.

Communication and Reasoning Exercises

53. Is it possible for a 2×3 matrix to equal a 3×2 matrix? Explain.

54. If A and B are 2×3 matrices and $A = B$, what can you say about $A - B$? Explain.

55. What does it mean when we say that $(A + B)_{ij} = A_{ij} + B_{ij}$?

56. What does it mean when we say that $(cA)_{ij} = c(A_{ij})$?

57. What would a 5×5 matrix A look like if $A_{ii} = 0$ for every i?

58. What would a matrix A look like if $A_{ij} = 0$ whenever $i \neq j$?

59. ▼ Give a formula for the ijth entry of the transpose of a matrix A.

60. ▼ A matrix is **symmetric** if it is equal to its transpose. Give an example of **(a)** a nonzero 2×2 symmetric matrix and **(b)** a nonzero 3×3 symmetric matrix.

61. ▼ A matrix is **skew-symmetric** or **antisymmetric** if it is equal to the negative of its transpose. Give an example of **(a)** a nonzero 2×2 skew-symmetric matrix and **(b)** a nonzero 3×3 skew-symmetric matrix.

62. ▼ Referring to Exercises 60 and 61, what can be said about a matrix that is both symmetric and skew-symmetric?

63. ▼ Why is matrix addition associative?

64. ▼ Is matrix subtraction associative? Explain.

65. Describe a scenario (possibly based on one of the preceding examples or exercises) in which you might wish to compute $A + B - C$ for certain matrices A, B, and C.

66. Describe a scenario (possibly based on one of the preceding examples or exercises) in which you might wish to compute $A - 2B$ for certain matrices A and B.

4.2   Matrix Multiplication

Multiplying Matrices

Suppose we download three movies at $10 each and five Chopin albums at $8 each. We calculate our total cost by computing each product's price × quantity and adding:

$$\text{Cost} = 10 \times 3 + 8 \times 5 = \$70.$$

Let us instead put the prices in a row vector,

$$P = \begin{bmatrix} 10 & 8 \end{bmatrix}, \qquad \text{The price matrix}$$

and the quantities purchased in a column vector,

$$Q = \begin{bmatrix} 3 \\ 5 \end{bmatrix}. \qquad \text{The quantity matrix}$$

Q: *Why do we use a row and a column instead of, say, two rows?*

A: It's rather a long story, but mathematicians found that it works best this way . . .

Because P represents the prices of the items we are purchasing and Q represents the quantities, it would be useful if the product PQ represented the total cost, a *single number* (which we can think of as a 1×1 matrix). For this to work, PQ should be calculated the same way we calculated the total cost:

$$PQ = \begin{bmatrix} 10 & 8 \end{bmatrix} \begin{bmatrix} 3 \\ 5 \end{bmatrix} = [10 \times 3 + 8 \times 5] = [70].$$

Notice that we obtain the answer by multiplying each entry in P (going from left to right) by the corresponding entry in Q (going from top to bottom) and then adding the results.

The Product *Row* × *Column*

The **product** AB of a row matrix A and a column matrix B is a 1×1 matrix. The length of the row in A must match the length of the column in B for the product to be defined. To find the product, multiply each entry in A (going from left to right) by the corresponding entry in B (going from top to bottom), and then add the results.

Visualizing Matrix Multiplication

$$\begin{bmatrix} 2 \\ 10 \\ -1 \end{bmatrix}$$

$\begin{bmatrix} 2 & 4 & 1 \end{bmatrix}$

2×2	$=$	4	Product of first entries $= 4$
4×10	$=$	40	Product of second entries $= 40$
$1 \times (-1)$	$=$	-1	Product of third entries $= -1$
		43	Sum of products $= 43$

Quick Examples

1. $\begin{bmatrix} 2 & 1 \end{bmatrix} \begin{bmatrix} -3 \\ 1 \end{bmatrix} = [2 \times (-3) + 1 \times 1] = [-6 + 1] = [-5]$

2. $\begin{bmatrix} 2 & 4 & 1 \end{bmatrix} \begin{bmatrix} 2 \\ 10 \\ -1 \end{bmatrix} = [2 \times 2 + 4 \times 10 + 1 \times (-1)]$

$$= [4 + 40 + (-1)] = [43]$$

Notes

1. In the discussion so far, *the row is on the left and the column is on the right* (RC again). (Later, we will consider products in which the column matrix is on the left and the row matrix is on the right.)

2. The row size has to match the column size. This means that, if we have a 1×3 row on the left, then the column on the right must be 3×1 in order for the product to make sense. For example, the product

$$\begin{bmatrix} a & b & c \end{bmatrix} \begin{bmatrix} x \\ y \end{bmatrix}$$

is not defined. ∎

EXAMPLE 1 Revenue

The *A-Plus* auto parts store mentioned in examples in Section 4.1 had the following sales in its Vancouver store:

	Vancouver
Wiper Blades	20
Cleaning Fluid (bottles)	10
Floor Mats	8

The store sells wiper blades for $7.00 each, cleaning fluid for $3.00 per bottle, and floor mats for $12.00 each. Use matrix multiplication to find the total revenue generated by sales of these items.

Solution We need to multiply each sales figure by the corresponding price and then add the resulting revenue figures. We represent the sales by a column vector, as suggested by the table:

$$Q = \begin{bmatrix} 20 \\ 10 \\ 8 \end{bmatrix}.$$

We put the selling prices in a row vector:

$$P = \begin{bmatrix} 7.00 & 3.00 & 12.00 \end{bmatrix}.$$

We can now compute the total revenue as the product

$$R = PQ = \begin{bmatrix} 7.00 & 3.00 & 12.00 \end{bmatrix} \begin{bmatrix} 20 \\ 10 \\ 8 \end{bmatrix}$$

$$= \begin{bmatrix} 140.00 + 30.00 + 96.00 \end{bmatrix} = \begin{bmatrix} 266.00 \end{bmatrix}.$$

So the sale of these items generated a total revenue of $266.00.

Note We could also have written the quantity sold as a row vector (which would be Q^T) and the prices as a column vector (which would be P^T) and then multiplied them in the opposite order $Q^T P^T$. Try this. ∎

EXAMPLE 2 Relationship with Linear Equations

a. Represent the matrix equation

$$\begin{bmatrix} 2 & -4 & 1 \end{bmatrix} \begin{bmatrix} x \\ y \\ z \end{bmatrix} = \begin{bmatrix} 5 \end{bmatrix}$$

as an ordinary equation.

b. Represent the linear equation $3x + y - z + 2w = 8$ as a matrix equation.

Solution

a. If we perform the multiplication on the left, we get the 1×1 matrix $[2x - 4y + z]$. Thus, the equation may be rewritten as

$$[2x - 4y + z] = [5]. \qquad \text{1} \times \text{1 matrix on the left} = \text{1} \times \text{1 matrix on the right}$$

Saying that these two 1×1 matrices are equal means that their entries are equal, so we get the equation

$$2x - 4y + z = 5.$$

b. This is the reverse of part (a):

$$\begin{bmatrix} 3 & 1 & -1 & 2 \end{bmatrix} \begin{bmatrix} x \\ y \\ z \\ w \end{bmatrix} = [8].$$

➡ **Before we go on ...** The row matrix $\begin{bmatrix} 3 & 1 & -1 & 2 \end{bmatrix}$ in Example 2 is the row of **coefficients** of the original equation. (See Section 3.1.) ∎

Now we turn to the general case of matrix multiplication.

The Product of Two Matrices: General Case

In general, for matrices A and B we can take the product AB only if the number of columns of A equals the number of rows of B (so that we can multiply the rows of A by the columns of B as above). The product AB is then obtained by taking its ijth entry to be

$$ij\text{th entry of } AB = \text{Row } i \text{ of } A \times \text{Column } j \text{ of } B. \qquad \text{As defined above}$$

Quick Examples

(R stands for row; C stands for column.)

3. $R_1 \rightarrow \begin{bmatrix} 2 & 0 & -1 & 3 \end{bmatrix} \begin{matrix} C_1 & C_2 & C_3 \\ \downarrow & \downarrow & \downarrow \\ \begin{bmatrix} 1 & 1 & -8 \\ 1 & -6 & 0 \\ 0 & 5 & 2 \\ -3 & 8 & 1 \end{bmatrix} \end{matrix} = \begin{bmatrix} R_1 \times C_1 & R_1 \times C_2 & R_1 \times C_3 \end{bmatrix}$

$$= \begin{bmatrix} -7 & 21 & -15 \end{bmatrix}$$

4. $\begin{matrix} R_1 \rightarrow \\ R_2 \rightarrow \end{matrix} \begin{bmatrix} 1 & -1 \\ 0 & 2 \end{bmatrix} \begin{matrix} C_1 & C_2 \\ \downarrow & \downarrow \\ \begin{bmatrix} 3 & 0 \\ 5 & -1 \end{bmatrix} \end{matrix} = \begin{bmatrix} R_1 \times C_1 & R_1 \times C_2 \\ R_2 \times C_1 & R_2 \times C_2 \end{bmatrix} = \begin{bmatrix} -2 & 1 \\ 10 & -2 \end{bmatrix}$

In matrix multiplication we always take

Rows on the left $\times$ Columns on the right.

Look at the dimensions in Quick Examples 3 and 4.

Match

$\downarrow \quad \downarrow$

$(1 \times 4)(4 \times 3) \rightarrow 1 \times 3$

Match

$\downarrow \quad \downarrow$

$(2 \times 2)(2 \times 2) \rightarrow 2 \times 2$

The fact that the number of columns in the left-hand matrix equals the number of rows in the right-hand matrix amounts to saying that the middle two numbers must match as above. If we "cancel" the middle matching numbers, we are left with the dimensions of the product.

Before continuing with examples, we state the rule for matrix multiplication formally.

Multiplication of Matrices: Formal Definition

If A is an $m \times n$ matrix and B is an $n \times k$ matrix, then the product AB is the $m \times k$ matrix whose ijth entry is the product

Row i of A $\times$ Column j of $B

$\downarrow \qquad\qquad \downarrow$

$$(AB)_{ij} = [a_{i1} \quad a_{i2} \quad a_{i3} \ldots a_{in}] \begin{bmatrix} b_{1j} \\ b_{2j} \\ b_{3j} \\ \vdots \\ b_{nj} \end{bmatrix} = a_{i1}b_{1j} + a_{i2}b_{2j} + a_{i3}b_{3j} + \cdots + a_{in}b_{nj}.$$

EXAMPLE 3   Matrix Product

Calculate:

a. $\begin{bmatrix} 2 & 0 & -1 & 3 \\ 1 & -1 & 2 & -2 \end{bmatrix} \begin{bmatrix} 1 & 1 & -8 \\ 1 & 0 & 0 \\ 0 & 5 & 2 \\ -2 & 8 & -1 \end{bmatrix}$ **b.** $\begin{bmatrix} -3 \\ 1 \end{bmatrix} [2 \quad 1]$

Solution

a. Before we start the calculation, we check that the dimensions of the matrices match up:

Match

$2 \times 4 \qquad 4 \times 3$

$$\begin{bmatrix} 2 & 0 & -1 & 3 \\ 1 & -1 & 2 & -2 \end{bmatrix} \begin{bmatrix} 1 & 1 & -8 \\ 1 & 0 & 0 \\ 0 & 5 & 2 \\ -2 & 8 & -1 \end{bmatrix}.$$

Using Technology

Technology can be used to multiply the matrices in Example 3. Here is an outline for part (a) (see the Technology Guides at the end of the chapter for additional details on using a TI-83/84 Plus or a spreadsheet):

TI-83/84 Plus

Enter the matrices [A] and [B] using MATRIX; EDIT
Home screen: [A] * [B]
[More details in the Technology Guide.]

Spreadsheet

Enter the matrices in convenient blocks of cells (e.g., A1–D2 and F1–H4).
Highlight a 2 × 3 block for the answer.
Type =MMULT(A1:D2,F1:H4)
Press Control+Shift+Enter
[More details in the Technology Guide.]

Ⓦ Website
www.WanerMath.com

→ Online Utilities
→ Matrix Algebra Tool
Enter matrices as shown:

```
Enter your matrices here.
A = [2, 0, -1, 3
1, -1, 2, -2]

B = [1, 1, -8
1, 0, 0
0, 5, 2
-2, 8, -1]
```

Type A*B in the formula box and press "Compute". If you try to multiply two matrices whose product is not defined, you will get an error alert box telling you that.

The product of the two matrices is defined, and the product will be a 2×3 matrix (we remove the matching 4s: $(2 \times 4)(4 \times 3) \to 2 \times 3$). To calculate the product, we follow the previous prescription:

$$\begin{matrix} & & C_1 \ \ C_2 \ \ \ C_3 \\ & & \downarrow \ \ \downarrow \ \ \ \downarrow \end{matrix}$$

$$\begin{matrix} R_1 \to \\ R_2 \to \end{matrix} \begin{bmatrix} 2 & 0 & -1 & 3 \\ 1 & -1 & 2 & -2 \end{bmatrix} \begin{bmatrix} 1 & 1 & -8 \\ 1 & 0 & 0 \\ 0 & 5 & 2 \\ -2 & 8 & -1 \end{bmatrix} = \begin{bmatrix} R_1 \times C_1 & R_1 \times C_2 & R_1 \times C_3 \\ R_2 \times C_1 & R_2 \times C_2 & R_2 \times C_3 \end{bmatrix}$$

$$= \begin{bmatrix} -4 & 21 & -21 \\ 4 & -5 & -2 \end{bmatrix}.$$

b. The dimensions of the two matrices given are 2×1 and 1×2. Because the 1s match, the product is defined, and the result will be a 2×2 matrix:

$$\begin{matrix} & & C_1 \ \ C_2 \\ & & \downarrow \ \ \downarrow \end{matrix}$$

$$\begin{matrix} R_1 \to \\ R_2 \to \end{matrix} \begin{bmatrix} -3 \\ 1 \end{bmatrix} \begin{bmatrix} 2 & 1 \end{bmatrix} = \begin{bmatrix} R_1 \times C_1 & R_1 \times C_2 \\ R_2 \times C_1 & R_2 \times C_2 \end{bmatrix} = \begin{bmatrix} -6 & -3 \\ 2 & 1 \end{bmatrix}.$$

Note In part (a) we *cannot* multiply the matrices in the opposite order because the middle dimensions do not match. We say simply that the product in the opposite order is **not defined**. In part (b) we *can* multiply the matrices in the opposite order, but we would get a 1×1 matrix if we did so. Thus, order is important when multiplying matrices. In general, if AB is defined, then BA need not even be defined. If BA is also defined, it may not have the same dimensions as AB. And even if AB and BA have the same dimensions, they may have different entries. (See the next example.) ∎

EXAMPLE 4   *AB* versus *BA*

Let $A = \begin{bmatrix} 1 & -1 \\ 0 & 2 \end{bmatrix}$ and $B = \begin{bmatrix} 3 & 0 \\ 5 & -1 \end{bmatrix}$. Find AB and BA.

Solution Note first that A and B are both 2×2 matrices, so the products AB and BA are both defined and are both 2×2 matrices—unlike the case in Example 3(b). We first calculate AB:

$$AB = \begin{bmatrix} 1 & -1 \\ 0 & 2 \end{bmatrix} \begin{bmatrix} 3 & 0 \\ 5 & -1 \end{bmatrix} = \begin{bmatrix} -2 & 1 \\ 10 & -2 \end{bmatrix}.$$

Now let's calculate BA:

$$BA = \begin{bmatrix} 3 & 0 \\ 5 & -1 \end{bmatrix} \begin{bmatrix} 1 & -1 \\ 0 & 2 \end{bmatrix} = \begin{bmatrix} 3 & -3 \\ 5 & -7 \end{bmatrix}.$$

Notice that BA has no resemblance to AB! Thus, we have discovered that, even for square matrices:

Matrix multiplication is not commutative.

In other words, $AB \neq BA$ in general, even when AB and BA both exist and have the same dimensions. (There are instances when $AB = BA$ for particular matrices A and B, but this is an exception, not the rule.)

EXAMPLE 5 Revenue

January sales at the *A-Plus* auto parts stores in Vancouver and Quebec are given in the following table:

	Vancouver	Quebec
Wiper Blades	20	15
Cleaning Fluid (bottles)	10	12
Floor Mats	8	4

The sale prices for these items are $7.00 each for wiper blades, $3.00 per bottle for cleaning fluid, and $12.00 each for floor mats. The costs to the company were $3.00 each for wiper blades, $1.00 per bottle for cleaning fluid, and $4.00 each for floor mats. Use matrix multiplication to compute the total revenue and total cost of the items listed at each store.

Solution We can do all of the requested calculations at once with a single matrix multiplication. Consider the following two labeled matrices:

$$Q = \begin{matrix} & \textbf{V} & \textbf{Q} \\ \textbf{Wb} \\ \textbf{Cf} \\ \textbf{Fm} \end{matrix} \begin{bmatrix} 20 & 15 \\ 10 & 12 \\ 8 & 4 \end{bmatrix}$$

$$P = \begin{matrix} & \textbf{Wb} & \textbf{Cf} & \textbf{Fm} \\ \textbf{Sale Prices} \\ \textbf{Cost Prices} \end{matrix} \begin{bmatrix} 7.00 & 3.00 & 12.00 \\ 3.00 & 1.00 & 4.00 \end{bmatrix}.$$

The first matrix records the quantities sold, while the second records the sales prices and cost prices. To compute the total revenue and costs at both stores, we calculate $T = PQ$:

$$T = PQ = \begin{bmatrix} 7.00 & 3.00 & 12.00 \\ 3.00 & 1.00 & 4.00 \end{bmatrix} \begin{bmatrix} 20 & 15 \\ 10 & 12 \\ 8 & 4 \end{bmatrix}$$

$$= \begin{bmatrix} 266.00 & 189.00 \\ 102.00 & 73.00 \end{bmatrix}.$$

We can label this matrix as follows:

$$R = \begin{matrix} & \textbf{V} & \textbf{Q} \\ \textbf{Sale Prices} \\ \textbf{Cost Prices} \end{matrix} \begin{bmatrix} 266.00 & 189.00 \\ 102.00 & 73.00 \end{bmatrix}.$$

In other words, the sales revenues were $266 for Vancouver and $189 for Quebec, while the total costs were $102 for Vancouver and $73 for Quebec.

➡ **Before we go on . . .** In Example 5 we were able to calculate PQ because the dimensions matched correctly: $(2 \times 3)(3 \times 2) \rightarrow 2 \times 2$. We could also have multiplied them in the opposite order and gotten a 3×3 matrix. Would the product QP be meaningful? In an application like this, not only do the dimensions have to match, but also the *labels* have to match for the result to be meaningful. The labels on the three columns of P are the parts that were sold, and these are also the labels on the three rows of Q. Therefore, we can "cancel labels" at the same time that we cancel the dimensions in the product. However, the labels on the two columns of Q do not match the labels on the two rows of P, and there is no useful interpretation of the product QP in this situation. ∎

There are very special square matrices of every size: 1×1, 2×2, 3×3, and so on, called the **identity** matrices.

Identity Matrix

The $n \times n$ identity matrix I is the matrix with 1s down the **main diagonal** (the diagonal starting at the top left) and 0s everywhere else. In symbols,

$$I_{ii} = 1, \quad \text{and}$$
$$I_{ij} = 0 \quad \text{if } i \neq j.$$

Quick Examples

5. 1×1 identity matrix $I = \begin{bmatrix} 1 \end{bmatrix}$

6. 2×2 identity matrix $I = \begin{bmatrix} 1 & 0 \\ 0 & 1 \end{bmatrix}$

7. 3×3 identity matrix $I = \begin{bmatrix} 1 & 0 & 0 \\ 0 & 1 & 0 \\ 0 & 0 & 1 \end{bmatrix}$

8. 4×4 identity matrix $I = \begin{bmatrix} 1 & 0 & 0 & 0 \\ 0 & 1 & 0 & 0 \\ 0 & 0 & 1 & 0 \\ 0 & 0 & 0 & 1 \end{bmatrix}$

Note Identity matrices are always square matrices, meaning that they have the same number of rows as columns. There is no such thing, for example, as the "2×4 identity matrix." ∎

The next example shows why I is interesting.

EXAMPLE 6 **Identity Matrix**

Evaluate the products AI and IA, where $A = \begin{bmatrix} a & b & c \\ d & e & f \\ g & h & i \end{bmatrix}$ and I is the 3×3 identity matrix.

Solution First notice that A is arbitrary; it could be any 3×3 matrix.

$$AI = \begin{bmatrix} a & b & c \\ d & e & f \\ g & h & i \end{bmatrix} \begin{bmatrix} 1 & 0 & 0 \\ 0 & 1 & 0 \\ 0 & 0 & 1 \end{bmatrix} = \begin{bmatrix} a & b & c \\ d & e & f \\ g & h & i \end{bmatrix}$$

and

$$IA = \begin{bmatrix} 1 & 0 & 0 \\ 0 & 1 & 0 \\ 0 & 0 & 1 \end{bmatrix} \begin{bmatrix} a & b & c \\ d & e & f \\ g & h & i \end{bmatrix} = \begin{bmatrix} a & b & c \\ d & e & f \\ g & h & i \end{bmatrix}.$$

In both cases the answer is the matrix A we started with. In symbols,

$$AI = A$$

and

$$IA = A$$

no matter which 3×3 matrix A you start with. Now this should remind you of a familiar fact from arithmetic:

$$a \cdot 1 = a$$

and

$$1 \cdot a = a.$$

That is why we call the matrix I the 3×3 *identity* matrix: because it appears to play the same role for 3×3 matrices that the identity 1 does for numbers.

➡ **Before we go on...** Try a similar calculation using 2×2 matrices: Let $A = \begin{bmatrix} a & b \\ c & d \end{bmatrix}$, let I be the 2×2 identity matrix, and check that $AI = IA = A$. In fact, the equation

$$AI = IA = A$$

works for square matrices of every dimension. It is also interesting to notice that $AI = A$ if I is the 2×2 identity matrix and A is any 3×2 matrix (try one). In fact, if I is any identity matrix, then $AI = A$ whenever the product is defined, and $IA = A$ whenever this product is defined. ■

We can now add to the list of properties we gave for matrix arithmetic at the end of Section 4.1 by writing down properties of matrix multiplication. In stating these properties, we shall assume that all matrix products we write are defined—that is, that the matrices have correctly matching dimensions. The first eight properties are the ones we've already seen; the rest are new.

Properties of Matrix Addition and Multiplication

If A, B, and C are matrices, if O is a zero matrix, and if I is an identity matrix, then the following hold:

$A + (B + C) = (A + B) + C$	*Additive associative law*
$A + B = B + A$	*Additive commutative law*
$A + O = O + A = A$	*Additive identity law*
$A + (-A) = O = (-A) + A$	*Additive inverse law*
$c(A + B) = cA + cB$	*Distributive law*
$(c + d)A = cA + dA$	*Distributive law*
$1A = A$	*Scalar unit*

$$
\begin{array}{ll}
0A = O & \textit{Scalar zero} \\
A(BC) = (AB)C & \textit{Multiplicative associative law} \\
c(AB) = (cA)B & \textit{Multiplicative associative law} \\
c(dA) = (cd)A & \textit{Multiplicative associative law} \\
AI = IA = A & \textit{Multiplicative identity law} \\
A(B + C) = AB + AC & \textit{Distributive law} \\
(A + B)C = AC + BC & \textit{Distributive law} \\
OA = AO = O & \textit{Multiplication by zero matrix}
\end{array}
$$

Note that we have not included a multiplicative commutative law for matrices, because the equation $AB = BA$ does not hold in general. In other words, matrix multiplication is *not* exactly like multiplication of numbers. (You have to be a little careful because it is easy to apply the commutative law without realizing it.)

We should also say a bit more about transposition. Transposition and multiplication have an interesting relationship. We write down the properties of transposition again, adding one new one.

Properties of Transposition

$$(A + B)^T = A^T + B^T$$
$$(cA)^T = c(A^T)$$
$$(AB)^T = B^T A^T$$

Notice the change in order in the last one. The order is crucial.

Quick Examples

9. $\left(\begin{bmatrix} 1 & -1 \\ 0 & 2 \end{bmatrix} \begin{bmatrix} 3 & 0 \\ 5 & -1 \end{bmatrix} \right)^T = \begin{bmatrix} -2 & 1 \\ 10 & -2 \end{bmatrix}^T = \begin{bmatrix} -2 & 10 \\ 1 & -2 \end{bmatrix}$ $(AB)^T$

10. $\begin{bmatrix} 3 & 0 \\ 5 & -1 \end{bmatrix}^T \begin{bmatrix} 1 & -1 \\ 0 & 2 \end{bmatrix}^T = \begin{bmatrix} 3 & 5 \\ 0 & -1 \end{bmatrix} \begin{bmatrix} 1 & 0 \\ -1 & 2 \end{bmatrix} = \begin{bmatrix} -2 & 10 \\ 1 & -2 \end{bmatrix}$ $B^T A^T$

11. $\begin{bmatrix} 1 & -1 \\ 0 & 2 \end{bmatrix}^T \begin{bmatrix} 3 & 0 \\ 5 & -1 \end{bmatrix}^T = \begin{bmatrix} 1 & 0 \\ -1 & 2 \end{bmatrix} \begin{bmatrix} 3 & 5 \\ 0 & -1 \end{bmatrix} = \begin{bmatrix} 3 & 5 \\ -3 & -7 \end{bmatrix}$ $A^T B^T$

These properties give you a glimpse of the field of mathematics known as **abstract algebra**. Algebraists study operations like these that resemble the operations on numbers but differ in some way, such as the lack of commutativity for multiplication seen here.

We end this section with more on the relationship between linear equations and matrix equations, which is one of the important applications of matrix multiplication.

EXAMPLE 7   Matrix Form of a System of Linear Equations

a. If

$$
A = \begin{bmatrix} 1 & -2 & 3 \\ 2 & 0 & -1 \\ -3 & 1 & 1 \end{bmatrix}, \quad X = \begin{bmatrix} x \\ y \\ z \end{bmatrix}, \quad \text{and} \quad B = \begin{bmatrix} 3 \\ -1 \\ 0 \end{bmatrix},
$$

rewrite the matrix equation $AX = B$ as a system of linear equations.

b. Express the following system of equations as a matrix equation of the form $AX = B$:

$$2x + y = 3$$
$$4x - y = -1.$$

Solution

a. The matrix equation $AX = B$ is

$$\begin{bmatrix} 1 & -2 & 3 \\ 2 & 0 & -1 \\ -3 & 1 & 1 \end{bmatrix} \begin{bmatrix} x \\ y \\ z \end{bmatrix} = \begin{bmatrix} 3 \\ -1 \\ 0 \end{bmatrix}.$$

As in Example 2(a), we first evaluate the left-hand side and then set it equal to the right-hand side:

$$\begin{bmatrix} 1 & -2 & 3 \\ 2 & 0 & -1 \\ -3 & 1 & 1 \end{bmatrix} \begin{bmatrix} x \\ y \\ z \end{bmatrix} = \begin{bmatrix} x - 2y + 3z \\ 2x - z \\ -3x + y + z \end{bmatrix}$$

$$\begin{bmatrix} x - 2y + 3z \\ 2x - z \\ -3x + y + z \end{bmatrix} = \begin{bmatrix} 3 \\ -1 \\ 0 \end{bmatrix}.$$

Because these two matrices are equal, their corresponding entries must be equal:

$$x - 2y + 3z = 3$$
$$2x \qquad - z = -1$$
$$-3x + y + z = 0.$$

In other words, the matrix equation $AX = B$ is equivalent to this system of linear equations. Notice that the coefficients of the left-hand sides of these equations are the entries of the matrix A. We call A the **coefficient matrix** of the system of equations. The entries of X are the unknowns, and the entries of B are the right-hand sides 3, -1, and 0.

b. As we saw in part (a), the coefficient matrix A has entries equal to the coefficients of the left-hand sides of the equations. Thus,

$$A = \begin{bmatrix} 2 & 1 \\ 4 & -1 \end{bmatrix}.$$

X is the column matrix consisting of the unknowns, while B is the column matrix consisting of the right-hand sides of the equations, so

$$X = \begin{bmatrix} x \\ y \end{bmatrix} \quad \text{and} \quad B = \begin{bmatrix} 3 \\ -1 \end{bmatrix}.$$

The system of equations can be rewritten as the matrix equation $AX = B$ with this A, X, and B.

This translation of systems of linear equations into matrix equations is really the first step in the method of solving linear equations discussed in Chapter 3. There, we worked with the **augmented matrix** of the system, which is simply A with B adjoined as an extra column.

Q : *When we write a system of equations as AX = B, couldn't we solve for the unknown X by dividing both sides by A?*

A : If we interpret division as multiplication by the inverse (for example, $2 \div 3 = 2 \times 3^{-1}$), we shall see in Section 4.3 that *certain* systems of the form $AX = B$ can be solved in this way, by multiplying both sides by A^{-1}. We first need to discuss what we mean by A^{-1} and how to calculate it.

4.2 EXERCISES

▼ more advanced   ◆ challenging
T indicates exercises that should be solved using technology

In Exercises 1–28, compute the products. Some of these may be undefined. Exercises marked **T** *should be done by using technology. The others should be done in two ways: by hand and by using technology where possible.* [**HINT:** *See Example 3.*]

1. $\begin{bmatrix} 1 & 3 & -1 \end{bmatrix} \begin{bmatrix} 9 \\ 1 \\ -1 \end{bmatrix}$

2. $\begin{bmatrix} 4 & 0 & -1 \end{bmatrix} \begin{bmatrix} -4 \\ 1 \\ 8 \end{bmatrix}$

3. $\begin{bmatrix} -1 & \frac{1}{2} \end{bmatrix} \begin{bmatrix} -\frac{1}{3} \\ 1 \end{bmatrix}$

4. $\begin{bmatrix} -1 & 1 \end{bmatrix} \begin{bmatrix} \frac{3}{4} \\ \frac{1}{4} \end{bmatrix}$

5. $\begin{bmatrix} 0 & -2 & 1 \end{bmatrix} \begin{bmatrix} x \\ y \\ z \end{bmatrix}$

6. $\begin{bmatrix} 4 & -1 & 1 \end{bmatrix} \begin{bmatrix} -x \\ x \\ y \end{bmatrix}$

7. $\begin{bmatrix} 1 & 3 & 2 \end{bmatrix} \begin{bmatrix} 1 \\ -1 \end{bmatrix}$

8. $\begin{bmatrix} 3 & 2 \end{bmatrix} \begin{bmatrix} 1 & -2 \end{bmatrix}$

9. $\begin{bmatrix} -1 & 1 \end{bmatrix} \begin{bmatrix} -3 & 1 & 4 & 3 \\ 0 & 1 & -2 & 1 \end{bmatrix}$

10. $\begin{bmatrix} 2 & -1 \end{bmatrix} \begin{bmatrix} -3 & 1 & 4 & 3 \\ 4 & 0 & 1 & 3 \end{bmatrix}$

11. $\begin{bmatrix} 1 & -1 & 2 & 3 \end{bmatrix} \begin{bmatrix} -1 & 2 & 0 \\ 2 & -1 & 0 \\ 0 & 5 & 2 \\ -1 & 8 & 1 \end{bmatrix}$

12. $\begin{bmatrix} 0 & 1 & -1 & 2 \end{bmatrix} \begin{bmatrix} 1 & -2 & 1 \\ 0 & 1 & 3 \\ 6 & 0 & 2 \\ -1 & -2 & 11 \end{bmatrix}$

13. $\begin{bmatrix} 1 & 0 & -1 \\ 1 & 1 & 2 \end{bmatrix} \begin{bmatrix} 0 & 1 & -1 \\ 1 & 0 & 1 \\ 4 & 8 & 0 \end{bmatrix}$

14. $\begin{bmatrix} 0 & 1 & -1 \\ 3 & 1 & -1 \end{bmatrix} \begin{bmatrix} 1 & 1 \\ 4 & 2 \\ 0 & 1 \end{bmatrix}$

15. $\begin{bmatrix} 1 & 0 \\ 1 & -1 \end{bmatrix} \begin{bmatrix} 0 & 1 \\ 0 & 1 \end{bmatrix}$

16. $\begin{bmatrix} 1 & -1 \\ 1 & -1 \end{bmatrix} \begin{bmatrix} 3 & -3 \\ 5 & -7 \end{bmatrix}$

17. $\begin{bmatrix} 0 & 1 \\ 0 & 1 \end{bmatrix} \begin{bmatrix} 1 & 0 \\ 1 & -1 \end{bmatrix}$

18. $\begin{bmatrix} 3 & -3 \\ 5 & -7 \end{bmatrix} \begin{bmatrix} 1 & -1 \\ 1 & -1 \end{bmatrix}$

19. $\begin{bmatrix} 1 & -1 \\ 1 & -1 \end{bmatrix} \begin{bmatrix} 2 & 3 \\ 2 & 3 \end{bmatrix}$

20. $\begin{bmatrix} 0 & 1 \\ 1 & 0 \end{bmatrix} \begin{bmatrix} 3 & -3 \\ 2 & -1 \end{bmatrix}$

21. $\begin{bmatrix} 1 & -1 \\ -1 & 1 \end{bmatrix} \begin{bmatrix} 2 & 3 \\ 2 & 3 \\ 1 & 1 \end{bmatrix}$

22. $\begin{bmatrix} 0 & 1 & -1 \\ 0 & -1 & 1 \end{bmatrix} \begin{bmatrix} 3 & -3 \\ 2 & -1 \end{bmatrix}$

23. $\begin{bmatrix} 1 & 0 & -1 \\ 2 & -2 & 1 \\ 0 & 0 & 1 \end{bmatrix} \begin{bmatrix} 1 & -1 & 4 \\ 1 & 1 & 0 \\ 0 & 4 & 1 \end{bmatrix}$

24. $\begin{bmatrix} 1 & 2 & 0 \\ 4 & -1 & 1 \\ 1 & 0 & 1 \end{bmatrix} \begin{bmatrix} 1 & 2 & -4 \\ 4 & 1 & 0 \\ 0 & -2 & 1 \end{bmatrix}$

25. $\begin{bmatrix} 1 & 0 & 1 & 0 \\ -1 & 1 & 0 & 1 \\ -2 & 0 & 1 & 4 \\ 0 & -1 & 0 & 1 \end{bmatrix} \begin{bmatrix} 1 \\ -3 \\ 2 \\ 0 \end{bmatrix}$

26. $\begin{bmatrix} 1 & 1 & -7 & 0 \\ -1 & 0 & 2 & 4 \\ -1 & 0 & -2 & 1 \\ 1 & -1 & 1 & 1 \end{bmatrix} \begin{bmatrix} 1 \\ -3 \\ 2 \\ 1 \end{bmatrix}$

27. **T** $\begin{bmatrix} 1.1 & 2.3 & 3.4 & -1.2 \\ 3.4 & 4.4 & 2.3 & 1.1 \\ 2.3 & 0 & -2.2 & 1.1 \\ 1.2 & 1.3 & 1.1 & 1.1 \end{bmatrix} \begin{bmatrix} -2.1 & 0 & -3.3 \\ -3.4 & -4.8 & -4.2 \\ 3.4 & 5.6 & 1 \\ 1 & 2.2 & 9.8 \end{bmatrix}$

28. **T** $\begin{bmatrix} 1.2 & 2.3 & 3.4 & 4.5 \\ 3.3 & 4.4 & 5.5 & 6.6 \\ 2.3 & -4.3 & -2.2 & 1.1 \\ 2.2 & -1.2 & -1 & 1.1 \end{bmatrix} \begin{bmatrix} 9.8 & 1 & -1.1 \\ 8.8 & 2 & -2.2 \\ 7.7 & 3 & -3.3 \\ 6.6 & 4 & -4.4 \end{bmatrix}$

29. Find[7] $A^2 = A \cdot A$, $A^3 = A \cdot A \cdot A$, A^4, and A^{100}, given that

$$A = \begin{bmatrix} 0 & 1 & 1 & 1 \\ 0 & 0 & 1 & 1 \\ 0 & 0 & 0 & 1 \\ 0 & 0 & 0 & 0 \end{bmatrix}.$$

30. Repeat Exercise 29 with $A = \begin{bmatrix} 0 & 2 & 0 & -1 \\ 0 & 0 & 2 & 0 \\ 0 & 0 & 0 & 2 \\ 0 & 0 & 0 & 0 \end{bmatrix}.$

Exercises 31–38 should be done in two ways: by hand and by using technology where possible.

Let

$$A = \begin{bmatrix} 0 & -1 & 0 & 1 \\ 10 & 0 & 1 & 0 \end{bmatrix}, B = \begin{bmatrix} 0 & -1 \\ 1 & 1 \\ -1 & 3 \\ 5 & 0 \end{bmatrix}, C = \begin{bmatrix} 1 & -1 \\ 1 & 1 \\ 1 & 1 \\ 1 & 1 \end{bmatrix}.$$

Evaluate:

31. AB **32.** AC **33.** $A(B - C)$ **34.** $(B - C)A$

Let $A = \begin{bmatrix} 1 & -1 \\ 0 & 2 \\ 0 & -2 \end{bmatrix}$, $B = \begin{bmatrix} 3 & 0 & -1 \\ 5 & -1 & 1 \end{bmatrix}$, $C = \begin{bmatrix} x & 1 & w \\ z & r & 4 \end{bmatrix}.$

Evaluate:

35. AB **36.** AC **37.** $A(B + C)$ **38.** $(B + C)A$

In Exercises 39–44, calculate (a) $P^2 = P \cdot P$, (b) $P^4 = P^2 \cdot P^2$, and (c) P^8. (Round all entries to four decimal places.)
(d) Without computing it explicitly, find $P^{1,000}$.

39. ▼ $P = \begin{bmatrix} 0.2 & 0.8 \\ 0.2 & 0.8 \end{bmatrix}$ **40.** ▼ $P = \begin{bmatrix} 0.1 & 0.1 \\ 0.9 & 0.9 \end{bmatrix}$

41. ▼ $P = \begin{bmatrix} 0.1 & 0.9 \\ 0 & 1 \end{bmatrix}$ **42.** ▼ $P = \begin{bmatrix} 1 & 0 \\ 0.8 & 0.2 \end{bmatrix}$

43. ▼ $P = \begin{bmatrix} 0.3 & 0.3 & 0.4 \\ 0.3 & 0.3 & 0.4 \\ 0.3 & 0.3 & 0.4 \end{bmatrix}$

What do you notice about the rows of P? Compare with Exercise 39, and state a general result about square matrices that these two exercises seem to suggest.

44. ▼ $P = \begin{bmatrix} -0.3 & -0.3 & -0.3 \\ 0.9 & 0.9 & 0.9 \\ 0.4 & 0.4 & 0.4 \end{bmatrix}$

What do you notice about the columns of P? Compare with Exercise 40, and state a general result about square matrices that these two exercises seem to suggest.

In Exercises 45–48, translate the given matrix equations into systems of linear equations. [HINT: See Example 7.]

45. $\begin{bmatrix} 2 & -1 & 4 \\ -4 & \frac{3}{4} & \frac{1}{3} \\ -3 & 0 & 0 \end{bmatrix} \begin{bmatrix} x \\ y \\ z \end{bmatrix} = \begin{bmatrix} 3 \\ -1 \\ 0 \end{bmatrix}$

46. $\begin{bmatrix} 1 & -1 & 4 \\ -\frac{1}{3} & -3 & \frac{1}{3} \\ 3 & 0 & 1 \end{bmatrix} \begin{bmatrix} x \\ y \\ z \end{bmatrix} = \begin{bmatrix} -3 \\ -1 \\ 2 \end{bmatrix}$

47. $\begin{bmatrix} 1 & -1 & 0 & 1 \\ 1 & 1 & 2 & 4 \end{bmatrix} \begin{bmatrix} x \\ y \\ z \\ w \end{bmatrix} = \begin{bmatrix} -1 \\ 2 \end{bmatrix}$

48. $\begin{bmatrix} 0 & 1 & 6 & 1 \\ 1 & -5 & 0 & 0 \end{bmatrix} \begin{bmatrix} x \\ y \\ z \\ w \end{bmatrix} = \begin{bmatrix} -2 \\ 9 \end{bmatrix}$

In Exercises 49–52, translate the given systems of equations into matrix form. [HINT: See Example 7.]

49. $\begin{aligned} x - y &= 4 \\ 2x - y &= 0 \end{aligned}$ **50.** $\begin{aligned} 2x + y &= 7 \\ -x &= 9 \end{aligned}$

51. $\begin{aligned} x + y - z &= 8 \\ 2x + y + z &= 4 \\ \frac{3x}{4} + \frac{z}{2} &= 1 \end{aligned}$ **52.** $\begin{aligned} x + y + 2z &= -2 \\ 4x + 2y - z &= -8 \\ \frac{x}{2} - \frac{y}{3} &= 4 \end{aligned}$

Applications

53. *Revenue* Your T-shirt operation is doing a booming trade. Last week you sold 50 tie-dyed shirts for \$15 each, 40 Suburban State University Crew shirts for \$10 each, and 30 Lacrosse T-shirts for \$12 each. Use matrix operations to calculate your total revenue for the week. [HINT: See Example 1.]

54. *Revenue* Karen Sandberg, your competitor in *Suburban State U's* T-shirt market, has apparently been undercutting your prices and outperforming you in sales. Last week she sold 100 tie-dyed shirts for \$10 each, 50 (low-quality) Crew shirts at \$5 apiece, and 70 Lacrosse T-shirts for \$8 each. Use matrix operations to calculate her total revenue for the week. [HINT: See Example 1.]

55. *Real Estate* The following table shows the cost of 1,000 square feet of luxury real estate in three cities in March 2014[8] together with the quantity your development company intends to purchase in each city:

---

[7] $A \cdot A \cdot A$ is $A(A \cdot A)$, or the equivalent $(A \cdot A)A$ by the associative law. Similarly, $A \cdot A \cdot A \cdot A = A(A \cdot A \cdot A) = (A \cdot A \cdot A)A = (A \cdot A)(A \cdot A)$; it doesn't matter where we place parentheses.

[8] Data rounded to the nearest 0.05. Source: "10 Most Expensive Markets for Real Estate," March 5, 2014 (www.cnbc.com).

	London	New York	Mumbai
Cost ($ million)	3.70	2.30	0.95
Quantity Purchased (thousands of square feet)	10	20	10

Use matrix multiplication to estimate the total cost of the purchase.

56. Real Estate Repeat Exercise 55 using the following table for Hong Kong, Paris, and Shanghai:[9]

	Hong Kong	Paris	Shanghai
Cost ($ million)	4.50	2.25	2.00
Quantity Purchased (thousands of square feet)	10	30	5

57. Revenue Recall the *Left Coast Bookstore* chain from Section 4.1. In January it sold 700 hardcover books, 1,300 softcover books, and 2,000 plastic books in San Francisco; it sold 400 hardcover, 300 softcover, and 500 plastic books in Los Angeles. Hardcover books sell for $30 each, softcover books sell for $10 each, and plastic books sell for $15 each. Write a column matrix with the price data, and show how matrix multiplication (using the sales and price data matrices) may be used to compute the total revenue at the two stores. [**HINT:** See Example 5.]

58. Profit Refer back to Exercise 57, and now suppose that each hardcover book costs the stores $10, each softcover book costs $5, and each plastic book costs $10. Use matrix operations to compute the total *profit* at each store in January. [**HINT:** See Example 5.]

▼ Income *Exercises 59–62 are based on the following spreadsheet, which shows the projected 2020 and 2030 U.S. male and female population in various age groups, as well as per capita incomes:[10]*

	A	B	C	D	E	F
1			2020 Population		2030 population	
2	Age	Mean Income ($1000)	Female (Millions)	Male (Millions)	Female (Millions)	Male (Millions)
3	15 to 24	14	23	24	25	26
4	25 to 44	42	43	43	45	46
5	45 to 64	48	43	41	42	41
6	65 to 84	29	31	20	40	31

59. Use matrix algebra to estimate the total income for females in 2020. (Round the answer to two significant digits.)

60. Use matrix algebra to estimate the total income for males in 2030. (Round the answer to two significant digits.)

61. Give a single matrix formula that expresses the difference in total income between males and females in 2020, and compute its value, rounded to two significant digits.

62. Give a single matrix formula that expresses the total income in 2030, and compute its value, rounded to two significant digits.

63. Consumption of Dairy Products The U.S. per capita consumption of yogurt and ice cream in 1983 and 2013 was as follows:[11]

	1983	2013
Yogurt (pounds)	3.2	14.9
Ice Cream (pounds)	26.3	20.1

Thinking of this table as a (labeled) 2×2 matrix P, compute the matrix product $\begin{bmatrix} -1 & 1 \end{bmatrix} P$. What does this product represent?

64. Consumption of Dairy Products The U.S. per capita consumption of cottage cheese and mozzarella in 1983 and 2013 was as follows:[12]

	1983	2013
Cottage Cheese (pounds)	4.1	2.1
Mozzarella (pounds)	3.7	10.8

Thinking of this table as a (labeled) 2×2 matrix P, compute the matrix product $P \begin{bmatrix} -1 \\ 1 \end{bmatrix}$. What does this product represent?

Foreclosure Crisis *Starting in 2010, on the heels of the 2007–2009 subprime mortgage crisis, the United States saw an epidemic of mortgage foreclosures, often initiated improperly by large financial institutions. Exercises 65–70 are based on the following table, which shows the numbers of foreclosures in three states during three months of 2011:[13]*

	June	July	Aug.
California	54,100	56,200	59,400
Florida	23,800	22,400	23,600
Texas	9,300	10,600	10,100

65. Each month, your law firm handled 10% of all foreclosures in California, 5% of all foreclosures in Florida, and 20% of all foreclosures in Texas. Use matrix multiplication to compute the total number of foreclosures handled by your firm in each of the months shown.

---

[9] See footnote for Exercise 55.

[10] The population figures are Census Bureau estimates, and the income figures are 2007 mean per capita incomes. All figures are approximate. Source: U.S. Census Bureau (www.census.gov).

[11] Figures are approximate. Source: United States Department of Agriculture (www.ers.usda.gov).

[12] *Ibid.*

[13] Figures are rounded. Source: www.realtytrac.com.

66. Your law firm handled 10% of all foreclosures in each state in June, 30% of all foreclosures in July, and 20% of all foreclosures in August 2011. Use matrix multiplication to compute the total number of foreclosures handled by your firm in each of the states shown.

67. Let A be the 3×3 matrix whose entries are the figures in the table, and let $B = \begin{bmatrix} 1 & 1 & 0 \end{bmatrix}$. What does the matrix BA represent?

68. Let A be the 3×3 matrix whose entries are the figures in the table, and let $B = \begin{bmatrix} 1 & 1 & 0 \end{bmatrix}^T$. What does the matrix AB represent?

69. ▼ Write a matrix product whose computation gives the total number by which the combined foreclosures for all three months in California and Texas exceeded the foreclosures in Florida. Calculate the product.

70. ▼ Write a matrix product whose computation gives the total number by which combined foreclosures in August exceeded foreclosures in June. Calculate the product.

71. ▼ **Costs** *Microbucks Computer Co.* makes two computers, the Pomegranate II and the Pomegranate Classic. The Pom II requires 2 processor chips, 16 memory chips, and 20 vacuum tubes, while the Pom Classic requires 1 processor chip, 4 memory chips, and 40 vacuum tubes. There are two companies that can supply these parts: *Motorel* can supply them at $100 per processor chip, $50 per memory chip, and $10 per vacuum tube, while *Intola* can supply them at $150 per processor chip, $40 per memory chip, and $15 per vacuum tube. Write down all of these data in two matrices: one showing the parts required for each model computer and the other showing the prices for each part from each supplier. Then show how matrix multiplication allows you to compute the total cost for parts for each model when parts are bought from either supplier.

72. ▼ **Profits** Refer back to Exercise 71. It actually costs *Motorel* only $25 to make each processor chip, $10 for each memory chip, and $5 for each vacuum tube. It costs *Intola* $50 per processor chip, $10 per memory chip, and $7 per vacuum tube. Use matrix operations to find the total profit Motorel and Intola would make on each model.

73. ▼ **Tourism in the 1990s** The following table gives the number of people (in thousands) who visited Australia and South Africa in 1998:[14]

	To	Australia	South Africa
From	**North America**	440	190
	Europe	950	950
	Asia	1,790	200

You estimate that 5% of all visitors to Australia and 4% of all visitors to South Africa decide to settle there permanently. Take A to be the 3×2 matrix whose entries are the 1998 tourism figures in the above table, and take

$$B = \begin{bmatrix} 0.05 \\ 0.04 \end{bmatrix} \quad \text{and} \quad C = \begin{bmatrix} 0.05 & 0 \\ 0 & 0.04 \end{bmatrix}.$$

Compute the products AB and AC. What do the entries in these matrices represent?

74. ▼ **Tourism in the 1990s** Referring to the tourism figures in Exercise 73, you estimate that from 1998 to 2018, tourism from North America to each of Australia and South Africa will have increased by 20%, tourism from Europe by 30%, and tourism from Asia by 10%. Take A to be the 3×2 matrix whose entries are the 1998 tourism figures, and take

$$B = \begin{bmatrix} 1.2 & 1.3 & 1.1 \end{bmatrix} \quad \text{and} \quad C = \begin{bmatrix} 1.2 & 0 & 0 \\ 0 & 1.3 & 0 \\ 0 & 0 & 1.1 \end{bmatrix}.$$

Compute the products BA and CA. What do the entries in these matrices represent?

75. 🅣 ▼ **Population Movement** In 2008 the population of the United States, broken down by regions, was 54.1 million in the Northeast, 65.7 million in the Midwest, 112.9 million in the South, and 70.3 million in the West. The table below shows the population movement during the period 2008–2009. (Thus, 99.23% of the population in the Northeast stayed there, while 0.16% of the population in the Northeast moved to the Midwest, and so on.)[15]

	To	Northeast	Midwest	South	West
From	**Northeast**	0.9923	0.0016	0.0042	0.0019
	Midwest	0.0018	0.9896	0.0047	0.0039
	South	0.0056	0.0059	0.9827	0.0058
	West	0.0024	0.0033	0.0044	0.9899

Set up the 2008 population figures as a row vector. Then use matrix multiplication to estimate the population in each region in 2009. (Round all answers to the nearest 0.1 million.)

76. 🅣 ▼ **Population Movement** Assume that the percentages given in Exercise 75 also describe the population movements from 2009 to 2010. Use two matrix multiplications to estimate, from the data in Exercise 75, the population in each region in 2010.

---

[14] Figures are rounded to the nearest 10,000. Sources: South African Dept. of Environmental Affairs and Tourism; Australia Tourist Commission/*New York Times*, January 15, 2000, p. C1.

[15] Note that this exercise ignores migration into or out of the country. Source: U.S. Census Bureau, Current Population Survey, 2009 Annual Social and Economic Supplement.

Communication and Reasoning Exercises

77. Give an example of two matrices A and B such that AB is defined but BA is not defined.

78. Give an example of two matrices A and B of different dimensions such that both AB and BA are defined.

79. Compare addition and multiplication of 1×1 matrices to the arithmetic of numbers.

80. In comparing the algebra of 1×1 matrices, as discussed so far, to the algebra of real numbers (see Exercise 79), what important difference do you find?

81. Comment on the following claim: Every matrix equation represents a system of equations.

82. When is it true that both AB and BA are defined, even though neither A nor B is a square matrix?

83. ▼ Find a scenario in which it would be useful to "multiply" two row vectors according to the rule

$$[a \quad b \quad c][d \quad e \quad f] = [ad \quad be \quad cf].$$

84. ▼ Make up an application whose solution reads as follows:

"Total revenue $= \begin{bmatrix} 10 & 100 & 30 \end{bmatrix} \begin{bmatrix} 10 & 0 & 3 \\ 1 & 2 & 0 \\ 0 & 1 & 40 \end{bmatrix}$."

85. ▼ What happens in a spreadsheet if, instead of using the function MMULT, you use "ordinary multiplication" as shown here?

	A	B	C	D	E	F	G
1	2	0	7		1	1	-8
2	1	-1	0		1	0	0
3	-2	1	1		0	5	2
4							
5	=A1:C3*E1:G3						
6							
7							

86. ▼ Define the *naïve product* $A \square B$ of two $m \times n$ matrices A and B by

$$(A \square B)_{ij} = A_{ij} B_{ij}.$$

(This is how someone who has never seen matrix multiplication before might think to multiply matrices.) Referring to Example 1 in this section, compute and comment on the meaning of $P \square (Q^T)$.

4.3 Matrix Inversion

Inverse of a Matrix

Now that we've discussed matrix addition, subtraction, and multiplication, you may well be wondering about matrix *division*. In the realm of real numbers, division can be thought of as a form of multiplication: Dividing 3 by 7 is the same as multiplying 3 by 1/7, the inverse of 7. In symbols, $3 \div 7 = 3 \times (1/7)$, or 3×7^{-1}. To imitate division of real numbers in the realm of matrices, we need to discuss the multiplicative **inverse**, A^{-1}, of a matrix A.

Note Because multiplication of real numbers is commutative, we can write, for example, $\frac{3}{7}$ as either 3×7^{-1} or $7^{-1} \times 3$. In the realm of matrices, multiplication is not commutative, so from now on we shall *never* talk about "division" of matrices. (By $\frac{B}{A}$, should we mean $A^{-1}B$ or BA^{-1}?) ∎

Before we try to find the inverse of a matrix, we must first know exactly what we *mean* by the inverse. Recall that the inverse of a number a is the number, often written a^{-1}, with the property that $a^{-1} \cdot a = a \cdot a^{-1} = 1$. For example, the inverse of 76 is the number $76^{-1} = 1/76$, because $(1/76) \cdot 76 = 76 \cdot (1/76) = 1$. This is the number calculated by the x^{-1} button found on many calculators. Not all numbers have an inverse. For example—and this is the only example—the number 0 has no inverse, because you cannot get 1 by multiplying 0 by anything.

The inverse of a matrix is defined similarly. To make life easier, we shall restrict attention to **square** matrices, which are matrices that have the same number of rows as columns.*

✱ Nonsquare matrices *cannot* have inverses in the sense that we shall be talking about. This is not a trivial fact to prove.

Inverse of a Matrix, Singular Matrix

The **inverse** of an $n \times n$ matrix A is the $n \times n$ matrix A^{-1} that, when multiplied by A on either side, yields the $n \times n$ identity matrix I. Thus,

$$AA^{-1} = A^{-1}A = I.$$

If A has an inverse, it is said to be **invertible**. Otherwise, it is said to be **singular**.

Quick Examples

1. The inverse of the 1×1 matrix $\begin{bmatrix} 3 \end{bmatrix}$ is $\begin{bmatrix} \frac{1}{3} \end{bmatrix}$, because $\begin{bmatrix} 3 \end{bmatrix}\begin{bmatrix} \frac{1}{3} \end{bmatrix} = \begin{bmatrix} 1 \end{bmatrix} = \begin{bmatrix} \frac{1}{3} \end{bmatrix}\begin{bmatrix} 3 \end{bmatrix}$.

2. The inverse of the $n \times n$ identity matrix I is I itself, because $II = I$. Thus, $I^{-1} = I$.

3. The inverse of the 2×2 matrix $A = \begin{bmatrix} 1 & -1 \\ -1 & -1 \end{bmatrix}$ is $A^{-1} = \begin{bmatrix} \frac{1}{2} & -\frac{1}{2} \\ -\frac{1}{2} & -\frac{1}{2} \end{bmatrix}$, because

$$\begin{bmatrix} 1 & -1 \\ -1 & -1 \end{bmatrix}\begin{bmatrix} \frac{1}{2} & -\frac{1}{2} \\ -\frac{1}{2} & -\frac{1}{2} \end{bmatrix} = \begin{bmatrix} 1 & 0 \\ 0 & 1 \end{bmatrix} \qquad AA^{-1} = I$$

and

$$\begin{bmatrix} \frac{1}{2} & -\frac{1}{2} \\ -\frac{1}{2} & -\frac{1}{2} \end{bmatrix}\begin{bmatrix} 1 & -1 \\ -1 & -1 \end{bmatrix} = \begin{bmatrix} 1 & 0 \\ 0 & 1 \end{bmatrix}. \qquad A^{-1}A = I$$

Notes

1. It is possible to show that if A and B are square matrices with $AB = I$, then it must also be true that $BA = I$. In other words, once we have checked that $AB = I$, we know that B is the inverse of A. The second check, that $BA = I$, is unnecessary.

2. If B is the inverse of A, then we can also say that A is the inverse of B (why?). Thus, we sometimes refer to such a pair of matrices as an **inverse pair** of matrices. ∎

EXAMPLE 1   Singular Matrix

Can $A = \begin{bmatrix} 1 & 1 \\ 0 & 0 \end{bmatrix}$ have an inverse?

Solution No. To see why not, notice that both entries in the second row of AB will be 0, no matter what B is. So AB cannot equal I, no matter what B is. Hence, A is singular.

➡ **Before we go on...** If you think about it, you can write down many similar examples of singular matrices. There is only one number with no multiplicative inverse (0), but there are many matrices that have no inverses. ∎

Finding the Inverse of a Square Matrix

Q: *In Quick Example 3, it was stated that the inverse of* $\begin{bmatrix} 1 & -1 \\ -1 & -1 \end{bmatrix}$ *is* $\begin{bmatrix} \frac{1}{2} & -\frac{1}{2} \\ -\frac{1}{2} & -\frac{1}{2} \end{bmatrix}$. *How was that obtained?*

A : We can think of the problem of finding A^{-1} as a problem of finding four unknowns, the four unknown entries of A^{-1}:

$$A^{-1} = \begin{bmatrix} x & y \\ z & w \end{bmatrix}.$$

These unknowns must satisfy the equation $AA^{-1} = I$, or

$$\begin{bmatrix} 1 & -1 \\ -1 & -1 \end{bmatrix} \begin{bmatrix} x & y \\ z & w \end{bmatrix} = \begin{bmatrix} 1 & 0 \\ 0 & 1 \end{bmatrix}.$$

If we were to try to find the first column of A^{-1}, consisting of x and z, we would have to solve

$$\begin{bmatrix} 1 & -1 \\ -1 & -1 \end{bmatrix} \begin{bmatrix} x \\ z \end{bmatrix} = \begin{bmatrix} 1 \\ 0 \end{bmatrix},$$

or

$$x - z = 1$$
$$-x - z = 0.$$

To solve this system by Gauss-Jordan reduction, we would row-reduce the augmented matrix, which is A with the column $\begin{bmatrix} 1 \\ 0 \end{bmatrix}$ adjoined:

$$\begin{bmatrix} 1 & -1 & | & 1 \\ -1 & -1 & | & 0 \end{bmatrix} \rightarrow \begin{bmatrix} 1 & 0 & | & x \\ 0 & 1 & | & z \end{bmatrix}.$$

To find the second column of A^{-1}, we would similarly row-reduce the augmented matrix obtained by tacking on to A the second column of the identity matrix:

$$\begin{bmatrix} 1 & -1 & | & 0 \\ -1 & -1 & | & 1 \end{bmatrix} \rightarrow \begin{bmatrix} 1 & 0 & | & y \\ 0 & 1 & | & w \end{bmatrix}.$$

The row operations used in doing these two reductions would be exactly the same. We could do both reductions simultaneously by "doubly augmenting" A, putting both columns of the identity matrix to the right of A:

$$\begin{bmatrix} 1 & -1 & | & 1 & 0 \\ -1 & -1 & | & 0 & 1 \end{bmatrix} \rightarrow \begin{bmatrix} 1 & 0 & | & x & y \\ 0 & 1 & | & z & w \end{bmatrix}.$$

We carry out this reduction in the following example.

EXAMPLE 2   **Computing the Inverse of a Matrix**

Find the inverse of each matrix.

a. $P = \begin{bmatrix} 1 & -1 \\ -1 & -1 \end{bmatrix}$ **b.** $Q = \begin{bmatrix} 1 & 0 & 1 \\ 2 & -2 & -1 \\ 3 & 0 & 0 \end{bmatrix}$

Solution

a. As described above, we put the matrix P on the left and the identity matrix I on the right to get a 2×4 matrix:

$$\begin{bmatrix} 1 & -1 & | & 1 & 0 \\ -1 & -1 & | & 0 & 1 \end{bmatrix}.$$
$$\qquad P \qquad\qquad I$$

We now row-reduce the whole matrix:

$$\begin{bmatrix} 1 & -1 & 1 & 0 \\ -1 & -1 & 0 & 1 \end{bmatrix} \begin{matrix} \\ R_2 + R_1 \end{matrix} \rightarrow \begin{bmatrix} 1 & -1 & 1 & 0 \\ 0 & -2 & 1 & 1 \end{bmatrix} \begin{matrix} 2R_1 - R_2 \\ \end{matrix} \rightarrow$$

$$\begin{bmatrix} 2 & 0 & 1 & -1 \\ 0 & -2 & 1 & 1 \end{bmatrix} \begin{matrix} \frac{1}{2}R_1 \\ -\frac{1}{2}R_2 \end{matrix} \rightarrow \begin{bmatrix} 1 & 0 & \frac{1}{2} & -\frac{1}{2} \\ 0 & 1 & -\frac{1}{2} & -\frac{1}{2} \end{bmatrix}.$$
$$\underset{I}{} \qquad \underset{P^{-1}}{}$$

We have now solved the systems of linear equations that define the entries of P^{-1}. Thus,

$$P^{-1} = \begin{bmatrix} \frac{1}{2} & -\frac{1}{2} \\ -\frac{1}{2} & -\frac{1}{2} \end{bmatrix}.$$

b. The procedure to find the inverse of a 3 × 3 matrix (or larger) is just the same as for a 2 × 2 matrix. We place Q on the left and the identity matrix (now 3 × 3) on the right, and reduce:

$$\overset{Q}{} \qquad \overset{I}{}$$
$$\begin{bmatrix} 1 & 0 & 1 & | & 1 & 0 & 0 \\ 2 & -2 & -1 & | & 0 & 1 & 0 \\ 3 & 0 & 0 & | & 0 & 0 & 1 \end{bmatrix} \begin{matrix} \\ R_2 - 2R_1 \\ R_3 - 3R_1 \end{matrix} \rightarrow \begin{bmatrix} 1 & 0 & 1 & | & 1 & 0 & 0 \\ 0 & -2 & -3 & | & -2 & 1 & 0 \\ 0 & 0 & -3 & | & -3 & 0 & 1 \end{bmatrix} \begin{matrix} 3R_1 + R_3 \\ R_2 - R_3 \end{matrix} \rightarrow$$

$$\begin{bmatrix} 3 & 0 & 0 & | & 0 & 0 & 1 \\ 0 & -2 & 0 & | & 1 & 1 & -1 \\ 0 & 0 & -3 & | & -3 & 0 & 1 \end{bmatrix} \begin{matrix} \frac{1}{3}R_1 \\ -\frac{1}{2}R_2 \\ -\frac{1}{3}R_3 \end{matrix} \rightarrow \begin{bmatrix} 1 & 0 & 0 & | & 0 & 0 & \frac{1}{3} \\ 0 & 1 & 0 & | & -\frac{1}{2} & -\frac{1}{2} & \frac{1}{2} \\ 0 & 0 & 1 & | & 1 & 0 & -\frac{1}{3} \end{bmatrix}.$$
$$\underset{I}{} \qquad \underset{Q^{-1}}{}$$

Thus,

$$Q^{-1} = \begin{bmatrix} 0 & 0 & \frac{1}{3} \\ -\frac{1}{2} & -\frac{1}{2} & \frac{1}{2} \\ 1 & 0 & -\frac{1}{3} \end{bmatrix}.$$

We have already checked that P^{-1} is the inverse of P. You should also check that Q^{-1} is the inverse of Q.

The method we used in Example 2 can be summarized as follows.

Inverting an $n \times n$ Matrix

To determine whether or not an $n \times n$ matrix A is invertible, and to find A^{-1} if it does exist, follow this procedure:

1. Write down the $n \times 2n$ matrix $[A \,|\, I]$. (This is A with the $n \times n$ identity matrix set next to it.)

2. Row-reduce $[A \,|\, I]$.

3. If the reduced form is $[I \,|\, B]$ (i.e., has the identity matrix in the left part), then A is invertible and $B = A^{-1}$. If row reduction does not result in I in the left part, then A is singular. (See Example 3.)

Although there is a general formula for the inverse of a matrix, it is not a simple one. In fact, using the formula for anything larger than a 3×3 matrix is so inefficient that the row-reduction procedure is the method of choice even for computer algorithms. However, the general formula is very simple for the special case of 2×2 matrices.

Formula for the Inverse of a 2 × 2 Matrix

The inverse of a 2×2 matrix is

$$\begin{bmatrix} a & b \\ c & d \end{bmatrix}^{-1} = \frac{1}{ad - bc} \begin{bmatrix} d & -b \\ -c & a \end{bmatrix}, \quad \text{provided that } ad - bc \neq 0.$$

If the quantity $ad - bc$ is zero, then the matrix is singular (non-invertible). The quantity $ad - bc$ is called the **determinant** of the matrix $\begin{bmatrix} a & b \\ c & d \end{bmatrix}$.

Quick Examples

4. $$\begin{bmatrix} 1 & 2 \\ 3 & 4 \end{bmatrix}^{-1} = \frac{1}{(1)(4) - (2)(3)} \begin{bmatrix} 4 & -2 \\ -3 & 1 \end{bmatrix} = -\frac{1}{2} \begin{bmatrix} 4 & -2 \\ -3 & 1 \end{bmatrix}$$
$$= \begin{bmatrix} -2 & 1 \\ \frac{3}{2} & -\frac{1}{2} \end{bmatrix}$$

5. $\begin{bmatrix} 1 & -1 \\ 2 & -2 \end{bmatrix}$ has determinant $ad - bc = (1)(-2) - (-1)(2) = 0$ and so is singular.

The formula for the inverse of a 2×2 matrix can be obtained by using the technique of row reduction. (See the Communication and Reasoning Exercises at the end of the section.)

As we mentioned earlier, not every square matrix has an inverse, as we see in the next example.

EXAMPLE 3 **Singular 3 × 3 Matrix**

Find the inverse of the matrix $S = \begin{bmatrix} 1 & 1 & 2 \\ -2 & 0 & 4 \\ 3 & 1 & -2 \end{bmatrix}$ if it exists.

Solution We proceed as before:

$$\begin{array}{c} S \qquad\qquad I \\ \left[\begin{array}{ccc|ccc} 1 & 1 & 2 & 1 & 0 & 0 \\ -2 & 0 & 4 & 0 & 1 & 0 \\ 3 & 1 & -2 & 0 & 0 & 1 \end{array}\right] \begin{array}{l} \\ R_2 + 2R_1 \\ R_3 - 3R_1 \end{array} \rightarrow \left[\begin{array}{ccc|ccc} 1 & 1 & 2 & 1 & 0 & 0 \\ 0 & 2 & 8 & 2 & 1 & 0 \\ 0 & -2 & -8 & -3 & 0 & 1 \end{array}\right] \begin{array}{l} 2R_1 - R_2 \\ \\ R_3 + R_2 \end{array} \end{array}$$

$$\rightarrow \left[\begin{array}{ccc|ccc} 2 & 0 & -4 & 0 & -1 & 0 \\ 0 & 2 & 8 & 2 & 1 & 0 \\ 0 & 0 & 0 & -1 & 1 & 1 \end{array}\right].$$

We stopped here, even though the reduction is incomplete, because there is *no hope* of getting the identity on the left-hand side. Completing the row reduction will not change the three zeros in the bottom row. So what is wrong? Nothing. As in Example 1, we have here a singular matrix. Any square matrix that, after row reduction, winds up with a row of zeros is singular. (See Exercise 77.)

➡ **Before we go on . . .** In practice, deciding whether a given matrix is invertible or singular is easy: Simply try to find its inverse. If the process works, then the matrix is invertible, and we get its inverse. If the process fails, then the matrix is singular. If you try to invert a singular matrix using a spreadsheet, calculator, or computer program, you should get an error. Sometimes, instead of an error, you will get a spurious answer due to round-off errors in the device. ∎

Using the Inverse to Solve a System of *n* Linear Equations in *n* Unknowns

Having used systems of equations and row reduction to find matrix inverses, we will now use matrix inverses to solve systems of equations. Recall that, at the end of Section 4.2, we saw that a system of linear equations could be written in the form

$$AX = B,$$

where A is the coefficient matrix, X is the column matrix of unknowns, and B is the column matrix of right-hand sides. Now suppose that there are as many unknowns as equations, so A is a square matrix, and suppose that A is invertible. The object is to solve for the matrix X of unknowns, so we multiply both sides of the equation by the inverse A^{-1} of A, getting

$$A^{-1}AX = A^{-1}B.$$

Notice that we put A^{-1} on the left on both sides of the equation. Order matters when multiplying matrices, so we have to be careful to do the same thing to both sides of the equation. But now $A^{-1}A = I$, so we can rewrite the last equation as

$$IX = A^{-1}B.$$

Also, $IX = X$ (I being the identity matrix), so we really have

$$X = A^{-1}B,$$

and we have solved for X!

Moreover, we have shown that, if A is invertible and $AX = B$, then the *only possible* solution is $X = A^{-1}B$. We should check that $A^{-1}B$ is actually a solution by substituting back into the original equation:

$$AX = A(A^{-1}B) = (AA^{-1})B = IB = B.$$

Thus, $X = A^{-1}B$ is a solution and is the only solution. Therefore, if A is invertible, $AX = B$ has exactly one solution.

On the other hand, if $AX = B$ has no solutions or has infinitely many solutions, we can conclude that A is not invertible (why?). To summarize, we have the following.

Solving the Matrix Equation $AX = B$

If A is an invertible matrix, then the matrix equation $AX = B$ has the unique solution

$$X = A^{-1}B.$$

Quick Example

6. The system of linear equations

$$
\begin{aligned}
2x \quad\;\;\; + z &= 9 \\
2x + y - z &= 6 \\
3x + y - z &= 9
\end{aligned}
$$

can be written as $AX = B$, where

$$
A = \begin{bmatrix} 2 & 0 & 1 \\ 2 & 1 & -1 \\ 3 & 1 & -1 \end{bmatrix}, \quad X = \begin{bmatrix} x \\ y \\ z \end{bmatrix}, \quad \text{and} \quad B = \begin{bmatrix} 9 \\ 6 \\ 9 \end{bmatrix}.
$$

The matrix A is invertible with inverse

$$
A^{-1} = \begin{bmatrix} 0 & -1 & 1 \\ 1 & 5 & -4 \\ 1 & 2 & -2 \end{bmatrix}. \qquad \text{You should check this.}
$$

Thus,

$$
X = A^{-1}B = \begin{bmatrix} 0 & -1 & 1 \\ 1 & 5 & -4 \\ 1 & 2 & -2 \end{bmatrix} \begin{bmatrix} 9 \\ 6 \\ 9 \end{bmatrix} = \begin{bmatrix} 3 \\ 3 \\ 3 \end{bmatrix},
$$

so $(x, y, z) = (3, 3, 3)$ is the (unique) solution to the system.

EXAMPLE 4   **Solving Systems of Equations Using an Inverse**

Solve the following three systems of equations:

a.
$$
\begin{aligned}
2x \quad\;\;\; + z &= 1 \\
2x + y - z &= 1 \\
3x + y - z &= 1
\end{aligned}
$$

b.
$$
\begin{aligned}
2x \quad\;\;\; + z &= 0 \\
2x + y - z &= 1 \\
3x + y - z &= 2
\end{aligned}
$$

c.
$$
\begin{aligned}
2x \quad\;\;\; + z &= 0 \\
2x + y - z &= 0 \\
3x + y - z &= 0
\end{aligned}
$$

Solution We *could* go ahead and row-reduce all three augmented matrices, as we did in Chapter 3, but this would require a lot of work. Notice that the coefficients are the same in all three systems. In other words, we can write the three systems in matrix form as follows:

a. $AX = B$ **b.** $AX = C$ **c.** $AX = D$

where the matrix A is the same in all three cases:

$$
A = \begin{bmatrix} 2 & 0 & 1 \\ 2 & 1 & -1 \\ 3 & 1 & -1 \end{bmatrix}.
$$

Now the solutions to these systems are

a. $X = A^{-1}B$ **b.** $X = A^{-1}C$ **c.** $X = A^{-1}D$

so the main work is the calculation of the single matrix A^{-1}, which we have already noted (see Quick Example 6) is

$$A^{-1} = \begin{bmatrix} 0 & -1 & 1 \\ 1 & 5 & -4 \\ 1 & 2 & -2 \end{bmatrix}.$$

Thus, the three solutions are as follows:

a. $X = A^{-1}B = \begin{bmatrix} 0 & -1 & 1 \\ 1 & 5 & -4 \\ 1 & 2 & -2 \end{bmatrix}\begin{bmatrix} 1 \\ 1 \\ 1 \end{bmatrix} = \begin{bmatrix} 0 \\ 2 \\ 1 \end{bmatrix}$

b. $X = A^{-1}C = \begin{bmatrix} 0 & -1 & 1 \\ 1 & 5 & -4 \\ 1 & 2 & -2 \end{bmatrix}\begin{bmatrix} 0 \\ 1 \\ 2 \end{bmatrix} = \begin{bmatrix} 1 \\ -3 \\ -2 \end{bmatrix}$

c. $X = A^{-1}D = \begin{bmatrix} 0 & -1 & 1 \\ 1 & 5 & -4 \\ 1 & 2 & -2 \end{bmatrix}\begin{bmatrix} 0 \\ 0 \\ 0 \end{bmatrix} = \begin{bmatrix} 0 \\ 0 \\ 0 \end{bmatrix}$

➡ **Before we go on . . .** We have been speaking of *the* inverse of a matrix A. Is there only one? It is not hard to prove that a matrix A cannot have more than one inverse. If B and C were both inverses of A, then

$B = BI$	Property of the identity
$\quad = B(AC)$	Because C is an inverse of A
$\quad = (BA)C$	Associative law
$\quad = IC$	Because B is an inverse of A
$\quad = C.$	Property of the identity

In other words, if B and C were both inverses of A, then B and C would have to be equal. ∎

FAQs

Which Method to Use in Solving a System

Q: *Now we have two methods to solve a system of linear equations $AX = B$: (1) Compute $X = A^{-1}B$, or (2) row-reduce the augmented matrix. Which is the better method?*

A: Each method has its advantages and disadvantages. Method (1), as we have seen, is very efficient when you must solve several systems of equations with the same coefficients, but it works only when the coefficient matrix is *square* (meaning that you have the same number of equations as unknowns) *and invertible* (meaning that there is a unique solution). The row-reduction method will work for all systems. Moreover, for all but the smallest systems the most efficient way to find A^{-1} is to use row reduction. Thus, in practice, the two methods are essentially the same when both apply.

4.3   EXERCISES

▼ more advanced   ◆ challenging
Ⓣ indicates exercises that should be solved using technology

In Exercises 1–6, determine whether or not the given pairs of matrices are inverse pairs. [**HINT:** See Quick Examples 1–3.]

1. $A = \begin{bmatrix} 0 & 1 \\ 1 & 0 \end{bmatrix}, B = \begin{bmatrix} 0 & 1 \\ 1 & 0 \end{bmatrix}$

2. $A = \begin{bmatrix} 2 & 0 \\ 0 & 3 \end{bmatrix}, B = \begin{bmatrix} \frac{1}{2} & 0 \\ 0 & \frac{1}{2} \end{bmatrix}$

3. $A = \begin{bmatrix} 2 & 1 & 1 \\ 0 & 1 & 1 \\ 0 & 0 & 1 \end{bmatrix}, B = \begin{bmatrix} \frac{1}{2} & -\frac{1}{2} & 0 \\ 0 & 1 & -1 \\ 0 & 0 & 1 \end{bmatrix}$

4. $A = \begin{bmatrix} 1 & 1 & 1 \\ 0 & 1 & 1 \\ 0 & 0 & 1 \end{bmatrix}, B = \begin{bmatrix} 1 & -1 & 0 \\ 0 & 1 & -1 \\ 0 & 0 & 1 \end{bmatrix}$

5. $A = \begin{bmatrix} a & 0 & 0 \\ 0 & b & 0 \\ 0 & 0 & 0 \end{bmatrix}, B = \begin{bmatrix} a^{-1} & 0 & 0 \\ 0 & b^{-1} & 0 \\ 0 & 0 & 0 \end{bmatrix}$ $(a, b \neq 0)$

6. $A = \begin{bmatrix} a & 0 & 0 \\ 0 & b & 0 \\ 0 & 0 & c \end{bmatrix}, B = \begin{bmatrix} a^{-1} & 0 & 0 \\ 0 & b^{-1} & 0 \\ 0 & 0 & c^{-1} \end{bmatrix}$ $(a, b, c \neq 0)$

In Exercises 7–26, use row reduction to find the inverses of the given matrices if they exist, and check your answers by multiplication. [**HINT:** See Example 2.]

7. $\begin{bmatrix} 1 & 1 \\ 2 & 1 \end{bmatrix}$ **8.** $\begin{bmatrix} 0 & 1 \\ 1 & 1 \end{bmatrix}$ **9.** $\begin{bmatrix} 0 & 1 \\ 1 & 0 \end{bmatrix}$ **10.** $\begin{bmatrix} 4 & 0 \\ 0 & 2 \end{bmatrix}$

11. $\begin{bmatrix} 2 & 1 \\ 1 & 1 \end{bmatrix}$ **12.** $\begin{bmatrix} 3 & 0 \\ 0 & \frac{1}{2} \end{bmatrix}$ **13.** $\begin{bmatrix} 2 & 1 \\ 4 & 2 \end{bmatrix}$ **14.** $\begin{bmatrix} 1 & 1 \\ 6 & 6 \end{bmatrix}$

15. $\begin{bmatrix} 1 & 1 & 1 \\ 0 & 1 & 1 \\ 0 & 0 & 1 \end{bmatrix}$ **16.** $\begin{bmatrix} 1 & 2 & 3 \\ 0 & 1 & 2 \\ 0 & 0 & 1 \end{bmatrix}$ **17.** $\begin{bmatrix} 1 & 1 & 1 \\ 1 & 0 & 2 \\ 1 & -1 & 1 \end{bmatrix}$

18. $\begin{bmatrix} 1 & 2 & 3 \\ 0 & 2 & 3 \\ 1 & 0 & 1 \end{bmatrix}$ **19.** $\begin{bmatrix} 1 & 1 & 1 \\ 1 & -1 & 0 \\ 1 & 2 & 3 \end{bmatrix}$ **20.** $\begin{bmatrix} 1 & -1 & 3 \\ 0 & 1 & 3 \\ 1 & 1 & 1 \end{bmatrix}$

21. $\begin{bmatrix} 1 & 1 & 1 \\ 1 & 0 & 1 \\ 1 & -1 & 1 \end{bmatrix}$ **22.** $\begin{bmatrix} 1 & 1 & 1 \\ 0 & 1 & 1 \\ 1 & 0 & 0 \end{bmatrix}$

23. $\begin{bmatrix} 1 & 0 & 1 & 0 \\ -1 & 1 & 0 & 1 \\ -1 & 0 & 0 & 1 \\ 0 & -1 & 0 & 1 \end{bmatrix}$ **24.** $\begin{bmatrix} 0 & 1 & 1 & 0 \\ -1 & 1 & 1 & 1 \\ -1 & 1 & 0 & 1 \\ 0 & -1 & 0 & 1 \end{bmatrix}$

25. $\begin{bmatrix} 1 & 2 & 3 & 4 \\ 0 & 1 & 2 & 3 \\ 0 & 0 & 1 & 2 \\ 0 & 0 & 0 & 1 \end{bmatrix}$ **26.** $\begin{bmatrix} 0 & 0 & 0 & 1 \\ 0 & 0 & 1 & 0 \\ 0 & 1 & 0 & 0 \\ 1 & 0 & 0 & 0 \end{bmatrix}$

In Exercises 27–34, compute the determinant of the given matrix. If the determinant is nonzero, use the formula for inverting a 2 × 2 matrix to calculate the inverse of the given matrix. [**HINT:** See Quick Examples 4 and 5.]

27. $\begin{bmatrix} 1 & 1 \\ 1 & -1 \end{bmatrix}$ **28.** $\begin{bmatrix} 4 & 1 \\ 0 & 2 \end{bmatrix}$ **29.** $\begin{bmatrix} 1 & 2 \\ 3 & 4 \end{bmatrix}$

30. $\begin{bmatrix} 1 & 0 \\ 0 & 1 \end{bmatrix}$ **31.** $\begin{bmatrix} \frac{1}{6} & -\frac{1}{6} \\ 0 & \frac{1}{6} \end{bmatrix}$ **32.** $\begin{bmatrix} 2 & 1 \\ 4 & 2 \end{bmatrix}$

33. $\begin{bmatrix} 1 & 0 \\ \frac{3}{4} & 0 \end{bmatrix}$ **34.** $\begin{bmatrix} 1 & 1 \\ 1 & 1 \end{bmatrix}$

Ⓣ *In Exercises 35–42, use technology to find the inverse of the given matrix (when it exists). Round all entries in your answer to two decimal places.* [Caution: Because of rounding errors, technology sometimes produces an "inverse" of a singular matrix. These often can be recognized by their huge entries.]

35. $\begin{bmatrix} 1.1 & 1.2 \\ 1.3 & -1 \end{bmatrix}$ **36.** $\begin{bmatrix} 0.1 & -3.2 \\ 0.1 & -1.5 \end{bmatrix}$

37. $\begin{bmatrix} 3.56 & 1.23 \\ -1.01 & 0 \end{bmatrix}$ **38.** $\begin{bmatrix} 9.09 & -5.01 \\ 1.01 & 2.20 \end{bmatrix}$

39. $\begin{bmatrix} 1.1 & 3.1 & 2.4 \\ 1.7 & 2.4 & 2.3 \\ 0.6 & -0.7 & -0.1 \end{bmatrix}$ **40.** $\begin{bmatrix} 2.1 & 2.4 & 3.5 \\ 6.1 & -0.1 & 2.3 \\ -0.3 & -1.2 & 0.1 \end{bmatrix}$

41. $\begin{bmatrix} 0.01 & 0.32 & 0 & 0.04 \\ -0.01 & 0 & 0 & 0.34 \\ 0 & 0.32 & -0.23 & 0.23 \\ 0 & 0.41 & 0 & 0.01 \end{bmatrix}$

42. $\begin{bmatrix} 0.01 & 0.32 & 0 & 0.04 \\ -0.01 & 0 & 0 & 0.34 \\ 0 & 0.32 & -0.23 & 0.23 \\ 0.01 & 0.96 & -0.23 & 0.65 \end{bmatrix}$

In Exercises 43–48, use matrix inversion to solve the given systems of linear equations. (You solved similar systems using row reduction in Chapter 3.) [**HINT:** See Quick Example 6.]

43. $x + y = 4$
$\quad\ x - y = 1$

44. $2x + y = 2$
$\quad\ 2x - 3y = 2$

45. $\dfrac{x}{3} + \dfrac{y}{2} = 0$

$\quad\ \dfrac{x}{2} + y = -1$

46. $\dfrac{2x}{3} - \dfrac{y}{2} = \dfrac{1}{6}$

$\quad\ \dfrac{x}{2} - \dfrac{y}{2} = -1$

47. $-x + 2y - z = 0$
$-x - y + 2z = 0$
$2x - z = 6$

48. $x + 2y = 4$
$y - z = 0$
$x + 3y - 2z = 5$

In Exercises 49 and 50, use matrix inversion to solve each collection of systems of linear equations. [HINT: See Example 4.]

49. a. $-x - 4y + 2z = 4$
$x + 2y - z = 3$
$x + y - z = 8$

b. $-x - 4y + 2z = 0$
$x + 2y - z = 3$
$x + y - z = 2$

c. $-x - 4y + 2z = 0$
$x + 2y - z = 0$
$x + y - z = 0$

50. a. $-x - 4y + 2z = 8$
$x - z = 3$
$x + y - z = 8$

b. $-x - 4y + 2z = 8$
$x - z = 3$
$x + y - z = 2$

c. $-x - 4y + 2z = 0$
$x - z = 0$
$x + y - z = 0$

Applications

Some of the following exercises are similar or identical to exercises and examples in Chapter 3. Use matrix inverses to find the solutions. We suggest that you invert some of the matrices by hand and others using technology.

51. Nutrition One serving of **Campbell Soup Company**'s Campbell's Pork & Beans contains 5 grams of protein and 21 grams of carbohydrates.[16] A typical slice of "lite" rye bread contains 4 grams of protein and 12 grams of carbohydrates.
   a. I am planning a meal of beans-on-toast, and I want it to supply 20 grams of protein and 80 grams of carbohydrates. How should I prepare my meal?
   b. If I require A grams of protein and B grams of carbohydrates, give a formula that tells me how many slices of bread and how many servings of Pork & Beans to use.

52. Nutrition According to the nutritional information on a package of **General Mills**' Honey Nut Cheerios brand cereal, each 1-ounce serving of Cheerios contains 3 grams of protein and 24 grams of carbohydrates.[17] Each half-cup serving of enriched skim milk contains 4 grams of protein and 6 grams of carbohydrates.
   a. I am planning a meal of cereal and milk, and I want it to supply 26 grams of protein and 78 grams of carbohydrates. How should I prepare my meal?
   b. If I require A grams of protein and B grams of carbohydrates, give a formula that tells me how many servings of milk and Honey Nut Cheerios to use.

53. Resource Allocation You manage an ice cream factory that makes three flavors: Creamy Vanilla, Continental Mocha, and Succulent Strawberry. Into each batch of Creamy Vanilla go two eggs, one cup of milk, and two cups of cream. Into each batch of Continental Mocha go one egg, one cup of milk, and two cups of cream. Into each batch of Succulent Strawberry go one egg, two cups of milk, and one cup of cream. Your stocks of eggs, milk, and cream vary from day to day. How many batches of each flavor should you make in order to use up all of your ingredients if you have the following amounts in stock?
   a. 350 eggs, 350 cups of milk, and 400 cups of cream
   b. 400 eggs, 500 cups of milk, and 400 cups of cream
   c. A eggs, B cups of milk, and C cups of cream

54. Resource Allocation The *Arctic Juice Company* makes three juice blends: PineOrange, using 2 quarts of pineapple juice and 2 quarts of orange juice per gallon; PineKiwi, using 3 quarts of pineapple juice and 1 quart of kiwi juice per gallon; and OrangeKiwi, using 3 quarts of orange juice and 1 quart of kiwi juice per gallon. The amount of each kind of juice the company has on hand varies from day to day. How many gallons of each blend can it make on a day with the following stocks?
   a. 800 quarts of pineapple juice, 650 quarts of orange juice, 350 quarts of kiwi juice.
   b. 650 quarts of pineapple juice, 800 quarts of orange juice, 350 quarts of kiwi juice.
   c. A quarts of pineapple juice, B quarts of orange juice, C quarts of kiwi juice.

Investing: Inverse ETFs (Exchange Traded Funds) Inverse ETFs, sometimes referred to as "bear market" or "short" funds, are designed to deliver the opposite of the performance of the index or category they track and so can be used by traders to bet against the stock market. Exercises 55–56 are based on the following table, which shows the performance of three such funds as of August 5, 2015:[18]

	Year-to-Date Loss (%)
MYY (ProShares Short Midcap 400)	6
SH (ProShares Short S&P 500)	5
REW (ProShares UltraShort Technology)	7

55. You invested a total of $9,000 in the three funds at the beginning of 2011, including an equal amount in SH and REW. Your year-to-date loss from the first two funds amounted to $400. How much did you invest in each of the three funds?

---

[16] According to the label information on a 16-ounce can.

[17] Actually, it is 23 grams of carbohydrates. We made it 24 grams to simplify the calculation.

[18] Based on prices at noon on August 5, 2015. YTD losses rounded to the nearest percentage point. Source: www.google.com/finance.

56. You invested a total of $6,000 in the three funds at the beginning of 2011, including an equal amount in MYY and SH. Your total year-to-date loss amounted to $360. How much did you invest in each of the three funds?

Investing: Lesser-Known Stocks *Exercises 57–58 are based on the following information about the stocks of* **Whitestone REIT, HCC Insurance Holdings, Inc.,** *and* **SanDisk Corporation:**[19]

	Price ($)	Dividend Yield (%)
WSR (WSR Whitestone REIT)	16	7
HCC (HCC Insurance Holdings, Inc.)	56	2
SNDK (SanDisk Corporation)	80	2

57. ▼ You invested a total of $8,400 in shares of the three stocks at the given prices and expected to earn $248 in annual dividends. If you purchased a total of 200 shares, how many shares of each stock did you purchase?

58. ▼ You invested a total of $11,200 in shares of the three stocks at the given prices and expected to earn $304 in annual dividends. If you purchased a total of 250 shares, how many shares of each stock did you purchase?

59. ⊤ ▼ *Population Movement* In 2009 the population of the United States, broken down by regions, was 54.6 million in the Northeast, 66.0 million in the Midwest, 111.8 million in the South, and 70.6 million in the West. The table below shows the population movement during the period 2008–2009. (Thus, 99.23% of the population in the Northeast stayed there, while 0.16% of the population in the Northeast moved to the Midwest, and so on.)[20]

To	Northeast	Midwest	South	West
From **Northeast**	0.9923	0.0016	0.0042	0.0019
Midwest	0.0018	0.9896	0.0047	0.0039
South	0.0056	0.0059	0.9827	0.0058
West	0.0024	0.0033	0.0044	0.9899

Set up the 2009 population figures as a row vector. Use matrix inversion and multiplication to estimate the population in each region in 2008. (Round all answers to the nearest 0.1 million.)

60. ▼ ⊤ *Population Movement* Assume that the percentages given in Exercise 59 also describe the population movements from 2007 to 2008. Use two matrix multiplications to estimate from the data in Exercise 59 the population in each region in 2007.

61. ◆ ⊤ *Rotations* If a point (x, y) in the plane is rotated counterclockwise about the origin through an angle of 45°, its new coordinates (x', y') are given by

$$\begin{bmatrix} x' \\ y' \end{bmatrix} = R \begin{bmatrix} x \\ y \end{bmatrix}$$

where R is the 2×2 matrix $\begin{bmatrix} a & -a \\ a & a \end{bmatrix}$ and $a = \sqrt{1/2} \approx 0.7071$.

  a. If the point $(2, 3)$ is rotated counterclockwise through an angle of 45°, what are its (approximate) new coordinates?
  b. Multiplication by what matrix would result in a counterclockwise rotation of 90°? 135°? (Express the matrices in terms of R.) [**HINT:** Think of a rotation through 90° as two successive rotations through 45°.]
  c. Multiplication by what matrix would result in a *clockwise* rotation of 45°?

62. ◆ ⊤ *Rotations* If a point (x, y) in the plane is rotated counterclockwise about the origin through an angle of 60°, its new coordinates (x', y') are given by

$$\begin{bmatrix} x' \\ y' \end{bmatrix} = S \begin{bmatrix} x \\ y \end{bmatrix}$$

where S is the 2×2 matrix $\begin{bmatrix} a & -b \\ b & a \end{bmatrix}$ and $a = 1/2$ and $b = \sqrt{3/4} \approx 0.8660$.

  a. If the point $(2, 3)$ is rotated counterclockwise through an angle of 60°, what are its (approximate) new coordinates?
  b. Referring to Exercise 61, multiplication by what matrix would result in a counterclockwise rotation of 105°? (Express the matrices in terms of S and the matrix R from Exercise 61.) [**HINT:** Think of a rotation through 105° as a rotation through 60° followed by a rotation through 45°.]
  c. Multiplication by what matrix would result in a *clockwise* rotation of 60°?

⊤ *Encryption* Matrices are commonly used to encrypt data. Here is a simple form such an encryption can take. First, we represent each letter in the alphabet by a number, so let us take <space> = 0, A = 1, B = 2 and so on. Thus, for example, "ABORT MISSION" becomes

$$[1 \quad 2 \quad 15 \quad 18 \quad 20 \quad 0 \quad 13 \quad 9 \quad 19 \quad 19 \quad 9 \quad 15 \quad 14].$$

To encrypt this coded phrase, we use an invertible matrix of any size with integer entries. For instance, let us take A to be the 2×2 matrix $\begin{bmatrix} 1 & 2 \\ 3 & 4 \end{bmatrix}$. *We can first arrange the coded sequence of numbers in the form of a matrix with two rows*

---

[19] Yields rounded to the nearest percentage point and stock prices at the close of the stock market on February 27, 2015, rounded to the nearest $1. Source: www.finance.yahoo.

[20] Note that this exercise ignores migration into or out of the country. Source: U.S. Census Bureau, Current Population Survey, 2009 Annual Social and Economic Supplement.

(using zero in the last place if we have an odd number of characters) and then multiply on the left by A:

$$\text{Encrypted matrix} = \begin{bmatrix} 1 & 2 \\ 3 & 4 \end{bmatrix} \begin{bmatrix} 1 & 15 & 20 & 13 & 19 & 9 & 14 \\ 2 & 18 & 0 & 9 & 19 & 15 & 0 \end{bmatrix}$$

$$= \begin{bmatrix} 5 & 51 & 20 & 31 & 57 & 39 & 14 \\ 11 & 117 & 60 & 75 & 133 & 87 & 42 \end{bmatrix},$$

which we can also write as

$$\begin{bmatrix} 5 & 11 & 51 & 117 & 20 & 60 & 31 & 75 & 57 & 133 & 39 & 87 & 14 & 42 \end{bmatrix}.$$

To decipher the encoded message, multiply the encrypted matrix by A^{-1}. Exercises 63–66 use the above matrix A for encoding and decoding.

63. ▼ Use the matrix A to encode the phrase "GO TO PLAN B".

64. ▼ Use the matrix A to encode the phrase "ABANDON SHIP".

65. ▼ Decode the following message, which was encrypted using the matrix A.

$$\begin{bmatrix} 33 & 69 & 54 & 126 & 11 & 27 & 20 & 60 & 29 & 59 & 65 & 149 & 41 & 87 \end{bmatrix}$$

66. ▼ Decode the following message, which was encrypted using the matrix A.

$$\begin{bmatrix} 59 & 141 & 43 & 101 & 7 & 21 & 29 & 59 & 65 & 149 & 41 & 87 \end{bmatrix}$$

Communication and Reasoning Exercises

67. Multiple choice: If A and B are square matrices with $AB = I$ and $BA = I$, then
(A) B is the inverse of A.
(B) A and B must be equal.
(C) A and B must both be singular.
(D) At least one of A and B is singular.

68. Multiple choice: If A is a square matrix with $A^3 = I$ then
(A) A must be the identity matrix.
(B) A is invertible.
(C) A is singular.
(D) A is both invertible and singular.

69. What can you say about the inverse of a 2×2 matrix of the form $\begin{bmatrix} a & b \\ a & b \end{bmatrix}$?

70. If you think of numbers as 1×1 matrices, which numbers are invertible 1×1 matrices?

71. ▼ Use matrix multiplication to check that the inverse of a general 2×2 matrix is given by

$$\begin{bmatrix} a & b \\ c & d \end{bmatrix}^{-1} = \frac{1}{ad - bc} \begin{bmatrix} d & -b \\ -c & a \end{bmatrix}$$

(provided that $ad - bc \neq 0$).

72. ◆ Derive the formula in Exercise 71 using row reduction. (Assume that $ad - bc \neq 0$.)

73. ▼ A **diagonal** matrix D has the following form:

$$D = \begin{bmatrix} d_1 & 0 & 0 & \cdots & 0 \\ 0 & d_2 & 0 & \cdots & 0 \\ 0 & 0 & d_3 & \cdots & 0 \\ \vdots & \vdots & \vdots & \ddots & \vdots \\ 0 & 0 & 0 & \cdots & d_n \end{bmatrix}.$$

When is D singular? Why?

74. ▼ If a square matrix A row-reduces to the identity matrix, must it be invertible? If so, say why. If not, give an example of such a (singular) matrix.

75. ▼ If A and B are invertible, check that $B^{-1}A^{-1}$ is the inverse of AB.

76. ▼ Solve the matrix equation $A(B + CX) = D$ for X. (You may assume that A and C are invertible square matrices.)

77. ◆ In Example 3 we said that, if a square matrix A row-reduces to a matrix with a row of zeros, then it is singular. Why?

78. ◆ Your friend has two square matrices A and B, neither of them the zero matrix, with the property that AB is the zero matrix. You immediately tell him that neither A nor B can possibly be invertible. How can you be so sure?

4.4 Game Theory

Two-Person Zero-Sum Games

It frequently happens that you are faced with making a decision or choosing a best strategy from several possible choices. For instance, you might need to decide whether to invest in stocks or bonds, whether to cut prices of the product you sell, or what offensive play to use in a football game. In these examples the result depends on something you cannot control: In the first case your success depends on the future behavior of the economy. In the second case it depends in part on whether your competitors also cut prices. In the third case it depends on the defensive strategy chosen by the opposing team.

In all three cases you are, in a sense, "playing a game" in which your "opponent" is the economy, your competitors, or the opposing team, respectively. The degree of success resulting from your decision can be measured numerically in each case—your investment profit, increased revenue, or yardage gained on the football field—and we call this your *payoff*. **Game theory** is the mathematical study of situations like these, which, because they involve two "players," are called **two-person games**. The simplest types of two-person games are **zero-sum** games: Each player's gain (or payoff) resulting from a decision is equal to the opponent's loss.[*]

Game theory is very new in comparison with most of the mathematics you learn. It was invented in the 1920s by the noted mathematicians Emile Borel (1871–1956) and John von Neumann (1903–1957). Game theory's connection with linear programming was discovered even more recently, in 1947, by von Neumann, and further advances were made by the mathematician John Nash (1928–2015),[†] for which he received the 1994 Nobel Prize for Economics.

The Payoff Matrix and Expected Payoff

In two-person zero-sum games we represent the various options and payoffs in a matrix, and we can then calculate the best strategy using matrix algebra and other techniques, as we now illustrate with a simple type of game.

We have probably all played the game "Rock, Paper, Scissors" at some time in our lives. It goes as follows: There are two players—let us call them A and B—and at each turn, each player produces, by a gesture of the hand, either paper, a pair of scissors, or a rock. Rock beats scissors (since a rock can crush scissors) but is beaten by paper (since a rock can be covered by paper), while scissors beat paper (since scissors can cut paper). The round is a draw if both A and B show the same item. We could turn this into a betting game if, at each turn, we require the loser to pay the winner 1¢. For instance, if A shows a rock and B shows paper, then A pays B 1¢.

Rock, Paper, Scissors is an example of a two-person zero-sum game because each player's loss is equal to the other player's gain. We can represent this game by a matrix, called the **payoff matrix**:

$$
\begin{array}{c}
 \\
\mathbf{A} \begin{array}{c} r \\ p \\ s \end{array}
\end{array}
\begin{array}{c}
\mathbf{B} \\
\begin{array}{ccc} r & p & s \end{array} \\
\begin{bmatrix} 0 & -1 & 1 \\ 1 & 0 & -1 \\ -1 & 1 & 0 \end{bmatrix}
\end{array}
\quad \text{or just} \quad P = \begin{bmatrix} 0 & -1 & 1 \\ 1 & 0 & -1 \\ -1 & 1 & 0 \end{bmatrix}
$$

if we choose to omit the labels. In the payoff matrix, Player A's options, or **moves**, are listed on the left, while Player B's options are listed on top. We think of A as playing the rows and B as playing the columns. The entries of the matrix are the **payoffs**. Positive payoffs indicate a win for the row player, while negative payoffs indicate a loss for the row player. Thus, for example, the p, s entry is the payoff if A plays p (paper) and B plays s (scissors). In this event, B wins, and the -1 payoff there indicates that A loses 1¢. (If that payoff were -2 instead, it would have meant that A loses 2¢.)

> ### Two-Person Zero-Sum Game, Strategies
>
> A **two-person zero-sum game** is one in which one player's loss equals the other's gain. We assume that the outcome is determined by each player's choice from among a fixed, finite set of moves. If Player A has m moves to choose from and Player B has n, we can represent the game using the **payoff matrix**, the $m \times n$ matrix showing Player A's payoff resulting from each possible pair of choices of moves.

---

[*] An example of a *non-zero-sum game* would be one in which the government taxed the earnings of the winner. In that case the winner's gain would be less than the loser's loss.

[†] Nash's turbulent life is the subject of the biography *A Beautiful Mind* by Sylvia Nasar (Simon & Schuster, 1998). The 2001 Academy Award–winning movie of the same title is a somewhat fictionalized account.

In each round of the game, the way a player chooses a move is called a **strategy**. A player using a **pure strategy** makes the same move each round of the game. For example, if a player in the above game chooses to play scissors at each turn, then that player is using the pure strategy s. A player using a **mixed strategy** chooses each move a certain percentage of the time in a random fashion; for instance, Player A might choose to play p 50% of the time and each of s and r 25% of the time.

Our ultimate goal is to be able to determine which strategy is best for each player to use. To do that, we need to know how to evaluate strategies. The fundamental calculation we need is that of the **expected payoff** resulting from the strategies used by the two players. Let's look at a simple example.

EXAMPLE 1   Expected Payoff

Consider the following game:

$$\mathbf{A} \begin{array}{c} \\ p \\ q \end{array} \overset{\begin{array}{cc} a & b \end{array}}{\begin{bmatrix} 3 & -1 \\ -2 & 3 \end{bmatrix}}.$$

The row player (Player A) decides to pick moves at random, choosing to play p 75% of the time and q 25% of the time. The column player (Player B) also picks moves at random, choosing a 20% of the time and b 80% of the time. On average, how much does A expect to win or lose?

Solution   Suppose they play the game 100 times. Each time they play, there are four possible outcomes:

Case 1: A plays p, B plays a.
Because A plays p only 75% of the time and B plays a only 20% of the time, we expect this case to occur $0.75 \times 0.20 = 0.15$, or 15% of the time, or 15 times out of 100. Each time this happens, A gains 3 points, so we get a contribution of $15 \times 3 = 45$ points to A's total winnings.

Case 2: A plays p, B plays b.
Because A plays p only 75% of the time and B plays b only 80% of the time, we expect this case to occur $0.75 \times 0.80 = 0.60$, or 60 times out of 100. Each time this happens, A loses 1 point, so we get a contribution of $60 \times -1 = -60$ to A's total winnings.

Case 3: A plays q, B plays a.
This case occurs $0.25 \times 0.20 = 0.05$, or 5 out of 100 times, with a loss of 2 points to A each time, giving a contribution of $5 \times -2 = -10$ to A's total winnings.

Case 4: A plays q, B plays b.
This case occurs $0.25 \times 0.80 = 0.20$, or 20 out of 100 times, with a gain of 3 points to A each time, giving a contribution of $20 \times 3 = 60$ to A's total winnings.

Summing to get A's total winnings and then dividing by the number of times the game is played, we get the average value of

$$(45 - 60 - 10 + 60)/100 = 0.35$$

so A can expect to win an average of 0.35 points per play of the game. We call 0.35 the **expected payoff** of the game resulting from these particular strategies for A and B.

This calculation was somewhat tedious, and it would only get worse if A and B had many moves to choose from. There is a far more convenient way of doing exactly the same calculation, using matrix multiplication: We start by representing the player's strategies as matrices. For reasons to become clear in a moment, we record A's strategy as a row matrix:

$$R = [0.75 \quad 0.25].$$

We record B's strategy as a column matrix:

$$C = \begin{bmatrix} 0.20 \\ 0.80 \end{bmatrix}.$$

(We will sometimes write column vectors using transpose notation, writing, for example, $[0.20 \quad 0.80]^T$ for the column above, to save space.) Now: *The expected payoff is the matrix product RPC, where P is the payoff matrix!*

$$\text{Expected payoff} = RPC = [0.75 \quad 0.25] \begin{bmatrix} 3 & -1 \\ -2 & 3 \end{bmatrix} \begin{bmatrix} 0.20 \\ 0.80 \end{bmatrix}$$

$$= [1.75 \quad 0] \begin{bmatrix} 0.20 \\ 0.80 \end{bmatrix} = [0.35].$$

Why does this work? Write out the arithmetic involved in the matrix product RPC to see what we calculated:

$$[0.75 \times 3 + 0.25 \times (-2)] \times 0.20 + [0.75 \times (-1) + 0.25 \times 3] \times 0.80$$
$$= 0.75 \times 3 \times 0.20 + 0.25 \times (-2) \times 0.20 + 0.75 \times (-1) \times 0.80 + 0.25 \times 3 \times 0.80$$
$$= \quad \text{Case 1} \quad + \quad \text{Case 3} \quad + \quad \text{Case 2} \quad + \quad \text{Case 4.}$$

So the matrix product does all at once the various cases we considered above.

Let's summarize what we just saw.

The Expected Payoff Resulting from Mixed Strategies *R* and *C*

The **expected payoff of a game resulting from given mixed strategies** is the average payoff that occurs if the game is played a large number of times with the row and column players using the given strategies.

To compute the expected payoff resulting from given mixed strategies:

1. Write the row player's mixed strategy as a row matrix R.

2. Write the column player's mixed strategy as a column matrix C.

3. Calculate the product RPC, where P is the payoff matrix. This product is a 1×1 matrix whose entry is the expected payoff e.

Quick Example

1. Consider a game of "Rock, Paper, Scissors" with a dollar payoff for the winner each round:

$$\begin{array}{c} \\ r \\ p \\ s \end{array} \begin{array}{c} \begin{array}{ccc} r & p & s \end{array} \\ \begin{bmatrix} 0 & -1 & 1 \\ 1 & 0 & -1 \\ -1 & 1 & 0 \end{bmatrix} \end{array}$$

Suppose that the row player plays *rock* half the time and each of the other two strategies a quarter of the time and that the column player always plays *paper*. We write

$$R = \begin{bmatrix} \frac{1}{2} & \frac{1}{4} & \frac{1}{4} \end{bmatrix} \quad \text{and} \quad C = \begin{bmatrix} 0 \\ 1 \\ 0 \end{bmatrix}.$$

So

$$e = RPC = \begin{bmatrix} \frac{1}{2} & \frac{1}{4} & \frac{1}{4} \end{bmatrix} \begin{bmatrix} 0 & -1 & 1 \\ 1 & 0 & -1 \\ -1 & 1 & 0 \end{bmatrix} \begin{bmatrix} 0 \\ 1 \\ 0 \end{bmatrix}$$

$$= \begin{bmatrix} \frac{1}{2} & \frac{1}{4} & \frac{1}{4} \end{bmatrix} \begin{bmatrix} -1 \\ 0 \\ 1 \end{bmatrix} = -\frac{1}{4}.$$

Thus, the row player can expect to lose, on average, $1 every four plays.

Solving a Game

Now that we know how to evaluate particular strategies, we want to find the *best* strategy. The next example takes us another step toward that goal.

EXAMPLE 2 Television Ratings Wars

Commercial TV station RTV and cultural station CTV are competing for viewers in the Tuesday prime-time 9–10 pm time slot. RTV is trying to decide whether to show a sitcom, a docudrama, a reality show, or a movie, while CTV is thinking about either a nature documentary, a symphony concert, a ballet, or an opera. A television rating company estimates the payoffs for the various alternatives as follows. (Each point indicates a shift of 1,000 viewers from one channel to the other; thus, for instance, -2 indicates a shift of 2,000 viewers from RTV to CTV.)

<table>
<thead>
<tr><th></th><th></th><th colspan="4">CTV</th></tr>
<tr><th></th><th></th><th>Nature Doc.</th><th>Symphony</th><th>Ballet</th><th>Opera</th></tr>
</thead>
<tbody>
<tr><td rowspan="4">RTV</td><td>Sitcom</td><td>2</td><td>1</td><td>-2</td><td>2</td></tr>
<tr><td>Docudrama</td><td>-1</td><td>1</td><td>-1</td><td>2</td></tr>
<tr><td>Reality Show</td><td>-2</td><td>0</td><td>0</td><td>1</td></tr>
<tr><td>Movie</td><td>3</td><td>1</td><td>-1</td><td>1</td></tr>
</tbody>
</table>

a. If RTV notices that CTV is showing nature documentaries half the time and symphonies the other half, what would RTV's best strategy be, and how many viewers would it gain if it followed this strategy?

b. If, on the other hand, CTV notices that RTV is showing docudramas half the time and reality shows the other half, what would CTV's best strategy be, and how many viewers would it gain or lose if it followed this strategy?

Solution

a. We are given the matrix of the game, P, in the table above, and we are given CTV's strategy $C = [0.50 \quad 0.50 \quad 0 \quad 0]^T$. We are not given RTV's strategy R. To say that RTV is looking for its best strategy is to say that it wants the resulting expected payoff $e = RPC$ to be as high as possible. So we take $R = [x \quad y \quad z \quad t]$ and look for values for x, y, z, and t that make RPC as high as possible. First, we calculate e in terms of these unknowns:

$$e = RPC = [x \quad y \quad z \quad t] \begin{bmatrix} 2 & 1 & -2 & 2 \\ -1 & 1 & -1 & 2 \\ -2 & 0 & 0 & 1 \\ 3 & 1 & -1 & 1 \end{bmatrix} \begin{bmatrix} 0.50 \\ 0.50 \\ 0 \\ 0 \end{bmatrix}$$

$$= [x \quad y \quad z \quad t] \begin{bmatrix} 1.5 \\ 0 \\ -1 \\ 2 \end{bmatrix} = 1.5x - z + 2t.$$

Now, the unknowns x, y, z, and t must be nonnegative and add up to 1 (why?). Also, because t has the largest coefficient, 2, we'll get the best result by making it as large as possible, namely, $t = 1$, leaving $x = y = z = 0$. Thus, RTV's best strategy is $R = [0 \quad 0 \quad 0 \quad 1]$. In other words, RTV should use the pure strategy of showing a movie every Tuesday evening. If it does so, the expected payoff will be

$$e = 1.5(0) - 0 + 2(1) = 2,$$

so RTV can expect to gain 2,000 viewers.

b. Here, we are given $R = [0 \quad 0.50 \quad 0.50 \quad 0]$ and are not given CTV's strategy C, so this time we take $C = [x \quad y \quad z \quad t]^T$ and calculate the resulting expected payoff e:

$$e = RPC = [0 \quad 0.50 \quad 0.50 \quad 0] \begin{bmatrix} 2 & 1 & -2 & 2 \\ -1 & 1 & -1 & 2 \\ -2 & 0 & 0 & 1 \\ 3 & 1 & -1 & 1 \end{bmatrix} \begin{bmatrix} x \\ y \\ z \\ t \end{bmatrix}$$

$$= [-1.5 \quad 0.5 \quad -0.5 \quad 1.5] \begin{bmatrix} x \\ y \\ z \\ t \end{bmatrix} = -1.5x + 0.5y - 0.5z + 1.5t.$$

Now, CTV wants e to be as *low* as possible (why?). Because x has the largest negative coefficient, CTV would like it to be as large as possible: $x = 1$, so the rest of the unknowns must be zero. Thus, CTV's best strategy is $C = [1 \quad 0 \quad 0 \quad 0]^T$; that is, show a nature documentary every night. If it does so, the expected payoff will be

$$e = -1.5(1) + 0.5(0) - 0.5(0) + 1.5(0) = -1.5.$$

So CTV can expect to gain 1,500 viewers.

Example 2 illustrates the fact that, no matter what mixed strategy one player selects, the other player can choose an appropriate *pure* counterstrategy to maximize its gain. How does this affect what decisions you should make as one of the players? If you were on the board of directors of RTV, you might reason as follows: Since for every mixed strategy you try, CTV can find a best counterstrategy (as in part (b)), it is in your company's best interest to select a mixed strategy that *minimizes* the effect of CTV's best counterstrategy. This is called the **minimax criterion**.

Minimax Criterion

A player using the **minimax criterion** chooses a strategy that, among all possible strategies, minimizes the effect of the other player's best counterstrategy. That is, an optimal (best) strategy according to the minimax criterion is one that minimizes the maximum damage the opponent can cause.

This criterion assumes that your opponent is determined to win. More precisely, it assumes the following.

Fundamental Principle of Game Theory

Each player tries to use its best possible strategy and assumes that the other player is doing the same.

This principle is not always followed by every player. For example, one of the players may be Nature and may choose its move at random, with no particular purpose in mind. In such a case, criteria other than the minimax criterion may be more appropriate. For example, there is the "maximax" criterion, which maximizes the maximum possible payoff (also known as the "reckless" strategy), or the criterion that seeks to minimize "regret" (the difference between the payoff you get and the payoff you *would have gotten* if you had known beforehand what was going to happen).[*] But we shall assume here the fundamental principle and try to find optimal strategies under the minimax criterion.

[*] See *Location in Space: Theoretical Perspectives in Economic Geography,* 3rd Edition, by Peter Dicken and Peter E. Lloyd, HarperCollins Publishers, 1990, p. 276.

Finding the optimal strategy is called **solving the game**. In general, solving a game can be done by using linear programming, as we shall see in Chapter 5. However, we can solve 2×2 games "by hand," as we shall see in the next example. First, we notice that some large games can be reduced to smaller games.

Consider the game in Example 2, which had the following matrix:

$$P = \begin{bmatrix} 2 & 1 & -2 & 2 \\ -1 & 1 & -1 & 2 \\ -2 & 0 & 0 & 1 \\ 3 & 1 & -1 & 1 \end{bmatrix}.$$

Compare the second and third columns through the eyes of the column player, CTV. Every payoff in the third column is, from CTV's point of view, as good as or better than the corresponding entry in the second column. Thus, no matter what RTV does, CTV will do better showing a ballet (third column) than a symphony (second column). We say that the third column **dominates** the second column. As far as CTV is concerned, we might as well forget about symphonies entirely, so we remove the second column. Similarly, the third column dominates the fourth, so we can remove the fourth column too. This gives us a smaller game to work with:

$$P = \begin{bmatrix} 2 & -2 \\ -1 & -1 \\ -2 & 0 \\ 3 & -1 \end{bmatrix}.$$

Now compare the first and last rows. Every payoff in the last row is larger than the corresponding payoff in the first row, so the last row is always better for RTV. Again, we say that the last row dominates the first row, and we can discard the first row. Similarly, the last row dominates the second row, so we discard the second row as well. This reduces us to the following game:

$$P = \begin{bmatrix} -2 & 0 \\ 3 & -1 \end{bmatrix}.$$

In this matrix, neither row dominates the other, and neither column dominates the other. So this is as far as we can go with this line of argument. We call this **reduction by dominance**.

Reduction by Dominance

One *row* **dominates** another if every entry in the former is greater than or equal to the corresponding entry in the latter. Put another way, one row dominates another if it is always at least as good for the row player.

One *column* dominates another if every entry in the former is less than or equal to the corresponding entry in the latter. Put another way, one column dominates another if it is always at least as good for the column player.

Procedure for Reducing by Dominance:

1. Check whether there is any row in the (remaining) matrix that is dominated by another row. Remove all dominated rows.

2. Check whether there is any column in the (remaining) matrix that is dominated by another column. Remove all dominated columns.

3. Repeat Steps 1 and 2 until there are no dominated rows or columns.*

---

*We can also reverse the order of Steps 1 and 2 as we did for the payoff matrix above: First remove dominated columns, then remove dominated rows, and repeat the two steps until there are no dominated columns or rows.

Let us now go back to the television ratings wars example and see how we can solve a game using the minimax criterion once we are down to a 2×2 payoff matrix.

EXAMPLE 3   **Solving a 2 × 2 Game**

Continuing from Example 2:

a. Find the optimal strategy for RTV.

b. Find the optimal strategy for CTV.

c. Find the expected payoff of the game if RTV and CTV use their optimal strategies.

Solution As in the text, we begin by reducing the game by dominance, which brings us down to the following 2×2 game:

		CTV	
		Nature Doc.	**Ballet**
RTV	**Reality Show**	-2	0
	Movie	3	-1

a. Now let's find RTV's optimal strategy. Because we don't yet know what it is, we write down a general strategy:

$$R = \begin{bmatrix} x & y \end{bmatrix}.$$

Because $x + y = 1$, we can replace y by $1 - x$:

$$R = \begin{bmatrix} x & 1 - x \end{bmatrix}.$$

We know that CTV's best counterstrategy to R will be a pure strategy (see the discussion after Example 2), so let's compute the expected payoff that results from each of CTV's possible pure strategies:

$$e = \begin{bmatrix} x & 1 - x \end{bmatrix} \begin{bmatrix} -2 & 0 \\ 3 & -1 \end{bmatrix} \begin{bmatrix} 1 \\ 0 \end{bmatrix}$$

$$= (-2)x + 3(1 - x) = -5x + 3$$

$$f = \begin{bmatrix} x & 1 - x \end{bmatrix} \begin{bmatrix} -2 & 0 \\ 3 & -1 \end{bmatrix} \begin{bmatrix} 0 \\ 1 \end{bmatrix}$$

$$= 0x - (1 - x) = x - 1.$$

Because both e and f depend on x, we can graph them as in Figure 1.

If, for instance, RTV happened to choose $x = 0.5$, then the expected payoffs resulting from CTV's two pure strategies are $e = -5(1/2) + 3 = 1/2$ and $f = 1/2 - 1 = -1/2$. The worst outcome for RTV is the lower of the two, f, and this will be true wherever the graph of f is below the graph of e. On the other hand, if RTV chose $x = 1$, the graph of e would be lower, and the worst possible expected value would be $e = -5(1) + 3 = -2$. Since RTV can choose x to be any value between 0 and 1, the worst possible outcomes are those shown by the colored portion of the graph in Figure 2.

Because RTV is trying to make the worst possible outcome as large as possible (that is, to minimize damages), it is seeking the point on the colored portion of the graph that is highest. This is the intersection point of the two lines. To calculate its coordinates, it's easiest to equate the two functions of x:

$$-5x + 3 = x - 1,$$

$$-6x = -4,$$

or

$$x = \frac{2}{3}.$$

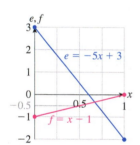

Figure 1

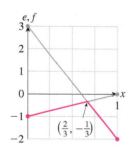

Figure 2

The e (or f) coordinate is then obtained by substituting $x = 2/3$ into the expression for e (or f), giving

$$e = -5\left(\frac{2}{3}\right) + 3$$

$$= -\frac{1}{3}.$$

We conclude that RTV's best strategy is to take $x = 2/3$, giving an expected value of $-1/3$. In other words, RTV's optimal mixed strategy is

$$R = \begin{bmatrix} \frac{2}{3} & \frac{1}{3} \end{bmatrix}.$$

Going back to the original game, RTV should show reality shows $2/3$ of the time and movies $1/3$ of the time. It should not bother showing any sitcoms or docudramas. It expects to lose, on average, 333 viewers to CTV, but all of its other options are worse.

b. To find CTV's optimal strategy, we must reverse roles and start by writing its unknown strategy as follows:

$$C = \begin{bmatrix} x \\ 1 - x \end{bmatrix}.$$

We calculate the expected payoffs for the two pure row strategies:

$$e = \begin{bmatrix} 1 & 0 \end{bmatrix} \begin{bmatrix} -2 & 0 \\ 3 & -1 \end{bmatrix} \begin{bmatrix} x \\ 1 - x \end{bmatrix}$$

$$= -2x$$

and

$$f = \begin{bmatrix} 0 & 1 \end{bmatrix} \begin{bmatrix} -2 & 0 \\ 3 & -1 \end{bmatrix} \begin{bmatrix} x \\ 1 - x \end{bmatrix}$$

$$= 3x - (1 - x)$$

$$= 4x - 1.$$

As with the row player, we know that the column player's best strategy will correspond to the intersection of the graphs of e and f (Figure 3). (Why is the upper edge colored, rather than the lower edge?) The graphs intersect when

$$-2x = 4x - 1$$

or

$$x = \frac{1}{6}.$$

The corresponding value of e (or f) is

$$e = -2\left(\frac{1}{6}\right) = -\frac{1}{3}.$$

Thus, CTV's optimal mixed strategy is $\begin{bmatrix} \frac{1}{6} & \frac{5}{6} \end{bmatrix}^T$, and the expected payoff is $-1/3$. So CTV should show nature documentaries $1/6$ of the time and ballets $5/6$ of the

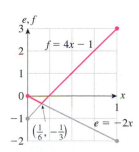

Figure 3

time. It should not bother to show symphonies or operas. It expects to gain, on average, 333 viewers from RTV.

c. We can now calculate the expected payoff as usual, using the optimal strategies we found in parts (a) and (b):

$$e = RPC$$

$$= \begin{bmatrix} \frac{2}{3} & \frac{1}{3} \end{bmatrix} \begin{bmatrix} -2 & 0 \\ 3 & -1 \end{bmatrix} \begin{bmatrix} \frac{1}{6} \\ \frac{5}{6} \end{bmatrix}$$

$$= -\frac{1}{3}.$$

➡ **Before we go on ...** In Example 3 it is no accident that the expected payoff resulting from the optimal strategies equals the expected payoff we found in parts (a) and (b). If we call the expected payoff resulting from the optimal strategies the **expected value of the game**, the row player's optimal strategy guarantees an expected payoff no smaller than the expected value, while the column player's optimal strategy guarantees an expected payoff no larger. Together they force the average payoff to be the expected value of the game. ∎

Expected Value of a Game

The **expected value of a game** is the expected payoff that results when the row and column players use their optimal (minimax) strategies. By using its optimal strategy, the row player guarantees an expected payoff no lower than the expected value of the game, no matter what the column player does. Similarly, by using its optimal strategy, the column player guarantees an expected payoff no higher than the expected value of the game, no matter what the row player does.

Q: *What about games that don't reduce to 2 × 2 matrices? Are these solved in a similar way?*

A: The method illustrated in Example 3 cannot easily be generalized to solve bigger games (i.e., games that cannot be reduced to 2 × 2 matrices); solving even a 2 × 3 game using this approach would require us to consider graphs in three dimensions. To be able to solve games of arbitrary size, we need to wait until the next chapter (Section 5.5), where we describe a method for solving a game, using the simplex method, that works for all payoff matrices.

Strictly Determined Games

Although we haven't yet discussed how to solve general $m \times n$ games, there are certain kinds of games that can be solved quite simply regardless of size, as illustrated by the following example.

| EXAMPLE 4 | Strictly Determined Game |

Solve the following game:

$$
\begin{array}{c}
 & \begin{array}{ccc} p & q & r \end{array} \\
\begin{array}{c} s \\ t \\ u \end{array}
\left[\begin{array}{rrr}
-4 & -3 & 3 \\
2 & -1 & -2 \\
1 & 0 & 2
\end{array} \right].
\end{array}
$$

Solution Call the row player A and the column player B. If we look carefully at this matrix, we see that no row dominates another and no column dominates another, so we can't reduce it. Nor do we know how to solve a 3 × 3 game, so it looks as if we're stuck. However, there is a way to understand this particular game. With the minimax criterion in mind, let's begin by considering the worst possible outcomes for the row player for each possible move. We do this by circling the smallest payoff in each row, the **row minima**:

$$
\begin{array}{c}
 & \begin{array}{ccc} p & q & r \end{array} & \text{Row minima} \\
\begin{array}{c} s \\ t \\ u \end{array}
\left[\begin{array}{rrr}
\boxed{-4} & -3 & 3 \\
2 & -1 & \boxed{-2} \\
1 & \boxed{0} & 2
\end{array} \right]
& \begin{array}{l} -4 \\ -2 \\ 0 \leftarrow \text{(largest)} \end{array}
\end{array}
$$

So, for example, if A plays move s, the worst possible outcome is to lose 4. Player A takes the least risk by using move u, which has the largest row minimum.

We do the same thing for the column player, remembering that smaller payoffs are better for B and larger payoffs are worse. We draw a box around the largest payoff in each column, the **column maxima**:

$$
\begin{array}{c}
 & \begin{array}{ccc} p & q & r \end{array} \\
\begin{array}{c} s \\ t \\ u \end{array}
\left[\begin{array}{rrr}
-4 & -3 & \boxed{3} \\
\boxed{2} & -1 & -2 \\
1 & \boxed{0} & 2
\end{array} \right].
\end{array}
$$

Column maxima $\quad 2 \qquad 0 \qquad 3$

$$\uparrow$$
(smallest)

Player B takes the least risk by using move q, which has the smallest column maximum.

Now put the circles and boxes together:

$$
\begin{array}{c}
 & \begin{array}{ccc} p & q & r \end{array} \\
\begin{array}{c} s \\ t \\ u \end{array}
\left[\begin{array}{rrr}
\boxed{-4} & -3 & \boxed{3} \\
\boxed{2} & -1 & \boxed{-2} \\
1 & \boxed{\boxed{0}} & 2
\end{array} \right].
\end{array}
$$

Notice that the uq entry is both circled and boxed: It is both a row minimum and a column maximum. We call such an entry a **saddle point**.

Now we claim that the optimal strategy for A is to always play u, while the optimal strategy for B is to always play q. By playing u, A guarantees that the payoff will be 0 or higher, no matter what B does, so the expected value of the game has to be *at least* 0. On the other hand, by playing q, B guarantees that the payoff will be 0 or less, so the expected value of the game has to be *no more than* 0. Combining these facts, we conclude that the expected value of the game must be exactly 0, and A and B have no strategies that could do any better for them than the pure strategies *u* and *q*.

➡️ **Before we go on . . .** You should consider what happens in an example like the television ratings wars game of Example 3. In that game, the largest row minimum is −1, while the smallest column maximum is 0; there is no saddle point. The row player can force a payoff of at least −1 by playing a pure strategy (always showing movies, for example) but can do better, forcing an expected payoff of −1/3, by playing a mixed strategy, as we saw in Example 3. Similarly, the column player can force the payoff to be 0 or less with a pure strategy, but can do better, forcing an expected payoff of −1/3, with a mixed strategy. Only when there is a saddle point will pure strategies be optimal. ▪

Strictly Determined Game

A **saddle point** is a payoff v that is simultaneously a row minimum and a column maximum (both boxed and circled in our approach). If a game has a saddle point, the corresponding row and column strategies are the optimal ones, the expected value of the game is the payoff v, and we say that the game is **strictly determined**.

If a game has two or more saddle points, then they all must have the same value, and all of the corresponding strategies are optimal.

FAQs

Solving a Game

Q: *We've seen several ways of trying to solve a game. What should I do and in what order?*

A: Here are the steps you should take when trying to solve a game:

1. Reduce by dominance. This should always be your first step.

2. If you were able to reduce to a 1×1 game, you're done. The optimal strategies are the corresponding pure strategies, as they dominate all the others.

3. Look for a saddle point in the reduced game. If it has one, the game is strictly determined, and the corresponding pure strategies are optimal.

4. If your reduced game is 2×2 and has no saddle point, use the method of Example 3 to find the optimal mixed strategies.

5. If your reduced game is larger than 2×2 and has no saddle point, you have to use linear programming to solve it, but that will have to wait until Chapter 5.

4.4 EXERCISES

▼ more advanced   ◆ challenging
T indicates exercises that should be solved using technology

In Exercises 1–4, calculate the expected payoff of the game with payoff matrix

$$P = \begin{bmatrix} 2 & 0 & -1 & 2 \\ -1 & 0 & 0 & -2 \\ -2 & 0 & 0 & 1 \\ 3 & 1 & -1 & 1 \end{bmatrix}$$

using the mixed strategies supplied. [HINT: See Example 1.]

1. $R = \begin{bmatrix} 0 & 1 & 0 & 0 \end{bmatrix}$, $C = \begin{bmatrix} 1 & 0 & 0 & 0 \end{bmatrix}^T$

2. $R = \begin{bmatrix} 0 & 0 & 0 & 1 \end{bmatrix}$, $C = \begin{bmatrix} 0 & 1 & 0 & 0 \end{bmatrix}^T$

3. $R = \begin{bmatrix} 0.5 & 0.5 & 0 & 0 \end{bmatrix}$, $C = \begin{bmatrix} 0 & 0 & 0.5 & 0.5 \end{bmatrix}^T$

4. $R = \begin{bmatrix} 0 & 0.5 & 0 & 0.5 \end{bmatrix}$, $C = \begin{bmatrix} 0.5 & 0.5 & 0 & 0 \end{bmatrix}^T$

In Exercises 5–8, either a mixed column or mixed row strategy is given. In each case, use

$$P = \begin{bmatrix} 0 & -1 & 5 \\ 2 & -2 & 4 \\ 0 & 3 & 0 \\ 1 & 0 & -5 \end{bmatrix},$$

and find the optimal pure strategy (or strategies) the other player should use. Express the answer as a row or column matrix. Also determine the resulting expected payoff. [HINT: See Example 2.]

5. $C = \begin{bmatrix} 0.25 & 0.75 & 0 \end{bmatrix}^T$

6. $C = \begin{bmatrix} \frac{1}{3} & \frac{1}{3} & \frac{1}{3} \end{bmatrix}^T$

7. $R = \begin{bmatrix} \frac{1}{2} & 0 & \frac{1}{4} & \frac{1}{4} \end{bmatrix}$

8. $R = \begin{bmatrix} 0.8 & 0.2 & 0 & 0 \end{bmatrix}$

In Exercises 9–14, reduce the given payoff matrix by dominance.

9.
$$\begin{array}{c c} & \begin{matrix} p & q & r \end{matrix} \\ \begin{matrix} a \\ b \end{matrix} & \begin{bmatrix} 1 & 1 & 10 \\ 2 & 3 & -4 \end{bmatrix} \end{array}$$

10.
$$\begin{array}{c c} & \begin{matrix} p & q & r \end{matrix} \\ \begin{matrix} a \\ b \end{matrix} & \begin{bmatrix} 2 & 0 & 10 \\ 15 & -4 & -5 \end{bmatrix} \end{array}$$

11.
$$\begin{array}{c c} & \begin{matrix} a & b & c \end{matrix} \\ \begin{matrix} 1 \\ 2 \\ 3 \end{matrix} & \begin{bmatrix} 2 & -4 & -9 \\ -1 & -2 & -3 \\ 5 & 0 & -1 \end{bmatrix} \end{array}$$

12.
$$\begin{array}{c c} & \begin{matrix} a & b & c \end{matrix} \\ \begin{matrix} 1 \\ 2 \\ 3 \end{matrix} & \begin{bmatrix} 0 & -1 & -5 \\ -3 & -10 & 10 \\ 2 & 3 & -4 \end{bmatrix} \end{array}$$

13.
$$\begin{array}{c c} & \begin{matrix} a & b & c \end{matrix} \\ \begin{matrix} p \\ q \\ r \\ s \end{matrix} & \begin{bmatrix} 1 & -1 & -5 \\ 4 & 0 & 2 \\ 3 & -3 & 10 \\ 3 & -5 & -4 \end{bmatrix} \end{array}$$

14.
$$\begin{array}{c c} & \begin{matrix} a & b & c \end{matrix} \\ \begin{matrix} p \\ q \\ r \\ s \end{matrix} & \begin{bmatrix} 2 & -4 & 9 \\ 1 & 1 & 0 \\ -1 & -2 & -3 \\ 1 & 1 & -1 \end{bmatrix} \end{array}$$

In Exercises 15–20, decide whether the game is strictly determined. If it is, give the players' optimal pure strategies and the value of the game. [HINT: See Example 4.]

15.
$$\begin{array}{c c} & \begin{matrix} p & q \end{matrix} \\ \begin{matrix} a \\ b \end{matrix} & \begin{bmatrix} 1 & 1 \\ 2 & -4 \end{bmatrix} \end{array}$$

16.
$$\begin{array}{c c} & \begin{matrix} p & q \end{matrix} \\ \begin{matrix} a \\ b \end{matrix} & \begin{bmatrix} -1 & 2 \\ 10 & -1 \end{bmatrix} \end{array}$$

17.
$$\begin{array}{c c} & \begin{matrix} p & q & r \end{matrix} \\ \begin{matrix} a \\ b \end{matrix} & \begin{bmatrix} 2 & 0 & -2 \\ -1 & 3 & 0 \end{bmatrix} \end{array}$$

18.
$$\begin{array}{c c} & \begin{matrix} p & q & r \end{matrix} \\ \begin{matrix} a \\ b \end{matrix} & \begin{bmatrix} -2 & 1 & -3 \\ -2 & 3 & -2 \end{bmatrix} \end{array}$$

19.
$$\begin{array}{c c} & \begin{matrix} a & b & c \end{matrix} \\ \begin{matrix} P \\ Q \\ R \\ S \end{matrix} & \begin{bmatrix} 1 & -1 & -5 \\ 4 & -4 & 2 \\ 3 & -3 & -10 \\ 5 & -5 & -4 \end{bmatrix} \end{array}$$

20.
$$\begin{array}{c c} & \begin{matrix} a & b & c \end{matrix} \\ \begin{matrix} P \\ Q \\ R \\ S \end{matrix} & \begin{bmatrix} -2 & -4 & 9 \\ 1 & 1 & 0 \\ -1 & -2 & -3 \\ 1 & 1 & -1 \end{bmatrix} \end{array}$$

In Exercises 21–24, find (a) the optimal mixed row strategy, (b) the optimal mixed column strategy, and (c) the expected value of the game. [HINT: See Example 3.]

21. $P = \begin{bmatrix} -1 & 2 \\ 0 & -1 \end{bmatrix}$

22. $P = \begin{bmatrix} -1 & 0 \\ 1 & -1 \end{bmatrix}$

23. $P = \begin{bmatrix} -1 & -2 \\ -2 & 1 \end{bmatrix}$

24. $P = \begin{bmatrix} -2 & -1 \\ -1 & -3 \end{bmatrix}$

Applications

In Exercises 25–32, set up the payoff matrix.

25. Games to Pass the Time You and your friend have come up with the following simple game to pass the time: In each round, you simultaneously call "heads" or "tails." If you have both called the same thing, your friend wins 1 point; if your calls differ, you win 1 point.

26. Games to Pass the Time Bored with the game in Exercise 25, you decide to use the following variation instead: If you both call "heads," your friend wins 2 points; if you both call "tails," your friend wins 1 point; if your calls differ, then you win 2 points if you called "heads" and 1 point if you called "tails."

27. War Games You are deciding whether to invade France, Sweden, or Norway, and your opponent is simultaneously deciding which of these three countries to defend. If you invade a country that your opponent is defending, you will be defeated (payoff: -1), but if you invade a country that your opponent is not defending, you will be successful (payoff: $+1$).

28. War Games You must decide whether to attack your opponent by sea or air, and your opponent must simultaneously decide whether to mount an all-out air defense, an all-out coastal defense (against an attack from the sea), or a combined air and coastal defense. If there is no defense for your mode of attack, you win 100 points. If your attack is met by a shared air and coastal defense, you win 50 points. If your attack is met by an all-out defense, you lose 200 points.

29. ▼ *Marketing* Your fast-food outlet, *Burger Queen*, has obtained a license to open branches in three closely situated South African cities: Brakpan, Nigel, and Springs. Your market surveys show that Brakpan and Nigel each provide a potential market of 2,000 burgers a day, while Springs provides a potential market of 1,000 burgers per day. Your company can finance an outlet in only one of those cities. Your main competitor, *Burger Princess*, has also obtained licenses for these cities and is similarly planning to open only one outlet. If you both happen to locate at the same city, you will share the total business from all three cities equally, but if you locate in different cities, you will each get all the business in the city in which you have located plus half the business in the third city. The payoff is the number of burgers you will sell per day minus the number of burgers your competitor will sell per day.

30. ▼ *Marketing* Repeat Exercise 29, given that the potential sales markets in the three cities are Brakpan: 2,500 per day, Nigel: 1,500 per day, and Springs: 1,200 per day.

31. ▼ *Betting* When you bet on a racehorse with odds of *m–n*, you stand to win *m* dollars for every bet of *n* dollars if your horse wins; for instance, if the horse you bet is running at 5–2 and wins, you will win $5 for every $2 you bet. (Thus, a $2 bet will return $7.) Here are some actual odds from a 1992 race at Belmont Park, New York.[21] The favorite at 5–2 was Pleasant Tap, the second choice was Thunder Rumble at 7–2, while the third choice was Strike the Gold at 4–1. Assume that you are making a $10 bet on one of these horses. The payoffs are your winnings. (If your horse does not win, you lose your entire bet. Of course, it is possible for none of your horses to win.)

32. ▼ *Betting* Referring to Exercise 31, suppose that just before the race, there has been frantic betting on Thunder Rumble with the result that the odds have dropped to 2–5 for that horse. The odds on the other two horses remain unchanged.

33. *Retail Discount Wars* Just one week after your *Abercrom B* men's fashion outlet has opened at a new location near *Burger Prince* in the Mall, your rival, *Abercrom A*, opens up directly across from you. You have been informed that Abercrom A is about to launch either a 30% off everything sale or a 50% off everything sale. You, on the other hand, have decided to either *increase* prices (to make your store seem more exclusive) or do absolutely nothing. You

construct the following payoff matrix, where the payoffs represent the number of customers your outlet can expect to gain from Abercrom A:

$$\begin{array}{cc} & \textbf{Abercrom A} \\ & \begin{array}{cc} \text{30\% Off} & \text{50\% Off} \end{array} \\ \textbf{Abercrom B} \begin{array}{c} \text{Do Nothing} \\ \text{Increase Prices} \end{array} & \begin{bmatrix} -60 & -40 \\ 30 & -50 \end{bmatrix}. \end{array}$$

There is a 20% chance that Abercrom A will opt for the "30% off" sale and an 80% chance that it will opt for the "50% Off" sale. Your sense from upper management at Abercrom B is that there is a 50% chance you will be given the go-ahead to raise prices. What is the expected resulting effect on your customer base?

34. *More Retail Discount Wars* Your *Abercrom B* men's fashion outlet has a 30% chance of launching an expensive new line of used auto-mechanic dungarees (complete with grease stains) and a 70% chance of staying instead with its traditional torn military-style dungarees. Your rival across from you in the mall, *Abercrom A*, appears to be deciding between a line of torn gym shirts and a more daring line of "empty shirts" (that is, empty shirt boxes). Your corporate spies reveal that there is a 20% chance that Abercrom A will opt for the empty shirt option. The following payoff matrix gives the number of customers your outlet can expect to gain from Abercrom A in each situation:

$$\begin{array}{cc} & \textbf{Abercrom A} \\ & \begin{array}{cc} \text{Torn Shirts} & \text{Empty Shirts} \end{array} \\ \textbf{Abercrom B} \begin{array}{c} \text{Mechanics} \\ \text{Military} \end{array} & \begin{bmatrix} 10 & -40 \\ -30 & 50 \end{bmatrix}. \end{array}$$

What is the expected resulting effect on your customer base?

35. *Factory Location*[22] A manufacturer of electrical machinery is located in a cramped, though low-rent, factory close to the center of a large city. The firm needs to expand, and it could do so in one of three ways: (1) Remain where it is and install new equipment, (2) move to a suburban site near the same city, or (3) relocate in a different part of the country where labor is cheaper. Its decision will be influenced by the fact that one of the following will happen: (I) The government may introduce a program of equipment grants, (II) a new suburban highway may be built, or (III) the government may institute a policy of financial help to companies who move into regions of high unemployment. The value to the company of each combination is given in the following table:

---

[21] Source: *New York Times*, September 18, 1992, p. B14.

[22] Adapted from an example in *Location in Space: Theoretical Perspectives in Economic Geography* by P. Dicken and P. E. Lloyd (Harper & Row, 1990).

Government's Options

		I	II	III
Manufacturer's Options	**1**	200	150	140
	2	130	220	130
	3	110	110	220

If the manufacturer judges that there is a 20% probability that the government will go with option I, a 50% probability that it will go with option II, and a 30% probability that it will go with option III, what is the manufacturer's best option?

36. *Crop Choice*[23] A farmer has a choice of growing wheat, barley, or rice. Her success will depend on the weather, which could be dry, average, or wet, as measured in the following table:

Weather

		Dry	Average	Wet
Crop Choices	**Wheat**	20	20	10
	Barley	10	15	20
	Rice	10	20	20

If the probability that the weather will be dry is 10%, the probability that it will be average is 60%, and the probability that it will be wet is 30%, what is the farmer's best choice of crop?

37. *Study Techniques* Your mathematics test is tomorrow and will cover the following topics: game theory, linear programming, and matrix algebra. You have decided to do an all-nighter and must determine how to allocate your 8 hours of study time among the three topics. If you were to spend the entire 8 hours on any one of these topics (thus using a pure strategy), you feel confident that you would earn a 90% score on that portion of the test but would not do so well on the other topics. You have come up with the following table, where the entries are your expected scores. (The fact that linear programming and matrix algebra are used in game theory is reflected in these numbers.)

Test

Your Strategies	Game Theory	Linear Programming	Matrix Algebra
Game Theory	90	70	70
Linear Programming	40	90	40
Matrix Algebra	60	40	90

You have been told that the test will be weighted as follows: game theory: 25%, linear programming: 50%, and matrix algebra: 25%.

---

[23] See footnote for Exercise 35.

a. If you spend 25% of the night on game theory, 50% on linear programming, and 25% on matrix algebra, what score do you expect to get on the test?

b. Is it possible to improve on this by altering your study schedule? If so, what is the highest score you can expect on the test?

c. If your study schedule is according to part (a) and your teacher decides to forget her promises about how the test will be weighted and instead bases it all on a single topic, which topic would be worst for you, and what score could you expect on the test?

38. *Study Techniques* Your friend Joe has been spending all of his time on fraternity activities and therefore knows absolutely nothing about any of the three topics on tomorrow's math test. (See Exercise 37.) Because you are recognized as an expert on the use of game theory to solve study problems, he has turned to you for advice as to how to spend his all-nighter. As the following table shows, his situation is not so rosy. (Since he knows no linear programming or matrix algebra, the table shows, for instance, that studying game theory all night will not be much use in preparing him for this topic.)

Test

Joe's Strategies	Game Theory	Linear Programming	Matrix Algebra
Game Theory	30	0	20
Linear Programming	0	70	0
Matrix Algebra	0	0	70

Assuming that the test will be weighted as described in Exercise 37, what are the answers to parts (a), (b), and (c) as they apply to Joe?

39. ▼ *Staff Cutbacks* Frank Tempest manages a large snowplow service in Manhattan, Kansas, and is alarmed by the recent weather trends; there have been no significant snowfalls in recent years. He is therefore contemplating laying off some of his workers but is unsure about whether to lay off 5, 10, or 15 of his 50 workers. Being very methodical, he estimates his annual net profits based on four possible annual snowfall figures: 0 inches, 20 inches, 40 inches, and 60 inches. (He takes into account the fact that, if he is running a small operation in the face of a large annual snowfall, he will lose business to his competitors because he will be unable to discount on volume.)

	0 Inches	20 Inches	40 Inches	60 Inches
5 Laid Off	−$500,000	−$200,000	$10,000	$200,000
10 Laid Off	−$200,000	$0	$0	$0
15 Laid Off	−$100,000	$10,000	−$200,000	−$300,000

a. During the past 10 years, the region has had 0 inches twice, 20 inches twice, 40 inches three times, and

60 inches three times. Based on this information, how many workers should Tempest lay off, and how much would it cost him?

b. There is a 50% chance that Tempest will lay off 5 workers and a 50% chance that he will lay off 15 workers. What is the worst thing Nature can do to him in terms of snowfall? How much would it cost him?

c. The Gods of Chaos (who control the weather) know that Tempest is planning to use the strategy in part (a) and are determined to hurt Tempest as much as possible. Tempest, being somewhat paranoid, suspects this. What should he do?

40. ▼ *Textbook Writing* You are writing a college-level textbook on finite mathematics and are trying to come up with the best combination of word problems. Over the years, you have accumulated a collection of amusing problems, serious applications, long complicated problems, and generic problems.[24] Before your book is published, it must be scrutinized by several reviewers, who, it seems, are never satisfied with the mix you use. You estimate that there are three kinds of reviewers: the "no-nonsense" types, who prefer applications and generic problems; the "dead serious" types, who feel that a college-level text should contain little or no humor and lots of long complicated problems; and the "laid-back" types, who believe that learning best takes place in a light-hearted atmosphere bordering on anarchy. You have drawn up the following chart, where the payoffs represent the reactions of reviewers on a scale of -10 (ballistic) to $+10$ (ecstatic):

Reviewers

		No-Nonsense	Dead Serious	Laid-Back
	Amusing	-5	-10	10
	Serious	5	3	0
You	Long	-5	5	3
	Generic	5	3	-10

a. Your first draft of the book contained no generic problems and equal numbers of the other categories. If half the reviewers of your book were "dead serious" and the rest were equally divided between the "no-nonsense" and "laid-back" types, what score would you expect?

b. In your second draft of the book, you tried to balance the content by including some generic problems and eliminating several amusing ones. You wound up with a mix of which one eighth were amusing, one quarter were serious, three eighths were long, and a quarter were generic. What kind of reviewer would be *least* impressed by this mix?

c. What kind of reviewer would be *most* impressed by the mix in your second draft?

41. *Price Wars* Computer Electronics, Inc. (CE) and the Gigantic Computer Store (GCS) are planning to discount the price they charge for the HAL laptop computer, of which they are the only distributors. Because Computer Electronics provides a free warranty service, it can generally afford to charge more. A market survey provides the following data on the gains to CE's market share that will result from different pricing decisions:

GCS

		$900	$1,000	$1,200
	$1,000	15%	60%	80%
CE	$1,200	15%	60%	60%
	$1,300	10%	20%	40%

a. Use reduction by dominance to determine how much each company should charge. What is the effect on CE's market share?

b. CE, which knows that GCS is planning to use reduction by dominance to determine its pricing policy, wants its own market share to be as large as possible. What effect, if any, would the information about GCS have on CE's best strategy?

42. *More Price Wars* (Refer to Exercise 41.) A new market survey results in the following revised data:

GCS

		$900	$1,000	$1,200
	$1,000	20%	60%	60%
CE	$1,200	15%	60%	60%
	$1,300	10%	20%	40%

a. Use reduction by dominance to determine how much each company should charge. What is the effect on CE's market share?

b. In general, why do price wars tend to force prices down?

43. *Wrestling Tournaments* City Community College (CCC) plans to host *Midtown Military Academy* (MMA) for a wrestling tournament. Each school has three wrestlers in the 190-pound weight class: CCC has Pablo, Sal, and Edison, while MMA has Carlos, Marcus, and Noto. Pablo can beat Carlos and Marcus, Marcus can beat Edison and Sal, Noto can beat Edison, while the other combinations will result in an even match. Set up a payoff matrix, and use reduction by dominance to decide which wrestler each team should choose as its champion. Does one school have an advantage over the other?

44. *Wrestling Tournaments* (Refer to Exercise 43.) One day before the wrestling tournament discussed in Exercise 43, Pablo sustains a hamstring injury and is replaced by Hans,

---

[24] Of the following type: "A certain company has three processing plants: A, B, and C, each of which uses three processes: P_1, P_2, and P_3. Process P_1 uses 100 units of chemical C_1, 50 units of C_2, . . . " and so on.

who (unfortunately for CCC) can be beaten by both Carlos and Marcus. Set up the payoff matrix, and use reduction by dominance to decide which wrestler each team should choose as its champion. Does one school have an advantage over the other?

45. *The Battle of Rabaul-Lae*[25] In the Second World War, during the struggle for New Guinea, intelligence reports revealed that the Japanese were planning to move a troop and supply convoy from the port of Rabaul at the eastern tip of New Britain to Lae, which lies just west of New Britain on New Guinea. The convoy could travel either via a northern route, which was plagued by poor visibility, or by a southern route, where the visibility was clear. General Kenney, who was the commander of the Allied Air Forces in the area, had the choice of concentrating reconnaissance aircraft on one route or the other and bombing the Japanese convoy once it was sighted. Kenney's staff drafted the following outcomes for his choices, where the payoffs are estimated days of bombing time:

		Japanese Commander's Strategies	
		Northern Route	Southern Route
Kenney's Strategies	Northern Route	2	2
	Southern Route	1	3

What would you have recommended to General Kenney? What would you have recommended to the Japanese commander?[26] How much bombing time results if these recommendations are followed?

46. *The Battle of Rabaul-Lae* Referring to Exercise 45, suppose that General Kenney had a third alternative: splitting his reconnaissance aircraft between the two routes, which would result in the following estimates:

		Japanese Commander's Strategies	
		Northern Route	Southern Route
Kenney's Strategies	Northern Route	2	2
	Split Reconnaissance	1.5	2.5
	Southern Route	1	3

What would you have recommended to General Kenney? What would you have recommended to the Japanese com-

mander? How much bombing time results if these recommendations are followed?

47. ▼ *The Prisoner's Dilemma* Slim Lefty and Joe Rap have been arrested for grand theft auto, having been caught red-handed driving away in a stolen 2012 Porsche. Although the police have more than enough evidence to convict them both, a confession would simplify the work of the prosecution. The police decide to interrogate the prisoners separately. Slim and Joe are both told of the following plea-bargaining arrangement: If both confess, they will each receive a 2-year sentence; if neither confesses, they will both receive 5-year sentences; and if only one confesses (and thus squeals on the other), he will receive a suspended sentence, while the other will receive a 10-year sentence. What should Slim do?

48. ▼ *More Prisoners' Dilemmas* Jane Good and Prudence Brown have been arrested for robbery, but the police lack sufficient evidence for a conviction and so decide to interrogate them separately in the hope of extracting a confession. Both Jane and Prudence are told the following: If they both confess, they will each receive a 5-year sentence; if neither confesses, they will be released; if one confesses, she will receive a suspended sentence, while the other will receive a 10-year sentence. What should Jane do?

49. ▼ *Campaign Strategies*[27] Florida and Ohio are "swing states" that have a large bounty of electoral votes and are therefore highly valued by presidential campaign strategists. Suppose that it is now the weekend before Election Day 2012, and each candidate (Romney and Obama) can visit only one more state. Further, to win the election, Romney needs to win both of these states. Currently, Romney has a 40% chance of winning Ohio and a 60% chance of winning Florida. Therefore, he has a $0.40 \times 0.60 = 0.24$, or 24%, chance of winning the election. Assume that each candidate can increase his probability of winning a state by 10% if he but not his opponent visits that state. If both candidates visit the same state, there is no effect.
   a. Set up a payoff matrix with Romney as the row player and Obama as the column player, where the payoff for a specific set of circumstances is the probability (expressed as a percentage) that Romney will win both states.
   b. Where should each candidate visit under the circumstances?

50. ▼ *Campaign Strategies* Repeat Exercise 49, this time assuming that Romney has an 80% chance of winning Ohio and a 90% chance of winning Florida.

51. *Advertising* The *Softex Shampoo Company* is considering how to split its advertising budget between ads on two radio stations: WISH and WASH. Its main competitor, *Splish Shampoo, Inc.*, has found out about this and is considering

---

[25] As discussed in *Games and Decisions* by R. D. Luce and H. Raiffa Section 11.3 (New York: Wiley, 1957). This is based on an article in the *Journal of the Operations Research Society of America* 2 (1954): 365–385.

[26] The correct answers to Exercises 45 and 46 correspond to the actual decisions both commanders made.

---

[27] Based on *Game Theory for Swingers: What states should the candidates visit before Election Day?* by Jordan Ellenberg. Source: www.slate.com.

countering Softex's ads with its own on the same radio stations. (Proposed jingle: "Softex, Shmoftex; Splash with Splish.") Softex has calculated that, were it to devote its entire advertising budget to ads on WISH, it would increase revenues in the coming month by $100,000 in the event that Splish was running all its ads on the less popular WASH, but Softex would lose $20,000 in revenues if Splish ran its ads on WISH. If, on the other hand, it devoted its entire budget to WASH ads, Softex would neither increase nor decrease revenues in the event that Splish was running all its ads on the more popular WISH and would in fact lose $20,000 in revenues if Splish ran its ads on WASH. What should Softex do, and what effect will this have on revenues?

52. *Labor Negotiations* The management team of the *Abstract Concrete Company* is negotiating a 3-year contract with the labor unions at one of its plants and is trying to decide on its offer for a salary increase. If it offers a 5% increase and the unions accept the offer, Abstract Concrete will gain $20 million in projected profits in the coming year, but if labor rejects the offer, the management team predicts that it will be forced to increase the offer to the union demand of 15%, thus halving the projected profits. If Abstract Concrete offers a 15% increase, the company will earn $10 million in profits over the coming year if the unions accept. If the unions reject, they will probably go out on strike (because management has set 15% as its upper limit), and management has decided that it can then in fact gain $12 million in profits by selling out the defunct plant in retaliation. What intermediate percentage should the company offer, and what profit should it project?

Communication and Reasoning Exercises

53. Why is a saddle point called a "saddle point"?

54. Can the payoff in a saddle point ever be larger than all other payoffs in a game? Explain.

55. ◆ One day, while browsing through an old *Statistical Abstract of the United States,* you come across the following data, which show the number of females employed (in

thousands) in various categories according to their educational attainment:[28]

	Managerial/ Professional	Technical/Sales/ Administrative	Service	Precision Production	Operators/ Fabricators
Less Than 4 Years of High School	260	1,080	2,020	260	1,400
4 Years of High School Only	2,430	9,510	3,600	570	2,130
1 to 3 Years of College	2,690	5,080	1,080	160	350
At Least 4 Years of College	7,210	2,760	380	70	110

Because you have been studying game theory that day, the first thing you do is to search for a saddle point. Having found one, you conclude that, as a female, your best strategy in the job market is to forget about a college career. Find the flaw in this reasoning.

56. ◆ Exercises 37 and 38 seem to suggest that studying a single topic before an exam is better than studying all the topics in that exam. Comment on this discrepancy between the game theory result and common sense.

57. ◆ Explain what is wrong with a decision to play the mixed strategy [0.5 0.5] by alternating the two strategies: Play the first strategy on the odd-numbered moves and the second strategy on the even-numbered moves. Illustrate your argument by devising a game in which your best strategy is [0.5 0.5].

58. ◆ Describe a situation in which a mixed strategy and a pure strategy are equally effective.

---

[28] Source: *Statistical Abstract of the United States 1991* (111th Ed.) U.S. Department of Commerce, Economics and Statistics Administration, and Bureau of the Census.

4.5 Input-Output Models

Sectors of the Economy

In this section we look at an application of matrix algebra developed by Wassily Leontief (1906–1999) in the middle of the twentieth century. In 1973 he won the Nobel Prize in Economics for this work. The application involves analyzing national and regional economies by looking at how various parts of the economy interrelate. We'll work out some of the details by looking at a simple scenario.

First, we can think of the economy of a country or a region as being composed of various **sectors**, or groups of one or more industries. Typical sectors are the manufacturing sector, the utilities sector, and the agricultural sector. To introduce the basic concepts, we shall consider two specific sectors: the coal-mining sector (Sector 1) and the electric utilities sector (Sector 2). Each produces a commodity: The coal-mining sector produces coal, and the electric utilities sector produces electricity. We measure these products by their dollar value. By **one unit** of a product, we mean $1 worth of that product.

Here is the scenario:

1. To produce one unit ($1 worth) of coal, assume that the coal-mining sector uses 50¢ worth of coal (to power mining machinery, say) and 10¢ worth of electricity.

2. To produce one unit ($1 worth) of electricity, assume that the electric utilities sector uses 25¢ worth of coal and 25¢ worth of electricity.

These are *internal* usage figures. In addition to this, assume that there is an *external* demand (from the rest of the economy) of 7,000 units ($7,000 worth) of coal and 14,000 units ($14,000 worth) of electricity over a specific time period (1 year, say). Our basic question is: How much should each of the two sectors supply to meet both internal and external demand?

The key to answering this question is to set up equations of the form

Total supply = Total demand.

The unknowns, the values we are seeking, are

x_1 = Total supply (in units) from Sector 1 (coal)

x_2 = Total supply (in units) from Sector 2 (electricity).

Our equations then take the following form:

Total supply from Sector 1 = Total demand for Sector 1 products

$$x_1 = 0.50x_1 \qquad + \qquad 0.25x_2 \qquad + \qquad 7,000$$

↑   ↑   ↑

Coal required by Sector 1   Coal required by Sector 2   External demand for coal

Total supply from Sector 2 = Total demand for Sector 2 products

$$x_2 = 0.10x_1 \qquad + \qquad 0.25x_2 \qquad + \qquad 14,000.$$

↑   ↑   ↑

Electricity required by Sector 1   Electricity required by Sector 2   External demand for electricity

This is a system of two linear equations in two unknowns:

$$x_1 = 0.50x_1 + 0.25x_2 + 7,000$$
$$x_2 = 0.10x_1 + 0.25x_2 + 14,000.$$

We can rewrite this system of equations in matrix form as follows:

$$\underbrace{\begin{bmatrix} x_1 \\ x_2 \end{bmatrix}}_{\text{Production}} = \underbrace{\begin{bmatrix} 0.50 & 0.25 \\ 0.10 & 0.25 \end{bmatrix} \begin{bmatrix} x_1 \\ x_2 \end{bmatrix}}_{\text{Internal demand}} + \underbrace{\begin{bmatrix} 7,000 \\ 14,000 \end{bmatrix}}_{\text{External demand}}.$$

In symbols,

$$X = AX + D.$$

Here,

$$X = \begin{bmatrix} x_1 \\ x_2 \end{bmatrix}$$

is called the **production vector**. Its entries are the amounts produced by the two sectors. The matrix

$$D = \begin{bmatrix} 7,000 \\ 14,000 \end{bmatrix}$$

is called the **external demand** vector, and

$$A = \begin{bmatrix} 0.50 & 0.25 \\ 0.10 & 0.25 \end{bmatrix} \qquad \text{Organization: } \begin{bmatrix} 1 \to 1 & 1 \to 2 \\ 2 \to 1 & 2 \to 2 \end{bmatrix}$$

is called the **technology matrix**. The entries of the technology matrix have the following meanings:

a_{11} = Units of Sector 1 needed to produce one unit of Sector 1
a_{12} = Units of Sector 1 needed to produce one unit of Sector 2
a_{21} = Units of Sector 2 needed to produce one unit of Sector 1
a_{22} = Units of Sector 2 needed to produce one unit of Sector 2.

Now that we have the matrix equation

$$X = AX + D,$$

we can solve it as follows. First, subtract AX from both sides:

$$X - AX = D.$$

Because $X = IX$, where I is the 2×2 identity matrix, we can rewrite this as

$$IX - AX = D.$$

Now factor out X:

$$(I - A)X = D.$$

If we multiply both sides by the inverse of $(I - A)$, we get the solution

$$X = (I - A)^{-1}D.$$

Input-Output Model

In an input-output model, an economy (or part of one) is divided into n **sectors**. We then record the $n \times n$ **technology matrix** A, whose ijth entry is the number of units from Sector i used in producing one unit from Sector j (in symbols, "$i \to j$"). To meet an **external demand** of D, the economy must produce X, where X is the **production vector**. These are related by the equations

$$X = AX + D$$

or

$$X = (I - A)^{-1}D. \qquad \text{Provided that } (I - A) \text{ is invertible}$$

Quick Example

1. In the scenario above,

$$A = \begin{bmatrix} 0.50 & 0.25 \\ 0.10 & 0.25 \end{bmatrix}, \quad X = \begin{bmatrix} x_1 \\ x_2 \end{bmatrix}, \quad \text{and} \quad D = \begin{bmatrix} 7,000 \\ 14,000 \end{bmatrix}.$$

The solution is

$$X = (I - A)^{-1}D$$

$$\begin{bmatrix} x_1 \\ x_2 \end{bmatrix} = \left(\begin{bmatrix} 1 & 0 \\ 0 & 1 \end{bmatrix} - \begin{bmatrix} 0.50 & 0.25 \\ 0.10 & 0.25 \end{bmatrix} \right)^{-1} \begin{bmatrix} 7,000 \\ 14,000 \end{bmatrix}$$

$$= \begin{bmatrix} 0.50 & -0.25 \\ -0.10 & 0.75 \end{bmatrix}^{-1} \begin{bmatrix} 7,000 \\ 14,000 \end{bmatrix}$$ Calculate $I - A$.

$$= \begin{bmatrix} \frac{15}{7} & \frac{5}{7} \\ \frac{2}{7} & \frac{10}{7} \end{bmatrix} \begin{bmatrix} 7,000 \\ 14,000 \end{bmatrix}$$ Calculate $(I - A)^{-1}$.

$$= \begin{bmatrix} 25,000 \\ 22,000 \end{bmatrix}.$$

In other words, to meet the demand, the economy must produce $25,000 worth of coal and $22,000 worth of electricity.

The next example uses actual data from the U.S. economy. (We have rounded the figures to make the computations less complicated.) It is rare to find input-output data already packaged for you as a technology matrix. Instead, the data commonly found in statistical sources come in the form of input-output tables, from which we will have to construct the technology matrix.

EXAMPLE 1   **Petroleum and Natural Gas**

Consider two sectors of the U.S. economy: crude petroleum and natural gas (*crude*) and petroleum refining and related industries (*refining*). According to government figures,[29] in 1998 the crude sector used $27,000 million worth of its own products and $750 million worth of the products of the refining sector to produce $87,000 million worth of goods (crude oil and natural gas). The refining sector in the same year used $59,000 million worth of the products of the crude sector and $15,000 million worth of its own products to produce $140,000 million worth of goods (refined oil and the like). What was the technology matrix for these two sectors? What was left over from each of these sectors for use by other parts of the economy or for export?

Solution First, for convenience, we record the given data in the form of a table, called the **input-output table**. (All figures are in millions of dollars.)

		To	
		Crude	**Refining**
From	**Crude**	27,000	59,000
	Refining	750	15,000
	Total Output	87,000	140,000

---

[29] The data have been rounded to two significant digits. Source: *Survey of Current Business*, December, 2001, U.S. Department of Commerce. The *Survey of Current Business* and the input-output tables themselves are available at the website of the Department of Commerce's Bureau of Economic Analysis (www.bea.gov).

The entries in the top portion are arranged in the same way as those of the technology matrix: The ijth entry represents the number of units of Sector i that went to Sector j. Thus, for instance, the 59,000 million entry in the 1, 2 position represents the number of units of Sector 1, crude, that were used by Sector 2, refining. ("From the side, to the top.")

We now construct the technology matrix. The technology matrix has entries a_{ij} = Units of Sector i used to produce *one* unit of Sector j. Thus,

a_{11} = Units of crude to produce one unit of crude. We are told that 27,000 million units of crude were used to produce 87,000 million units of crude. Thus, to produce *one* unit of crude, $27{,}000/87{,}000 \approx 0.31$ units of crude were used, so $a_{11} \approx 0.31$. (We have rounded this value to two significant digits; further digits are not reliable because of rounding of the original data.)

a_{12} = Units of crude to produce one unit of refined: $a_{12} = 59{,}000/140{,}000 \approx 0.42$

a_{21} = Units of refined to produce one unit of crude: $a_{21} = 750/87{,}000 \approx 0.0086$

a_{22} = Units of refined to produce one unit of refined: $a_{22} = 15{,}000/140{,}000 \approx 0.11$.

This gives the technology matrix

$$A = \begin{bmatrix} 0.31 & 0.42 \\ 0.0086 & 0.11 \end{bmatrix}. \qquad \text{Technology matrix}$$

In short, *we obtained the technology matrix from the input-output table by dividing the Sector 1 column by the Sector 1 total, and the Sector 2 column by the Sector 2 total.*

Now we also know the total output from each sector, so *we have already been given the production vector:*

$$X = \begin{bmatrix} 87{,}000 \\ 140{,}000 \end{bmatrix}. \qquad \text{Production vector}$$

What we are asked for is the external demand vector D, the amount available for the outside economy. To find D, we use the equation

$$X = AX + D, \qquad \text{Relationship of } X, A, \text{ and } D$$

where, this time, we are given A and X and we must solve for D. Solving for D gives

$$D = X - AX$$

$$= \begin{bmatrix} 87{,}000 \\ 140{,}000 \end{bmatrix} - \begin{bmatrix} 0.31 & 0.42 \\ 0.0086 & 0.11 \end{bmatrix} \begin{bmatrix} 87{,}000 \\ 140{,}000 \end{bmatrix}$$

*Why?

$$\approx \begin{bmatrix} 87{,}000 \\ 140{,}000 \end{bmatrix} - \begin{bmatrix} 86{,}000 \\ 16{,}000 \end{bmatrix} = \begin{bmatrix} 1{,}000 \\ 124{,}000 \end{bmatrix}. \qquad \text{We rounded to two significant digits.*}$$

The first number, $1,000 million, is the amount produced by the crude sector that is available to be used by other parts of the economy or to be exported. (In fact, because something has to happen to all that crude petroleum and natural gas, this is the amount that is actually used or exported, where use can include stockpiling.) The second number, $124,000 million, represents the amount produced by the refining sector that is available to be used by other parts of the economy or to be exported.

Note that we could have calculated D more simply from the input-output table. The internal use of units from the crude sector was the sum of the outputs from that sector:

$$27,000 + 59,000 = 86,000.$$

Because 87,000 units were actually produced by the sector, that left a surplus of $87,000 - 86,000 = 1,000$ units for export. We could compute the surplus from the refining sector similarly. (The two calculations actually come out slightly differently because we rounded the intermediate results.) The calculation in Example 2 below cannot be done as trivially, however.

Using Technology

See the Technology Guides at the end of the chapter to see how to compute the technology matrix and the external demand vector in Example 1 using a TI-83/84 Plus or a spreadsheet.

Input-Output Table

National economic data are often given in the form of an **input-output table**. The ijth entry in the top portion of the table is the number of units that go from Sector i to Sector j. The "Total outputs" are the total numbers of units produced by each sector. We obtain the technology matrix from the input-output table by dividing the Sector 1 column by the Sector 1 total, the Sector 2 column by the Sector 2 total, and so on.

Quick Example

2. Input-output table:

		To	
		Skateboards	**Wood**
From	**Skateboards**	20,000*	0
	Wood	100,000	500,000
	Total Output	200,000	5,000,000

* The production of skateboards required skateboards because skateboard workers tend to commute to work on (what else?) skateboards!

Technology matrix:

$$A = \begin{bmatrix} \frac{20,000}{200,000} & \frac{0}{5,000,000} \\ \frac{100,000}{200,000} & \frac{500,000}{5,000,000} \end{bmatrix} = \begin{bmatrix} 0.1 & 0 \\ 0.5 & 0.1 \end{bmatrix}.$$

EXAMPLE 2 **Rising Demand**

Suppose that external demand for refined petroleum rises to $200,000 million, but the demand for crude remains $1,000 million (as in Example 1). How do the production levels of the two sectors considered in Example 1 have to change?

Solution We are being told that now

$$D = \begin{bmatrix} 1,000 \\ 200,000 \end{bmatrix},$$

and we are asked to find X. Remember that we can calculate X from the formula

$$X = (I - A)^{-1}D.$$

Now

$$I - A = \begin{bmatrix} 1 & 0 \\ 0 & 1 \end{bmatrix} - \begin{bmatrix} 0.31 & 0.42 \\ 0.0086 & 0.11 \end{bmatrix} = \begin{bmatrix} 0.69 & -0.42 \\ -0.0086 & 0.89 \end{bmatrix}.$$

We take the inverse using our favorite technique and find that, to four significant digits,*

$$(I - A)^{-1} \approx \begin{bmatrix} 1.458 & 0.6880 \\ 0.01409 & 1.130 \end{bmatrix}.$$

Now we can compute X:

$$X = (I - A)^{-1}D = \begin{bmatrix} 1.458 & 0.6880 \\ 0.01409 & 1.130 \end{bmatrix}\begin{bmatrix} 1,000 \\ 200,000 \end{bmatrix} \approx \begin{bmatrix} 140,000 \\ 230,000 \end{bmatrix}.$$

(As in Example 1, we have rounded all the entries in the answer to two significant digits.) Comparing this vector to the production vector used in Example 1, we see that production in the crude sector has to increase from $87,000 million to $140,000 million, while production in the refining sector has to increase from $140,000 million to $230,000 million.

* Because A is accurate to two digits, we should use more than two significant digits in intermediate calculations so as not to lose additional accuracy. We must, of course, round the final answer to two digits.

➡ **Before we go on ...** Using the matrix $(I - A)^{-1}$, we have a slightly different way of solving Example 2. We are asking for the effect on production of a *change* in the final demand of 0 for crude and $200,000 - 124,000 = \$76,000$ million for refined products. If we multiply $(I - A)^{-1}$ by the matrix representing this *change,* we obtain

$$\begin{bmatrix} 1.458 & 0.6880 \\ 0.01409 & 1.130 \end{bmatrix}\begin{bmatrix} 0 \\ 76,000 \end{bmatrix} \approx \begin{bmatrix} 53,000 \\ 90,000 \end{bmatrix}.$$

$(I - A)^{-1} \times$ Change in demand $=$ Change in production

We see the changes required in production: an increase of $53,000 million in the crude sector and an increase of $90,000 million in the refining sector.

Notice that the increase in external demand for the products of the refining sector requires the crude sector to increase production as well, even though there is no increase in the *external* demand for its products. The reason is that, to increase production, the refining sector needs to use more crude oil, so the *internal* demand for crude oil goes up. The inverse matrix $(I - A)^{-1}$ takes these **indirect effects** into account in a nice way.

By replacing the $76,000 by $1 in the computation we just did, we see that a $1 increase in external demand for refined products will require an increase in production of $0.6880 in the crude sector as well as an increase in production of $1.130 in the refining sector. This is how we interpret the entries in $(I - A)^{-1}$, and this is why it is useful to look at this matrix inverse rather than just solve $(I - A)X = D$ for X using, say, Gauss-Jordan reduction. Looking at $(I - A)^{-1}$, we can also find the effects of an increase of $1 in external demand for crude: an increase in production of $1.458 in the crude sector and an increase of $0.01409 in the refining sector.

Here are some questions to think about: Why are the diagonal entries of $(I - A)^{-1}$ (slightly) larger than 1? Why is the entry in the lower left so small in comparison to the others? ∎

Interpreting $(I - A)^{-1}$: Indirect Effects

If A is the technology matrix, then the ijth entry of $(I - A)^{-1}$ is the change in the number of units Sector i must produce to meet a one-unit increase in external demand for Sector j products. To meet a rising external demand, the necessary change in production for each sector is given by

$$\text{Change in production} = (I - A)^{-1}D^+,$$

where D^+ is the change in external demand.

Quick Example

3. Take Sector 1 to be skateboards and Sector 2 to be wood, and assume that

$$(I - A)^{-1} = \begin{bmatrix} 1.1 & 0 \\ 0.6 & 1.1 \end{bmatrix}.$$

Then

$a_{11} = 1.1 =$ Number of additional units of skateboards that must be produced to meet a one-unit increase in the demand for skateboards (Why is this number larger than 1?)

$a_{12} = 0 =$ Number of additional units of skateboards that must be produced to meet a one-unit increase in the demand for wood (Why is this number 0?)

$a_{21} = 0.6 =$ Number of additional units of wood that must be produced to meet a one-unit increase in the demand for skateboards

$a_{22} = 1.1 =$ Number of additional units of wood that must be produced to meet a one-unit increase in the demand for wood.

To meet an increase in external demand of 100 skateboards and 400 units of wood, the necessary change in production is

$$(I - A)^{-1}D^+ = \begin{bmatrix} 1.1 & 0 \\ 0.6 & 1.1 \end{bmatrix}\begin{bmatrix} 100 \\ 400 \end{bmatrix} = \begin{bmatrix} 110 \\ 500 \end{bmatrix},$$

so 110 additional skateboards and 500 additional units of wood will need to be produced.

In the preceding examples we used only two sectors of the economy. The data used in Examples 1 and 2 were taken from an input-output table published by the U.S. Department of Commerce, in which the whole U.S. economy was broken down into 85 sectors. This in turn was a simplified version of a model in which the economy was broken into about 500 sectors. Obviously, computers are required to make a realistic input-output analysis possible. Many governments collect and publish input-output data as part of their national planning. The United Nations collects these data and publishes collections of national statistics. These and other useful statistics can be found at the website of the United Nations Statistics Division: http://unstats .un.org/unsd.

EXAMPLE 3 Kenya Economy

Consider four sectors of the economy of Kenya: (1) the traditional economy, (2) agriculture, (3) manufacture of metal products and machinery, and (4) wholesale and retail trade.[30] The input-output table for these four sectors for 1976 looks like this (all numbers are thousands of K£):

		To			
		1	**2**	**3**	**4**
	1	8,600	0	0	0
From	**2**	0	20,000	24	0
	3	1,500	530	15,000	660
	4	810	8,500	5,800	2,900
	Total Output	87,000	530,000	110,000	180,000

Suppose that external demand for agriculture increased by K£50,000,000 and that external demand for metal products and machinery increased by K£10,000,000. How would production in these four sectors have to change to meet this rising demand?

Solution To find the change in production necessary to meet the rising demand, we need to use the formula

$$\text{Change in production} = (I - A)^{-1}D^+,$$

where A is the technology matrix and D^+ is the change in demand:

$$D^+ = \begin{bmatrix} 0 \\ 50,000 \\ 10,000 \\ 0 \end{bmatrix}.$$

With entries shown rounded to two significant digits, the matrix A is

$$A = \begin{bmatrix} 0.099 & 0 & 0 & 0 \\ 0 & 0.038 & 0.00022 & 0 \\ 0.017 & 0.001 & 0.14 & 0.0037 \\ 0.0093 & 0.016 & 0.053 & 0.016 \end{bmatrix}.$$ Entries shown are rounded to two significant digits.

The next calculation is best done by using technology:

$$\text{Change in production} = (I - A)^{-1}D^+ = \begin{bmatrix} 0 \\ 52,000 \\ 12,000 \\ 1,500 \end{bmatrix}.$$ Entries shown are rounded to two significant digits.

[30] Figures are rounded. Source: *Input-Output Tables for Kenya 1976,* Central Bureau of Statistics of the Ministry of Economic Planning and Community Affairs, Kenya.

Looking at this result, we see that the changes in external demand will leave the traditional economy unaffected, production in agriculture will rise by K£52 million, production in the manufacture of metal products and machinery will rise by K£12 million, and activity in wholesale and retail trade will rise by K£1.5 million.

➡ **Before we go on . . .** Can you see why the traditional economy was unaffected in Example 3? Although it takes inputs from other parts of the economy, the traditional economy is not itself an input to any other part. In other words, there is no intermediate demand for the products of the traditional economy coming from any other part of the economy, so an increase in production in any other sector of the economy will require no increase from the traditional economy. On the other hand, the wholesale and retail trade sector does provide input to the agriculture and manufacturing sectors, so increases in those sectors do require an increase in the trade sector.

One more point: If you calculate $(I - A)^{-1}$, you will notice how small the off-diagonal entries are. This tells us that increases in each sector have relatively small effects on the other sectors. We say that these sectors are **loosely coupled**. Regional economies, where many products are destined to be shipped out to the rest of the country, tend to show this phenomenon even more strongly. Notice in Example 2 that those two sectors are **strongly coupled**, because a rise in demand for refined products requires a comparable rise in the production of crude. ∎

4.5   EXERCISES

▼ more advanced   ◆ challenging

Ⓣ indicates exercises that should be solved using technology

1. Let A be the technology matrix $A = \begin{bmatrix} 0.2 & 0.05 \\ 0.8 & 0.01 \end{bmatrix}$, where Sector 1 is paper and Sector 2 is wood. Fill in the missing quantities.

a. ___ units of wood are needed to produce one unit of paper.

b. ___ units of paper are used in the production of one unit of paper.

c. The production of each unit of wood requires the use of ___ units of paper.

2. Let A be the technology matrix $A = \begin{bmatrix} 0.01 & 0.001 \\ 0.2 & 0.004 \end{bmatrix}$, where Sector 1 is processor chips and Sector 2 is silicon. Fill in the missing quantities.

a. ___ units of silicon are required in the production of one unit of silicon.

b. ___ units of processor chips are used in the production of one unit of silicon.

c. The production of each unit of processor chips requires the use of ___ units of silicon.

3. Each unit of television news requires 0.2 units of television news and 0.5 units of radio news. Each unit of radio news requires 0.1 units of television news and no radio news. With Sector 1 as television news and Sector 2 as radio news, set up the technology matrix A.

4. Production of one unit of cologne requires no cologne and 0.5 units of perfume. Into one unit of perfume go 0.1 units of cologne and 0.3 units of perfume. With Sector 1 as cologne and Sector 2 as perfume, set up the technology matrix A.

In Exercises 5–12, you are given a technology matrix A and an external demand vector D. Find the corresponding production vector X. [**HINT:** See Quick Example 1.]

5. $A = \begin{bmatrix} 0.5 & 0.4 \\ 0 & 0.5 \end{bmatrix}, D = \begin{bmatrix} 10,000 \\ 20,000 \end{bmatrix}$

6. $A = \begin{bmatrix} 0.5 & 0.4 \\ 0 & 0.5 \end{bmatrix}, D = \begin{bmatrix} 20,000 \\ 10,000 \end{bmatrix}$

7. $A = \begin{bmatrix} 0.1 & 0.4 \\ 0.2 & 0.5 \end{bmatrix}, D = \begin{bmatrix} 25,000 \\ 15,000 \end{bmatrix}$

8. $A = \begin{bmatrix} 0.1 & 0.2 \\ 0.4 & 0.5 \end{bmatrix}, D = \begin{bmatrix} 24,000 \\ 14,000 \end{bmatrix}$

9. $A = \begin{bmatrix} 0.5 & 0.1 & 0 \\ 0 & 0.5 & 0.1 \\ 0 & 0 & 0.5 \end{bmatrix}, D = \begin{bmatrix} 1,000 \\ 1,000 \\ 2,000 \end{bmatrix}$

10. $A = \begin{bmatrix} 0.5 & 0.1 & 0 \\ 0 & 0.5 & 0.1 \\ 0 & 0 & 0.5 \end{bmatrix}, D = \begin{bmatrix} 3,000 \\ 3,800 \\ 2,000 \end{bmatrix}$

11. $A = \begin{bmatrix} 0.2 & 0.2 & 0 \\ 0.2 & 0.4 & 0.2 \\ 0 & 0.2 & 0.2 \end{bmatrix}, D = \begin{bmatrix} 16{,}000 \\ 8{,}000 \\ 8{,}000 \end{bmatrix}$

12. $A = \begin{bmatrix} 0.2 & 0.2 & 0.2 \\ 0.2 & 0.4 & 0.2 \\ 0.2 & 0.2 & 0.2 \end{bmatrix}, D = \begin{bmatrix} 7{,}000 \\ 14{,}000 \\ 7{,}000 \end{bmatrix}$

13. Given $A = \begin{bmatrix} 0.1 & 0.4 \\ 0.2 & 0.5 \end{bmatrix}$, find the changes in production required to meet an increase in demand of 50 units of Sector 1 products and 30 units of Sector 2 products.

14. Given $A = \begin{bmatrix} 0.5 & 0.4 \\ 0 & 0.5 \end{bmatrix}$, find the changes in production required to meet an increase in demand of 20 units of Sector 1 products and 10 units of Sector 2 products.

15. Let $(I - A)^{-1} = \begin{bmatrix} 1.5 & 0.1 & 0 \\ 0.2 & 1.2 & 0.1 \\ 0.1 & 0.7 & 1.6 \end{bmatrix}$, and assume that the external demand for the products in Sector 1 increases by one unit. By how many units should each sector increase production? What do the columns of the matrix $(I - A)^{-1}$ tell you? [**HINT:** See Quick Example 3.]

16. Let $(I - A)^{-1} = \begin{bmatrix} 1.5 & 0.1 & 0 \\ 0.1 & 1.1 & 0.1 \\ 0 & 0 & 1.3 \end{bmatrix}$, and assume that the external demand for the products in each of the sectors increases by one unit. By how many units should each sector increase production? [**HINT:** See Quick Example 3.]

In Exercises 17 and 18, obtain the technology matrix from the given input-output table. [**HINT:** See Example 1.]

17.

		To		
		A	**B**	**C**
From	**A**	1,000	2,000	3,000
	B	0	4,000	0
	C	0	1,000	3,000
	Total Output	5,000	5,000	6,000

18.

		To		
		A	**B**	**C**
From	**A**	0	100	300
	B	500	400	300
	C	0	0	600
	Total Output	1,000	2,000	3,000

Applications

19. *Campus Food* The two campus cafeterias, the Main Dining Room and Bits & Bytes, typically use each other's food

in doing business on campus. One weekend, the input-output table was as follows:[31]

		To	
		Main DR	**Bits & Bytes**
From	**Main DR**	$10,000	$20,000
	Bits & Bytes	5,000	0
	Total Output	50,000	40,000

Given that the demand for food on campus last weekend was $45,000 from the Main Dining Room and $30,000 from Bits & Bytes, how much did the two cafeterias have to produce to meet the demand last weekend?

20. *Plagiarism* Two student groups at *Enormous State University*, the Choral Society and the Football Club, maintain files of term papers that they write and offer to students for research purposes. Some of these papers they use themselves in generating more papers. To avoid suspicion of plagiarism by faculty members (who seem to have astute memories), each paper is given to students or used by the clubs only once. (No copies are kept.) The number of papers that were used in the production of new papers last year is shown in the following input-output table:

		To	
		Choral Soc.	**Football Club**
From	**Choral Soc.**	20	10
	Football Club	10	30
	Total Output	100	200

Given that 270 Choral Society papers and 810 Football Club papers will be used by students outside of these two clubs next year, how many new papers do the two clubs need to write?

21. ⊤ *Communication Equipment* Two sectors of the U.S. economy are (1) audio, video, and communication equipment and (2) electronic components and accessories. In 1998 the input-output table involving these two sectors was as follows. (All figures are in millions of dollars):[32]

		To	
		Equipment	**Components**
From	**Equipment**	6,000	500
	Components	24,000	30,000
	Total Output	90,000	140,000

---

[31] For some reason, the Main Dining Room consumes a lot of its own food!

[32] The data have been rounded. Source: *Survey of Current Business*, December 2001, U.S. Department of Commerce.

Determine the production levels necessary in these two sectors to meet an external demand for $80,000 million of communication equipment and $90,000 million of electronic components. Round answers to two significant digits.

22. **Ⓣ Wood and Paper** Two sectors of the U.S. economy are (1) lumber and wood products and (2) paper and allied products. In 1998 the input-output table involving these two sectors was as follows. (All figures are in millions of dollars.)[33]

		To	
		Wood	**Paper**
From	**Wood**	36,000	7,000
	Paper	100	17,000
	Total Output	120,000	120,000

If external demand for lumber and wood products rises by $10,000 million and external demand for paper and allied products rises by $20,000 million, what increase in output of these two sectors is necessary? Round answers to two significant digits.

23. **Australia Economy** Two sectors of the Australian economy are (1) textiles and (2) clothing and footwear. The 1977 input-output table[34] involving these two sectors results in the following value for $(I - A)^{-1}$:

$$(I - A)^{-1} = \begin{bmatrix} 1.228 & 0.182 \\ 0.006 & 1.1676 \end{bmatrix}.$$

Complete the following sentences.
a. ____ additional dollars worth of clothing and footwear must be produced to meet a $1 increase in the demand for textiles.
b. 0.182 additional dollars worth of ____ must be produced to meet a $1 increase in the demand for ____.

24. **Australia Economy** Two sectors of the Australian economy are (1) community services and (2) recreation services. The 1978–79 input-output table[35] involving these two sectors results in the following value for $(I - A)^{-1}$:

$$(I - A)^{-1} = \begin{bmatrix} 1.0066 & 0.00576 \\ 0.00496 & 1.04206 \end{bmatrix}.$$

Complete the following sentences.
a. 0.00496 additional dollars worth of ____ must be produced to meet a $1 increase in the demand for ____.
b. ____ additional dollars worth of community services must be produced to meet a $1 increase in the demand for community services.

Mexico Economy *Economists generally divide a country's economy into three broad sectors: primary, secondary, and tertiary. The primary sector is the sector using natural resources to produce raw materials, and includes oil extraction, agriculture, mining, and fishing. The secondary or industrial sector is the sector that uses raw materials to produce manufactured goods, and includes oil refining, textiles, and electronics. The tertiary or services sector provides services, and includes tourism, financial services, and health care. Exercises 25–30 are based on the following technology matrix for Mexico in 2008. (The sectors are in the order primary, secondary, and tertiary, and entries are rounded to two decimal places.)[36]*

$$A = \begin{bmatrix} 0.09 & 0.03 & 0.00 \\ 0.14 & 0.23 & 0.08 \\ 0.07 & 0.12 & 0.15 \end{bmatrix}$$

25. In 2008, Mexico produced around 590 billion pesos of raw materials, 11,000 billion pesos of manufactured goods, and 9,600 billion pesos of services.[37] Use these data to construct the input-output table for Mexico in 2008. [**HINT:** See Example 1.]

26. Suppose, in a particular year, Mexico had produced around 860 billion pesos of raw materials, 23,000 billion pesos of manufactured goods, and 8,500 billion pesos of services. What would the input-output table for the Mexican economy have been? [**HINT:** See Example 1.]

27. Use the technology matrix and the production data in Exercise 25 to calculate the amounts exported out of the country from each sector of the Mexico economy in 2008. (Round answers to two significant digits.)

28. Use the technology matrix and the production data in Exercise 26 to construct the amounts that would have been exported out of the country from each sector of the Mexico economy. (Round answers to two significant digits.)

29. **Ⓣ** Determine how the three sectors of the Mexico economy would react to a decrease in demand for tourism (tertiary sector) of 1,000 billion pesos and an increase in the demand for raw materials of 2,000 billion pesos.

30. **Ⓣ** Determine how the three sectors of the Mexico economy would react to an increase in demand for tourism (tertiary sector) of 1,000 billion pesos and a decrease in the other two sectors of 1,000 billion pesos each.

Ⓣ *Exercises 31–34 require the use of technology.*

31. **Ⓣ United States Input-Output Table** Four sectors of the U.S. economy are (1) livestock and livestock products, (2) other agricultural products, (3) forestry and fishery products, and (4) agricultural, forestry, and fishery services.

---

[33] See footnote for Exercise 21.

[34] Source: *Australian National Accounts and Input-Output Tables 1978–1979*, Australian Bureau of Statistics.

[35] *Ibid.*

---

[36] Source for data: Instituto Nacional de Estadística y Geografía (www.inegi.org.mx).

[37] *Ibid.*

In 1977 the input-output table involving these four sectors was as follows. (All figures are in millions of dollars.)[38]

		To			
		1	**2**	**3**	**4**
From	**1**	11,937	9	109	855
	2	26,649	4,285	0	4,744
	3	0	0	439	61
	4	5,423	10,952	3,002	216
	Total Output	97,795	120,594	14,642	47,473

Determine how these four sectors would react to a simultaneous increase in demand of $1,000 million in every sector. (Round answers to four significant digits.)

32. ⊤ *United States Input-Output Table* Four sectors of the U.S. economy are (1) motor vehicles, (2) truck and bus bodies, trailers, and motor vehicle parts, (3) aircraft and parts, and (4) other transportation equipment. In 1998 the input-output table involving these four sectors was as follows. (All figures in millions of dollars.)[39]

		To			
		1	**2**	**3**	**4**
From	**1**	75	1,092	0	1,207
	2	64,858	13,081	7	1,070
	3	0	0	21,782	0
	4	0	0	0	1,375
	Total Output	230,676	135,108	129,376	44,133

Determine how these four sectors would react to a simultaneous increase in demand of $1,000 million in every sector. (Round answers to four significant digits.)

33. ⊤ ▼ *Australia Input-Output Table* Four sectors of the Australian economy are (1) agriculture, (2) forestry, fishing, and hunting, (3) meat and milk products, and (4) other food products. In 1978–79 the input-output table involving these four sectors was as follows. (All figures are in millions of Australian dollars.)[40]

		To			
		1	**2**	**3**	**4**
From	**1**	678.4	3.7	3,341.5	1,023.5
	2	15.5	6.9	17.1	124.5
	3	47.3	4.3	893.1	145.8
	4	312.5	22.1	83.2	693.5
	Total Output	9,401.3	685.8	6,997.3	4,818.3

a. How much additional production by the meat and milk sector is necessary to accommodate a $100 increase in the demand for agriculture?

b. Which sector requires the most of its own product in order to meet a $1 increase in external demand for that product?

34. ⊤ ▼ *Australia Input-Output Table* Four sectors of the Australian economy are (1) petroleum and coal products, (2) nonmetallic mineral products, (3) basic metals and products, and (4) fabricated metal products. In 1978–79 the input-output table involving these four sectors was as follows. (All figures are in millions of Australian dollars.)[41]

		To			
		1	**2**	**3**	**4**
From	**1**	174.1	30.5	120.3	14.2
	2	0	190.1	55.8	12.6
	3	2.1	40.2	1,418.7	1,242.0
	4	0.1	7.3	40.4	326.0
	Total Output	3,278.0	2,188.8	6,541.7	4,065.8

a. How much additional production by the petroleum and coal products sector is necessary to accommodate a $1,000 increase in the demand for fabricated metal products?

b. Which sector requires the most of the product of some other sector in order to meet a $1 increase in external demand for that product?

Communication and Reasoning Exercises

35. What would it mean if the technology matrix A were the zero matrix?

36. Can an external demand be met by an economy whose technology matrix A is the identity matrix? Explain.

37. ▼ What would it mean if the total output figure for a particular sector of an input-output table were equal to the sum of the figures in the row for that sector?

38. ▼ What would it mean if the total output figure for a particular sector of an input-output table were less than the sum of the figures in the row for that sector?

39. ▼ What does it mean if an entry in the matrix $(I - A)^{-1}$ is zero?

40. ▼ Why do we expect the diagonal entries in the matrix $(I - A)^{-1}$ to be slightly larger than 1?

41. ▼ Why do we expect the off-diagonal entries of $(I - A)^{-1}$ to be less than 1?

42. ▼ Why do we expect all the entries of $(I - A)^{-1}$ to be nonnegative?

---

[38] Source: *Survey of Current Business,* December 2001, U.S. Department of Commerce.

[39] *Ibid.*

[40] Source: *Australian National Accounts and Input-Output Tables 1978–1979,* Australian Bureau of Statistics.

[41] *Ibid.*

CHAPTER 4 REVIEW

KEY CONCEPTS

 www.WanerMath.com
Go to the Website to find a comprehensive and interactive Web-based summary of Chapter 4.

4.1 Matrix Addition and Scalar Multiplication

$m \times n$ matrix, dimensions, entries [p. 254]
Referring to the entries of a matrix [p. 254]
Matrix equality [p. 255]
Row, column, and square matrices [p. 255]
Addition and subtraction of matrices [p. 256]
Scalar multiplication [p. 258]
Properties of matrix addition and scalar multiplication [p. 259]
The transpose of a matrix [p. 260]
Properties of transposition [p. 261]

4.2 Matrix Multiplication

Multiplying a row by a column [p. 265]
Linear equation as a matrix equation [p. 266]
The product of two matrices: general case [p. 267]

Identity matrix [p. 271]
Properties of matrix addition and multiplication [p. 272]
Properties of transposition and multiplication [p. 273]
A system of linear equations can be written as a single matrix equation [p. 273]

4.3 Matrix Inversion

The inverse of a matrix, singular matrix [p. 280]
Procedure for finding the inverse of a matrix [p. 282]
Formula for the inverse of a 2×2 matrix; determinant of a 2×2 matrix [p. 283]
Using an inverse matrix to solve a system of equations [p. 285]

4.4 Game Theory

Two-person zero-sum game, payoff matrix [p. 291]
A strategy specifies how a player chooses a move [p. 291]
The expected payoff of a game for given mixed strategies R and C [p. 293]
An optimal strategy, according to the minimax criterion, is one that

minimizes the maximum damage your opponent can cause you. [p. 296]
The Fundamental Principle of Game Theory [p. 296]
Procedure for reducing by dominance [p. 297]
Procedure for solving a 2×2 game [p. 297]
The expected value of a game is its expected payoff when the players use their optimal strategies [p. 300]
A strictly determined game is one with a saddle point [p. 302]
Steps to follow in solving a game [p. 302]

4.5 Input-Output Models

An input-output model divides an economy into sectors. The technology matrix records the interactions of these sectors and allows us to relate external demand to the production vector. [p. 310]
Procedure for finding a technology matrix from an input-output table [p. 313]
The entries of $(I - A)^{-1}$ [p. 315]

REVIEW EXERCISES

For Exercises 1–10, let

$$A = \begin{bmatrix} 1 & 2 & 3 \\ 4 & 5 & 6 \end{bmatrix}, \quad B = \begin{bmatrix} 1 & -1 \\ 0 & 1 \end{bmatrix},$$

$$C = \begin{bmatrix} -1 & 0 \\ 1 & 1 \\ 0 & 1 \end{bmatrix}, \quad and \quad D = \begin{bmatrix} -3 & -2 & -1 \\ 1 & 2 & 3 \end{bmatrix}.$$

Determine whether each expression is defined. If it is, evaluate it.

1. $A + B$

2. $A - D$

3. $2A^T + C$

4. AB

5. $A^T B$

6. A^2

7. B^2

8. B^3

9. $AC + B$

10. $CD + B$

In Exercises 11–16, find the inverse of the given matrix, or determine that the matrix is singular.

11. $\begin{bmatrix} 1 & -1 \\ 0 & 1 \end{bmatrix}$

12. $\begin{bmatrix} 1 & 2 \\ 0 & 0 \end{bmatrix}$

13. $\begin{bmatrix} 1 & 2 & 3 \\ 0 & 4 & 1 \\ 0 & 0 & 1 \end{bmatrix}$

14. $\begin{bmatrix} 1 & 2 & 3 & 4 \\ 1 & 3 & 4 & 2 \\ 0 & 1 & 2 & 3 \\ 0 & 0 & 1 & 2 \end{bmatrix}$

15. $\begin{bmatrix} 1 & 2 & 3 & 4 \\ 2 & 3 & 3 & 3 \\ 0 & 1 & 2 & 3 \\ 0 & 0 & 1 & 2 \end{bmatrix}$

16. $\begin{bmatrix} 0 & 1 & 0 & 0 \\ 1 & 0 & 0 & 0 \\ 0 & 0 & 0 & 1 \\ 0 & 0 & 1 & 0 \end{bmatrix}$

In Exercises 17–20, write the given system of linear equations as a matrix equation, and solve by inverting the coefficient matrix.

17. $x + 2y = 0$
$3x + 4y = 2$

18. $x + y + z = 3$
$y + 2z = 4$
$y - z = 1$

19. $x + y + z = 2$
$x + 2y + z = 3$
$x + y + 2z = 1$

20. $x + y = 0$
$y + z = 1$
$z + w = 0$
$x - w = 3$

In Exercises 21–24, solve the game with the given payoff matrix, and give the expected value of the game.

21. $P = \begin{bmatrix} 2 & 1 & 3 & 2 \\ -1 & 0 & -2 & 1 \\ 2 & 0 & 1 & 3 \end{bmatrix}$
22. $P = \begin{bmatrix} 3 & -3 & -2 \\ -1 & 3 & 0 \\ 2 & 2 & 1 \end{bmatrix}$

23. $P = \begin{bmatrix} -1 & -3 & -2 \\ -1 & 3 & 0 \\ 3 & 3 & -1 \end{bmatrix}$

24. $P = \begin{bmatrix} 1 & 4 & 3 & 3 \\ 0 & -1 & 2 & 3 \\ 2 & 0 & -1 & 2 \end{bmatrix}$

In Exercises 25–28, find the production vector X corresponding to the given technology matrix A and external demand vector D.

25. $A = \begin{bmatrix} 0.3 & 0.1 \\ 0 & 0.3 \end{bmatrix}, D = \begin{bmatrix} 700 \\ 490 \end{bmatrix}$

26. $A = \begin{bmatrix} 0.7 & 0.1 \\ 0.1 & 0.7 \end{bmatrix}, D = \begin{bmatrix} 1,000 \\ 2,000 \end{bmatrix}$

27. $A = \begin{bmatrix} 0.2 & 0.2 & 0.2 \\ 0 & 0.2 & 0.2 \\ 0 & 0 & 0.2 \end{bmatrix}, D = \begin{bmatrix} 32,000 \\ 16,000 \\ 8,000 \end{bmatrix}$

28. $A = \begin{bmatrix} 0.5 & 0.1 & 0 \\ 0.1 & 0.5 & 0.1 \\ 0 & 0.1 & 0.5 \end{bmatrix}, D = \begin{bmatrix} 23,000 \\ 46,000 \\ 23,000 \end{bmatrix}$

Applications: OHaganBooks.com
[Try the game at www.OHaganBooks.com]

It is now July 1, and online sales of romance, science fiction, and horror novels at OHaganBooks.com were disappointingly slow over the past month. Exercises 29–34 are based on the following tables.

Inventory of books in stock on June 1 at the OHaganBooks .com warehouses in Texas and Nevada:

Books in Stock (June 1)

	Romance	Sci Fi	Horror
Texas	2,500	4,000	3,000
Nevada	1,500	3,000	1,000

Online sales during June:

June Sales

	Romance	Sci Fi	Horror
Texas	300	500	100
Nevada	100	600	200

New books purchased each month:

Monthly Purchases

	Romance	Sci Fi	Horror
Texas	400	400	300
Nevada	200	400	300

July Sales (Projected)

	Romance	Sci Fi	Horror
Texas	280	550	100
Nevada	50	500	120

29. *Inventory* Use matrix algebra to compute the inventory at each warehouse at the end of June.

30. *Inventory* Use matrix algebra to compute the change in inventory at each warehouse during June.

31. *Inventory* Assuming that sales continue at the level projected for July for the next few months, write down a matrix equation showing the inventory N at each warehouse x months after July 1. How many months from now will OHaganBooks.com run out of Sci Fi novels at the Nevada warehouse?

32. *Inventory* Assuming that sales continue at the level projected for July for the next few months, write down a matrix equation showing the change in inventory N at each warehouse x months after July 1. How many months from now will OHaganBooks.com have 1,000 more horror novels in stock in Texas than currently?

33. *Revenue* It is now the end of July, and OHaganBooks .com's e-commerce manager bursts into the CEO's office. "I thought you might want to know, John, that our sales figures are exactly what I projected a month ago. Is that good market analysis or what?" OHaganBooks.com has charged an average of $5 for romance novels, $6 for science fiction novels, and $5.50 for horror novels. Use the projected July sales figures from above and matrix arithmetic to compute the total revenue OHaganBooks.com earned at each warehouse in July.

34. *Cost* OHaganBooks.com pays an average of $2 for romance novels, $3.50 for science fiction novels, and $1.50 for horror novels. Use this information together with the monthly purchasing information to compute the monthly purchasing cost.

Acting on a "tip" from Marjory Duffin, John O'Hagan decided that his company should invest a significant sum in shares of Duffin House Publishers (DHP) and Duffin Subprime Ventures (DSV). Exercises 35–38 are based on the following table, which shows what information John was able to piece together later, after some of the records had been deleted by an angry student intern.

Date	Number of Shares: DHP	Price per Share: DHP	Number of Shares: DSV	Price per Share: DSV
July 1	?	$20	?	$10
August 1	?	$10	?	$20
September 1	?	$5	?	$40
Total	5,000		7,000	

35. **Investments** Over the 3 months shown, the company invested a total of $50,000 in DHP stock and, on August 15, was paid dividends of 10¢ per share held on that date, for a total of $300. Use matrix inversion to determine how many shares of DHP OHaganBooks.com purchased on each of the three dates shown.

36. **Investments** Over the 3 months shown, the company invested a total of $150,000 in DSV stock and, on July 15, was paid dividends of 20¢ per share held on that date, for a total of $600. Use matrix inversion to determine how many shares of DSV OHaganBooks.com purchased on each of the three dates shown.

37. **Investments** (Refer to Exercise 35.) On October 1 the shares of DHP purchased on July 1 were sold at $3 per share. The remaining shares were sold 1 month later at $1 per share. Use matrix algebra to determine the total loss (taking into account the dividends paid on August 15) incurred as a result of the Duffin stock debacle.

38. **Investments** (Refer to Exercise 36.) On September 15, DSV announced a two-for-one stock split, so each share originally purchased was converted into two shares. On October 1 the company paid an additional special dividend of 10¢ per share. On October 15, OHaganBooks.com sold 3,000 shares at $20 per share. One week later the subprime market crashed, *Duffin Subprime Ventures* declared bankruptcy (after awarding its fund manager a $10 million bonus), and DSV stock became worthless. Use matrix algebra to determine the total loss (taking into account the dividends paid on July 15) incurred as a result of the Duffin stock debacle.

OHaganBooks.com has two main competitors—JungleBooks.com and FarmerBooks.com—and no other competitors of any significance on the horizon. Exercises 39–42 are based on the following table, which shows the movement of customers during July.[42] (Thus, for instance, the first row tells us that

80% of OHaganBooks.com's customers remained loyal, 10% of them went to JungleBooks.com, and the remaining 10% went to FarmerBooks.com.)

	To OHagan	To Jungle	To Farmer
From OHagan	0.8	0.1	0.1
From Jungle	0.4	0.6	0
From Farmer	0.2	0	0.8

At the beginning of July, OHaganBooks.com had an estimated 2,000 customers, while its two competitors had 4,000 each.

39. **Competition** Set up the July 1 customer numbers in a row matrix, and use matrix arithmetic to estimate the number of customers each company has at the end of July.

40. **Competition** Assuming that the July trends continue in August, predict the number of customers each company will have at the end of August.

41. **Competition** Assuming that the July trends continue in August, why is it not possible for a customer of FarmerBooks.com on July 1 to have ended up as a JungleBooks.com customer 2 months later without having ever been an OHaganBooks.com customer?

42. **Competition** Name one or more important factors that the model we have used does not take into account.

Publisher Marjory Duffin reveals that JungleBooks.com may be launching a promotional scheme in which it will offer either two books for the price of one, or three books for the price of two. (Marjory can't quite seem to remember which and is not certain whether they will go with the scheme at all.) John O'Hagan's marketing advisers Flood and O'Lara seem to have different ideas as to how to respond. Flood suggests that the company counter by offering three books for the price of one, while O'Lara suggests that it offer instead a free copy of the Finite Mathematics Student Solutions Manual with every purchase. After a careful analysis, O'Hagan comes up with the following payoff matrix, where the payoffs represent the number of customers, in thousands, he expects to gain from JungleBooks.com:

	JungleBooks No Promo	2 for Price of 1	3 for Price of 2
No Promo	0	−60	−40
O'Hagan 3 for Price of 1	30	20	10
Finite Math	20	0	15

Use the above information in Exercises 43–48.

43. **Competition** After a very expensive dinner at an exclusive restaurant, Marjory suddenly "remembers" that the JungleBooks.com CEO mentioned to her (at a less expensive restaurant) that there is only a 20% chance that JungleBooks.com will launch a "2 for the price of 1" promotion and a 40% chance that it will launch a "3 for the price of 2" promotion. What should OHaganBooks.com do in view of this information, and what will the expected effect be on its customer base?

---

[42] By a "customer" of one of the three e-commerce sites, we mean someone who purchases more at that site than at either of the two competitors.

44. *Competition* JungleBooks.com CEO François Dubois has been told by someone with personal ties to OHaganBooks.com staff that OHaganBooks.com is in fact 80% certain to opt for the "3 for the price of 1" option and will certainly go with one of the two possible promos. What should JungleBooks.com do in view of this information, and what will the expected effect be on its customer base?

45. *Corporate Spies* One of John O'Hagan's trusted marketing advisers has, without knowing it, accidentally sent him a copy of the following e-mail:

> To: René, JungleBooks.com Marketing Department
>
> From: O'Lara
>
> Hey René somehow the CEO here says he has learned that there is a 20% chance that you will opt for the "2 for the price of 1" promotion, and a 40% chance that you will opt for the "3 for the price of 2" promotion. Thought you might want to know. So when do I get my "commission"?—Jim

What will each company do in view of this new information, and what will the expected effect be on its customer base?

46. *More Corporate Spies* The next day, everything changes: OHaganBooks.com's mole at JungleBooks.com, Davíde DuPont, is found unconscious in the coffee room at JungleBooks.com headquarters, clutching in his hand the following correspondence, which he had apparently received moments earlier:

> To: Davíde
> From: John O'Hagan
> Subject: Re: Urgent Information
> Davíde: This information is much appreciated and definitely changes my plans—J
> >
> >To: John O
> >From: Davíde
> >Subject: Urgent Information
> >
> >Thought you might want to know that JungleBooks.com
> >thinks you are 80% likely to opt for 3 for 1 and 20%
> >likely to opt for Finite Math.
> >
> >—D

What will each company do in view of this information, and what will the expected effect be on its customer base?

47. *Things Unravel* John O'Hagan is about to go with the option chosen in Exercise 45 when he hears word about an exposé in the *Publisher Enquirer* on corporate spying in the two companies. Each company now knows that no information about the other's intentions can be trusted. Now what should OHaganBooks.com do, and how many customers should it expect to gain or lose?

48. *Competition* As a result of the *Publisher Enquirer* exposé it is now apparent that, not only can each company make no assumptions about the strategies the other might be using, but the payoff matrix they have been using is wrong. A crack investigative reporter at the *Enquirer* publishes the following revised matrix:

$$P = \begin{bmatrix} 0 & -60 & -40 \\ 30 & 20 & 10 \\ 20 & 0 & 20 \end{bmatrix}.$$

Now what should JungleBooks.com do, and how many customers should it expect to gain or lose?

Some of the books sold by OHaganBooks.com are printed at Bruno Mills, Inc., a combined paper mill and printing company. Exercises 49–52 are based on the following typical monthly input-output table for Bruno Mills's paper and book printing sectors.

		To	
		Paper	**Books**
From	**Paper**	$20,000	$50,000
	Books	2,000	5,000
	Total Output	200,000	100,000

49. *Production* Find the technology matrix for Bruno Mills's paper and book printing sectors.

50. *Production* Compute $(I - A)^{-1}$. What is the significance of the $(1, 2)$ entry?

51. *Production* Approximately $1,700 worth of the books sold each month by OHaganBooks.com are printed at Bruno Mills, and OHaganBooks.com uses approximately $170 worth of Bruno Mills's paper products each month. What is the total value of paper and books that must be produced by Bruno Mills to meet demand from OHaganBooks.com?

52. *Production* Currently, Bruno Mills has a monthly capacity of $500,000 of paper products and $200,000 of books. What level of external demand would cause Bruno Mills to meet the capacity for both products?

Projecting Market Share

You are the sales director at *Selular,* a cellphone provider, and things are not looking good for your company: *iClone,* a recently launched competitor, is beginning to chip away at Selular's market share. Particularly disturbing are rumors of fierce brand loyalty by iClone customers, with several bloggers suggesting that iClone retains close to 100% of their customers. Worse, you will shortly be presenting a sales report to the board of directors, and the CEO has "suggested" that your report include 2-, 5-, and 10-year projections of Selular's market share given the recent impact on the market by iClone. The CEO also wants a worst-case scenario projecting what would happen if iClone customers were so loyal that none of them ever switch services.

The sales department has conducted two market surveys, taken one quarter apart, of the major cellphone providers, which are *iClone, Selular, AB&C,* and some smaller companies lumped together as "Other," and has given you the data shown in Figure 4, which shows the percentages of subscribers who switched from one service to another during the quarter.*

* If you go on to study probability theory in Chapter 7, you will see how this scenario can be interpreted as a Markov system, and you will revisit the analysis below in that context. The percentages in the figure and market shares below reflect actual data for several cellphone services in the United States during a single quarter of 2003. (See the exercises for Section 7.7.)

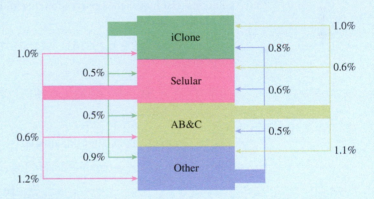

Figure 4

For example, by the end of the quarter, 0.5% of iClone's customers had switched to Selular, 0.5% to AB&C, and 0.9% to Other. The current market shares are as follows: iClone: 29.7%, Selular: 19.3%, AB&C: 18.1%, Other: 32.9%.

"This is simple," you tell yourself. "Since I know the current market shares and percentage movements over one quarter, I can easily calculate the market shares next quarter (assuming that the percentages that switch services remain the same), then repeat the calculation for the following quarter, and so on until I get the long-term prediction I am seeking." So you begin your calculations by computing the market shares next quarter. First you note that, since a total of $0.5 + 0.5 + 0.9 = 1.9\%$ of iClone users switched to other brands, the rest, 98.1%, stayed with iClone. Similarly, 97.2% of Selular users, 97.3% of AB&C, and 98.1% of Other stayed with their respective brands. Then you compute

iClone share after one quarter

$= 98.1\%$ of iClone $+ 1.0\%$ of Selular $+ 1.0\%$ of AB&C $+ 0.8\%$ of Other

$= (0.981)(0.297) + (0.010)(0.193) + (0.010)(0.181) + (0.008)(0.329)$

$= 0.297729$, or 29.7729%.

Similarly,

Selular share:

$$= (0.005)(0.297) + (0.972)(0.193) + (0.006)(0.181) + (0.006)(0.329)$$
$$= 0.192141$$

AB&C share:

$$= (0.005)(0.297) + (0.006)(0.193) + (0.973)(0.181) + (0.005)(0.329)$$
$$= 0.180401$$

Other share:

$$= (0.009)(0.297) + (0.012)(0.193) + (0.011)(0.181) + (0.981)(0.329)$$
$$= 0.329729.$$

So you project the market shares after one quarter to be iClone: 29.7729%, Selular: 19.2141%, AB&C: 18.0401%, Other: 32.9729%. You now begin to realize that repeating this kind of calculation for the large number of quarters required for long-term projections will be tedious. You call in your student intern (who happens to be a mathematics major) to find out whether she can help. After taking one look at the calculations, she makes the observation that all you have really done is compute the product of two matrices:

$$[0.297729 \quad 0.192141 \quad 0.180401 \quad 0.329729]$$

$$= [0.297 \quad 0.193 \quad 0.181 \quad 0.329] \begin{bmatrix} .981 & .005 & .005 & .009 \\ .010 & .972 & .006 & .012 \\ .010 & .006 & .973 & .011 \\ .008 & .006 & .005 & .981 \end{bmatrix}$$

Market shares after one quarter = Market shares at start of quarter $\times$ A.

The 4×4 matrix A is organized as follows:

		To			
		iClone	Selular	AB&C	Other
	iClone	.981	.005	.005	.009
From	**Selular**	.010	.972	.006	.012
	AB&C	.010	.006	.973	.011
	Other	.008	.006	.005	.981

Since the market shares one quarter later can be obtained from the shares at the start of the quarter, you realize that you can now obtain the shares *two* quarters later by multiplying the result by A, and at this point, you start using technology (such as the Matrix Algebra Tool at www.WanerMath.com) to continue the calculation:

Market shares after two quarters = Market shares after one quarter $\times$ A

$$= [0.297729 \quad 0.192141 \quad 0.180401 \quad 0.329729] \begin{bmatrix} .981 & .005 & .005 & .009 \\ .010 & .972 & .006 & .012 \\ .010 & .006 & .973 & .011 \\ .008 & .006 & .005 & .981 \end{bmatrix}$$

$$\approx [0.298435 \quad 0.191310 \quad 0.179820 \quad 0.330434].$$

Although the use of matrices has simplified your work, multiplying the result by A over and over again to get the market shares for successive months is still tedious. Would it not be possible to get, say, the market share after 10 years (40 quarters) with a single calculation? To explore this, you decide to use symbols for the various market shares:

$$m_0 = \text{Starting market shares} = [0.297 \quad 0.193 \quad 0.181 \quad 0.329]$$

$$m_1 = \text{Market shares after 1 quarter}$$

$$= [0.297729 \quad 0.192141 \quad 0.180401 \quad 0.329729]$$

$$m_2 = \text{Market shares after 2 quarters}$$
$$\vdots$$
$$m_n = \text{Market share after } n \text{ quarters.}$$

You then rewrite the relationships above as

$$m_1 = m_0 A \quad \text{Shares after one quarter} = \text{Shares at start of quarter} \times A$$

$$m_2 = m_1 A \quad \text{Shares after two quarters} = \text{Shares after one quarter} \times A.$$

On an impulse, you substitute the first equation in the second:

$$m_2 = m_1 A = (m_0 A)A = m_0 A^2.$$

Continuing, you get

$$m_3 = m_2 A = (m_0 A^2)A = m_0 A^3$$
$$\vdots$$
$$m_n = m_0 A^n,$$

which is exactly the formula you need! You can now obtain the 2-, 5-, and 10-year projections each in a single step (with the aid of technology):

2-year projection: $m_8 = m_0 A^8$

$$= [0.297 \quad 0.193 \quad 0.181 \quad 0.329] \begin{bmatrix} .981 & .005 & .005 & .009 \\ .010 & .972 & .006 & .012 \\ .010 & .006 & .973 & .011 \\ .008 & .006 & .005 & .981 \end{bmatrix}^8$$

$$\approx [0.302237 \quad 0.186875 \quad 0.176691 \quad 0.334198]$$

5-year projection: $m_{20} = m_0 A^{20} \approx [0.307992 \quad 0.180290 \quad 0.171938 \quad 0.33978]$

10-year projection: $m_{40} = m_0 A^{40} \approx [0.313857 \quad 0.173809 \quad 0.167074 \quad 0.34526]$.

In particular, Selular's market shares are projected to decline slightly: 2-year projection: 18.7%, 5-year projection: 18.0%, 10-year projection: 17.4%.

You now move on to the worst-case scenario, which you represent by assuming 100% loyalty by iClone users with the remaining percentages staying the same (Figure 5).

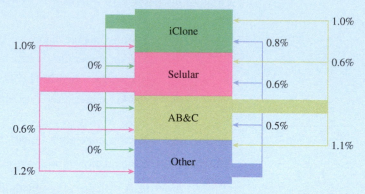

Figure 5

The matrix A corresponding to this diagram is $\begin{bmatrix} 1 & 0 & 0 & 0 \\ .010 & .972 & .006 & .012 \\ .010 & .006 & .973 & .011 \\ .008 & .006 & .005 & .981 \end{bmatrix}$, and you find

2-year projection: $m_8 = m_0 A^8 \approx [0.346335 \quad 0.175352 \quad 0.165200 \quad 0.313113]$

5-year projection: $m_{20} = m_0 A^{20} \approx [0.413747 \quad 0.152940 \quad 0.144776 \quad 0.288536]$

10-year projection: $m_{40} = m_0 A^{40} \approx [0.510719 \quad 0.123553 \quad 0.117449 \quad 0.248279]$.

Thus, in the worst-case scenario, Selular's market shares are projected to decline more rapidly: 2-year projection: 17.5%, 5-year projection: 15.3%, 10-year projection: 12.4%.

EXERCISES

1. Project Selular's market share 20 and 30 years from now based on the original data shown in Figure 4. (Round all figures to the nearest 0.1%.)

2. Using the original data, what can you say about each company's share in 60 and 80 years (assuming that current trends continue)? (Round all figures to the nearest 0.1%.)

3. Obtain a sequence of projections several hundreds of years into the future using the scenarios in both Figure 4 and Figure 5. What do you notice?

4. Compute a sequence of larger and larger powers of the matrix A for both scenarios with all figures rounded to four decimal places. What do you notice?

5. Suppose the trends noted in the market survey were in place for several quarters *before* the current quarter. Using the scenario in Figure 4, determine what the companies' market shares were one quarter before the present and one year before the present (to the nearest 0.1%). Do the same for the scenario in Figure 5.

6. If A is the matrix from the scenario in Figure 4, compute A^{-1}. In light of the preceding exercise, what do the entries in A^{-1} mean?

Section 4.1

Example 2 (page 257) The *A-Plus* auto parts store chain has two outlets, one in Vancouver and one in Quebec. Among other things, it sells wiper blades, windshield cleaning fluid, and floor mats. The monthly sales of these items at the two stores for 2 months are given in the following tables:

January Sales

	Vancouver	Quebec
Wiper Blades	20	15
Cleaning Fluid (bottles)	10	12
Floor Mats	8	4

February Sales

	Vancouver	Quebec
Wiper Blades	23	12
Cleaning Fluid (bottles)	8	12
Floor Mats	4	5

Use matrix arithmetic to calculate the change in sales of each product in each store from January to February.

Solution

On the TI-83/84 Plus, matrices are referred to as [A], [B], and so on through [J]. To enter a matrix, press MATRIX to bring up the matrix menu, select EDIT, select a matrix, and press ENTER. Then enter the dimensions of the matrix followed by its entries. When you want to use a matrix, press MATRIX, select the matrix, and press ENTER.

On the TI-83/84 Plus, adding matrices is similar to adding numbers. The sum of the matrices [A] and [B] is [A] + [B]; their difference, of course, is [A] - [B]. As in the text, for this example,

1. Create two matrices, [J] and [F].

2. Compute their difference, [F] - [J] using [F] - [J] → [D].

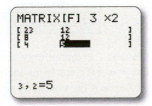

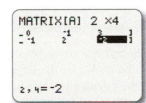

Note that we have stored the difference in the matrix [D] in case we need it for later use.

Section 4.2

Example 3(a) (page 268) Calculate

$$\begin{bmatrix} 2 & 0 & -1 & 3 \\ 1 & -1 & 2 & -2 \end{bmatrix} \begin{bmatrix} 1 & 1 & -8 \\ 1 & 0 & 0 \\ 0 & 5 & 2 \\ -2 & 8 & -1 \end{bmatrix}.$$

Solution

On the TI-83/84 Plus the format for multiplying matrices is the same as for multiplying numbers: [A] [B] or [A] * [B] will give the product. We enter the matrices and then multiply them. (Note that, while editing, you can see only three columns of [A] at a time.)

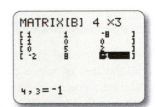

Note that if you try to multiply two matrices whose product is not defined, you will get the error "DIM MISMATCH" (dimension mismatch).

Example 6 (page 271)—Identity Matrix On the TI-83/84 Plus, the function `identity(n)` (in the MATRIX MATH menu) returns the $n \times n$ identity matrix.

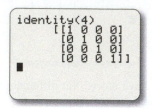

Section 4.3

Example 2(b) (page 281) Find the inverse of

$$Q = \begin{bmatrix} 1 & 0 & 1 \\ 2 & -2 & -1 \\ 3 & 0 & 0 \end{bmatrix}.$$

Solution

On a TI-83/84 Plus, you can invert the square matrix [A] by entering [A] x^{-1} ENTER.

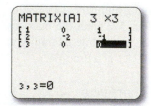
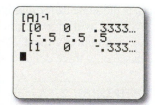

(Note that you can use the right and left arrow keys to scroll the answer, which cannot be shown on the screen all at once.) You could also use the calculator to help you go through the row reduction, as described in Chapter 3.

Example 4 (page 285) Solve the following three systems of equations:

a. $2x \qquad + z = 1$
$2x + y - z = 1$
$3x + y - z = 1$

b. $2x \qquad + z = 0$
$2x + y - z = 1$
$3x + y - z = 2$

c. $2x \qquad + z = 0$
$2x + y - z = 0$
$3x + y - z = 0$

Solution

1. Enter the four matrices A, B, C, and D:

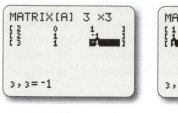
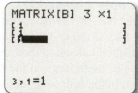

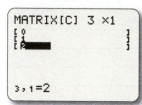
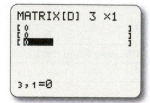

2. Compute the solutions $A^{-1}B$, $A^{-1}C$, and $A^{-1}D$:

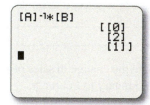
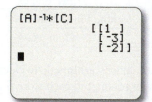

Section 4.5

Example 1 (page 311) Recall that the input-output table in Example 1 looks like this:

		To	
		Crude	**Refining**
From	**Crude**	27,000	59,000
	Refining	750	15,000
	Total Output	87,000	140,000

What was the technology matrix for these two sectors? What was left over from each of these sectors for use by other parts of the economy or for export?

Solution

There are several ways to use these data to create the technology matrix in your TI-83/84 Plus. For small matrices like this the most straightforward is to use the matrix editor, where you can give each entry as the appropriate quotient:

and so on. Once we have the technology matrix [A] and the production vector [B] (remember that we can't use [X] as a matrix name), we can calculate the external demand vector:

$$[B]-[A]*[B]$$
$$[[1000\ \ ]$$
$$[124250]]$$

Of course, when interpreting these numbers, we must remember to round to two significant digits, because our original data were accurate to only that many digits.

Here is an alternative way to calculate the technology matrix that may be better for examples with more sectors:

1. Begin by entering the columns of the input-output table as lists in the list editor ([STAT] EDIT) (see below left).

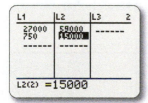   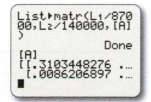

2. We now want to divide each column by the total output of its sector and assemble the results into a matrix. We can do this using the List ▶ matr function (under the [MATRIX] MATH menu) shown on the right above.

Example 3 (page 316)   Consider four sectors of the economy of Kenya: (1) the traditional economy, (2) agriculture, (3) manufacture of metal products and machinery, and (4) wholesale and retail trade. The input-output table for these four sectors for 1976 looks like this (all numbers are thousands of K£):

		To			
		1	**2**	**3**	**4**
From	**1**	8,600	0	0	0
	2	0	20,000	24	0
	3	1,500	530	15,000	660
	4	810	8,500	5,800	2,900
	Total Output	87,000	530,000	110,000	180,000

Suppose that external demand for agriculture increased by K£50,000,000 and that external demand for metal products and machinery increased by K£10,000,000. How would production in these four sectors have to change to meet this rising demand?

Solution

1. Enter the technology matrix A as [A] and D^+ as [D] using one of the techniques above:

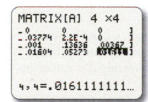   

2. You can then compute the change in production with the formula $\texttt{(identity(4)-[A])}^{-1}\texttt{[D]}$.

$$\texttt{(identity(4)-[A]}$$
$$\texttt{)}^{-1}\texttt{[D]}$$
$$[[0\qquad\quad]$$
$$[51963.42476]$$
$$[11645.36129]$$
$$[1471.104956]]$$

Section 4.1

Example 2 (page 257) The *A-Plus* auto parts store chain has two outlets, one in Vancouver and one in Quebec. Among other things, it sells wiper blades, windshield cleaning fluid, and floor mats. The monthly sales of these items at the two stores for 2 months are given in the following tables:

January Sales

	Vancouver	Quebec
Wiper Blades	20	15
Cleaning Fluid (bottles)	10	12
Floor Mats	8	4

February Sales

	Vancouver	Quebec
Wiper Blades	23	12
Cleaning Fluid (bottles)	8	12
Floor Mats	4	5

Use matrix arithmetic to calculate the change in sales of each product in each store from January to February.

Solution

To enter a matrix in a spreadsheet, we put its entries in any convenient block of cells. For example, the matrix *A* in Quick Example 1 of this section might look like this:

	A	B	C
1	2	0	1
2	33	-22	0

Spreadsheets refer to such blocks of data as **arrays**, which it can handle in much the same way as it handles single cells of data. For instance, in typing a formula, just as clicking on a cell creates a reference to that cell, selecting a whole array of cells will create a reference to that array. An array is referred to using an **array range** consisting of the top left and bottom right cell coordinates, separated by a colon. For example, the array range A1:C2 refers to the 2×3 matrix above, with top left corner A1 and bottom right corner C2.

1. To add or subtract two matrices in a spreadsheet, first input their entries in two separate arrays in the spreadsheet. (We have also added labels as in the previous tables, which you might do if you wanted to save the spreadsheet for later use.)

	A	B	C
1	January Sales		
2		Vancouver	Quebec
3	wiper blades	20	15
4	cleaning fluid (bottles)	10	12
5	floor mats	8	4
6			
7	February Sales		
8		Vancouver	Quebec
9	wiper blades	23	12
10	cleaning fluid (bottles)	8	12
11	floor mats	4	5

2. Select (highlight) a block of the same size (3×2 in this case) where you would like the answer, $F - J$, to appear, enter the formula =B9:C11-B3:C5, and then type Control+Shift+Enter. The easiest way to do this is as follows:

- Highlight cells B15:C17. *Where you want the answer to appear*
- Type "=".
- Highlight the matrix *F*. *Cells B9 through C11*
- Type "-".
- Highlight the matrix *J*. *Cells B3 through C5*
- Press Control+Shift+Enter. *Not just Enter*

	A	B	C
1	January Sales		
2		Vancouver	Quebec
3	wiper blades	20	15
4	cleaning fluid (bottles)	10	12
5	floor mats	8	4
6			
7	February Sales		
8		Vancouver	Quebec
9	wiper blades	23	12
10	cleaning fluid (bottles)	8	12
11	floor mats	4	5
12			
13	Change in Sales		
14		Vancouver	Quebec
15	wiper blades	=B9:C11-B3:C5	
16	cleaning fluid (bottles)		
17	floor mats		

Typing Control+Shift+Enter (instead of Enter) tells the spreadsheet that your formula is an *array formula,* one that returns a matrix rather than a single number.[43] Once entered, the formula bar will show the formula you entered enclosed in "curly braces," indicating that it is an array formula. Note that you must use Control+Shift+Enter to delete any array you create: Select the block you wish to delete and press Delete followed by Control+Shift+Enter.

Section 4.2

Example 3(a) (page 268)   Calculate the product

$$\begin{bmatrix} 2 & 0 & -1 & 3 \\ 1 & -1 & 2 & -2 \end{bmatrix} \begin{bmatrix} 1 & 1 & -8 \\ 1 & 0 & 0 \\ 0 & 5 & 2 \\ -2 & 8 & -1 \end{bmatrix}.$$

Solution

In a spreadsheet the function we use for matrix multiplication is MMULT. (Ordinary multiplication, *, will *not* work.)

1. Enter the two matrices as shown in the spreadsheet, and highlight a block where you want the answer to appear. (Note that it should have the correct dimensions for the product: 2 × 3.)

2. Enter the formula =MMULT(A1:D2,F1:H4) (using the mouse to avoid typing the array ranges if you like), and press Control+Shift+Enter. The product will appear in the region you highlighted. If you try to multiply two matrices whose product is not defined, you will get the error "#VALUE!".

Example 6—Identity Matrix (page 271)   There is no spreadsheet function that returns an identity matrix. If you need a small identity matrix, it's simplest to just enter

the 1s and 0s by hand. If you need a large identity matrix, here is one way to get it quickly.

1. Say we want a 4 × 4 identity matrix in the cells B1:E4. Enter the following formula in cell B1:

```
=IF(ROW(B1)-ROW($B$1)
=COLUMN(B1)-COLUMN($B$1),1,0)
```

	A	B	C	D	E
1		=IF(ROW(B1)-ROW($B$1)=COLUMN(B1)-COLUMN($B$1),1,0)			
2					
3					
4					

2. Press Enter, then copy cell B1 to cells B1:E4. The formula will return 1s along the diagonal of the matrix and 0s elsewhere, giving you the identity matrix. Why does this formula work?

	A	B	C	D	E
1		1	0	0	0
2		0	1	0	0
3		0	0	1	0
4		0	0	0	1

Section 4.3

Example 2(b) (page 281)   Find the inverse of

$$Q = \begin{bmatrix} 1 & 0 & 1 \\ 2 & -2 & -1 \\ 3 & 0 & 0 \end{bmatrix}.$$

Solution

In a spreadsheet the function MINVERSE computes the inverse of a matrix.

1. Enter Q somewhere convenient, for example, in cells A1:C3.

2. Choose the block where you would like the inverse to appear, highlight the whole block.

3. Enter the formula =MINVERSE(A1:C3) and press Control+Shift+Enter.

The inverse will appear in the region you highlighted. (To convert the answer to fractions, format the cells as fractions.)

---

[43] Note that on a Mac, Command+Enter has the same effect as Control+Shift+Enter.

If a matrix is singular, a spreadsheet will register an error by showing #NUM! in each cell.

	A	B	C	D	E	F	G
1	1	0	1		0	0	0.3333333
2	2	-2	-1		-0.5	-0.5	0.5
3	3	0	0		1	0	-0.3333333

Although the spreadsheet appears to invert the matrix in one step, it is going through the procedure in the text or some variation of it to find the inverse. Of course, you could also use the spreadsheet to help you go through the row reduction, just as in Chapter 3.

Example 4 (page 285) Solve the following three systems of equations:

a. $2x \quad + z = 1$
$2x + y - z = 1$
$3x + y - z = 1$

b. $2x \quad + z = 0$
$2x + y - z = 1$
$3x + y - z = 2$

c. $2x \quad + z = 0$
$2x + y - z = 0$
$3x + y - z = 0$

Solution

Spreadsheets instantly update calculated results every time the contents of a cell are changed. We can take advantage of this to solve the three systems of equations given above using the same worksheet as follows:

1. Enter the matrices A and B from the matrix equation $AX = B$.

2. Select a 3×1 block of cells for the matrix X.

3. The Excel formula that we can use to calculate X is

$$\texttt{=MMULT(MINVERSE(A1:C3),E1:E3)} \qquad A^{-1}B$$

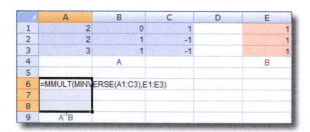

(As usual, use the mouse to select the ranges for A and B while typing the formula, and don't forget to press Control+Shift+Enter.) Having obtained the solution to part (a), you can now simply modify the entries in column E to see the solutions for parts (b) and (c).

Note Your spreadsheet for part (a) may look like this:

	A	B	C	D	E
1	2	0	1		1
2	2	1	-1		1
3	3	1	-1		1
4		A			B
5					
6	1.11022E-16				
7	2				
8	1				
9	A⁻¹B				

What is that strange number doing in cell A6? "E-16" represents "$\times 10^{-16}$", so the entry is really

$$1.11022 \times 10^{-16} = 0.000\,000\,000\,000\,000\,111022 \approx 0.$$

Mathematically, it is supposed to be *exactly* zero (see the solution to part (a) in the text), but Excel made a small error in computing the inverse of A, resulting in this spurious value. Note, however, that it is accurate (agrees with zero) to 15 decimal places! In practice, when we see numbers arise in matrix calculations that are far smaller than all the other entries, we can often assume that they are supposed to be zero. ■

Section 4.5

Example 1 (page 311) Recall that the input-output table in Example 1 looks like this:

		To	
		Crude	**Refining**
From	**Crude**	27,000	59,000
	Refining	750	15,000
	Total Output	87,000	140,000

What was the technology matrix for these two sectors? What was left over from each of these sectors for use by other parts of the economy or for export?

Solution

1. Enter the input-output table in a spreadsheet:

	A	B	C	D
1			To	
2			Crude	Refining
3	From	Crude	27000	59000
4		Refining	750	15000
5		Total Output	87000	140000

2. To obtain the technology matrix, we divide each column by the total output of its sector:

	A	B	C	D
1			To	
2			Crude	Refining
3	From	Crude	27000	59000
4		Refining	750	15000
5		Total Output	87000	140000
6				
7			=C3/C$5	
8				

	A	B	C	D
1			To	
2			Crude	Refining
3	From	Crude	27000	59000
4		Refining	750	15000
5		Total Output	87000	140000
6				
7			0.3103448	0.4214286
8			0.0086207	0.1071429

The formula =C3/C$5 is copied into the shaded 2 × 2 block shown above. (The $ sign in front of the 5 forces the program to always divide by the total in Row 5 even when the formula is copied from Row 7 to Row 8.) The result is the technology matrix shown in the bottom screenshot above.

3. Using the techniques discussed in Section 4.2, we can now compute $D = X - AX$ to find the demand vector.

Example 3 (page 316) Consider four sectors of the economy of Kenya: (1) the traditional economy, (2) agriculture, (3) manufacture of metal products and machinery, and (4) wholesale and retail trade. The input-output table for these four sectors for 1976 looks like this (all numbers are thousands of K£):

		To			
		1	2	3	4
From	1	8,600	0	0	0
	2	0	20,000	24	0
	3	1,500	530	15,000	660
	4	810	8,500	5,800	2,900
	Total Output	87,000	530,000	110,000	180,000

Suppose that external demand for agriculture increased by K£50,000,000 and that external demand for metal products and machinery increased by K£10,000,000.

How would production in these four sectors have to change to meet this rising demand?

Solution

1. Enter the input-output table in the spreadsheet.

2. Compute the technology matrix by dividing each column by the column total.

3. Insert the identity matrix I in preparation for the next step.

4. To see how each sector reacts to rising external demand, you must calculate the inverse matrix $(I - A)^{-1}$, as shown below. (Remember to use Control+Shift+Enter each time.)

5. To compute $(I - A)^{-1}D^{+}$, enter D^{+} as a column, and use the MMULT operation. (See Example 3 in Section 4.2.)

	A	B	C	D	E	F	G	H	I
13		(I − A)					(I − A)⁻¹		
14	0.9011494	0	0	0		1.1096939	0	0	0
15	0	0.9622642	-0.0002182	0		5.034E-06	1.039216	0.0002626	9.786E-07
16	-0.0172414	-0.001	0.8636364	-0.0036667		0.0222032	0.0012755	1.1581566	0.0043181
17	-0.0093103	-0.0160377	-0.0527273	0.9838889		0.0116908	0.0170079	0.0620708	1.0166062
18									
19	D		(I − A)⁻¹D						
20	0		0						
21	50000		51963.425						
22	10000		11845.361						
23	0		1471.105						

Note Here is one of the beauties of spreadsheet programs: Once you are done with the calculation, you can use the spreadsheet as a template for any 4 × 4 input-output table by just changing the entries of the input-output matrix and/or D^{+}. The rest of the computation will then be done automatically as the spreadsheet is updated. In other words, you can use it to do your homework! ∎

5

LINEAR PROGRAMMING

CASE STUDY

The Diet Problem

The *Galaxy Nutrition* health-food mega-store chain provides free online nutritional advice and support to its customers. As website technical consultant, you are planning to construct an interactive web page to assist customers in preparing a diet tailored to their nutritional and budgetary requirements. Ideally, the customer would select foods to consider and specify nutritional and/or budgetary constraints, and the tool should return the optimal diet meeting those requirements. You would also like the web page to allow the customer to decide whether, for instance, to find the cheapest possible diet meeting the requirements, the diet with the lowest number of calories, or the diet with the least total carbohydrates.

How do you go about constructing such a web page?

Image Studios/UpperCut Images/Getty Images

Introduction

In this chapter we begin to look at one of the most important types of problems for business and the sciences: finding the largest or smallest possible value of some quantity (such as profit or cost) under certain constraints (such as limited resources). We call such problems **optimization** problems because we are trying to find the best, or optimum, value. The optimization problems we look at in this chapter involve linear functions only and are known as **linear programming** (LP) problems. One of the main purposes of calculus, which you may study later, is to solve nonlinear optimization problems.

Linear programming problems involving only two unknowns can usually be solved by a graphical method that we discuss in Sections 5.1 and 5.2. When there are three or more unknowns, we must use an algebraic method, as we had to do for systems of linear equations. The method we use is called the **simplex method**. Invented in 1947 by George B. Dantzig* (1914–2005), the simplex method is still the most commonly used technique to solve LP problems in real applications, from finance to the computation of trajectories for guided missiles.

The simplex method can be used for hand calculations when the numbers are fairly small and the unknowns are few. Practical problems often involve large numbers and many unknowns, however. Problems such as routing telephone calls or airplane flights or allocating resources in a manufacturing process can involve tens of thousands of unknowns. Solving such problems by hand is obviously impractical, so computers are regularly used. Although computer programs most often use the simplex method, mathematicians are always seeking faster methods. The first radically different method of solving LP problems was the **ellipsoid algorithm** published in 1979 by the Soviet mathematician Leonid G. Khachiyan[2] (1952–2005). In 1984, Narendra Karmarkar (1957–), a researcher at Bell Labs, created a more efficient method, now known as **Karmarkar's algorithm**. Although these methods (and others since developed) can be shown to be faster than the simplex method in the worst cases, it seems to be true that the simplex method is still the fastest in the applications that arise in practice.

Calculators and spreadsheets are very useful aids in the simplex method. In practice, software packages do most of the work, so you can think of what we teach you here as a peek inside a "black box." What the software cannot do for you is convert a real situation into a mathematical problem, so the most important lessons to get out of this chapter are (1) how to recognize and set up a linear programming problem and (2) how to interpret the results.

* Dantzig is the real-life source of the story of the student who, walking in late to a math class, copies down two problems on the board, thinking they're homework. After much hard work he hands in the solutions, only to discover that he has just solved two famous unsolved problems. This actually happened to Dantzig in graduate school in 1939.[1]

5.1 Graphing Linear Inequalities

Inequalities

By the end of the next section we will be solving linear programming (LP) problems with two unknowns. We use inequalities to describe the *constraints* in such problems, so we start by reviewing some basic notation for inequalities.

---

[1] Sources: D. J. Albers, and C. Reid, "An Interview of George B. Dantzig: The Father of Linear Programming," *College Mathematics Journal,* v. 17 (1986), pp. 293–314. The article is quoted and discussed in the context of the urban legends it inspired at www.snopes.com/college/homework/unsolvable.asp.

[2] Dantzig and Khachiyan died approximately two weeks apart in 2005. The *New York Times* ran their obituaries together on May 23, 2005.

Strict Inequalities	Quick Examples
$a < b$ means that a **is less than** b.	$3 < 99$, $\;-2 < -1$, $\;0 < 3$
$a > b$ means that a **is greater than** b.	$4 > 3$, $\;1.78 > 1.76$, $\;\dfrac{1}{3} > \dfrac{1}{4}$
Non-Strict Inequalities[*]	Quick Examples
$a \leq b$ means that a **is less than or equal to** b.	$3 \leq 99$, $\;-2 \leq -2$, $\;0 \leq 3$
$a \geq b$ means that a **is greater than or equal to** b.	$3 \geq 3$, $\;1.78 \geq 1.76$, $\;\dfrac{1}{3} \geq \dfrac{1}{4}$

[*] In this chapter we focus on the non-strict inequalities.

Following are some of the basic rules for manipulating inequalities. Although we illustrate all of them with the inequality $\leq$, they apply equally well to inequalities with $\geq$ and to the strict inequalities $<$ and $>$.

Rules for Manipulating Inequalities	Quick Examples
1. The same quantity can be added to or subtracted from both sides of an inequality: If $x \leq y$, then $x + a \leq y + a$ for any real number a.	$x \leq y$ implies $x - 4 \leq y - 4$
2. Both sides of an inequality can be multiplied or divided by a positive constant: If $x \leq y$ and a is positive, then $ax \leq ay$.	$x \leq y$ implies $3x \leq 3y$
3. Both sides of an inequality can be multiplied or divided by a negative constant if the inequality is *reversed*: If $x \leq y$ and a is negative, then $ax \geq ay$.	$x \leq y$ implies $-3x \geq -3y$
4. Switching the left and right sides *reverses* the inequality: If $x \leq y$, then $y \geq x$; if $y \geq x$, then $x \leq y$.	$3x \geq 5y$ implies $5y \leq 3x$

Here are the particular kinds of inequalities in which we're interested, again stated in terms of non-strict inequalities.

Linear Inequalities and Solving Inequalities

An **inequality in the unknown x** is the statement that one expression involving x is less than or equal to (or greater than or equal to) another. Similarly, we can have an **inequality in x and y**, which involves expressions that contain x and y; an **inequality in x, y, and z**; and so on. A **linear inequality** in one or more unknowns is an inequality of the form

$$ax \leq b \quad \text{or} \quad ax \geq b \qquad (a \text{ and } b \text{ real constants})$$
$$ax + by \leq c \quad \text{or} \quad ax + by \geq c \qquad (a, b, \text{ and } c \text{ real constants})$$
$$ax + by + cz \leq d \qquad (a, b, c, \text{ and } d \text{ real constants})$$
$$ax + by + cz + dw \leq e \qquad (a, b, c, d, \text{ and } e \text{ real constants})$$

and so on.

Quick Examples

1. $2x + 8 \geq 89$ Linear inequality in x

2. $2x^3 \leq x^3 + y$ Nonlinear inequality in x and y

3. $3x - 2y \geq 8$ Linear inequality in x and y

4. $x^2 + y^2 \leq 19z$ Nonlinear inequality in x, y, and z

5. $3x - 2y + 4z \leq 0$ Linear inequality in x, y, and z

A **solution** of an inequality in the unknown x is a value for x that makes the inequality true. For example, $2x + 8 \geq 89$ has a solution $x = 50$ because $2(50) + 8 \geq 89$. Of course, it has many other solutions as well. Similarly, a solution of an inequality in x and y is a pair of values (x, y) making the inequality true. For example, $(5, 1)$ is a solution of $3x - 2y \geq 8$ because $3(5) - 2(1) \geq 8$. To **solve** an inequality is to find the set of *all* solutions.

Solving Linear Inequalities in Two Variables

Our first goal is to solve linear inequalities in two variables—that is, inequalities of the form $ax + by \leq c$. As an example, let's solve

$$2x + 3y \leq 6.$$

We already know how to solve the *equation* $2x + 3y = 6$. As we saw in Chapter 1, the solution of this equation may be pictured as the set of all points (x, y) on the straight-line graph of the equation. This straight line has x-intercept 3 (obtained by putting $y = 0$ in the equation) and y-intercept 2 (obtained by putting $x = 0$ in the equation) and is shown in Figure 1.

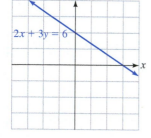

Notice that, if (x, y) is any point on the line, then x and y not only satisfy the *equation* $2x + 3y = 6$, but also satisfy the *inequality* $2x + 3y \leq 6$, because being equal to 6 qualifies as being less than or equal to 6.

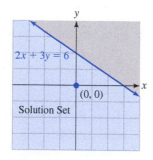

Figure 1

Figure 2

Solution Set

Q: *Do the points on the line give all possible solutions to the inequality?*

A: No. For example, try the origin, $(0, 0)$. Because $2(0) + 3(0) = 0 \leq 6$, the point $(0, 0)$ is a solution that does not lie on the line. In fact, here is a possibly surprising fact: The solution to any linear inequality in two unknowns is represented by an entire **half plane**: the set of all points on one side of the line (including the line itself). Thus, because $(0, 0)$ is a solution of $2x + 3y \leq 6$ and is not on the line, every point on the same side of the line as $(0, 0)$ is a solution as well. (The colored region below the line in Figure 2 shows which half plane constitutes the solution set.)

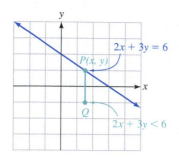

Figure 3

To see why the solution set of $2x + 3y \leq 6$ is the entire half plane shown in Figure 2, start with any point P on the line $2x + 3y = 6$. We already know that P is a solution of $2x + 3y \leq 6$. If we choose any point Q directly below P, the x-coordinate of Q will be the same as that of P, and the y-coordinate will be smaller. So the value of $2x + 3y$ at Q will be smaller than the value at P, which is 6. Thus, $2x + 3y < 6$ at Q, so Q is another solution of the inequality. (See Figure 3.) In other words, *every point beneath the line is a solution of* $2x + 3y \leq 6$. On the other hand, any point above the line is directly above a point on the line, so $2x + 3y > 6$ for such a point. Thus, *no point above the line is a solution of* $2x + 3y \leq 6$.

The same kind of argument can be used to show that the solution set of every inequality of the form $ax + by \leq c$ or $ax + by \geq c$ consists of the half plane above

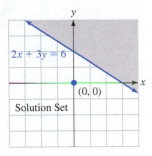

$2x + 3y = 6$

$(0, 0)$

Solution Set

Figure 4

or below the line $ax + by = c$. The "test-point" procedure we describe below gives us an easy method for deciding whether the solution set includes the region above or below the corresponding line.

Now we are going to do something that will appear backward at first (but makes it simpler to sketch sets of solutions of *systems* of linear inequalities). For our standard drawing of the region of solutions of $2x + 3y \leq 6$, we are going to *shade only the part that we do not want and leave the solution region blank*. Think of covering over or "blocking out" the unwanted points, leaving those that we do want in full view (but remember that the points on the boundary line are also points that we want). The result is Figure 4. The reason we do this should become clear in Example 2.

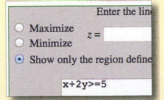
Sketching the Region Represented by a Linear Inequality in Two Variables

1. Sketch the straight line obtained by replacing the given inequality with an equality.

2. Choose a test point that is not on the line; $(0, 0)$ is a good choice if the line does not pass through the origin.

3. If the test point satisfies the inequality, then the set of solutions is the line plus the entire region on the same side of the line as the test point. Otherwise, it is the line plus the region on the other side of the line. In either case, shade (block out) the side that does *not* contain the solutions, leaving the solution set unshaded.

Quick Example

6. Here are the three steps used to graph the inequality $x + 2y \geq 5$:

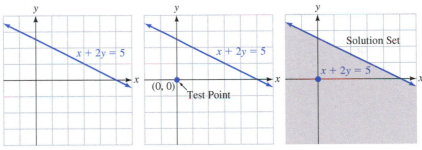

1. Sketch the line $x + 2y = 5$.

2. Test the point $(0, 0)$. $0 + 2(0) \not\geq 5$. The inequality is not satisfied.

3. Because the inequality is not satisfied, shade the region containing the test point.

EXAMPLE 1 **Graphing Single Inequalities**

Sketch the regions determined by each of the following inequalities:

a. $3x - 2y \leq 6$ **b.** $6x \leq 12 + 4y$ **c.** $x \leq -1$ **d.** $y \geq 0$ **e.** $x \geq 3y$

Solution

*Remember here and also in parts (b)–(e) that the solution set also includes the boundary line.

a. The boundary line $3x - 2y = 6$ has x-intercept 2 and y-intercept -3 (Figure 5). We use $(0, 0)$ as a test point (because it is not on the line). Because $3(0) - 2(0) \leq 6$, the inequality is satisfied by the test point $(0, 0)$, so it lies inside the solution set. The solution set is shown in Figure 5.*

b. The given inequality, $6x \leq 12 + 4y$, can be rewritten in the form $ax + by \leq c$ by subtracting $4y$ from both sides:

$$6x - 4y \leq 12.$$

Dividing both sides by 2 gives the inequality $3x - 2y \leq 6$, which we considered in part (a). Now, *applying the rules for manipulating inequalities does not affect the set of solutions.* Thus, the inequality $6x \leq 12 + 4y$ has the same set of solutions as $3x - 2y \leq 6$. (See Figure 5.)

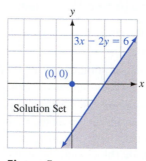

Figure 5   **Figure 6**

c. The region $x \leq -1$ has as boundary the vertical line $x = -1$. The test point $(0, 0)$ is not in the solution set, as shown in Figure 6.

d. The region $y \geq 0$ has as boundary the horizontal line $y = 0$ (that is, the x-axis). We cannot use $(0, 0)$ for the test point because it lies on the boundary line. Instead, we choose a convenient point that is not on the line $y = 0$—say, $(0, 1)$. Because $1 \geq 0$, this point is in the solution set, giving us the region shown in Figure 7.

e. The line $x \geq 3y$ has as boundary the line $x = 3y$ or, solving for y,

$$y = \frac{1}{3}x.$$

This line passes through the origin with slope $1/3$, so again we cannot choose the origin as a test point. Instead, we choose $(0, 1)$. Substituting these coordinates in $x \geq 3y$ gives $0 \geq 3(1)$, which is false, so $(0, 1)$ is not in the solution set, as shown in Figure 8.

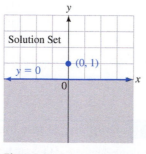 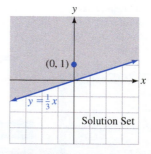

Figure 7   **Figure 8**

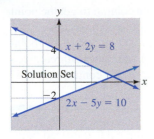

Figure 9

*Although these graphs are quite easy to do by hand, the more lines we have to graph, the more difficult it becomes to get everything in the right place, and this is where graphing technology can become important. This is especially true when, for instance, three or more lines intersect at points that are very close together and hard to distinguish in hand-drawn graphs.

EXAMPLE 2 **Graphing Simultaneous Inequalities**

Sketch the region of points that satisfy both inequalities:

$$2x - 5y \leq 10$$
$$x + 2y \leq 8.$$

Solution Each inequality has a solution set that is a half plane. If a point is to satisfy *both* inequalities, it must lie in both sets of solutions. Put another way, if we cover the points that are not solutions to $2x - 5y \leq 10$ and then also cover the points that are not solutions to $x + 2y \leq 8$, the points that remain uncovered must be the points we want, those that are solutions to both inequalities. The result is shown in Figure 9, where the unshaded region (including its boundary) is the set of solutions.*

As a check, we can look at points in various regions in Figure 9. For example, our graph shows that $(0, 0)$ should satisfy both inequalities, and it does:

$$2(0) - 5(0) = 0 \leq 10 \quad ✔$$
$$0 + 2(0) = 0 \leq 8. \quad ✔$$

On the other hand, $(0, 5)$ should fail to satisfy one of the inequalities:

$$2(0) - 5(5) = -25 \leq 10 \quad ✔$$
$$0 + 2(5) = 10 > 8. \quad ✗$$

One more: $(5, -1)$ should fail one of the inequalities:

$$2(5) - 5(-1) = 15 > 10 \quad ✗$$
$$5 + 2(-1) = 3 \leq 8. \quad ✔$$

EXAMPLE 3 **Corner Points**

Sketch the region of solutions of the following system of inequalities, and list the coordinates of all the corner points.

$$3x - y \leq 6$$
$$x + y \geq 6$$
$$y \leq 6$$

Solution Shading the regions that we do not want leaves us with the triangle shown in Figure 10. We label the corner points A, B, and C as shown.

Each of these corner points lies at the intersection of two of the bounding lines. So to find the coordinates of each corner point, we need to solve the system of equations given by the two lines. To do this systematically, we make the following table:

Figure 10

Point	Lines through Point	Coordinates
A	$y = 6$ $x + y = 6$	$(0, 6)$
B	$y = 6$ $3x - y = 6$	$(4, 6)$
C	$x + y = 6$ $3x - y = 6$	$(3, 3)$

* Technology Note Using the trace feature makes it easy to locate corner points graphically. Remember to zoom in for additional accuracy when appropriate. Of course, you can also use technology to help solve the systems of equations, as we discussed in Chapter 3.

Here, we have solved each system of equations in the middle column to get the point on the right, using the techniques of Chapter 3. You should do this for practice.*

As a partial check that we have drawn the correct region, let us choose any point in its interior—say, $(3, 5)$. We can easily check that $(3, 5)$ satisfies all three given inequalities. It follows that all of the points in the triangular region containing $(3, 5)$ are also solutions.

➡️ **Before we go on . . .** What if the constraint $y \leq 6$ in Example 3 had instead been, say, $y \leq 2$? We would need to adjust Figure 10 by moving the horizontal line at $y = 6$ down to $y = 2$ (below the point C) and then graying out everything above it. What we would find is that everything had been grayed out (see Figure 11)! This means that there would be nothing at all in the solution set. Put another way, the solution set of the system $3x - y \leq 6$, $x + y \geq 6$, $y \leq 2$ is **empty**.

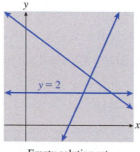

Empty solution set

Figure 11

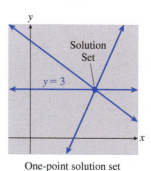

One-point solution set

Figure 12

If, instead, the last constraint had been $y \leq 3$, then all three lines would intersect at the single point $(3, 3)$. Even though everything would still be grayed out (see Figure 12), the solution set would consist of that single point $(3, 3)$, because that point is on the boundary of each of the three regions $3x - y \leq 6$, $x + y \geq 6$, and $y \leq 3$ and hence is in each of their solution sets.

Take another look at the regions of solutions in Examples 2 and 3 (Figures 13 and 14).

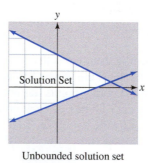

Unbounded solution set

Figure 13

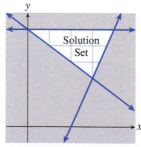

Bounded solution set

Figure 14

Notice that the solution set in Figure 13 extends infinitely far to the left, whereas the one in Figure 14 is completely enclosed by a boundary. Sets that are completely enclosed are called **bounded**, and sets that extend infinitely in one or more directions are **unbounded**. For example, all the solution sets in Example 1 are unbounded. ∎

| EXAMPLE 4 | Resource Allocation |

Socaccio Pistachio, Inc. makes two types of pistachio nuts: Dazzling Red and Organic. Pistachio nuts require food color and salt, and the following table shows the amount of food color and salt required for a 1-kilogram batch of pistachios as well as the total amount of these ingredients available each day:

	Dazzling Red	Organic	Total Available
Food Color (grams)	2	1	20
Salt (grams)	10	20	220

Use a graph to show the possible numbers of batches of each type of pistachio Socaccio can produce each day. This region (the solution set of a system of inequalities) is called the **feasible region**.

Solution As we did in Chapter 3, we start by identifying the unknowns: Let x be the number of batches of Dazzling Red manufactured per day, and let y be the number of batches of Organic manufactured each day.

Now, because of our experience with systems of linear equations, we are tempted to say: For food color, $2x + y = 20$, and for salt, $10x + 20y = 220$. However, no one is saying that Socaccio has to use all available ingredients; the company might choose to use fewer than the total available amounts if this proves more profitable. Thus, $2x + y$ can be anything *up to a total of* 20. In other words,

$$2x + y \leq 20.$$

Similarly,

$$10x + 20y \leq 220.$$

There are two more restrictions that are not explicitly mentioned: Neither x nor y can be negative. (The company cannot produce a negative number of batches of nuts.) Therefore, we have the additional restrictions

$$x \geq 0 \quad \text{and} \quad y \geq 0.$$

These two inequalities tell us that the feasible region (solution set) is restricted to the first quadrant, because in the other quadrants, either x or y or both x and y are negative. So instead of shading out all other quadrants, we can simply restrict our drawing to the first quadrant.

The (bounded) feasible region shown in Figure 15 is a graphical representation of the limitations the company faces.

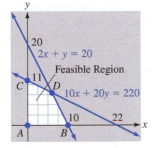

Figure 15

➡ **Before we go on...** Every point in the feasible region in Example 4 represents a value for x and a value for y that do not violate any of the company's restrictions. For example, the point $(5, 6)$ lies well inside the region, so the company can produce five batches of Dazzling Red nuts and six batches of Organic without exceeding the limitations on ingredients [that is, $2(5) + 6 = 16 \leq 20$ and $10(5) + 20(6) = 170 \leq 220$]. The corner points A, B, C, and D are significant if the company wishes to realize the greatest profit, as we will see in Section 5.2. We can find the corners as in the following table:

Point	Lines through Point	Coordinates
A		$(0, 0)$
B		$(10, 0)$
C		$(0, 11)$
D	$2x + y = 20$ $10x + 20y = 220$	$(6, 8)$

(We have not listed the lines through the first three corners because their coordinates can be read easily from the graph.) Points on the line segment *DB* represent use of all the food color (because the segment lies on the line $2x + y = 20$), and points on the line segment *CD* represent use of all the salt (because the segment lies on the line $10x + 20y = 220$). Note that the point *D* is the only solution that uses all of both ingredients. ∎

FAQs

Recognizing Whether to Use a Linear Inequality or a Linear Equation

Q : *How do I know whether to model a situation by a linear inequality such as* $3x + 2y \leq 10$ *or by a linear equation such as* $3x + 2y = 10$?

A : Here are some key phrases to look for: *at most, up to, no more than, at least, or more, exactly.* Suppose, for instance, that nuts cost 3¢, bolts cost 2¢, *x* is the number of nuts you can buy, and *y* is the number of bolts you can buy.

- If you have *up to* 10¢ to spend, then $3x + 2y \leq 10$.
- If you must spend *exactly* 10¢, then $3x + 2y = 10$.
- If you plan to spend *at least* 10¢, then $3x + 2y \geq 10$.

The use of inequalities to model a situation is often more realistic than the use of equations; for instance, one cannot always expect to exactly fill all orders, spend the exact amount of one's budget, or keep a plant operating at exactly 100% capacity.

5.1 EXERCISES

▼ more advanced   ◆ challenging

T indicates exercises that should be solved using technology

In Exercises 1–26, sketch the region that corresponds to the given inequalities, say whether the region is bounded or unbounded, and find the coordinates of all corner points (if any). [**HINT:** See Examples 1, 2, and 3.]

1. $2x + y \leq 10$

2. $4x - y \leq 12$

3. $-x - 2y \leq 8$

4. $-x + 2y \geq 4$

5. $3x + 2y \geq 5$

6. $2x - 3y \leq 7$

7. $x \leq 3y$

8. $y \geq 3x$

9. $\dfrac{3x}{4} - \dfrac{y}{4} \leq 1$

10. $\dfrac{x}{3} + \dfrac{2y}{3} \geq 2$

11. $x \geq -5$

12. $y \leq -4$

13. $4x - y \leq 8$
$x + 2y \leq 2$

14. $2x + y \leq 4$
$x - 2y \geq 2$

15. $3x + 2y \geq 6$
$3x - 2y \leq 6$
$x \geq 0$

16. $3x + 2y \leq 6$
$3x - 2y \geq 6$
$-y \geq 2$

17. $x + y \geq 5$
$x \leq 10$
$y \leq 8$
$x \geq 0, y \geq 0$

18. $2x + 4y \geq 12$
$x \leq 5$
$y \leq 3$
$x \geq 0, y \geq 0$

19. $20x + 10y \leq 100$
$10x + 20y \leq 100$
$10x + 10y \leq 60$
$x \geq 0, y \geq 0$

20. $30x + 20y \leq 600$
$10x + 40y \leq 400$
$20x + 30y \leq 450$
$x \geq 0, y \geq 0$

21. $20x + 10y \geq 100$
$10x + 20y \geq 100$
$10x + 10y \geq 80$
$x \geq 0, y \geq 0$

22. $30x + 20y \geq 600$
$10x + 40y \geq 400$
$20x + 30y \geq 600$
$x \geq 0, y \geq 0$

23. $-3x + 2y \leq 5$
$3x - 2y \leq 6$
$x \leq 2y$
$x \geq 0, y \geq 0$

24. $-3x + 2y \leq 5$
$3x - 2y \geq 6$
$y \leq x/2$
$x \geq 0, y \geq 0$

25. $2x - y \geq 0$
$x - 3y \leq 0$
$x \geq 0, y \geq 0$

26. $-x + y \geq 0$
$4x - 3y \geq 0$
$x \geq 0, y \geq 0$

T *In Exercises 27–32, we suggest that you use technology. Graph the region corresponding to the inequalities, and find the coordinates of all corner points (if any) to two decimal places.*

27. $2.1x - 4.3y \geq 9.7$

28. $-4.3x + 4.6y \geq 7.1$

29. $-0.2x + 0.7y \geq 3.3$
$1.1x + 3.4y \geq 0$

30. $0.2x + 0.3y \geq 7.2$
$2.5x - 6.7y \leq 0$

31. $4.1x - 4.3y \leq 4.4$
$7.5x - 4.4y \leq 5.7$
$4.3x + 8.5y \leq 10$

32. $2.3x - 2.4y \leq 2.5$
$4.0x - 5.1y \leq 4.4$
$6.1x + 6.7y \leq 9.6$

Applications

33. *Resource Allocation* You manage an ice cream factory that makes two flavors: Creamy Vanilla and Continental Mocha. Into each quart of Creamy Vanilla go 2 eggs and 3 cups of cream. Into each quart of Continental Mocha go 1 egg and 3 cups of cream. You have in stock 500 eggs and 900 cups of cream. Draw the feasible region showing the number of quarts of vanilla and number of quarts of mocha that can be produced. Find the corner points of the region. [**HINT:** See Example 4.]

34. *Resource Allocation* Podunk Institute of Technology's Math Department offers two courses: Finite Math and Applied Calculus. Each section of Finite Math has 60 students, and each section of Applied Calculus has 50. The department is allowed to offer a total of up to 110 sections. Furthermore, no more than 6,000 students want to take a math course. (No student will take more than one math course.) Draw the feasible region that shows the number of sections of each class that can be offered. Find the corner points of the region. [**HINT:** See Example 4.]

35. *Nutrition* Ruff, Inc. makes dog food out of chicken and grain. Chicken has 10 grams of protein and 5 grams of fat

per ounce, and grain has 2 grams of protein and 2 grams of fat per ounce. A bag of dog food must contain at least 200 grams of protein and at least 150 grams of fat. Draw the feasible region that shows the number of ounces of chicken and number of ounces of grain Ruff can mix into each bag of dog food. Find the corner points of the region.

36. *Purchasing* Enormous State University's Business School is buying computers. The school has two models to choose from: the Pomegranate and the iZac. Each Pomegranate comes with 400 GB of memory and 80 TB of disk space, and each iZac has 300 GB of memory and 100 TB of disk space. For reasons related to its accreditation, the school would like to be able to say that it has a total of at least 48,000 GB of memory and at least 12,800 TB of disk space. Draw the feasible region that shows the number of each kind of computer it can buy. Find the corner points of the region.

37. *Nutrition* **Gerber Products**' Gerber Mixed Cereal for Baby contains, in each serving, 60 calories and 11 grams of carbohydrates. Gerber Mango Tropical Fruit Dessert contains, in each serving, 80 calories and 21 grams of carbohydrates.[3] You want to provide your child with at least 140 calories and at least 32 grams of carbohydrates. Draw the feasible region that shows the number of servings of cereal and number of servings of dessert that you can give your child. Find the corner points of the region.

38. *Nutrition* **Gerber Products**' Gerber Mixed Cereal for Baby contains, in each serving, 60 calories, 11 grams of carbohydrates, and no vitamin C. Gerber Apple Banana Juice contains, in each serving, 60 calories, 15 grams of carbohydrates, and 120% of the U.S. Recommended Daily Allowance (RDA) of vitamin C for infants.[4] You want to provide your child with at least 120 calories, at least 26 grams of carbohydrates, and at least 50% of the U.S. RDA of vitamin C for infants. Draw the feasible region that shows the number of servings of cereal and number of servings of juice that you can give your child. Find the corner points of the region.

39. *Municipal Bond Funds* The **Pioneer Investment Management** Municipal High Income fund (MHI) and the **BlackRock Advisors** BlackRock Municipal fund (BKK) are tax-exempt municipal bond funds. In 2015 the Pioneer fund was expected to yield 6%, while the BlackRock fund was expected to yield 5%.[5] You would like to invest a total of up to $80,000 and earn at least $4,200 in interest in the coming year (based on the given yields). Draw the feasible region that shows how much money you can invest in each fund. Find the corner points of the region.

---

[3] Source: Nutrition information supplied with the products.

[4] *Ibid.*

[5] Expected yields based on returns as of April 2015. Source: CEF Connect (www.cefconnect.com).

40. *Mutual Funds* In 2015, the Phoenix/Zweig Advisors Zweig Total Return fund (ZTR) was expected to yield 5%, and the Madison Asset Management Madison Strategic Sector Premium fund (MSP) was expected to yield 7%.[6] You would like to invest a total of up to $60,000 and earn at least $3,500 in interest. Draw the feasible region that shows how much money you can invest in each fund (based on the given yields). Find the corner points of the region.

41. ▼ *Investments: Financial Stocks* (Compare Exercise 51 in Section 3.1.) During the first quarter of 2015, Toronto Dominion Bank (TD) stock cost $45 per share and was expected to yield 4% per year in dividends, while CNA Financial Corp. (CNA) stock cost $40 per share and was expected to yield 2.5% per year in dividends.[7] You have up to $25,000 to invest in these stocks and would like to earn at least $760 in dividends over the course of a year. (Assume the dividend to be unchanged for the year.) Draw the feasible region that shows how many shares in each company you can buy. Find the corner points of the region. (Round each coordinate to the nearest whole number.)

42. ▼ *Investments: High-Dividend Stocks* (Compare Exercise 52 in Section 3.1.) During the first quarter of 2015, Plains All American Pipeline L.P. (PAA) stock cost $50 per share and was expected to yield 5% per year in dividends, while Total SA (TOT) stock cost $50 per share and was expected to yield 6% per year in dividends.[8] You have up to $45,000 to invest in these stocks and would like to earn at least $2,400 in dividends over the course of a year. (Assume the dividend to be unchanged for the year.) Draw the feasible region that shows how many shares in each company you can buy. Find the corner points of the region. (Round each coordinate to the nearest whole number.)

43. ▼ *Advertising* You are the marketing director for a company that manufactures bodybuilding supplements, and you are planning to run ads in Sports Illustrated and GQ Magazine. Based on readership data, you estimate that each one-page ad in *Sports Illustrated* will be read by 650,000 people in your target group, while each one-page ad in *GQ* will be read by 150,000.[9] You would like your ads to be read by at least 3 million people in the target group, and, to ensure the broadest possible audience, you would like to place at least three full-page ads in each magazine. Draw the feasible region that shows how many pages you can purchase in each magazine. Find the corner points of the region. (Round each coordinate to the nearest whole number.)

44. ▼ *Advertising* You are the marketing director for a company that manufactures bodybuilding supplements and you are planning to run ads in Sports Illustrated and Muscle and Fitness. Based on readership data, you estimate that each one-page ad in *Sports Illustrated* will be read by 650,000 people in your target group, while each one-page ad in *Muscle and Fitness* will be read by 250,000 people in your target group.[10] You would like your ads to be read by at least 4 million people in the target group, and, to ensure the broadest possible audience, you would like to place at least three full-page ads in each magazine during the year. Draw the feasible region showing how many pages you can purchase in each magazine. Find the corner points of the region. (Round each coordinate to the nearest whole number.)

Communication and Reasoning Exercises

45. Find a system of inequalities whose solution set is unbounded.

46. Find a system of inequalities whose solution set is empty.

47. How would you use linear inequalities to describe the triangle with corner points $(0, 0)$, $(2, 0)$, and $(0, 1)$?

48. Explain the advantage of shading the region of points that do not satisfy the given inequalities. Illustrate with an example.

49. Describe at least one drawback to the method of finding the corner points of a feasible region by drawing its graph, when the feasible region arises from real-life constraints.

50. Draw several bounded regions described by linear inequalities. For each region you draw, find the point that gives the greatest possible value of $x + y$. What do you notice?

In Exercises 51–54, you are mixing x grams of ingredient A and y grams of ingredient B. Choose the equation or inequality that models the given requirement.

51. There should be at least 3 more grams of ingredient A than ingredient B.
   (A) $3x - y \le 0$ **(B)** $x - 3y \ge 0$
   (C) $x - y \ge 3$ **(D)** $3x - y \ge 0$

52. The mixture should contain at least 25% of ingredient A by weight.
   (A) $4x - y \le 0$ **(B)** $x - 4y \ge 0$
   (C) $x - y \ge 4$ **(D)** $3x - y \ge 0$

53. ▼ There should be at least 3 parts (by weight) of ingredient A to 2 parts of ingredient B.
   (A) $3x - 2y \ge 0$ **(B)** $2x - 3y \ge 0$
   (C) $3x + 2y \ge 0$ **(D)** $2x + 3y \ge 0$

[6] See footnote for Exercise 39.

[7] Stock prices and yields are approximate. Source: http://finance.yahoo.com.

[8] *Ibid.*

[9] The readership data for *Sports Illustrated* is based, in part, on the results of a readership survey taken in March 2000. The readership data for *GQ* is fictitious. Source: Mediamark Research Inc./*New York Times*, May 29, 2000, p. C1.

[10] The readership data for both magazines are based on the results of a readership survey taken in March 2000. Source: Mediamark Research Inc./*New York Times*, May 29, 2000, p. C1.

54. ▼ There should be no more of ingredient A (by weight) than ingredient B.
(A) $x - y = 0$ (B) $x - y \leq 0$
(C) $x - y \geq 0$ (D) $x + y \geq y$

55. ▼ You are setting up a system of inequalities in the unknowns x and y. The inequalities represent constraints faced by *Fly-by-Night Airlines*, where x represents the number of first-class tickets it should issue for a specific flight and y represents the number of business-class tickets it should issue for that flight. You find that the feasible region is empty. How do you interpret this?

56. ▼ In the situation described in Exercise 55, is it possible instead for the feasible region to be unbounded? Explain your answer.

57. ▼ Create an interesting scenario that leads to the following system of inequalities:

$$20x + 40y \leq 1{,}000$$
$$30x + 20y \leq 1{,}200$$
$$x \geq 0, y \geq 0.$$

58. ▼ Create an interesting scenario that leads to the following system of inequalities:

$$20x + 40y \geq 1{,}000$$
$$30x + 20y \geq 1{,}200$$
$$x \geq 0, y \geq 0.$$

5.2   Solving Linear Programming Problems Graphically

Fundamentals

As we saw in Example 4 in Section 5.1, in some scenarios the possibilities are restricted by a system of linear inequalities. In that example it would also be natural to ask which of the various possibilities gives the company the largest profit. This is a kind of problem known as a *linear programming problem* (commonly referred to as an LP problem).

Linear Programming Problems in Two Unknowns

A **linear programming (LP) problem** in two unknowns x and y is one in which we are to find the maximum or minimum value of a linear expression

$$ax + by,$$

called the **objective function**, subject to a number of linear **constraints** of the form

$$cx + dy \leq e \quad \text{or} \quad cx + dy \geq e.$$

The solution set of points (x, y) satisfying all constraints is called the **feasible region** for the problem, the largest or smallest value of the objective function is called the **optimal value**, and a pair of values of x and y that gives the optimal value constitutes an **optimal solution**.

Quick Example

1. Maximize $p = x + y$ Objective function
    subject to $x + 2y \leq 12$
            $2x + y \leq 12$ Constraints
            $x \geq 0, y \geq 0.$

See Example 1 for a method of solving this LP problem (that is, finding an optimal solution and value).

To solve LP problems, we use the following result.

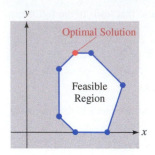

Figure 16

Fundamental Theorem of Linear Programming

(a) Linear programming problems with bounded, nonempty feasible regions always have optimal solutions.

(b) If an LP problem has optimal solutions, then at least one of these solutions occurs at a corner point of the feasible region (Figure 16).

Let's use this to solve an LP problem, and then we'll discuss why it's true.

EXAMPLE 1   **Solving a Linear Programming Problem**

Maximize $p = x + y$

subject to $x + 2y \leq 12$

$2x + y \leq 12$

$x \geq 0, y \geq 0.$

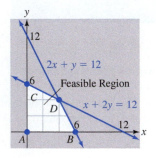

Figure 17

Solution We begin by drawing the feasible region for the problem. We do this using the techniques of Section 5.1, and we get Figure 17. The feasible region is bounded and nonempty, so the first part of the Fundamental Theorem of Linear Programming tells us that the problem has an optimal solution.

Each **feasible point** (point in the feasible region) gives an x and a y satisfying the constraints. The question now is, which of these points gives the largest value of the objective function $p = x + y$? The second part of the Fundamental Theorem of Linear Programming tells us that the largest value must occur at one (or more) of the corners of the feasible region. In the following table, we list the coordinates of each corner point, and we compute the value of the objective function at each corner:

Corner Point	Lines through Point	Coordinates	$p = x + y$
A		$(0, 0)$	0
B		$(6, 0)$	6
C		$(0, 6)$	6
D	$x + 2y = 12$ $2x + y = 12$	$(4, 4)$	8

Now we simply pick the one that gives the largest value for p, which is D. Therefore, the optimal value of p is 8, and an optimal solution is $(4, 4)$.

Now we owe you an explanation of why one of the corner points should be an optimal solution. The question is, which point in the feasible region gives the largest possible value of $p = x + y$?

Consider first an easier question: Which points result in a *particular value* of p? For example, which points result in $p = 2$? These would be the points on the line $x + y = 2$, which is the line labeled $p = 2$ in Figure 18.

Now suppose we want to know which points make $p = 4$: These would be the points on the line $x + y = 4$, which is the line labeled $p = 4$ in Figure 18. Notice that this line is parallel to but higher than the line $p = 2$. (If p represented profit in an application, we would call these **isoprofit lines**, or **constant-profit lines**.) Imagine moving this line up or down in the picture. As we move the line down, we see smaller values of p, and as we move it up, we see larger values. Several more of these lines are drawn in Figure 18. Look, in particular, at the line labeled $p = 10$. This line does not meet the feasible region, meaning that no feasible point makes p as large as 10. Starting with the line $p = 2$, as we move the line up, increasing p, there will be a last line that meets the feasible region. In the figure, it is clear that this is the line $p = 8$, and this meets the feasible region in only one point, which is the corner point D. Therefore, D gives the greatest value of p of all feasible points.

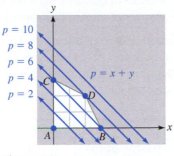

Figure 18   **Figure 19**

If we had been asked to maximize some other objective function, such as $p = x + 3y$, then the optimal solution might be different. Figure 19 shows some of the isoprofit lines for this objective function. This time, the last point that is hit as p increases is C, not D. This tells us that the optimal solution is $(0, 6)$, giving the optimal value $p = 18$.

This discussion should convince you that the optimal value in an LP problem will always occur at one of the corner points. By the way, it is possible for the optimal value to occur at *two* corner points and at all points along an edge connecting them. (Do you see why?) We will see this in Example 3(b).

Here is a summary of the method we have just been using.

Graphical Method for Solving Linear Programming Problems in Two Unknowns (Bounded Feasible Regions)

1. Graph the feasible region, and check that it is bounded.

2. Compute the coordinates of the corner points.

3. Substitute the coordinates of the corner points into the objective function to see which gives the maximum (or minimum) value of the objective function.

4. Any such corner point is an optimal solution.

Note If the feasible region is unbounded, this method will work only if there are optimal solutions; otherwise, it will not work. We will show you a method for deciding this after Example 3. ∎

Applications: Bounded Feasible Regions

EXAMPLE 2   **Resource Allocation**

Acme Baby Foods mixes two strengths of apple juice. One quart of Beginner's juice is made from 30 fluid ounces of water and 2 fluid ounces of apple juice concentrate. One quart of Advanced juice is made from 20 fluid ounces of water and 12 fluid ounces of concentrate. Every day Acme has available 30,000 fluid ounces of water and 3,600 fluid ounces of concentrate. Acme makes a profit of 20¢ on each quart of Beginner's juice and 30¢ on each quart of Advanced juice. How many quarts of each should Acme make each day to get the largest profit? How would this change if Acme made a profit of 40¢ on Beginner's juice and 20¢ on Advanced juice?

Solution The first question we are asked gives unknown quantities as

$$x = \text{number of quarts of Beginner's juice made each day}$$
$$y = \text{number of quarts of Advanced juice made each day.}$$

(In this context, x and y are often called the **decision variables**, because we must decide what their values should be in order to get the largest profit.) We can write down the data given in the form of a table. (The numbers in the first two columns are amounts per quart of juice.)

	Beginner's, x	Advanced, y	Available
Water (ounces)	30	20	30,000
Concentrate (ounces)	2	12	3,600
Profit (¢)	20	30	

Because nothing in the problem says that Acme must use up all the water or concentrate, just that it can use no more than what is available, the first two rows of the table give us two inequalities:

$$30x + 20y \le 30,000$$
$$2x + 12y \le 3,600.$$

Dividing the first inequality by 10 and the second by 2 gives

$$3x + 2y \le 3,000$$
$$x + 6y \le 1,800.$$

We also have that $x \ge 0$ and $y \ge 0$ because Acme can't make a negative amount of juice. To finish setting up the problem, we are asked to maximize the profit, which is

$$p = 20x + 30y. \quad \text{Expressed in cents}$$

This gives us our LP problem:

$$\text{Maximize} \quad p = 20x + 30y$$
$$\text{subject to} \quad 3x + 2y \le 3,000$$
$$x + 6y \le 1,800$$
$$x \ge 0, y \ge 0.$$

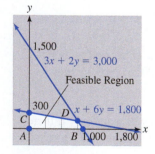

Figure 20

The (bounded) feasible region is shown in Figure 20.

The corners and the values of the objective function are listed in the following table:

Point	Lines through Point	Coordinates	$p = 20x + 30y$
A		$(0, 0)$	0
B		$(1,000, 0)$	20,000
C		$(0, 300)$	9,000
D	$3x + 2y = 3,000$ $x + 6y = 1,800$	$(900, 150)$	22,500

We are seeking to maximize the objective function p, so we look for corner points that give the maximum value for p. Because the maximum occurs at the point D, we conclude that the (only) optimal solution occurs at D. Thus, the company should make 900 quarts of Beginner's juice and 150 quarts of Advanced juice, for a largest possible profit of 22,500¢, or $225.

If, instead, the company made a profit of 40¢ on each quart of Beginner's juice and 20¢ on each quart of Advanced juice, then we would have $p = 40x + 20y$. This gives the following table:

Point	Lines through Point	Coordinates	$p = 40x + 20y$
A		$(0, 0)$	0
B		$(1,000, 0)$	40,000
C		$(0, 300)$	6,000
D	$3x + 2y = 3,000$ $x + 6y = 1,800$	$(900, 150)$	39,000

We can see that, in this case, Acme should make 1,000 quarts of Beginner's juice and no Advanced juice, for a largest possible profit of 40,000¢, or $400.

➡ **Before we go on ...** Notice that, in the first version of the problem in Example 2, the company used all the water and juice concentrate:

Water: $30(900) + 20(150) = 30,000$

Concentrate: $2(900) + 12(150) = 3,600.$

In the second version the company used all the water but not all the concentrate:

Water: $30(1,000) + 20(0) = 30,000$

Concentrate: $2(1,000) + 12(0) = 2,000 < 3,600.$ ■

EXAMPLE 3   Investments

The *Solid Trust Savings & Loan Company* has set aside $25 million for loans to home buyers. Its policy is to allocate at least $10 million annually for luxury condominiums. A government housing development grant that the company receives requires, however, that at least one third of its total loans be allocated to low-income housing.

a. Solid Trust's return on condominiums is 12%, and its return on low-income housing is 10%. How much should the company allocate for each type of housing to maximize its total return?

b. Redo part (a), assuming that the return is 12% on both condominiums and low-income housing.

Solution

a. We first identify the unknowns: Let x be the annual amount (in millions of dollars) allocated to luxury condominiums, and let y be the annual amount allocated to low-income housing.

We now look at the constraints. The first constraint is mentioned in the first sentence: The total the company can invest is $25 million. Thus,

$$x + y \leq 25.$$

(The company is not required to invest all of the $25 million; rather, it can invest *up to* $25 million.) Next, the company has allocated at least $10 million to condos. Rephrasing this in terms of the unknowns, we get

The amount allocated to condos is at least $10 million.

The phrase "is at least" means $\geq$. Thus, we obtain a second constraint:

$$x \geq 10.$$

The third constraint is that at least one third of the total financing must be for low-income housing. Rephrasing this, we say:

The amount allocated to low-income housing is at least one third of the total.

Because the total investment will be $x + y$, we get

$$y \geq \frac{1}{3}(x + y).$$

We put this in the standard form of a linear inequality as follows:

$$3y \geq x + y \qquad \text{Multiply both sides by 3.}$$
$$-x + 2y \geq 0. \qquad \text{Subtract } x + y \text{ from both sides.}$$

There are no further constraints.

Now, what about the return on these investments? According to the data, the annual return is given by

$$p = 0.12x + 0.10y.$$

We want to make this quantity p as large as possible. In other words, we want to

$$\text{Maximize} \quad p = 0.12x + 0.10y$$
$$\text{subject to} \quad x + y \leq 25$$
$$x \geq 10$$
$$-x + 2y \geq 0$$
$$x \geq 0, y \geq 0.$$

(Do you see why the inequalities $x \geq 0$ and $y \geq 0$ are slipped in here?) The feasible region is shown in Figure 21.

We now make a table that gives the (approximate) return on investment at each corner point:

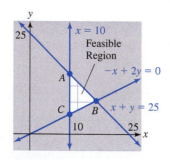

Figure 21

Point	Lines through Point	Coordinates	$p = 0.12x + 0.10y$
A	$x = 10$ $x + y = 25$	$(10, 15)$	2.7
B	$x + y = 25$ $-x + 2y = 0$	$(50/3, 25/3)$	2.833
C	$x = 10$ $-x + 2y = 0$	$(10, 5)$	1.7

From the table we see that the values of x and y that maximize the return are $x = 50/3$ and $y = 25/3$, which give a total return of about $2.833 million. In other words, the most profitable course of action is to invest about $16.667 million in loans for condominiums and $8.333 million in loans for low-income housing, giving a maximum annual return of about $2.833 million.

b. The LP problem is the same as that for part (a) except for the objective function:

$$\text{Maximize} \quad p = 0.12x + 0.12y$$
$$\text{subject to} \quad x + y \le 25$$
$$x \ge 10$$
$$-x + 2y \ge 0$$
$$x \ge 0, y \ge 0.$$

Here are the values of p at the three corners:

Point	Coordinates	$p = 0.12x + 0.12y$
A	$(10, 15)$	3
B	$(50/3, 25/3)$	3
C	$(10, 5)$	1.8

Looking at the table, we see that a curious thing has happened: We get the same maximum annual return at both A and B. Thus, we could choose either option to maximize the annual return. In fact, any point along the line segment AB will yield an annual return of \$3 million. For example, the point $(12, 13)$ lies on the line segment AB and also yields an annual revenue of \$3 million. This happens because the "isoreturn" lines are parallel to that edge.

➡ **Before we go on . . .** What breakdowns of investments would lead to the *lowest* return for parts (a) and (b)? ∎

Unbounded Feasible Regions

The preceding examples all had bounded feasible regions. If the feasible region is unbounded, then, *provided that there are optimal solutions,* the fundamental theorem of linear programming guarantees that the above method will work. The following procedure determines whether or not optimal solutions exist and finds them when they do.

Graphical Method for Solving Linear Programming Problems in Two Unknowns (Unbounded Feasible Regions)

If the feasible region of an LP problem is unbounded, proceed as follows:

1. Draw a rectangle large enough that all the corner points are inside the rectangle (and not on its boundary):

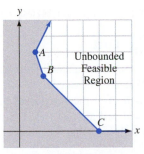

Corner points: A, B, C

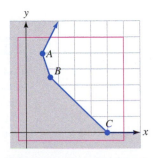

Corner points inside the rectangle

2. Shade the outside of the rectangle so as to define a new bounded feasible region, and locate the new corner points:

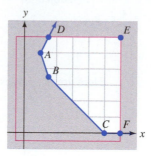

New corner points: D, E, and F

3. Obtain the optimal solutions using this bounded feasible region.

4. If any optimal solutions occur at one of the original corner points (A, B, and C in the figure), then the LP problem has that corner point as an optimal solution. Otherwise, the LP problem has no optimal solutions. When the latter occurs, we say that the **objective function is unbounded**, because it can assume arbitrarily large (positive or negative) values.

Q: *Do I always have to add a bounding rectangle to deal with unbounded regions?*

A: Not always; under some circumstances you can tell right away whether optimal solutions exist when the feasible region is unbounded. *Note that the following apply only when we have the constraints $x \geq 0$ and $y \geq 0$:*

    a. If you are minimizing $c = ax + by$ with a and b nonnegative, then optimal solutions always exist.

    b. If you are maximizing $p = ax + by$ with a and b both positive, then there is no optimal solution.

    c. If you are maximizing $p = ax + by$ with $a \leq 0$ and $b \leq 0$, then optimal solutions always exist.

    d. If you are minimizing $c = ax + by$ with a and b both negative, then there is no optimal solution.

Do you see why these statements are true? (Think about what happens in (4) above in each case.)

Applications: Unbounded Feasible Regions

EXAMPLE 4   Cost

You are the manager of a small store that specializes in hats, sunglasses, and other accessories. You are considering a sales promotion of a new line of hats and sunglasses. You will offer the sunglasses only to customers who purchase two or more hats, so you will sell at least twice as many hats as pairs of sunglasses. Moreover, your supplier tells you that, because of seasonal demand, your order of sunglasses cannot exceed 100 pairs. To ensure that the sale items fill out the large display you have set aside, you estimate that you should order at least 210 items in all.

a. Assume that you will lose $3 on every hat and $2 on every pair of sunglasses sold. Given the constraints above, how many hats and pairs of sunglasses should you order to lose the least amount of money in the sales promotion?

b. Suppose instead that you lose $1 on every hat sold but make a profit of $5 on every pair of sunglasses sold. How many hats and pairs of sunglasses should you order to make the largest profit in the sales promotion?

c. Now suppose that you make a profit of $1 on every hat sold but lose $5 on every pair of sunglasses sold. How many hats and pairs of sunglasses should you order to make the largest profit in the sales promotion?

Solution

a. The unknowns are

$$x = \text{number of hats you order}$$
$$y = \text{number of pairs of sunglasses you order.}$$

The objective is to minimize the total loss:

$$c = 3x + 2y.$$

Now for the constraints. The requirement that you will sell at least twice as many hats as sunglasses can be rephrased as

The number of hats is at least twice the number of pairs of sunglasses,

or

$$x \geq 2y,$$

which, in standard form, is

$$x - 2y \geq 0.$$

Next, your order of sunglasses cannot exceed 100 pairs, so

$$y \leq 100.$$

Finally, you would like to sell at least 210 items in all, giving

$$x + y \geq 210.$$

Thus, the LP problem is the following:

$$
\begin{aligned}
\text{Minimize} \quad & c = 3x + 2y \\
\text{subject to} \quad & x - 2y \geq 0 \\
& y \leq 100 \\
& x + y \geq 210 \\
& x \geq 0, y \geq 0.
\end{aligned}
$$

The feasible region is shown in Figure 22. This region is unbounded, so there is no guarantee that there are any optimal solutions. However, the objective is to minimize $c = 3x + 2y$, which has the form $c = ax + by$ with a and b nonnegative, and we do have the constraints $x \geq 0$ and $y \geq 0$. Thus, by part (a) of the Q&A above, this LP problem does have optimal solutions (and there is no need to draw

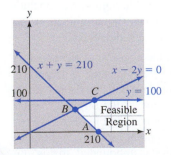

Figure 22

a bounding rectangle). We obtain the solution as usual, by listing the corners of the feasible region along with the corresponding values of the objective function c:

Point	Lines through Point	Coordinates	$c = 3x + 2y$
A		$(210, 0)$	630
B	$x + y = 210$ $x - 2y = 0$	$(140, 70)$	560
C	$x - 2y = 0$ $y = 100$	$(200, 100)$	800

The corner point that gives the minimum value of the objective function c is B. Thus, the combination that gives the smallest loss, \$560, is 140 hats and 70 pairs of sunglasses.

b. The LP problem is the following:

$$\text{Maximize} \quad p = -x + 5y$$
$$\text{subject to} \quad x - 2y \geq 0$$
$$y \leq 100$$
$$x + y \geq 210$$
$$x \geq 0, y \geq 0.$$

This time, the objective function $p = -x + 5y$ does not have nonnegative coefficients, so we must follow the procedure described above: We enclose the corner points in a rectangle as shown in Figure 23. (There are infinitely many possible rectangles we could have used. We chose one that gives convenient coordinates for the new corners.)

We now list all the corners of this bounded region along with the corresponding values of the objective function c:

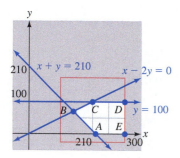

Figure 23

Point	Lines through Point	Coordinates	$p = -x + 5y$
A		$(210, 0)$	-210
B	$x + y = 210$ $x - 2y = 0$	$(140, 70)$	210
C	$x - 2y = 0$ $y = 100$	$(200, 100)$	300
D		$(300, 100)$	200
E		$(300, 0)$	-300

The corner point that gives the maximum value of the objective function p is C. Because C is one of the corner points of the original feasible region, we conclude that our LP problem has an optimal solution at C. Thus, the combination that gives the largest profit, \$300, is 200 hats and 100 pairs of sunglasses.

c. The objective function is now $p = x - 5y$, which is the negative of the objective function used in part (b). Thus, the table of values of p is the same as in part (b) except that it has opposite signs in the p column. This time we find that the maximum value of p occurs at E. However, E is not a corner point of the original feasible region, so the LP problem has no optimal solution. Referring to Figure 22, we can make the objective p as large as we like by choosing a point far to the right in the unbounded feasible region. Thus, the objective function is unbounded; that is, it is possible to make an arbitrarily large profit.

EXAMPLE 5   **Resource Allocation**

You are composing a very avant-garde ballade for violins and bassoons. In your ballade, each violinist plays a total of two notes, and each bassoonist plays only one note. To make your ballade long enough, you decide that it should contain at least 200 instrumental notes. Furthermore, after playing the requisite two notes, each violinist will sing one soprano note, while each bassoonist will sing three soprano notes.* To make the ballade sufficiently interesting, you have decided on a minimum of 300 soprano notes. To give your composition a sense of balance, you wish to have no more than three times as many bassoonists as violinists. Violinists charge $200 per performance, and bassoonists charge $400 per performance. How many of each should your ballade call for in order to minimize personnel costs?

* Whether or not these musicians are capable of singing decent soprano notes will be left to chance. You reason that a few bad notes will add character to the ballade.

Solution First, the unknowns are x = number of violinists and y = number of bassoonists. The constraint on the number of instrumental notes implies that

$$2x + y \geq 200$$

because the total number is to be *at least* 200. Similarly, the constraint on the number of soprano notes is

$$x + 3y \geq 300.$$

The next one is a little tricky. As usual, we reword it in terms of the quantities x and y.

The number of bassoonists should be no more than three times the number of violinists.

Thus, $y \leq 3x$

or $3x - y \geq 0.$

Finally, the total cost per performance will be

$$c = 200x + 400y.$$

We wish to minimize total cost. So our linear programming problem is as follows:

$$
\begin{aligned}
\text{Minimize} \quad & c = 200x + 400y \\
\text{subject to} \quad & 2x + y \geq 200 \\
& x + 3y \geq 300 \\
& 3x - y \geq 0 \\
& x \geq 0, y \geq 0.
\end{aligned}
$$

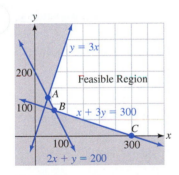

Figure 24

We get the feasible region shown in Figure 24.† Although the feasible region is unbounded, we can appeal again to part (a) of the Q&A before Example 4 and conclude that there are optimal solutions without the need to draw a bounding rectangle.

† In Figure 24 you can see how graphing technology would help in determining the corner points: Unless you are very confident in the accuracy of your sketch, how do you know that the line $y = 3x$ falls to the left of the point B? If it were to fall to the right, then B would not be a corner point, and the solution would be different. You could (and should) check that B satisfies the inequality $3x - y \geq 0$ so that the line falls to the left of B as shown. However, if you use a graphing calculator or computer, you can be fairly confident of the picture that is produced without doing further calculations.

Point	Lines through Point	Coordinates	$c = 200x + 400y$
A	$2x + y = 200$ $3x - y = 0$	$(40, 120)$	56,000
B	$2x + y = 200$ $x + 3y = 300$	$(60, 80)$	44,000
C		$(300, 0)$	60,000

From the table we see that the minimum cost occurs at B, so the minimum cost is $44,000 per performance, employing 60 violinists and 80 bassoonists. (Quite a wasteful ballade, one might say.)

FAQs

Recognizing a Linear Programming Problem, Setting Up Inequalities, and Dealing with Unbounded Regions

Q : *How do I recognize when an application leads to an LP problem as opposed to a system of linear equations?*

A : Here are some cues that suggest an LP problem:

- Key phrases suggesting inequalities rather than equalities, such as *at most, up to, no more than, at least,* and *or more.*
- A quantity that is being maximized or minimized (this will be the objective). Key phrases are *maximum, minimum, most, least, largest, greatest, smallest, as large as possible,* and *as small as possible.*

Q : *How do I deal with tricky phrases such as "there should be no more than twice as many nuts as bolts" or "at least 50% of the total should be bolts"?*

A : The easiest way to deal with phrases like this is to use the technique we discussed in Chapter 3: Reword the phrases using "the number of . . . ", as in

The number of nuts (x) is no more than twice the number of bolts (y): $x \leq 2y$.
The number of bolts is at least 50% of the total: $y \geq 0.50(x + y)$.

Q : *When do I not need to add a bounding rectangle to solve an LP problem with an unbounded feasible region?*

A : When you have the constraints $x \geq 0$ and $y \geq 0$, there is no need to add a bounding rectangle if:

- you are minimizing $c = ax + by$ with a and b nonnegative, in which case optimal solutions always exist (Examples 4(a) and 5 are of this type.);
- you are maximizing $p = ax + by$ with a and b both positive, in which case there is no optimal solution;
- you are maximizing $p = ax + by$ with $a \leq 0$ and $b \leq 0$, in which case optimal solutions always exist; or
- you are minimizing $c = ax + by$ with a and b both negative, in which case there is no optimal solution.

Using Technology

 Website

www.WanerMath.com
For an online utility that does everything (solves linear programming problems with two unknowns and even draws the feasible region), go to the Website and follow

→ Online Utilities
→ Linear Programming Grapher.

5.2   EXERCISES

▼ more advanced   ◆ challenging
T indicates exercises that should be solved using technology

In Exercises 1–24, solve the given LP problem. If no optimal solution exists, indicate whether the feasible region is empty or the objective function is unbounded. [**HINT:** See Example 1.]

1. Maximize $p = x + y$
subject to $x + 2y \leq 9$
$2x + y \leq 9$
$x \geq 0, y \geq 0.$

2. Maximize $p = x + 2y$
subject to $x + 3y \leq 24$
$2x + y \leq 18$
$x \geq 0, y \geq 0.$

3. Minimize $c = x + y$
subject to $x + 2y \geq 6$
$2x + y \geq 6$
$x \geq 0, y \geq 0.$

4. Minimize $c = x + 2y$
subject to $x + 3y \geq 30$
$2x + y \geq 30$
$x \geq 0, y \geq 0.$

5. Maximize $p = 3x + y$
subject to $3x - 7y \leq 0$
$7x - 3y \geq 0$
$x + y \leq 10$
$x \geq 0, y \geq 0.$

6. Maximize $p = x - 2y$
subject to $x + 2y \leq 8$
$x - 6y \leq 0$
$3x - 2y \geq 0$
$x \geq 0, y \geq 0.$

7. Maximize $p = 3x + 2y$
   subject to $0.2x + 0.1y \leq 1$
   $0.15x + 0.3y \leq 1.5$
   $10x + 10y \leq 60$
   $x \geq 0, y \geq 0.$

8. Maximize $p = x + 2y$
   subject to $30x + 20y \leq 600$
   $0.1x + 0.4y \leq 4$
   $0.2x + 0.3y \leq 4.5$
   $x \geq 0, y \geq 0.$

9. Minimize $c = 0.2x + 0.3y$
   subject to $0.2x + 0.1y \geq 1$
   $0.15x + 0.3y \geq 1.5$
   $10x + 10y \geq 80$
   $x \geq 0, y \geq 0.$

10. Minimize $c = 0.4x + 0.1y$
    subject to $30x + 20y \geq 600$
    $0.1x + 0.4y \geq 4$
    $0.2x + 0.3y \geq 4.5$
    $x \geq 0, y \geq 0.$

11. Maximize and minimize $p = x + 2y$
    subject to $x + y \geq 2$
    $x + y \leq 10$
    $x - y \leq 2$
    $x - y \geq -2.$

12. Maximize and minimize $p = 2x - y$
    subject to $x + y \geq 2$
    $x - y \leq 2$
    $x - y \geq -2$
    $x \leq 10, y \leq 10.$

13. Maximize $p = 2x - y$
    subject to $x + 2y \geq 6$
    $x \qquad \leq 8$
    $x \geq 0, y \geq 0.$

14. Maximize $p = x - 3y$
    subject to $2x + y \geq 4$
    $y \leq 5$
    $x \geq 0, y \geq 0.$

15. Maximize $p = 2x + 3y$
    subject to $0.1x + 0.2y \geq 1$
    $2x + y \geq 10$
    $x \geq 0, y \geq 0.$

16. Maximize $p = 3x + 2y$
    subject to $0.1x + 0.1y \geq 0.2$
    $y \leq 10$
    $x \geq 0, y \geq 0.$

17. Minimize $c = x - 3y$
    subject to $3x + y \geq 5$
    $2x - y \geq 0$
    $x - 3y \leq 0$
    $x \geq 0, y \geq 0.$

18. Minimize $c = 3x - y$
    subject to $2x - y \geq 3$
    $x - y \geq 0$
    $x - 2y \leq 0$
    $x \geq 0, y \geq 0.$

19. Minimize $c = 2x + 4y$
    subject to $0.1x + 0.1y \geq 1$
    $x + 2y \geq 14$
    $x \geq 0, y \geq 0.$

20. Maximize $p = 2x + 3y$
    subject to $-x + y \geq 10$
    $x + 2y \leq 12$
    $x \geq 0, y \geq 0.$

21. Minimize $c = 3x - 3y$
    subject to $\dfrac{x}{4} \leq y$
    $y \leq \dfrac{2x}{3}$
    $x + y \geq 5$
    $x + 2y \leq 10$
    $x \geq 0, y \geq 0.$

22. Minimize $c = -x + 2y$
    subject to $y \leq \dfrac{2x}{3}$
    $x \leq 3y$
    $y \geq 4$
    $x \geq 6$
    $x + y \leq 16.$

23. Maximize $p = x + y$
    subject to $x + 2y \geq 10$
    $2x + 2y \leq 10$
    $2x + y \geq 10$
    $x \geq 0, y \geq 0.$

24. Minimize $c = 3x + y$
    subject to $10x + 20y \geq 100$
    $0.3x + 0.1y \geq 1$
    $x \geq 0, y \geq 0.$

Applications

25. Resource Allocation You manage an ice cream factory that makes two flavors: Creamy Vanilla and Continental Mocha. Into each quart of Creamy Vanilla go 2 eggs and 3 cups of cream. Into each quart of Continental Mocha go 1 egg and 3 cups of cream. You have in stock 500 eggs and 900 cups of cream. You make a profit of $3 on each quart of Creamy Vanilla and $2 on each quart of Continental Mocha. How many quarts of each flavor should you make to earn the largest profit? [**HINT:** See Example 2.]

26. Resource Allocation *Podunk Institute of Technology's* Math Department offers two courses: Finite Math and Applied Calculus. Each section of Finite Math has 60 students, and each section of Applied Calculus has 50. The department is allowed to offer a total of up to 110 sections. Furthermore, no more than 6,000 students want to take a math course. (No student will take more than one math course.) Suppose the university makes a profit of $100,000 on each section of Finite Math and $50,000 on each section of Applied Calculus. (The profit is the difference between what the students are charged and what the professors are paid.) How many sections of each course should the department offer to make the largest profit? [**HINT:** See Example 2.]

27. Nutrition *Ruff, Inc.* makes dog food out of chicken and grain. Chicken has 10 grams of protein and 5 grams of fat per ounce, and grain has 2 grams of protein and 2 grams of fat per ounce. A bag of dog food must contain at least 200 grams of protein and at least 150 grams of fat. If chicken costs 10¢ per ounce and grain costs 1¢ per ounce, how many ounces of each should Ruff use in each bag of dog food to minimize cost? [**HINT:** See Example 4.]

28. Purchasing *Enormous State University's* Business School is buying computers. The school has two models from which to choose, the Pomegranate and the iZac. Each Pomegranate comes with 400 GB of memory and 80 TB of disk space; each iZac has 300 GB of memory and 100 TB of disk space. For reasons related to its accreditation the school would like to be able to say that it has a total of at least 48,000 GB of memory and at least 12,800 TB of disk space. If the Pomegranate and the iZac cost $2,000 each, how many of each should the school buy to keep the cost as low as possible? [**HINT:** See Example 4.]

29. Nutrition **Gerber Products'** Gerber Mixed Cereal for Baby contains, in each serving, 60 calories and 11 grams of carbohydrates. Gerber Mango Tropical Fruit Dessert contains, in each serving, 80 calories and 21 grams of carbohydrates.[11] If the cereal costs 30¢ per serving and the dessert costs 50¢ per serving, and you want to provide your child with at least 140 calories and at least 32 grams of carbohydrates, how can you do so at the least cost? (Fractions of servings are permitted.)

30. Nutrition **Gerber Products'** Gerber Mixed Cereal for Baby contains, in each serving, 60 calories, 10 grams of carbohydrates, and no vitamin C. Gerber Apple Banana Juice contains, in each serving, 60 calories, 15 grams of carbohydrates, and 120% of the U.S. Recommended Daily Allowance (RDA) of vitamin C for infants.[12] The cereal costs 10¢ per serving, and the juice costs 30¢ per serving. If you want to provide your child with at least 120 calories, at least 25 grams of carbohydrates, and at least 60% of the U.S. RDA of vitamin C for infants, how can you do so at the least cost? (Fractions of servings are permitted.)

31. Energy Efficiency You are thinking of making your home more energy efficient by replacing some of the light bulbs with compact fluorescent bulbs and insulating part or all of your exterior walls. Each compact fluorescent light bulb costs $4 and saves you an average of $2 per year in energy costs, and each square foot of wall insulation costs $1 and saves you an average of $0.20 per year in energy costs.[13] Your home has 60 light fittings and 1,100 square feet of uninsulated exterior wall. You can spend no more than $1,200 and would like to save as much per year in energy costs as possible. How many compact fluorescent light bulbs and how many square feet of insulation should you purchase? How much will you save in energy costs per year?

32. Energy Efficiency (Compare with Exercise 31.) You are thinking of making your mansion more energy efficient by replacing some of the light bulbs with compact fluorescent bulbs and insulating part or all of your exterior walls. Each compact fluorescent light bulb costs $4 and saves you an average of $2 per year in energy costs, and each square foot of wall insulation costs $1 and saves you an average of $0.20 per year in energy costs.[14] Your mansion has 200 light fittings and 3,000 square feet of uninsulated exterior wall. To impress your friends, you would like to spend as much as possible but save no more than $800 per year in energy costs. (You are proud of your large utility bills.) How many compact fluorescent light bulbs and how many square feet of insulation should you purchase? How much will you save in energy costs per year?

Bodybuilding Supplements *Exercises 33–36 are based on the following data on four bodybuilding supplements. (Figures shown correspond to a single serving.)*[15]

---

[11] Source: Nutrition information supplied with the products.

[12] *Ibid.*

[13] Source: American Council for an Energy-Efficient Economy/*New York Times*, December 1, 2003, p. C6.

[14] *Ibid.*

[15] Source: Nutritional information supplied by the manufacturers/ www.bodybuilding.com. Cost per serving is approximate and varies considerably. "BCAAs" refers to the branched-chain amino acids leucine, isoleucine, and valine in the optimal 2:1:1 ratio.

	Creatine (grams)	L-Glutamine (grams)	BCAAs (grams)	Cost ($)
Xtend (SciVation)	0	2.5	7	1.00
Gainz (MP Hardcore)	2	3	6	1.10
Strongevity (Bill Phillips)	2.5	1	0	1.20
Muscle Physique (EAS)	2	2	0	1.00

33. Your personal trainer suggests that you supplement with at least 4 grams of creatine, 36 grams of L-glutamine, and 84 grams of BCAAs each week. You are thinking of combining Xtend and Gainz to provide you with the required nutrients. How many servings of each should you combine to obtain a week's supply that meets your trainer's specifications at the least cost?

34. Your friend's personal trainer suggests that she supplement with at least 30 grams of creatine, 42 grams of L-glutamine, and 36 grams of BCAAs each week. Your friend is thinking of combining Gainz and Muscle Physique to provide her with the required nutrients. How many servings of each should she combine to obtain a week's supply that meets her trainer's specifications at the least cost?

35. Your new personal trainer suggests that you supplement with at least 40 grams of creatine and 38 grams of L-glutamine but no more than 90 grams of BCAAs each week. You are thinking of combining Gainz and Strongevity to create a week's supply that meets your new trainer's specifications.
    a. Can you combine the products in such a way that the number of servings of Gainz exceeds that of Strongevity by as much as possible? If so, how many servings of each should you combine? If not, explain why not.
    b. Can you combine the products in such a way that the number of servings of Strongevity exceeds that of Gainz by as much as possible? If so, how many servings of each should you combine? If not, explain why not.

36. Your friend's new personal trainer suggests that she supplement with no more than 20 grams of creatine but at least 20 grams of L-glutamine and 42 grams of BCAAs each week. She is thinking of combining Xtend and Strongevity to create a week's supply that meets her new trainer's specifications.
    a. Can she combine the products in such a way that the number of servings of Xtend exceeds that of Strongevity by as much as possible? If so, how many servings of each should she combine? If not, explain why not.
    b. Can she combine the products in such a way that the number of servings of Strongevity exceeds that of Xtend by as much as possible? If so, how many servings of each should she combine? If not, explain why not.

37. *Resource Allocation* Your salami manufacturing plant can order up to 1,000 pounds of pork and 2,400 pounds of beef per day for use in manufacturing its two specialties: *Count Dracula Salami* and *Frankenstein Sausage*. Production of the Count Dracula variety requires 1 pound of pork and 3 pounds of beef for each salami, while the Frankenstein variety requires 2 pounds of pork and 2 pounds of beef for every sausage. In view of your heavy investment in advertising Count Dracula Salami, you have decided that at least one third of the total production should be Count Dracula. On the other hand, because of the health-conscious consumer climate, your Frankenstein Sausage (sold as having less beef) is earning your company a profit of $3 per sausage, while sales of the Count Dracula variety are down and it is earning your company only $1 per salami. Given these restrictions, how many of each kind of sausage should you produce to maximize profits, and what is the maximum possible profit? [**HINT:** See Example 3.]

38. *Project Design* The *Megabuck Hospital Corporation* is to build a state-subsidized nursing home serving homeless patients as well as high-income patients. State regulations require that every subsidized nursing home must house a minimum of 1,000 homeless patients and no more than 750 high-income patients in order to qualify for state subsidies. The overall capacity of the nursing home is to be 2,100 patients. The board of directors, under pressure from a neighborhood group, insists that the number of homeless patients should not exceed twice the number of high-income patients. Because of the state subsidy, the nursing home will make an average profit of $10,000 per month for every homeless patient it houses, whereas the profit per high-income patient is estimated at $8,000 per month. How many of each type of patient should the nursing home house to maximize profit? [**HINT:** See Example 3.]

39. *Television Advertising* On Monday evenings in April 2015, each episode of *The Big Bang Theory* was typically watched by 1.8 million viewers, while each episode of *American Dad* was typically watched by 1.5 million viewers.[16] Your marketing services firm has been hired to promote *Bald No More*'s hair replacement process by buying a total of at least 30 commercial spots during episodes of *The Big Bang Theory* and *American Dad*. You have been quoted a price of $3,000 per spot for *The Big Bang Theory* and $1,000 per spot for *American Dad*. Bald No More's advertising budget for TV commercials is $120,000, and, because of the company president's fondness for physics, it would like no more than 50% of the total number of spots to appear on *American Dad*. How many spots should you purchase on each show to maximize exposure? [**HINT:** Calculate exposure as Number of ads × Number of viewers.]

---

[16] Ratings are for April 4, 2015. Source: Nielsen Media Research/www.tvbythenumbers.com.

40. *Television Advertising* On Monday evenings in April 2015, each episode of *WWE Entertainment* was typically watched by 3.8 million viewers, while each episode of *American Dad* was typically watched by 1.5 million viewers.[17] Your marketing services firm has been hired to promote *Gauss Jordan* sneakers by buying at least 40 commercial spots during episodes of *WWE Entertainment* and *American Dad*. You have been quoted a price of $4,000 per spot for *WWE Entertainment* and $1,000 per spot for *American Dad*. Gauss Jordan, Inc.'s advertising budget for TV commercials is $260,000, and it would like at least 75% of the total number of spots to appear on *WWE Entertainment*. How many spots should you purchase on each show to maximize exposure? [**HINT:** Calculate exposure as Number of ads × Number of viewers.]

Investing *Exercises 41 and 42 are based on the following data on four stocks:*[18]

	Price ($)	Dividend Yield (%)	52-Week Price Change ($)
OCR (**Omnicare**)	90	1	28
RCKY (**Rocky Brands**)	20	2	6
GCO (**Genesco**)	70	0	−5
ATVI (**Activision Blizzard**)	25	1	5

41. ▼ You are planning to invest up to $10,000 in OCR and RCKY shares. You want your investment to yield at least $120 in dividends, and, for tax reasons, you want to minimize the 52-week gain in the total value of the shares. How many shares of each company should you purchase? (Fractions of shares are permitted.)

42. ▼ You are planning to invest up to $43,000 in GCO and ATVI shares. For tax reasons you want your investment to yield no more than $10 in dividends. You want to minimize the 52-week gain (or maximize the loss) in the total value of the shares. How many shares of each company should you purchase? (Fractions of shares are permitted.)

43. ▼ *Investments: Financial Stocks* (Compare Exercise 41 in Section 5.1.) During the first quarter of 2015, **Toronto Dominion Bank** (TD) stock cost $45 per share, was expected to yield 4% per year in dividends, and had a risk index of 3.0 per share, while **CNA Financial Corp.** (CNA) stock cost $40 per share, was expected to yield 2.5% per year in dividends, and had a risk index of 2.0 per share.[19] You have up to $25,000 to invest in these stocks and would like to earn at least $760 in dividends over the course of a year. (Assume the dividend to be unchanged for the year.) How many shares (to the nearest tenth of a unit) of each stock should you purchase to meet your requirements and minimize the total risk index for your portfolio? What is the minimum total risk index?

44. ▼ *Investments: High-Dividend Stocks* (Compare Exercise 42 in Section 5.1.) During the first quarter of 2015, **Plains All American Pipeline L.P.** (PAA) stock cost $50 per share, was expected to yield 5% per year in dividends, and had a risk index of 2.0, while **Total SA** (TOT) stock cost $50 per share, was expected to yield 6% per year in dividends, and had a risk index of 3.0.[20] You have up to $45,000 to invest in these stocks and would like to earn at least $2,400 in dividends over the course of a year. (Assume the dividend to be unchanged for the year.) How many shares of each stock should you purchase to meet your requirements and minimize the total risk index for your portfolio? What is the minimum total risk index?

45. ▼ *Planning* My friends: I, the mighty Brutus, have decided to prepare for retirement by instructing young warriors in the arts of battle and diplomacy. For each hour spent in battle instruction, I have decided to charge 50 ducats. For each hour spent in diplomacy instruction, I shall charge 40 ducats. Because of my advancing years, I can spend no more than 50 hours per week instructing the youths, although the great Jove knows that they are sorely in need of instruction! Because of my fondness for physical pursuits, I have decided to spend no more than one third of the total time in diplomatic instruction. However, the present border crisis with the Gauls is a sore indication of our poor abilities as diplomats. As a result, I have decided to spend at least 10 hours per week instructing in diplomacy. Finally, to complicate things further, there is the matter of Scarlet Brew: I have estimated that each hour of battle instruction will require 10 gallons of Scarlet Brew to quench my students' thirst and that each hour of diplomacy instruction, being less physically demanding, requires half that amount. Because my harvest of red berries has far exceeded my expectations, I estimate that I'll have to use at least 400 gallons per week in order to avoid storing the fine brew at great expense. Given all these restrictions, how many hours per week should I spend in each type of instruction to maximize my income?

46. ▼ *Planning* Repeat Exercise 45 with the following changes: I would like to spend no more than half the total time in diplomatic instruction, and I must use at least 600 gallons of Scarlet Brew.

47. ▼ *Resource Allocation* One day, Gillian the magician summoned the wisest of her women. "Devoted sisters of the Coven," she began, "I have a quandary: As you well know, I possess great expertise in sleep spells and shock spells, but

---

[17] See footnote for Exercise 39.

[18] Approximate price as of May 13, 2015. 52-week price changes are approximate. Source: http://finance.google.com.

[19] Stock prices and yields are approximate, and risk indices are fictitious. Source: http://finance.google.com.

[20] *Ibid.*

unfortunately, these can be a drain on my aural energy resources, and I would like my net expenditure of aural energy to be a minimum yet still meet my commitments in protecting the Sisterhood from the ever-present threat of trolls. Specifically, I have estimated that each sleep spell keeps us safe for an average of 3 hours, while every shock spell protects us for only 1 hour. We certainly require enough protection to last 24 hours of each day and possibly more, just to be safe. At the same time, I have noticed that each of my sleep spells can immobilize two trolls at once, whereas one of my powerful shock spells can immobilize four trolls at once. We are faced, my sisters, with an onslaught of as many as 26 trolls per day! Finally, as you are no doubt aware, the Bylaws of the Coven dictate that for a magician to remain in good standing, she should cast no more shock spells than sleep spells, whereas—and I quote from Bylaw 33c—"The number of sleep spells shall never exceed thrice that of shock spells by more than three." What do I do, oh Wise Ones?" How would they respond if:

a. Each sleep spell uses 50 therms of aural energy and each shock spell uses 20 therms?

b. Each sleep spell uses 40 therms of aural energy whereas each shock spell *boosts* aural energy by 10 therms?

c. Each sleep spell uses 10 therms of aural energy whereas each shock spell boosts aural energy by 40 therms?

[**HINT:** See Example 4.]

48. ▼ *Risk Management* The Grand Vizier of the Kingdom of Um is being blackmailed by nine individuals and is having a very difficult time keeping them from going public. He has been keeping them at bay with two kinds of payoff: gold from the Royal Treasury and political favors. Through long experience, he has learned that each gold payoff gives him peace for an average of about 1 month and has an exposure risk index of -5, while each political favor earns him about a month and a half of reprieve but has an exposure risk index of $+1$. To maintain his flawless reputation in the Court, he feels that he cannot afford any revelations about his tainted past to come to light within the next year. So it is imperative that he make at least nine payoffs this year, that his blackmailers be kept at bay for 12 months, and that he maintain a total exposure risk index of no more than 3. Furthermore, he would like to keep the number of gold payoffs at no more than 60% of the combined number of payoffs because the outward flow of gold bars might arouse suspicion on the part of the Royal Treasurer. The gold payoffs and political favors tend to affect his travel budget (he frequently travels to the Himalayas for vizier-related reasons). He would like to maintain his flawless reputation in the Court in such a way that the net loss to his travel budget is a minimum. What is he to do, and what is the effect on his travel budget if:

a. Each gold bar removed from the treasury depletes his travel budget by 2 Orbs, and, as a result of administrative costs, each political favor depletes his travel budget by 4 Orbs.

b. Each gold bar removed from the treasury somehow *adds* two Orbs to his travel budget, but each political favor depletes it by one Orb.

c. Each gold bar removed from the treasury depletes his travel budget by 6 Orbs, but, for reasons too complicated to explain, each political favor *adds* an Orb to the budget.

[**HINT:** See Example 4.]

49. ◆ *Management*[21] You are the service manager for a supplier of closed-circuit television systems. Your company can provide up to 160 hours per week of technical service for your customers, but the demand for technical service far exceeds this amount. As a result, you have been asked to develop a model to allocate service technicians' time between new customers (those still covered by service contracts) and old customers (whose service contracts have expired). To ensure that new customers are satisfied with your company's service, the sales department has instituted a policy that at least 100 hours per week be allocated to servicing new customers. At the same time, your superiors have informed you that the company expects your department to generate at least $1,200 per week in revenues. Technical service time for new customers generates an average of $10 per hour (because much of the service is still under warranty), and that for old customers generates $30 per hour. How many hours per week should you allocate to each type of customer to generate the most revenue?

50. ◆ *Scheduling*[22] The *Scottsville Textile Mill* produces several different fabrics on eight dobby looms that operate 24 hours per day and are scheduled for 30 days in the coming month. The mill will produce only Fabric 1 and Fabric 2 during the coming month. Each dobby loom can turn out 4.63 yards of either fabric per hour. Assume that there is a monthly demand of 16,000 yards of Fabric 1 and 12,000 yards of Fabric 2. Profits are calculated as 33¢ per yard for each fabric produced on the dobby looms.

a. Will it be possible to satisfy total demand?

b. In the event that total demand is not satisfied, the Scottsville Textile Mill will need to purchase the fabrics from another mill to make up the shortfall. Its profits on resold fabrics ordered from another mill amount to 20¢ per yard for Fabric 1 and 16¢ per yard for Fabric 2. How many yards of each fabric should it produce to maximize profits?

---

[21] Loosely based on a similiar problem in *An Introduction to Management Science* (6th Ed.) by D. R. Anderson, D. J. Sweeney, and T. A. Williams (West, 1991).

[22] Adapted from *The Calhoun Textile Mill Case* by J. D. Camm, P. M. Dearing, and S. K. Tadisina as presented for case study in *An Introduction to Management Science* (6th Ed.) by D. R. Anderson, D. J. Sweeney, and T. A. Williams (West, 1991). Our exercise uses a subset of the data given in the cited study.

Communication and Reasoning Exercises

51. If a linear programming problem has a bounded, nonempty feasible region, then optimal solutions
(A) must exist.   (B) may or may not exist.
(C) cannot exist.

52. If a linear programming problem has an unbounded, non-empty feasible region, then optimal solutions
(A) must exist.   (B) may or may not exist.
(C) cannot exist.

53. What can you say if the optimal value occurs at two adjacent corner points?

54. Describe at least one drawback to using the graphical method to solve a linear programming problem arising from a real-life situation.

55. The feasible region of your LP problem is unbounded, and two of your constraints are $x \geq 0$ and $y \geq 0$. Decide in each case whether a bounding rectangle is necessary to decide whether an optimal solution exists. If it is not necessary, state whether the LP problem does or does not have a solution.
a. You are minimizing $c = 4x + y$.
b. You are maximizing $p = 2x$.
c. You are maximizing $p = 4x - y$.
d. You are maximizing $p = 2x + y$.

56. The feasible region of your LP problem is unbounded, and two of your constraints are $x \geq 0$ and $y \geq 0$. Decide in each case whether a bounding rectangle is necessary to decide whether an optimal solution exists. If it is not necessary, state whether the LP problem does or does not have a solution.
a. You are minimizing $c = 4x - y$.
b. You are minimizing $c = -x - y$.
c. You are minimizing $c = -5y$.
d. You are maximizing $p = -2x - y$.

57. Create a linear programming problem in two variables that has no optimal solution.

58. Create a linear programming problem in two variables that has more than one optimal solution.

59. Create an interesting scenario leading to the following linear programming problem:

$$\text{Maximize} \quad p = 10x + 10y$$
$$\text{subject to} \quad 20x + 40y \leq 1,000$$
$$30x + 20y \leq 1,200$$
$$x \geq 0, y \geq 0.$$

60. Create an interesting scenario leading to the following linear programming problem:

$$\text{Minimize} \quad c = 10x + 10y$$
$$\text{subject to} \quad 20x + 40y \geq 1,000$$
$$30x + 20y \geq 1,200$$
$$x \geq 0, y \geq 0.$$

61. ▼ Use an example to show why there may be no optimal solution to a linear programming problem if the feasible region is unbounded.

62. ▼ Use an example to illustrate why, in the event that an optimal solution does occur despite an unbounded feasible region, that solution corresponds to a corner point of the feasible region.

63. ▼ You are setting up an LP problem for Fly-by-Night Airlines with the unknowns x and y, where x represents the number of first-class tickets it should issue for a specific flight and y represents the number of business-class tickets it should issue for that flight, and the problem is to maximize profit. You find that there are two different corner points that maximize the profit. How do you interpret this?

64. ▼ In the situation described in Exercise 63, you find that there are no optimal solutions. How do you interpret this?

65. ◆ Consider the following example of a *nonlinear* programming problem: Maximize $p = xy$ subject to $x \geq 0, y \geq 0$, $x + y \leq 2$. Show that p is zero on every corner point but is greater than zero at many non-corner points.

66. ◆ Solve the nonlinear programming problem in Exercise 65.

5.3   The Simplex Method: Solving Standard Maximization Problems

Standard Maximization Problems and Slack Variables

The method discussed in Section 5.2 works quite well for LP problems in two unknowns, but what about three or more unknowns? Because we need an axis for each unknown, we would need to draw graphs in three dimensions (where we have x-, y-, and z-coordinates) to deal with problems in three unknowns, and we would have to draw in hyperspace to answer questions involving four or more unknowns. Given the state of technology as this book is being written, we can't easily do this. So we need another method for solving LP problems that will work for any number of unknowns. One such method, called the **simplex method**, has been the method of choice since it was invented by George Dantzig in 1947. (See the Introduction to this chapter for more about Dantzig.) To illustrate it best, we first use it to solve only so-called standard maximization problems.

General Linear Programming Problem

A **linear programming problem in n unknowns** $x_1, x_2, \ldots, x_n$ is one in which we are to find the maximum or minimum value of a linear **objective function**

$$a_1x_1 + a_2x_2 + \cdots + a_nx_n,$$

where $a_1, a_2, \ldots, a_n$ are numbers, subject to a number of linear **constraints** of the form

$$b_1x_1 + b_2x_2 + \cdots + b_nx_n \leq c \quad \text{or} \quad b_1x_1 + b_2x_2 + \cdots + b_nx_n \geq c,$$

where $b_1, b_2, \ldots, b_n$, and c are numbers.

Standard Maximization Problem

A **standard maximization problem** is an LP problem in which we are required to *maximize* (not minimize) an objective function of the form

$$p = a_1x_1 + a_2x_2 + \cdots + a_nx_n$$

subject to the constraints

$$x_1 \geq 0, x_2 \geq 0, \ldots, x_n \geq 0$$

and further constraints of the form

$$b_1x_1 + b_2x_2 + \cdots + b_nx_n \leq c$$

with c *nonnegative*. It is important that the relation here be $\leq$, *not* $=$ or $\geq$.

Note As in Chapter 3, we will almost always use $x, y, z, \ldots$ for the unknowns. Subscripted variables $x_1, x_2, \ldots$ are very useful names when you start running out of letters of the alphabet, but we should not find ourselves in that predicament. ■

Quick Examples

1. Maximize $p = 2x - 3y + 3z$
   subject to $2x \quad\quad + z \leq 7$
   $-x + 3y - 6z \leq 6$
   $x \geq 0, y \geq 0, z \geq 0.$

   This is a standard maximization problem.

2. Maximize $p = 2x_1 + x_2 - x_3 + x_4$
   subject to $x_1 - 2x_2 \quad\quad + x_4 \leq 0$
   $3x_1 \quad\quad\quad\quad \leq 1$
   $\quad\quad x_2 + x_3 \quad\quad \leq 2$
   $x_1 \geq 0, x_2 \geq 0, x_3 \geq 0, x_4 \geq 0.$

   This is a standard maximization problem.

3. Maximize $\quad p = 2x - 3y + 3z$
subject to $\quad 2x \quad\;\; + \;\; z \geq 7$
$-x + 3y - 6z \leq 6 \quad$ This is *not* a standard maximization problem.
$x \geq 0,\, y \geq 0,\, z \geq 0.$

The inequality $2x + z \geq 7$ cannot be written in the required form. If we reverse the inequality by multiplying both sides by -1, we get $-2x - z \leq -7$, but a negative value on the right side is not allowed.

Consider the following standard maximization problem (which we will actually solve in Example 1 below):

Maximize $\quad p = 3x + 2y + z$
subject to $\quad 2x + 2y + \;\; z \leq 10$
$x + 2y + 3z \leq 15$
$x \geq 0,\, y \geq 0,\, z \geq 0.$

The constraints are linear inequalities, but the simplex method is a matrix-based method that works by finding nonnegative solutions of a related system of linear *equations* rather than inequalities. Because the solutions that come out of the simplex method will always be nonnegative, we can assume the inequalities $x \geq 0$, $y \geq 0$, and $z \geq 0$ listed at the end and pay them no more attention. This leaves us with the first two inequalities, which we need to convert somehow to linear equations.

Look at the first inequality. It says that the left-hand side, $2x + 2y + z$, must have some positive number (or zero) *added to it* if it is to equal 10. Because we don't yet know what x, y, and z are, we are not yet sure what number to add to the left-hand side. So we invent a new unknown, $s \geq 0$, called a **slack variable**, to "take up the slack," so that

$$2x + 2y + z + s = 10.$$

Turning to the next inequality, $x + 2y + 3z \leq 15$, we now add a slack variable to its left-hand side to get it up to the value of the right-hand side. We might have to add a different number than we did the last time, so we use a new slack variable, $t \geq 0$, and obtain

$$x + 2y + 3z + t = 15. \quad \text{Use a different slack variable for each constraint.}$$

Now we have a system of three linear equations (including the one that defines the objective function), which we can write in standard form as follows:

$$\begin{aligned}
2x + 2y + \;\; z + s \quad\quad &= 10 \\
x + 2y + 3z \quad\;\; + t \quad\;\; &= 15 \\
-3x - 2y - \;\; z \quad\quad\;\; + p &= 0.
\end{aligned}$$

Note three things: First, all the variables are neatly aligned in columns, as they were in Chapter 3. Second, in rewriting the objective function $p = 3x + 2y + z$, we have left the coefficient of p as $+1$ and brought the other variables over to the same side of the equation as p. This will be our standard procedure from now on. *Don't* write $3x + 2y + z - p = 0$ (even though it means the same thing) because the negative coefficients will be important in the simplex method. Third, the above system of equations has fewer equations than unknowns and hence cannot have a unique solution.[*]

[*] That is what we might expect, however. The (possibly infinitely many) solutions that result (with all the variables nonnegative) turn out to correspond to points in the feasible region of the LP problem.

Equation Form of a Standard Maximization Problem

Given a standard maximization problem,

$$\text{Maximize} \quad p = a_1x_1 + a_2x_2 + \cdots + a_nx_n$$
$$\text{subject to} \quad b_1x_1 + b_2x_2 + \cdots + b_nx_n \leq c$$
$$b_1'x_1 + b_2'x_2 + \cdots + b_n'x_n \leq c'$$
$$\vdots$$
$$x_1 \geq 0, x_2 \geq 0, \ldots, x_n \geq 0,$$

its **equation form** is the system of linear equations

$$b_1x_1 + b_2x_2 + \cdots + b_nx_n \quad + s_1 \qquad\qquad = c$$
$$b_1'x_1 + b_2'x_2 + \cdots + b_n'x_n \qquad\quad + s_2 \qquad = c'$$
$$\vdots$$
$$-a_1x_1 - a_2x_2 - \cdots - a_nx_n \qquad\qquad\qquad + p = 0$$

where $s_1, s_2, \ldots$ are called **slack variables**.

Quick Example

4. The standard LP problem

$$\text{Maximize} \quad p = 2x - 3y + 3z$$
$$\text{subject to} \quad 2x \qquad\quad + z \leq 7$$
$$-x + 3y - 6z \leq 6$$
$$x + y + z \leq 15$$
$$x \geq 0, y \geq 0, z \geq 0$$

has the equation form

$$2x \qquad\quad + z + s \qquad\qquad\qquad = 7$$
$$-x + 3y - 6z \qquad\quad + t \qquad\qquad = 6$$
$$x + y + z \qquad\qquad\quad + u \qquad = 15$$
$$-2x + 3y - 3z \qquad\qquad\qquad\quad + p = 0.$$

The Simplex Method

The idea behind the simplex method is this: In any linear programming problem, there is a feasible region. If there are only two unknowns, we can draw the region; if there are three unknowns, it is a solid region in space; and if there are four or more unknowns, it is an abstract higher-dimensional region. But it is a faceted region with corners (think of a diamond), and it is at one of these corners that we will find the optimal solution. Geometrically, what the simplex method does is to start at the corner where all the unknowns are 0 (possible because we are talking of standard maximization problems) and then walk around the region, from corner to adjacent corner, always increasing the value of the objective function, until the best corner is found. In practice, we will visit only a small number of the corners before finding the right one. Algebraically, as we are about to see, this walking around is accomplished by matrix manipulations of the same sort as those used in the chapter on systems of linear equations.

We describe the method while working through an example.

EXAMPLE 1   Meet the Simplex Method

Maximize $p = 3x + 2y + z$

subject to $2x + 2y + z \le 10$

$x + 2y + 3z \le 15$

$x \ge 0, y \ge 0, z \ge 0.$

Solution

Step 1 *Write the LP problem in equation form.* (We already did this above.)

$$
\begin{aligned}
2x + 2y + z + s \quad\quad\quad &= 10 \\
x + 2y + 3z \quad + t \quad\quad &= 15 \\
-3x - 2y - z \quad\quad + p &= 0
\end{aligned}
$$

Step 2 *Set up the initial tableau.* We represent our system of equations by the following table (which is simply the augmented matrix in disguise), called **the initial tableau**:

	x	*y*	*z*	*s*	*t*	*p*	
	2	2	1	1	0	0	10
	1	2	3	0	1	0	15
	−3	−2	−1	0	0	1	0

The labels along the top keep track of which columns belong to which variables.

Now notice a peculiar thing. If we rewrite the matrix using the variables s, t, and p first, we get the matrix

$$
\begin{array}{cccccc}
s & t & p & x & y & z \\
\left[\begin{array}{cccccc|c}
1 & 0 & 0 & 2 & 2 & 1 & 10 \\
0 & 1 & 0 & 1 & 2 & 3 & 15 \\
0 & 0 & 1 & -3 & -2 & -1 & 0
\end{array}\right]
\end{array},
$$

Matrix with s, t, and p columns first

which is already in reduced form. We can therefore read off the general solution (see Section 3.2) to our system of equations as

$$
\begin{aligned}
s &= 10 - 2x - 2y - z \\
t &= 15 - x - 2y - 3z \\
p &= 0 + 3x + 2y + z
\end{aligned}
$$

x, y, z arbitrary.

Thus, we get a whole family of solutions, one for each choice of x, y, and z. One possible choice is to set x, y, and z all equal to 0. This gives the particular solution

$$s = 10, \quad t = 15, \quad p = 0, \quad x = 0, \quad y = 0, \quad z = 0. \qquad \text{Set } x = y = z = 0 \text{ above.}$$

This solution is called the **basic solution** associated with the tableau. The variables s and t are called the **active** variables, and x, y, and z are the **inactive** variables. (Other terms used are **basic** and **nonbasic** variables.)*

We can obtain the basic solution directly from the tableau as follows:

• The active variables correspond to the cleared columns (columns with only one nonzero entry).

---

* In the language of Chapter 3, x, y, and z are the *parameters* of the general solution. But in the context of the simplex method, they are always chosen to be zero, hence the term *inactive*.

- The values of the active variables are calculated as shown below.
- All other variables are inactive and are set equal to zero.

Inactive	Inactive	Inactive	Active	Active	Active	
$x = 0$	$y = 0$	$z = 0$	$s = \frac{10}{1}$	$t = \frac{15}{1}$	$p = \frac{0}{1}$	
x	y	z	s	t	p	
2	2	1	1	0	0	10
1	2	3	0	1	0	15
-3	-2	-1	0	0	1	0

As an additional aid to recognizing which variables are active and which are inactive, we label each row with the name of the corresponding active variable. Thus, the complete initial tableau looks like this:

	x	y	z	s	t	p	
s	2	2	1	1	0	0	10
t	1	2	3	0	1	0	15
p	-3	-2	-1	0	0	1	0

This basic solution represents our starting position $x = y = z = 0$ in the feasible region in xyz-space.

We now need to move to another corner point. To do so, we choose a pivot* in one of the first three columns of the tableau and clear its column. Then we will get a different basic solution, which corresponds to another corner point. Thus, to move from corner point to corner point, all we have to do is choose suitable pivots and clear columns in the usual manner.

The next two steps give the procedure for choosing the pivot.

Step 3 *Select the pivot column* (the column that contains the pivot we are seeking).

> ### Selecting the Pivot Column
>
> Choose the negative number with the largest magnitude on the left-hand side of the bottom row (that is, don't consider the last number in the bottom row). Its column is the pivot column. (If there are two or more candidates, choose any one.) If all the numbers on the left-hand side of the bottom row are zero or positive, then we are done, and the basic solution is the optimal solution.

Simple enough. The most negative number in the bottom row is -3, so we choose the x column as the pivot column:

	x	y	z	s	t	p	
s	2	2	1	1	0	0	10
t	1	2	3	0	1	0	15
p	-3	-2	-1	0	0	1	0

↑
Pivot column

* See Section 3.2 for a discussion of pivots and pivoting.

Q: *Why choose the pivot column this way?*

A: The variable labeling the pivot column is going to be increased from 0 to something positive. In the equation $p = 3x + 2y + z$, the fastest way to increase p is to increase x because p would increase by 3 units for every 1-unit increase in x. (If we chose to increase y, then p would increase by only 2 units for every 1-unit increase in y; and if we increased z instead, p would grow even more slowly.) In short, choosing the pivot column this way makes it likely that we'll increase p as much as possible.

Step 4 *Select the pivot in the pivot column.*

Selecting the Pivot

1. The pivot must always be a positive number. (This rules out zeros and negative numbers, such as the -3 in the bottom row.)

2. For each positive entry b in the pivot column, compute the ratio a/b, where a is the number in the rightmost column in that row. We call this a **test ratio**.

3. Of these ratios, choose the smallest one. (If there are two or more candidates, choose any one.) The corresponding number b is the pivot.

In our example the test ratio in the first row is $10/2 = 5$, and the test ratio in the second row is $15/1 = 15$. Here, 5 is the smallest, so the 2 in the upper left is our pivot.

	x	y	z	s	t	p		Test ratios
s	$\boxed{2}$	2	1	1	0	0	10	$10/2 = 5$
t	1	2	3	0	1	0	15	$15/1 = 15$
p	-3	-2	-1	0	0	1	0	

Q: *Why select the pivot this way?*

A: The rule given above guarantees that, after pivoting, all variables will be nonnegative in the basic solution. In other words, it guarantees that we will remain in the feasible region. We will explain further after finishing this example.

Step 5 *Use the pivot to clear the column in the normal manner and then relabel the pivot row with the label from the pivot column.* It is important to follow the exact prescription described in Section 3.2 for formulating the row operations:

$$aR_c \pm bR_p. \qquad \text{\textit{a and b both positive}}$$

Row to change   Pivot row

All entries in the last column should remain nonnegative after pivoting. Furthermore, because the x column (and no longer the s column) will be cleared, x will become an active variable. In other words, the s on the left of the pivot will be replaced by x. We call s the **departing**, or **exiting variable** and x the **entering variable** for this step.

Entering variable
↓

		x	y	z	s	t	p		
Departing variable →	s	$\boxed{2}$	2	1	1	0	0	10	
	t	1	2	3	0	1	0	15	$2R_2 - R_1$
	p	-3	-2	-1	0	0	1	0	$2R_3 + 3R_1$

This gives

	x	y	z	s	t	p	
x	2	2	1	1	0	0	10
t	0	2	5	-1	2	0	20
p	0	2	1	3	0	2	30

This is the second tableau.

Step 6 *Go to Step 3.* But wait! According to Step 3, we are finished because there are no negative numbers in the bottom row. Thus, we can read off the answer.

Remember, though, that the solution for x, the first active variable, is not just $x = 10$ but is $x = 10/2 = 5$ because the pivot has not been reduced to a 1. Similarly, $t = 20/2 = 10$ and $p = 30/2 = 15$. All the other variables are zero because they are inactive. Thus, the solution is as follows: p has a maximum value of 15, and this occurs when $x = 5$, $y = 0$, and $z = 0$. (The slack variables then have the values $s = 0$ and $t = 10$.)

Q : *Why can we stop when there are no negative numbers in the bottom row? Why does this tableau give an optimal solution?*

A : The bottom row corresponds to the equation $2y + z + 3s + 2p = 30$, or

$$p = 15 - y - \frac{1}{2}z - \frac{3}{2}s.$$

Think of this as part of the general solution to our original system of equations, with y, z, and s as the parameters. Because these variables must be nonnegative, *the largest possible value of p in any feasible solution of the system comes when all three of the parameters are 0.* Thus, the current basic solution must be an optimal solution.[*]

[*] Calculators or spreadsheets could obviously be a big help in the calculations here, just as in Chapter 3. We'll say more about that after the next couple of examples.

We owe some further explanation for Step 4 of the simplex method. After Step 3, we knew that x would be the entering variable, and we needed to choose the departing variable. In the next basic solution, x was to have some positive value, and we wanted this value to be as large as possible (to make p as large as possible) without making any other variables negative. Look again at the equations written in Step 2:

$$s = 10 - 2x - 2y - z$$
$$t = 15 - x - 2y - 3z.$$

We needed to make either s or t into an inactive variable and hence zero. Also, y and z were to remain inactive. If we had made s inactive, then we would have had $0 = 10 - 2x$, so $x = 10/2 = 5$. This would have made $t = 15 - 5 = 10$, which would be fine. On the other hand, if we had made t inactive, then we would have had $0 = 15 - x$, so $x = 15$, and this would have made $s = 10 - 2 \cdot 15 = -20$, which would *not* be fine, because slack variables must be nonnegative. In other words, we had a choice of making $x = 10/2 = 5$ or $x = 15/1 = 15$, but making x larger than 5 would have made another variable negative. We were thus compelled to choose the smaller ratio, 5, and make s the departing variable. Of course, we do not have to think it through this way every time. We just use the rule stated in Step 4. (For a graphical explanation, see Example 3.)

EXAMPLE 2   **Simplex Method**

Find the maximum value of $p = 12x + 15y + 5z$, subject to the constraints

$$2x + 2y + z \le 8$$
$$x + 4y - 3z \le 12$$
$$x \ge 0, y \ge 0, z \ge 0.$$

Solution Following Step 1, we introduce slack variables to write the LP problem in equation form:

$$2x + 2y + z + s = 8$$
$$x + 4y - 3z + t = 12$$
$$-12x - 15y - 5z + p = 0.$$

We now follow with Step 2, setting up the initial tableau:

	x	y	z	s	t	p	
s	2	2	1	1	0	0	8
t	1	4	-3	0	1	0	12
p	-12	-15	-5	0	0	1	0

For Step 3 we select the column over the negative number with the largest magnitude in the bottom row, which is the y column. For Step 4, finding the pivot, we see that the test ratios are $8/2$ and $12/4$, the smallest being $12/4 = 3$. So we select the pivot in the t row and clear its column:

	x	y	z	s	t	p		
s	2	2	1	1	0	0	8	$2R_1 - R_2$
t	1	$\boxed{4}$	-3	0	1	0	12	
p	-12	-15	-5	0	0	1	0	$4R_3 + 15R_2$

The departing variable is t, and the entering variable is y. This gives the second tableau:

	x	y	z	s	t	p	
s	3	0	5	2	-1	0	4
y	1	4	-3	0	1	0	12
p	-33	0	-65	0	15	4	180

We now go back to Step 3. Because we still have negative numbers in the bottom row, we choose the one with the largest magnitude (which is -65), and thus our pivot column is the z column. Because negative numbers can't be pivots, the only possible choice for the pivot is the 5. (We need not compute the test ratios because there would be only one from which to choose.) We now clear this column, remembering to take care of the departing and entering variables:

	x	y	z	s	t	p		
s	3	0	$\boxed{5}$	2	-1	0	4	
y	1	4	-3	0	1	0	12	$5R_2 + 3R_1$
p	-33	0	-65	0	15	4	180	$R_3 + 13R_1$

This gives

	x	y	z	s	t	p	
z	3	0	5	2	-1	0	4
y	14	20	0	6	2	0	72
p	6	0	0	26	2	4	232

Notice how the value of p keeps climbing: It started at 0 in the first tableau, went up to $180/4 = 45$ in the second, and is currently at $232/4 = 58$. Because there are no more negative numbers in the bottom row, we are done and can write down the solution: p has a maximum value of $232/4 = 58$, and this occurs when

$$x = 0$$
$$y = \frac{72}{20} = \frac{18}{5} \quad \text{and}$$
$$z = \frac{4}{5}.$$

The slack variables are both zero.

As a partial check on our answer, we can substitute these values into the objective function and the constraints:

$$58 = 12(0) + 15(18/5) + 5(4/5) \qquad ✔$$
$$2(0) + 2(18/5) + (4/5) = 8 \le 8 \qquad ✔$$
$$0 + 4(18/5) - 3(4/5) = 12 \le 12. \qquad ✔$$

We say that this is only a partial check because it shows only that our solution is feasible and that we have correctly calculated p. It does not show that we have the optimal solution. This check will *usually* catch any arithmetic mistakes we make, but it is not foolproof.

Applications

In the next example (further exploits of *Acme Baby Foods*—compare Example 2 in Section 5.2) we show how the simplex method relates to the graphical method.

EXAMPLE 3 Resource Allocation

Acme Baby Foods makes two puddings, vanilla and chocolate. Each serving of vanilla pudding requires 2 teaspoons of sugar and 25 fluid ounces of water, and each serving of chocolate pudding requires 3 teaspoons of sugar and 15 fluid ounces of water. Acme has available each day 3,600 teaspoons of sugar and 22,500 fluid ounces of water. Acme makes no more than 600 servings of vanilla pudding because that is all that it can sell each day. If Acme makes a profit of 10¢ on each serving of vanilla pudding and 7¢ on each serving of chocolate, how many servings of each should it make to maximize its profit?

Solution We first identify the unknowns. Let

$$x = \text{the number of servings of vanilla pudding}$$
$$y = \text{the number of servings of chocolate pudding.}$$

The objective function is the profit $p = 10x + 7y$, which we need to maximize. For the constraints, we start with the fact that Acme will make no more than 600 servings of vanilla: $x \leq 600$. We can put the remaining data in a table as follows:

	Vanilla	Chocolate	Total Available
Sugar (teaspoons)	2	3	3,600
Water (ounces)	25	15	22,500

Because Acme can use no more sugar and water than is available, we get the following two constraints:

$$2x + 3y \leq 3,600$$
$$25x + 15y \leq 22,500. \quad \text{Note that all the terms are divisible by 5.}$$

Thus, our linear programming problem is this:

Maximize $p = 10x + 7y$ subject to

$$x \leq 600$$
$$2x + 3y \leq 3,600$$
$$5x + 3y \leq 4,500 \quad \text{We divided } 25x + 15y \leq 22,500 \text{ by 5.}$$
$$x \geq 0, y \geq 0.$$

Next, we introduce the slack variables and set up the initial tableau:

$$x \quad + s \qquad\qquad = 600$$
$$2x + 3y \quad + t \qquad\quad = 3,600$$
$$5x + 3y \qquad + u \qquad = 4,500$$
$$-10x - 7y \qquad\qquad + p = 0.$$

Note that we have had to introduce a third slack variable, u. There need to be as many slack variables as there are constraints (other than those of the $x \geq 0$ variety).

Q : *What do the slack variables say about Acme puddings?*

A : The first slack variable, *s*, represents the number you must add to the number of servings of vanilla pudding actually made to obtain the maximum of 600 servings. The second slack variable, *t*, represents the amount of sugar that is left over once the puddings are made, and (5 ×) *u* represents the amount of water left over.

We now use the simplex method to solve the problem:

	x	y	s	t	u	p		
s	$\boxed{1}$	0	1	0	0	0	600	
t	2	3	0	1	0	0	3,600	$R_2 - 2R_1$
u	5	3	0	0	1	0	4,500	$R_3 - 5R_1$
p	-10	-7	0	0	0	1	0	$R_4 + 10R_1$

	x	y	s	t	u	p		
x	1	0	1	0	0	0	600	
t	0	3	-2	1	0	0	2,400	$R_2 - R_3$
u	0	$\boxed{3}$	-5	0	1	0	1,500	
p	0	-7	10	0	0	1	6,000	$3R_4 + 7R_3$

	x	y	s	t	u	p		
x	1	0	1	0	0	0	600	$3R_1 - R_2$
t	0	0	$\boxed{3}$	1	-1	0	900	
y	0	3	-5	0	1	0	1,500	$3R_3 + 5R_2$
p	0	0	-5	0	7	3	28,500	$3R_4 + 5R_2$

	x	y	s	t	u	p	
x	3	0	0	-1	1	0	900
s	0	0	3	1	-1	0	900
y	0	9	0	5	-2	0	9,000
p	0	0	0	5	16	9	90,000

Using Technology

See the Technology Guides at the end of the chapter for a discussion of using a TI-83/84 Plus to help with the simplex method in Example 3. Or, go to the Website at www.WanerMath.com and follow the path

→ Online Utilities
→ Pivot and Gauss-Jordan Tool

to find a utility that allows you to avoid doing the calculations in each pivot step: Just highlight the entry you wish to use as a pivot, and press "Pivot on Selection".

Thus, the solution is as follows: The maximum value of p is $90,000/9 = 10,000¢ = \$100$, which occurs when $x = 900/3 = 300$, and $y = 9,000/9 = 1,000$. (The slack variables are $s = 900/3 = 300$ and $t = u = 0$.)

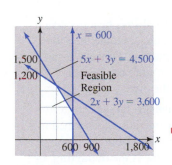

Figure 25

➡ **Before we go on...** Because the problem in Example 3 had only two variables, we could have solved it graphically. It is interesting to think about the relationship between the two methods. Figure 25 shows the feasible region. Each tableau in the simplex method corresponds to a corner of the feasible region, given by the corresponding basic solution. In this example the sequence of basic solutions is

$$(x, y) = (0, 0), (600, 0), (600, 500), (300, 1,000).$$

This is the sequence of corners shown in Figure 26. In general, we can think of the simplex method as walking from corner to corner of the feasible region until we locate the optimal solution. In problems with many variables and many constraints, the simplex method usually visits only a small fraction of the total number of corners.

We can also explain again, in a different way, the reason we use the test ratios when choosing the pivot. For example, when choosing the first pivot, we had to choose among the test ratios 600, 1,800, and 900. (Look at the first tableau.) In

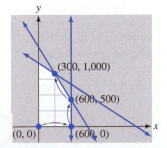

Figure 26

Figure 25, you can see that those are the three x-intercepts of the lines that bound the feasible region. If we had chosen 1,800 or 900, we would have jumped along the x-axis to a point outside of the feasible region, which we do not want to do. In general, the test ratios measure the distance from the current corner to the constraint lines, and we must choose the smallest such distance to avoid crossing any of them into the unfeasible region.

It is also interesting in an application like this to think about the values of the slack variables. We said above that s is the difference between the maximum 600 servings of vanilla that might be made and the number that is actually made. In the optimal solution, $s = 300$, which says that 300 fewer servings of vanilla were made than the maximum possible. Similarly, t was the amount of sugar left over. In the optimal solution, $t = 0$, which tells us that all of the available sugar is used. Finally, $u = 0$, so all of the available water is used as well. ∎

Summary: The Simplex Method for Standard Maximization Problems

To solve a standard maximization problem using the simplex method, we take the following steps:

1. Convert to a system of equations by introducing **slack variables** to turn the constraints into equations and by rewriting the objective function in standard form.

2. Write down the initial **tableau**.

3. Select the pivot column: Choose the negative number with the largest magnitude in the left-hand side of the bottom row. Its column is the pivot column. (If there are two or more candidates, choose any one.) If all the numbers in the left-hand side of the bottom row are zero or positive, then we are finished, and the basic solution maximizes the objective function. (See below for the basic solution.)

4. Select the pivot in the pivot column: The pivot must always be a positive number. For each positive entry b in the pivot column, compute the ratio a/b, where a is the number in the last column in that row. Of these **test ratios**, choose the smallest one. (If there are two or more candidates, choose any one.) The corresponding number b is the pivot.

5. Use the pivot to clear the column in the normal manner (taking care to follow the exact prescription for formulating the row operations described in Chapter 3), and then relabel the pivot row with the label from the pivot column. The variable originally labeling the pivot row is the **departing**, or **exiting, variable**, and the variable labeling the column is the **entering variable**.

6. Go to Step 3.

To get the **basic solution** corresponding to any tableau in the simplex method, set to zero all variables that do not appear as row labels. The value of a variable that does appear as a row label (an **active variable**) is the number in the rightmost column in that row divided by the number in that row in the column labeled by the same variable.

FAQs

Troubleshooting the Simplex Method

Q : *What if there is no candidate for the pivot in the pivot column? For example, what do we do with a tableau like the following?*

	x	y	z	s	t	p	
z	0	0	5	2	0	0	4
y	-8	20	0	6	5	0	72
p	-20	0	0	26	15	4	232

A : Here, the pivot column is the *x* column, but there is no suitable entry for a pivot (because zeros and negative numbers can't be pivots). This happens when the feasible region is unbounded and there is also no optimal solution. In other words, *p* can be made as large as we like without violating the constraints.

Q : *What should we do if there is a negative number in the rightmost column?*

A : A negative number will not appear above the bottom row in the rightmost column if we follow the procedure correctly. (The bottom right entry is allowed to be negative if the objective takes on negative values as in a negative profit, or loss.) Following are the most likely errors leading to this situation:

- The pivot was chosen incorrectly. (Don't forget to choose the *smallest* test ratio.) When this mistake is made, one or more of the variables will be negative in the corresponding basic solution.
- The row operation instruction was written backward or performed backward (for example, instead of $R_2 - R_1$, it was $R_1 - R_2$). This mistake can be corrected by multiplying the row by -1.
- An arithmetic error occurred. (We all make those annoying errors from time to time.)

Q : *What about zeros in the rightmost column?*

A : Zeros are permissible in the rightmost column. For example, the constraint $x - y \leq 0$ will lead to a zero in the rightmost column.*

Q : *What happens if we choose a pivot column other than the one with the most negative number in the bottom row?*

A : There is no harm in doing this as long as we choose the pivot in that column using the smallest test ratio. All it might do is slow the whole calculation by adding extra steps.

* When there are zeros in the rightmost column there is a potential problem of *cycling,* where a sequence of pivots brings you back to a tableau you already considered, with no change in the objective function. You can usually break out of a cycle by choosing a different pivot. This problem should not arise in the exercises.

One last suggestion: If it is possible to do a simplification step (dividing a row by a positive number) *at any stage*, we should do so. As we saw in Chapter 3, this can help to prevent the numbers from getting out of hand.

5.3 EXERCISES

▼ more advanced   ◆ challenging
T indicates exercises that should be solved using technology

1. Maximize $p = 2x + y$
subject to $x + 2y \le 6$
$-x + y \le 4$
$x + y \le 4$
$x \ge 0, y \ge 0.$

[**HINT:** See Examples 1 and 2.]

2. Maximize $p = x$
subject to $x - y \le 4$
$-x + 3y \le 4$
$x \ge 0, y \ge 0.$

[**HINT:** See Examples 1 and 2.]

3. Maximize $p = x - y$
subject to $5x - 5y \le 20$
$2x - 10y \le 40$
$x \ge 0, y \ge 0.$

4. Maximize $p = 2x + 3y$
subject to $3x + 8y \le 24$
$6x + 4y \le 30$
$x \ge 0, y \ge 0.$

5. Maximize $p = 5x - 4y + 3z$
subject to $5x + 5z \le 100$
$5y - 5z \le 50$
$5x - 5y \le 50$
$x \ge 0, y \ge 0, z \ge 0.$

6. Maximize $p = 6x + y + 3z$
subject to $3x + y \le 15$
$2x + 2y + 2z \le 20$
$x \ge 0, y \ge 0, z \ge 0.$

7. Maximize $p = 7x + 5y + 6z$
subject to $x + y - z \le 3$
$x + 2y + z \le 8$
$x + y \le 5$
$x \ge 0, y \ge 0, z \ge 0.$

8. Maximize $p = 3x + 4y + 2z$
subject to $3x + y + z \le 5$
$x + 2y + z \le 5$
$x + y + z \le 4$
$x \ge 0, y \ge 0, z \ge 0.$

9. Maximize $z = 3x_1 + 7x_2 + 8x_3$
subject to $5x_1 - x_2 + x_3 \le 1,500$
$2x_1 + 2x_2 + x_3 \le 2,500$
$4x_1 + 2x_2 + x_3 \le 2,000$
$x_1 \ge 0, x_2 \ge 0, x_3 \ge 0.$

10. Maximize $z = 3x_1 + 4x_2 + 6x_3$
subject to $5x_1 - x_2 + x_3 \le 1,500$
$2x_1 + 2x_2 + x_3 \le 2,500$
$4x_1 + 2x_2 + x_3 \le 2,000$
$x_1 \ge 0, x_2 \ge 0, x_3 \ge 0.$

11. Maximize $p = x + y + z + w$
subject to $x + y + z \le 3$
$y + z + w \le 4$
$x + z + w \le 5$
$x + y + w \le 6$
$x \ge 0, y \ge 0, z \ge 0, w \ge 0.$

12. Maximize $p = x - y + z + w$
subject to $x + y + z \le 3$
$y + z + w \le 3$
$x + z + w \le 4$
$x + y + w \le 4$
$x \ge 0, y \ge 0, z \ge 0, w \ge 0.$

13. ▼ Maximize $p = x + y + z + w + v$
subject to $x + y \le 1$
$y + z \le 2$
$z + w \le 3$
$w + v \le 4$
$x \ge 0, y \ge 0, z \ge 0, w \ge 0, v \ge 0.$

14. ▼ Maximize $p = x + 2y + z + 2w + v$
subject to $x + y \le 1$
$y + z \le 2$
$z + w \le 3$
$w + v \le 4$
$x \ge 0, y \ge 0, z \ge 0, w \ge 0, v \ge 0.$

T *In Exercises 15–20 we suggest the use of technology. Round all answers to two decimal places.*

15. Maximize $p = 2.5x + 4.2y + 2z$
subject to $0.1x + y - 2.2z \le 4.5$
$2.1x + y + z \le 8$
$x + 2.2y \le 5$
$x \ge 0, y \ge 0, z \ge 0.$

16. Maximize $p = 2.1x + 4.1y + 2z$
subject to $3.1x + 1.2y + z \le 5.5$
$x + 2.3y + z \le 5.5$
$2.1x + y + 2.3z \le 5.2$
$x \ge 0, y \ge 0, z \ge 0.$

17. Maximize $p = x + 2y + 3z + w$
subject to $x + 2y + 3z \le 3$
$ y + z + 2.2w \le 4$
$x + z + 2.2w \le 5$
$x + y + 2.2w \le 6$
$x \ge 0, y \ge 0, z \ge 0, w \ge 0.$

18. Maximize $p = 1.1x - 2.1y + z + w$
subject to $x + 1.3y + z \le 3$
$ 1.3y + z + w \le 3$
$x + z + w \le 4.1$
$x + 1.3y + w \le 4.1$
$x \ge 0, y \ge 0, z \ge 0, w \ge 0.$

19. Maximize $p = x - y + z - w + v$
subject to $x + y \le 1.1$
$y + z \le 2.2$
$z + w \le 3.3$
$w + v \le 4.4$
$x \ge 0, y \ge 0, z \ge 0, w \ge 0, v \ge 0.$

20. Maximize $p = x - 2y + z - 2w + v$
subject to $x + y \le 1.1$
$y + z \le 2.2$
$z + w \le 3.3$
$w + v \le 4.4$
$x \ge 0, y \ge 0, z \ge 0, w \ge 0, v \ge 0.$

Applications

21. *Purchasing* You are in charge of purchases at the student-run used-book supply program at your college, and you must decide how many introductory calculus, history, and marketing texts should be purchased from students for resale. Because of budget limitations, you cannot purchase more than 650 of these textbooks each semester. There are also shelf-space limitations: Calculus texts occupy 2 units of shelf space each, history books 1 unit each, and marketing texts 3 units each, and you can spare at most 1,000 units of shelf space for the texts. If the used-book program makes a profit of $10 on each calculus text, $4 on each history text, and $8 on each marketing text, how many of each type of text should you purchase to maximize profit? What is the maximum profit the program can make in a semester? [**HINT:** See Example 3.]

22. *Sales* The Marketing Club at your college has decided to raise funds by selling three types of T-shirts: one with a single-color "ordinary" design, one with a two-color "fancy" design, and one with a three-color "very fancy" design. The club feels that it can sell up to 300 T-shirts. "Ordinary" T-shirts will cost the club $6 each, "fancy" T-shirts $8 each, and "very fancy" T-shirts $10 each, and the club has a total purchasing budget of $3,000. It will sell "ordinary" T-shirts at a profit of $4 each, "fancy" T-shirts at a profit of $5 each, and "very fancy" T-shirts at a profit of $4 each. How many of each kind of T-shirt should the club order to maximize profit? What is the maximum profit the club can make? [**HINT:** See Example 3.]

23. *Resource Allocation* Arctic Juice Company makes three juice blends: PineOrange, using 2 portions of pineapple juice and 2 portions of orange juice per gallon; PineKiwi, using 3 portions of pineapple juice and 1 portion of kiwi juice per gallon; and OrangeKiwi, using 3 portions of orange juice and 1 portion of kiwi juice per gallon. Each day the company has 800 portions of pineapple juice, 650 portions of orange juice, and 350 portions of kiwi juice available. Its profit on PineOrange is $1 per gallon, its profit on PineKiwi is $2 per gallon, and its profit on OrangeKiwi is $1 per gallon. How many gallons of each blend should it make each day to maximize profit? What is the largest possible profit the company can make?

24. *Purchasing* Trans Global Tractor Trailers has decided to spend up to $1,500,000 on a fleet of new trucks, and it is considering three models: the Gigahaul, which has a capacity of 6,000 cubic feet and is priced at $60,000; the Megahaul, with a capacity of 5,000 cubic feet, priced at $50,000; and the Picohaul, with a capacity of 2,000 cubic feet, priced at $40,000. The anticipated annual revenues are $500,000 for each new truck purchased (regardless of size). Trans Global would like a total capacity of up to 130,000 cubic feet and feels that it cannot provide drivers and maintenance for more than 30 trucks. How many of each should it purchase to maximize annual revenue? What is the largest possible revenue it can make?

25. *Resource Allocation* The *Enormous State University* History Department offers three courses—Ancient, Medieval, and Modern History—and the department chairperson is trying to decide how many sections of each to offer this semester. The department may offer up to 45 sections total, up to 5,000 students would like to take a course, and there are 60 professors to teach them. (No student will take more than one history course, and no professor will teach more than one section.) Sections of Ancient History have 100 students each, sections of Medieval History have 50 students each, and sections of Modern History have 200 students each. Modern History sections are taught by a team of two professors, while Ancient History and Medieval History need only one professor per section. Ancient History nets the university $10,000 per section, Medieval nets $20,000 per section, and Modern History nets $30,000 per section. How many sections of each course should the department offer in order to generate the largest profit? What is the largest profit possible? Will there be any unused time slots, any students who did not get into classes, or any professors without anything to teach?

26. *Resource Allocation* You manage an ice cream factory that makes three flavors: Creamy Vanilla, Continental Mocha, and Succulent Strawberry. Into each batch of Creamy Vanilla go 2 eggs, 1 cup of milk, and 2 cups of cream. Into each batch of Continental Mocha go 1 egg, 1 cup of milk, and 2 cups of cream. Into each batch of Succulent Strawberry go 1 egg, 2 cups of milk, and 2 cups of cream. You have in stock 200 eggs, 120 cups of milk, and 200 cups of cream. You make a profit of $3 on each batch of Creamy Vanilla, $2 on each batch of Continental Mocha, and $4 on each batch of Succulent Strawberry.

  a. How many batches of each flavor should you make to maximize your profit?

  b. In your answer to part (a), have you used all the ingredients?

  c. Because of the poor strawberry harvest this year, you cannot make more than 10 batches of Succulent Strawberry. Does this affect your maximum profit?

27. *Agriculture* Your small farm encompasses 100 acres, and you are planning to grow tomatoes, lettuce, and carrots in the coming planting season. Fertilizer costs per acre are $5 for tomatoes, $4 for lettuce, and $2 for carrots. Based on past experience, you estimate that each acre of tomatoes will require an average of 4 hours of labor per week, while tending to lettuce and carrots will each require an average of 2 hours per week. You estimate a profit of $2,000 for each acre of tomatoes, $1,500 for each acre of lettuce, and $500 for each acre of carrots. You can afford to spend no more than $400 on fertilizer, and your farm laborers can supply up to 500 hours per week. How many acres of each crop should you plant to maximize total profits? In this event, will you be using all 100 acres of your farm?

28. *Agriculture* Your farm encompasses 500 acres, and you are planning to grow soybeans, corn, and wheat in the coming planting season. Fertilizer costs per acre are $5 for soybeans, $2 for corn, and $1 for wheat. You estimate that each acre of soybeans will require an average of 5 hours of labor per week, while tending to corn and wheat will each require an average of 2 hours per week. On the basis of past yields and current market prices, you estimate a profit of $3,000 for each acre of soybeans, $2,000 for each acre of corn, and $1,000 for each acre of wheat. You can afford to spend no more than $3,000 on fertilizer, and your farm laborers can supply 3,000 hours per week. How many acres of each crop should you plant to maximize total profits? In this event, will you be using all the available labor?

29. *Resource Allocation* (Compare Exercise 36 in Chapter 3 Review) The *Enormous State University* Choral Society is planning its annual Song Festival, when it will serve three kinds of delicacies: granola treats, nutty granola treats, and nuttiest granola treats. The following table shows some of the ingredients required for a single serving of each delicacy as well as the total amount of each ingredient available:

	Granola	Nutty Granola	Nuttiest Granola	Total Available
Toasted Oats (ounces)	1	1	5	1,500
Almonds (ounces)	4	8	8	10,000
Raisins (ounces)	2	4	8	4,000

The society makes a profit of $6 on each serving of granola, $8 on each serving of nutty granola, and $3 on each serving of nuttiest granola. Assuming that the Choral Society can sell all that it makes, how many servings of each will maximize profits? How much of each ingredient will be left over?

30. *Resource Allocation* Repeat Exercise 29, but this time assume that the Choral Society makes a $3 profit on each of its delicacies.

Gaming *Execises 31 and 32 are based on the following table, which shows some parameters of various weapons used in role-playing gaming:*[23]

Weapon	Cost (gold pieces)	Damage to Medium Targets	Critical Damage	Weight (pounds)
Axe (Throwing)	8	6	12	2
Javelin	1	6	12	2
Longsword	15	8	32	4
Mace (Light)	5	6	12	4
Spear	2	8	24	6

31. *Orcs* The Orc leader Achlúk has up to 50,000 gold pieces to spend on an arsenal of axes, maces, and spears for his army of orcs for a planned assault on Hobshire, in which he would like to inflict as much damage on medium targets (like humans and hobbits) as possible. To avoid excessive transportation costs, Achlúk needs to limit the total weight of the arsenal to 40,000 pounds or less, and, as his orcs are particularly fond of axes but not particularly skilled at spear-throwing, he would like to include at least as many axes as spears in the arsenal. What should his weapons arsenal look like, and how much damage on medium targets can be inflicted?

32. *Elves* The Elf leader Galandir has up to 30,000 gold pieces to spend on an arsenal of javelins, longswords, and spears

[23] Source: Dungeons and Dragons Wiki (www.dandwiki.com). (Critical Damage is a weighted measure of "critical hit damage" as defined there.) See also Paul Tozour's blog at www.gamasutra.com/blogs/PaulTozour/20130707/195718/ for a discussion of similar scenarios in the context of arming a video game battle tank.

for her band of elves for a planned assault on Mordrúk, in which she would like to inflict as much critical damage as possible. For the sake of swiftness the total weight of Galandir's arsenal cannot exceed 3,000 pounds, and, as the elves are particularly skilled at javelin-throwing, she would like to include at least half as many javelins as swords. What should her weapons arsenal look like, and how much critical damage can be inflicted?

33. *Recycling* **Safety-Kleen** operates the world's largest oil re-refinery in Elgin, Illinois. You have been hired by the company to determine how to allocate its intake of up to 50 million gallons of used oil to its three refinery processes: A, B, and C. You are told that electricity costs for process A amount to $150,000 per million gallons treated, while for processes B and C, the costs are $100,000 and $50,000, respectively, per million gallons treated. Process A can recover 60% of the used oil, process B can recover 55%, and process C can recover only 50%. Assuming a revenue of $4 million per million gallons of recovered oil and an annual electrical budget of $3 million, how much used oil would you allocate to each process to maximize total revenues?[24]

34. *Recycling* Repeat Exercise 33, but this time assume that process C can handle only up to 20 million gallons per year.

Bodybuilding Supplements *Exercises 35 and 36 are based on the following data on four popular bodybuilding supplements. (Figures shown correspond to a single serving.)*[25]

	Creatine (grams)	L-Glutamine (grams)	BCAAs (grams)
Xtend (**SciVation**)	0	2.5	7
Gainz (**MP Hardcore**)	2	3	6
Strongevity (**Bill Phillips**)	2.5	1	0
Muscle Physique (**EAS**)	2	2	0

35. Your personal trainer suggests that you supplement with as much BCAAs as possible but with no more than 40 grams of creatine and 60 grams of L-glutamine per week. You are thinking of combining Xtend, Gainz, and Strongevity to provide you with the required nutrients. How many

servings of each should you combine to obtain a week's supply that meets your trainer's specifications and also includes at least as many servings of Strongevity as Xtend? How much BCAAs will you obtain?

36. Your friend's personal trainer suggests that she supplement with as much L-glutamine as possible but with no more than 60 grams of creatine and 60 grams of BCAAs per week. She is thinking of combining Gainz, Strongevity, and Muscle Physique to provide her with the required nutrients. How many servings of each should she combine to obtain a week's supply that meets her trainer's specifications and also includes no more servings of Gainz than of Muscle Physique? How much L-glutamine will she obtain?

Investing *Exercises 37 and 38 are based on the following data on three stocks:*[26]

	Price ($)	Dividend Yield (%)	52-Week Price Change ($)
DUK (**Duke Energy Corp**)	80	4	4
DTV (**DIRECTV**)	100	0	10
OCR (**Omnicare, Inc.**)	90	1	30

37. ▼ You are planning to invest up to $90,000 in DUK, DTV, and OCR shares. You desire to maximize the 52-week gain but, for tax reasons, want to earn no more than $900 in dividends. Your broker suggests that because DTV stock pays no dividends, you should invest everything in DTV. Is she right?

38. ▼ Repeat Exercise 37 under the assumption that the 52-week change in DTV stock is $30 but its price is unchanged.

39. ▣ ▼ *Loan Planning*[27] *Enormous State University's* employee credit union has $5 million available for loans in the coming year. As VP in charge of finances, you must decide how much capital to allocate to each of four different kinds of loans, as shown in the following table:

Type of Loan	Annual Rate of Return (%)
Automobile	8
Furniture	10
Signature	12
Other secured	10

---

[24] These figures are realistic: Safety-Kleen's actual 1993 capacity was 50 million gallons, its recycled oil sold for approximately $4 per gallon, its recycling process could recover approximately 55% of the used oil, and its electrical bill was $3 million. Source: Oil Recycler Greases Rusty City's Economy, *Chicago Tribune*, May 30, 1993, Section 7, p.1.

[25] Source: Nutritional information supplied by the manufacturers/ www.bodybuilding.com. "BCAAs" refers to the branched-chain amino acids leucine, isoleucine, and valine in the optimal 2:1:1 ratio.

[26] Approximate price during May 2014; 52-week price changes and dividend yields are approximate. Source: http://finance.google.com.

[27] Adapted from an exercise in *An Introduction to Management Science* (6th. ed.) by D. R. Anderson, D. J. Sweeney, and T. A. Williams (West, 1991).

State laws and credit union policies impose the following restrictions:

- Signature loans may not exceed 10% of the total investment of funds.
- Furniture loans plus other secured loans may not exceed automobile loans.
- Other secured loans may not exceed 200% of automobile loans.

How much should you allocate to each type of loan to maximize the annual return?

40. **T ▼ Investments** You have $100,000 that you are considering investing in three dividend-yielding bank stocks: **Banco Santander Brasil**, **Bank of Hawaii**, and **Banco Santander Chile**. You have the following data:[28]

Stock	Yield (%)
BSBR (Banco Santander Brasil)	7
BOH (Bank of Hawaii)	5
SAN (Banco Santander Chile)	4

Your broker has made the following suggestions:

- At least 50% of your total investment should be in SAN.
- No more than 10% of your total investment should be in BSBR.

How much should you invest in each stock to maximize your anticipated dividends while following your broker's advice?

41. **▼ Portfolio Management** If x dollars are invested in a company that controls, say, 30% of the market with five brand names, then $0.30x$ is a measure of market exposure, and $5x$ is a measure of brand-name exposure. Now suppose you are a broker at a large securities firm, and one of your clients would like to invest up to $100,000 in recording industry stocks. You decide to recommend a combination of stocks in four of the world's largest recording companies: **Warner Music**, **Universal Music**, **Sony**, and **EMI**. (See the table.)[29]

	Warner Music	Universal Music	Sony	EMI
Market Share (%)	12	20	20	15
Number of Labels (brands)	8	20	10	15

You would like your client's brand-name exposure to be as large as possible but his total market exposure to be $15,000 or less. (This would reflect an average of 15%.) Furthermore, you would like at least 20% of the investment to be in **Universal** because you feel that its control of the DGG and Phillips labels is advantageous for its classical music operations. How much should you advise your client to invest in each company?

42. **▼ Portfolio Management** Referring to Exercise 41, suppose instead that you wanted your client to maximize his total market exposure but limit his brand-name exposure to 1.5 million or less (representing an average of 15 labels or fewer per company), and still invest at least 20% of the total in **Universal**. How much should you advise your client to invest in each company?

43. **T ▼ Transportation Scheduling** (This exercise is almost identical to Exercise 26 in Section 3.3 but is more realistic; one cannot always expect to fill all orders exactly and keep all plants operating at 100 percent capacity.) The *Tubular Ride Boogie Board Company* has manufacturing plants in Tucson, Arizona, and Toronto, Ontario. You have been given the job of coordinating distribution of the latest model, the Gladiator, to their outlets in Honolulu and Venice Beach. The Tucson plant, when operating at full capacity, can manufacture 620 Gladiator boards per week, while the Toronto plant, beset by labor disputes, can produce only 410 boards per week. The outlet in Honolulu orders 500 Gladiator boards per week, while the Venice Beach outlet orders 530 boards per week. Transportation costs are as follows: Tucson to Honolulu: $10 per board; Tucson to Venice Beach: $5 per board; Toronto to Honolulu: $20 per board; Toronto to Venice Beach: $10 per board. Your manager has informed you that the company's total transportation budget is $6,550. You realize that it may not be possible to fill all the orders, but you would like the total number of boogie boards shipped to be as large as possible. Given this, how many Gladiator boards should you order shipped from each manufacturing plant to each distribution outlet?

44. **T ▼ Transportation Scheduling** Repeat Exercise 43, but use a transportation budget of $5,050.

45. **T ▼ Transportation Scheduling** Your publishing company is about to start a promotional blitz for its new book, *Advanced Quantum Mechanics for the Liberal Arts*. You have 20 salespeople stationed in Chicago and 10 in Denver. You would like to fly at most 10 salespeople into Los Angeles and at most 15 into New York. A round-trip plane flight from Chicago to Los Angeles costs $195;[30] one from Chicago to New York costs $182; one from Denver to Los

---

[28] Yields are as of September 2011. Source: www.google.com/finance.

[29] The number of labels includes only major labels. Market shares are approximate and represent the period 2000–2002. Sources: various, including www.emigroup.com, http://finance.vivendi.com/discover/financial, and http://business2.com, March 2002.

---

[30] Prices from Travelocity, at www.travelocity.com, for the week of June 3, 2002, as of May 5, 2002.

Angeles costs \$395; and one from Denver to New York costs \$166. You want to spend at most \$4,520 on plane flights. How many salespeople should you fly from each of Chicago and Denver to each of Los Angeles and New York to have the most salespeople on the road?

46. **T** ▼ *Transportation Scheduling* Repeat Exercise 45, but this time, spend at most \$5,770.

Communication and Reasoning Exercises

47. Can the following linear programming problem be stated as a standard maximization problem? If so, do it; if not, explain why.

$$\text{Maximize} \quad p = 3x - 2y$$
$$\text{subject to} \quad x - y + z \geq 0$$
$$x - y - z \leq 6$$
$$x \geq 0, y \geq 0, z \geq 0.$$

48. Can the following linear programming problem be stated as a standard maximization problem? If so, do it; if not, explain why.

$$\text{Maximize} \quad p = -3x - 2y$$
$$\text{subject to} \quad x - y + z \geq 0$$
$$x - y - z \geq -6$$
$$x \geq 0, y \geq 0, z \geq 0.$$

49. Why is the simplex method useful? (After all, we do have the graphical method for solving LP problems.)

50. Are there any types of linear programming problems that cannot be solved with the methods of this section but that can be solved by using the methods of Section 5.2? Explain.

51. ▼ Your friend Janet is going around telling everyone that if there are only two constraints in a linear programming problem, then, in any optimal basic solution, at most two unknowns (other than the objective) will be nonzero. Is she correct? Explain.

52. ▼ Your other friend Jason is going around telling everyone that if there is only one constraint in a standard linear programming problem, then you will have to pivot at most once to obtain an optimal solution. Is he correct? Explain.

53. ▼ What is a "basic solution"? How might one find a basic solution of a given system of linear equations?

54. ▼ In a typical simplex method tableau, there are more unknowns than equations, and we know from Chapter 3 that this typically implies the existence of infinitely many solutions. How are the following types of solutions interpreted in the simplex method?
   a. Solutions in which all the variables are positive.
   b. Solutions in which some variables are negative.
   c. Solutions in which the inactive variables are zero.

55. ◆ Can the value of the objective function decrease in passing from one tableau to the next? Explain.

56. ◆ Can the value of the objective function remain unchanged in passing from one tableau to the next? Explain.

5.4 The Simplex Method: Solving General Linear Programming Problems

Standard and Nonstandard Linear Programming Problems

We saw in Section 5.3 that a general LP problem may or may not be a standard maximization problem. Recall that, for an LP problem to be standard, it needs to satisfy two requirements:

1. We are *maximizing* (not minimizing) an objective function.

2. The constraints (apart from the requirement that each variable be nonnegative) are all ≤ constraints, with the right-hand sides nonnegative.

General linear programming problems can have constraints such as $2x + 3y \geq 4$, or $2x + 3y = 4$, which violate (2), or we might want to minimize, rather than maximize, the objective function, violating (1). Nonstandard problems like these are almost as easy to deal with as the standard kind, and we use a slight modification of the simplex method to handle them. The best way to illustrate their solution is by means of examples.

First, we discuss nonstandard maximization problems: LP problems that violate (2).

Nonstandard Maximization Problems

EXAMPLE 1   Maximizing with Mixed Constraints

Maximize $p = 4x + 12y + 6z$

subject to $x + y + z \leq 100$

$4x + 10y + 7z \leq 480$

$x + y + z \geq 60$

$x \geq 0, y \geq 0, z \geq 0.$

Solution We begin by turning the first two inequalities into equations as usual because they have the standard form. We get

$$x + y + z + s = 100$$
$$4x + 10y + 7z + t = 480.$$

We are tempted to use a slack variable for the third inequality, $x + y + z \geq 60$, but *adding* something positive to the left-hand side will not make it equal to the right: It will get even bigger. To make it equal to 60, we must *subtract* some nonnegative number. We will call this number u (because we have already used s and t) and refer to u as a **surplus variable** rather than a slack variable. Thus, we write

$$x + y + z - u = 60.$$

Continuing with the setup, we have

$$x + y + z + s = 100$$
$$4x + 10y + 7z + t = 480$$
$$x + y + z - u = 60$$
$$-4x - 12y - 6z + p = 0.$$

This leads to the initial tableau:

	x	y	z	s	t	u	p	
s	1	1	1	1	0	0	0	100
t	4	10	7	0	1	0	0	480
*u	1	1	1	0	0	-1	0	60
p	-4	-12	-6	0	0	0	1	0

We put a star next to the third row because the basic solution corresponding to this tableau is

$$x = y = z = 0, \quad s = 100, \quad t = 480, \quad u = 60/(-1) = -60.$$

Several things are wrong here. First, the values $x = y = z = 0$ do not satisfy the third inequality, $x + y + z \geq 60$. Thus, this basic solution is *not feasible*. Second—and this is really the same problem—the surplus variable u is negative, whereas we said that it should be nonnegative. The star next to the row labeled u alerts us to the fact that the present basic solution is not feasible and that the problem is located in the starred row, where the active variable u is negative.

Whenever an active variable is negative, we star the corresponding row.

In setting up the initial tableau, we star those rows coming from $\geq$ inequalities.

The simplex method as described in Section 5.3 assumed that we began in the feasible region, but now we do not. Our first task is to get ourselves into the feasible region. In practice, we can think of this as getting rid of the stars on the rows. Once we get into the feasible region, we go back to the method of Section 5.3.

There are several ways to get into the feasible region. The method we have chosen is one of the simplest to state and carry out. (We will see why this method works at the end of the example.)

The Simplex Method for General Linear Programming Problems

Star all rows that give a negative value for the associated active variable (except for the objective variable, which is allowed to be negative). If there are starred rows, you will need to begin with Phase I.

Phase I: Getting into the Feasible Region (Getting Rid of the Stars)
In the first starred row, find the largest positive number. Use test ratios as in Section 5.3 to find the pivot in that column (exclude the bottom row), and then pivot on that entry. (If the lowest ratio occurs in both a starred row and an unstarred row, pivot in a starred row rather than the unstarred one.) Check to see which rows should now be starred. Repeat until no starred rows remain, and then go on to Phase II.

Phase II: Use the Simplex Method for Standard Maximization Problems
If there are any negative entries on the left side of the bottom row after Phase I, use the method described in the preceding section. Otherwise, there is nothing to do in Phase II, and you are done.

Because there is a starred row, we need to use Phase I. The largest positive number in the starred row is 1, which occurs three times. Arbitrarily select the first, which is in the first column. In that column the smallest test ratio happens to be given by the 1 in the u row, so this is our first pivot:

Pivot column
↓

	x	y	z	s	t	u	p		
s	1	1	1	1	0	0	0	100	$R_1 - R_3$
t	4	10	7	0	1	0	0	480	$R_2 - 4R_3$
*u	[1]	1	1	0	0	-1	0	60	
p	-4	-12	-6	0	0	0	1	0	$R_4 + 4R_3$

This gives

	x	y	z	s	t	u	p	
s	0	0	0	1	0	1	0	40
t	0	6	3	0	1	4	0	240
x	1	1	1	0	0	-1	0	60
p	0	-8	-2	0	0	-4	1	240

Notice that we removed the star from Row 3. To see why, look at the basic solution given by this tableau:

$$x = 60, \quad y = 0, \quad z = 0, \quad s = 40, \quad t = 240, \quad u = 0.$$

None of the variables is negative anymore, so there are no rows to star. The basic solution is therefore feasible—it satisfies all the constraints.

Now that there are no more stars, we have completed Phase I, so we proceed to Phase II, which is just the method of Section 5.3:

	x	y	z	s	t	u	p	
s	0	0	0	1	0	1	0	40
t	0	6	3	0	1	4	0	240
x	1	1	1	0	0	-1	0	60
p	0	-8	-2	0	0	-4	1	240

$6R_3 - R_2$

$3R_4 + 4R_2$

	x	y	z	s	t	u	p	
s	0	0	0	1	0	1	0	40
y	0	6	3	0	1	4	0	240
x	6	0	3	0	-1	-10	0	120
p	0	0	6	0	4	4	3	1,680

And we are finished. Thus, the solution is

$$p = 1,680/3 = 560, \quad x = 120/6 = 20, \quad y = 240/6 = 40, \quad z = 0.$$

The slack and surplus variables are

$$s = 40, \quad t = 0, \quad u = 0.$$

➡ **Before we go on . . .** We owe you an explanation of why this method works. When we perform a pivot in Phase I, one of two things will happen. As in Example 1, we may pivot in a starred row. In that case, the negative active variable in that row will become inactive (hence zero), and some other variable will be made active with a positive value because we are pivoting on a positive entry. Thus, at least one star will be eliminated. (We will not introduce any new stars because pivoting on the entry with the smallest test ratio will keep all nonnegative variables nonnegative.[*])

The second possibility is that we may pivot on some row other than a starred row. Choosing the pivot via test ratios again guarantees that no new starred rows are created. A little bit of algebra shows that the value of the negative variable in the first starred row must increase toward zero. (Choosing the *largest* positive entry in the starred row will make it a little more likely that we will increase the value of that variable as much as possible; the rationale for choosing the largest entry is the same as that for choosing the most negative entry in the bottom row during Phase II.) Repeating this procedure as necessary, the value of the variable must eventually become zero or positive, assuming that there are feasible solutions to begin with.

So one way or the other, we can eventually get rid of all of the stars. ∎

Here is an example that begins with two starred rows.

[*] except on rare occasions when the rightmost entry in a previously starred row is zero. We can prevent this from happening by a slight modification in the way that Phase 1 handles rows whose rightmost entry is zero (as implemented in the online Simplex Method Tool at the Website). However, as with cycling, we will not encounter this issue in either the examples or exercises.

EXAMPLE 2   **More Mixed Constraints**

Maximize $p = 2x + y$

subject to $x + y \geq 35$

$x + 2y \leq 60$

$2x + y \geq 60$

$x \leq 25$

$x \geq 0, y \geq 0.$

Solution We introduce slack and surplus variables, and we write down the initial tableau:

$$x + y - s = 35$$
$$x + 2y + t = 60$$
$$2x + y - u = 60$$
$$x + v = 25$$
$$-2x - y + p = 0.$$

	x	y	s	t	u	v	p	
*s	1	1	−1	0	0	0	0	35
t	1	2	0	1	0	0	0	60
*u	2	1	0	0	−1	0	0	60
v	1	0	0	0	0	1	0	25
p	−2	−1	0	0	0	0	1	0

We locate the largest positive entry in the first starred row (Row 1). There are two to choose from (both 1s); let's choose the one in the x column. The entry with the smallest test ratio in that column is the 1 in the v row, so that is the entry we use as the pivot:

Pivot column
↓

	x	y	s	t	u	v	p		
*s	1	1	−1	0	0	0	0	35	$R_1 - R_4$
t	1	2	0	1	0	0	0	60	$R_2 - R_4$
*u	2	1	0	0	−1	0	0	60	$R_3 - 2R_4$
v	1	0	0	0	0	1	0	25	
p	−2	−1	0	0	0	0	1	0	$R_5 + 2R_4$

	x	y	s	t	u	v	p	
*s	0	1	−1	0	0	−1	0	10
t	0	2	0	1	0	−1	0	35
*u	0	1	0	0	−1	−2	0	10
x	1	0	0	0	0	1	0	25
p	0	−1	0	0	0	2	1	50

Notice that both stars are still there because the basic solutions for s and u remain negative (but less so). The only positive entry in the first starred row is the 1 in the y column, and that entry also has the smallest test ratio in its column. (Actually, it is tied with the 1 in the u column, so we could choose either one.)

	x	y	s	t	u	v	p		
*s	0	1	−1	0	0	−1	0	10	
t	0	2	0	1	0	−1	0	35	$R_2 - 2R_1$
*u	0	1	0	0	−1	−2	0	10	$R_3 - R_1$
x	1	0	0	0	0	1	0	25	
p	0	−1	0	0	0	2	1	50	$R_5 + R_1$

	x	y	s	t	u	v	p	
y	0	1	−1	0	0	−1	0	10
t	0	0	2	1	0	1	0	15
u	0	0	1	0	−1	−1	0	0
x	1	0	0	0	0	1	0	25
p	0	0	−1	0	0	1	1	60

The basic solution is $x = 25$, $y = 10$, $s = 0$, $t = 15$, $u = 0/(-1) = 0$, and $v = 0$. Because there are no negative variables left (even u has become 0), we are in the feasible region, so we can go on to Phase II, shown next. (Filling in the instructions for the row operations is an exercise.)

	x	y	s	t	u	v	p	
y	0	1	−1	0	0	−1	0	10
t	0	0	2	1	0	1	0	15
u	0	0	1	0	−1	−1	0	0
x	1	0	0	0	0	1	0	25
p	0	0	−1	0	0	1	1	60

	x	y	s	t	u	v	p	
y	0	1	0	0	−1	−2	0	10
t	0	0	0	1	2	3	0	15
s	0	0	1	0	−1	−1	0	0
x	1	0	0	0	0	1	0	25
p	0	0	0	0	−1	0	1	60

	x	y	s	t	u	v	p	
y	0	2	0	1	0	−1	0	35
u	0	0	0	1	2	3	0	15
s	0	0	2	1	0	1	0	15
x	1	0	0	0	0	1	0	25
p	0	0	0	1	0	3	2	135

The optimal solution is

$$x = 25, \quad y = 35/2 = 17.5, \quad p = 135/2 = 67.5 \qquad (s = 7.5, t = 0, u = 7.5).$$

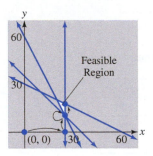

Figure 27

➡ **Before we go on . . .** Because Example 2 had only two unknowns, we can picture the sequence of basic solutions on the graph of the feasible region. This is shown in Figure 27.

You can see that there was no way to jump from $(0, 0)$ in the initial tableau directly into the feasible region because the first jump must be along an axis. (Why?) Also notice that the third jump did not move at all. To which step of the simplex method does this correspond? ∎

Minimization Problems

Now that we know how to deal with nonstandard constraints, we consider **minimization** problems, problems in which we have to minimize, rather than maximize, the objective function. The idea is to *convert a minimization problem into a maximization problem*, which we can then solve as usual.

Suppose, for instance, that we want to minimize $c = 10x - 30y$ subject to some constraints. The technique is as follows: Define a new variable p by taking p to be the negative of c so that $p = -c$. Then, the larger we make p, the smaller c becomes. For example, if we can make p increase from -10 to -5, then c will decrease from 10 to 5. So if we are looking for the smallest value of c, we might as well look for the largest value of p instead. More concisely,

Minimizing c is the same as maximizing p = −c.

Now because $c = 10x - 30y$, we have $p = -10x + 30y$, and the requirement that we "minimize $c = 10x - 30y$" is now replaced by "maximize $p = -10x + 30y$."

Minimization Problems

We convert a minimization problem into a maximization problem by taking the negative of the objective function. All the constraints remain unchanged.

Quick Example

Minimization Problem	⟶	Maximization Problem
Minimize $c = 10x - 30y$		Maximize $p = -10x + 30y$
subject to $2x + y \leq 160$		subject to $2x + y \leq 160$
$x + 3y \geq 120$		$x + 3y \geq 120$
$x \geq 0, y \geq 0.$		$x \geq 0, y \geq 0.$

EXAMPLE 3   **Purchasing**

You are in charge of ordering furniture for your company's new headquarters. You need to buy at least 200 tables, 500 chairs, and 300 computer desks. *Wall-to-Wall Furniture* (WWF) is offering a package of 20 tables, 25 chairs, and 18 computer desks for $2,000, whereas rival *Acme Furniture* (AF) is offering a package of 10 tables, 50 chairs, and 24 computer desks for $3,000. How many packages should you order from each company to minimize your total cost?

Solution The unknowns here are

x = number of packages ordered from WWF
y = number of packages ordered from AF.

We can put the information about the various kinds of furniture in a table:

	WWF	AF	Needed
Tables	20	10	200
Chairs	25	50	500
Computer Desks	18	24	300
Cost ($)	2,000	3,000	

From this table we get the following LP problem:

Minimize $c = 2,000x + 3,000y$
subject to $20x + 10y \geq 200$
$25x + 50y \geq 500$
$18x + 24y \geq 300$
$x \geq 0, y \geq 0.$

Before we start solving this problem, notice that all the inequalities may be simplified. The first is divisible by 10, the second by 25, and the third by 6. (However, this affects the meaning of the surplus variables; see the "Before we go on" discussion below.) Dividing gives the following simpler problem:

Minimize $c = 2,000x + 3,000y$
subject to $2x + y \geq 20$
$x + 2y \geq 20$
$3x + 4y \geq 50$
$x \geq 0, y \geq 0.$

Following the discussion that preceded this example, we convert to a maximization problem:

Maximize $p = -2,000x - 3,000y$
subject to $2x + y \geq 20$
$x + 2y \geq 20$
$3x + 4y \geq 50$
$x \geq 0, y \geq 0.$

We introduce surplus variables:

$$2x + y - s \qquad\qquad = 20$$
$$x + 2y \qquad - t \qquad = 20$$
$$3x + 4y \qquad\qquad - u \qquad = 50$$
$$2,000x + 3,000y \qquad\qquad + p = 0.$$

The initial tableau is then

	x	y	s	t	u	p	
*s	2	1	−1	0	0	0	20
*t	1	2	0	−1	0	0	20
*u	3	4	0	0	−1	0	50
p	2,000	3,000	0	0	0	1	0

The largest entry in the first starred row is the 2 in the upper left, which happens to give the smallest test ratio in its column.

	x	y	s	t	u	p		
*s	2	1	-1	0	0	0	20	
*t	1	2	0	-1	0	0	20	$2R_2 - R_1$
*u	3	4	0	0	-1	0	50	$2R_3 - 3R_1$
p	2,000	3,000	0	0	0	1	0	$R_4 - 1,000R_1$

	x	y	s	t	u	p		
x	2	1	-1	0	0	0	20	$3R_1 - R_2$
*t	0	3	1	-2	0	0	20	
*u	0	5	3	0	-2	0	40	$3R_3 - 5R_2$
p	0	2,000	1,000	0	0	1	$-20,000$	$3R_4 - 2,000R_2$

	x	y	s	t	u	p		
x	6	0	-4	2	0	0	40	$5R_1 - R_3$
y	0	3	1	-2	0	0	20	$5R_2 + R_3$
*u	0	0	4	10	-6	0	20	
p	0	0	1,000	4,000	0	3	$-100,000$	$R_4 - 400R_3$

	x	y	s	t	u	p		
x	30	0	-24	0	6	0	180	$R_1/6$
y	0	15	9	0	-6	0	120	$R_2/3$
t	0	0	4	10	-6	0	20	$R_3/2$
p	0	0	-600	0	2,400	3	$-108,000$	$R_4/3$

This completes Phase I. We are not yet at the optimal solution, so after performing the simplifications indicated, we proceed with Phase II.

	x	y	s	t	u	p		
x	5	0	-4	0	1	0	30	$R_1 + 2R_3$
y	0	5	3	0	-2	0	40	$2R_2 - 3R_3$
t	0	0	2	5	-3	0	10	
p	0	0	-200	0	800	1	$-36,000$	$R_4 + 100R_3$

	x	y	s	t	u	p	
x	5	0	0	10	-5	0	50
y	0	10	0	-15	5	0	50
s	0	0	2	5	-3	0	10
p	0	0	0	500	500	1	$-35,000$

The optimal solution is

$$x = 50/5 = 10, \quad y = 50/10 = 5, \quad p = -35,000,$$
$$\text{so} \quad c = 35,000 \quad (s = 5, t = 0, u = 0).$$

You should buy 10 packages from Wall-to-Wall Furniture and 5 from Acme Furniture, for a minimum cost of $35,000.

➡ **Before we go on ...** The surplus variables in Example 3 represent pieces of furniture over and above the minimum requirements. The order you place will give you 50 extra tables ($s = 5$, but s was introduced after we divided the first inequality by 10, so the actual surplus is $10 \times 5 = 50$), the correct number of chairs ($t = 0$), and the correct number of computer desks ($u = 0$). ∎

The preceding LP problem is an example of a **standard minimization problem**—in a sense the opposite of a standard maximization problem: We are *minimizing* an objective function, where all the constraints have the form $Ax + By + Cz + \cdots \geq N$ with N nonnegative. We will discuss standard minimization problems more fully in Section 5.5, as well as another method of solving them.

FAQs

When to Switch to Phase II, Equality Constraints, and Troubleshooting

Q: *How do I know when to switch to Phase II?*

A: After each step, check the basic solution for starred rows. You are not ready to proceed with Phase II until all the stars are gone.

Q: *How do I deal with an equality constraint, such as $2x + 7y - z = 90$?*

A: Although we haven't given examples of equality constraints, they can be treated by the following trick: Replace an equality by two inequalities. For example, replace the equality $2x + 7y - z = 90$ by the two inequalities $2x + 7y - z \leq 90$ and $2x + 7y - z \geq 90$. A little thought will convince you that these two inequalities amount to the same thing as the original equality!

Q: *What happens if it is impossible to choose a pivot using the instructions in Phase I?*

A: In that case, the LP problem has no solution. In fact, the feasible region is empty. If it is impossible to choose a pivot in Phase II, then the feasible region is unbounded, and there is no optimal solution.

5.4   EXERCISES

▼ more advanced   ◆ challenging
T indicates exercises that should be solved using technology

1. Maximize $\quad p = x + y$
subject to $\qquad x + 2y \geq 6$
$\qquad\qquad -x + y \leq 4$
$\qquad\qquad 2x + y \leq 8$
$\qquad\qquad x \geq 0, y \geq 0.$ [**HINT:** See Examples 1 and 2.]

2. Maximize $\quad p = 3x + 2y$
subject to $\qquad x + 3y \geq 6$
$\qquad\qquad -x + y \leq 4$
$\qquad\qquad 2x + y \leq 8$
$\qquad\qquad x \geq 0, y \geq 0.$ [**HINT:** See Examples 1 and 2.]

3. Maximize $p = 12x + 10y$
  subject to $x + y \leq 25$
          $x \geq 10$
          $-x + 2y \geq 0$
          $x \geq 0, y \geq 0.$

4. Maximize $p = x + 2y$
  subject to $x + y \leq 25$
           $y \geq 10$
          $2x - y \geq 0$
          $x \geq 0, y \geq 0.$

5. Maximize $p = 2x + 5y + 3z$
  subject to $x + y + z \leq 150$
          $x + y + z \geq 100$
          $x \geq 0, y \geq 0, z \geq 0.$

6. Maximize $p = 3x + 2y + 2z$
  subject to $x + y + 2z \leq 38$
          $2x + y + z \geq 24$
          $x \geq 0, y \geq 0, z \geq 0.$

7. Maximize $p = 10x + 20y + 15z$
  subject to $x + 2y + z \leq 40$
          $2y - z \geq 10$
          $2x - y + z \geq 20$
          $x \geq 0, y \geq 0, z \geq 0.$

8. Maximize $p = 10x + 10y + 15z$
  subject to $x - y + z \leq 12$
          $2x - 2y + z \geq 15$
          $-y + z \geq 3$
          $x \geq 0, y \geq 0, z \geq 0.$

9. Maximize $p = x + y + 3z + w$
  subject to $x + y + z + w \leq 40$
          $2x + y - z - w \geq 10$
          $x + y + z + w \geq 10$
          $x \geq 0, y \geq 0, z \geq 0, w \geq 0.$

10. Maximize $p = x + y + 4z + 2w$
  subject to $x + y + z + w \leq 50$
          $2x + y - z - w \geq 10$
          $x + y + z + w \geq 20$
          $x \geq 0, y \geq 0, z \geq 0, w \geq 0.$

11. Minimize $c = 6x + 6y$
  subject to $x + 2y \geq 20$
          $2x + y \geq 20$
          $x \geq 0, y \geq 0.$ [**HINT:** See Example 3.]

12. Minimize $c = 3x + 2y$
  subject to $x + 2y \geq 20$
          $2x + y \geq 10$
          $x \geq 0, y \geq 0.$ [**HINT:** See Example 3.]

13. Minimize $c = 2x + y + 3z$
  subject to $x + y + z \geq 100$
          $2x + y \geq 50$
          $y + z \geq 50$
          $x \geq 0, y \geq 0, z \geq 0.$

14. Minimize $c = 2x + 2y + 3z$
  subject to $x + z \geq 100$
          $2x + y \geq 50$
          $y + z \geq 50$
          $x \geq 0, y \geq 0, z \geq 0.$

15. Minimize $c = 50x + 50y + 11z$
  subject to $2x + z \geq 3$
          $2x + y - z \geq 2$
          $3x + y - z \leq 3$
          $x \geq 0, y \geq 0, z \geq 0.$

16. Minimize $c = 50x + 11y + 50z$
  subject to $3x + z \geq 8$
          $3x + y - z \geq 6$
          $4x + y - z \leq 8$
          $x \geq 0, y \geq 0, z \geq 0.$

17. Minimize $c = x + y + z + w$
  subject to $5x - y + w \geq 1{,}000$
          $z + w \leq 2{,}000$
          $x + y \leq 500$
          $x \geq 0, y \geq 0, z \geq 0, w \geq 0.$

18. Minimize $c = 5x + y + z + w$
  subject to $5x - y + w \geq 1{,}000$
          $z + w \leq 2{,}000$
          $x + y \leq 500$
          $x \geq 0, y \geq 0, z \geq 0, w \geq 0.$

T *In Exercises 19–24, we suggest the use of technology. Round all answers to two decimal places.*

19. Maximize $p = 2x + 3y + 1.1z + 4w$
  subject to $1.2x + y + z + w \leq 40.5$
          $2.2x + y - z - w \geq 10$
          $1.2x + y + z + 1.2w \geq 10.5$
          $x \geq 0, y \geq 0, z \geq 0, w \geq 0.$

20. Maximize $p = 2.2x + 2y + 1.1z + 2w$
  subject to $x + 1.5y + 1.5z + w \leq 50.5$
          $2x + 1.5y - z - w \geq 10$
          $x + 1.5y + z + 1.5w \geq 21$
          $x \geq 0, y \geq 0, z \geq 0, w \geq 0.$

21. Minimize $c = 2.2x + y + 3.3z$
  subject to $x + 1.5y + 1.2z \geq 100$
          $2x + 1.5y \geq 50$
          $1.5y + 1.1z \geq 50$
          $x \geq 0, y \geq 0, z \geq 0.$

22. Minimize $c = 50.3x + 10.5y + 50.3z$
subject to
$$3.1x \quad\;\; + 1.1z \geq 28$$
$$3.1x + y - 1.1z \geq 23$$
$$4.2x + y - 1.1z \geq 28$$
$$x \geq 0, y \geq 0, z \geq 0.$$

23. Minimize $c = 1.1x + y + 1.5z - w$
subject to
$$5.12x - y \quad\;\; + w \leq 1,000$$
$$z + w \geq 2,000$$
$$1.22x + y \quad\quad\quad \leq 500$$
$$x \geq 0, y \geq 0, z \geq 0, w \geq 0.$$

24. Minimize $c = 5.45x + y + 1.5z + w$
subject to
$$5.12x - y \quad\;\; + w \geq 1,000$$
$$z + w \geq 2,000$$
$$1.12x + y \quad\quad\quad \leq 500$$
$$x \geq 0, y \geq 0, z \geq 0, w \geq 0.$$

Applications

25. *Agriculture* (Compare Exercise 27 in Section 5.3.) Your small farm encompasses 100 acres, and you are planning to grow tomatoes, lettuce, and carrots in the coming planting season. Fertilizer costs per acre are $5 for tomatoes, $4 for lettuce, and $2 for carrots. Based on past experience, you estimate that each acre of tomatoes will require an average of 4 hours of labor per week, while tending to lettuce and carrots will each require an average of 2 hours per week. You estimate a profit of $2,000 for each acre of tomatoes, $1,500 for each acre of lettuce, and $500 for each acre of carrots. You would like to spend at least $400 on fertilizer (your niece owns the company that manufactures it), and your farm laborers can supply up to 500 hours per week. How many acres of each crop should you plant to maximize total profits? In this event, will you be using all 100 acres of your farm? [**HINT:** See Example 3.]

26. *Agriculture* (Compare Exercise 28 in Section 5.3.) Your farm encompasses 900 acres, and you are planning to grow soybeans, corn, and wheat in the coming planting season. Fertilizer costs per acre are $5 for soybeans, $2 for corn, and $1 for wheat. You estimate that each acre of soybeans will require an average of 5 hours of labor per week, while tending to corn and wheat will each require an average of 2 hours per week. On the basis of past yields and current market prices, you estimate a profit of $3,000 for each acre of soybeans, $2,000 for each acre of corn, and $1,000 for each acre of wheat. You can afford to spend no more than $3,000 on fertilizer, but your labor union contract stipulates at least 2,000 hours per week of labor. How many acres of each crop should you plant to maximize total profits? In this event, will you be using more than 2,000 hours of labor? [**HINT:** See Example 3.]

27. *Politics* The political pollster *Canter* is preparing for a national election. It would like to poll at least 1,500 Democrats and 1,500 Republicans. Each mailing to the East Coast gets responses from 100 Democrats and 50 Republicans. Each mailing to the Midwest gets responses from 100 Democrats and 100 Republicans. And each mailing to the West Coast gets responses from 50 Democrats and 100 Republicans. Mailings to the East Coast cost $40 each to produce and mail, mailings to the Midwest cost $60 each, and mailings to the West Coast cost $50 each. How many mailings should Canter send to each area of the country to get the responses it needs at the least possible cost? What will it cost?

28. *Purchasing* *Bingo's Copy Center* needs to buy white paper and yellow paper. Bingo's can buy from three suppliers. *Harvard Paper* sells a package of 20 reams of white and 10 reams of yellow for $60, *Yale Paper* sells a package of 10 reams of white and 10 reams of yellow for $40, and *Dartmouth Paper* sells a package of 10 reams of white and 20 reams of yellow for $50. If Bingo's needs 350 reams of white and 400 reams of yellow, how many packages should it buy from each supplier to minimize the cost? What is the least possible cost?

29. *Resource Allocation* *Succulent Citrus* produces orange juice and orange concentrate. This year the company anticipates a demand of at least 10,000 quarts of orange juice and 1,000 quarts of orange concentrate. Each quart of orange juice requires 10 oranges, and each quart of concentrate requires 50 oranges. The company also anticipates using at least 200,000 oranges for these products. Each quart of orange juice costs the company 50¢ to produce, and each quart of concentrate costs $2.00 to produce. How many quarts of each product should Succulent Citrus produce to meet the demand and minimize total costs?

30. *Resource Allocation* *Fancy Pineapple* produces pineapple juice and canned pineapple rings. This year the company anticipates a demand of at least 10,000 pints of pineapple juice and 1,000 cans of pineapple rings. Each pint of pineapple juice requires 2 pineapples, and each can of pineapple rings requires 1 pineapple. The company anticipates using at least 20,000 pineapples for these products. Each pint of pineapple juice costs the company 20¢ to produce, and each can of pineapple rings costs 50¢ to produce. How many pints of pineapple juice and cans of pineapple rings should Fancy Pineapple produce to meet the demand and minimize total costs?

31. *Latin Music Sales (Digital)* You are about to go live with a Latin music download service called *iYayay* that will compete head-to-head with **Apple**'s iTunes. (Good luck!) You will be selling digital albums of regional (Mexican/Tejano) music for $5 each, pop/rock albums for $4 each, and tropical (salsa/merengue/cumbia/bachata) albums for $6 each. Your servers can handle up to 40,000 downloaded albums per day, and you anticipate on the basis of national sales[31] that sales

[31] In 2013, total U.S. revenues from regional music were about four times those from tropical music. Source: Recording Industry Association of America http://riaa.org.

of regional music will be at least four times those from tropical music. You also anticipate that you will sell at least 10,000 pop/rock albums per day as a result of the very attractive $4 price. On the basis of these assumptions, how many of each type of album should you sell for a maximum daily revenue, and what will your daily revenue be?

32. *Latin Music Sales (Digital)* It seems that your biggest rival, Lupita Pelogrande, has learned about your music store and has decided to launch *iRico*, a competitor Latin music download service that will also offer regional (Mexican/Tejano) music albums for $5 each and pop/rock albums for $4 each. However, instead of tropical music, *iRico* will offer reggaeton albums at only $3 per album. Lupita's servers are more robust than yours and can handle up to 60,000 downloaded albums per day. She anticipates that revenues from regional music can total to up to double those from reggaeton and that *iRico* can sell at least 50,000 albums per day, not even counting reggaeton. On the basis of these assumptions, how many of each type of album should *iRico* sell for a maximum daily revenue, and what will the daily revenue be?

Gaming *Exercises 33–36 are based on the following table, which shows some parameters of various weapons used in role-playing gaming:*[32]

Weapon	Cost (gold pieces)	Damage to Medium Targets	Critical Damage	Weight (pounds)
Axe (Throwing)	8	6	12	2
Javelin	1	6	12	2
Longsword	15	8	32	4
Mace (Light)	5	6	12	4
Spear	2	8	24	6

33. *Orcs* (Compare Exercise 31 in Section 5.3.) The Orc leader Achlúk has up to 600 gold pieces to spend on an arsenal of axes, maces, and spears for his army of orcs for a planned assault on Hobshire, in which he would like to inflict as much damage on medium targets (such as humans and hobbits) as possible and a total of at least 2,400 units of critical damage. As his orcs are particularly fond of maces but not particularly skilled at spear-throwing, he would like to include at least twice as many maces as spears in the arsenal. What should his weapons arsenal look like, and how much damage on medium targets can be inflicted?

34. *Elves* (Compare Exercise 32 in Section 5.3.) The Elf leader Galandir is assembling an arsenal of up to 6,000 pounds in weight of javelins, longswords, and spears for her band of elves in a planned assault on Mordrúk. She needs to inflict as much critical damage as possible and at least 2,000 units of damage on medium targets. As the elves in her band prefer spears to longswords, she would like to ensure that there are at least as many spears as longswords. What should her weapons arsenal look like, and how much critical damage can be inflicted?

35. *Orcs* Having suffered an embarrassing defeat in Hobshire, Achlúk is rearming his army of orcs with axes, maces, and spears. This time, his budget is tight, and he wants to spend as little as possible but still inflict at least 2,000 units of damage on medium targets and 2,400 units of critical damage while keeping the total weight to no more than 1,000 pounds. What should his weapons arsenal look like, and how much will it cost?

36. *Elves* The Elf leader Galandir is having second thoughts about how to arm her band of elves with javelins, longswords, and spears for the planned assault on Mordrúk. For the sake of swiftness she wants the arsenal to weigh as little as possible but still inflict at least 24,000 units of critical damage and 2,000 units of damage on medium targets at a total cost of no more than 15,000 gold pieces. What should her weapons arsenal look like, and how much will it weigh?

37. ▌*Nutrition* **Gerber Products'** Gerber Mixed Cereal for Baby contains, in each serving, 60 calories and no vitamin C. Gerber Mango Tropical Fruit Dessert contains, in each serving, 80 calories and 45% of the U.S. Recommended Daily Allowance (RDA) of vitamin C for infants. Gerber Apple Banana Juice contains, in each serving, 60 calories and 120% of the RDA of vitamin C for infants.[33] The cereal costs 10¢ per serving, the dessert costs 53¢ per serving, and the juice costs 27¢ per serving. If you want to provide your child with at least 120 calories and at least 120% of the RDA of vitamin C, how can you do so at the least cost?

38. ▌*Nutrition* **Gerber Products'** Gerber Mixed Cereal for Baby contains, in each serving, 60 calories, no vitamin C, and 11 grams of carbohydrates. Gerber Mango Tropical Fruit Dessert contains, in each serving, 80 calories, 45% of the RDA of vitamin C for infants, and 21 grams of carbohydrates. Gerber Apple Banana Juice contains, in each serving, 60 calories, 120% of the RDA of vitamin C for infants, and 15 grams of carbohydrates.[34] Assume that the cereal costs 11¢ per serving, the dessert costs 50¢ per serving, and the juice costs 30¢ per serving. If you want to provide your child with at least 180 calories, at least 120% of the RDA of vitamin C, and at least 37 grams of carbohydrates, how can you do so at the least cost?

---

[32] Source: Dungeons and Dragons Wiki (www.dandwiki.com.) (Critical Damage is a weighted measure of "critical hit damage" as defined there.) See also Paul Tozour's blog at www.gamasutra.com/blogs/PaulTozour/20130707/195718/ for a discussion of similar scenarios in the context of arming a video game battle tank.

[33] Source: Nutrition information supplied with the products.
[34] *Ibid.*

39. ▮ *Purchasing Cheapskate Electronics Store* needs to update its inventory of stereos, TVs, and DVD players. There are three suppliers it can buy from: *Nadir* offers a bundle consisting of 5 stereos, 10 TVs, and 15 DVD players for $3,000. *Blunt* offers a bundle consisting of 10 stereos, 10 TVs, and 10 DVD players for $4,000. *Sonny* offers a bundle consisting of 15 stereos, 10 TVs, and 10 DVD players for $5,000. Cheapskate Electronics needs at least 150 stereos, 200 TVs, and 150 DVD players. How can it update its inventory at the least possible cost? What is the least possible cost?

40. ▮ *Purchasing Federal Rent-a-Car* is putting together a new fleet. It is considering package offers from three car manufacturers. *Fred Motors* is offering 5 small cars, 5 medium cars, and 10 large cars for $500,000. *Admiral Motors* is offering 5 small, 10 medium, and 5 large cars for $400,000. *Chrysalis* is offering 10 small, 5 medium, and 5 large cars for $300,000. Federal would like to buy at least 550 small cars, at least 500 medium cars, and at least 550 large cars. How many packages should it buy from each car maker to keep the total cost as small as possible? What will be the total cost?

▮ *Bodybuilding Supplements Exercises 41 and 42 are based on the following data on four bodybuilding supplements. (Figures shown correspond to a single serving.)*[35]

	Creatine (grams)	L-Glutamine (grams)	BCAAs (grams)	Approximate Cost ($)
Xtend (SciVation)	0	2.5	7	1.00
Gainz (MP Hardcore)	2	3	6	1.10
Strongevity (Bill Phillips)	2.5	1	0	1.30
Muscle Physique (EAS)	2	2	0	1.00

41. Your personal trainer suggests that you supplement with at least 70 grams of creatine, 50 grams of L-glutamine, and 60 grams of BCAAs per week. You are thinking of combining Xtend, Gainz, and Strongevity to provide you with the required nutrients. How many servings of each should you combine to obtain a week's supply that meets your trainer's specifications at the least cost? How much will the week's supply cost? [**HINT:** Use the Excel pivot and Gauss-Jordan tool in decimal mode to do the pivoting.]

42. Your friend's personal trainer suggests that she supplement with at least 60 grams of each of creatine, L-glutamine, and

BCAAs per week. She is thinking of combining Gainz, Strongevity, and Muscle Physique to provide her with the required nutrients. How many servings of each should she combine to obtain a week's supply that meets her trainer's specifications at the least cost? How much will the week's supply cost? [**HINT:** Use the Excel pivot and Gauss-Jordan tool in decimal mode to do the pivoting.]

43. ▼ *Subsidies* The Miami Beach City Council has offered to subsidize hotel development in Miami Beach, and it is hoping for at least two hotels with a total capacity of at least 1,400. Suppose that you are a developer interested in taking advantage of this offer by building a small group of hotels in Miami Beach. You are thinking of three prototypes: a convention-style hotel with 500 rooms costing $100 million, a vacation-style hotel with 200 rooms costing $20 million, and a small motel with 50 rooms costing $4 million. The City Council will approve your plans, provided that you build at least one convention-style hotel and no more than two small motels.
   a. How many of each type of hotel should you build to satisfy the city council's wishes and stipulations while minimizing your total cost?
   b. Now assume that the city council will give developers 20% of the cost of building new hotels in Miami Beach, up to $50 million.[36] Will the city's $50 million subsidy be sufficient to cover 20% of your total costs?

44. ▼ *Subsidies* Refer back to Exercise 43. You are about to begin the financial arrangements for your new hotels when the city council informs you that it has changed its mind and now requires at least two vacation-style hotels and no more than four small motels.
   a. How many of each type of hotel should you build to satisfy the city council's wishes and stipulations while minimizing your total costs?
   b. Will the city's $50 million subsidy limit still be sufficient to cover 20% of your total costs?

45. ▼ *Transportation Scheduling* We return to your exploits coordinating distribution for the *Tubular Ride Boogie Board Company*.[37] You will recall that the company has manufacturing plants in Tucson, Arizona, and Toronto, Ontario, and you have been given the job of coordinating distribution of their latest model, the Gladiator, to their outlets in Honolulu and Venice Beach. The Tucson plant can manufacture up to 620 boards per week, while the Toronto plant, beset by labor disputes, can produce no more than 410 Gladiator boards per week. The outlet in Honolulu orders 500 Gladiator boards per week, while the Venice Beach outlet orders 530 boards per week. Transportation

[35] Source: Nutritional information supplied by the manufacturers/ www.bodybuilding.com. Cost per serving is approximate and varies considerably. "BCAAs" refers to the branched-chain amino acids leucine, isoleucine, and valine in the optimal 2:1:1 ratio.

[36] The Miami Beach City Council made such an offer in 1993. (*Chicago Tribune*, June 20, 1993, Section 7, p. 8).

[37] See Exercise 26 in Section 3.3 and Exercise 43 in Section 5.3. This time, we will use the simplex method to solve the version of this problem we first considered in Section 3.3.

costs are as follows: Tucson to Honolulu: $10 per board; Tucson to Venice Beach: $5 per board; Toronto to Honolulu: $20 per board; Toronto to Venice Beach: $10 per board. Your manager has said that you are to be sure to fill all orders and ship the boogie boards at a minimum total transportation cost. How will you do it?

46. ▼ *Transportation Scheduling* In the situation described in Exercise 45, you have just been notified that workers at the Toronto boogie board plant have gone on strike, resulting in a total work stoppage. You are to come up with a revised delivery schedule by tomorrow with the understanding that the Tucson plant can push production to a maximum of 1,000 boards per week. What should you do?

47. ▼ *Finance* Senator Porkbarrel habitually overdraws his three bank accounts: at the *Congressional Integrity Bank, Citizens' Trust,* and *Checks R Us.* There are no penalties because the overdrafts are subsidized by the taxpayer. The Senate Ethics Committee tends to let slide irregular banking activities as long as they are not flagrant. At the moment (because of Congress's preoccupation with a Supreme Court nominee), a total overdraft of up to $10,000 will be overlooked. Porkbarrel's conscience makes him hesitate to overdraw accounts at banks whose names include expressions such as "integrity" and "citizens' trust." The effect is that his overdrafts at the first two banks combined amount to no more than one quarter of the total. On the other hand, the financial officers at Integrity Bank, aware that Senator Porkbarrel is a member of the Senate Banking Committee, "suggest" that he overdraw at least $2,500 from their bank. Find the amount he should overdraw from each bank to avoid investigation by the Ethics Committee and overdraw his account at Integrity by as much as his sense of guilt will allow.

48. ▼ *Scheduling* Because Joe Slim's brother was recently elected to the State Senate, Joe's financial advisement concern, *Inside Information Inc.,* has been doing a booming trade, even though the financial counseling he offers is quite worthless. (None of his seasoned clients pays the slightest attention to his advice.) Slim charges different hourly rates to different categories of individuals: $5,000 per hour for private citizens, $50,000 per hour for corporate executives, and $10,000 per hour for presidents of universities. Because of his taste for leisure, he feels that he can spend no more than 40 hours per week in consultation. On the other hand, Slim feels that it would be best for his intellect were he to devote at least 10 hours of consultation each week to university presidents. However, Slim always feels somewhat uncomfortable dealing with academics, so he would prefer to spend no more than half his consultation time with university presidents. Furthermore, he likes to think of himself as representing the interests of the common citizen, so he wishes to offer at least 2 more hours of his time each week to private citizens than to corporate executives and university presidents combined. Given all these restrictions, how many hours each week should he spend with each type of client to maximize his income?

49. ▼ *Transportation Scheduling* Your publishing company is about to start a promotional blitz for its new book, *Advanced Quantum Mechanics for the Liberal Arts.* You have 20 salespeople stationed in Chicago and 10 in Denver. You would like to fly at least 10 salespeople to Los Angeles and at least 15 to New York. A round-trip plane flight from Chicago to Los Angeles costs $200; from Chicago to New York costs $125; from Denver to Los Angeles costs $225; and from Denver to New York costs $280.[38] How many salespeople should you fly from each of Chicago and Denver to each of Los Angeles and New York to spend the least amount on plane flights?

50. ▼ *Transportation Scheduling* Repeat Exercise 49, but now suppose that you would like at least 15 salespeople in Los Angeles.

51. **T** ▼ *Hospital Staffing* As the staff director of a new hospital, you are planning to hire cardiologists, rehabilitation specialists, and infectious disease specialists. According to recent data, each cardiology case averages $12,000 in revenue, each physical rehabilitation case averages $19,000, and each infectious disease case averages $14,000.[39] You judge that each specialist you employ will expand the hospital caseload by about 10 patients per week. You already have 3 cardiologists on staff, and the hospital is equipped to admit up to 200 patients per week. According to past experience, each cardiologist and rehabilitation specialist brings in one government research grant per year, while each infectious disease specialist brings in three. Your board of directors would like to see a total of at least 30 grants per year and would like your weekly revenue to be as large as possible. How many of each kind of specialist should you hire?

52. **T** ▼ *Hospital Staffing* Referring to Exercise 51, you completely misjudged the number of patients each type of specialist would bring to the hospital per week. It turned out that each cardiologist brought in 120 new patients per year, each rehabilitation specialist brought in 90 per year, and each infectious disease specialist brought in 70 per year.[40] It also turned out that your hospital could deal with no more than 1,960 new patients per year. Repeat Exercise 51 in light of this corrected data.

Communication and Reasoning Exercises

53. Explain the need for Phase I in a nonstandard LP problem.

54. Explain the need for Phase II in a nonstandard LP problem.

---

[38] Approximate prices advertised on various websites in September 2011.

[39] These (rounded) figures are based on an Illinois survey of 1.3 million hospital admissions (*Chicago Tribune,* March 29, 1993, Section 4, p. 1). Source: Lutheran General Health System, Argus Associates, Inc.

[40] These (rounded) figures were obtained from the survey referenced in Exercise 51 by dividing the average hospital revenue per physician by the revenue per case.

55. Explain briefly why we would need to use Phase I in solving a linear programming problem with the constraint $x + 2y - z \geq 3$.

56. Which rows do we star, and why?

57. Consider the following linear programming problem:

$$\text{Maximize} \quad p = x + y$$
$$\text{subject to} \quad x - 2y \geq 0$$
$$2x + y \leq 10$$
$$x \geq 0, y \geq 0.$$

This problem

(A) must be solved using the techniques of Section 5.4.
(B) must be solved using the techniques of Section 5.3.
(C) can be solved using the techniques of either section.

58. Consider the following linear programming problem:

$$\text{Maximize} \quad p = x + y$$
$$\text{subject to} \quad x - 2y \geq 1$$
$$2x + y \leq 10$$
$$x \geq 0, y \geq 0.$$

This problem

(A) must be solved using the techniques of Section 5.4.
(B) must be solved using the techniques of Section 5.3.
(C) can be solved using the techniques of either section.

59. ▼ Find a linear programming problem in three variables that requires one pivot in Phase I.

60. ▼ Find a linear programming problem in three variables that requires two pivots in Phase I.

61. ▼ Find a linear programming problem in two or three variables with no optimal solution, and show what happens when you try to solve it using the simplex method.

62. ▼ Find a linear programming problem in two or three variables with more than one optimal solution, and investigate which solution is found by the simplex method.

5.5 The Simplex Method and Duality

Dual Linear Programming Problems

We mentioned **standard minimization problems** in Section 5.4. These problems have the following form.

Standard Minimization Problem

A **standard minimization problem** is an LP problem in which we are required to *minimize* (not maximize) a linear objective function

$$c = as + bt + cu + \cdots$$

of the variables $s, t, u, \ldots$ (in this section we will always use the letters $s, t, u, \ldots$ for the unknowns in a standard minimization problem) subject to the constraints

$$s \geq 0, \quad t \geq 0, \quad u \geq 0, \ldots$$

and further constraints of the form

$$As + Bt + Cu + \cdots \geq N,$$

where $A, B, C, \ldots$, and N are numbers with N nonnegative.

A **standard linear programming problem** is an LP problem that is either a standard maximization problem or a standard minimization problem. An LP problem satisfies the **nonnegative objective condition** if all the coefficients in the objective function are nonnegative.

> ### Quick Examples
>
> **Standard Minimization and Maximization Problems**
>
> 1. Minimize $c = 2s + 3t + 3u$
>   subject to
> $$2s \quad\;\; + \;\; u \geq 10$$
> $$s + 3t - 6u \geq 5$$
> $$s \geq 0,\, t \geq 0,\, u \geq 0.$$
>   This is a standard minimization problem satisfying the nonnegative objective condition.
>
> 2. Maximize $p = 2x + 3y + 3z$
>   subject to
> $$2x \quad\;\; + \;\; z \leq 7$$
> $$x + 3y - 6z \leq 6$$
> $$x \geq 0,\, y \geq 0,\, z \geq 0.$$
>   This is a standard maximization problem satisfying the nonnegative objective condition.
>
> 3. Minimize $c = 2s - 3t + 3u$
>   subject to
> $$2s \quad\;\; + \;\; u \geq 10$$
> $$s + 3t - 6u \geq 5$$
> $$s \geq 0,\, t \geq 0,\, u \geq 0.$$
>   This is a standard minimization problem that does *not* satisfy the nonnegative objective condition.

We saw a way of solving minimization problems in Section 5.4, but a mathematically elegant relationship between maximization and minimization problems gives us another way of solving minimization problems that satisfy the nonnegative objective condition. This relationship is called **duality**.

To describe duality, we must first represent an LP problem by a matrix. This matrix is *not* the first tableau but something simpler: Pretend you forgot all about slack variables and also forgot to change the signs of the objective function.[*] As an example, consider the following two standard[†] problems:

* Forgetting these things is exactly what happens to many students under test conditions!

† Although duality does not require the problems to be standard, it does require them to be written in so-called *standard form:* In the case of a maximization problem, all constraints need to be (re)written using ≤, while for a minimization problem, all constraints need to be (re)written using ≥. Note that this means that the right-hand side of a constraint may wind up negative. It is least confusing to stick with standard problems, which is what we will do in this section.

Problem 1

$$\text{Maximize} \quad p = 20x + 20y + 50z$$
$$\text{subject to} \quad 2x + \;\; y + 3z \leq 2{,}000$$
$$x + 2y + 4z \leq 3{,}000$$
$$x \geq 0,\, y \geq 0,\, z \geq 0.$$

We represent this problem by the matrix

$$\begin{bmatrix} 2 & 1 & 3 & | & 2{,}000 \\ 1 & 2 & 4 & | & 3{,}000 \\ 20 & 20 & 50 & | & 0 \end{bmatrix} \quad \begin{matrix} \text{Constraint 1} \\ \text{Constraint 2} \\ \text{Objective} \end{matrix}$$

Notice that the coefficients of the objective function go in the bottom row, and we place a zero in the bottom right corner.

Problem 2 (from Example 3 in Section 5.4)

$$\text{Minimize} \quad c = 2{,}000s + 3{,}000t$$
$$\text{subject to} \quad 2s + \;\; t \geq 20$$
$$s + 2t \geq 20$$
$$3s + 4t \geq 50$$
$$s \geq 0,\, t \geq 0.$$

Problem 2 is represented by

$$\begin{bmatrix} 2 & 1 & 20 \\ 1 & 2 & 20 \\ 3 & 4 & 50 \\ 2{,}000 & 3{,}000 & 0 \end{bmatrix} \begin{matrix} \text{Constraint 1} \\ \text{Constraint 2} \\ \text{Constraint 3} \\ \text{Objective} \end{matrix}$$

These two problems are related: The matrix for Problem 1 is the transpose of the matrix for Problem 2. (Recall that the transpose of a matrix is obtained by writing its columns as rows; see Section 4.1.) When we have a pair of LP problems related in this way, we say that the two are *dual* LP problems.

Dual Linear Programming Problems

Two LP problems, one a maximization and one a minimization problem, are **dual** if the matrix that represents one is the transpose of the matrix that represents the other.

Finding the Dual of a Given Problem
Given an LP problem, we find its dual as follows:

1. Represent the problem as a matrix (see above).
2. Take the transpose of the matrix.
3. Write down the dual, which is the LP problem corresponding to the new matrix. If the original problem was a maximization problem, its dual will be a minimization problem, and vice versa.

The original problem is called the **primal problem**, and its dual is referred to as the **dual problem**.

Quick Example

Primal problem
4. Minimize $\quad c = s + 2t$
   subject to $\quad 5s + 2t \geq 60$
   $\qquad\qquad 3s + 4t \geq 80$
   $\qquad\qquad s + t \geq 20$
   $\qquad\qquad s \geq 0, t \geq 0.$

$\overset{1}{\rightarrow}$

$$\begin{bmatrix} 5 & 2 & 60 \\ 3 & 4 & 80 \\ 1 & 1 & 20 \\ 1 & 2 & 0 \end{bmatrix}$$

$\overset{2}{\rightarrow}$

$$\begin{bmatrix} 5 & 3 & 1 & 1 \\ 2 & 4 & 1 & 2 \\ 60 & 80 & 20 & 0 \end{bmatrix}$$

$\overset{3}{\rightarrow}$

Dual problem
Maximize $\quad p = 60x + 80y + 20z$
subject to $\quad 5x + 3y + z \leq 1$
$\qquad\qquad 2x + 4y + z \leq 2$
$\qquad\qquad x \geq 0, y \geq 0, z \geq 0.$

The following theorem justifies what we have been doing, and says that solving the dual problem of an LP problem is equivalent to solving the original problem.

Fundamental Theorem of Duality

(a) If an LP problem has an optimal solution, then so does its dual. Moreover, the primal problem and the dual problem have the same optimal value for their objective functions.

(**b**) Contained in the final tableau of the simplex method applied to an LP problem is the solution to its dual problem: It is given by the bottom entries in the columns associated with the slack variables, divided by the entry under the objective variable.

* The proof of the theorem is beyond the scope of this book but can be found in a textbook devoted to linear programming, such as *Linear Programming* by Vašek Chvátal (San Francisco: W. H. Freeman and Co., 1983), which has a particularly well-motivated discussion.

The theorem* gives us an alternative way of solving minimization problems that satisfy the nonnegative objective condition. Let's illustrate by solving Problem 2 above.

EXAMPLE 1 **Solving by Duality**

Minimize $c = 2{,}000s + 3{,}000t$
subject to $2s + \ t \geq 20$
$s + 2t \geq 20$
$3s + 4t \geq 50$
$s \geq 0, t \geq 0.$

Solution

Step 1 *Find the dual problem.* Write the primal problem in matrix form and take the transpose:

$$
\begin{bmatrix}
2 & 1 & 20 \\
1 & 2 & 20 \\
3 & 4 & 50 \\
2{,}000 & 3{,}000 & 0
\end{bmatrix}
\rightarrow
\begin{bmatrix}
2 & 1 & 3 & 2{,}000 \\
1 & 2 & 4 & 3{,}000 \\
20 & 20 & 50 & 0
\end{bmatrix}.
$$

The dual problem is

Maximize $p = 20x + 20y + 50z$
subject to $2x + \ y + 3z \leq 2{,}000$
$x + 2y + 4z \leq 3{,}000$
$x \geq 0, y \geq 0, z \geq 0.$

Step 2 *Use the simplex method to solve the dual problem.* Because we have a standard maximization problem, we do not have to worry about Phase I but go straight to Phase II.

	x	y	z	s	t	p	
s	2	1	3	1	0	0	2,000
t	1	2	4	0	1	0	3,000
p	−20	−20	−50	0	0	1	0

	x	y	z	s	t	p	
z	2	1	3	1	0	0	2,000
t	−5	2	0	−4	3	0	1,000
p	40	−10	0	50	0	3	100,000

	x	y	z	s	t	p	
z	9	0	6	6	−3	0	3,000
y	−5	2	0	−4	3	0	1,000
p	15	0	0	30	15	3	105,000

Note that the maximum value of the objective function is $p = 105,000/3 = 35,000$. By the theorem this is also the optimal value of c in the primal problem!

Step 3 *Read off the solution to the primal problem by dividing the bottom entries in the columns associated with the slack variables by the entry in the p column.* Here is the final tableau again with the entries in question highlighted:

	x	y	z	s	t	p	
z	9	0	6	6	-3	0	3,000
y	-5	2	0	-4	3	0	1,000
p	15	0	0	30	15	3	105,000

The solution to the primal problem is

$$s = 30/3 = 10, \quad t = 15/3 = 5, \quad c = 105,000/3 = 35,000.$$

(Compare this with the method we used to solve Example 3 in Section 5.4. Which method seems more efficient?)

➡ **Before we go on . . .** Can you now see the reason for using the variable names $s, t, u, \ldots$ in standard minimization problems? ■

Q: *Is the theorem also useful for solving problems that do not satisfy the nonnegative objective condition?*

A: Consider a standard minimization problem that does not satisfy the nonnegative objective condition, such as

Minimize $c = 2s - t$
subject to $2s - 3t \geq 1$
              $5s + 6t \geq 7$
              $s \geq 0, t \geq 0.$

Its dual would be

Maximize $p = x + 7y$
subject to $2x + 5y \leq 2$
            $-3x + 6y \leq -1$
            $x \geq 0, y \geq 0.$

This is not a standard maximization problem because the right-hand side of the second constraint is negative. (Multiplying both sides of the second constraint by -1 to make the right-hand side nonnegative results in a $\geq$ type inequality.) To solve the dual by the simplex method will require using Phase I as well as Phase II, and we may as well just solve the primal problem that way to begin with. Thus, duality helps us to solve problems only when the primal problem satisfies the nonnegative objective condition.

In general, if a problem does not satisfy the nonnegative objective condition, its dual is not standard.

Shadow Costs

A common kind of economic application is one that leads to a standard minimization problem involving minimizing cost. The minimum cost that is obtained depends on the constraints, which, because it is a standard minimization problem, are all of the form

$$As + Bt + Cu + \cdots \geq N.$$

N is typically the number of units of some required quantity, such as protein, and increasing that requirement will likely raise the minimum cost. In a standard minimization problem it turns out that the minimum cost increases by a fixed amount for each additional unit of N (up to a certain "allowable" maximum), and this fixed amount is called the **shadow cost** of the associated requirement.[*]

    In the following example we see how the solution to the dual problem also gives us the shadow costs associated with all the requirements.

[*] You might recognize the shadow cost as an associated *marginal cost*: the increase in minimum cost per additional unit of the required quantity.

EXAMPLE 2   **Shadow Costs**

You are trying to decide how many vitamin pills to take. SuperV brand vitamin pills each contain 2 milligrams of vitamin X, 1 milligram of vitamin Y, and 1 milligram of vitamin Z. Topper brand vitamin pills each contain 1 milligram of vitamin X, 1 milligram of vitamin Y, and 2 milligrams of vitamin Z. You want to take enough pills daily to get at least 12 milligrams of vitamin X, 10 milligrams of vitamin Y, and 12 milligrams of vitamin Z. However, SuperV pills cost 4¢ each, and Toppers cost 3¢ each, and you would like to minimize the total cost of your daily dosage.

a. How many of each brand of pill should you take?

b. Determine the shadow costs of each vitamin. That is, if you raised the daily requirements of any vitamin, by how much would the cost increase for each additional milligram added to the requirement?

Solution

a. This is a straightforward minimization problem. The unknowns are

$$s = \text{number of SuperV brand pills}$$
$$t = \text{number of Topper brand pills}.$$

The linear programming problem is

$$
\begin{aligned}
\text{Minimize} \quad & c = 4s + 3t \\
\text{subject to} \quad & 2s + t \geq 12 \\
& s + t \geq 10 \\
& s + 2t \geq 12 \\
& s \geq 0, t \geq 0.
\end{aligned}
$$

We solve this problem by using the simplex method on its dual, which is

$$
\begin{aligned}
\text{Maximize} \quad & p = 12x + 10y + 12z \\
\text{subject to} \quad & 2x + y + z \leq 4 \\
& x + y + 2z \leq 3 \\
& x \geq 0, y \geq 0, z \geq 0.
\end{aligned}
$$

After pivoting three times, we arrive at the final tableau:

	x	y	z	s	t	p	
x	6	0	−6	6	−6	0	6
y	0	1	3	−1	2	0	2
p	0	0	6	2	8	1	32

Therefore, the answer to the original problem is that you should take two SuperV vitamin pills and eight Toppers at a cost of 32¢ per day.

b. Now, the key to finding the shadow costs, which determine how changing your daily vitamin requirements would affect your minimum cost, is to look at the solution to the dual problem. From the tableau we see that $x = 1$, $y = 2$, and $z = 0$. To see what x, y, and z might tell us about the original problem, let's look at their units. In the inequality $2x + y + z \leq 4$, the coefficient 2 of x has units "milligrams of vitamin X per SuperV pill," and the 4 on the right-hand side has units "cents per SuperV pill." For $2x$ to have the same units as the 4 on the right-hand side, x must have units "cents per milligram of vitamin X." Similarly, y must have units "cents per milligram of vitamin Y," and z must have units "cents per milligram of vitamin Z."

These are exactly the units that the shadow costs we are seeking should have: cost per additional milligram of vitamin. One can show (although we will not do it here) that this is no coincidence: x is in fact the shadow cost of vitamin X. That is, x is the amount that would be added to the minimum cost for each increase[*] of 1 milligram of vitamin X in our daily requirement. For example, if we were to increase our requirement from 12 milligrams to 14 milligrams, an increase of 2 milligrams, the minimum cost would change by $2x = 2$¢, from 32¢ to 34¢. (Try it; you'll end up taking four SuperV pills and six Toppers.)

Similarly, $y = 2$ is the shadow cost of vitamin Y; each increase of 1 milligram in the requirement for vitamin Y would increase the cost by 2¢. The same holds for (small) decreases: Each decrease of 1 milligram in the requirement for vitamin Y would decrease the cost by 2¢ (and similarly for vitamin X).

What about $z = 0$? The shadow cost of vitamin Z is 0¢ per milligram, meaning that you can increase your requirement of vitamin Z without changing your cost. In fact, the solution $s = 2$ and $t = 8$ provides you with 18 milligrams of vitamin Z, so you can increase the required amount of vitamin Z up to 18 milligrams (or decrease it to 0) without changing the solution at all.

We can also interpret the shadow costs as the effective cost to you of each milligram of each vitamin in the optimal solution. You are paying 1¢ per milligram of vitamin X, paying 2¢ per milligram of vitamin Y, and getting the vitamin Z for free. This gives a total cost of $1 \times 12 + 2 \times 10 + 0 \times 12 = 32$¢, as we know. Again, if you change your requirements slightly, these are the amounts you will pay per milligram of each vitamin.

[*] To be scrupulously correct, this works only for changes within a certain range, not necessarily for very large changes.

Application to Game Theory

We return to a topic we discussed in Section 4.4: solving two-person zero-sum games. In that section we described how to solve games that could be reduced to 2×2 games or smaller. It turns out that we can solve larger games using linear programming and duality. We summarize the procedure, work through an example, and then discuss why it works.

Solving a Matrix Game

Step 1 Reduce the payoff matrix by dominance.

Step 2 Add a fixed number k to each of the entries so that they all become nonnegative and no column is all zero.

Step 3 Write 1s to the right of and below the matrix, and then write down the associated standard maximization problem. Solve this primal problem using the simplex method.

Step 4 Find the optimal strategies and the expected value as follows:

Column Strategy

1. Express the solution to the primal problem as a column vector.

2. Normalize by dividing each entry of the solution vector by p (which is also the sum of all the entries).

3. Insert zeros in positions corresponding to the columns deleted during reduction by dominance.

Row Strategy

1. Express the solution to the dual problem as a row vector.

2. Normalize by dividing each entry by p, which will once again be the sum of all the entries.

3. Insert zeros in positions corresponding to the rows deleted during reduction by dominance.

Value of the Game

$$e = \frac{1}{p} - k$$

EXAMPLE 3   **Restaurant Inspector**

You manage two restaurants, *Tender Steaks Inn* (TSI) and *Break for a Steak* (BFS). Even though you run the establishments impeccably, the Department of Health has been sending inspectors to your restaurants on a daily basis and fining you for minor infractions. You've found that you can head off a fine if you're present, but you can cover only one restaurant at a time. The Department of Health, on the other hand, has two inspectors, who sometimes visit the same restaurant and sometimes split up, one to each restaurant. The average fines you have been getting are shown in the following matrix:

Health Inspectors

		Both at BFS	Both at TSI	One at Each
You go to	**TSI**	$8,000	0	$2,000
	BFS	0	$10,000	$4,000

How should you choose which restaurant to visit to minimize your expected fine?

Solution This matrix is not quite the payoff matrix because fines, being penalties, should be negative payoffs. Thus, the payoff matrix is the following:

$$P = \begin{bmatrix} -8,000 & 0 & -2,000 \\ 0 & -10,000 & -4,000 \end{bmatrix}.$$

We follow the steps above to solve the game using the simplex method.

Step 1 There are no dominated rows or columns, so this game does not reduce.

Step 2 We add $k = 10,000$ to each entry so that none are negative, getting the following new matrix (with no zero column):

$$\begin{bmatrix} 2,000 & 10,000 & 8,000 \\ 10,000 & 0 & 6,000 \end{bmatrix}.$$

Step 3 We write 1s to the right and below this matrix:

$$\begin{bmatrix} 2,000 & 10,000 & 8,000 & 1 \\ 10,000 & 0 & 6,000 & 1 \\ 1 & 1 & 1 & 0 \end{bmatrix}.$$

The corresponding standard maximization problem is the following:

Maximize $p = x + y + z$

subject to $2,000x + 10,000y + 8,000z \le 1$

$10,000x \qquad\qquad + 6,000z \le 1$

$x \ge 0, y \ge 0, z \ge 0.$

Step 4 We use the simplex method to solve this problem. After pivoting twice, we arrive at the final tableau:

	x	y	z	s	t	p	
y	0	50,000	34,000	5	-1	0	4
x	10,000	0	6,000	0	1	0	1
p	0	0	14,000	5	4	50,000	9

Column Strategy The solution to the primal problem is

$$\begin{bmatrix} x \\ y \\ z \end{bmatrix} = \begin{bmatrix} \frac{1}{10,000} \\ \frac{4}{50,000} \\ 0 \end{bmatrix}.$$

We divide each entry by $p = 9/50,000$, which is also the sum of the entries. This gives the optimal column strategy:

$$C = \begin{bmatrix} \frac{5}{9} \\ \frac{4}{9} \\ 0 \end{bmatrix}.$$

Thus, the inspectors' optimal strategy is to stick together, visiting BFS with probability 5/9 and TSI with probability 4/9.

Row Strategy The solution to the dual problem is

$$\begin{bmatrix} s & t \end{bmatrix} = \begin{bmatrix} \frac{5}{50,000} & \frac{4}{50,000} \end{bmatrix}.$$

Once again, we divide by $p = 9/50{,}000$ to find the optimal row strategy:

$$R = \begin{bmatrix} \frac{5}{9} & \frac{4}{9} \end{bmatrix}.$$

Thus, you should visit TSI with probability 5/9 and BFS with probability 4/9.

Value of the Game Your expected average fine is

$$e = \frac{1}{p} - k = \frac{50{,}000}{9} - 10{,}000 = -\frac{40{,}000}{9} \approx -\$4{,}444.$$

➡ **Before we go on...** We owe you an explanation of why the procedure we used in Example 3 works. The main point is to understand how we turn a game into a linear programming problem. It's not hard to see that adding a fixed number k to all the payoffs will change only the payoff, increasing it by k, and will not change the optimal strategies. So let's pick up Example 3 from the point at which we were considering the following game:

$$P = \begin{bmatrix} 2{,}000 & 10{,}000 & 8{,}000 \\ 10{,}000 & 0 & 6{,}000 \end{bmatrix}.$$

We are looking for the optimal strategies R and C for the row and column players, respectively; if e is the value of the game, we will have $e = RPC$. Let's concentrate first on the column player's strategy $C = \begin{bmatrix} u & v & w \end{bmatrix}^T$, where u, v, and w are the unknowns we want to find. Because e is the value of the game, if the column player uses the optimal strategy C and the row player uses any old strategy S, the expected value with these strategies has to be e or better for the column player, so $SPC \le e$. Let's write that out for two particular choices of S. First, consider $S = \begin{bmatrix} 1 & 0 \end{bmatrix}$:

$$\begin{bmatrix} 1 & 0 \end{bmatrix} \begin{bmatrix} 2{,}000 & 10{,}000 & 8{,}000 \\ 10{,}000 & 0 & 6{,}000 \end{bmatrix} \begin{bmatrix} u \\ v \\ w \end{bmatrix} \le e.$$

Multiplied out, this gives

$$2{,}000u + 10{,}000v + 8{,}000w \le e.$$

Next, do the same thing for $S = \begin{bmatrix} 0 & 1 \end{bmatrix}$:

$$\begin{bmatrix} 0 & 1 \end{bmatrix} \begin{bmatrix} 2{,}000 & 10{,}000 & 8{,}000 \\ 10{,}000 & 0 & 6{,}000 \end{bmatrix} \begin{bmatrix} u \\ v \\ w \end{bmatrix} \le e$$

$$10{,}000u + 6{,}000w \le e.$$

It turns out that if these two inequalities are true, then $SPC \le e$ for any S at all, which is what the column player wants. These are starting to look like constraints in a linear programming problem, but the variable e appearing on the right is in the way. We get around this by dividing by e, which we know to be positive because all of the payoffs are nonnegative and no column is all zero (so the column player can't force the value of the game to be 0; here is where we need these assumptions). We get the following inequalities:

$$2{,}000\left(\frac{u}{e}\right) + 10{,}000\left(\frac{v}{e}\right) + 8{,}000\left(\frac{w}{e}\right) \le 1$$

$$10{,}000\left(\frac{u}{e}\right) \qquad\qquad + 6{,}000\left(\frac{w}{e}\right) \le 1.$$

Now we're getting somewhere. To make these look even more like linear constraints, we replace our unknowns u, v and w with new unknowns, $x = u/e$, $y = v/e$, and $z = w/e$. Our inequalities then become

$$2{,}000x + 10{,}000y + 8{,}000z \leq 1$$
$$10{,}000x \qquad\qquad + 6{,}000z \leq 1.$$

What about an objective function? From the point of view of the column player, the objective is to find a strategy that will minimize the expected value e. To write e in terms of our new variables x, y, and z, we use the fact that our original variables, being the entries in the column strategy, have to add up to 1: $u + v + w = 1$. Dividing by e gives

$$\frac{u}{e} + \frac{v}{e} + \frac{w}{e} = \frac{1}{e}$$

or

$$x + y + z = \frac{1}{e}.$$

Now we notice that, if we *maximize* $p = x + y + z = 1/e$, it will have the effect of minimizing e, which is what we want. So we get the following linear programming problem:

$$\text{Maximize} \quad p = x + y + z$$
$$\text{subject to} \quad 2{,}000x + 10{,}000y + 8{,}000z \leq 1$$
$$10{,}000x \qquad\qquad + 6{,}000z \leq 1$$
$$x \geq 0, y \geq 0, z \geq 0.$$

Why can we say that x, y, and z should all be nonnegative? Because the unknowns u, v, w, and e must all be nonnegative.

So now, if we solve this linear programming problem to find x, y, z, and p, we can find the column player's optimal strategy by computing $u = xe = x/p$, $v = y/p$, and $w = z/p$. Moreover, the value of the game is $e = 1/p$. (If we added k to all the payoffs, we should now adjust by subtracting k again to find the correct value of the game.)

Turning now to the row player's strategy, if we repeat the above type of argument from the row player's viewpoint, we'll end up with the following linear programming problem to solve:

$$\text{Minimize} \quad c = s + t$$
$$\text{subject to} \quad 2{,}000s + 10{,}000t \geq 1$$
$$10{,}000s \qquad\qquad \geq 1$$
$$8{,}000s + 6{,}000t \geq 1$$
$$s \geq 0, t \geq 0.$$

This is, of course, the dual to the problem we solved to find the column player's strategy, so we know that we can read its solution off of the same final tableau. The optimal value of c will be the same as the value of p, so $c = 1/e$ also. The entries in the optimal row strategy will be s/c and t/c. ∎

FAQs

When to Use Duality

Q : *Given a minimization problem, when should I use duality, and when should I use the two-phase method in Section 5.4?*

A : If the original problem satisfies the nonnegative objective condition (none of the coefficients in the objective function are negative), then you can use duality to convert the problem to a standard maximization one, which can be solved with the one-phase method. If the original problem does not satisfy the nonnegative objective condition, then dualizing results in a nonstandard LP problem, so dualizing may not be worthwhile.

Q : *When is it absolutely necessary to use duality?*

A : Never. Duality gives us an efficient but not necessary alternative for solving standard minimization problems.

5.5 EXERCISES

▼ more advanced   ◆ challenging
T indicates exercises that should be solved using technology

In Exercises 1–8, write down (without solving) the dual LP problem. [**HINT:** See Quick Example 4.]

1. Maximize $p = 2x + y$
subject to $x + 2y \leq 6$
$-x + y \leq 2$
$x \geq 0, y \geq 0.$

2. Maximize $p = x + 5y$
subject to $x + y \leq 6$
$-x + 3y \leq 4$
$x \geq 0, y \geq 0.$

3. Minimize $c = 2s + t + 3u$
subject to $s + t + u \geq 100$
$2s + t \geq 50$
$s \geq 0, t \geq 0, u \geq 0.$

4. Minimize $c = 2s + 2t + 3u$
subject to $s + u \geq 100$
$2s + t \geq 50$
$s \geq 0, t \geq 0, u \geq 0.$

5. Maximize $p = x + y + z + w$
subject to $x + y + z \leq 3$
$y + z + w \leq 4$
$x + z + w \leq 5$
$x + y + w \leq 6$
$x \geq 0, y \geq 0, z \geq 0, w \geq 0.$

6. Maximize $p = x + y + z + w$
subject to $x + y + z \leq 3$
$y + z + w \leq 3$
$x + z + w \leq 4$
$x + y + w \leq 4$
$x \geq 0, y \geq 0, z \geq 0, w \geq 0.$

7. Minimize $c = s + 3t + u$
subject to $5s - t + v \geq 1{,}000$
$u - v \geq 2{,}000$
$s + t \geq 500$
$s \geq 0, t \geq 0, u \geq 0, v \geq 0.$

8. Minimize $c = 5s + 2u + v$
subject to $s - t + 2v \geq 2{,}000$
$u + v \geq 3{,}000$
$s + t \geq 500$
$s \geq 0, t \geq 0, u \geq 0, v \geq 0.$

In Exercises 9–22, solve the given standard minimization problem using duality. (You may already have seen some of these in earlier sections, but now you will be solving them using a different method.) [**HINT:** See Example 1.]

9. Minimize $c = s + t$
subject to $s + 2t \geq 6$
$2s + t \geq 6$
$s \geq 0, t \geq 0.$

10. Minimize $c = s + 2t$

subject to $s + 3t \geq 30$

$2s + t \geq 30$

$s \geq 0, t \geq 0.$

11. Minimize $c = 6s + 6t$

subject to $s + 2t \geq 20$

$2s + t \geq 20$

$s \geq 0, t \geq 0.$

12. Minimize $c = 3s + 2t$

subject to $s + 2t \geq 20$

$2s + t \geq 10$

$s \geq 0, t \geq 0.$

13. Minimize $c = 0.2s + 0.3t$

subject to $2s + t \geq 10$

$s + 2t \geq 10$

$s + t \geq 8$

$s \geq 0, t \geq 0.$

14. Minimize $c = 0.4s + 0.1t$

subject to $3s + 2t \geq 60$

$s + 2t \geq 40$

$2s + 3t \geq 45$

$s \geq 0, t \geq 0.$

15. Minimize $c = 2s + t$

subject to $3s + t \geq 30$

$s + t \geq 20$

$s + 3t \geq 30$

$s \geq 0, t \geq 0.$

16. Minimize $c = s + 2t$

subject to $4s + t \geq 100$

$2s + t \geq 80$

$s + 3t \geq 150$

$s \geq 0, t \geq 0.$

17. Minimize $c = s + 2t + 3u$

subject to $3s + 2t + u \geq 60$

$2s + t + 3u \geq 60$

$s \geq 0, t \geq 0, u \geq 0.$

18. Minimize $c = s + t + 2u$

subject to $s + 2t + 2u \geq 60$

$2s + t + 3u \geq 60$

$s \geq 0, t \geq 0, u \geq 0.$

19. Minimize $c = 2s + t + 3u$

subject to $s + t + u \geq 100$

$2s + t \geq 50$

$t + u \geq 50$

$s \geq 0, t \geq 0, u \geq 0.$

20. Minimize $c = 2s + 2t + 3u$

subject to $s + u \geq 100$

$2s + t \geq 50$

$t + u \geq 50$

$s \geq 0, t \geq 0, u \geq 0.$

21. Minimize $c = s + t + u$

subject to $3s + 2t + u \geq 60$

$2s + t + 3u \geq 60$

$s + 3t + 2u \geq 60$

$s \geq 0, t \geq 0, u \geq 0.$

22. Minimize $c = s + t + 2u$

subject to $s + 2t + 2u \geq 60$

$2s + t + 3u \geq 60$

$s + 3t + 6u \geq 60$

$s \geq 0, t \geq 0, u \geq 0.$

In Exercises 23–28, solve the game with the given payoff matrix. [**HINT:** *See Example 3.*]

23. $P = \begin{bmatrix} -1 & 1 & 2 \\ 2 & -1 & -2 \end{bmatrix}$ **24.** $P = \begin{bmatrix} 1 & -1 & 2 \\ 1 & 2 & 0 \end{bmatrix}$

25. $P = \begin{bmatrix} -1 & 1 & 2 \\ 2 & -1 & -2 \\ 1 & 2 & 0 \end{bmatrix}$ **26.** $P = \begin{bmatrix} 1 & -1 & 2 \\ 1 & 2 & 0 \\ 0 & 1 & 1 \end{bmatrix}$

27. $\boxed{\text{T}}\ P = \begin{bmatrix} -1 & 1 & 2 & -1 \\ 2 & -1 & -2 & -3 \\ 1 & 2 & 0 & 1 \\ 0 & 2 & 3 & 3 \end{bmatrix}$

28. $\boxed{\text{T}}\ P = \begin{bmatrix} 1 & -1 & 2 & 0 \\ 1 & 2 & 0 & 1 \\ 0 & 1 & 1 & 0 \\ 2 & 0 & -2 & 2 \end{bmatrix}$

Applications

Many of Exercises 29–40 are similar or identical to ones in preceding exercise sets. Use duality to answer them.

29. *Nutrition* Meow makes cat food out of fish and cornmeal. Fish has 8 grams of protein and 4 grams of fat per ounce, and cornmeal has 4 grams of protein and 8 grams of fat. A jumbo can of cat food must contain at least 48 grams of protein and 48 grams of fat. If fish and cornmeal both cost

5¢ per ounce, how many ounces of each should Meow use in each can of cat food to minimize costs? What are the shadow costs of protein and of fat? [**HINT:** See Example 2.]

30. Nutrition *Oz* makes lion food out of giraffe and gazelle meat. Giraffe meat has 18 grams of protein and 36 grams of fat per pound, while gazelle meat has 36 grams of protein and 18 grams of fat per pound. A batch of lion food must contain at least 36,000 grams of protein and 54,000 grams of fat. Giraffe meat costs $2 per pound and gazelle meat costs $4 per pound. How many pounds of each should go into each batch of lion food to minimize costs? What are the shadow costs of protein and fat? [**HINT:** See Example 2.]

31. Nutrition *Ruff* makes dog food out of chicken and grain. Chicken has 10 grams of protein and 5 grams of fat per ounce, and grain has 2 grams of protein and 2 grams of fat per ounce. A bag of dog food must contain at least 200 grams of protein and at least 150 grams of fat. If chicken costs 10¢ per ounce and grain costs 1¢ per ounce, how many ounces of each should Ruff use in each bag of dog food to minimize cost? What are the shadow costs of protein and fat?

32. Purchasing The *Enormous State University*'s Business School is buying computers. The school has two models to choose from, the Pomegranate and the iZac. Each Pomegranate comes with 400 GB of memory and 80 TB of disk space, while each iZac has 300 GB of memory and 100 TB of disk space. For reasons related to its accreditation the school would like to be able to say that it has a total of at least 48,000 GB of memory and at least 12,800 TB of disk space. If both the Pomegranate and the iZac cost $2,000 each, how many of each should the school buy to keep the cost as low as possible? What are the shadow costs of memory and disk space?

33. Nutrition Gerber Products' Gerber Mixed Cereal for Baby contains, in each serving, 60 calories and no vitamin C. Each serving of Gerber Mango Tropical Fruit Dessert contains 80 calories and 45% of the U.S. Recommended Daily Allowance (RDA) of vitamin C for infants. Each serving of Gerber Apple Banana Juice contains 60 calories and 120% of the U.S. RDA of vitamin C for infants.[41] The cereal costs 10¢ per serving, the dessert costs 53¢ per serving, and the juice costs 27¢ per serving. If you want to provide your child with at least 120 calories and at least 120% of the U.S. RDA of vitamin C, how can you do so at the least cost? What are your shadow costs for calories and vitamin C?

34. Nutrition Gerber Products' Gerber Mixed Cereal for Baby contains, in each serving, 60 calories, no vitamin C, and 11 grams of carbohydrates. Each serving of Gerber Mango Tropical Fruit Dessert contains 80 calories, 45% of the U.S. Recommended Daily Allowance (RDA) of vitamin C for infants, and 21 grams of carbohydrates. Each serving of

Gerber Apple Banana Juice contains 60 calories, 120% of the U.S. RDA of vitamin C for infants, and 15 grams of carbohydrates.[42] Assume that the cereal costs 11¢ per serving, the dessert costs 50¢ per serving, and the juice costs 30¢ per serving. If you want to provide your child with at least 180 calories, at least 120% of the U.S. RDA of vitamin C, and at least 37 grams of carbohydrates, how can you do so at the least cost? What are your shadow costs for calories, vitamin C, and carbohydrates?

35. Politics The political pollster *Canter* is preparing for a national election. It would like to poll at least 1,500 Democrats and 1,500 Republicans. Each mailing to the East Coast gets responses from 100 Democrats and 50 Republicans. Each mailing to the Midwest gets responses from 100 Democrats and 100 Republicans. Each mailing to the West Coast gets responses from 50 Democrats and 100 Republicans. Mailings to the East Coast cost $40 each to produce and mail, mailings to the Midwest cost $60 each, and mailings to the West Coast cost $50 each. How many mailings should Canter send to each area of the country to get the responses it needs at the least possible cost? What will it cost? What are the shadow costs of a Democratic response and of a Republican response?

36. Purchasing *Bingo's Copy Center* needs to buy white paper and yellow paper. Bingo's can buy from three suppliers. *Harvard Paper* sells a package of 20 reams of white and 10 reams of yellow for $60; *Yale Paper* sells a package of 10 reams of white and 10 reams of yellow for $40, and *Dartmouth Paper* sells a package of 10 reams of white and 20 reams of yellow for $50. If Bingo's needs 350 reams of white and 400 reams of yellow, how many packages should it buy from each supplier to minimize the cost? What is the lowest possible cost? What are the shadow costs of white paper and yellow paper?

37. ▼ **Advertising** You are the marketing director for a company that manufactures bodybuilding supplements, and you are planning to run ads in **Sports Illustrated** and **GQ Magazine**. On the basis of readership data, you estimate that each ad in *Sports Illustrated* will be read by 600,000 people in your target group, while each ad in *GQ* will be read by 150,000.[43] You would like your ads to be read by at least 9 million people in the target group, and you plan to place at least 6 ads in *Sports Illustrated* and at least 8 ads in *GQ* during the next year. *Sports Illustrated* quotes you $2,000 per ad, while GQ quotes you $1,000 per ad. How many ads should be placed in each magazine to satisfy your requirements at a minimum cost?

---

[42] *Ibid.*

[43] The readership data for *Sports Illustrated* is roughly based on the results of a readership survey taken in March 2000. The readership data for *GQ* is fictitious. Source: Mediamark Research Inc./*New York Times*, May 29, 2000, p. C1.

[41] Source: Nutrition information supplied with the products.

38. ▼ *Advertising* You are the marketing director for a company that manufactures bodybuilding supplements, and you are planning to run ads in **Sports Illustrated** and **Muscle and Fitness**. On the basis of readership data, you estimate that each ad in *Sports Illustrated* will be read by 600,000 people in your target group, while each ad in *Muscle and Fitness* will be read by 300,000 people in your target group.[44] You would like your ads to be read by at least 9 million people in the target group, and you plan to place at least 3 ads in *Sports Illustrated* and at least 4 ads in *Muscle and Fitness* during the next year. *Sports Illustrated* quotes you $3,000 per ad, while *Muscle and Fitness* quotes you $2,000 per ad. How many ads should be placed in each magazine to satisfy your requirements at a minimum cost?

39. ▼ *Resource Allocation* One day, Gillian the Magician summoned the wisest of her women. "Devoted sisters of the Coven," she began, "I have a quandary: As you well know, I possess great expertise in sleep spells and shock spells, but unfortunately, these are proving to be a drain on my aural energy resources; each sleep spell costs me 50 therms of aural energy, while each shock spell requires 20 therms. Clearly, I would like to hold my overall expenditure of aural energy to a minimum and still meet my commitments in protecting the Sisterhood from the ever-present threat of trolls. Specifically, I have estimated that each sleep spell keeps us safe for an average of 1.5 hours, while every shock spell protects us for only one half hour. We certainly require enough protection to last 24 hours of each day and possibly more, just to be safe. At the same time, I have noticed that each of my sleep spells can immobilize 3 trolls at once, while one of my typical shock spells (having a narrower range) can immobilize only 2 trolls at once. We are faced, my sisters, with an onslaught of as many as 52 trolls per day! Finally, as you are no doubt aware, the Bylaws of the Coven dictate that for a magician to remain in good standing, she should cast at least as many shock spells as sleep spells. What do I do, oh Wise Ones?"

40. ▼ *Risk Management* The Grand Vizier of the Kingdom of Um is being blackmailed by numerous individuals and is having a very difficult time keeping his blackmailers from going public. He has been keeping them at bay with two kinds of payoff: gold from the Royal Treasury and political favors. Through bitter experience, he has learned that each gold payoff gives him peace for an average of about 1 month, and each political favor seems to earn him about a month and a half of reprieve. To maintain his flawless reputation in the court, he feels that he cannot afford any revelations about his tainted past to come to light within the next year. Thus, it is imperative that his blackmailers be kept at

bay for at least 12 months. Furthermore, he would like to keep the number of gold payoffs at no more than one quarter of the combined number of payoffs because the outward flow of gold bars might arouse suspicion on the part of the Royal Treasurer. The gold payoffs tend to deplete the Grand Vizier's travel budget. (The treasury has been subsidizing his numerous trips to the Himalayas.) He estimates that each gold bar removed from the treasury will cost him four trips. On the other hand, because the administering of political favors tends to cost him valuable travel time, he suspects that each political favor will cost him about two trips. Now, he would obviously like to keep his blackmailers silenced and lose as few trips as possible. What is he to do? How many trips will he lose in the next year?

41. ▼ *Game Theory: Politics* Incumbent Tax N. Spend and challenger Trick L. Down are running for county executive, and polls show them to be in a dead heat. The election hinges on three cities: Littleville, Metropolis, and Urbantown. The candidates have decided to spend the last weeks before the election campaigning in those three cities; each day each candidate will decide in which city to spend the day. Pollsters have determined the following payoff matrix, where the payoff represents the number of votes gained or lost for each 1-day campaign trip:

T. N. Spend

		Littleville	Metropolis	Urbantown
T. L. Down	**Littleville**	−200	−300	300
	Metropolis	−500	500	−100
	Urbantown	−500	0	0

What percentage of time should each candidate spend in each city to maximize votes gained? If both candidates use their optimal strategies, what is the expected vote?

42. ▼ *Game Theory: Marketing* Your company's new portable phone/music player/browser/bottle washer, the *Run-Man*, will compete against the established market leader, the *iNod*, in a saturated market. (Thus, for each device you sell, one fewer iNod is sold.) You are planning to launch the RunMan with a traveling road show, concentrating on two cities: New York and Boston. The makers of the iNod will do the same to try to maintain their sales. If, on a given day, you both go to New York, you will lose 1,000 units in sales to the iNod. If you both go to Boston, you will lose 750 units in sales. On the other hand, if you go to New York and your competitor to Boston, you will gain 1,500 units in sales from them. If you go to Boston and they to New York, you will gain 500 units in sales. What percentage of time should you spend in New York and what percentage in Boston, and how do you expect your sales to be affected?

43. ▼ *Game Theory: Morra Games* A three-finger Morra game is a game in which two players simultaneously show

[44] The readership data for both magazines are roughly based on the results of a readership survey taken in March 2000. Source: Mediamark Research Inc./*New York Times*, May 29, 2000, p. C1.

one, two, or three fingers at each round. The outcome depends on a predetermined set of rules. Here is an interesting example: If the numbers of fingers shown by A and B differ by 1, then A loses one point. If they differ by more than 1, the round is a draw. If they show the same number of fingers, A wins an amount equal to the sum of the fingers shown. Determine the optimal strategy for each player and the expected value of the game.

44. **T ▼ *Game Theory: Morra Games*** Referring to Exercise 43, consider the following rules for a three-finger Morra game: If the sum of the fingers shown is odd, then A wins an amount equal to that sum. If the sum is even, B wins the sum. Determine the optimal strategy for each player and the expected value of the game. [**HINT:** Use technology to do the pivoting in the associated linear programming problem.]

45. **T ◆ *Game Theory: Military Strategy*** Colonel Blotto is a well-known game in military strategy.[45] Here is a version of this game: Colonel Blotto has four regiments under his command, while his opponent, Captain Kije, has three. The armies are to try to occupy two locations, and each commander must decide how many regiments to send to each location. The army that sends more regiments to a location captures that location as well as the other army's regiments. If both armies send the same number of regiments to a location, then there is a draw. The payoffs are one point for each location captured and one point for each regiment captured. Find the optimum strategy for each commander and also the value of the game.

46. **T ◆ *Game Theory: Military Strategy*** Referring to Exercise 45, consider the version of Colonel Blotto with the same payoffs given there except that Captain Kije earns two points for each location captured, while Colonel Blotto continues to earn only one point. Find the optimum strategy for each commander and also the value of the game. Round all figures to two decimal places.

---

[45] See Samuel Karlin, *Mathematical Methods and Theory in Games, Programming and Economics* (Addison-Wesley, 1959).

Communication and Reasoning Exercises

47. A minimization problem has three variables and two constraints (other than those of the form $s \geq 0$, $t \geq 0$, . . .). How many variables and constraints (other than those of the form $x \geq 0$, $y \geq 0$, . . .) does the dual problem have? Why?

48. A minimization problem has more constraints (other than those of the form $s \geq 0$, $t \geq 0$, . . .) than variables. What can you say about the dual problem? Why?

49. Give one possible advantage of using duality to solve a standard minimization problem.

50. To ensure that the dual of a minimization problem will result in a standard maximization problem,
    (A) the primal problem should satisfy the nonnegative objective condition.
    (B) the primal problem should be a standard minimization problem.
    (C) the primal problem should not satisfy the nonnegative objective condition.

51. Give an example of a standard minimization problem whose dual is *not* a standard maximization problem. How would you go about solving your problem?

52. Give an example of a nonstandard minimization problem whose dual is a standard maximization problem.

53. If the primal problem is a standard minimization problem not satisfying the nonnegative objective condition, what can you say about the dual problem? Why?

54. If the primal problem is a nonstandard maximization problem, what can you say about the objective function of the dual problem? Why?

55. ▼ Given a minimization problem, when would you solve it by applying the simplex method to its dual, and when would you apply the simplex method to the minimization problem itself?

56. ▼ Create an interesting application that leads to a standard maximization problem. Solve it using the simplex method, and note the solution to its dual problem. What does the solution to the dual tell you about your application?

KEY CONCEPTS

5.1 Graphing Linear Inequalities

Inequalities, strict and nonstrict
[p. 338]
Linear inequalities [p. 339]
Solution of an inequality [p. 340]
Sketching the region represented by a
linear inequality in two variables
[p. 341]
Bounded and unbounded regions
[p. 344]
Feasible region [p. 345]

5.2 Solving Linear Programming Problems Graphically

Linear programming (LP) problem in
two unknowns; objective function;
constraints; feasible region; optimal
value; optimal solution [p. 349]
Fundamental Theorem of Linear
Programming [p. 350]

Graphical method for solving an
LP problem [p. 351]
Decision variables [p. 352]
Graphical method for solving an LP
problem with an unbounded feasible
region [p. 355]

5.3 The Simplex Method: Solving Standard Maximization Problems

General linear programming problem in
n unknowns [p. 367]
Standard maximization problem [p. 367]
Slack variable [p. 368]
Equation form of a standard maximiza-
tion problem [p. 369]
Tableau [p. 370]
Active (or basic) variables; inactive
(or nonbasic) variables; basic
solution [p. 370]
Rules for selecting the pivot
column [p. 371]
Rules for selecting the pivot; test
ratios [p. 372]
Departing or exiting variable, entering
variable [p. 373]

5.4 The Simplex Method: Solving General Linear Programming Problems

Surplus variable [p. 386]
Phase I and Phase II for solving general
LP problems [p. 387]
Using the simplex method to solve a
minimization problem [p. 391]

5.5 The Simplex Method and Duality

Standard minimization problem
[p. 400]
Standard LP problem [p. 400]
Nonnegative objective condition
[p. 400]
Dual LP problems; primal problem;
dual problem [p. 402]
Fundamental Theorem of Duality
[p. 402]
Shadow costs [p. 405]
Game theory: The LP problem
associated with a two-person
zero-sum game [p. 406]

REVIEW EXERCISES

In Exercises 1–4, sketch the region corresponding to the given inequalities, say whether it is bounded, and give the coordinates of all corner points.

1. $2x - 3y \leq 12$

2. $x \leq 2y$

3. $x + 2y \leq 20$
$3x + 2y \leq 30$
$x \geq 0, y \geq 0$

4. $3x + 2y \geq 6$
$2x - 3y \leq 6$
$3x - 2y \geq 0$
$x \geq 0, y \geq 0$

In Exercises 5–8, solve the given linear programming problem graphically.

5. Maximize $p = 2x + y$
subject to $3x + y \leq 30$
$x + y \leq 12$
$x + 3y \leq 30$
$x \geq 0, y \geq 0.$

6. Maximize $p = 2x + 3y$
subject to $x + y \geq 10$
$2x + y \geq 12$
$x + y \leq 20$
$x \geq 0, y \geq 0.$

7. Minimize $c = 2x + y$
subject to $3x + y \geq 30$
$x + 2y \geq 20$
$2x - y \geq 0$
$x \geq 0, y \geq 0.$

8. Minimize $c = 3x + y$
subject to $3x + 2y \geq 6$
$2x - 3y \leq 0$
$3x - 2y \geq 0$
$x \geq 0, y \geq 0.$

In Exercises 9–18, solve the given linear programming problem using the simplex method. If no optimal solution exists, indicate whether the feasible region is empty or the objective function is unbounded.

9. Maximize $p = x + y + 2z$
subject to $x + 2y + 2z \leq 60$
$2x + y + 3z \leq 60$
$x \geq 0, y \geq 0, z \geq 0.$

10. Maximize $p = x + y + 2z$
subject to $x + 2y + 2z \leq 60$
$2x + y + 3z \leq 60$
$x + 3y + 6z \leq 60$
$x \geq 0, y \geq 0, z \geq 0.$

11. Maximize $p = x + y + 3z$
subject to $x + y + z \geq 100$
$y + z \leq 80$
$x \quad + z \leq 80$
$x \geq 0, y \geq 0, z \geq 0.$

12. Maximize $p = 2x + y$
subject to $x + 2y \geq 12$
$2x + y \leq 12$
$x + y \leq 5$
$x \geq 0, y \geq 0.$

13. Minimize $c = x + 2y + 3z$
subject to $3x + 2y + z \geq 60$
$2x + y + 3z \geq 60$
$x \geq 0, y \geq 0, z \geq 0.$

14. Minimize $c = 5x + 4y + 3z$
subject to $x + y + 4z \geq 30$
$2x + y + 3z \geq 60$
$x \geq 0, y \geq 0, z \geq 0.$

15. ⊤ Minimize $c = x - 2y + 4z$
subject to $3x + 2y - z \geq 10$
$2x + y + 3z \geq 20$
$x + 3y - 2z \geq 30$
$x \geq 0, y \geq 0, z \geq 0.$

16. ⊤ Minimize $c = x + y - z$
subject to $3x + 2y + z \geq 60$
$2x + y + 3z \geq 60$
$x + 3y + 2z \geq 60$
$x \geq 0, y \geq 0, z \geq 0.$

17. Minimize $c = x + y + z + w$
subject to $x + y \geq 30$
$x + z \geq 20$
$x + y - w \leq 10$
$y + z - w \leq 10$
$x \geq 0, y \geq 0, z \geq 0, w \geq 0.$

18. Minimize $c = 4x + y + z + w$
subject to $x + y \geq 30$
$y - z \leq 20$
$z - w \leq 10$
$x \geq 0, y \geq 0, z \geq 0, w \geq 0.$

In Exercises 19–22, solve the given linear programming problem using duality.

19. Minimize $c = 2x + y$
subject to $3x + 2y \geq 60$
$2x + y \geq 60$
$x + 3y \geq 60$
$x \geq 0, y \geq 0.$

20. Minimize $c = 2x + y + 2z$
subject to $3x + 2y + z \geq 100$
$2x + y + 3z \geq 200$
$x \geq 0, y \geq 0, z \geq 0.$

21. Minimize $c = 2x + y$
subject to $3x + 2y \geq 10$
$2x - y \leq 30$
$x + 3y \geq 60$
$x \geq 0, y \geq 0.$

22. Minimize $c = 2x + y + 2z$
subject to $3x - 2y + z \geq 100$
$2x + y - 3z \leq 200$
$x \geq 0, y \geq 0, z \geq 0.$

In Exercises 23–26, solve the game with the given payoff matrix.

23. $P = \begin{bmatrix} -1 & 2 & -1 \\ 1 & -2 & 1 \\ 3 & -1 & 0 \end{bmatrix}$ **24.** $P = \begin{bmatrix} -3 & 0 & 1 \\ -4 & 0 & 0 \\ 0 & -1 & -2 \end{bmatrix}$

25. $P = \begin{bmatrix} -3 & -2 & 3 \\ 1 & 0 & 0 \\ -2 & 2 & 1 \end{bmatrix}$ **26.** $P = \begin{bmatrix} -4 & -2 & -3 \\ 1 & -3 & -2 \\ -3 & 1 & -4 \end{bmatrix}$

Exercises 27–30 are adapted from the Actuarial Exam on Operations Research.

27. You are given the following linear programming problem:

$$\text{Minimize} \quad c = x + 2y$$
$$\text{subject to} \quad -2x + y \geq 1$$
$$x - 2y \geq 1$$
$$x \geq 0, y \geq 0.$$

Which of the following is true?
(A) The problem has no feasible solutions.
(B) The objective function is unbounded.
(C) The problem has optimal solutions.

28. Repeat Exercise 27 with the following linear programming problem:

$$\text{Maximize} \quad p = x + y$$
$$\text{subject to} \quad -2x + y \leq 1$$
$$x - 2y \leq 2$$
$$x \geq 0, y \geq 0.$$

29. Determine the optimal value of the objective function. You are given the following linear programming problem.

$$\text{Maximize} \quad Z = x_1 + 4x_2 + 2x_3 - 10$$
$$\text{subject to} \quad 4x_1 + x_2 + x_3 \leq 45$$
$$-x_1 + x_2 + 2x_3 \leq 0$$
$$x_1, x_2, x_3 \geq 0.$$

30. Determine the optimal value of the objective function. You are given the following linear programming problem.

$$\text{Minimize} \quad Z = x_1 + 4x_2 + 2x_3 + x_4 + 40$$
$$\text{subject to} \quad 4x_1 + x_2 + x_3 \leq 45$$
$$-x_1 + 2x_2 + x_4 \geq 40$$
$$x_1, x_2, x_3 \geq 0.$$

Applications: OHaganBooks.com
[Try the game at www.OHaganBooks.com]

In Exercises 31–34, you are the buyer for OHaganBooks.com and are considering increasing stocks of romance and horror novels at the new OHaganBooks.com warehouse in Texas. You have offers from several publishers: Duffin House, Higgins Press, McPhearson Imprints, *and* O'Conell Books. *Duffin offers a package of 5 horror novels and 5 romance novels for $50, Higgins offers a package of 5 horror and 10 romance novels for $80, McPhearson offers a package of 10 horror novels and 5 romance novels for $80, and O'Conell offers a package of 10 horror novels and 10 romance novels for $90.*

31. How many packages should you purchase from Duffin House and Higgins Press to obtain at least 4,000 horror novels and 6,000 romance novels at minimum cost? What is the minimum cost?

32. How many packages should you purchase from McPhearson Imprints and O'Conell Books to obtain at least 5,000 horror novels and 4,000 romance novels at minimum cost? What is the minimum cost?

33. Refer to the scenario in Exercise 31. As it turns out, John O'Hagan promised Marjory Duffin that OHaganBooks .com would buy at least 20% more packages from Duffin as from Higgins, but you still want to obtain at least 4,000 horror novels and 6,000 romance novels at minimum cost.
   a. Referring to your solution of Exercise 31, say which of the following statements are possible *without solving the problem*:
   (A) The cost will stay the same.
   (B) The cost will increase.
   (C) The cost will decrease.
   (D) It will be impossible to meet all the conditions.
   (E) The cost will become unbounded.
   b. If you wish to meet all the requirements at minimum cost, how many packages should you purchase from each publisher? What is the minimum cost?

34. Refer to Exercise 32. You are about to place the order meeting the requirements of Exercise 32 when you are told that you can order no more than a total of 500 packages and that at least half of the packages should be from McPhearson. Explain why this is impossible by referring to the feasible region for Exercise 32.

35. Investments Marjory Duffin's portfolio manager has suggested two high-yielding stocks: European Emerald Emporium (EEE) and Royal Ruby Retailers (RRR).[46] EEE shares cost $50, yield 4.5% in dividends, and have a risk index of 2.0 per share. RRR shares cost $55, yield 5% in dividends, and have a risk index of 3.0 per share. Marjory has up to $12,100 to invest and would like to earn at least $550 in

dividends. How many shares of each stock should she purchase to meet her requirements and minimize the total risk index for her portfolio? What is the minimum total risk index?

36. Investments Marjory Duffin's other portfolio manager has suggested another two high-yielding stocks: Countrynarrow Mortgages (CNM) and Scotland Subprime (SS).[47] CNM shares cost $40, yield 5.5% in dividends, and have a risk index of 1.0 per share. SS shares cost $25, yield 7.5% in dividends, and have a risk index of 1.5 per share. Marjory can invest up to $30,000 in these stocks and would like to earn at least $1,650 in dividends. How many shares of each stock should she purchase to meet her requirements and minimize the total risk index for her portfolio?

37. Resource Allocation Billy-Sean O'Hagan has joined the Physics Society at *Suburban State University*, and the group is planning to raise money to support the dying space program by making and selling umbrellas. The society intends to make three models: the Sprinkle, the Storm, and the Hurricane. The amounts of cloth, metal, and wood used in making each model are given in this table:

	Sprinkle	Storm	Hurricane	Total Available
Cloth (square yards)	1	2	2	600
Metal (pounds)	2	1	3	600
Wood (pounds)	1	3	6	600
Profit ($)	1	1	2	

The table also shows the amounts of each material available in a given day and the profits to be made from each model. How many of each model should the society make to maximize its profit?

38. Profit *Duffin House*, which is now the largest publisher of books sold at the OHaganBooks.com site, prints three kinds of books: paperback, quality paperback, and hardcover. The amounts of paper, ink, and time on the presses required for each kind of book are given in this table:

	Paperback	Quality Paperback	Hardcover	Total Available
Paper (pounds)	3	2	1	6,000
Ink (gallons)	2	1	3	6,000
Time (minutes)	10	10	10	22,000
Profit ($)	1	2	3	

The table also lists the total amounts of paper, ink, and time available in a given day and the profits made on each kind

---

[46] RRR and EEE happen to be, respectively, the ticker symbols of RSC Holdings (an equipment rental provider) and Evergreen Energy Inc. (an environmentally friendly energy technology company) and thus have nothing to do with rubies and emeralds.

---

[47] CNM is actually the ticker symbol of Carnegie Wave, whereas SS is not the ticker symbol of any U.S.-based company we are aware of.

of book. How many of each kind of book should Duffin print to maximize profit?

39. *Purchases* You are just about to place book orders from *Duffin House* and *Higgins Press* (see Exercise 31) when everything changes: Duffin House informs you that, because of a global romance crisis, its packages now each will contain 5 horror novels but only 2 romance novels and still cost $50 per package. Packages from Higgins Press will now contain 10 of each type of novel, but now cost $150 per package. *Ewing Books* enters the fray and offers its own package of 5 horror and 5 romance novels for $100. The sales manager now tells you that at least 50% of the packages must come from Higgins Press, and, as before, you want to obtain at least 4,000 horror novels and 6,000 romance novels at minimum cost. Taking all of this into account, how many packages should you purchase from each publisher? What is the minimum cost?

40. *Purchases* You are about to place book orders from *McPhearson Imprints* and *O'Conell Books* (see Exercise 32) when you get an e-mail from McPhearson Imprints saying sorry, but they have stopped publishing romance novels because of the global romance crisis and can now offer only packages of 10 horror novels for $50. O'Conell is still offering packages of 10 horror novels and 10 romance novels for $90, and now the U.S. Treasury, in an attempt to bolster the floundering romance industry, is offering its own package of 20 romance novels for $120. Furthermore, Congress, in approving this measure, has passed legislation dictating that at least two thirds of the packages in every order must come from the U.S. Treasury. As before, you wish to obtain at least 5,000 horror novels and 4,000 romance novels at minimum cost. Taking all of this into account, how many packages should you purchase from each supplier? What is the minimum cost?

41. *Degree Requirements* During his lunch break, John O'Hagan decides to devote some time to assisting his son Billy-Sean, who continues to have a terrible time planning his college course schedule. The latest *Bulletin of Suburban State University* claims to have added new flexibility to its course requirements, but it remains as complicated as ever. It reads as follows:

> *All candidates for the degree of Bachelor of Arts at SSU must take at least 120 credits from the Sciences, Fine Arts, Liberal Arts, and Mathematics combined, including at least as many Science credits as Fine Arts credits, and at most twice as many Mathematics credits as Science credits, but with Liberal Arts credits exceeding Mathematics credits by no more than one third of the number of Fine Arts credits.*

Science and fine arts credits cost $300 each, and liberal arts and mathematics credits cost $200 each. John would like to have Billy-Sean meet all the requirements at a minimum total cost.

a. Set up (without solving) the associated linear programming problem.

b. ⊤ Use technology to determine how many of each type of credit Billy-Sean should take. What will the total cost be?

42. *Degree Requirements* No sooner had the "new and flexible" course requirement been released than the English Department again pressured the University Senate to include their vaunted "Verbal Expression" component in place of the fine arts requirement in all programs (including the sciences):

> *All candidates for the degree of Bachelor of Science at SSU must take at least 120 credits from the Liberal Arts, Sciences, Verbal Expression, and Mathematics, including at most as many Science credits as Liberal Arts credits, and at least twice as many Verbal Expression credits as Science credits and Liberal Arts credits combined, with Liberal Arts credits exceeding Mathematics credits by at least a quarter of the number of Verbal Expression credits.*

Science credits cost $300 each, while each credit in the remaining subjects now costs $400. John would like to have Billy-Sean meet all the requirements at a minimum total cost.

a. Set up (without solving) the associated linear programming problem.

b. ⊤ Use technology to determine how many of each type of credit Billy-Sean should take. What will the total cost be?

43. *Shipping* On the same day that the sales department at *Duffin House* received an order for 600 packages from the OHaganBooks.com Texas headquarters, it received an additional order for 200 packages from FantasyBooks.com, based in California. Duffin House has warehouses in New York and Illinois. The New York warehouse has 600 packages in stock, but the Illinois warehouse is closing down and has only 300 packages in stock. Shipping costs per package of books are as follows: New York to Texas: $20; New York to California: $50; Illinois to Texas: $30; Illinois to California: $40. What is the lowest total shipping cost for which Duffin House can fill the orders? How many packages should be sent from each warehouse to each online bookstore at a minimum shipping cost?

44. *Transportation Scheduling* *Duffin House* is about to start a promotional blitz for its new book, *Advanced String Theory for the Liberal Arts*. The company has 25 salespeople stationed in Austin and 10 in San Diego, and would like to fly at least 15 to sales fairs in each of Houston and Cleveland. A round-trip plane flight from Austin to Houston costs $200; from Austin to Cleveland costs $150; from San Diego to Houston costs $400; and from San Diego to Cleveland costs $200. How many salespeople should the company fly from each of Austin and San Diego to each of Houston and Cleveland for the lowest total cost in airfare?

45. *Marketing* Marjory Duffin, head of *Duffin House*, reveals to John O'Hagan that FantasyBooks.com is considering several promotional schemes: It may offer two books for the price of one, three books for the price of two, or possibly a free copy of *Brain Surgery for Klutzes* with each order. OHaganBooks.com's marketing advisers Floody and O'Lara seem to have different ideas as to how to respond. Floody suggests offering *three* books for the price of one, while O'Lara suggests instead offering a free copy of the *Finite Mathematics Student Solutions Manual* with every purchase. After a careful analysis, O'Hagan comes up with the following payoff matrix, where the payoffs represent the number of customers, in thousands, O'Hagan expects to gain from FantasyBooks.com:

		FantasyBooks		
	No Promo	2 for Price of 1	3 for Price of 2	Brain Surgery
OHagan No Promo	0	−60	−40	10
3 for Price of 1	30	20	10	15
Finite Math	20	0	15	10

Find the optimal strategies for both companies and the expected shift in customers.

46. *Study Techniques* Billy-Sean's friend Pat from college has been spending all of his time in fraternity activities and therefore knows absolutely nothing about any of the three topics on tomorrow's math test. He has turned to Billy-Sean for advice as to how to spend his all-nighter. The table below shows the scores Pat could expect to earn if the entire test were to be in a specific subject. (Because he knows no linear programming or matrix algebra, the table shows, for instance, that studying game theory all night will not be much use in preparing him for this topic.)

		Test		
		Game Theory	Linear Programming	Matrix Algebra
Pat's Strategies	Study Game Theory	30	0	20
	Study Linear Programming	0	70	0
	Study Matrix Algebra	0	0	70

What percentage of the night should Pat spend on each topic, assuming the principles of game theory, and what score can he expect to get?

The Diet Problem

The *Galaxy Nutrition* health-food mega-store chain provides free online nutritional advice and support to its customers. As website technical consultant, you are planning to construct an interactive web page to assist customers in preparing a diet tailored to their nutritional and budgetary requirements. Ideally, the customer would select foods to consider and specify nutritional and/or budgetary constraints, and the tool should return the optimal diet meeting those requirements. You would also like the web page to allow the customer to decide whether, for instance, to find the cheapest possible diet meeting the requirements, the diet with the lowest number of calories, or the diet with the least total carbohydrates.

After doing a little research, you notice that the kind of problem you are trying to solve is quite well known and referred to as the *diet problem*, and that solving the diet problem is a famous example of linear programming. Indeed, there are already some online pages that solve versions of the problem that minimize total cost, so you have adequate information to assist you as you plan the page.*

You decide to start on a relatively small scale with a program that uses a list of 10 foods and minimizes either total caloric intake or total cost and satisfies a small list of requirements. Following is a small part of a table of nutritional information from the demo at the NEOS Wiki (all the values shown are for a single serving) as well as approximate minimum daily requirements:

*See, for instance, the Diet Problem Solver at the NEOS Guide: www.neos-guide.org/content/diet-problem-solver.

Image Studios/UpperCut Images/Getty Images

	Price per Serving	Calories	Total Fat (grams)	Carbs (grams)	Dietary Fiber (grams)	Protein (grams)	Vit C (IU)
Tofu	$0.31	88.2	5.5	2.2	1.4	9.4	0.1
Roast Chicken	$0.84	277.4	10.8	0	0	42.2	0
Spaghetti with Sauce	$0.78	358.2	12.3	58.3	11.6	8.2	27.9
Tomato	$0.27	25.8	0.4	5.7	1.4	1.0	23.5
Oranges	$0.15	61.6	0.2	15.4	3.1	1.2	69.7
Wheat Bread	$0.05	65.0	1.0	12.4	1.3	2.2	0
Cheddar Cheese	$0.25	112.7	9.3	0.4	0	7.0	0
Oatmeal	$0.82	145.1	2.3	25.3	4.0	6.1	0
Peanut Butter	$0.07	188.5	16.0	6.9	2.1	7.7	0
White Tuna in Water	$0.69	115.6	2.1	0	0	22.7	0
Minimum Requirements		2,200	20	80	25	60	90

Source: www.neos-guide.org/content/diet-problem-solver.

Now you get to work. As always, you start by identifying the unknowns. Since the output of the web page will consist of a recommended diet, the unknowns should logically be the number of servings of each item of food selected by the user. In your first trial run, you decide to include all the 10 food items listed, so you take

x_1 = Number of servings of tofu

x_2 = Number of servings of roast chicken

$\vdots$

x_{10} = Number of servings of white tuna in water.

You now set up a linear programming problem for two sample scenarios.

Scenario 1 (Minimum Cost): Satisfy all minimum nutritional requirements at a minimum cost. Here the linear programming problem is

Minimize

$c = 0.31x_1 + 0.84x_2 + 0.78x_3 + 0.27x_4 + 0.15x_5 + 0.05x_6 + 0.25x_7$
$\quad + 0.82x_8 + 0.07x_9 + 0.69x_{10}$

subject to

$88.2x_1 + 277.4x_2 + 358.2x_3 + 25.8x_4 + 61.6x_5 + 65x_6 + 112.7x_7$
$\quad + 145.1x_8 + 188.5x_9 + 115.6x_{10} \geq 2{,}200$

$5.5x_1 + 10.8x_2 + 12.3x_3 + 0.4x_4 + 0.2x_5 + 1x_6 + 9.3x_7 + 2.3x_8$
$\quad + 16x_9 + 2.1x_{10} \geq 20$

$2.2x_1 + 58.3x_3 + 5.7x_4 + 15.4x_5 + 12.4x_6 + 0.4x_7 + 25.3x_8 + 6.9x_9 \geq 80$

$1.4x_1 + 11.6x_3 + 1.4x_4 + 3.1x_5 + 1.3x_6 + 4x_8 + 2.1x_9 \geq 25$

$9.4x_1 + 42.2x_2 + 8.2x_3 + 1x_4 + 1.2x_5 + 2.2x_6 + 7x_7 + 6.1x_8 + 7.7x_9$
$\quad + 22.7x_{10} \geq 60$

$0.1x_1 + 27.9x_3 + 23.5x_4 + 69.7x_5 \geq 90.$

This is clearly the kind of linear programming problem no one in their right mind would like to do by hand (solving it requires 16 tableaus!), so you decide to use the online simplex method tool at the Website (Website → Online Utilities → Simplex Method Tool).

Here is a picture of the input, entered almost exactly as written above. You need to enter each constraint on a new line, and

$$\texttt{Minimize c = 0.31x1 + ... Subject to}$$

must be typed on a single line.

```
   Type your linear programming problem below. (Press "Example" to see how to set it up.)
Minimize   c =
0.31x1+0.84x2+0.78x3+0.27x4+0.15x5+0.05x6+0.25x7+0.82x8+0.07x9+0.69x10 Subject to
88.2x1+277.4x2+358.2x3+25.8x4+61.6x5+65x6+112.7x7+145.1x8+188.5x9+115.6x10 >= 2200
5.5x1+10.8x2+12.3x3+0.4x4+0.2x5+1x6+9.3x7+2.3x8+16x9+2.1x10 >= 20
2.2x1+58.3x3+5.7x4+15.4x5+12.4x6+0.4x7+25.3x8+6.9x9 >= 80
1.4x1+11.6x3+1.4x4+3.1x5+1.3x6+4x8+2.1x9 >= 25
9.4x1+42.2x2+8.2x3+1x4+1.2x5+2.2x6+7x7+6.1x8+7.7x9+22.7x10 >= 60
0.1x1+0x2+27.9x3+23.5x4+69.7x5 >= 90
```

Clicking "Solve" results in the following solution:

$$c = 0.981126; \quad x_1 = 0, \quad x_2 = 0, \quad x_3 = 0, \quad x_4 = 0, \quad x_5 = 1.29125,$$
$$x_6 = 0, \quad x_7 = 0, \quad x_8 = 0, \quad x_9 = 11.2491, \quad x_{10} = 0.$$

This means that you can satisfy all the daily requirements for less than $1 on a diet of 1.3 servings of oranges and 11.2 servings of peanut butter. Although you enjoy peanut butter, 11.2 servings seems a little over the top, so you modify the LP problem by adding a new constraint (which also suggests to you that some kind of flexibility needs to be built into the site to allow users to set limits on the number of servings of any one item):

$$x_9 \leq 3.$$

This new constraint results in the following solution:

$$c = 1.59981; \quad x_1 = 0, \quad x_2 = 0, \quad x_3 = 0, \quad x_4 = 0, \quad x_5 = 1.29125,$$
$$x_6 = 23.9224, \quad x_7 = 0, \quad x_8 = 0, \quad x_9 = 3, \quad x_{10} = 0.$$

Because wheat bread is cheap and, in large enough quantities, supplies ample protein, the program has now substituted 23.9 servings of wheat bread for the missing peanut butter for a total cost of $1.60.

Unfettered, you now add

$$x_6 \leq 4$$

and obtain the following spaghetti, bread, and peanut butter diet for $3.40 per day:

$$c = 3.40305; \quad x_1 = 0, \quad x_2 = 0, \quad x_3 = 3.83724, \quad x_4 = 0, \quad x_5 = 0,$$
$$x_6 = 4, \quad x_7 = 0, \quad x_8 = 0, \quad x_9 = 3, \quad x_{10} = 0.$$

Scenario 2 (Minimum Calories): Minimize total calories and satisfy all minimum nutritional requirements (except for caloric intake).

Here, the linear programming problem is

Minimize

$$c = 88.2x_1 + 277.4x_2 + 358.2x_3 + 25.8x_4 + 61.6x_5 + 65x_6 + 112.7x_7$$
$$+ 145.1x_8 + 188.5x_9 + 115.6x_{10}$$

subject to

$$5.5x_1 + 10.8x_2 + 12.3x_3 + 0.4x_4 + 0.2x_5 + 1x_6 + 9.3x_7 + 2.3x_8 + 16x_9$$
$$+ 2.1x_{10} \geq 20$$
$$2.2x_1 + 58.3x_3 + 5.7x_4 + 15.4x_5 + 12.4x_6 + 0.4x_7 + 25.3x_8 + 6.9x_9 \geq 80$$
$$1.4x_1 + 11.6x_3 + 1.4x_4 + 3.1x_5 + 1.3x_6 + 4x_8 + 2.1x_9 \geq 25$$
$$9.4x_1 + 42.2x_2 + 8.2x_3 + 1x_4 + 1.2x_5 + 2.2x_6 + 7x_7 + 6.1x_8 + 7.7x_9$$
$$+ 22.7x_{10} \geq 60$$
$$0.1x_1 + 27.9x_3 + 23.5x_4 + 69.7x_5 \geq 90.$$

You obtain the following 716-calorie tofu, tomato, and tuna diet:

$$x_1 = 2.07232, \quad x_2 = 0, \quad x_3 = 0, \quad x_4 = 15.7848, \quad x_5 = 0, \quad x_6 = 0,$$
$$x_7 = 0, \quad x_8 = 0, \quad x_9 = 0, \quad x_{10} = 1.08966.$$

As 16 servings of tomatoes seems a little over the top, you add the new constraint $x_4 \leq 3$ and obtain a 783-calorie tofu, tomato, orange, and tuna diet:

$$x_1 = 2.81682, \quad x_2 = 0, \quad x_3 = 0, \quad x_4 = 3, \quad x_5 = 5.43756, \quad x_6 = 0,$$
$$x_7 = 0, \quad x_8 = 0, \quad x_9 = 0, \quad x_{10} = 1.05713.$$

What the trial runs have shown you is that your website will need to allow the user to set reasonable upper bounds for the number of servings of each kind of food considered. You now get to work writing the algorithm, which appears here:

Website → Online Utilities → Diet Problem Solver

EXERCISES

1. Briefly explain why roast chicken, which supplies protein more cheaply than either tofu or tuna, does not appear in the optimal solution in either scenario.

2. Consider the optimal solution obtained in Scenario 1 when peanut butter and bread were restricted. Experiment on the Simplex Method Tool by increasing the protein requirement 10 grams at a time until chicken appears in the optimal diet. At what level of protein does the addition of chicken first become necessary?

3. What constraints would you add for a person who wants to eat at most two servings of chicken a day and is allergic to tomatoes and peanut butter? What is the resulting diet for Scenario 2?

4. What is the linear programming problem for someone who wants as much protein as possible at a cost of no more than $6 per day with no more than 50 grams of carbohydrates per day, assuming that they want to satisfy the minimum requirements for all the remaining nutrients? What is the resulting diet?

5. Is it possible to obtain a diet with no bread or peanut butter in Scenario 1 costing less than $4 per day?

Section 5.1

Some calculators, including the TI-83/84 Plus, will shade one side of a graph, but you need to tell the calculator which side to shade. For instance, to obtain the solution set of $2x + 3y \le 6$ shown in Figure 4:

1. Solve the corresponding equation $2x + 3y = 6$ for y, and use the input shown below:

2. The icon to the left of "Y_1" tells the calculator to shade above the line. You can cycle through the various shading options by positioning the cursor to the left of Y_1 and pressing $\boxed{\text{ENTER}}$ until you see the one you want. Here's what the graph will look like:

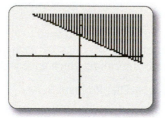

Section 5.3

Example 3 (page 376) The *Acme Baby Foods* example in the text leads to the following linear programming problem:

$$\begin{aligned}
\text{Maximize} \quad & p = 10x + 7y \\
\text{subject to} \quad & x \qquad\quad \le 600 \\
& 2x + 3y \le 3{,}600 \\
& 5x + 3y \le 4{,}500 \\
& x \ge 0, y \ge 0.
\end{aligned}$$

Solve it using technology.

Solution

When we introduce slack variables, we get the following system of equations:

$$\begin{aligned}
x \qquad\quad + s \qquad\qquad\quad &= 600 \\
2x + 3y \qquad + t \qquad\quad &= 3{,}600 \\
5x + 3y \qquad\qquad + u \quad &= 4{,}500 \\
-10x - 7y \qquad\qquad\quad + p &= 0.
\end{aligned}$$

We use the PIVOT program for the TI-83/84 Plus to help with the simplex method. This program is available at the Website by following

Everything for Finite Math → Math Tools for Chapter 5.

Because the calculator handles decimals as easily as integers, there is no need to avoid them, except perhaps to save limited screen space. If we don't need to avoid decimals, we can use the traditional Gauss-Jordan method (see the discussion at the end of Section 3.2): After selecting your pivot and before clearing the pivot column, *divide the pivot row by the value of the pivot, thereby turning the pivot into a 1.*

The main drawback to using the TI-83/84 Plus is that we can't label the rows and columns. We can mentally label them as we go, but we can do without labels entirely if we wish. We begin by entering the initial tableau as the matrix [A]. (Another drawback to using the TI-83/84 Plus is that it can't show the whole tableau at once. Here and below, we show tableaux across several screens. Use the TI-83/84 Plus's arrow keys to scroll a matrix left and right so you can see all of it.)

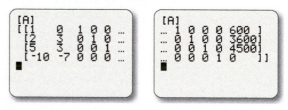

The following is the sequence of tableaux we get while using the simplex method with the help of the PIVOT program:

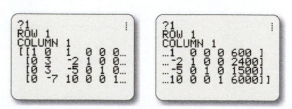

After determining that the next pivot is in the third row and second column, we divide the third row by the pivot, 3, and then pivot:

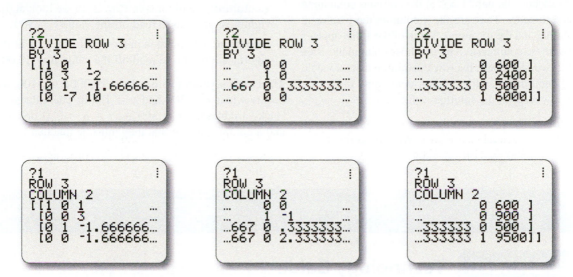

The next pivot is the 3 in the second row, third column. We divide its row by 3 and pivot:

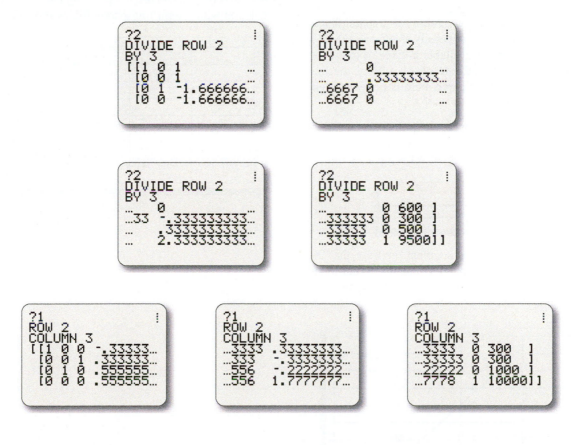

There are no negative numbers in the bottom row, so we're finished. How do we read off the optimal solutions if we don't have labels, though? Look at the columns containing one 1 and three 0s. They are the x column, the y column, the s column, and the p column. Think of the 1 that appears in each of these columns as a pivot whose column has been cleared. If we had labels, the row containing a pivot would have the same label as the column containing that pivot. We can now read off the solution as follows:

x column: The pivot is in the first row, so Row 1 would have been labeled with x. We look at the rightmost column to read off the value $x = 300$.

y column: The pivot is in Row 3, so we look at the rightmost column to read off the value $y = 1,000$.

s column: The pivot is in Row 2, so we look at the rightmost column to read off the value $s = 300$.

p column: The pivot is in Row 3, so we look at the rightmost column to read off the value $p = 10,000$.

Thus, the maximum value of p is $10,000¢ = \$100$, which occurs when $x = 300$ and $y = 1,000$. The values of the slack variables are $s = 300$ and $t = u = 0$. (Look at the t and u columns to see that they must be inactive.)

Spreadsheet Technology Guide

Section 5.1

Excel is not a particularly good tool for graphing linear inequalities because it cannot easily shade one side of a line. One solution that is available in Excel is to use the "error bar" feature to indicate which side of the line *should* be shaded. For example, here is how we might graph the inequality $2x + 3y \leq 6$.

1. Create a scatter graph using two points to construct a line segment (as in Chapter 1). (Notice that we had to solve the equation $2x + 3y = 6$ for y.)

2. Double-click on the line segment, and use the "X-Error Bars" feature to obtain a diagram similar to the one below, where the error bars indicate the direction of shading.

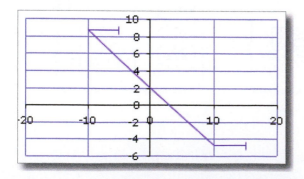

Alternatively, you can use the Drawing Palette to create a polygon with a semi-transparent fill, as shown below.

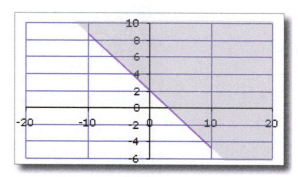

6

SETS AND COUNTING

CASE STUDY

Designing a Puzzle

As Product Design Manager for *Cerebral Toys, Inc.*, you are constantly on the lookout for ideas for intellectually stimulating yet inexpensive toys. Your design team recently came up with an idea for a puzzle consisting of a number of plastic cubes. Each cube will have two faces colored red, two colored white, and two colored blue, and there will be exactly two cubes with each possible configuration of colors. The goal of the puzzle is to seek out the matching pairs, thereby enhancing a child's geometric intuition and three-dimensional manipulation skills.

If the kit is to include every possible configuration of colors, how many cubes will the kit contain?

Image Source/Getty Images

www.WanerMath.com

At the Website, in addition to the resources listed in the Preface, you will find:

• A utility to compute factorials, permutations, and combinations

427

Introduction

The theory of sets is the foundation for most of mathematics. It also has direct applications—for example, in searching computer databases. We will use set theory extensively in the chapter on probability, so much of this chapter revolves around the idea of a **set of outcomes** of a procedure such as rolling a pair of dice or choosing names from a database. Also important in probability is the theory of **counting** the number of elements in a set, which is called **combinatorics**.

Counting elements is not a trivial proposition; for example, the betting game Lotto (used in many state lotteries) has you pick six numbers from some range—say, 1–55. If your six numbers match the six numbers chosen in the official drawing, you win the top prize. How many Lotto tickets would you need to buy to guarantee that you will win? That is, how many Lotto tickets are possible? By the end of this chapter we will be able to answer these questions.

6.1   Sets and Set Operations

In this section we introduce some of the basic ideas of set theory. Some of the examples and applications we see here are derived from the theory of probability and will recur throughout the rest of this chapter and the next.

Sets

Visualizing a Set

Set

Elements of the set

Sets and Elements

A **set** is a collection of items, referred to as the **elements** of the set.

Quick Examples

We usually use a capital letter to name a set and braces to enclose the elements of a set.

$W = \{\text{Amazon, eBay, Apple}\}$
$N = \{1, 2, 3, \ldots\}$

$x \in A$ means that x **is an element of** the set A. If x is not an element of A, we write $x \notin A$.

Amazon $\in W$ (W as above)
Microsoft $\notin W$ $2 \in N$

$B = A$ means that A and B have the same elements. The order in which the elements are listed does not matter.

$\{5, -9, 1, 3\} = \{-9, 1, 3, 5\}$
$\{1, 2, 3, 4\} \neq \{1, 2, 3, 6\}$

$B \subseteq A$ means that B is a **subset** of A; every element of B is also an element of A.

$\{\text{eBay, Apple}\} \subseteq W$
$\{1, 2, 3, 4\} \subseteq \{1, 2, 3, 4\}$

$B \subset A$ means that B is a **proper subset** of A: $B \subseteq A$, but $B \neq A$.

$\{\text{eBay, Apple}\} \subset W$
$\{1, 2, 3\} \subset \{1, 2, 3, 4\}$
$\{1, 2, 3\} \subset N$ (N as above)

$\varnothing$ is the **empty set**, the set containing no elements. It is a subset of every set.

$\varnothing \subseteq W$
$\varnothing \subset W$

A **finite** set has finitely many elements. An **infinite** set does not have finitely many elements.

$W = \{\text{Amazon, eBay, Apple}\}$
is a finite set.
$N = \{1, 2, 3, \ldots\}$ is an infinite set.

One type of set we'll use often is the **set of outcomes** of some activity or experiment. For example, if we toss a coin and observe which side faces up, there are two possible outcomes: heads (H) and tails (T). The set of outcomes of tossing a coin once can be written

$$S = \{H, T\}.$$

As another example, suppose we roll a die that has faces numbered 1 through 6, as usual, and observe which number faces up. The set of outcomes *could* be represented as

$$S = \left\{ \boxdot, \boxdot, \boxdot, \boxdot, \boxdot, \boxdot \right\}.$$

However, we can much more easily write

$$S = \{1, 2, 3, 4, 5, 6\}.$$

EXAMPLE 1   Two Dice: Distinguishable versus Indistinguishable

a. Suppose we have two dice that we can distinguish in some way—say, one is green and one is red. If we roll both dice, what is the set of outcomes?

b. Describe the set of outcomes if the dice are indistinguishable.

Solution

a. A systematic way of laying out the set of outcomes for a distinguishable pair of dice is shown in Figure 1.

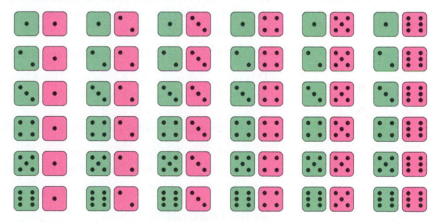

Figure 1

In the first row all the green dice show a 1, in the second row a 2, in the third row a 3, and so on. Similarly, in the first column all the red dice show a 1, in the second column a 2, and so on. The diagonal pairs (top left to bottom right) show all the "doubles." Using the picture as a guide, we can write the set of 36 outcomes as follows:

$$S = \begin{Bmatrix} (1, 1), & (1, 2), & (1, 3), & (1, 4), & (1, 5), & (1, 6), \\ (2, 1), & (2, 2), & (2, 3), & (2, 4), & (2, 5), & (2, 6), \\ (3, 1), & (3, 2), & (3, 3), & (3, 4), & (3, 5), & (3, 6), \\ (4, 1), & (4, 2), & (4, 3), & (4, 4), & (4, 5), & (4, 6), \\ (5, 1), & (5, 2), & (5, 3), & (5, 4), & (5, 5), & (5, 6), \\ (6, 1), & (6, 2), & (6, 3), & (6, 4), & (6, 5), & (6, 6) \end{Bmatrix}. \qquad \text{Distinguishable dice}$$

Notice that S is also the set of outcomes when we roll a single die twice, if we take the first number in each pair to be the outcome of the first roll and the second number to be the outcome of the second roll.

b. If the dice are truly indistinguishable, we will have no way of knowing which die is which once they are rolled. Think of placing two identical dice in a closed box and then shaking the box. When we look inside afterward, there is no way to tell which die is which. (If we make a small marking on one of the dice or somehow keep track of it as it bounces around, we are *distinguishing* the dice.) We regard two dice as **indistinguishable** if we make no attempt to distinguish them. Thus, for example, the two different outcomes $(1, 3)$ and $(3, 1)$ from part (a) would represent the same outcome in part (b) (one die shows a 3 and the other a 1). Because the set of outcomes should contain each outcome only once, we can remove $(3, 1)$. Following this approach gives the following smaller set of outcomes:

$$S = \left\{ \begin{array}{l} (1, 1), \ (1, 2), \ (1, 3), \ (1, 4), \ (1, 5), \ (1, 6), \\ \qquad (2, 2), \ (2, 3), \ (2, 4), \ (2, 5), \ (2, 6), \\ \qquad\qquad (3, 3), \ (3, 4), \ (3, 5), \ (3, 6), \\ \qquad\qquad\qquad (4, 4), \ (4, 5), \ (4, 6), \\ \qquad\qquad\qquad\qquad (5, 5), \ (5, 6), \\ \qquad\qquad\qquad\qquad\qquad (6, 6) \end{array} \right\}. \qquad \text{\textcolor{blue}{Indistinguishable dice}}$$

EXAMPLE 2   **Set-Builder Notation**

Let $B = \{0, 2, 4, 6, 8\}$. B is the set of all nonnegative* even integers less than 10. If we don't want to list the individual elements of B, we can instead use set-builder notation and write

$$B = \{n \,|\, n \text{ is a nonnegative even integer less than 10}\}.$$

This is read "*B is the set of all n such that n is a nonnegative even integer less than 10.*" Here is the correspondence between the words and the symbols:

B is the set of all n such that n is a nonnegative even integer less than 10.

$$B = \{n \,|\, n \text{ is a nonnegative even integer less than 10}\}.$$

Venn Diagrams

We can visualize sets and relations between sets using **Venn diagrams**. In a Venn diagram we represent a set as a region, often a disk (Figure 2).

The elements of A are the points inside the region. The following Venn diagrams illustrate the relations we've discussed so far.

Figure 2

Venn Diagrams for Set Relations

$x \in A$

$x \notin A$

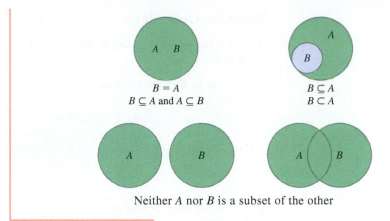

Neither A nor B is a subset of the other

Note Although the diagram for $B \subseteq A$ suggests a proper subset, it is customary to use the same diagram for both subsets and proper subsets. ∎

EXAMPLE 3 **Customer Interests**

NobelBooks.com (a fierce competitor of OHaganBooks.com) maintains a database of customers and the types of books they have purchased. In the company's database is the set of customers

$$S = \{\text{Einstein, Bohr, Millikan, Heisenberg, Schrödinger, Dirac}\}.$$

A search of the database for customers who have purchased cookbooks yields the subset

$$A = \{\text{Einstein, Bohr, Heisenberg, Dirac}\}.$$

Another search, this time for customers who have purchased mysteries, yields the subset

$$B = \{\text{Bohr, Heisenberg, Schrödinger}\}.$$

NobelBooks.com wants to promote a new combination mystery/cookbook and wants to target two subsets of customers: those who have purchased either cookbooks or mysteries (or both) and, for additional promotions, those who have purchased both cookbooks and mysteries. Name the customers in each of these subsets.

Solution We can picture the database and the two subsets using the Venn diagram in Figure 3.

The set of customers who have purchased either cookbooks or mysteries (or both) consists of the customers who are in A or B or both: Einstein, Bohr, Heisenberg, Schrödinger, and Dirac. The set of customers who have purchased both cookbooks and mysteries consists of the customers in the overlap of A and B, Bohr and Heisenberg.

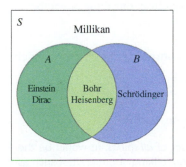

Figure 3

Set Operations

Set operations produce new sets from old ones, just as operations on numbers produce new numbers from old. Following is a list of some important examples.

Some Set Operations

$A \cup B$ is the **union** of A and B, the set of all elements that are either in A or in B (or in both).

$$A \cup B = \{x \mid x \in A \text{ or } x \in B\}$$

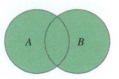

$A \cap B$ is the **intersection** of A and B, the set of all elements that are common to A and B.

$$A \cap B = \{x \mid x \in A \text{ and } x \in B\}$$

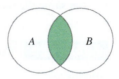

Logical Equivalents

Union: For an element to be in $A \cup B$, it must be in A **or** in B.
Intersection: For an element to be in $A \cap B$, it must be in A **and** in B.

Quick Examples

If $A = \{a, b, c, d\}$ and $B = \{c, d, e, f\}$, then

1. $A \cup B = \{a, b, c, d, e, f\}$
2. $A \cap B = \{c, d\}$.

Note Mathematicians always use "or" in its *inclusive* sense: one thing or another *or both.* ∎

There is one other operation we use, called the **complement** of a set A, which, roughly speaking, is the set of things *not* in A.

Q: *Why only "roughly"? Why not just form the set of things not in A?*

A: This would amount to assuming that there is a set of *all things.* (It would be the complement of the empty set.) Although tempting, talking about entities such as the "set of all things" leads to paradoxes.[*] Instead, we first need to fix a set S of all *objects under consideration,* or the *universe of discourse,* which we generally call the **universal set** for the discussion. For example, when we search the web, we take S to be the set of all web pages. When talking about integers, we take S to be the set of all integers. In other words, our choice of universal set depends on the context. The complement of a set $A \subseteq S$ is then the set of *elements of S* that are not in A.

---

[*] The most famous such paradox is called Russell's Paradox, after the mathematical logician (and philosopher and pacifist) Bertrand Russell. It goes like this: If there were a set of all things, then there would also be a (smaller) set of all sets. Call it S. Now, because S is the *set of all* sets, it must contain itself as a member. In other words, $S \in S$. Let P be the subset of S consisting of all sets that are *not* members of themselves. Now we pose the following question: Is P a member of itself? If it is, then, because it is the set of all sets that are *not* members of themselves, it is not a member of itself. On the other hand, if it is *not* a member of itself, then it qualifies as an element of P. In other words, it *is* a member of itself! Because neither can be true, something is wrong. What is wrong is the assumption that there is such a thing as the set of all sets or the set of all things.

Complement

If S is the universal set and $A \subseteq S$, then A' is the **complement** of A (in S), the set of all elements of S not in A.

$$A' = \{x \in S \mid x \notin A\}$$ = Green region below

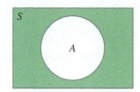

Logical Equivalent

For an element to be in A', it must be in S but **not** in A.

> **Quick Example**
>
> **3.** If $S = \{a, b, c, d, e, f, g\}$ and $A = \{a, b, c, d\}$, then
>
> $$A' = \{e, f, g\}.$$

In the following example we use set operations to describe the sets we found in Example 3, as well as some others.

EXAMPLE 4   **Customer Interests**

NobelBooks.com maintains a database of customers and the types of books they have purchased. In the company's database is the set of customers

$$S = \{\text{Einstein, Bohr, Millikan, Heisenberg, Schrödinger, Dirac}\}.$$

A search of the database for customers who have purchased cookbooks yields the subset

$$A = \{\text{Einstein, Bohr, Heisenberg, Dirac}\}.$$

Another search, this time for customers who have purchased mysteries, yields the subset

$$B = \{\text{Bohr, Heisenberg, Schrödinger}\}.$$

A third search, for customers who have registered with the site but have not used their first-time customer discount, yields the subset

$$C = \{\text{Millikan}\}.$$

Use set operations to describe the following subsets:

a. The subset of customers who have purchased either cookbooks or mysteries

b. The subset of customers who have purchased both cookbooks and mysteries

c. The subset of customers who have not purchased cookbooks

d. The subset of customers who have purchased cookbooks but have not used their first-time customer discount

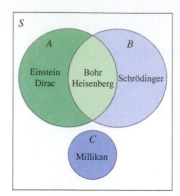

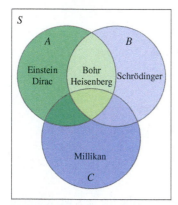

Figure 4

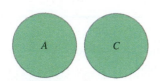

Figure 5

*People new to set theory sometimes find it strange to consider the empty set a valid set. Here is one of the times when it is very useful to do so. If we did not, we would have to say that $A \cap C$ was defined only when A and C had something in common. Having to deal with the fact that this set operation was not always defined would quickly get tiresome.

Visualizing $A \times B$

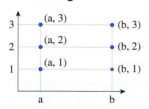

Solution Figure 4 shows two alternative Venn diagram representations of the database. Although the second version shows C overlapping A and B, the placement of the names inside shows that there are no customers in those overlaps.

a. The subset of customers who have bought either cookbooks *or* mysteries is

$$A \cup B = \{\text{Einstein, Bohr, Heisenberg, Schrödinger, Dirac}\}.$$

b. The subset of customers who have bought both cookbooks *and* mysteries is

$$A \cap B = \{\text{Bohr, Heisenberg}\}.$$

c. The subset of customers who have *not* bought cookbooks is

$$A' = \{\text{Millikan, Schrödinger}\}.$$

Note that, for the universal set, we are using the set S of all customers in the database.

d. The subset of customers who have bought cookbooks but have not used their first-time purchase discount is the empty set:

$$A \cap C = \varnothing.$$

When the intersection of two sets is empty, we say that the two sets are **disjoint**. In a Venn diagram, disjoint sets are drawn as regions that don't overlap, as in Figure 5.*

➡ **Before we go on . . .** Computer databases and the web can be searched using so-called Boolean searches. These are search requests using "and," "or," and "not." Using "and" gives the intersection of separate searches, using "or" gives the union, and using "not" gives the complement. In the next section we'll see how web search engines allow such searches. ∎

Cartesian Product

There is one more set operation we need to discuss.

> **Cartesian Product**
>
> The **Cartesian product** of two sets, A and B, is the set of all ordered pairs (a, b) with $a \in A$ and $b \in B$.
>
> $$A \times B = \{(a, b) \mid a \in A \text{ and } b \in B\}$$
>
> In words, $A \times B$ is the set of all ordered pairs whose first component is in A and whose second component is in B.
>
> **Quick Examples**
>
> **4.** If $A = \{a, b\}$ and $B = \{1, 2, 3\}$, then
>
> $$A \times B = \{(a, 1), (a, 2), (a, 3), (b, 1), (b, 2), (b, 3)\}. \quad \text{See the figure in the margin.}$$
>
> **5.** If $S = \{H, T\}$, then
>
> $$S \times S = \{(H, H), (H, T), (T, H), (T, T)\}.$$
>
> In other words, if S is the set of outcomes of tossing a coin once, then $S \times S$ is the set of outcomes of tossing a coin twice.

6. If $S = \{1, 2, 3, 4, 5, 6\}$, then

$$S \times S = \left\{\begin{array}{llllll}(1,1), & (1,2), & (1,3), & (1,4), & (1,5), & (1,6),\\(2,1), & (2,2), & (2,3), & (2,4), & (2,5), & (2,6),\\(3,1), & (3,2), & (3,3), & (3,4), & (3,5), & (3,6),\\(4,1), & (4,2), & (4,3), & (4,4), & (4,5), & (4,6),\\(5,1), & (5,2), & (5,3), & (5,4), & (5,5), & (5,6),\\(6,1), & (6,2), & (6,3), & (6,4), & (6,5), & (6,6)\end{array}\right\}.$$

In other words, if S is the set of outcomes of rolling a die once, then $S \times S$ is the set of outcomes of rolling a die twice (or rolling two distinguishable dice).

7. If $A = \{$red, yellow$\}$ and $B = \{$Mustang, Firebird$\}$, then

$A \times B = \{$(red, Mustang), (red, Firebird), (yellow, Mustang), (yellow, Firebird)$\}$ which we might also write as

$A \times B = \{$red Mustang, red Firebird, yellow Mustang, yellow Firebird$\}$.

EXAMPLE 5   Representing Cartesian Products

The manager of an automobile dealership has collected data on the number of pre-owned **Acura**, **Infiniti**, **Lexus**, and **Mercedes** cars the dealership has from the 2009, 2010, and 2011 model years. In entering this information on a spreadsheet, the manager would like to have each spreadsheet cell represent a particular year and make. Describe this set of cells.

Solution Because each cell represents a year and a make, we can think of the cell as a pair (year, make), as in (2009, Acura). Thus, the set of cells can be thought of as a Cartesian product:

$Y = \{2009, 2010, 2011\}$ Year of car

$M = \{$Acura, Infiniti, Lexus, Mercedes$\}$ Make of car

$$Y \times M = \left\{\begin{array}{llll}(2009, \text{Acura}), & (2009, \text{Infiniti}), & (2009, \text{Lexus}), & (2009, \text{Mercedes}),\\(2010, \text{Acura}), & (2010, \text{Infiniti}), & (2010, \text{Lexus}), & (2010, \text{Mercedes}),\\(2011, \text{Acura}), & (2011, \text{Infiniti}), & (2011, \text{Lexus}), & (2011, \text{Mercedes})\end{array}\right\}.$$ Cells

Thus, the manager might arrange the spreadsheet as follows:

	A	B	C	D	E
1		**Acura**	**Infiniti**	**Lexus**	**Mercedes**
2	**2009**	(2009 Acura)	(2009 Infiniti)	(2009 Lexus)	(2009 Mercedes)
3	**2010**	(2010 Acura)	(2010 Infiniti)	(2010 Lexus)	(2010 Mercedes)
4	**2011**	(2011 Acura)	(2011 Infiniti)	(2011 Lexus)	(2011 Mercedes)

The highlighting shows the 12 cells to be filled in, representing the numbers of cars of each year and make. For example, in cell B2 should go the number of 2009 Acuras the dealership has.

➡ **Before we go on . . .** The arrangement in the spreadsheet in Example 5 is consistent with the matrix notation in Chapter 4. We could also have used the elements of Y as column labels along the top and the elements of M as row labels down the side. Along those lines, we can also visualize the Cartesian product $Y \times M$ as a set of points in the xy-plane ("Cartesian plane") as shown in Figure 6.

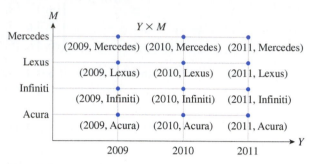

Figure 6

FAQs

The Many Meanings of "And"

Q : *Suppose A is the set of actors and B is the set of all baseball players. Then the set of all actors and baseball players is A ∩ B—right?*

A : Wrong. The fact that the word "and" appears in the description of a set does not always mean that the set is an intersection; the word "and" can mean different things in different contexts. $A \cap B$ refers to the set of elements that are in both A and B, hence to actors who are also baseball players. On the other hand, the set of all actors and baseball players is the set of people who are either actors or baseball players (or both), which is $A \cup B$. We can use the word "and" to describe both sets:

$A \cap B = \{$people who are both actors *and* baseball players$\}$

$A \cup B = \{$people who are actors *or* baseball players$\}$

$\quad\quad = \{$all actors *and* baseball players$\}$.

6.1 EXERCISES

▼ more advanced ◆ challenging
🅣 indicates exercises that should be solved using technology

In Exercises 1–16, list the elements in the given set.

1. The set F consisting of the four seasons

2. The set A consisting of the authors of this book

3. The set I of all positive integers no greater than 6

4. The set N of all negative integers greater than -3

5. $A = \{n \mid n$ is a positive integer and $0 \leq n \leq 3\}$
[**HINT:** See Example 2.]

6. $A = \{n \mid n$ is a positive integer and $0 < n < 8\}$
[**HINT:** See Example 2.]

7. $B = \{n \mid n$ is an even positive integer and $0 \leq n \leq 8\}$

8. $B = \{n \mid n$ is an odd positive integer and $0 \leq n \leq 8\}$

9. The set of all outcomes (**a**) of tossing a pair of distinguishable coins and (**b**) of tossing a pair of indistinguishable coins
[**HINT:** See Example 1.]

10. The set of outcomes (**a**) of tossing three distinguishable coins and (**b**) of tossing three indistinguishable coins
[**HINT:** See Example 1.]

11. The set of all outcomes of rolling two distinguishable dice such that the numbers add to 6

12. The set of all outcomes of rolling two distinguishable dice such that the numbers add to 8

13. The set of all outcomes of rolling two indistinguishable dice such that the numbers add to 6

14. The set of all outcomes of rolling two indistinguishable dice such that the numbers add to 8

15. The set of all outcomes of rolling two distinguishable dice such that the numbers add to 13

16. The set of all outcomes of rolling two distinguishable dice such that the numbers add to 1

In Exercises 17–20, draw a Venn diagram that illustrates the relationships among the given sets. [HINT: See Example 3.]

17. S = {eBay, Google, Amazon, OHaganBooks, Hotmail}, A = {Amazon, OHaganBooks}, B = {eBay, Amazon}, C = {Amazon, Hotmail}

18. S = {Apple, Dell, Gateway, Pomegranate, Compaq}, A = {Gateway, Pomegranate, Compaq}, B = {Dell, Gateway, Pomegranate, Compaq}, C = {Apple, Dell, Compaq}

19. S = {eBay, Google, Amazon, OHaganBooks, Hotmail}, A = {Amazon, Hotmail}, B = {eBay, Google, Amazon, Hotmail}, C = {Amazon, Hotmail}

20. S = {Apple, Dell, Gateway, Pomegranate, Compaq}, A = {Apple, Dell, Pomegranate, Compaq}, B = {Pomegranate}, C = {Pomegranate}

Let A = {June, Janet, Jill, Justin, Jeffrey, Jello}, B = {Janet, Jello, Justin}, and C = {Sally, Solly, Molly, Jolly, Jello}. In Exercises 21–34, find the given set. [HINT: See Quick Examples 1 and 2.]

21. $A \cup B$ **22.** $A \cup C$

23. $A \cup \varnothing$ **24.** $B \cup \varnothing$

25. $A \cup (B \cup C)$ **26.** $(A \cup B) \cup C$

27. $C \cap B$ **28.** $C \cap A$

29. $A \cap \varnothing$ **30.** $\varnothing \cap B$

31. $(A \cap B) \cap C$ **32.** $A \cap (B \cap C)$

33. $(A \cap B) \cup C$ **34.** $A \cup (B \cap C)$

In Exercises 35–42, A = {small, medium, large}, B = {blue, green}, and C = {triangle, square}. [HINT: See Quick Examples 4–7.]

35. List the elements of $A \times C$.

36. List the elements of $B \times C$.

37. List the elements of $A \times B$.

38. The elements of $A \times B \times C$ are the ordered triples (a, b, c) with $a \in A$, $b \in B$, and $c \in C$. List all the elements of $A \times B \times C$.

39. ▯ Represent $B \times C$ as cells in a spreadsheet. [HINT: See Example 5.]

40. ▯ Represent $A \times C$ as cells in a spreadsheet. [HINT: See Example 5.]

41. ▯ Represent $A \times B$ as cells in a spreadsheet.

42. ▯ Represent $A \times A$ as cells in a spreadsheet.

Let A = {H, T} be the set of outcomes when a coin is tossed, and let B = {1, 2, 3, 4, 5, 6} be the set of outcomes when a die is rolled. In Exercises 43–46, write the given set in terms of A and/or B, and list its elements.

43. The set of outcomes when a die is rolled and then a coin tossed

44. The set of outcomes when a coin is tossed twice

45. The set of outcomes when a coin is tossed three times

46. The set of outcomes when a coin is tossed twice and then a die is rolled

Let S be the set of outcomes when two distinguishable dice are rolled, let E be the subset of outcomes in which at least one die shows an even number, and let F be the subset of outcomes in which at least one die shows an odd number. In Exercises 47–52, list the elements in the given subset.

47. E' **48.** F'

49. $(E \cup F)'$ **50.** $(E \cap F)'$

51. $E' \cup F'$ **52.** $E' \cap F'$

In Exercises 53–60, use Venn diagrams to illustrate the given identity for subsets A, B, and C of S.

53. ▼ $(A \cup B)' = A' \cap B'$ DeMorgan's law

54. ▼ $(A \cap B)' = A' \cup B'$ DeMorgan's law

55. ▼ $(A \cap B) \cap C = A \cap (B \cap C)$ Associative law

56. ▼ $(A \cup B) \cup C = A \cup (B \cup C)$ Associative law

57. ▼ $A \cup (B \cap C) = (A \cup B) \cap (A \cup C)$ Distributive law

58. ▼ $A \cap (B \cup C) = (A \cap B) \cup (A \cap C)$ Distributive law

59. ▼ $S' = \varnothing$ **60.** ▼ $\varnothing' = S$

Applications

Databases *A freelance computer consultant keeps a database of her clients, which contains the names*

   S = {Acme, Brothers, Crafts, Dion, Effigy, Floyd, Global, Hilbert}.

The following clients owe her money:

   A = {Acme, Crafts, Effigy, Global}.

The following clients have done at least $10,000 worth of business with her:

   B = {Acme, Brothers, Crafts, Dion}.

The following clients have employed her in the last year:

   C = {Acme, Crafts, Dion, Effigy, Global, Hilbert}.

In Exercises 61–68, a subset of clients is described that the consultant could find using her database. Write the subset in terms of A, B, and C, and list the clients in that subset. [HINT: See Example 4.]

61. The clients who owe her money and have done at least $10,000 worth of business with her

62. The clients who owe her money or have done at least $10,000 worth of business with her

63. The clients who have done at least $10,000 worth of business with her or have employed her in the last year

64. The clients who have done at least $10,000 worth of business with her and have employed her in the last year

65. The clients who do not owe her money and have employed her in the last year

66. The clients who do not owe her money or have employed her in the last year

67. ▼ The clients who owe her money, have not done at least $10,000 worth of business with her, and have not employed her in the last year

68. ▼ The clients who either do not owe her money, have done at least $10,000 worth of business with her, or have employed her in the last year

69. Ⓣ *Boat Sales* You are given data on revenues from sales of sailboats, motor boats, and yachts for each of the years 2003 through 2006. How would you represent these data in a spreadsheet? The cells in your spreadsheet represent elements of which set?

70. Ⓣ *Health-Care Spending* Spending in most categories of health care in the United States increased dramatically in the last 30 years of the 1900s.[1] You are given data showing total spending on prescription drugs, nursing homes, hospital care, and professional services for each of the last three decades of the 1900s. How would you represent these data in a spreadsheet? The cells in your spreadsheet represent elements of which set?

Communication and Reasoning Exercises

71. You sell iPads and *jPads*. Let *I* be the set of all iPads you sold last year, and let *J* be the set of all jPads you sold last year. What set represents the collection of all iPads and jPads you sold combined?

72. You sell two models of music players: the *yoVaina Grandote* and the *yoVaina Minúsculito,* and each comes in three colors: Infraroja, Ultravioleta, and Radiografía. Let *M* be the set of models, and let *C* be the set of colors. What set represents the different choices a customer can make?

73. You are searching online for techno music that is neither European nor Dutch. In set notation, which set of music files are you searching for?
(A) Techno ∩ (European ∩ Dutch)′
(B) Techno ∩ (European ∪ Dutch)′
(C) Techno ∪ (European ∩ Dutch)′
(D) Techno ∪ (European ∪ Dutch)′

74. You would like to see either a World War II movie, or one that is based on a comic book character but does not feature aliens. Which set of movies are you interested in seeing?
(A) WWII ∩ (Comix ∩ Aliens′)
(B) WWII ∩ (Comix ∪ Aliens′)
(C) WWII ∪ (Comix ∩ Aliens′)
(D) WWII ∪ (Comix ∪ Aliens′)

75. ▼ Explain, illustrating by means of an example, why $(A \cap B) \cup C \neq A \cap (B \cup C)$.

76. ▼ Explain, making reference to operations on sets, why the statement "He plays soccer or rugby and cricket" is ambiguous.

77. ▼ Explain the meaning of a universal set, and give two different universal sets that could be used in a discussion about sets of positive integers.

78. ▼ Is the set of outcomes when two indistinguishable dice are rolled (Example 1) a Cartesian product of two sets? If so, which two sets? If not, why not?

79. ▼ Design a database scenario that leads to the following statement: To keep the factory operating at maximum capacity, the plant manager should select the suppliers in $A \cap (B \cup C')$.

80. ▼ Design a database scenario that leads to the following statement: To keep her customers happy, the bookstore owner should stock periodicals in $A \cup (B \cap C')$.

81. ▼ Rewrite in set notation: She prefers movies that are not violent, are shorter than 2 hours, and have neither a tragic ending nor an unexpected ending.

82. ▼ Rewrite in set notation: He will cater for any event as long as there are no more than 1,000 people, it lasts for at least 3 hours, and it is within a 50-mile radius of Toronto.

83. ▼ When this book was being written, the copy editor wanted to delete the comma in the following sentence (see Exercise 74): "You would like to see either a World War II movie, or one that is based on a comic book character but does not feature aliens." Explain why this would have resulted in an ambiguity.

84. ▼ When an older version of this book was being written, the copy editor wanted to delete the comma in the following sentence: "You would like to see a World War II movie, based on a comic book character but not featuring aliens." Explain why removing the comma would have had no effect on the meaning of the sentence.

---

[1] Source: Department of Health and Human Services/*New York Times*, January 8, 2002, p. A14.

Cardinality

In this section we begin to look at a deceptively simple idea: the size of a set, which we call its **cardinality**.

Visualizing Cardinality

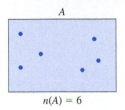

$n(A) = 6$

Cardinality

If A is a finite set, then its **cardinality** is

$$n(A) = \text{number of elements in } A.$$

> **Quick Examples**
>
> 1. Let $S = \{a, b, c\}$. Then $n(S) = 3$.
> 2. Let S be the set of outcomes when two distinguishable dice are rolled. Then $n(S) = 36$ (see Example 1 in Section 6.1).
> 3. $n(\varnothing) = 0$ because the empty set has no elements.

Counting the elements in a small, simple set is straightforward. To count the elements in a large, complicated set, we try to describe the set as built of simpler sets using the set operations. We then need to know how to calculate the number of elements in, for example, a union, based on the number of elements in the simpler sets whose union we are taking.

The Cardinality of a Union

How can we calculate $n(A \cup B)$ if we know $n(A)$ and $n(B)$? Our first guess might be that $n(A \cup B)$ is $n(A) + n(B)$. But consider a simple example. Let

$$A = \{a, b, c\}$$

and

$$B = \{b, c, d\}.$$

Then $A \cup B = \{a, b, c, d\}$, so $n(A \cup B) = 4$, but $n(A) + n(B) = 3 + 3 = 6$. The calculation $n(A) + n(B)$ gives the wrong answer because the elements b and c are counted twice: once for being in A and again for being in B. To correct for this overcounting, we need to subtract the number of elements that get counted twice, which is the number of elements that A and B have in common, or $n(A \cap B) = 2$ in this case. So we get the right number for $n(A \cup B)$ from the following calculation:

$$n(A) + n(B) - n(A \cap B) = 3 + 3 - 2 = 4.$$

This argument leads to the following general formula.

Cardinality of a Union

If A and B are finite sets, then

$$n(A \cup B) = n(A) + n(B) - n(A \cap B).$$

Visualizing Cardinality of a Union

Disjoint sets

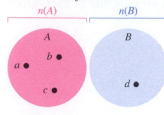

$$n(A \cup B) = n(A) + n(B)$$
$$= 3 + 1$$

Not disjoint

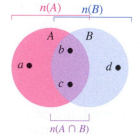

$$n(A \cup B) = n(A) + n(B) - n(A \cap B)$$
$$= 3 + 3 - 2$$

denniszv/Shutterstock.com

* We confirmed this number by a search for "asteroid threat" OR "William Burrows."

† For example, in September 2015 a search on Google gave the following results: "Romulan mind probe": 6 results, "Abraham Lincoln is born on Earth": 4 results, "Romulan mind probe" OR "Abraham Lincoln is born on Earth": 3,840 results!

In particular, if A and B are disjoint (meaning that $A \cap B = \varnothing$), then

$$n(A \cup B) = n(A) + n(B).$$

(When A and B are disjoint, we say that $A \cup B$ is a **disjoint union**.)

Quick Examples

4. If $A = \{a, b, c, d\}$ and $B = \{b, c, d, e, f\}$, then

$$n(A \cup B) = n(A) + n(B) - n(A \cap B) = 4 + 5 - 3 = 6.$$

In fact, $A \cup B = \{a, b, c, d, e, f\}$.

5. If $A = \{a, b, c\}$ and $B = \{d, e, f\}$, then $A \cap B = \varnothing$, so

$$n(A \cup B) = n(A) + n(B) = 3 + 3 = 6.$$

EXAMPLE 1 Wikipedia Searches

In September 2015 a search on Wikipedia for the phrase "asteroid threat" yielded 9 articles containing that phrase, and a search for "William Burrows" yielded 30 articles. A search for articles containing both phrases yielded 2 articles. How many articles contained either "asteroid threat," "William Burrows," or both?

Solution Let A be the set of sites containing "asteroid threat," and let B be the set of sites containing "William Burrows." We are told that

$$n(A) = 9$$
$$n(B) = 30$$
$$n(A \cap B) = 2. \quad \text{"asteroid threat" AND "William Burrows"}$$

The formula for the cardinality of the union tells us that

$$n(A \cup B) = n(A) + n(B) - n(A \cap B) = 9 + 30 - 2 = 37.\text{*}$$

So 37 articles in the Wikipedia database contained one or both of the phrases "asteroid threat" and "William Burrows."

➡ **Before we go on...** Most search engines use "OR" for union and "AND" for intersection, as does that of Wikipedia. Note that the popular web search engines like Google and Bing do not adhere to the search rule you enter but instead present results that *they think* you want. So although the formula

$$n(A \cup B) = n(A) + n(B) - n(A \cap B)$$

always holds mathematically, you will usually find that, in Google or Bing searches, the numbers don't add up.† ∎

Q : *Is there a similar formula for $n(A \cap B)$?*

A : The formula for the cardinality of a union can also be thought of as a formula for the cardinality of an intersection. We can solve for $n(A \cap B)$ to get

$$n(A \cap B) = n(A) + n(B) - n(A \cup B).$$

In fact, we can think of this formula as an equation relating four quantities. If we know any three of them, we can use the equation to find the fourth. (See Example 2 below.)

Q: *Is there a similar formula for $n(A')$?*

A: We can get a formula for the cardinality of a complement as follows: If S is our universal set and $A \subseteq S$, then S is the disjoint union of A and its complement. That is,

$$S = A \cup A' \quad \text{and} \quad A \cap A' = \varnothing.$$

Applying the cardinality formula for a disjoint union, we get

$$n(S) = n(A) + n(A').$$

We can then solve for $n(A')$ or for $n(A)$ to get the formulas shown below.

Visualizing Cardinality of a Complement

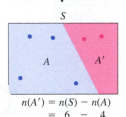

$$n(A') = n(S) - n(A)$$
$$= 6 - 4$$

Cardinality of a Complement

If S is a finite universal set and A is a subset of S, then

$$n(A') = n(S) - n(A)$$

and

$$n(A) = n(S) - n(A').$$

Quick Example

6. If $S = \{a, b, c, d, e, f\}$ and $A = \{a, b, c, d\}$, then

$$n(A') = n(S) - n(A) = 6 - 4 = 2.$$

In fact, $A' = \{e, f\}$.

EXAMPLE 2 Database Searches

In September 2015 a search on **Amazon.com** found 65,000 cookbooks.[2] Of these, 14,000 were on regional cooking, 5,000 were on vegetarian cooking, and 17,000 were on either regional or vegetarian cooking (or both). How many of these books were not on both regional and vegetarian cooking?

Solution Let S be the set of all 65,000 books on cooking, let A be the set of books on regional cooking, and let B be the set of books on vegetarian cooking. We wish to find the size of the complement of the set of books on both regional and vegetarian cooking—that is, $n((A \cap B)')$. Using the formula for the cardinality of a complement, we have

$$n((A \cap B)') = n(S) - n(A \cap B) = 65,000 - n(A \cap B).$$

To find $n(A \cap B)$, we use the formula for the cardinality of a union:

$$n(A \cup B) = n(A) + n(B) - n(A \cap B).$$

---

[2] Precisely, it found that number of books under the subject "Cookbooks, Food & Wine." Regional cooking falls under "Regional & International," and vegetarian cooking falls under "Vegetarian & Vegan." Figures are rounded to the nearest 1,000.

Substituting the values we were given, we find

$$17{,}000 = 14{,}000 + 5{,}000 - n(A \cap B),$$

which we can solve to get

$$n(A \cap B) = 2{,}000.$$

Therefore,

$$n((A \cap B)') = 65{,}000 - n(A \cap B) = 65{,}000 - 2{,}000 = 63{,}000.$$

So 63,000 of the cookbooks were not on both regional and vegetarian cooking.

EXAMPLE 3   Apple Sales

The following table shows sales, in millions of items, of Macs, iPhones, and iPads sold by **Apple** in 2012, 2013, and 2014:[3]

	Macs (A)	iPhones (B)	iPads (C)	Total
2012 (U)	18	125	58	201
2013 (V)	16	150	71	238
2014 (W)	19	169	68	256
Total	53	444	197	695

Let S be the set of all these items sold, and label the sets representing the sales in each row and column as shown (so that, for example, A is the set of all Macs sold during the 3 years). Describe the following sets and compute their cardinality:

a. U' **b.** $A \cap U'$ **c.** $(A \cap U)'$ **d.** $C \cup U$

Solution   Before answering parts (a)–(d), first notice that each number in the table is the cardinality of a specific set; for instance, the 18 million Macs sold in 2012 is the cardinality of the set $A \cap U$ of items that were Macs and also sold in 2012, the 125 million iPhones sold in 2012 is the cardinality of $B \cap U$, and so on:

	Macs (A)	iPhones (B)	iPads (C)	Total
2012 (U)	$n(A \cap U)$	$n(B \cap U)$	$n(C \cap U)$	$n(U)$
2013 (V)	$n(A \cap V)$	$n(B \cap V)$	$n(C \cap V)$	$n(V)$
2014 (W)	$n(A \cap W)$	$n(B \cap W)$	$n(C \cap W)$	$n(W)$
Total	$n(A)$	$n(B)$	$n(C)$	$n(S)$

Now let us answer the specific questions: Because all the figures are stated in millions of items, we'll give our calculations and results in millions of items as well.

a. U' is the set of all items not sold in 2012. To compute its cardinality, we could add the totals for all the other years listed in the rightmost column:

$$n(U') = n(V) + n(W) = 238 + 256 = 494 \text{ million items.}$$

---

[3] Figures are rounded. Source: Apple quarterly press releases, www.investor.apple.com.

Alternatively, we can use the formula for the cardinality of a complement (referring again to the totals in the table):

$$n(U') = n(S) - n(U) = 695 - 201 = 494 \text{ million items.}$$

b. $A \cap U'$ is the intersection of the set of all Macs and the set of all items not sold in 2012. In other words, it is the set of all Macs not sold in 2012. Here is the table with the corresponding sets A and U' shaded ($A \cap U'$ is the overlap):

	Macs (A)	iPhones (B)	iPads (C)	Total
2012 (U)	18	125	58	201
2013 (V)	16	150	71	238
2014 (W)	19	169	68	256
Total	53	444	197	695

From the table,

$$n(A \cap U') = 16 + 19 = 35 \text{ million items.}$$

c. $A \cap U$ is the set of all Macs sold in 2012, so $(A \cap U)'$ is the set of all items remaining if we exclude Macs sold in 2012:

	Macs (A)	iPhones (B)	iPads (C)	Total
2012 (U)	18	125	58	201
2013 (V)	16	150	71	238
2014 (W)	19	169	68	256
Total	53	444	197	695

From the formula for the cardinality of a complement,

$$n((A \cap U)') = n(S) - n(A \cap U)$$
$$= 695 - 18 = 677 \text{ million items.}$$

d. $C \cup U$ is the set of items that either were iPads or sold in 2012:

	Macs (A)	iPhones (B)	iPads (C)	Total
2012 (U)	18	125	58	201
2013 (V)	6	150	71	238
2014 (W)	19	169	68	256
Total	53	444	197	695

To compute it, we can use the formula for the cardinality of a union:

$$n(C \cup U) = n(C) + n(U) - n(C \cap U)$$
$$= 197 + 201 - 58 = 340 \text{ million items.}$$

To determine the cardinality of a union of three or more sets, for example, $n(A \cup B \cup C)$, we can think of $A \cup B \cup C$ as a union of two sets, $(A \cup B)$ and C, and then analyze each piece using the techniques we already have. Alternatively, there are formulas for the cardinalities of unions of any number of sets, but these formulas get more and more complicated as the number of sets grows. In many applications, such as the following example, we can use Venn diagrams instead.

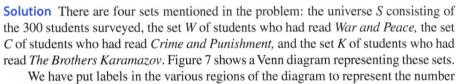

EXAMPLE 4   Reading Lists

A survey of 300 college students found that 100 had read *War and Peace,* 120 had read *Crime and Punishment,* and 100 had read *The Brothers Karamazov.* It also found that 40 had read only *War and Peace,* 70 had read *War and Peace* but not *The Brothers Karamazov,* and 80 had read *The Brothers Karamazov* but not *Crime and Punishment.* Only 10 had read all three novels. How many had read none of these three novels?

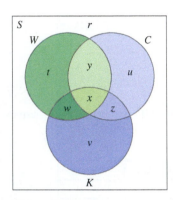

Figure 7

Solution There are four sets mentioned in the problem: the universe S consisting of the 300 students surveyed, the set W of students who had read *War and Peace,* the set C of students who had read *Crime and Punishment,* and the set K of students who had read *The Brothers Karamazov.* Figure 7 shows a Venn diagram representing these sets.

We have put labels in the various regions of the diagram to represent the number of students in each region. For example, x represents the number of students in $W \cap C \cap K$, which is the number of students who have read all three novels. We are told that this number is 10, so

$$x = 10.$$

(You should draw the diagram for yourself and fill in the numbers as we go along.) We are also told that 40 students had read only *War and Peace,* so

$$t = 40.$$

We are given none of the remaining regions directly. However, because 70 had read *War and Peace* but not *The Brothers Karamazov,* we see that t and y must add up to 70. Because we already know that $t = 40$, it follows that $y = 30$. Further, because a total of 100 students had read *War and Peace,* we have

$$x + y + t + w = 100.$$

Substituting the known values of x, y, and t gives

$$10 + 30 + 40 + w = 100,$$

so $w = 20$. Because 80 students had read *The Brothers Karamazov* but not *Crime and Punishment,* we see that $v + w = 80$, so $v = 60$ (because we know that $w = 20$). We can now calculate z using the fact that a total of 100 students had read *The Brothers Karamazov:*

$$60 + 20 + 10 + z = 100,$$

giving $z = 10$. Similarly, we can now get u using the fact that 120 students had read *Crime and Punishment:*

$$10 + 30 + 10 + u = 120,$$

giving $u = 70$. Of the 300 students surveyed, we've now found $x + y + z + w + t + u + v = 240$. This leaves

$$r = 60$$

who had read none of the three novels.

The Cardinality of a Cartesian Product

We've covered all the operations except Cartesian product. To find a formula for $n(A \times B)$, consider the following simple example:

$$A = \{H, T\}$$
$$B = \{1, 2, 3, 4, 5, 6\},$$

so

$$A \times B = \{H1, H2, H3, H4, H5, H6, T1, T2, T3, T4, T5, T6\}.$$

As we saw in Example 5 in Section 6.1, the elements of $A \times B$ can be arranged in a table or spreadsheet with $n(A) = 2$ rows and $n(B) = 6$ elements in each row:

	A	B	C	D	E	F	G
1		1	2	3	4	5	6
2	H	H1	H2	H3	H4	H5	H6
3	T	T1	T2	T3	T4	T5	T6

In a region with two rows and six columns, there are $2 \times 6 = 12$ cells. So

$$n(A \times B) = n(A)n(B)$$

in this case. There is nothing particularly special about this example, however, and that formula holds true in general.

Cardinality of a Cartesian Product

If A and B are finite sets, then

$$n(A \times B) = n(A)n(B).$$

Quick Example

7. If $A = \{a, b, c\}$ and $B = \{x, y, z, w\}$, then

$$n(A \times B) = n(A)n(B) = 3 \times 4 = 12.$$

EXAMPLE 5   Coin Tosses

a. If we toss a coin twice and observe the sequence of heads and tails, how many possible outcomes are there?

b. If we toss a coin three times, how many possible outcomes are there?

c. If we toss a coin ten times, how many possible outcomes are there?

Solution

a. Let $A = \{H, T\}$ be the set of possible outcomes when a coin is tossed once. The set of outcomes when a coin is tossed twice is $A \times A$, which has

$$n(A \times A) = n(A)n(A) = 2 \times 2 = 4$$

possible outcomes.

b. When a coin is tossed three times, we can think of the set of outcomes as the product of the set of outcomes for the first two tosses, which is $A \times A$, and the set of outcomes for the third toss, which is just A. The set of outcomes for the three tosses is then $(A \times A) \times A$, which we usually write as $A \times A \times A$ or A^3. The number of outcomes is

$$n((A \times A) \times A) = n(A \times A)n(A) = (2 \times 2) \times 2 = 8.$$

c. Considering the result of part (b), we can easily see that the set of outcomes here is $A^{10} = A \times A \times \cdots \times A$ (10 copies of A), or the set of ordered sequences of ten Hs and Ts. It's also easy to see that

$$n(A^{10}) = [n(A)]^{10} = 2^{10} = 1{,}024.$$

➡ **Before we go on . . .** We can start to see the power of these formulas for cardinality. In Example 5 we were able to calculate that there are 1,024 possible outcomes when we toss a coin 10 times without writing out all 1,024 possibilities and counting them. ∎

6.2 EXERCISES

▼ more advanced   ◆ challenging
🅣 indicates exercises that should be solved using technology

Let $A = \{$Dirk, Johan, Frans, Sarie$\}$, $B = \{$Frans, Sarie, Tina, Klaas, Henrika$\}$, $C = \{$Hans, Frans$\}$. *Find the numbers indicated in Exercises 1–6.* [**HINT:** See Quick Examples 1–5.]

1. $n(A) + n(B)$ **2.** $n(A) + n(C)$

3. $n(A \cup B)$ **4.** $n(A \cup C)$

5. $n(A \cup (B \cap C))$ **6.** $n(A \cap (B \cup C))$

7. Verify that $n(A \cup B) = n(A) + n(B) - n(A \cap B)$ with A and B as above.

8. Verify that $n(A \cup C) = n(A) + n(C) - n(A \cap C)$ with A and C as above.

Let $A = \{$H, T$\}$, $B = \{1, 2, 3, 4, 5, 6\}$, and $C = \{$red, green, blue$\}$. *Find the numbers indicated in Exercises 9–14.* [**HINT:** See Example 5.]

9. $n(A \times A)$ **10.** $n(B \times B)$

11. $n(B \times C)$ **12.** $n(A \times C)$

13. $n(A \times B \times B)$ **14.** $n(A \times B \times C)$

15. If $n(A) = 43, n(B) = 20$, and $n(A \cap B) = 3$, find $n(A \cup B)$.

16. If $n(A) = 60, n(B) = 20$, and $n(A \cap B) = 1$, find $n(A \cup B)$.

17. If $n(A \cup B) = 100$ and $n(A) = n(B) = 60$, find $n(A \cap B)$.

18. If $n(A) = 100, n(A \cup B) = 150$, and $n(A \cap B) = 40$, find $n(B)$.

Let $S = \{$Barnsley, Manchester United, Southend, Sheffield United, Liverpool, Maroka Swallows, Witbank Aces, Royal Tigers, Dundee United, Lyon$\}$ *be a universal set,* $A = \{$Southend, Liverpool, Maroka Swallows, Royal Tigers$\}$, *and* $B = \{$Barnsley, Manchester United, Southend$\}$. *Find the numbers indicated in Exercises 19–24.* [**HINT:** See Quick Example 6.]

19. $n(A')$ **20.** $n(B')$ **21.** $n((A \cap B)')$

22. $n((A \cup B)')$ **23.** $n(A' \cap B')$ **24.** $n(A' \cup B')$

25. With S, A, and B as above, verify that $n((A \cap B)') = n(A') + n(B') - n((A \cup B)')$.

26. With S, A, and B as above, verify that $n(A' \cap B') + n(A \cup B) = n(S)$.

In Exercises 27–30, use the given information to complete the solution of each partially solved Venn diagram. [**HINT:** See Example 4.]

27.

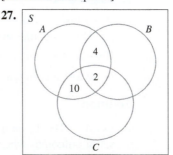

$n(A) = 20, n(B) = 20, n(C) = 28,$
$n(B \cap C) = 8, n(S) = 50$

28.

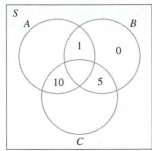

$n(A) = 16, n(B) = 11, n(C) = 30, n(S) = 40$

29. ▼

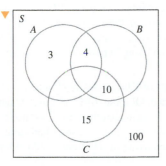

$n(A) = 10, n(B) = 19, n(S) = 140$

30. ▼

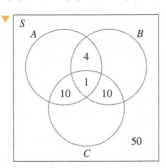

$n(A \cup B) = 30, n(B \cup C) = 30, n(A \cup C) = 35$

Applications

31. Web Searches In November 2011 a search using the web search engine Bing for "asteroid" yielded 25.0 million websites containing that word. A search for "comet" yielded 93.5 million sites. A search for sites containing both words yielded 3.1 million sites.[4] How many websites contained either "asteroid" or "comet" or both? [**HINT:** See Example 1.]

32. Web Searches In November 2011 a search using the web search engine Bing for "tea party" yielded 58.0 million websites containing that phrase. A search for "coffee party" yielded 0.6 million sites. A search for sites containing both phrases yielded 0.1 million sites.[5] How many websites

[4] Back in 2011, Bing (but not Google) could still be relied on to adhere to the search rule you entered (see the "Before we go on" discussion following Example 1). Figures are rounded to the nearest 0.1 million.

[5] *Ibid.*

contained either "tea party" or "coffee party" or both? [**HINT:** See Example 1.]

33. Amusement On a particularly boring transatlantic flight, one of the authors amused himself by counting the heads of the people in the seats in front of him. He noticed that all 37 of them either had black hair or had a whole row to themselves (or both). Of this total, 33 had black hair, and 6 were fortunate enough to have a whole row of seats to themselves. How many of the black-haired people had whole rows to themselves?

34. Restaurant Menus While scanning through the dessert menu of your favorite restaurant, you notice that it lists 14 desserts that include yogurt, fruit, or both. Of these, 8 include yogurt, and 9 include fruit. How many of the desserts with yogurt also include fruit?

35. Mobile Gamers Of a total of 132 million mobile gamers (people who use smartphones or tablets or both for gaming) in India in 2014, 123 million used smartphones, and 44 million used tablets.[6] How many used only tablets?

36. Mobile Gamers Of a total of 208 million mobile gamers (people who use smartphones or tablets or both for gaming) in India in 2016, 194 million used smartphones, and 73 million used tablets.[7] How many did not use tablets?

Publishing *Exercises 37–42 are based on the following table, which shows the results of a survey of authors by a fictitious publishing company:*

	New Authors	Established Authors	Total
Successful	5	25	30
Unsuccessful	15	55	70
Total	20	80	100

Consider the following subsets of the set S of all authors represented in the table: C, the set of successful authors; U, the set of unsuccessful authors; N, the set of new authors; and E, the set of established authors. [**HINT:** See Example 3.]

37. Describe the sets $C \cap N$ and $C \cup N$ in words. Use the table to compute $n(C), n(N), n(C \cap N)$, and $n(C \cup N)$. Verify that $n(C \cup N) = n(C) + n(N) - n(C \cap N)$.

38. Describe the sets $N \cap U$ and $N \cup U$ in words. Use the table to compute $n(N), n(U), n(N \cap U)$, and $n(N \cup U)$. Verify that $n(N \cup U) = n(N) + n(U) - n(N \cap U)$.

39. Describe the set $C \cap N'$ in words, and find the number of elements it contains.

40. Describe the set $U \cup E'$ in words, and find the number of elements it contains.

41. ▼ What percentage of established authors are successful? What percentage of successful authors are established?

[6] Figures are rounded. Source for data: www.emarketer.com.

[7] Figures are estimates; the last two by www.emarketer.com.

42. ▼ What percentage of new authors are unsuccessful? What percentage of unsuccessful authors are new?

Housing Starts *Exercises 43–48 are based on the following table, which shows the number of new (single-unit) houses started, in thousands, in the different regions of the United States during 2012–2014.*[8] *Take S to be the set of all housing starts represented in the table, and label the sets representing the housing starts in each row and column as shown (so, for example, N is the set of all housing starts in the Northeast during 2012–2014).*

	Northeast (N)	Midwest (M)	South (T)	West (W)	Total
2012 (A)	50	90	280	110	530
2013 (B)	60	100	330	130	620
2014 (C)	50	110	350	150	660
Total	160	300	960	390	1,810

In each exercise, use symbols to describe the given set, and compute its cardinality.

43. The set of housing starts in the Midwest in 2014

44. The set of housing starts either in the West or in 2012

45. The set of housing starts in 2013 excluding housing starts in the South

46. The set of housing starts in the Northeast after 2012

47. ▼ The set of housing starts in 2013 in the West and Midwest

48. ▼ The set of housing starts in the South and West in years other than 2012

Stocks *Exercises 49–54 are based on the following table, which shows the stock market performance of 40 industries from five sectors of the U.S. economy as of noon on September 11, 2015.*[9] *(Take S to be the set of all 40 industries represented in the table.)*

	Increased (X)	Decreased (Y)	Unchanged (Z)	Totals
Financials (F)	3	4	1	8
Manufacturing (M)	8	3	3	14
Information Technology (T)	6	1	0	7
Health Care (H)	4	1	1	6
Utilities (U)	3	1	1	5
Totals	24	10	6	40

49. Use symbols to describe the set of non-manufacturing industries that increased. How many elements are in this set?

50. Use symbols to describe the set of industries in the manufacturing sector that did not increase. How many elements are in this set?

51. Compute $n(H' \cup Z)$. What does this number represent?

52. Compute $n(H \cup Z')$. What does this number represent?

53. Calculate $\dfrac{n(T \cap Y)}{n(Y)}$. What does the answer represent?

54. Calculate $\dfrac{n(U \cap X)}{n(U)}$. What does the answer represent?

55. ▼ ***Medicine*** In a study of Tibetan children,[10] a total of 1,556 children were examined. Of these, 1,024 had rickets. Of the 243 urban children in the study, 93 had rickets.
  a. How many children living in nonurban areas had rickets?
  b. How many children living in nonurban areas did not have rickets?

56. ▼ ***Medicine*** In a study of Tibetan children,[11] a total of 1,556 children were examined. Of these, 615 had caries (cavities). Of the 1,313 children living in nonurban areas, 504 had caries.
  a. How many children living in urban areas had caries?
  b. How many children living in urban areas did not have caries?

57. ***Entertainment*** According to a survey of 100 people regarding their movie attendance in the last year, 40 had seen a science fiction movie, 55 had seen an adventure movie, and 35 had seen a horror movie. Moreover, 25 had seen a science fiction movie and an adventure movie, 5 had seen an adventure movie and a horror movie, and 15 had seen a science fiction movie and a horror movie. Only 5 people had seen a movie from all three categories.
  a. Use the given information to set up a Venn diagram and solve it. [**HINT:** See Example 4.]
  b. Complete the following sentence: The survey suggests that __ % of science fiction movie fans are also horror movie fans.

58. ***Athletics*** Of the 4,700 students at *Medium Suburban College*, 50 play collegiate soccer, 60 play collegiate lacrosse, and 96 play collegiate football. Only 4 students play both collegiate soccer and lacrosse, 6 play collegiate soccer and football, and 16 play collegiate lacrosse and football. No students play all three sports.
  a. Use the given information to set up a Venn diagram and solve it. [**HINT:** See Example 4.]
  b. Complete the following sentence: __ % of the college soccer players also play one of the other two sports at the collegiate level.

59. ***Entertainment*** In a survey of 100 *Enormous State University* students, 21 enjoyed classical music, 22 enjoyed rock music, and 27 enjoyed house music. Five of the students

---

[8] Figures are rounded. Source: www.census.gov.

[9] Unchanged" includes industries that moved by less than 0.1%. Source for data: Fidelity https://eresearch.fidelity.com.

[10] Source: N. S. Harris et al., "Nutritional and Health Status of Tibetan Children Living at High Altitudes," *New England Journal of Medicine*, 344(5), February 1, 2001, pp. 341–347.

[11] *Ibid.*

enjoyed both classical and rock. How many of those who enjoyed rock did not enjoy classical music?

60. *Entertainment* Refer back to Exercise 59. You are also told that 5 students enjoyed all three kinds of music while 53 enjoyed music in none of these categories. How many students enjoyed both classical and rock but disliked house music?

Communication and Reasoning Exercises

61. If A and B are finite sets with $A \subset B$, how are $n(A)$ and $n(B)$ related?

62. If A and B are subsets of the finite set S with $A \subset B$, how are $n(A')$ and $n(B')$ related?

63. Why is the Cartesian product referred to as a "product"? [**HINT:** Think about cardinality.]

64. Refer back to your answer to Exercise 63. What set operation could you use to represent the *sum* of two disjoint sets A and B? Why?

65. Formulate an interesting application whose answer is $n(A \cap B) = 20$.

66. Formulate an interesting application whose answer is $n(A \times B) = 120$.

67. ▼ When is $n(A \cup B) \neq n(A) + n(B)$?

68. ▼ When is $n(A \times B) = n(A)$?

69. ▼ When is $n(A \cup B) = n(A)$?

70. ▼ When is $n(A \cap B) = n(A)$?

71. ◆ Use a Venn diagram or some other method to obtain a formula for $n(A \cup B \cup C)$ in terms of $n(A)$, $n(B)$, $n(C)$, $n(A \cap B)$, $n(A \cap C)$, $n(B \cap C)$, and $n(A \cap B \cap C)$.

72. ◆ Suppose that A and B are sets with $A \subset B$ and $n(A)$ at least 2. Arrange the following numbers from smallest to largest (if two numbers are equal, say so): $n(A)$, $n(A \times B)$, $n(A \cap B)$, $n(A \cup B)$, $n(B \times A)$, $n(B)$, $n(B \times B)$.

6.3 Decision Algorithms: The Addition and Multiplication Principles

The Addition and Multiplication Principles

Let's start with a really simple example. You walk into an ice cream parlor and find that you can choose between ice cream, of which there are 15 flavors, and frozen yogurt, of which there are 5 flavors. If you want a single scoop of one of these, how many different selections can you make? Clearly, you have $15 + 5 = 20$ different desserts from which to choose. Mathematically, this is an example of the formula for the cardinality of a disjoint union: If we let A be the set of ice creams you can choose from and let B be the set of frozen yogurts, then $A \cap B = \varnothing$ and we want $n(A \cup B)$. But the formula for the cardinality of a disjoint union is $n(A \cup B) = n(A) + n(B)$, which gives $15 + 5 = 20$ in this case.

This example illustrates a very useful general principle.

Addition Principle

When choosing among r disjoint alternatives, suppose that

alternative 1 has n_1 possible outcomes,

alternative 2 has n_2 possible outcomes,

⋮

alternative r has n_r possible outcomes,

with no two of these outcomes the same. Then there are a total of $n_1 + n_2 + \cdots + n_r$ possible outcomes.

Quick Example

1. At a restaurant you can choose among 8 chicken dishes, 10 beef dishes, 4 seafood dishes, and 12 vegetarian dishes. This gives a total of $8 + 10 + 4 + 12 = 34$ different dishes to choose from.

Here is another simple example. In that ice cream parlor, not only can you choose from 15 flavors of ice cream, but you can also choose from 3 different sizes of cone. How many different ice cream cones can you select from? This time, we want to choose both a flavor and a size, or, in other words, a pair (flavor, size). Therefore, if we let A again be the set of ice cream flavors and now let C be the set of cone sizes, the pair we want to choose is an element of $A \times C$, the Cartesian product. To find the number of choices we have, we use the formula for the cardinality of a Cartesian product: $n(A \times C) = n(A)n(C)$. In this case we get $15 \times 3 = 45$ different ice cream cones we can select.

This example illustrates another general principle.

Multiplication Principle

When making a sequence of choices with r steps, suppose that

> step 1 has n_1 possible outcomes
>
> step 2 has n_2 possible outcomes
>
> $\vdots$
>
> step r has n_r possible outcomes

* See Example 3 for a case in which different sequences of choices can lead to the same outcome, with the result that the multiplication principle does not apply.

and that each sequence of choices results in a distinct outcome.* Then there are a total of $n_1 \times n_2 \times \cdots \times n_r$ possible outcomes.

Quick Example

2. At a restaurant you can choose among 5 appetizers, 34 main dishes, and 10 desserts. This gives a total of $5 \times 34 \times 10 = 1{,}700$ different meals (each including one appetizer, one main dish, and one dessert) from which you can choose.

Things get more interesting when we have to use the addition and multiplication principles in tandem.

EXAMPLE 1   Desserts

You walk into an ice cream parlor and find that you can choose between ice cream, of which there are 15 flavors, and frozen yogurt, of which there are 5 flavors. In addition, you can choose among 3 different sizes of cones for your ice cream or 2 different sizes of cups for your yogurt. If you want only a single item, how many different desserts can you choose from?

Solution   It helps to think about a definite procedure for deciding which dessert you will choose. Here is one we can use:

Alternative 1:   An ice cream cone
   Step 1 Choose a flavor.
   Step 2 Choose a size.

Alternative 2:   A cup of frozen yogurt
   Step 1 Choose a flavor.
   Step 2 Choose a size.

*An algorithm is a procedure with
definite rules for what to do at
every step.

That is, we can choose between alternative 1 and alternative 2. If we choose alternative 1, we have a sequence of two choices to make: flavor and size. The same is true of alternative 2. We shall call a procedure in which we make a sequence of decisions a **decision algorithm**.* Once we have a decision algorithm, we can use the addition and multiplication principles to count the number of possible outcomes.

Alternative 1: An ice cream cone
   Step 1 Choose a flavor: 15 choices.
   Step 2 Choose a size: 3 choices.
   There are $15 \times 3 = 45$ possible choices in alternative 1.   Multiplication principle

Alternative 2: A cup of frozen yogurt
   Step 1 Choose a flavor: 5 choices.
   Step 2 Choose a size: 2 choices.
   There are $5 \times 2 = 10$ possible choices in alternative 2.   Multiplication principle

So there are $45 + 10 = 55$ possible choices of desserts.   Addition principle

➡ **Before we go on . . .** Decision algorithms can be illustrated by **decision trees**. To simplify the picture, suppose we had fewer choices in Example 1—say, only two choices of ice cream flavor: vanilla and chocolate, and two choices of yogurt flavor: banana and raspberry. This would give us a total of $2 \times 3 + 2 \times 2 = 10$ possible desserts. We can illustrate the decisions we need to make when choosing what to buy in the diagram in Figure 8, called a *decision tree.*

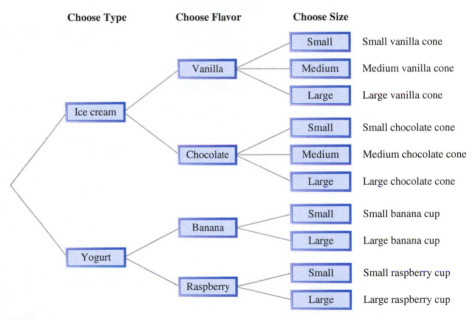

Figure 8   ■

Decision Algorithms and Decision Trees

We referred to the sequence of decisions we made in Example 1 as an example of a *decision algorithm.* Let's look at these gadgets a little more closely.

Decision Algorithm

A **decision algorithm** is a procedure in which we make a sequence of decisions. We can use decision algorithms to determine the number of possible items of a specified type (for example, ice cream cones) by pretending that we are *designing* such an item and listing the decisions or choices we should make at each stage of the process.

To count the number of possible gadgets, pretend you are *designing* a gadget, and list the decisions to be made at each stage.

Quick Example

3. Your local **Apple** store has iPads in two sizes: the larger Air and the smaller Mini. The Air is available in two colors (blue, green), and the Mini is available in four colors (blue, green, pink, purple). A decision algorithm for "designing" an iPad is as follows:

> *Alternative 1:* Select an Air:
> **Step 1** Choose a color: Two choices.
> (So there are two choices for alternative 1.)
>
> *Alternative 2:* Select a Mini:
> **Step 1** Choose a color: Four choices.
> (So there are four choices for alternative 2.)

Thus, there are $2 + 4 = 6$ possible choices of iPads.

Decision Tree

A decision algorithm can be illustrated by a **decision tree** in which the choices we make when we follow the algorithm are represented by branches.

Quick Example

4. The following tree illustrates the decision algorithm in Quick Example 3 to select an iPad:

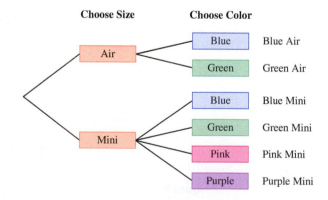

We do not use decision trees much in this chapter because, while they provide a good way of thinking about decision algorithms, they're not really practical for counting large sets. Similar diagrams will be very useful, however, in the chapter on probability.

Caution
For a decision algorithm to give the correct number of possible items, there must be a one-to-one correspondence between sequences of choices and resulting items. So it is necessary that each sequence of choices results in a distinct item. In other words, *changing one or more choices must result in a different item.* (See Example 3.)

EXAMPLE 2 **Exams**

An exam is broken into two parts, Part A and Part B, both of which you are required to do. In Part A you can choose between answering 10 true-false questions and answering 4 multiple-choice questions, each of which has 5 answers to choose from. In Part B you can choose between answering 8 true-false questions and answering 5 multiple-choice questions, each of which has 4 answers to choose from. How many different collections of answers are possible?

Solution While deciding what answers to write down, we use the following decision algorithm:

Step 1 Do Part A.
  Alternative 1: Answer the 10 true-false questions.
    Steps 1–10 Choose true or false for each question: 2 choices each.
    There are $2 \times 2 \times \cdots \times 2 = 2^{10} = 1{,}024$ choices in alternative 1.

  Alternative 2: Answer the 4 multiple-choice questions.
    Steps 1–4 Choose one answer for each question: 5 choices each.
    There are $5 \times 5 \times 5 \times 5 = 5^4 = 625$ choices in alternative 2.
  $1{,}024 + 625 = 1{,}649$ choices in step 1

Step 2 Do Part B.
  Alternative 1: Answer the 8 true-false questions: 2 choices each.
    $2^8 = 256$ choices in alternative 1

  Alternative 2: Answer the 5 multiple-choice questions, 4 choices each.
    $4^5 = 1{,}024$ choices in alternative 2
  $256 + 1{,}024 = 1{,}280$ choices in step 2

There are $1{,}649 \times 1{,}280 = 2{,}110{,}720$ different collections of answers possible.

The next example illustrates the need to select your decision algorithm with care.

EXAMPLE 3 **Scrabble**

You are playing Scrabble and have the following letters to work with: k, e, r, e. Because you are losing the game, you would like to use all your letters to make a single word, but you can't think of any four-letter words that use all these letters. In desperation, you decide to list *all* the four-letter sequences possible to see whether there are any valid words among them. How large is your list?

Solution It may first occur to you to try the following decision algorithm.

Step 1 Select the first letter: 4 choices.
Step 2 Select the second letter: 3 choices.
Step 3 Select the third letter: 2 choices.
Step 4 Select the last letter: 1 choice.

This gives $4 \times 3 \times 2 \times 1 = 24$ choices. However, something is wrong with the algorithm.

Q: *What is wrong with this decision algorithm?*

A: We didn't take into account the fact that there are two "e"s;* different decisions in Steps 1–4 can produce the same sequence. Suppose, for example, that we selected the first "e" in step 1, the second "e" in step 2, and then the "k" and the "r." This would produce the sequence "eekr." If we selected the *second* "e" in step 1, the *first* "e" in step 2, and then the "k" and "r," we would obtain the *same* sequence: "eekr." In other words, the decision algorithm produces two copies of the sequence "eekr" in the associated decision tree. (In fact, it produces two copies of each possible sequence of the letters.) In short, different sequences of choices produce the same result, violating the requirement in the note of caution that precedes Example 2:

For a decision algorithm to be valid, each sequence of choices must produce a different result.

* Consider the following extreme case: If all four letters were "e," then there would be only a single sequence—"eeee"—and not the 24 predicted by the decision algorithm.

Because our original algorithm is not valid, we need a new one. Here is a strategy that works nicely for this example. Imagine, as before, that we are going to construct a sequence of four letters. This time we are going to imagine that we have a sequence of four empty slots: ▢▢▢▢. Instead of selecting letters to fill the slots from left to right, we are going to select *slots* in which to place each of the letters. Remember that we have to use the letters k, e, r, e. We proceed as follows, leaving the "e"s until last:

Step 1 Select an empty slot for the k: 4 choices. (e.g., ▢▢k▢)
Step 2 Select an empty slot for the r: 3 choices. (e.g., r▢k▢)
Step 3 Place the "e"s in the remaining two slots: 1 choice!

Thus, the multiplication principle yields $4 \times 3 \times 1 = 12$ choices.

➡ **Before we go on ...** You should try constructing a decision tree for Example 3. You will see that each sequence of four letters is produced exactly once when we use the correct (second) decision algorithm. ∎

FAQs

Creating and Testing a Decision Algorithm

Q: *How do I set up a decision algorithm to count how many items there are in a given scenario?*

A: Pretend that you are *designing* such an item (for example, pretend that you are designing an ice cream cone), and come up with a systematic procedure for doing so, listing the decisions you should make at each stage.

Q: *In my decision algorithm, where do I use alternatives, and where do I use steps?*

A: Think of your procedure in terms of "or" and "and": If your procedure includes a list of consecutive instructions ("Do this *and* that *and* that *and* that . . ."), then use steps for that part of the decision algorithm; if your procedure includes a list of different possibilities ("Do this *or* that *or* that *or* that . . ."), then use alternatives for that part of the decision algorithm.

Q : *Once I have my decision algorithm, how do I check whether it is valid?*

A : Ask yourself the following question: "Is it possible to get the same item (the exact same ice cream cone, say) by making different decisions when applying the algorithm?" If the answer is "yes," then your decision algorithm is invalid. Otherwise, it is valid.

6.3 EXERCISES

▼ more advanced   ◆ challenging

T indicates exercises that should be solved by using technology

1. An experiment requires a choice among three initial setups. The first setup can result in two possible outcomes, the second in three possible outcomes, and the third in five possible outcomes. What is the total number of outcomes possible? [**HINT:** See Quick Example 1.]

2. A surgical procedure requires choosing among four alternative methodologies. The first can result in four possible outcomes, the second can result in three possible outcomes, and the remaining methodologies can each result in two possible outcomes. What is the total number of outcomes possible? [**HINT:** See Quick Example 1.]

3. An experiment requires a sequence of three steps. The first step can result in two possible outcomes, the second in three possible outcomes, and the third in five possible outcomes. What is the total number of outcomes possible? [**HINT:** See Quick Example 2.]

4. A surgical procedure requires four steps. The first can result in four possible outcomes, the second can result in three possible outcomes, and the remaining two can each result in two possible outcomes. What is the total number of outcomes possible? [**HINT:** See Quick Example 2.]

For the decision algorithms in Exercises 5–12, find how many outcomes are possible. [**HINT:** See Example 1.]

5. Alternative 1:
    Step 1: 1 outcome
    Step 2: 2 outcomes

    Alternative 2:
    Step 1: 2 outcomes
    Step 2: 2 outcomes
    Step 3: 1 outcome

6. Alternative 1:
    Step 1: 1 outcome
    Step 2: 2 outcomes
    Step 3: 2 outcomes

    Alternative 2:
    Step 1: 2 outcomes
    Step 2: 2 outcomes

7. Step 1:
    Alternative 1: 1 outcome
    Alternative 2: 2 outcomes

    Step 2:
    Alternative 1: 2 outcomes
    Alternative 2: 2 outcomes
    Alternative 3: 1 outcome

8. Step 1:
    Alternative 1: 1 outcome
    Alternative 2: 2 outcomes
    Alternative 3: 2 outcomes

    Step 2:
    Alternative 1: 2 outcomes
    Alternative 2: 2 outcomes

9. Alternative 1:
    Step 1:
       Alternative 1: 3 outcomes
       Alternative 2: 1 outcome
    Step 2: 2 outcomes

    Alternative 2: 5 outcomes

10. Alternative 1: 2 outcomes

    Alternative 2:
    Step 1:
       Alternative 1: 4 outcomes
       Alternative 2: 1 outcome
    Step 2: 2 outcomes

11. Step 1:
    Alternative 1:
       Step 1: 3 outcomes
       Step 2: 1 outcome
    Alternative 2: 2 outcomes

    Step 2: 5 outcomes

12. Step 1: 2 outcomes

    Step 2:
    Alternative 1:
       Step 1: 4 outcomes
       Step 2: 1 outcome
    Alternative 2: 2 outcomes

13. How many different four-letter sequences can be formed from the letters a, a, a, b? [**HINT:** See Example 3.]

14. How many different five-letter sequences can be formed from the letters a, a, a, b, c? [**HINT:** See Example 3.]

Applications

15. *Ice Cream* When **Baskin-Robbins** was founded in 1945, it made 31 different flavors of ice cream.[12] If you had a choice of having a single flavor of ice cream in a cone, a cup, or a sundae, how many different desserts could you have?

16. *Ice Cream* At the beginning of 2002, **Baskin-Robbins** claimed to have "nearly 1,000 different ice cream flavors."[13] Assuming that you could choose from 1,000 different flavors, that you could have a single flavor of ice cream in a cone, a cup, or a sundae, and that you could choose from a dozen different toppings, how many different desserts could you have?

17. *Binary Codes* A binary digit, or "bit," is either 0 or 1. A nybble is a four-bit sequence. How many different nybbles are possible?

---

[12] Source: Company website (www.baskinrobbins.com).

[13] *Ibid.*

18. **Ternary Codes** A ternary digit is either 0, 1, or 2. How many sequences of six ternary digits are possible?

19. **Ternary Codes** A ternary digit is either 0, 1, or 2. How many sequences of six ternary digits are possible containing a single 1 and a single 2?

20. **Binary Codes** A binary digit, or "bit," is either 0 or 1. A nybble is a four-bit sequence. How many different nybbles containing a single 1 are possible?

21. **Reward** While selecting candy for students in his class, Professor Murphy must choose between gummy candy and licorice nibs. Gummy candy packets come in three sizes, while packets of licorice nibs come in two. If he chooses gummy candy, he must select gummy bears, gummy worms, or gummy dinos. If he chooses licorice nibs, he must choose between red and black. How many choices does he have? [**HINT:** See Example 2.]

22. **Productivity** Professor Oger must choose between an extra writing assignment and an extra reading assignment for the upcoming spring break. For the writing assignment there are two essay topics to choose from and three different mandatory lengths (30 pages, 35 pages, or 40 pages). The reading topic would consist of one scholarly biography combined with one volume of essays. There are five biographies and two volumes of essays to choose from. How many choices does she have? [**HINT:** See Example 2.]

23. **DVD Discs** DVD discs at your local computer store are available in two types (DVD-R and DVD-RW), packaged singly, in spindles of 50, or in spindles of 100. When purchasing singly, you can choose from five colors; when purchasing in spindles of 50 or 100, you have two choices: silver or an assortment of colors. If you are purchasing DVD discs, how many possibilities do you have to choose from?

24. **Radar Detectors** Radar detectors are either powered by their own battery or plug into the cigarette lighter socket. All radar detectors come in two models: no-frills and fancy. In addition, detectors powered by their own batteries detect either radar or laser or both, whereas the plug-in types come in models that detect either radar or laser, but not both. How many different radar detectors can you buy?

25. **Multiple-Choice Tests** Professor Easy's final examination has 10 true-false questions followed by 2 multiple-choice questions. In each of the multiple-choice questions, you must select the correct answer from a list of five. How many answer sheets are possible?

26. **Multiple-Choice Tests** Professor Tough's final examination has 20 true-false questions followed by 3 multiple-choice questions. In each of the multiple-choice questions, you must select the correct answer from a list of six. How many answer sheets are possible?

27. **Tests** A test requires that you answer either Part A or Part B. Part A consists of 8 true-false questions, and Part B consists of 5 multiple-choice questions with one correct answer out of five. How many different completed answer sheets are possible?

28. **Tests** A test requires that you answer first Part A and then either Part B or Part C. Part A consists of 4 true-false questions, Part B consists of 4 multiple-choice questions with one correct answer out of five, and Part C consists of 3 multiple-choice questions with one correct answer out of six. How many different completed answer sheets are possible?

29. **Stock Portfolios** Your broker has suggested that you diversify your investments by splitting your portfolio among mutual funds, municipal bond funds, stocks, and precious metals. She suggests four good mutual funds, three municipal bond funds, eight stocks, and three precious metals (gold, silver, and platinum).
   a. Assuming that your portfolio is to contain one of each type of investment, how many different portfolios are possible?
   b. Assuming that your portfolio is to contain three mutual funds, two municipal bond funds, one stock, and two precious metals, how many different portfolios are possible?

30. **Menus** The local diner offers a meal combination consisting of an appetizer, a soup, a main course, and a dessert. There are five appetizers, two soups, four main courses, and five desserts. Your diet restricts you to choosing between a dessert and an appetizer. (You cannot have both.) Given this restriction, how many three-course meals are possible?

31. **Computer Codes** A computer byte consists of eight bits, each bit being either a 0 or a 1. If characters are represented using a code that uses a byte for each character, how many different characters can be represented?

32. **Computer Codes** Some written languages, such as Chinese and Japanese, use tens of thousands of different characters. If a language uses roughly 50,000 characters, a computer code for this language would have to use how many bytes per character? (See Exercise 31.)

33. **Symmetries of a Five-Pointed Star** A five-pointed star will appear unchanged if it is rotated through any one of the angles 0°, 72°, 144°, 216°, or 288°. It will also appear unchanged if it is flipped about the axis shown in the figure. A *symmetry* of the five-pointed star consists of either a rotation, or a rotation followed by a flip. How many different symmetries are there altogether?

34. *Symmetries of a Six-Pointed Star* A six-pointed star will appear unchanged if it is rotated through any one of the angles 0°, 60°, 120°, 180°, 240°, or 300°. It will also appear unchanged if it is flipped about the axis shown in the figure. A *symmetry* of the six-pointed star consists of either a rotation, or a rotation followed by a flip. How many different symmetries are there altogether?

35. *Variables in Visual Basic* A variable name in the programming language Visual Basic must begin with a letter (uppercase or lowercase) possibly followed by letters, digits (0–9), and various special characters. How many different Visual Basic variable names of length up to three are possible that consist of letters and digits and end in a digit?

36. *Employee IDs* A company assigns to each of its employees an ID code that consists of one, two, or three uppercase letters followed by a digit from 0 through 9. How many employee codes does the company have available?

37. *Tournaments* How many ways are there of filling in the blanks for the following (fictitious) soccer tournament?

```
North Carolina ──┐
                 ├──┐
Central Connecticut ─┘  │
                        ├───────
Virginia ────────┐      │
                 ├──┘
            Syracuse
```

38. *Tournaments* How many ways are there of filling in the blanks for a (fictitious) soccer tournament involving the four teams San Diego State, De Paul, Colgate, and Hofstra?

```
San Diego State ──┐
                  ├──┐
──────────────────┘  │
                     ├──── De Paul
──────────────────┐  │
                  ├──┘
Colgate ──────────┘
```

39. ▼ *Telephone Numbers* Suppose a telephone number consists of a sequence of seven digits not starting with 0 or 1.
  a. How many telephone numbers are possible?
  b. How many of them begin with either 463, 460, or 400?
  c. How many telephone numbers are possible if no two adjacent digits are the same? (For example, 235-9350 is permitted, but 223-6789 is not.)

40. ▼ *Social Security Numbers* A Social Security Number is a sequence of nine digits.
  a. How many Social Security Numbers are possible?
  b. How many of them begin with either 023 or 003?
  c. How many Social Security Numbers are possible if no two adjacent digits are the same? (For example, 235-93-2345 is permitted, but 126-67-8189 is not.)

41. ▼ *Credit Card Numbers* The vast majority of **Visa** and **Discover Card** credit cards have 16-digit card numbers. The first digit is either 4 (Visa) or 6 (Discover), digits 2 through 6 identify the issuer, digits 7 through 15 identify the customer, and the last digit is a "check digit" determined by the digits that precede it.[14] Your company, *CreditXplosion, Inc.,* issues both Visa and Discover credit cards.
  a. How many different CreditXplosion cards are possible?
  b. How many different CreditXplosion Discover card numbers are possible in which the check digit is wrong?

42. ▼ *Credit Card Numbers* Credit cards issued by **American Express** have 15-digit card numbers. The first two digits are either 34 or 37, digits 3 and 4 identify the currency, digits 5 through 11 identify the account, digits 12 through 14 identify the card within the account, and the last digit is a "check digit" determined by the digits that precede it.[15]
  a. Your company, *BuyersXplosion, Inc.,* has an account at American Express. How many different cards issued to BuyersXplosion's account are possible based on U.S. dollars and Mexican pesos?
  b. How many different U.S. dollar–denominated American Express cards (issued to anybody's account, not just BuyersXplosion's) are possible with a wrong check digit?

43. ▼ *DNA Chains: Life in Nature* DNA (deoxyribonucleic acid) is the basic building block of reproduction in living things. A DNA chain is a sequence of chemicals called *bases*. There are four possible bases: thymine (T), cytosine (C), adenine (A), and guanine (G).
  a. How many three-element DNA chains are possible?
  b. How many n-element DNA chains are possible?
  c. A human DNA chain has 2.1×10^{10} elements. How many human DNA chains are possible?

44. ▼ *DNA Chains: Synthetic Life* (Refer to Exercise 43.) In 2014, scientists announced the creation of the first reproducing organisms using an expanded DNA alphabet with two new bases ("X" and "Y") in addition to the four found in nature.[16]
  a. How many four-element expanded DNA chains are possible?
  b. How many n-element expanded DNA chains are possible?

---

[14] Source: wikipedia.org.

[15] *Ibid.*

[16] Source: D. A. Malyshev et al., "A Semi-synthetic Organism with an Expanded Genetic Alphabet," *Nature,* 509, 2014, pp. 385–388 (http://dx.doi.org/10.1038/nature13314).

c. A "super-human" DNA chain would, like that of natural humans, have 2.1×10^{10} elements but using the expanded alphabet. How many super-human DNA chains are possible?

45. ▼ *HTML* Colors in HTML (the language in which many web pages are written) can be represented by six-digit hexadecimal codes: sequences of six integers ranging from 0 to 15 (represented as 0, . . . , 9, A, B, . . . , F).
  a. How many different colors can be represented?
  b. Some monitors can display only colors encoded with pairs of repeating digits (such as 44DD88). How many colors can these monitors display?
  c. Grayscale shades are represented by sequences *xyxyxy* consisting of a repeated pair of digits. How many grayscale shades are possible?
  d. The pure colors are pure red: *xy*0000; pure green: 00*xy*00; and pure blue: 0000*xy*. (*xy* = *FF* gives the brightest pure color, while *xy* = 00 gives the darkest: black.) How many pure colors are possible?

46. ▼ *Telephone Numbers* In the past, a local telephone number in the United States consisted of a sequence of two letters followed by five digits. Three letters were associated with each number from 2 to 9 (just as in the standard telephone layout shown in the figure) so that each telephone number corresponded to a sequence of seven digits. How many different sequences of seven digits were possible?

47. ▼ *Romeo and Juliet* Here is a list of the main characters in Shakespeare's *Romeo and Juliet*. The first seven characters are male, and the last four are female.

Escalus, *prince of Verona*
Paris, *kinsman to the prince*
Romeo, *of Montague Household*
Mercutio, *friend of Romeo*
Benvolio, *friend of Romeo*
Tybalt, *nephew to Lady Capulet*
Friar Lawrence, *a Franciscan*
Lady Montague, *of Montague Household*
Lady Capulet, *of Capulet Household*
Juliet, *of Capulet Household*
Juliet's Nurse

A total of 10 male and 8 female actors are available to play these roles. How many possible casts are there? (All roles are to be played by actors of the same gender as the character.)

48. ▼ *Swan Lake* The *Enormous State University*'s Accounting Society has decided to produce a version of the ballet *Swan Lake* in which all the female roles (including all of the swans) will be danced by men and vice versa. Here are the main characters:

Prince Siegfried
Prince Siegfried's Mother
Princess Odette, *the White Swan*
The Evil Duke Rotbart
Odile, *the Black Swan*
Cygnet #1, *young swan*
Cygnet #2, *young swan*
Cygnet #3, *young swan*

The ESU Accounting Society has on hand a total of 4 female dancers and 12 male dancers who are to be considered for the main roles. How many possible casts are there?

49. ▼ *License Plates* Many U.S. license plates display a sequence of three letters followed by three digits.
  a. How many such license plates are possible?
  b. To avoid confusion of letters with digits, some states do not issue standard plates with the last letter an I, O, or Q. How many license plates are still possible?
  c. Assuming that the letter combinations VET, MDZ, and DPZ are reserved for disabled veterans, medical practitioners, and disabled persons, respectively, how many license plates are possible for other vehicles, also taking the restriction in part (b) into account?

50. ▼ *License Plates*[17] License plates in Montana have a sequence consisting of (1) a digit from 1 to 9, (2) a letter, (3) a dot, (4) a letter, and (5) a four-digit number.
  a. How many different license plates are possible?
  b. If numbers that end with 0 are reserved for official state vehicles, how many license plates are possible for other vehicles?

51. ▼ *Mazes*
  a. How many four-letter sequences are possible that contain only the letters R and D, with D occurring only once?
  b. Use part (a) to calculate the number of possible routes from Start to Finish in the maze shown in the figure, where each move is either to the right or down.

  c. Comment on what would happen if we also allowed left and up moves.

---

[17] Source: The License Plates of the World website (http://servo.oit .gatech.edu/~mk5/).

52. ▼ *Mazes*
    a. How many six-letter sequences are possible that contain only the letters R and D, with D occurring only once?
    b. Use part (a) to calculate the number of possible routes from Start to Finish in the maze shown in the figure, where each move is either to the right or down.

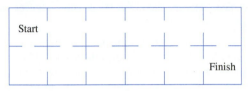

    c. Comment on what would happen if we also allowed left and up moves.

53. ▼ *Car Engines*[18] In a six-cylinder V6 engine, the even-numbered cylinders are on the left, and the odd-numbered cylinders are on the right. A good firing order is a sequence of the numbers 1 through 6 in which right and left sides alternate.
    a. How many possible good firing sequences are there?
    b. How many good firing sequences are there that start with a cylinder on the left?

54. ▼ *Car Engines* Repeat Exercise 53 for an eight-cylinder V8 engine.

55. ▼ *Minimalist Art* You are exhibiting your collection of minimalist paintings. Art critics have raved about your paintings, each of which consists of 10 vertical colored lines set against a white background. You have used the following rule to produce your paintings: Every second line, starting with the first, is to be either blue or gray, while the remaining five lines are to be either all light blue, all red, or all purple. Your collection is complete: Every possible combination that satisfies the rules occurs. How many paintings are you exhibiting?

56. ▼ *Combination Locks* Dripping wet after your shower, you have clean forgotten the combination of your lock. It is one of those "standard" combination locks, which uses a three-number combination with each number in the range 0 through 39. All you remember is that the second number is either 27 or 37, while the third number ends in a 5. In desperation you decide to go through all possible combinations

using the information you remember. Assuming that it takes about 10 seconds to try each combination, what is the longest possible time you may have to stand dripping in front of your locker?

57. ▼ *Product Design* Your company has patented an electronic digital padlock that a user can program with his or her own four-digit code. (Each digit can be 0 through 9.) The padlock is designed to open if either the correct code is keyed in or—and this is helpful for forgetful people—if exactly one of the digits is incorrect.
    a. How many incorrect codes will open a programmed padlock?
    b. How many codes will open a programmed padlock?

58. ▼ *Product Design* Your company has patented an electronic digital padlock that has a telephone-style keypad. Each digit from 2 through 9 corresponds to three letters of the alphabet (see the figure for Exercise 46). How many different four-letter sequences correspond to a single four-digit sequence using digits in the range 2 through 9?

59. ▼ *Calendars* The *World Almanac*[19] features a "perpetual calendar," a collection of 14 possible calendars. Why does this suffice to ensure that there is a calendar for every conceivable year?

60. ▼ *Calendars* How many possible calendars are there that have February 12 falling on a Sunday, Monday, or Tuesday?

61. ▼ *Programming in Visual Basic* (Some programming knowledge is assumed for this exercise.) How many iterations will be carried out in the following routine?

```
For i = 1 to 10
    For j = 2 to 20
        For k = 1 to 10
            Print i, j, k
        Next k
    Next j
Next i
```

62. ▼ *Programming in JavaScript* (Some programming knowledge is assumed for this exercise.) How many iterations will be carried out in the following routine?

```
for (i = 1; i <= 2; i++) {
    for (j = 1; j <= 2; j++) {
        for (k = 1; k <= 2; k++)
            sum += i+j+k;
    }
}
```

63. ◆ *Building Blocks* Use a decision algorithm to show that a rectangular solid with dimensions $m \times n \times r$ can be

[18] Adapted from an exercise in *Basic Techniques of Combinatorial Theory* by D. I. A. Cohen (New York: John Wiley, 1978).

[19] Source: *The World Almanac and Book of Facts* 1992 (New York: Pharos Books, 1992).

constructed with $m \cdot n \cdot r$ cubical $1 \times 1 \times 1$ blocks. (See the figure.)

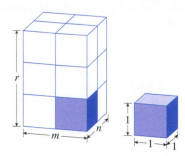

Rectangular Solid Made Up of $1 \times 1 \times 1$ Cubes

64. ◆ *Matrices* (Some knowledge of matrices is assumed for this exercise.) Use a decision algorithm to show that an $m \times n$ matrix must have $m \cdot n$ entries.

65. ◆ *Morse Code* In Morse code, each letter of the alphabet is encoded by a different sequence of dots and dashes. Different letters may have sequences of different lengths. How long should the longest sequence be to allow for every possible letter of the alphabet?

66. ◆ *Numbers* How many odd numbers between 10 and 99 have distinct digits?

Communication and Reasoning Exercises

67. Complete the following sentence: The multiplication principle is based on the cardinality of the _____ of two sets.

68. Complete the following sentence: The addition principle is based on the cardinality of the _____ of two disjoint sets.

69. You are packing for a short trip and want to take 2 of the 10 shirts you have hanging in your closet. Critique the following decision algorithm and calculation of how many different ways you can choose two shirts to pack: Step 1, choose one shirt, 10 choices. Step 2, choose another shirt, 9 choices. Hence, there are 90 possible choices of two shirts.

70. You are designing an advertising logo that consists of a tower of five squares. Three are yellow, one is blue, and one is green. Critique the following decision algorithm and calculation of the number of different five-square sequences: Step 1: Choose the first square, 5 choices. Step 2: Choose the second square, 4 choices. Step 3: Choose the third square: 3 choices. Step 4: Choose the fourth square, 2 choices. Step 5: Choose the last square: 1 choice. Hence, there are 120 possible five-square sequences.

71. ▼ Construct a decision algorithm that gives the correct number of five-square sequences in Exercise 70.

72. ▼ Find an interesting application that requires a decision algorithm with two steps in which each step has two alternatives.

6.4   **Permutations and Combinations**

Certain classes of counting problems come up frequently, and it is useful to develop formulas to deal with them without having to invoke decision algorithms.

Permutations

EXAMPLE 1   **Casting**

Ms. Birkitt, the English teacher at Brakpan Girls High School, wanted to stage a production of R. B. Sheridan's play *The School for Scandal*. The casting was going well until she was left with five unfilled characters and five seniors who were yet to be assigned roles. The characters were Lady Sneerwell, Lady Teazle, Mrs. Candour, Maria, and Snake; the unassigned seniors were April, May, June, Julia, and Augusta. How many possible assignments are there?

Solution To decide on a specific assignment, we use the following algorithm:

Step 1 Choose a senior to play Lady Sneerwell: 5 choices.

Step 2 Choose one of the remaining seniors to play Lady Teazle: 4 choices.

Step 3 Choose one of the now remaining seniors to play Mrs. Candour: 3 choices.

Step 4 Choose one of the now remaining seniors to play Maria: 2 choices.

Step 5 Choose the remaining senior to play Snake: 1 choice.

Thus, there are $5 \times 4 \times 3 \times 2 \times 1 = 120$ possible assignments of seniors to roles.

What the situation in Example 1 has in common with many others is that we start with a set—here, the set of seniors—and we want to know how many ways we can put the elements of that set in order in a list. In this example, an ordered list of the five seniors, for instance,

1. May
2. Augusta
3. June
4. Julia
5. April

corresponds to a particular casting:

Cast

Lady Sneerwell	May
Lady Teazle	Augusta
Mrs. Candour	June
Maria	Julia
Snake	April

We call an ordered list of items a **permutation** of those items.

If we have n items, how many permutations of those items are possible? We can use a decision algorithm similar to the one we used in Example 1 to select a permutation.

Step 1 Select the first item: n choices.

Step 2 Select the second item: $n - 1$ choices.

Step 3 Select the third item: $n - 2$ choices.

⋮

Step $n - 1$ Select the next-to-last item: 2 choices.

Step n Select the last item: 1 choice.

Thus, there are $n \times (n - 1) \times (n - 2) \times \cdots \times 2 \times 1$ possible permutations. We call this number n **factorial**, which we write as $n!$.

Permutations

A **permutation of n items** is an ordered list of those items. The number of possible permutations of n items is given by n **factorial**, which is

$$n! = n \times (n - 1) \times (n - 2) \times \cdots \times 2 \times 1$$

for n a positive integer, and

$$0! = 1.$$

Quick Examples

1. The number of permutations of five items is
   $5! = 5 \times 4 \times 3 \times 2 \times 1 = 120.$

Visualizing Permutations

Permutations of three colors in a flag:

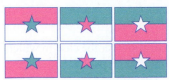

$3! = 3 \times 2 \times 1 = 6$ possible flags

Scott Barrow, Inc./SuperStock

2. The number of ways four CDs can be played in sequence is
$$4! = 4 \times 3 \times 2 \times 1 = 24.$$

3. The number of ways three cars can be matched with three drivers is
$$3! = 6.$$

Sometimes, instead of constructing an ordered list of *all* the items of a set, we might want to construct a list of only *some* of the items, as in the next example.

EXAMPLE 2 Corporations[20]

In the second quarter of 2015 the 10 largest publicly traded companies (by market capitalization) were, in order of ranking, **Apple**, **Microsoft**, **Exxon Mobil**, **Berkshire Hathaway**, **Google**, **Petro China**, **ICBC**, **Wells Fargo**, **Johnson & Johnson**, and **General Electric**. You would like to apply to six of these companies for a job, and you would like to list them in order of job preference. How many such ordered lists are possible?

Solution We want to count ordered lists, but we can't use the permutation formula because we don't want all 10 companies in the list, just 6 of them. So we fall back to a decision algorithm:

Step 1 Choose the first company: 10 choices.

Step 2 Choose the second company: 9 choices.

Step 3 Choose the third one: 8 choices.

Step 4 Choose the fourth one: 7 choices.

Step 5 Choose the fifth one: 6 choices.

Step 6 Choose the sixth one: 5 choices.

Thus, there are $10 \times 9 \times 8 \times 7 \times 6 \times 5 = 151{,}200$ possible lists of 6. We call this number the **number of permutations of 6 items chosen from 10**, or the **number of permutations of 10 items taken 6 at a time**.

➡ **Before we go on . . .** We wrote the answer as the product $10 \times 9 \times 8 \times 7 \times 6 \times 5$. But it is useful to notice that we can write this number in a more compact way:

$$10 \times 9 \times 8 \times 7 \times 6 \times 5 = \frac{10 \times 9 \times 8 \times 7 \times 6 \times 5 \times 4 \times 3 \times 2 \times 1}{4 \times 3 \times 2 \times 1}$$

$$= \frac{10!}{4!} = \frac{10!}{(10-6)!}$$ ∎

So we can generalize our definition of permutation to allow for the case in which we use only some of the items, not all. Check that, if $r = n$ below, this is the same definition we gave above.

---

[20] Based on the Financial Times Global 500 rankings. Source: www.wikipedia.org.

Permutations of *n* items taken *r* at a time

A **permutation of *n* items taken *r* at a time** is an ordered list of *r* items chosen from a set of *n* items. The number of permutations of *n* items taken *r* at a time is given by

$$P(n, r) = n \times (n - 1) \times (n - 2) \times \cdots \times (n - r + 1).$$

(There are *r* terms multiplied together.) We can also write

$$P(n, r) = \frac{n!}{(n - r)!}.$$

Quick Example

4. The number of permutations of six items taken two at a time is

$$P(6, 2) = 6 \times 5 = 30,$$

which we could also calculate as

$$P(6, 2) = \frac{6!}{(6 - 2)!} = \frac{6!}{4!} = \frac{720}{24} = 30.$$

Combinations

What if we don't care about the order of the items we're choosing? Consider the following example.

EXAMPLE 3   Corporations

Suppose we simply wanted to pick two of the 10 companies listed in Example 2 to apply to, without regard to order. How many possible choices do we have? What if we wanted to choose six to apply to, without regard to order?

Solution To answer the first question, our first guess might be $P(10, 2) = 10 \times 9 = 90$. However, that is the number of *ordered lists* of two companies. We said that we don't care which is first and which is second. For example, we consider the list

1. **Microsoft**
2. **Johnson & Johnson**

to be the same as

1. **Johnson & Johnson**
2. **Microsoft**.

Because every set of two companies occurs twice in the 90 lists, once in one order and again in the reverse order, we would count every set of two twice. Thus, there are $90/2 = 45$ possible choices of two companies.

Now, if we wish to pick six companies, again we might start with $P(10, 6) = 151,200$. But now, every set of six companies appears as many times as there are different orders in which they could be listed. Six things can be listed in $6! = 720$ different orders, so the number of ways of choosing six companies is $151,200/720 = 210$.

In Example 3 we were concerned with counting not the number of ordered lists, but the number of *unordered sets* of companies. For ordered lists we used the word *permutation*; for unordered sets we use the word **combination**.

Visualizing Permutation versus Combination

1.
2.
3.

Permutation   Combination

Permutations and Combinations

A **permutation** of n items taken r at a time is an *ordered list* of r items chosen from n. A **combination** of n items taken r at a time is an *unordered set* of r items chosen from n.

Note Because lists are usually understood to be ordered, when we refer to a list of items, we will always mean an *ordered* list. Similarly, because sets are understood to be unordered, when we refer to a set of items, we will always mean an *unordered* set. In short:

Lists are ordered. Sets are unordered. ■

Quick Example

5. There are six permutations of the three letters a, b, c taken two at a time:

   1. a, b;   2. b, a;   3. a, c;   4. c, a;   5. b, c;   6. c, b.
   There are six lists containing two of the letters a, b, c.

   There are three combinations of the three letters a, b, c taken two at a time:

   1. {a, b};   2. {a, c};   3. {b, c}.
   There are three sets containing two of the letters a, b, c.

How do we count the number of possible combinations of n items taken r at a time? We generalize the calculation done in Example 3. The number of permutations is $P(n, r)$, but each set of r items occurs $r!$ times because this is the number of ways in which those r items can be ordered. So the number of combinations is $P(n, r)/r!$.

Combinations of n items taken r at a time

The number of **combinations of n items taken r at a time** is given by

$$C(n, r) = \frac{P(n, r)}{r!} = \frac{n \times (n - 1) \times (n - 2) \times \cdots \times (n - r + 1)}{r!}.$$

We can also write

$$C(n, r) = \frac{n!}{r!(n - r)!}.$$

Quick Examples

6. The number of combinations of six items taken two at a time is

$$C(6, 2) = \frac{6 \times 5}{2 \times 1} = 15,$$

which we can also calculate as

$$C(6, 2) = \frac{6!}{2!(6 - 2)!} = \frac{6!}{2!4!} = \frac{720}{2 \times 24} = 15.$$

7. The number of sets of four marbles chosen from six is

$$C(6, 4) = \frac{6 \times 5 \times 4 \times 3}{4 \times 3 \times 2 \times 1} = 15.$$

Note There are other common notations for $C(n, r)$. Calculators often have $_nC_r$. In mathematics we often write $\binom{n}{r}$, which is also known as a **binomial coefficient**. Because $C(n, r)$ is the number of ways of choosing a set of r items from n, it is often read "n choose r." ∎

EXAMPLE 4   Calculating Combinations

Calculate: **a.** $C(11, 3)$ **b.** $C(11, 8)$

Solution The easiest way to calculate $C(n, r)$ by hand is to use the first formula above:

$$\begin{aligned} C(n, r) &= \frac{P(n, r)}{r!} \\ &= \frac{n \times (n - 1) \times (n - 2) \times \cdots \times (n - r + 1)}{r \times (r - 1) \times (r - 2) \times \cdots \times 1}. \end{aligned}$$

Both the numerator and the denominator have r factors, so we can begin with n/r and then continue multiplying by decreasing numbers on the top and the bottom until we hit 1 in the denominator. When calculating, it helps to cancel common factors from the numerator and denominator before doing the multiplication in either one.

a. $C(11, 3) = \dfrac{11 \times 10 \times 9}{3 \times 2 \times 1} = \dfrac{11 \times \overset{5}{\cancel{10}} \times \overset{3}{\cancel{9}}}{\cancel{3} \times \cancel{2} \times 1} = 165$

b. $C(11, 8) = \dfrac{11 \times 10 \times 9 \times 8 \times 7 \times 6 \times 5 \times 4}{8 \times 7 \times 6 \times 5 \times 4 \times 3 \times 2 \times 1}$

$\qquad\qquad = \dfrac{11 \times \overset{5}{\cancel{10}} \times \overset{3}{\cancel{9}}}{\cancel{3} \times \cancel{2} \times 1} = 165$

➡ **Before we go on . . .** It is no coincidence that the answers for parts (a) and (b) of Example 4, and also for Quick Examples 6 and 7, are the same. Consider what each represents. $C(11, 3)$ is the number of ways of choosing 3 items from 11—for example, electing 3 trustees from a slate of 11. Electing those 3 is the same as choosing the 8 who *do not* get elected. Thus, there are exactly as many ways to choose 3 items from 11 as there are ways to choose 8 items from 11. So $C(11, 3) = C(11, 8)$. In general,

$$C(n, r) = C(n, n - r).$$

We can also see this equality by using the formula

$$C(n, r) = \frac{n!}{r!(n-r)!}.$$

If we substitute $n - r$ for r, we get exactly the same formula.

Use the equality $C(n, r) = C(n, n - r)$ to make your calculations easier. Choose the one with the smaller denominator to begin with. ■

EXAMPLE 5   Calculating Combinations

Calculate: **a.** $C(11, 11)$ **b.** $C(11, 0)$

Solution

a. $C(11, 11) = \dfrac{11 \times 10 \times 9 \times 8 \times 7 \times 6 \times 5 \times 4 \times 3 \times 2 \times 1}{11 \times 10 \times 9 \times 8 \times 7 \times 6 \times 5 \times 4 \times 3 \times 2 \times 1} = 1$

b. What do we do with that 0? What does it mean to multiply 0 numbers together? We know from above that $C(11, 0) = C(11, 11)$, so we must have $C(11, 0) = 1$. How does this fit with the formulas? Go back to the calculation of $C(11, 11)$:

$$1 = C(11, 11) = \frac{11!}{11!(11-11)!} = \frac{11!}{11!0!}.$$

This equality is true only if we agree that $0! = 1$, which we do. Then

$$C(11, 0) = \frac{11!}{0!11!} = 1.$$

➡ **Before we go on . . .** There is nothing special about 11 in the calculation in Example 5. In general,

$$C(n, n) = C(n, 0) = 1.$$

After all, there is only one way to choose n items out of n: Choose them all. Similarly, there is only one way to choose 0 items out of n: Choose none of them. ■

Now for a few more complicated examples that illustrate the applications of the counting techniques we've discussed.

EXAMPLE 6   Lotto

In the betting game Lotto, used in many state lotteries, you choose six different numbers in the range 1–55 (the upper number varies). The order in which you choose them is irrelevant. If your six numbers match the six numbers chosen in the official drawing, you win the top prize. If Lotto tickets cost $1 for two sets of numbers and you decide to buy tickets that cover every possible combination, thereby guaranteeing that you will win the top prize, how much money will you have to spend?

Solution We first need to know how many sets of numbers are possible. Because order does not matter, we are asking for the number of combinations of 55 numbers taken 6 at a time. This is

$$C(55, 6) = \frac{55 \times 54 \times 53 \times 52 \times 51 \times 50}{6 \times 5 \times 4 \times 3 \times 2 \times 1} = 28{,}989{,}675.$$

Because $1 buys you two of these, you need to spend $28,989,675/2 = \$14,494,838$ (rounding up to the nearest dollar) to be assured of a win!

➡ **Before we go on ...** The calculation in Example 6 shows that you should not bother buying all these tickets if the winning prize is less than about $14.5 million. Even if the prize is higher, you need to account for the fact that many people will play and the prize may end up split among several winners, not to mention the impracticality of filling out millions of betting slips. ◼

EXAMPLE 7   **Marbles**

A bag contains three red marbles, three blue ones, three green ones, and two yellow ones (all distinguishable from one another).

a. How many sets of four marbles are possible?

b. How many sets of four are there such that each one is a different color?

c. How many sets of four are there in which at least two are red?

d. How many sets of four are there in which none are red but at least one is green?

Solution

a. We simply need to find the number of ways of choosing 4 marbles out of 11, which is

$$C(11, 4) = 330 \text{ possible sets of 4 marbles.}$$

b. We use a decision algorithm for choosing such a set of marbles:

Step 1 Choose one red one from the three red ones: $C(3, 1) = 3$ choices.
Step 2 Choose one blue one from the three blue ones: $C(3, 1) = 3$ choices.
Step 3 Choose one green one from the three green ones: $C(3, 1) = 3$ choices.
Step 4 Choose one yellow one from the two yellow ones: $C(2, 1) = 2$ choices.

This gives a total of $3 \times 3 \times 3 \times 2 = 54$ possible sets.

c. We need another decision algorithm. To say that at least two marbles must be red means that either two are red or three are red (with a total of three red ones). In other words, we have two *alternatives*.

Alternative 1: Exactly two red marbles
    Step 1 Choose two red ones: $C(3, 2) = 3$ choices.
    Step 2 Choose two nonred ones. There are eight of these, so we get $C(8, 2) = 28$ possible choices.
Thus, the total number of choices for this alternative is $3 \times 28 = 84$.

Alternative 2: Exactly three red marbles.
    Step 1 Choose the three red ones: $C(3, 3) = 1$ choice.
    Step 2 Choose one nonred one: $C(8, 1) = 8$ choices.
Thus, the total number of choices for this alternative is $1 \times 8 = 8$.

By the addition principle, we get a total of $84 + 8 = 92$ sets.

d. The phrase "at least one green" tells us that we again have some alternatives:

Alternative 1: One green marble
    Step 1 Choose one green marble from the three: $C(3, 1) = 3$ choices.
    Step 2 Choose three nongreen, nonred marbles: $C(5, 3) = 10$ choices.
Thus, the total number of choices for alternative 1 is $3 \times 10 = 30$.

Alternative 2: Two green marbles

Step 1 Choose two green marbles from the three: $C(3, 2) = 3$ choices.
Step 2 Choose two nongreen, nonred marbles: $C(5, 2) = 10$ choices.
Thus, the total number of choices for alternative 2 is $3 \times 10 = 30$.

Alternative 3: Three green marbles

Step 1 Choose three green marbles from the three: $C(3, 3) = 1$ choice.
Step 2 Choose one nongreen, nonred marble: $C(5, 1) = 5$ choices.
Thus, the total number of choices for alternative 3 is $1 \times 5 = 5$.

The addition principle now tells us that the number of sets of four marbles with none red but at least one green is $30 + 30 + 5 = 65$.

➡ **Before we go on . . .** Here is an easier way to answer Example 7(d). First, the total number of sets having *no* red marbles is $C(8, 4) = 70$. Next, of those, the number containing no green marbles is $C(5, 4) = 5$. This leaves $70 - 5 = 65$ sets that contain no red marbles but have at least one green marble. (We have really used here the formula for the cardinality of the complement of a set.) ■

The last example concerns poker hands. For those unfamiliar with playing cards, here is a short description. A standard deck consists of 52 playing cards. Each card is in one of 13 denominations: ace (A), 2, 3, 4, 5, 6, 7, 8, 9, 10, jack (J), queen (Q), and king (K), and in one of four suits: hearts (♥), diamonds (♦), clubs (♣), and spades (♠). Thus, for instance, the jack of spades, J♠, refers to the denomination of jack in the suit of spades. The entire deck of cards is thus

A♥	2♥	3♥	4♥	5♥	6♥	7♥	8♥	9♥	10♥	J♥	Q♥	K♥
A♦	2♦	3♦	4♦	5♦	6♦	7♦	8♦	9♦	10♦	J♦	Q♦	K♦
A♣	2♣	3♣	4♣	5♣	6♣	7♣	8♣	9♣	10♣	J♣	Q♣	K♣
A♠	2♠	3♠	4♠	5♠	6♠	7♠	8♠	9♠	10♠	J♠	Q♠	K♠

EXAMPLE 8 **Poker Hands**

In the card game poker, a hand consists of a set of 5 cards from a standard deck of 52. A **full house** is a hand consisting of three cards of one denomination ("three of a kind"—e.g., three 10s) and two of another ("two of a kind"—e.g., two queens). Here is an example of a full house: 10♣, 10♦, 10♠, Q♥, Q♣.

a. How many different poker hands are there?

b. How many different full houses are there that contain three 10s and two queens?

c. How many different full houses are there altogether?

Solution

a. Because the order of the cards doesn't matter, we simply need to know the number of ways of choosing a set of 5 cards out of 52, which is

$$C(52, 5) = 2,598,960 \text{ hands.}$$

b. Here is a decision algorithm for choosing a full house with three 10s and two queens:

Step 1 Choose three 10s. Because there are four 10s to choose from, we have $C(4, 3) = 4$ choices.
Step 2 Choose two queens: $C(4, 2) = 6$ choices.

Thus, there are $4 \times 6 = 24$ possible full houses with three 10s and two queens.

c. Here is a decision algorithm for choosing a full house:

Step 1 Choose a denomination for the three of a kind; 13 choices.

Step 2 Choose three cards of that denomination. Because there are four cards of each denomination (one for each suit), we get $C(4, 3) = 4$ choices.

Step 3 Choose a different denomination for the two of a kind. There are only 12 denominations left, so we have 12 choices.

Step 4 Choose two of that denomination: $C(4, 2) = 6$ choices.

Thus, by the multiplication principle there are a total of $13 \times 4 \times 12 \times 6 = 3{,}744$ possible full houses.

FAQs

Recognizing When to Use Permutations or Combinations

Q : *How can I tell whether a given application calls for permutations or combinations?*

A : Decide whether the application calls for ordered lists (as in situations in which order is implied) or for unordered sets (as in situations in which order is not relevant). Ordered lists are permutations, whereas unordered sets are combinations.

6.4 EXERCISES

▼ more advanced   ◆ challenging
T indicates exercises that should be solved using technology

In Exercises 1–16, evaluate the number. [**HINT:** See Quick Examples 4–7.]

1. 6!

2. 7!

3. 8!/6!

4. 10!/8!

5. $P(6, 4)$

6. $P(8, 3)$

7. $P(6, 4)/4!$

8. $P(8, 3)/3!$

9. $C(3, 2)$

10. $C(4, 3)$

11. $C(10, 8)$

12. $C(11, 9)$

13. $C(20, 1)$

14. $C(30, 1)$

15. $C(100, 98)$

16. $C(100, 97)$

17. How many ordered lists are there of four items chosen from six?

18. How many ordered sequences are possible that contain three objects chosen from seven?

19. How many unordered sets are possible that contain three objects chosen from seven?

20. How many unordered sets are there of four items chosen from six?

21. How many five-letter sequences are possible that use the letters b, o, g, e, y once each?

22. How many six-letter sequences are possible that use the letters q, u, a, k, e, s once each?

23. How many three-letter sequences are possible that use the letters q, u, a, k, e, s at most once each?

24. How many three-letter sequences are possible that use the letters b, o, g, e, y at most once each?

25. How many three-letter (unordered) sets are possible that use the letters q, u, a, k, e, s at most once each?

26. How many three-letter (unordered) sets are possible that use the letters b, o, g, e, y at most once each?

27. ▼ How many six-letter sequences are possible that use the letters a, u, a, a, u, k? [**HINT:** Use the decision algorithm discussed in Example 3 of Section 6.3.]

28. ▼ How many six-letter sequences are possible that use the letters f, f, a, a, f, f? [**HINT:** See the hint for Exercise 27.]

Marbles For Exercises 29–42, a bag contains three red marbles, two green ones, one lavender one, two yellows, and two orange marbles. [**HINT:** See Example 7.]

29. How many possible sets of four marbles are there?

30. How many possible sets of three marbles are there?

31. How many sets of four marbles include all the red ones?

32. How many sets of three marbles include all the yellow ones?

33. How many sets of four marbles include none of the red ones?

34. How many sets of three marbles include none of the yellow ones?

35. How many sets of four marbles include one of each color other than lavender?

36. How many sets of five marbles include one of each color?

37. How many sets of five marbles include at least two red ones?

38. How many sets of five marbles include at least one yellow one?

39. How many sets of five marbles include at most one of the yellow ones?

40. How many sets of five marbles include at most one of the red ones?

41. ▼ How many sets of five marbles include either the lavender one or exactly one yellow one but not both colors?

42. ▼ How many sets of five marbles include at least one yellow one but no green ones?

Dice If a die is rolled 30 times, there are 6^{30} different sequences possible. Exercises 43–46 ask how many of these sequences satisfy certain conditions. [HINT: Use the decision algorithm discussed in Example 3 of Section 6.3.]

43. ▼ What fraction of these sequences have exactly five 1s?

44. ▼ What fraction of these sequences have exactly five 1s and five 2s?

45. ▼ What fraction of these sequences have exactly 15 even numbers?

46. ▼ What fraction of these sequences have exactly 10 numbers less than or equal to 2?

In Exercises 47–52, calculate how many different sequences can be formed that use the letters of each given word. Leave your answer as a product of terms of the form $C(n, r)$. [HINT: Decide where, for example, all the s's will go rather than what will go in each position.]

47. ▼ mississippi

48. ▼ mesopotamia

49. ▼ megalomania

50. ▼ schizophrenia

51. ▼ casablanca

52. ▼ desmorelda

Applications

53. *Itineraries* Your international diplomacy trip requires stops in Thailand, Singapore, Hong Kong, and Bali. How many possible itineraries are there? [HINT: See Examples 1 and 2.]

54. *Itineraries* Refer back to Exercise 53. How many possible itineraries are there in which the last stop is Thailand?

Poker Hands A poker hand consists of 5 cards from a standard deck of 52. (See the chart preceding Example 8.) In Exercises 55–60, find the number of different poker hands of the specified type. [HINT: See Example 8.]

55. Two pairs (two of one denomination, two of another denomination, and one of a third)

56. Three of a kind (three of one denomination, one of another denomination, and one of a third)

57. Two of a kind (two of one denomination and three of different denominations)

58. Four of a kind (all four of one denomination and one of another)

59. ▼ Straight (five cards of consecutive denominations: A, 2, 3, 4, 5 up through 10, J, Q, K, A, not all of the same suit) (Note that the ace counts either as a 1 or as the denomination above king.)

60. ▼ Flush (five cards all of the same suit but not consecutive denominations)

Dogs of the Dow The "Dogs of the Dow" are the stocks listed on the Dow with the highest dividend yield. Exercises 61 and 62 are based on the following table, which shows the top 10 stocks of the "Dogs of the Dow" list for 2015, based on their performance the preceding year.[21]

Symbol	Company	Price ($)	Yield
T	AT&T	33.59	5.48%
VZ	Verizon	46.78	4.70%
CVX	Chevron	112.18	3.82%
MCD	McDonald's	93.70	3.63%
PFE	Pfizer	31.15	3.60%
GE	General Electric	25.27	3.48%
MRK	Merck	56.79	3.17%
CAT	Caterpillar	91.53	3.06%
XOM	ExxonMobil	92.45	2.99%
KO	Coca-Cola	42.22	2.89%

61. You decide to make a small portfolio consisting of a collection of 5 of the top 10 Dogs of the Dow.
   a. How many portfolios are possible?
   b. How many of these portfolios contain VZ and MCD but neither KO nor XOM?
   c. How many of these portfolios contain at least four stocks with yields above 3.5%?

62. You decide to make a small portfolio consisting of a collection of 6 of the top 10 Dogs of the Dow.
   a. How many portfolios are possible?
   b. How many of these portfolios contain MRK but not PFE?
   c. How many of these portfolios contain at most one stock priced above $60?

---

[21] Source: www.dogsofthedow.com.

Day Trading *Day traders typically buy and sell stocks (or other investment instruments) during the trading day and sell all investments by the end of the day. Exercises 63 and 64 are based on the following table, which shows the closing prices on September 22, 2015, of 12 stocks selected by your broker, Prudence Swift, as well as the change that day.*[22]

Tech Stocks	Close	Change
AAPL (Apple)	$113.40	−1.81
ADBE (Adobe Systems)	$84.66	1.34
EBAY (eBay)	$25.61	−0.31
MSFT (Microsoft)	$3.90	−0.21
S (Sprint)	$4.40	0.02
WIFI (Boingo Wireless)	$8.51	0.56
Non-Tech Stocks		
ANF (Abercrombie & Fitch)	$21.81	−0.02
B (Boeing)	$133.99	−2.03
F (Ford Motor Co.)	$13.91	−0.40
GE (General Electric)	$25.10	0.01
GIS (General Mills)	$57.12	0.33
JNJ (Johnson & Johnson)	$93.26	0.13

63. On the morning of September 22, 2015, Swift advised you to purchase a collection of three tech stocks and two non-tech stocks, all chosen at random from those listed in the table. You were to sell all the stocks at the end of the trading day.
    a. How many possible collections are possible?
    b. You tend to have bad luck with stocks—they usually start going down the moment you buy them. How many of the collections in part (a) consist entirely of stocks that declined in value by the end of the day?
    c. Using the answers to parts (a) and (b), what would you say your chances were of choosing a collection consisting entirely of stocks that declined in value by the end of the day?

64. On the morning of September 22, 2015, Swift advised your friend to purchase a collection of three stocks chosen at random from those listed in the table. Your friend was to sell all the stocks at the end of the trading day.
    a. How many possible collections are possible?
    b. How many of the collections in part (a) included exactly two tech stocks that increased in value by the end of the day?
    c. Using the answers to parts (a) and (b), what would you say the chances were that your friend chose a collection that included exactly two tech stocks that increased in value by the end of the day?

Elimination Tournaments *In an elimination tournament the teams are arranged in opponent pairs for the first round, and the winner of each round goes on to the next round until the champion emerges. The following diagram illustrates a 16-team tournament bracket, in which the 16 participating teams are arranged on the left under Round 1 and the winners of each round are added as the tournament progresses. The top team in each game is considered the "home" team, so the top-to-bottom order matters.*

To seed a tournament means to select which teams to play each other in the first round according to their preliminary ranking. For instance, in professional tennis and NCAA basketball the seeding is set up in the following order based on the preliminary rankings: 1 versus 16, 8 versus 9, 5 versus 12, 4 versus 13, 6 versus 11, 3 versus 14, 7 versus 10, and 2 versus 15.[23] *Exercises 65–68 are based on various types of elimination tournaments. (Leave each answer as a formula.)*

65. a. How many different seedings of a 16-team tournament are possible? (Express the answer as a formula.)
    b. In how many seedings will the top-ranked team play the bottom-ranked team, the second-ranked team play the second-lowest-ranked team, and so on?

66. a. How many different seedings of an 8–team tournament are possible? (Express the answer as a formula.)
    b. In how many seedings will each team play a team with adjacent ranking?

67. ▼ In 2014, after the NCAA basketball 64-team tournament had already been seeded, **Quicken Loans**, backed by investor Warren Buffett, offered a billion dollar prize for picking

---

all the winners.[24] An *upset* occurs when a team beats a higher-ranked team. How many configurations (filling in of all the winners in all the rounds) were possible in which there were exactly 15 upsets in the first four rounds?

68. ▼ Refer back to Exercise 67. In the 2013 NCAA playoffs there were 10 upsets in the first round, 4 in the second round, and 3 in each of the third and fourth rounds.[25] How many configurations (filling in of all the winners in all the rounds) of this type were possible? (In the last two rounds, rankings are not taken into consideration.)

Popular Movies in 2011 *Exercises 69 and 70 are based on the following list of top DVD rentals (based on revenue) for the weekend ending November 6, 2011:*[26]

Title	Rank
Captain America: The First Avenger	1
Bad Teacher	2
Fast Five	3
Cars 2	4
Trespass	5
Bridesmaids	6
Zookeeper	7
Transformers: Dark of the Moon	8
Scream	9
Thor	10

69. ▼ Rather than studying for math, you and your buddies decide to get together for a marathon movie-watching, popcorn-guzzling event on Saturday night. You decide to watch four movies selected at random from the above list.
  a. How many sets of four movies are possible?
  b. Your best friends, the Lara twins, refuse to see *Bridesmaids* on the grounds that it is "for girlie men" and also insist that at least one of *Captain America* or *Thor* be among the movies selected. How many of the possible groups of four will satisfy the twins?
  c. Comparing the answers in parts (a) and (b), would you say that the Lara twins are more likely than not to be satisfied with your random selection?

70. ▼ Rather than studying for astrophysics, you and your friends decide to get together for a marathon movie-watching, gummy-bear-munching event on Saturday night. You decide to watch three movies selected at random from the above list.
  a. How many sets of three movies are possible?
  b. Your best friends, the Pelogrande twins, refuse to see either *Cars 2* or *Zookeeper* on the grounds that they are "for idiots" and also insist that no more than one of *Bad Teacher* and *Fast Five* should be among the movies selected. How many of the possible groups of three will satisfy the twins?
  c. Comparing the answers in parts (a) and (b), would you say that the Pelogrande twins are more likely than not to be satisfied with your random selection?

71. ▼ *Traveling Salesperson* Suppose you are a salesperson who must visit the following 23 cities: Dallas, Tampa, Orlando, Fairbanks, Seattle, Detroit, Chicago, Houston, Arlington, Grand Rapids, Urbana, San Diego, Aspen, Little Rock, Tuscaloosa, Honolulu, New York, Ithaca, Charlottesville, Lynchville, Raleigh, Anchorage, and Los Angeles. Leave all your answers in factorial form.
  a. How many possible itineraries are there that visit each city exactly once?
  b. Repeat part (a) in the event that the first five stops have already been determined.
  c. Repeat part (a) in the event that your itinerary must include the sequence Anchorage, Fairbanks, Seattle, Chicago, and Detroit, in that order.

72. ▼ *Traveling Salesperson* Refer back to Exercise 71 (and leave all your answers in factorial form).
  a. How many possible itineraries are there that start and end at Detroit and visit every other city exactly once?
  b. How many possible itineraries are there that start and end at Detroit and visit Chicago twice and every other city once?
  c. Repeat part (a) in the event that your itinerary must include the sequence Anchorage, Fairbanks, Seattle, Chicago, and New York, in that order.

73. ▼ (*From the GMAT*) Ben and Ann are among seven contestants from which four semifinalists are to be selected. Of the different possible selections, how many contain neither Ben nor Ann?
  (**A**) 5   (**B**) 6   (**C**) 7   (**D**) 14   (**E**) 21

74. ▼ (*Based on a question from the GMAT*) Ben and Ann are among seven contestants from which four semifinalists are to be selected. Of the different possible selections, how many contain Ben but not Ann?
  (**A**) 5   (**B**) 8   (**C**) 9   (**D**) 10   (**E**) 20

---

[24] No one won the prize, and the offer was scrapped the following year as a result of a series of lawsuits and countersuits by Yahoo, SCA Promotions (a sweepstakes company), and Berkshire-Hathaway. Sources: http://abcnews.go.com/Sports/warren-buffet-backs-billion-dollar-march-madness-challenge/story?id=21615743, http://money.cnn.com/2015/03/12/news/buffett-ncaa-bracket-bet/index.html.

[25] Source: *Washington Post,* March 16, 2014 (www.washingtonpost.com).

[26] Source: Home Media Magazine (www.imdb.com/Charts/videolast).

75. ◆ (*From the GMAT exam*) If 10 persons meet at a reunion and each person shakes hands exactly once with each of the others, what is the total number of handshakes?
(**A**) $10 \cdot 9 \cdot 8 \cdot 7 \cdot 6 \cdot 5 \cdot 4 \cdot 3 \cdot 2 \cdot 1$ (**B**) $10 \cdot 10$
(**C**) $10 \cdot 9$ (**D**) 45 (**E**) 36

76. ◆ (*Based on a question from the GMAT exam*) If 12 businesspeople have a meeting and each pair exchanges business cards, how many business cards, total, get exchanged?
(**A**) $12 \cdot 11 \cdot 10 \cdot 9 \cdot 8 \cdot 7 \cdot 6 \cdot 5 \cdot 4 \cdot 3 \cdot 2 \cdot 1$ (**B**) $12 \cdot 12$
(**C**) $12 \cdot 11$ (**D**) 66 (**E**) 72

77. ◆ *Product Design* The Honest Lock Company plans to introduce what it refers to as the "true combination lock." The lock will open if the correct set of three numbers from 0 through 39 is entered in any order.
a. How many different combinations of three different numbers are possible?
b. If it is allowed that a number appear twice (but not three times), how many more possibilities are created?
c. If it is allowed that any or all of the numbers may be the same, what is the total number of combinations possible?

78. ◆ *Product Design* Repeat Exercise 77 for a lock based on selecting from the numbers 0 through 19.

79. ◆ *Theory of Linear Programming* (Some familiarity with linear programming is assumed for this exercise.) Suppose you have a linear programming problem with two unknowns and 20 constraints. You decide that graphing the feasible region would take a lot of work, but then you recall that corner points are obtained by solving a system of two equations in two unknowns obtained from two of the constraints. Thus, you decide that it might pay instead to locate all the possible corner points by solving all possible combinations of two equations and then checking whether each solution is a feasible point.
a. How many systems of two equations in two unknowns will you be required to solve?
b. Generalize this to n constraints.

80. ◆ *More Theory of Linear Programming* (Some familiarity with linear programming is assumed for this exercise.) Before the advent of the simplex method for solving linear programming problems, the following method was used: Suppose you have a linear programming problem with three unknowns and 20 constraints. You locate corner points as follows: Selecting three of the constraints, you turn them into equations (by replacing the inequalities with equalities), solve the resulting system of three equations in three unknowns, and then check to see whether the solution is feasible.
a. How many systems of three equations in three unknowns will you be required to solve?
b. Generalize this to n constraints.

Communication and Reasoning Exercises

81. If you were hard pressed to study for an exam on counting and had only enough time to study one topic, would you choose the formula for the number of permutations or the multiplication principle? Give reasons for your choice.

82. The formula for $C(n, r)$ is written as a ratio of two whole numbers. Can $C(n, r)$ ever be a fraction and not a whole number? Explain (without actually discussing the formula itself).

83. Which of the following represent permutations?
(**A**) An arrangement of books on a shelf
(**B**) A group of 10 people in a bus
(**C**) A committee of 5 senators chosen from 100
(**D**) A presidential cabinet of 5 portfolios chosen from 20

84. Which of the following represent combinations?
(**A**) A portfolio of five stocks chosen from the S&P Top Ten
(**B**) A group of 5 tenors for a choir chosen from 12 singers
(**C**) A new company CEO and a new CFO chosen from five candidates
(**D**) The *New York Times* Top Ten Bestseller list

85. When you click on "Get Driving Directions" on Mapquest .com, do you get a permutation or a combination of driving instructions? Explain, and give a simple example to illustrate why the answer is important.

86. If you want to know how many possible lists, arranged in alphabetical order, there are of five students selected from your class, you would use the formula for permutations— right? (Explain your answer.)

87. You are tutoring your friend for a test on sets and counting, and she asks the question "How do I know what formula to use for a given problem?" What is a good way to respond?

88. Complete the following: If a counting procedure has five alternatives, each of which has four steps of two choices each, then there are ___ outcomes. On the other hand, if there are five steps, each of which has four alternatives of two choices each, then there are ___ outcomes.

89. ▼ A textbook has the following exercise: "Three students from a class of 50 are selected to take part in a play. How many casts are possible?" Comment on this exercise.

90. ▼ Explain why the coefficient of a $a^2 b^4$ in $(a + b)^6$ is $C(6, 2)$. (This is a consequence of the **binomial theorem**.) [**HINT:** In the product $(a + b)(a + b) \cdots (a + b)$ (six times), in how many different ways can you pick two a's and four b's to multiply together?]

CHAPTER 6 REVIEW

KEY CONCEPTS

www.WanerMath.com
Go to the Website to find a comprehensive and interactive Web-based summary of Chapter 6.

6.1 Sets and Set Operations

Sets, elements, subsets, proper subsets, empty set, finite and infinite sets [p. 428]

Visualizing sets and relations between sets using Venn diagrams [p. 430]

Union:
$A \cup B = \{x \,|\, x \in A \text{ or } x \in B\}$ [p. 432]

Intersection: $A \cap B = \{x \,|\, x \in A \text{ and } x \in B\}$ [p. 432]

Universal sets, complements [p. 433]

Disjoint sets: $A \cap B = \varnothing$ [p. 434]

Cartesian product: $A \times B = \{(a, b) \,|\, a \in A \text{ and } b \in B\}$ [p. 434]

6.2 Cardinality

Cardinality: $n(A) =$ number of elements in A. [p. 439]

If A and B are finite sets, then
$n(A \cup B) = n(A) + n(B) - n(A \cap B)$. [p. 439]

If A and B are disjoint finite sets, then
$n(A \cup B) = n(A) + n(B)$. In this case we say that $A \cup B$ is a **disjoint union**. [p. 440]

If S is a finite universal set and A is a subset of S, then
$n(A') = n(S) - n(A)$ and
$n(A) = n(S) - n(A')$. [p. 441]

If A and B are finite sets, then
$n(A \times B) = n(A)n(B)$. [p. 445]

6.3 Decision Algorithms: The Addition and Multiplication Principles

Addition principle [p. 449]

Multiplication principle [p. 450]

Decision algorithm: a procedure for making a sequence of decisions to choose an element of a set [p. 452]

6.4 Permutations and Combinations

n factorial:
$n! = n \times (n - 1) \times (n - 2) \times \cdots \times 2 \times 1$ [p. 461]

Permutation of n items taken r at a time:

$$P(n, r) = \frac{n!}{(n - r)!} \quad \text{[p. 463]}$$

Combination of n items taken r at a time:

$$C(n, r) = \frac{P(n, r)}{r!} = \frac{n!}{r!(n - r)!}$$

[p. 464]

Using the equality
$C(n, r) = C(n, n - r)$ to simplify calculations of combinations [p. 465]

REVIEW EXERCISES

In Exercises 1–5, list the elements of the given set.

1. The set N of all negative integers greater than or equal to -3

2. The set of all outcomes of tossing a coin five times

3. The set of all outcomes of tossing two distinguishable dice such that the numbers are different

4. The sets $(A \cap B) \cup C$ and $A \cap (B \cup C)$, where $A = \{1, 2, 3, 4, 5\}$, $B = \{3, 4, 5\}$, and $C = \{1, 2, 5, 6, 7\}$

5. The sets $A \cup B'$ and $A \times B'$, where $A = \{a, b\}$, $B = \{b, c\}$, and $S = \{a, b, c, d\}$

In Exercises 6–10, write the indicated set in terms of the given sets.

6. S: the set of all customers; A: the set of all customers who owe money; B: the set of all customers who owe at least $1,000. The set of all customers who owe money but owe less than $1,000

7. A: the set of outcomes when a day in August is selected; B: the set of outcomes when a time of day is selected. The set of outcomes when a day in August and a time of that day are selected

8. S: the set of outcomes when two dice are rolled; E: those outcomes in which at most one die shows an even number, F: those outcomes in which the sum of the numbers is 7. The set of outcomes in which both dice show an even number or sum to seven

9. S: the set of all integers; E: the set of all even integers; Q: the set of all integers that are perfect squares ($Q = \{0, 1, 4, 9, 16, 25, \ldots\}$). The set of all integers that are not odd perfect squares

10. S: the set of all integers; N: the set of all negative integers; E: the set of all even integers; T: the set of all integers that are multiples of 3 ($T = \{0, 3, -3, 6, -6, 9, -9, \ldots\}$). The set of all even integers that are neither negative nor multiples of three

In Exercises 11–14, give a formula for the cardinality rule or rules needed to answer the question, and then give the solution.

11. You have read 150 of the 400 novels in your home, but your sister Roslyn has read 200, of which only 50 are novels you have read as well. How many have neither of you read?

12. There are 32 students in categories A and B combined; 24 are in A, and 24 are in B. How many are in both A and B?

13. You roll two dice, one red and one green. Losing combinations are doubles (both dice show the same number) and outcomes in which the green die shows an odd number and the red die shows an even number. The other combinations are winning ones. How many winning combinations are there?

14. The **Apple** iMac used to come in three models, each with five colors to choose from. How many combinations were possible?

Recall that a poker hand consists of 5 cards from a standard deck of 52. In Exercises 15–18, find the number of different poker hands of the specified type. Leave your answer in terms of combinations.

15. Two of a kind with no aces

16. A full house with either two kings and three queens or two queens and three kings

17. Straight flush (five cards of the same suit with consecutive denominations: A, 2, 3, 4, 5 up through 10, J, Q, K, A)

18. Three of a kind with no aces

In Exercises 19–24, consider a bag containing four red marbles, two green ones, one transparent one, three yellow ones, and two orange ones.

19. How many possible sets of five marbles are there in which all of them are red or green?

20. How many possible sets of five marbles are there in which none of them are red or green?

21. How many sets of five marbles include all the red ones?

22. How many sets of five marbles do not include all the red ones?

23. How many sets of five marbles include at least two yellow ones?

24. How many sets of five marbles include at most one of the red ones but no yellow ones?

Applications: OHaganBooks.com
[Try the game at www.OHaganBooks.com]

Inventories OHaganBooks.com currently operates three warehouses: one in Washington, one in California, and the new one in Texas. Exercises 25–30 are based on the following table, which shows the book inventories at each warehouse:

	Sci Fi	Horror	Romance	Other	Total
Washington	10,000	12,000	12,000	30,000	64,000
California	8,000	12,000	6,000	16,000	42,000
Texas	15,000	15,000	20,000	44,000	94,000
Total	33,000	39,000	38,000	90,000	200,000

Take the first letter of each category to represent the corresponding set of books; for instance, S is the set of sci fi books in stock, W is the set of books in the Washington warehouse, and so on. In each exercise, describe the given set in words, and compute its cardinality.

25. $S \cup T$

26. $H \cap C$

27. $C \cup S'$

28. $(R \cap T) \cup H$

29. $R \cap (T \cup H)$

30. $(S \cap W) \cup (H \cap C')$

Customers OHaganBooks.com has two main competitors: JungleBooks.com and FarmerBooks.com. At the beginning of August, OHaganBooks.com had 3,500 customers. Of these, a total of 2,000 customers were shared with JungleBooks.com, and 1,500 were shared with FarmerBooks.com. Furthermore, 1,000 customers were shared with both. JungleBooks.com has a total of 3,600 customers, FarmerBooks.com has 3,400, and they share 1,100 customers between them. Use these data for Exercises 31–36.

31. How many of all these customers are exclusive OHaganBooks.com customers?

32. How many customers of the other two companies are not customers of OHaganBooks.com?

33. Which of the three companies has the largest number of exclusive customers?

34. Which of the three companies has the smallest number of exclusive customers?

35. OHaganBooks.com is interested in merging with one of its two competitors. Which merger would give it the largest combined customer base, and how large would that be?

36. Referring to Exercise 35, which merger would give OHaganBooks.com the largest *exclusive* customer base, and how large would that be?

Online IDs As the customer base at OHaganBooks.com grows, IT manager Ruth Nabarro is thinking of introducing identity codes for all the customers, using capital letters and/or numbers. Help her answer the questions in Exercises 37–40.

37. If she uses three-letter codes, how many different customers can be identified?

38. If she uses codes with three different letters, how many different customers can be identified?

39. It appears that Nabarro has finally settled on codes consisting of two letters followed by two digits. For technical reasons, the letters must be different, and the first digit cannot be a zero. How many different customers can be identified?

40. O'Hagan sends Nabarro the following memo:

> To: Ruth Nabarro, Software Manager
> From: John O'Hagan, CEO
> Subject: Customer Identity Codes
>
> I have read your proposal for the customer ID codes. However, due to our ambitious expansion plans, I would like our system software to allow for at least 500,000 customers. Please adjust your proposal accordingly.

Nabarro is determined to have a sequence of letters followed by some digits, and, for reasons too complicated to explain, there cannot be more than two letters, the letters must be different, the digits must all be different, and the first digit cannot be a zero. What is the form of the shortest code she can use to satisfy the CEO, and how many different customers can be identified?

Degree Requirements *After an exhausting day at the office, John O'Hagan returns home and finds himself having to assist his son Billy-Sean, who continues to have a terrible time planning his first-year college course schedule. The latest* Bulletin of Suburban State University *reads as follows:*

> *All candidates for the degree of Bachelor of Arts at SSU must take, in their first year, at least 10 courses in the Sciences, Fine Arts, Liberal Arts, and Mathematics combined, of which at least 2 must be in each of the Sciences and Fine Arts, and exactly 3 must be in each of the Liberal Arts and Mathematics.*

Help him with the answers to Exercises 41–43.

41. If the bulletin lists exactly five first-year-level science courses and six first-year-level courses in each of the other categories, how many course combinations are possible that meet the minimum requirements?

42. Reading through the course descriptions in the bulletin a second time, John O'Hagan notices that Calculus I (listed as one of the mathematics courses) is a required course for many of the other courses and so decides that it would be best if Billy-Sean included Calculus I. Further, two of the Fine Arts courses cannot both be taken in the first year. How many course combinations are possible that meet the minimum requirements and include Calculus I?

43. To complicate things further, in addition to the requirement in Exercise 42, Physics II has Physics I as a prerequisite. (Both are listed as first-year science courses, but it is not necessary to take both.) How many course combinations are possible that include Calculus I and meet the minimum requirements?

Designing a Puzzle

As Product Design Manager for *Cerebral Toys, Inc.,* you are constantly on the lookout for ideas for intellectually stimulating yet inexpensive toys. You recently received the following memo from Felix Frost, the developmental psychologist on your design team.

To: Felicia
From: Felix
Subject: Crazy Cubes

We've hit on an excellent idea for a new educational puzzle (which we are calling "Crazy Cubes" until Marketing comes up with a better name). Basically, Crazy Cubes will consist of a set of plastic cubes. Two faces of each cube will be colored red, two will be colored blue, and two will be colored white, and there will be exactly two cubes with each possible configuration of colors. The goal of the puzzle is to seek out the matching pairs, thereby enhancing a child's geometric intuition and three-dimensional manipulation skills. The kit will include every possible configuration of colors. We are, however, a little stumped on the following question: How many cubes will the kit contain? In other words, how many possible ways can one color the faces of a cube so that two faces are red, two are blue, and two are white?

Looking at the problem, you reason that the following three-step decision algorithm ought to suffice:

Step 1 Choose a pair of faces to color red; $C(6, 2) = 15$ choices.

Step 2 Choose a pair of faces to color blue; $C(4, 2) = 6$ choices.

Step 3 Choose a pair of faces to color white; $C(2, 2) = 1$ choice.

This algorithm appears to give a total of $15 \times 6 \times 1 = 90$ possible cubes. However, before sending your reply to Felix, you realize that something is wrong, because there are different choices that result in the same cube. To describe some

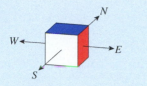

Figure 9

Figure 10

Figure 11

Rotate

Blue →   ← White

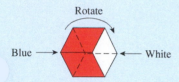

Figure 12

A tetrahedron

✱ There is a beautiful way of calculating this and similar numbers, called Pólya enumeration, but it requires a discussion of topics well outside the scope of this book. Take this as a hint that counting techniques can use some of the most sophisticated mathematics.

of these choices, imagine a cube oriented so that four of its faces are facing the four compass directions (Figure 9). Consider choice 1, with the top and bottom faces blue, north and south faces white, and east and west faces red; and choice 2, with the top and bottom faces blue, north and south faces red, and east and west faces white. These cubes are actually the same, as you see by rotating the second cube 90 degrees (Figure 10).

You therefore decide that you need a more sophisticated decision algorithm. Here is one that works:

Alternative 1: Faces with the same color opposite each other. Place one of the blue faces down. Then the top face is also blue. The cube must look like the one drawn in Figure 10. Thus, there is only one choice here.

Alternative 2: Red faces opposite each other and the other colors on adjacent pairs of faces. Again there is only one choice, as you can see by putting the red faces on the top and bottom and then rotating.

Alternative 3: White faces opposite each other and the other colors on adjacent pairs of faces; one possibility.

Alternative 4: Blue faces opposite each other and the other colors on adjacent pairs of faces; one possibility.

Alternative 5: Faces with the same color adjacent to each other. Look at the cube so that the edge common to the two red faces is facing you and horizontal (Figure 11). Then the faces on the left and right must be of different colors because they are opposite each other. Assume that the face on the right is white. (If it's blue, then rotate the die with the red edge still facing you to move it there, as in Figure 12.) This leaves two choices for the other white face, on the upper or the lower of the two back faces. This alternative gives two choices.

It follows that there are $1 + 1 + 1 + 1 + 2 = 6$ choices. Because the Crazy Cubes kit will feature two of each cube, the kit will require 12 different cubes.✱

EXERCISES

In all of the following exercises, there are three colors to choose from: red, white, and blue.

1. To enlarge the kit, Felix suggests including two each of two-colored cubes (using two of the colors red, white, and blue) with three faces one color and three another. How many additional cubes will be required?

2. If Felix now suggests adding two each of cubes with two faces one color, one face another color, and three faces the third color, how many additional cubes will be required?

3. Felix changes his mind and suggests that the kit use tetrahedral blocks with two colors instead (see the figure). How many of these would be required?

4. Once Felix finds the answer to Exercise 3, he decides to go back to the cube idea, but this time he insists that all possible combinations of up to three colors should be included. (For instance, some cubes will be all one color, others will be two colors.) How many cubes should the kit contain?

7

PROBABILITY

CASE STUDY

The Monty Hall Problem

On the game show *Let's Make a Deal*, you are shown three doors—A, B, and C—and behind one of them is the Big Prize. After you select one of them—say, door A—to make things more interesting, the host (Monty Hall) opens one of the other doors— say, door B—revealing that the Big Prize is not there. He then offers you the opportunity to change your selection to the remaining door, door C. Should you switch or stick with your original guess?

Does it make any difference?

Everett Collection, Inc.

Introduction

What is the probability of winning the lottery twice? What are the chances that a college athlete whose drug test is positive for steroid use is actually using steroids? You are playing poker and have been dealt two jacks. What is the likelihood that one of the next three cards you are dealt will also be a jack? These are all questions about probability.

Understanding probability is important in many fields, ranging from risk management in business through hypothesis testing in psychology to quantum mechanics in physics. Historically, the theory of probability arose in the sixteenth and seventeenth centuries from attempts by mathematicians such as Gerolamo Cardano, Pierre de Fermat, Blaise Pascal, and Christiaan Huygens to understand games of chance. Andrey Nikolaevich Kolmogorov set forth the foundations of modern probability theory in his 1933 book *Foundations of the Theory of Probability.*

The goal of this chapter is to familiarize you with the basic concepts of modern probability theory and to give you a working knowledge that you can apply in a variety of situations. In the first two sections the emphasis is on translating real-life situations into the language of sample spaces, events, and probability. Once we have mastered the language of probability, we spend the rest of the chapter studying some of its theory and applications. The last section gives an interesting application of both probability and matrix arithmetic.

7.1   Sample Spaces and Events

Sample Spaces

At the beginning of a football game, to ensure fairness, the referee tosses a coin to decide who will get the ball first. When the ref tosses the coin and observes which side faces up, there are two possible results: heads (H) and tails (T). These are the *only* possible results, ignoring the (remote) possibility that the coin lands on its edge. The act of tossing the coin is an example of an **experiment**. The two possible results, H and T, are possible **outcomes** of the experiment, and the set $S = \{H, T\}$ of all possible outcomes is the **sample space** for the experiment.

Experiments, Outcomes, and Sample Spaces

An **experiment** is an occurrence with a result, or **outcome**, that is uncertain before the experiment takes place. The set of all possible outcomes is called the **sample space** for the experiment.

Quick Examples

1. **Experiment:** Flip a coin, and observe the side facing up.
   Outcomes: H, T
   Sample Space: $S = \{H, T\}$

2. **Experiment:** Select a student in your class.
   Outcomes: The students in your class
   Sample Space: The set of students in your class

3. **Experiment:** Select a student in your class, and observe the color of his or her hair.
   Outcomes: red, black, brown, blond, green, . . .
   Sample Space: {red, black, brown, blond, green, . . .}

4. **Experiment:** Cast a die, and observe the number facing up.
   Outcomes: 1, 2, 3, 4, 5, 6
   Sample Space: $S = \{1, 2, 3, 4, 5, 6\}$

5. **Experiment:** Cast two distinguishable dice (see Example 1(a) of Section 6.1), and observe the numbers facing up.
   Outcomes: $(1, 1), (1, 2), \ldots, (6, 6)$ (36 outcomes)

$$\text{Sample Space:} \quad S = \begin{Bmatrix} (1,1), & (1,2), & (1,3), & (1,4), & (1,5), & (1,6), \\ (2,1), & (2,2), & (2,3), & (2,4), & (2,5), & (2,6), \\ (3,1), & (3,2), & (3,3), & (3,4), & (3,5), & (3,6), \\ (4,1), & (4,2), & (4,3), & (4,4), & (4,5), & (4,6), \\ (5,1), & (5,2), & (5,3), & (5,4), & (5,5), & (5,6), \\ (6,1), & (6,2), & (6,3), & (6,4), & (6,5), & (6,6) \end{Bmatrix}$$

$n(S) = $ the number of outcomes in $S = 36$

6. **Experiment:** Cast two indistinguishable dice (see Example 1(b) of Section 6.1), and observe the numbers facing up.
   Outcomes: $(1, 1), (1, 2), \ldots, (6, 6)$ (21 outcomes)

$$\text{Sample Space:} \quad S = \begin{Bmatrix} (1,1), & (1,2), & (1,3), & (1,4), & (1,5), & (1,6), \\ & (2,2), & (2,3), & (2,4), & (2,5), & (2,6), \\ & & (3,3), & (3,4), & (3,5), & (3,6), \\ & & & (4,4), & (4,5), & (4,6), \\ & & & & (5,5), & (5,6), \\ & & & & & (6,6) \end{Bmatrix}$$

$n(S) = 21$

7. **Experiment:** Cast two dice, and observe the *sum* of the numbers facing up.
   Outcomes: 2, 3, 4, 5, 6, 7, 8, 9, 10, 11, 12
   Sample Space: $S = \{2, 3, 4, 5, 6, 7, 8, 9, 10, 11, 12\}$

8. **Experiment:** Choose 2 cars (without regard to order) at random from a fleet of 10.
   Outcomes: Collections of 2 cars chosen from 10
   Sample Space: The set of all collections of 2 cars chosen from 10

$n(S) = C(10, 2) = 45$

The following example introduces a sample space that we'll use in several other examples.

EXAMPLE 1 **School and Work**

In a survey conducted by the Bureau of Labor Statistics,[1] the high school graduating class of 2010 was divided into those who went on to college and those who did not. Those who went on to college were further divided into those who went to 2-year colleges and those who went to 4-year colleges. All graduates were also asked whether they were working or not. Find the sample space for the experiment "Select a member of the high school graduating class of 2010, and classify his or her subsequent school and work activity."

Solution The tree in Figure 1 shows the various possibilities.

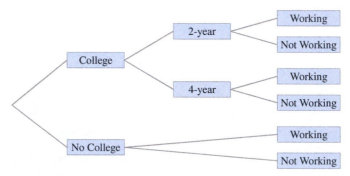

Figure 1

The sample space is

$$S = \{\text{2-year college \& working, 2-year college \& not working,}$$
$$\text{4-year college \& working, 4-year college \& not working,}$$
$$\text{no college \& working, no college \& not working}\}.$$

Events

In Example 1, suppose we are interested in the event that a 2010 high school graduate was working. In mathematical language we are interested in the *subset* of the sample space consisting of all outcomes in which the graduate was working.

Visualizing an Event
In the following figure, the favorable outcomes (events in E) are shown in green.

Sample Space S

Events

Given a sample space S, an **event** E is a subset of S. The outcomes in E are called the **favorable** outcomes. We say that E **occurs** in a particular experiment if the outcome of that experiment is one of the elements of E—that is, if the outcome of the experiment is favorable.

[1] "College Enrollment and Work Activity of High School Graduates," U.S. Bureau of Labor Statistics (www.bls.gov/news.release/hsgec.htm)

Quick Examples

9. **Experiment:** Roll a die, and observe the number facing up.

   $$S = \{1, 2, 3, 4, 5, 6\}$$

   Event: E: The number observed is odd.

   $$E = \{1, 3, 5\}$$

10. **Experiment:** Roll two distinguishable dice, and observe the numbers facing up.

    $$S = \{(1, 1), (1, 2), \ldots, (6, 6)\}$$

    Event: F: The dice show the same number.

    $$F = \{(1, 1), (2, 2), (3, 3), (4, 4), (5, 5), (6, 6)\}$$

11. **Experiment:** Roll two distinguishable dice, and observe the numbers facing up.

    $$S = \{(1, 1), (1, 2), \ldots, (6, 6)\}$$

    Event: G: The sum of the numbers is 1.

    $$G = \varnothing \qquad \text{There are no favorable outcomes.}$$

12. **Experiment:** Select a city beginning with "J."
    Event: E: The city is Johannesburg.

    $$E = \{\text{Johannesburg}\} \qquad \text{An event can consist of a single outcome.}$$

13. **Experiment:** Roll a die, and observe the number facing up.
    Event: E: The number observed is either even or odd.

    $$E = S = \{1, 2, 3, 4, 5, 6\} \qquad \text{An event can consist of all possible outcomes.}$$

14. **Experiment:** Select a student in your class.
    Event: E: The student has red hair.

    $$E = \{\text{red-haired students in your class}\}$$

15. **Experiment:** Draw a hand of 2 cards from a deck of 52.
    Event: H: Both cards are diamonds.

    H is the set of all hands of 2 cards chosen from 52 such that both cards are diamonds.

Here are some more examples of events.

EXAMPLE 2 **Dice**

We roll a red die and a green die and observe the numbers facing up. Describe the following events as subsets of the sample space.

a. E: The sum of the numbers showing is 6.

b. F: The sum of the numbers showing is 2.

Solution Here (again) is the sample space for the experiment of throwing two dice:

$$S = \begin{Bmatrix} (1,1), & (1,2), & (1,3), & (1,4), & (1,5), & (1,6), \\ (2,1), & (2,2), & (2,3), & (2,4), & (2,5), & (2,6), \\ (3,1), & (3,2), & (3,3), & (3,4), & (3,5), & (3,6), \\ (4,1), & (4,2), & (4,3), & (4,4), & (4,5), & (4,6), \\ (5,1), & (5,2), & (5,3), & (5,4), & (5,5), & (5,6), \\ (6,1), & (6,2), & (6,3), & (6,4), & (6,5), & (6,6) \end{Bmatrix}.$$

a. In mathematical language, E is the subset of S that consists of all those outcomes in which the sum of the numbers showing is 6. Here is the sample space once again, with the outcomes in question shown in color:

$$S = \begin{Bmatrix} (1,1), & (1,2), & (1,3), & (1,4), & (1,5), & (1,6), \\ (2,1), & (2,2), & (2,3), & (2,4), & (2,5), & (2,6), \\ (3,1), & (3,2), & (3,3), & (3,4), & (3,5), & (3,6), \\ (4,1), & (4,2), & (4,3), & (4,4), & (4,5), & (4,6), \\ (5,1), & (5,2), & (5,3), & (5,4), & (5,5), & (5,6), \\ (6,1), & (6,2), & (6,3), & (6,4), & (6,5), & (6,6) \end{Bmatrix}.$$

Thus, $E = \{(1,5), (2,4), (3,3), (4,2), (5,1)\}$.

b. The only outcome in which the numbers showing add to 2 is $(1,1)$. Thus,

$$F = \{(1,1)\}.$$

EXAMPLE 3 School and Work

Let S be the sample space of Example 1. List the elements in the following events:

a. The event E that a 2010 high school graduate was working.

b. The event F that a 2010 high school graduate was not going to a 2-year college.

Solution

a. We had this sample space:

$S = \{$2-year college & working, two-year college & not working, 4-year college & working, four-year college & not working, no college & working, no college & not working$\}$.

We are asked for the event that a graduate was working. Whenever we encounter a phrase involving "the event that . . . ," we mentally translate this into mathematical language by changing the wording.

Replace the phrase "the event that . . ." by the phrase "the subset of the sample space consisting of all outcomes in which . . . ."

Thus, we are interested in the subset of the sample space consisting of all outcomes in which the graduate was working. This gives

$E = \{$2-year college & working, 4-year college & working, no college & working$\}$.

The outcomes in E are illustrated by the shaded cells in Figure 2.

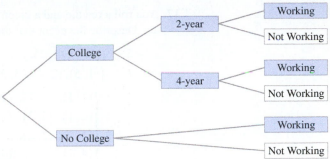

Figure 2

b. We are looking for the event that a graduate was not going to a 2-year college; that is, the subset of the sample space consisting of all outcomes in which the graduate was not going to a 2-year college. Thus,

$F = \{$4-year college & working, 4-year college & not working,
no college & working, no college & not working$\}$.

The outcomes in F are illustrated by the shaded cells in Figure 3.

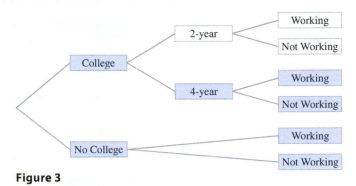

Figure 3

Complement, Union, and Intersection of Events

Events may often be described in terms of other events, using set operations such as complement, union, and intersection.

Visualizing the Complement

Sample Space S

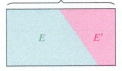

Complement of an Event

The **complement** of an event E is the set of outcomes not in E. Thus, the complement of E represents the event that E *does not occur*.

Quick Examples

16. You take four shots at the goal during a soccer game and record the number of times you score. Describe the event that you score at least twice, and also describe its complement.

$S = \{0, 1, 2, 3, 4\}$ Set of outcomes
$E = \{2, 3, 4\}$ Event that you score at least twice
$E' = \{0, 1\}$ Event that you do not score at least twice

17. You roll a red die and a green die and observe the two numbers facing up. Describe the event that the sum of the numbers is not 6.

$$S = \{(1, 1), (1, 2), \ldots, (6, 6)\}$$

$$F = \{(1, 5), (2, 4), (3, 3), (4, 2), (5, 1)\} \qquad \text{Sum of numbers is 6.}$$

$$F' = \begin{cases} (1, 1), \ (1, 2), \ (1, 3), \ (1, 4), \qquad\quad (1, 6), \\ (2, 1), \ (2, 2), \ (2, 3), \qquad\quad (2, 5), \ (2, 6), \\ (3, 1), \ (3, 2), \qquad\quad (3, 4), \ (3, 5), \ (3, 6), \\ (4, 1), \qquad\quad (4, 3), \ (4, 4), \ (4, 5), \ (4, 6), \\ \qquad\quad (5, 2), \ (5, 3), \ (5, 4), \ (5, 5), \ (5, 6), \\ (6, 1), \ (6, 2), \ (6, 3), \ (6, 4), \ (6, 5), \ (6, 6) \end{cases}$$

Sum of numbers is not 6.

Union of Events

The **union** of the events E and F is the set of all outcomes in E or F (or both). Thus, $E \cup F$ represents the event that E occurs *or* F occurs (or both).*

*As in Chapter 6, when we use the word "or," we agree to mean one or the other *or both*. This is called the **inclusive or**, and mathematicians have agreed to take this as the meaning of *or* to avoid confusion.

Quick Example

18. Roll a die.

E: The outcome is a 5; $E = \{5\}$.

F: The outcome is an even number; $F = \{2, 4, 6\}$.

$E \cup F$: The outcome is either a 5 *or* an even number:
$E \cup F = \{2, 4, 5, 6\}$.

Intersection of Events

The **intersection** of the events E and F is the set of all outcomes common to E and F. Thus, $E \cap F$ represents the event that both E *and* F occur.

Quick Example

19. Roll two dice: one red and one green.

E: The red die is 2.

F: The green die is odd.

$E \cap F$: The red die is 2, and the green die is odd:
$E \cap F = \{(2, 1), (2, 3), (2, 5)\}$.

EXAMPLE 4   **Weather**

Let R be the event that it will rain tomorrow, let P be the event that it will be pleasant, let C be the event that it will be cold, and let H be the event that it will be hot.

a. Express in words: $R \cap P'$, $R \cup (P \cap C)$.

b. Express in symbols: Tomorrow will be either a pleasant day or a cold and rainy day; it will not, however, be hot.

Solution The key here is to remember that intersection corresponds to *and* and union to *or*.

a. $R \cap P'$ is the event that it will rain *and* it will not be pleasant.

$R \cup (P \cap C)$ is the event that either it will rain, or it will be pleasant and cold.

b. If we rephrase the given statement using *and* and *or* we get "Tomorrow will be either a pleasant day or a cold and rainy day, and it will not be hot."

$$[P \cup (C \cap R)] \cap H'$$ Pleasant, or cold and rainy, and not hot.

The nuances of the English language play an important role in this formulation. For instance, the effect of the pause (comma) after "rainy day" suggests placing the preceding clause $P \cup (C \cap R)$ in brackets. In addition, the phrase "cold and rainy" suggests that C and R should be grouped together in their own parentheses.

The next example is essentially Example 3 in Section 6.2, but translated here into the language of events.

EXAMPLE 5   **Apple Sales**

The following table shows sales, in millions of items, of Macs, iPhones, and iPads sold by **Apple** in 2012, 2013, and 2014.[2]

	Macs (A)	iPhones (B)	iPads (C)	Total
2012 (U)	18	125	58	201
2013 (V)	16	150	71	238
2014 (W)	19	169	68	256
Total	53	444	197	695

Consider the experiment in which a device is selected at random from the 695 million devices represented in the table. Label the events representing the items in each row and column as shown (so that, for example, A is the event that the device selected was a Mac). Describe the following events and compute their cardinality:

a. U' **b.** $A \cap U'$ **c.** $(A \cap U)'$ **d.** $C \cup U$

Solution Before answering parts (a)–(d), first notice that the sample space S is the set of all items represented in the table, so S has a total of 695 million outcomes. Also, each number in the table is the cardinality of a specific event; for instance, the 18 million Macs sold in 2012 is the cardinality of the event $A \cap U$ that the device selected was a Mac and also sold in 2012, the 125 million iPhones sold in 2012 is the cardinality of the event $B \cap U$, and so on:

	Macs (A)	iPhones (B)	iPads (C)	Total
2012 (U)	$n(A \cap U)$	$n(B \cap U)$	$n(C \cap U)$	$n(U)$
2013 (V)	$n(A \cap V)$	$n(B \cap V)$	$n(C \cap V)$	$n(V)$
2014 (W)	$n(A \cap W)$	$n(B \cap W)$	$n(C \cap W)$	$n(W)$
Total	$n(A)$	$n(B)$	$n(C)$	$n(S)$

----

[2] Figures are rounded. Source: Apple quarterly press releases, www.investor.apple.com.

Now let us answer the specific questions: Because all the figures are stated in millions of items, we'll give our calculations and results in millions of items as well.

a. U' is the event that the device was not sold in 2012. Its cardinality is

$$n(U') = n(S) - n(U) = 695 - 201 = 494 \text{ million items.}$$

b. $A \cap U'$ is the event that the device was a Mac not sold in 2012:

	Macs (*A*)	iPhones (*B*)	iPads (*C*)	Total
2012 (*U*)	18	125	58	201
2013 (*V*)	16	150	71	238
2014 (*W*)	19	169	68	256
Total	53	444	197	695

From the table,

$$n(A \cap U') = 16 + 19 = 35 \text{ million items.}$$

c. $A \cap U$ is the event that the device was a Mac sold in 2012, so $(A \cap U)'$ is the event that the device was not a Mac sold in 2012:

	Macs (*A*)	iPhones (*B*)	iPads (*C*)	Total
2012 (*U*)	18	125	58	201
2013 (*V*)	16	150	71	238
2014 (*W*)	19	169	68	256
Total	53	444	197	695

From the formula for the cardinality of the complement of an event, its cardinality is

$$n((A \cap U)') = n(S) - n(A \cap U)$$
$$= 695 - 18 = 677 \text{ million items.}$$

d. $C \cup U$ is the event that the device was either an iPad or sold in 2012:

	Macs (*A*)	iPhones (*B*)	iPads (*C*)	Total
2012 (*U*)	18	125	58	201
2013 (*V*)	16	150	71	238
2014 (*W*)	19	169	68	256
Total	53	444	197	695

From the formula for the cardinality of a union, we have

$$n(C \cup U) = n(C) + n(U) - n(C \cap U)$$
$$= 197 + 201 - 58 = 340 \text{ million items.}$$

The case in which $E \cap F$ is empty is interesting, and we give it a name.

**Visualizing Mutually
Exclusive Events**

Sample Space S

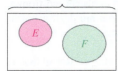

Mutually Exclusive Events

If E and F are events, then E and F are said to be **disjoint** or **mutually exclusive**
if $E \cap F$ is empty. (Hence, they have no outcomes in common.)

Interpretation

It is impossible for mutually exclusive events to occur simultaneously.

Quick Examples

In each of the following examples, E and F are mutually exclusive events.

20. Roll a die, and observe the number facing up. E: The outcome is even;
F: The outcome is odd.

$$E = \{2, 4, 6\}, F = \{1, 3, 5\}$$

21. Toss a coin three times, and record the sequence of heads and tails. E:
All three tosses land the same way up, F: One toss shows heads, and
the other two show tails.

$$E = \{HHH, TTT\}, F = \{HTT, THT, TTH\}$$

22. Observe the weather at 10 am tomorrow. E: It is raining; F: There is
not a cloud in the sky.

FAQs

Specifying the Sample Space

Q: *How do I determine the sample space in a given application?*

A: Strictly speaking, an experiment should include a description of what kinds of objects
are in the sample space, as in:

> *Cast a die, and observe the number facing up.*
> Sample space: the possible numbers facing up, $\{1, 2, 3, 4, 5, 6\}$.

> *Choose a person at random, and record her Social Security number and
> whether she is blonde.*
> Sample space: pairs (nine-digit number, Y/N).

However, in many of the scenarios discussed in this chapter and the next, an experiment is specified more vaguely, as in "Select a student in your class." In cases like
this, the nature of the sample space should be determined from the context. For
example, if the discussion is about grade-point averages and gender, the sample
space can be taken to consist of pairs (grade-point average, M/F).

7.1   EXERCISES

▼ more advanced   ◆ challenging
 indicates exercises that should be solved using technology

*In Exercises 1–18, describe the sample space S of the experi-
ment, and list the elements of the given event. (Assume that the
coins are distinguishable and that what is observed are the
faces or numbers that face up.)* [**HINT:** See Examples 1–3.]

1. Two coins are tossed; the result is at most one tail.

2. Two coins are tossed; the result is one or more heads.

3. Three coins are tossed; the result is at most one head.

4. Three coins are tossed; the result is more tails than heads.

5. Two distinguishable dice are rolled; the numbers add to 5.

6. Two distinguishable dice are rolled; the numbers add to 9.

7. Two indistinguishable dice are rolled; the numbers add to 4.

8. Two indistinguishable dice are rolled; one of the numbers is even and the other is odd.

9. Two indistinguishable dice are rolled; both numbers are prime.[3]

10. Two indistinguishable dice are rolled; neither number is prime.

11. A letter is chosen at random from those in the word *Mozart*; the letter is a vowel.

12. A letter is chosen at random from those in the word *Mozart*; the letter is neither *a* nor *m*.

13. A sequence of two different letters is randomly chosen from those of the word *sore*; the first letter is a vowel.

14. A sequence of two different letters is randomly chosen from those of the word *hear*; the second letter is not a vowel.

15. A sequence of two different digits is randomly chosen from the digits 0–4; the first digit is larger than the second.

16. A sequence of two different digits is randomly chosen from the digits 0–4; the first digit is twice the second.

17. You are considering purchasing either a domestic car, an imported car, a van, an antique car, or an antique truck; you do not buy a car.

18. You are deciding whether to enroll for Psychology 1, Psychology 2, Economics 1, General Economics, or Math for Poets; you decide to avoid economics.

19. A packet of gummy candy contains four strawberry gums, four lime gums, two black currant gums, and two orange gums. April May sticks her hand in and selects four at random. Complete the following sentences:
    a. The sample space is the set of . . . .
    b. April is particularly fond of combinations of two strawberry and two black currant gums. The event that April will get the combination she desires is the set of . . . .

20. A bag contains three red marbles, two blue ones, and four yellow ones. Alexandra Great pulls out three of them at random. Complete the following sentences:
    a. The sample space is the set of . . . .
    b. The event that Alexandra gets one of each color is the set of . . . .

21. ▼ President Barack H. Obama's first cabinet consisted of the Secretaries of Agriculture, Commerce, Defense, Education, Energy, Health and Human Services, Homeland Security, Housing and Urban Development, Interior, Labor, State, Transportation, Treasury, Veterans Affairs, and the Attorney General.[4] Assuming that President Obama had 20 candidates, including Hillary Clinton, to fill these posts

(and wished to assign no one to more than one post), complete the following sentences:
    a. The sample space is the set of . . . .
    b. The event that Hillary Clinton is the Secretary of State is the set of . . . .

22. ▼ A poker hand consists of a set of 5 cards chosen from a standard deck of 52 playing cards. You are dealt a poker hand. Complete the following sentences:
    a. The sample space is the set of . . . .
    b. The event "a full house" is the set of . . . . (Recall that a full house is three cards of one denomination and two of another.)

Suppose two dice (one red, one green) are rolled. Consider the following events. A: the red die shows 1; B: the numbers add to 4; C: at least one of the numbers is 1; and D: the numbers do not add to 11. In Exercises 23–30, express the stated event in symbolic form and say how many elements it contains. [**HINT:** See Example 5.]

23. The red die shows 1, and the numbers add to 4.

24. The red die shows 1, but the numbers do not add to 11.

25. The numbers do not add to 4.

26. The numbers add to 11.

27. The numbers do not add to 4, but they do add to 11.

28. Either the numbers add to 11 or the red die shows a 1.

29. At least one of the numbers is 1, or the numbers add to 4.

30. Either the numbers add to 4, or they add to 11, or at least one of them is 1.

Let W be the event that you will use the book's Website tonight, let I be the event that your math grade will improve, and let E be the event that you will use the Website every night. In Exercises 31–38, express the given event in symbols.

31. You will use the Website tonight, and your math grade will improve.

32. You will use the Website tonight, or your math grade will not improve.

33. Either you will use the Website every night or your math grade will not improve.

34. Your math grade will not improve even though you use the Website every night.

35. ▼ Either your math grade will improve, or you will use the Website tonight but not every night.

36. ▼ You will use the Website either tonight or every night, and your grade will improve.

37. ▼ (Compare Exercise 35.) Either your math grade will improve or you will use the Website tonight, but you will not use it every night.

38. ▼ (Compare Exercise 36.) Either you will use the Website tonight, or you will use it every night and your grade will improve.

---

[3] A positive integer is **prime** if it is neither 1 nor a product of smaller integers.

[4] Source: The White House website (www.whitehouse.gov).

In Exercises 39–42, Pablo randomly picks three marbles from a bag of eight marbles (four red ones, two green ones, and two yellow ones).

39. How many outcomes are there in the sample space? How many outcomes are there in the event that none of the marbles he picks are red?

40. How many outcomes are there in the sample space? How many outcomes are there in the event that all of the marbles he picks are red?

41. ▼ How many outcomes are there in the event that Pablo picks one marble of each color?

42. ▼ How many outcomes are there in the event that the marbles Pablo picks are not all the same color?

Applications

Housing Prices *Exercises 43–48 are based on the map below, which shows the percentage change in housing prices from June 2014 to June 2015 in each of nine regions (U.S. Census divisions).*[5]

43. You are choosing a region of the country to move to. Describe the event *E* that the region you choose saw an increase in housing prices of 6% or more.

44. You are choosing a region of the country to move to. Describe the event *E* that the region you choose saw an increase in housing prices of less than 4%.

45. You are choosing a region of the country to move to. Let *E* be the event that the region you choose saw an increase in housing prices of 6% or more, and let *F* be the event that the region you choose is on the east coast. Describe the events *E* ∪ *F* and *E* ∩ *F* both in words and by listing the outcomes of each.

46. You are choosing a region of the country to move to. Let *E* be the event that the region you choose saw an increase in housing prices of less than 4%, and let *F* be the event that the region you choose is not on the east coast. Describe the events *E* ∪ *F* and *E* ∩ *F* both in words and by listing the outcomes of each.

47. ▼ You are choosing a region of the country to move to. Which of the following pairs of events are mutually exclusive?

a. *E*: You choose a region from among the two with the highest percentage increase in housing prices.
F: You choose a region that is not on the east or west coast.

b. *E*: You choose a region from among the two with the highest percentage increase in housing prices.
F: You choose a region that is on the east coast.

48. ▼ You are choosing a region of the country to move to. Which of the following pairs of events are mutually exclusive?

a. *E*: You choose a region from among the three with the least increase in housing prices.
F: You choose a region from among the central divisions.

b. *E*: You choose a region from among the three with the least increase in housing prices.
F: You choose a region on the east or west coast.

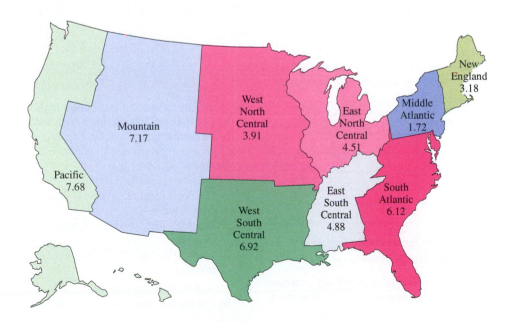

---

[5] Source: Federal Housing Finance Agency (www.fhfa.gov).

Publishing *Exercises 49–56 are based on the accompanying table, which shows the results of a survey of authors by a (fictitious) publishing company.* [**HINT:** See Example 5.]

	New Authors	Established Authors	Total
Successful	5	25	30
Unsuccessful	15	55	70
Total	20	80	100

Consider the following events: S: An author is successful; U: An author is unsuccessful; N: An author is new; and E: An author is established.

49. Describe the events $S \cap N$ and $S \cup N$ in words. Use the table to compute $n(S \cap N)$ and $n(S \cup N)$.

50. Describe the events $N \cap U$ and $N \cup U$ in words. Use the table to compute $n(N \cap U)$ and $n(N \cup U)$.

51. Which of the following pairs of events are mutually exclusive: *N* and *E*; *N* and *S*; *S* and *E*?

52. Which of the following pairs of events are mutually exclusive: *U* and *E*; *U* and *S*; *S* and *N*?

53. Describe the event $S \cap N'$ in words, and find the number of elements it contains.

54. Describe the event $U \cup E'$ in words, and find the number of elements it contains.

55. ▼ What percentage of established authors are successful? What percentage of successful authors are established?

56. ▼ What percentage of new authors are unsuccessful? What percentage of unsuccessful authors are new?

Stocks *Exercises 57–62 are based on the following table, which shows the stock market performance of 40 industries from five sectors of the U.S. economy as of noon on September 11, 2015.*[6] *(Take S to be the set of all 40 industries represented in the table.)*

	Increased (*X*)	Decreased (*Y*)	Unchanged (*Z*)	Totals
Financials (*F*)	3	4	1	8
Manufacturing (*M*)	8	3	3	14
Information Technology (*T*)	6	1	0	7
Health Care (*H*)	4	1	1	6
Utilities (*U*)	3	1	1	5
Totals	24	10	6	40

57. Use symbols to describe the event that an industry increased in value but was not in the manufacturing sector. How many elements are in this event?

58. Use symbols to describe the event that an industry was in the manufacturing sector and did not increase in value. How many elements are in this event?

59. Describe the event $H' \cup Z$ in words, and compute $n(H' \cup Z)$.

60. Describe the event $H \cup Z'$ in words, and compute $n(H \cup Z')$.

61. Find all pairs of mutually exclusive events among the events *F, M, T, X, Y,* and *Z*.

62. Find all pairs of events that are not mutually exclusive among the events *M, T, Y* and *Z*.

Animal Psychology *Exercises 63–68 concern the following chart, which shows the way in which a dog moves its facial muscles when torn between the drives of fight and flight.*[7] *The "fight" drive increases from left to right; the "flight" drive increases from top to bottom. (Notice that an increase in the "fight" drive causes the dog's upper lip to lift, while an increase in the "flight" drive draws its ears downward.)*

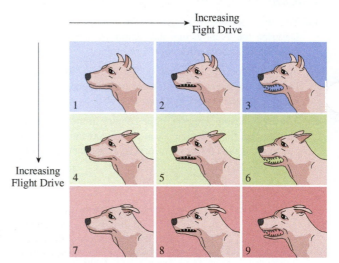

63. ▼ Let *E* be the event that the dog's flight drive is strongest, let *F* be the event that the dog's flight drive is weakest, let *G* be the event that the dog's fight drive is strongest, and let *H* be the event that the dog's fight drive is weakest. Describe the following events in terms of *E, F, G,* and *H* using the symbols ∩, ∪, and ′.
   a. The dog's flight drive is not strongest, and its fight drive is weakest.
   b. The dog's flight drive is strongest, or its fight drive is weakest.
   c. Neither the dog's flight drive nor its fight drive is strongest.

64. ▼ Let *E* be the event that the dog's flight drive is strongest, let *F* be the event that the dog's flight drive is weakest, let *G*

---

[6] "Unchanged" includes industries that moved by less than 0.1%.
Source for data: Fidelity (https://eresearch.fidelity.com).

[7] Source: *On Aggression* by Konrad Lorenz (Fakenham, Norfolk: University Paperback Edition, Cox & Wyman Limited, 1967).

be the event that the dog's fight drive is strongest, and let H be the event that the dog's fight drive is weakest. Describe the following events in terms of E, F, G, and H using the symbols $\cap$, $\cup$, and $'$.

a. The dog's flight drive is weakest, and its fight drive is not weakest.

b. The dog's flight drive is not strongest, or its fight drive is weakest.

c. Either the dog's flight drive or its fight drive fails to be strongest.

65. ▼ Describe the following events explicitly (as subsets of the sample space):

a. The dog's fight and flight drives are both strongest.

b. The dog's fight drive is strongest, but its flight drive is neither weakest nor strongest.

66. ▼ Describe the following events explicitly (as subsets of the sample space):

a. Neither the dog's fight drive nor its flight drive is strongest.

b. The dog's fight drive is weakest, but its flight drive is neither weakest nor strongest.

67. ▼ Describe the following events in words:
   a. $\{1, 4, 7\}$ b. $\{1, 9\}$ c. $\{3, 6, 7, 8, 9\}$

68. ▼ Describe the following events in words:
   a. $\{7, 8, 9\}$ b. $\{3, 7\}$ c. $\{1, 2, 3, 4, 7\}$

Exercises 69–72 use counting arguments from the preceding chapter.

69. ▼ *Gummy Bears* A bag contains six gummy bears. Noel picks four at random. How many possible outcomes are there? If one of the gummy bears is raspberry, how many of these outcomes include the raspberry gummy bear?

70. ▼ *Chocolates* My couch potato friend enjoys sitting in front of the TV and grabbing handfuls of 5 chocolates at random from his snack jar. Unbeknownst to him, I have replaced one of the 20 chocolates in his jar with a cashew. (He hates cashews with a passion.) How many possible outcomes are there the first time he grabs 5 chocolates? How many of these include the cashew?

71. ▼ *Horse Races* The seven contenders in the fifth horse race at Aqueduct on February 18, 2002, were Pipe Bomb, Expect a Ship, All That Magic, Electoral College, Celera, Cliff Glider, and Inca Halo.[8] You are interested in the first three places (winner, second place, and third place) for the race.

a. Find the cardinality $n(S)$ of the sample space S of all possible finishes of the race. (A finish for the race consists of a first, a second, and a third place winner.)

b. Let E be the event that Electoral College is in second or third place, and let F be the event that Celera is the winner. Express the event $E \cap F$ in words, and find its cardinality.

72. ▼ *Intramurals* The following five teams will be participating in *Urban University*'s hockey intramural tournament: the Independent Wildcats, the Phi Chi Bulldogs, the Gate Crashers, the Slide Rule Nerds, and the City Slickers. Prizes will be awarded for the winner and runner-up.

a. Find the cardinality $n(S)$ of the sample space S of all possible outcomes of the tournament. (An outcome of the tournament consists of a winner and a runner-up.)

b. Let E be the event that the City Slickers are runners-up, and let F be the event that the Independent Wildcats are neither the winners nor runners-up. Express the event $E \cup F$ in words, and find its cardinality.

Communication and Reasoning Exercises

73. Complete the following sentence. An event is a ____.

74. Complete the following sentence. Two events E and F are mutually exclusive if their intersection is ____.

75. If E and F are events, then $(E \cap F)'$ is the event that ____.

76. If E and F are events, then $(E' \cap F')$ is the event that ____.

77. Let E be the event that you meet a tall, dark stranger. Which of the following could reasonably represent the experiment and sample space in question?

(A) You go on vacation and lie in the sun; S is the set of cloudy days.

(B) You go on vacation and spend an evening at the local dance club; S is the set of people you meet.

(C) You go on vacation and spend an evening at the local dance club; S is the set of people you do not meet.

78. Let E be the event that you buy a Porsche. Which of the following could reasonably represent the experiment and sample space in question?

(A) You go to an auto dealership and select a Mustang; S is the set of colors available.

(B) You go to an auto dealership and select a red car; S is the set of cars you decide not to buy.

(C) You go to an auto dealership and select a red car; S is the set of car models available.

79. ▼ True or false? Every set S is the sample space for some experiment. Explain.

80. ▼ True or false? Every sample space S is a finite set. Explain.

81. ▼ Describe an experiment in which a die is cast and the set of outcomes is $\{0, 1\}$.

82. ▼ Describe an experiment in which two coins are flipped and the set of outcomes is $\{0, 1, 2\}$.

83. ▼ Two distinguishable dice are rolled. Could there be two mutually exclusive events that both contain outcomes in which the numbers facing up add to 7?

84. ▼ Describe an experiment in which two dice are rolled, and describe two mutually exclusive events that both contain outcomes in which both dice show a 1.

---

[8] Source: *Newsday*, Feb. 18, 2002, p. A36.

7.2 Relative Frequency

Fundamentals

Suppose you have a coin that you think is not fair and you would like to determine the likelihood that heads will come up when it is tossed. You could estimate this likelihood by tossing the coin a large number of times and counting the number of times heads comes up. Suppose, for instance, that in 100 tosses of the coin, heads comes up 58 times. The fraction of times that heads comes up, $58/100 = .58$, is the **relative frequency**, or **estimated probability** of heads coming up when the coin is tossed. In other words, saying that the relative frequency of heads coming up is .58 is the same as saying that heads came up 58% of the time in your series of experiments.

Now let's think about this example in terms of sample spaces and events. First of all, there is an experiment that has been repeated $N = 100$ times: Toss the coin, and observe the side facing up. The sample space for this experiment is $S = \{H, T\}$. Also, there is an event E in which we are interested: the event that heads comes up, which is $E = \{H\}$. The number of times E has occurred, or the **frequency** of E, is $fr(E) = 58$. The relative frequency of the event E is then

$$P(E) = \frac{fr(E)}{N} \qquad \begin{array}{l}\text{Frequency of event } E \\ \hline \text{Number of repetitions } N\end{array}$$

$$= \frac{58}{100} = .58.$$

Notes

1. The relative frequency gives us an *estimate* of the likelihood that heads will come up when that particular coin is tossed. This is why statisticians often use the alternative term *estimated probability* to describe it.

2. The larger the number of times the experiment is performed, the more accurate an estimate we expect this estimated probability to be. ∎

Visualizing Relative Frequency

$$P(E) = \frac{fr(E)}{N} = \frac{4}{10} = .4$$

Relative Frequency

When an experiment is performed a number of times, the **relative frequency** or **estimated probability** of an event E is the fraction of times that the event E occurs. If the experiment is performed N times and the event E occurs $fr(E)$ times, then the relative frequency is given by

$$P(E) = \frac{fr(E)}{N}. \qquad \text{Fraction of times } E \text{ occurs}$$

The number $fr(E)$ is called the **frequency** of E. N, the number of times that the experiment is performed, is called the number of **trials** or the **sample size**. If E consists of a single outcome s, then we refer to $P(E)$ as the relative frequency or estimated probability of the outcome s, and we write $P(s)$.

The collection of the estimated probabilities of *all* the outcomes is the **relative frequency distribution** or **estimated probability distribution**.

Quick Examples

1. **Experiment:** Roll a pair of dice, and add the numbers that face up.
   Event: E: The sum is 5.

If the experiment is repeated 100 times and E occurs on 10 of the rolls, then the relative frequency of E is

$$P(E) = \frac{fr(E)}{N} = \frac{10}{100} = .10.$$

2. If 10 rolls of a single die resulted in the outcomes 2, 1, 4, 4, 5, 6, 1, 2, 2, 1, then the associated relative frequency distribution is shown in the following table:

Outcome	1	2	3	4	5	6
Rel. Frequency	.3	.3	0	.2	.1	.1

3. **Experiment:** Note the cloud conditions on a particular day at noon.

If the experiment is repeated a number of times and the sky is clear 20% of those times, partly cloudy 30% of those times, and overcast the rest of those times, then the relative frequency distribution is as follows:

Outcome	Clear	Partly Cloudy	Overcast
Rel. Frequency	.20	.30	.50

---

*The official plural form of Prius, according to Toyota.

EXAMPLE 1 **Sales of Hybrid Vehicles**

In a survey of 250 hybrid vehicles sold in the United States, 125 were Toyota Prii,* 30 were Honda Civics, 20 were Toyota Camrys, 15 were Ford Escapes, and the rest were other makes.[9] What is the relative frequency that a hybrid vehicle sold in the United States is not a Toyota Camry?

Solution The experiment consists of choosing a hybrid vehicle sold in the United States and determining its make. The sample space suggested by the information given is

$$S = \{\text{Toyota Prius, Honda Civic, Toyota Camry, Ford Escape, Other}\},$$

and we are interested in the event

$$E = \{\text{Toyota Prius, Honda Civic, Ford Escape, Other}\}.$$

The sample size is $N = 250$, of which 20 were Toyota Camrys. Thus, the frequency of E is $fr(E) = 250 - 20 = 230$, and the relative frequency of E is

$$P(E) = \frac{fr(E)}{N} = \frac{230}{250} = .92.$$

➡ **Before we go on . . .** In Example 1 you might ask how accurate the estimate of .92 is or how well it reflects *all* of the hybrid vehicles sold in the United States absent any information about national sales figures. The field of statistics provides the tools needed to say to what extent this estimated probability can be trusted. ∎

---

[9] The proportions are based on approximate actual cumulative sales through October 2011 (www.wikipedia.com).

EXAMPLE 2   **Auctions on eBay**

The following chart shows the results of a survey of the bid prices for 50 paintings on **eBay** with the highest number of bids:[10]

Bid Price	$0–$9.99	$10–$49.99	$50–$99.99	≥$100
Frequency	6	23	15	6

Consider the experiment in which a painting is chosen and the bid price is observed.

a. Find the relative frequency distribution.

b. Find the relative frequency that a painting in the survey had a bid price of less than $50.

Solution

a. The following table shows the relative frequency of each outcome, which we find by dividing each frequency by the sum $N = 50$:

Bid Price	$0–$9.99	$10–$49.99	$50–$99.99	≥$100
Rel. Frequency	$\dfrac{6}{50} = .12$	$\dfrac{23}{50} = .46$	$\dfrac{15}{50} = .30$	$\dfrac{6}{50} = .12$

b. Method 1: Computing Directly

$$E = \{\$0–\$9.99, \$10–\$49.99\}$$

Thus,

$$P(E) = \frac{fr(E)}{N} = \frac{6 + 23}{50} = \frac{29}{50} = .58.$$

Method 2: Using the Relative Frequency Distribution
Notice that we can obtain the same answer from the distribution in part (a) by simply adding the relative frequencies of the outcomes in E:

$$P(E) = .12 + .46 = .58.$$

Q: *Why did we get the same result in Example 2(b) by simply adding the relative frequencies of the outcomes in E?*

A: The reason can be seen by doing the calculation in the first method a slightly different way:

$$P(E) = \frac{fr(E)}{N} = \frac{6 + 23}{50}$$

$$= \frac{6}{50} + \frac{23}{50}. \qquad \text{Sum of relative frequencies of the individual outcomes}$$

This property of relative frequency distributions is discussed below.

---

[10] In the category "Art—Direct from Artist" on November 14, 2011 (www.eBay.com).

Following are some important properties of estimated probability that we can observe in Example 2.

Some Properties of Relative Frequency Distributions

Let $S = \{s_1, s_2, \ldots, s_n\}$ be a sample space, and let $P(s_i)$ be the relative frequency of the event $\{s_i\}$. Then

1. $0 \leq P(s_i) \leq 1$

2. $P(s_1) + P(s_2) + \cdots + P(s_n) = 1$

3. If $E = \{e_1, e_2, \ldots, e_r\}$, then $P(E) = P(e_1) + P(e_2) + \cdots + P(e_r)$.

In words:

1. The relative frequency of each outcome is a number between 0 and 1 (inclusive).

2. The relative frequencies of all the outcomes add up to 1.

3. The relative frequency of an event E is the sum of the relative frequencies of the individual outcomes in E.

Relative Frequency and Increasing Sample Size

A "fair" coin is one that is as likely to come up heads as it is to come up tails. In other words, we expect heads to come up 50% of the time if we toss such a coin many times. Put more precisely, we expect the relative frequency to approach .5 as the number of trials gets larger. Figure 4 shows how the relative frequency behaved for one sequence of coin tosses. For each N we have plotted what fraction of times the coin came up heads in the first N tosses.

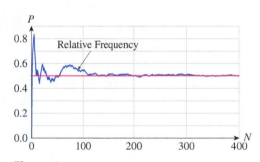

Figure 4

* This can be made more precise by the concept of a "limit" used in calculus.

Notice that the relative frequency graph meanders as N increases, sometimes getting closer to .5 and sometimes drifting away. However, the graph tends to meander within smaller and smaller distances of .5 as N increases.*

In general, this is how relative frequency seems to behave: As N gets large, the relative frequency appears to approach some fixed value. Some refer to this value as the "actual" probability, whereas others point out that there are difficulties with this notion. For instance, how can we actually determine this limit to any accuracy by experiment? How exactly is the experiment conducted? Technical and philosophical issues aside, the relative frequencies do approach a fixed value, and in the next section we will talk about how we use probability models to predict this limiting value.

Using Technology

See the Technology Guides at the end of the chapter to see how to use a TI-83/84 Plus or a spreadsheet to simulate experiments.

7.2 EXERCISES

▼ more advanced   ◆ challenging
T indicates exercises that should be solved using technology

In Exercises 1–6, calculate the relative frequency P(E) using the given information.

1. $N = 100$, $fr(E) = 40$ **2.** $N = 500$, $fr(E) = 300$

3. Eight hundred adults are polled, and 640 of them support universal health-care coverage. E is the event that an adult supports universal health-care coverage. [**HINT:** See Example 1.]

4. Eight hundred adults are polled, and 640 of them support universal health-care coverage. E is the event that an adult does not support universal health-care coverage. [**HINT:** See Example 1.]

5. A die is rolled 60 times with the following result: 1, 2, and 3 each come up 8 times, and 4, 5, and 6 each come up 12 times. E is the event that the number that comes up is at most 4.

6. A die is rolled 90 times with the following result: 1 and 2 never come up, 3 and 4 each come up 30 times, and 5 and 6 each come up 15 times. E is the event that the number that comes up is at least 4.

Exercises 7–12 are based on the following table, which shows the frequency of outcomes when two distinguishable coins were tossed 4,000 times and the uppermost faces were observed. [**HINT:** See Example 2.]

Outcome	HH	HT	TH	TT
Frequency	1,100	950	1,200	750

7. Determine the relative frequency distribution.

8. What is the relative frequency that heads comes up at least once?

9. What is the relative frequency that the second coin lands with heads up?

10. What is the relative frequency that the first coin lands with heads up?

11. Would you judge the second coin to be fair? Give a reason for your answer.

12. Would you judge the first coin to be fair? Give a reason for your answer.

In Exercises 13–18, say whether the given distribution can be a relative frequency distribution. If your answer is no, indicate why not. [**HINT:** See the properties of relative frequency distributions.]

13.

Outcome	1	2	3	5
Rel. Frequency	.4	.6	0	0

14.

Outcome	A	B	C	D
Rel. Frequency	.2	.1	.2	.1

15.

Outcome	HH	HT	TH	TT
Rel. Frequency	.5	.4	.5	− .4

16.

Outcome	2	4	6	8
Rel. Frequency	25	25	25	25

17.

Outcome	−3	−2	−1	0
Rel. Frequency	.2	.3	.2	.3

18.

Outcome	HH	HT	TH	TT
Rel. Frequency	0	0	0	1

In Exercises 19 and 20, complete the given relative frequency distribution and compute the stated relative frequencies. [**HINT:** See the properties of relative frequency distributions.]

19.

Outcome	1	2	3	4	5
Rel. Frequency	.2	.3	.1	.1	

    a. $P(\{1, 3, 5\})$ **b.** $P(E')$ where $E = \{1, 2, 3\}$

20.

Outcome	1	2	3	4	5
Rel. Frequency	.4		.3	.1	.1

    a. $P(\{2, 3, 4\})$ **b.** $P(E')$ where $E = \{3, 4\}$

T *Exercises 21–24 require the use of a calculator or computer with a random number generator.*

21. Simulate 100 tosses of a fair coin, and compute the estimated probability that heads comes up.

22. Simulate 100 throws of a fair die, and calculate the estimated probability that the result is a 6.

23. Simulate 50 tosses of two coins, and compute the estimated probability that the outcome is one head and one tail (in any order).

24. Simulate 100 throws of two fair dice, and calculate the estimated probability that the result is a double 6.

Applications

25. *Latin Music Sales: 2013* In a survey of 500 Latin music downloads in 2013, 270 were regional (Mexican/Tejano), 150 were pop-rock, 70 were tropical (salsa/merengue/

cumbia/bachata), and 10 were urban (reggaeton).[11] Calculate the following relative frequencies:

a. That a music download was regional

b. That a music download was either tropical or urban

c. That a music download was not urban [**HINT:** See Example 1.]

26. _Latin Music Sales: 2009_ In a survey of 400 Latin music downloads in 2009, 200 were regional (Mexican/Tejano), 130 were pop-rock, 45 were tropical (salsa/merengue/cumbia/bachata), and 25 were urban (reggaeton).[12] Calculate the following relative frequencies:

a. That a music download was pop-rock

b. That a music download was neither tropical nor regional

c. That a music download was not regional [**HINT:** See Example 1.]

27. _Subprime Mortgages during the 2000–2008 Housing Bubble_ The following chart shows the results of a survey of the status of subprime home mortgages in Texas in November 2008:[13]

Mortgage Status	Current	Past Due	In Foreclosure	Repossessed
Frequency	134	52	9	5

(The four categories are mutually exclusive; for instance, "Past Due" refers to a mortgage whose payment status is past due but is not in foreclosure, and "In Foreclosure" refers to a mortgage that is in the process of being foreclosed but not yet repossessed.)

a. Find the relative frequency distribution for the experiment of randomly selecting a subprime mortgage in Texas and determining its status.

b. What is the relative frequency that a randomly selected subprime mortgage in Texas was not current? [**HINT:** See Example 2.]

28. _Subprime Mortgages during the 2000–2008 Housing Bubble_ The following chart shows the results of a survey of the status of subprime home mortgages in Florida in November 2008:[14]

Mortgage Status	Current	Past Due	In Foreclosure	Repossessed
Frequency	110	65	60	15

(The four categories are mutually exclusive; for instance, "Past Due" refers to a mortgage whose payment status is

past due but is not in foreclosure, and "In Foreclosure" refers to a mortgage that is in the process of being foreclosed but not yet repossessed.)

a. Find the relative frequency distribution for the experiment of randomly selecting a subprime mortgage in Florida and determining its status.

b. What is the relative frequency that a randomly selected subprime mortgage in Florida was neither in foreclosure nor repossessed? [**HINT:** See Example 2.]

29. _Population Age in Mexico_ The following table shows the results of a survey of randomly selected residents of Mexico:[15]

Age	0–14	15–29	30–64	>64
Percentage	30	27	37	6

a. Find the associated relative frequency distribution. [**HINT:** See Quick Example 3.]

b. Find the relative frequency that a resident of Mexico is _not_ from 15 to 64 years old.

30. _Population Age in the United States_ The following table shows the results of a survey of randomly selected U.S. residents:[16]

Age	0–14	15–29	30–64	>64
Percentage	20	21	46	13

a. Find the associated relative frequency distribution. [**HINT:** See Quick Example 3.]

b. Find the relative frequency that a resident of the United States is 30 years old or older.

31. _Motor Vehicle Safety_ The following table shows crashworthiness ratings for 10 small SUVs.[17] (3 = Good, 2 = Acceptable, 1 = Marginal, 0 = Poor)

Frontal Crash Test Rating	3	2	1	0
Frequency	1	4	4	1

a. Find the relative frequency distribution for the experiment of choosing a small SUV at random and determining its frontal crash rating.

b. What is the relative frequency that a randomly selected small SUV will have a crash test rating of "Acceptable" or better?

---

[11] Based on digital revenues in 2013. Source: The Recording Industry Association of America (www.riaa.com).

[12] Based on digital revenues in 2009. Source: _Ibid._

[13] Based on actual data in 2008. Source: Federal Reserve Bank of New York (www.newyorkfed.org/regional/subprime.html).

[14] _Ibid._

[15] Based on population distribution in 2010. Source: Instituto Nacional de Estadística y Geografía (www.inegi.org.mx).

[16] _Ibid._ (The data for the United States was also provided by the Instituto Nacional de Estadística y Geografía.)

[17] Ratings by the Insurance Institute for Highway Safety. Sources: Oak Ridge National Laboratory: "An Analysis of the Impact of Sport Utility Vehicles in the United States," Stacy C. Davis, Lorena F. Truett (August 2000) Insurance Institute for Highway Safety (www-cta.ornl.gov/Publications/Final SUV report.pdf).

32. *Motor Vehicle Safety* The following table shows crash-worthiness ratings for 16 small cars.[18] (3 = Good, 2 = Acceptable, 1 = Marginal, 0 = Poor)

Frontal Crash Test Rating	3	2	1	0
Frequency	1	11	2	2

a. Find the relative frequency distribution for the experiment of choosing a small car at random and determining its frontal crash rating.

b. What is the relative frequency that a randomly selected small car will have a crash test rating of "Marginal" or worse?

33. *Internet Connections* The following pie chart shows the relative frequency distribution resulting from a survey of 2,000 U.S. households with Internet connections back in 2003:[19]

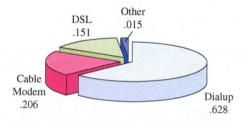

Determine the **frequency distribution**, that is, the total number of households with each type of Internet connection in the survey.

34. *Internet Connections* The following pie chart shows the relative frequency distribution resulting from a survey of 3,000 U.S. rural households with Internet connections back in 2003:[20]

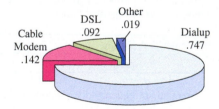

Determine the **frequency distribution**, that is, the total number of households with each type of Internet connection in the survey.

35. ▼ *Stock Market Gyrations* The following chart shows the day-by-day change in the Dow Jones Industrial Average

during 20 successive business days in October 2008 during the 2008 financial crisis:[21]

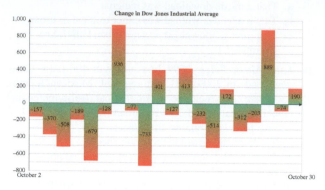

Use the chart to construct the relative frequency distribution using the following three outcomes. Surge: The Dow was up by more than 300 points; Plunge: The Dow was down by more than 300 points; Steady: The Dow changed by 300 points or less.

36. ▼ *Stock Market Gyrations* Repeat Exercise 35 using the following chart for November–December 2008:[22]

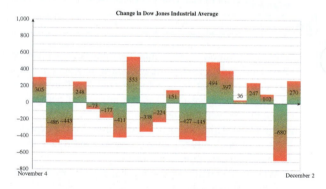

Publishing *Exercises 37–46 are based on the following table, which shows the results of a survey of 100 authors by a (fictitious) publishing company:*

	New Authors	Established Authors	Total
Successful	5	25	30
Unsuccessful	15	55	70
Total	20	80	100

Compute the relative frequencies of the given events if an author as specified is chosen at random.

37. ▼ An author is established and successful.

38. ▼ An author is unsuccessful and new.

---

[18] See footnote for Exercise 31.

[19] Based on a 2003 survey. Source: "A Nation Online: Entering the Broadband Age," U.S. Department of Commerce, September 2004 (www.ntia.doc.gov/reports/anol/index.html).

[20] *Ibid.*

[21] Source: http://finance.google.com.

[22] *Ibid.*

39. ▼ An author is a new author.

40. ▼ An author is successful.

41. ▼ An author is unsuccessful.

42. ▼ An author is established.

43. ▼ A successful author is established.

44. ▼ An unsuccessful author is established.

45. ▼ An established author is successful.

46. ▼ A new author is unsuccessful.

47. ▼ *Public Health* A random sampling of chicken in supermarkets revealed that approximately 80% was contaminated with the organism *Campylobacter*.[23] Of the contaminated chicken, 20% had the strain that is resistant to antibiotics. Construct a relative frequency distribution showing the following outcomes when chicken is purchased at a supermarket: *U:* the chicken is not infected with *Campylobacter*; *C:* the chicken is infected with nonresistant *Campylobacter*; *R:* the chicken is infected with resistant *Campylobacter.*

48. ▼ *Public Health* A random sampling of turkey in supermarkets found 58% to be contaminated with *Campylobacter*, and 84% of those to be contaminated with the strain that is resistant to antibiotics.[24] Construct a relative frequency distribution showing the following outcomes when turkey is purchased at a supermarket: *U:* the turkey is not infected with *Campylobacter*; *C:* the turkey is infected with nonresistant *Campylobacter*; and *R:* the turkey is infected with resistant *Campylobacter.*

49. ▼ *Organic Produce* A 2001 Agriculture Department study of more than 94,000 samples from more than 20 crops showed that 73% of conventionally grown foods had residues from at least one pesticide. Moreover, conventionally grown foods were six times as likely to contain multiple pesticides as organic foods. Of the organic foods tested, 23% had pesticide residues, which includes 10% with multiple pesticide residues.[25] Compute two estimated probability distributions: one for conventional produce and one for organic produce, showing the relative frequencies that a randomly selected product has no pesticide residues, has residues from a single pesticide, and has residues from multiple pesticides.

50. ▼ *Organic Produce* Repeat Exercise 49 using the following information for produce from California: 31% of conventional food and 6.5% of organic food had residues from at least one pesticide. Assume that, as in Exercise 49, conventionally grown foods are six times as likely to contain multiple pesticides as organic foods. Also assume that 3% of the organic food has residues from multiple pesticides.

51. ▼ *Steroids Testing* A pharmaceutical company is running trials on a new test for anabolic steroids. The company uses the test on 400 athletes known to be using steroids and 200 athletes known not to be using steroids. Of those using steroids, the new test is positive for 390 and negative for 10. Of those not using steroids, the test is positive for 10 and negative for 190. What is the relative frequency of a **false negative** result (the probability that an athlete using steroids will test negative)? What is the relative frequency of a **false positive** result (the probability that an athlete not using steroids will test positive)?

52. ▼ *Lie Detectors* A manufacturer of lie detectors is testing its newest design. It asks 300 subjects to lie deliberately and another 500 to tell the truth. Of those who lied, the lie detector caught 200. Of those who told the truth, the lie detector accused 200 of lying. What is the relative frequency of the machine wrongly letting a liar go, and what is the probability that it will falsely accuse someone who is telling the truth?

53. 🅣 ◆ *Public Health* Refer back to Exercise 47. Simulate the experiment of selecting chicken at a supermarket and determining the following outcomes: *U:* The chicken is not infected with *Campylobacter; C:* The chicken is infected with nonresistant *Campylobacter; R:* The chicken is infected with resistant *Campylobacter.* [**HINT:** Generate integers in the range 1–100. The outcome is determined by the range. For instance, if the number is in the range 1–20, regard the outcome as *U*, etc.]

54. 🅣 ◆ *Public Health* Repeat Exercise 53, but use turkeys and the data given in Exercise 48.

Communication and Reasoning Exercises

55. Complete the following. The relative frequency of an event *E* is defined to be _____.

56. If two people each flip a coin 100 times and compute the relative frequency that heads comes up, they will both obtain the same result—right?

57. How many different answers are possible if you flip a coin 100 times and compute the relative frequency that heads comes up? What are the possible answers?

58. Interpret the popularity rating of the student council president as a relative frequency by specifying an appropriate experiment and also what is observed.

59. ▼ Ruth tells you that when you roll a pair of fair dice, the probability of obtaining a pair of matching numbers is 1/6. To test this claim, you roll a pair of fair dice 20 times and never once get a pair of matching numbers. This proves that either Ruth is wrong or the dice are not fair—right?

---

[23] *Campylobacter* is one of the leading causes of food poisoning in humans. Thoroughly cooking the meat kills the bacteria. Source: *New York Times*, October 20, 1997, p. A1. Publication of this article first brought *Campylobacter* to the attention of a wide audience.

[24] *Ibid.*

[25] The 10% figure is an estimate. Source: *New York Times*, May 8, 2002, p. A29.

60. ▼ Juan tells you that when you roll a pair of fair dice, the probability that the numbers add up to 7 is 1/6. To test this claim, you roll a pair of fair dice 24 times, and the numbers add up to 7 exactly four times. This proves that Juan is right—right?

61. ▼ How would you measure the relative frequency that the weather service accurately predicts the next day's high temperature?

62. ▼ Suppose that you toss a coin 100 times and get 70 heads. If you continue tossing the coin, the estimated probability of heads overall should approach 50% if the coin is fair. Will you have to get more tails than heads in subsequent tosses to "correct" for the 70 heads you got in the first 100 tosses?

7.3 Probability and Probability Models

What Is Probability?

It is understandable if you are a little uncomfortable with using relative frequency as the estimated probability because it does not always agree with what you intuitively think to be true. For instance, if you toss a fair coin (one as likely to come up heads as tails) 100 times and heads happen to come up 62 times, the experiment seems to suggest that the probability of heads is .62, even though you *know* that the "actual" probability is .50 (because the coin is fair).

Q : *So what do we mean by "actual" probability?*

A : There are various philosophical views as to exactly what we should mean by "actual" probability. For example, (finite) *frequentists* say that there is no such thing as "actual probability"—all we should really talk about is what we can actually measure: the relative frequency. *Propensitists* say that the actual probability p of an event is a property of the event that makes its relative frequency tend to p in the long run; that is, p will be the limiting value of the relative frequency as the number of trials in a repeated experiment gets larger and larger. (See Figure 4 in Section 7.2.) *Bayesians,* on the other hand, argue that the actual probability of an event is the degree to which we *expect* it to occur, given our knowledge about the nature of the experiment. These and other viewpoints have been debated in considerable depth in the literature.*

*The interested reader should consult references in the philosophy of probability. For an online summary, see, for example, the Stanford Encyclopedia of Philosophy (http://plato.stanford.edu/contents.html).

Mathematicians tend to avoid the whole debate and talk instead about *abstract* probability, or **probability distributions**, based purely on the properties of relative frequency listed in Section 7.2. Specific probability distributions can then be used as *models* in real-life situations such as flipping a coin or tossing a die, to predict (or model) relative frequency.

> ### Probability Distribution; Probability
>
> (Compare with the properties of relative frequency distributions in Section 7.2.) A (finite) **probability distribution** is an assignment of a number $P(s_i)$, the **probability of s_i**, to each outcome of a finite sample space $S = \{s_1, s_2, \ldots, s_n\}$. The probabilities must satisfy
>
> **1.** $0 \le P(s_i) \le 1$
>
> and
>
> **2.** $P(s_1) + P(s_2) + \cdots + P(s_n) = 1$.

We find the **probability of an event** E, written $P(E)$, by adding up the probabilities of the outcomes in E.

If $P(E) = 0$, we call E an **impossible event**. The empty event $\varnothing$ is always impossible, since *something* must happen.

Quick Examples

1. All the examples of estimated probability distributions in Section 7.2 are examples of probability distributions. (See the Quick Examples in that section.)

2. Let us take $S = \{H, T\}$ and make the assignments $P(H) = .5$, $P(T) = .5$. Because these numbers are between 0 and 1 and add to 1, they specify a probability distribution.

3. In Quick Example 2, we can instead make the assignments $P(H) = .2$, $P(T) = .8$. Because these numbers are between 0 and 1 and add to 1, they, too, specify a probability distribution.

4. With $S = \{H, T\}$ again, we could also take $P(H) = 1$, $P(T) = 0$, so $\{T\}$ is an impossible event.

5. The following table gives a probability distribution for the sample space $S = \{1, 2, 3, 4, 5, 6\}$:

Outcome	1	2	3	4	5	6
Probability	.3	.3	0	.1	.2	.1

It follows that

$$P(\{1, 6\}) = .3 + .1 = .4$$
$$P(\{2, 3\}) = .3 + 0 = .3$$
$$P(3) = 0. \qquad \text{\{3\} is an impossible event.}$$

The above Quick Examples included models for the experiments of flipping fair and unfair coins. In general, we have the following.

Probability Models

A **probability model** for a particular experiment is a probability distribution that predicts the relative frequency of each outcome if the experiment is performed a large number of times. (See Figure 4 at the end of Section 7.2).* Just as we think of relative frequency as *estimated probability*, we can think of modeled probability as *theoretical probability*.

Quick Examples

6. **Fair Coin Model:** (See Quick Example 2.) Flip a fair coin, and observe the side that faces up. Because we expect that heads is as likely to come up as tails, we model this experiment with the probability distribution specified by $S = \{H, T\}$, $P(H) = .5$, $P(T) = .5$. Figure 4 in Section 7.2 suggests that the relative frequency of heads

* Just how large is a "large number of times"? That depends on the nature of the experiment. For example, if you toss a fair coin 100 times, then the relative frequency of heads will be between .45 and .55 about 73% of the time. If an outcome is extremely unlikely (such as winning the lottery), you might need to repeat the experiment billions or trillions of times before the relative frequency approaches any specific number.

approaches .5 as the number of coin tosses gets large, so the fair coin model predicts the relative frequency for a large number of coin tosses quite well.

7. **Unfair Coin Model:** (See Quick Example 3.) Take $S = \{H, T\}$ and $P(H) = .2$, $P(T) = .8$. We can think of this distribution as a model for the experiment of flipping an unfair coin that is four times as likely to land with tails uppermost than heads.

8. **Fair Die Model:** Roll a fair die, and observe the uppermost number. Because we expect to roll each specific number one sixth of the time, we model the experiment with the probability distribution specified by $S = \{1, 2, 3, 4, 5, 6\}$, $P(1) = 1/6$, $P(2) = 1/6, \ldots, P(6) = 1/6$. This model predicts, for example, that the relative frequency of throwing a 5 approaches $1/6$ as the number of times you roll the die gets large.

9. Roll a pair of fair dice. (Recall that there are a total of 36 outcomes if the dice are distinguishable.) Then an appropriate model of the experiment has

$$S = \left\{\begin{array}{llllll}(1,1), & (1,2), & (1,3), & (1,4), & (1,5), & (1,6), \\ (2,1), & (2,2), & (2,3), & (2,4), & (2,5), & (2,6), \\ (3,1), & (3,2), & (3,3), & (3,4), & (3,5), & (3,6), \\ (4,1), & (4,2), & (4,3), & (4,4), & (4,5), & (4,6), \\ (5,1), & (5,2), & (5,3), & (5,4), & (5,5), & (5,6), \\ (6,1), & (6,2), & (6,3), & (6,4), & (6,5), & (6,6) \end{array}\right\},$$

with each outcome being assigned a probability of $1/36$.

10. In the experiment in Quick Example 9, take E to be the event that the sum of the numbers that face up is 5, so

$$E = \{(1,4), (2,3), (3,2), (4,1)\}.$$

By the properties of probability distributions,

$$P(E) = \frac{1}{36} + \frac{1}{36} + \frac{1}{36} + \frac{1}{36} = \frac{4}{36} = \frac{1}{9}.$$

Notice that, in all of the Quick Examples above except Quick Example 7, all the outcomes are equally likely, and each outcome s has a probability of

$$P(s) = \frac{1}{\text{Total number of outcomes}} = \frac{1}{n(S)}.$$

More generally, in Quick Example 10 we saw that adding the probabilities of the individual outcomes in an event E amounted to computing the ratio (Number of favorable outcomes)/(Total number of outcomes):

$$P(E) = \frac{\text{Number of favorable outcomes}}{\text{Total number of outcomes}} = \frac{n(E)}{n(S)}.$$

Visualizing Probability for Equally Likely Outcomes

Sample Space S

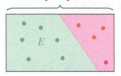

$$P(E) = \frac{n(E)}{n(S)} = \frac{6}{10} = .6$$

Probability Model for Equally Likely Outcomes

In an experiment in which all outcomes are equally likely, we model the experiment by taking the probability of an event E to be

$$P(E) = \frac{\text{Number of favorable outcomes}}{\text{Total number of outcomes}} = \frac{n(E)}{n(S)}.$$

Note Remember that this formula will work *only* when the outcomes are equally likely. If, for example, a die is *weighted*, then the outcomes may not be equally likely, and the formula above will not give an appropriate probability model. ∎

Quick Examples

11. Toss a fair coin three times. In this case, $S = \{$HHH, HHT, HTH, HTT, THH, THT, TTH, TTT$\}$. The probability that we throw exactly two heads is

    $$P(E) = \frac{n(E)}{n(S)} = \frac{3}{8}.$$ There are eight equally likely outcomes, and $E = \{$HHT, HTH, THH$\}$.

12. Roll a pair of fair dice. The probability that we roll a double (both dice show the same number) is

    $$P(E) = \frac{n(E)}{n(S)} = \frac{6}{36} = \frac{1}{6}.$$ $E = \{(1, 1), (2, 2), (3, 3), (4, 4), (5, 5), (6, 6)\}$

13. Randomly choose a person from a class of 40, in which 6 have red hair. If E is the event that a randomly selected person in the class has red hair, then

    $$P(E) = \frac{n(E)}{n(S)} = \frac{6}{40} = .15$$

EXAMPLE 1 Sales of Hybrid Vehicles

(Compare Example 1 in Section 7.2.) A total of 1.9 million hybrid vehicles had been sold in the United States through October of 2011. Of these, 955,000 were Toyota Prii, 205,000 were Honda Civics, 170,000 were Toyota Camrys, 105,000 were Ford Escapes, and the rest were other makes.[26]

a. What is the probability that a randomly selected hybrid vehicle sold in the United States was either a Toyota Prius or a Honda Civic?

b. What is the probability that a randomly selected hybrid vehicle sold in the United States was not a Toyota Camry?

---

[26] Source for sales data: www.wikipedia.com.

Solution

a. The experiment suggested by the question consists of randomly choosing a hybrid vehicle sold in the United States and determining its make. We are interested in the event E that the hybrid vehicle was either a Toyota Prius or a Honda Civic. So

$$S = \text{the set of hybrid vehicles sold; } n(S) = 1,900,000$$

$$E = \text{the set of Toyota Prii and Honda Civics sold;}$$
$$n(E) = 955,000 + 205,000 = 1,160,000.$$

Are the outcomes equally likely in this experiment? Yes, because we are as likely to choose one vehicle as another. Thus,

$$P(E) = \frac{n(E)}{n(S)} = \frac{1,160,000}{1,900,000} \approx .61.$$

b. Let the event F consist of those hybrid vehicles sold that were not Toyota Camrys:

$$n(F) = 1,900,000 - 170,000 = 1,730,000.$$

Hence,

$$P(F) = \frac{n(F)}{n(S)} = \frac{1,730,000}{1,900,000} \approx .91.$$

Q: *In Example 1 of Section 7.2 we had a similar example about hybrid vehicles, but we called the probabilities calculated there relative frequencies. Here, they are probabilities. What is the difference?*

A: In Example 1 of Section 7.2, the data were based on the results of a survey, or sample, of only 250 hybrid vehicles (out of a total of about 1.9 million sold in the United States) and were therefore incomplete. (A statistician would say that we were given *sample data.*) It follows that any inference we draw from the 250 surveyed, such as the probability that a hybrid vehicle sold in the United States is not a Toyota Camry, is uncertain, and this is the cue that tells us that we are working with relative frequency, or estimated probability. Think of the survey as an experiment (choosing a hybrid vehicle) repeated 250 times—exactly the setting for estimated probability.

In Example 1 above, on the other hand, the data do not describe how *some* hybrid vehicle sales are broken down into the categories described; they describe how *all 1.9 million* hybrid vehicle sales in the United States are broken down. (The statistician would say that we were given *population data* in this case, because the data describe the entire "population" of hybrid vehicles sold in the United States.)

EXAMPLE 2 Indistinguishable Dice

We recall from Section 7.1 that the sample space when we roll a pair of indistinguishable dice is

$$S = \begin{cases} (1, 1), \ (1, 2), \ (1, 3), \ (1, 4), \ (1, 5), \ (1, 6), \\ (2, 2), \ (2, 3), \ (2, 4), \ (2, 5), \ (2, 6), \\ (3, 3), \ (3, 4), \ (3, 5), \ (3, 6), \\ (4, 4), \ (4, 5), \ (4, 6), \\ (5, 5), \ (5, 6), \\ (6, 6) \end{cases}.$$

Construct a probability model for this experiment.

Solution Because there are 21 outcomes, it is tempting to say that the probability of each outcome should be taken to be 1/21. However, the outcomes are not all equally likely. For instance, the outcome (2, 3) is twice as likely as (2, 2), because (2, 3) can occur in two ways. (It corresponds to the event {(2, 3), (3, 2)} for distinguishable dice.) For purposes of calculating probability, it is easiest to use calculations for distinguishable dice.* Here are some examples.

* Note that any pair of real dice can be distinguished in principle because they possess slight differences, although we may regard them as indistinguishable by not attempting to distinguish them. Thus, the probabilities of events must be the same as for the corresponding events for distinguishable dice.

Outcome (Indistinguishable Dice)	(1, 1)	(1, 2)	(2, 2)	(1, 3)	(2, 3)	(3, 3)
Corresponding Event (Distinguishable Dice)	{(1, 1)}	{(1, 2), (2, 1)}	{(2, 2)}	{(1, 3), (3, 1)}	{(2, 3), (3, 2)}	{(3, 3)}
Probability	$\frac{1}{36}$	$\frac{2}{36} = \frac{1}{18}$	$\frac{1}{36}$	$\frac{2}{36} = \frac{1}{18}$	$\frac{2}{36} = \frac{1}{18}$	$\frac{1}{36}$

If we continue this process for all 21 outcomes, we will find that they add to 1. Figure 5 illustrates the complete probability distribution:

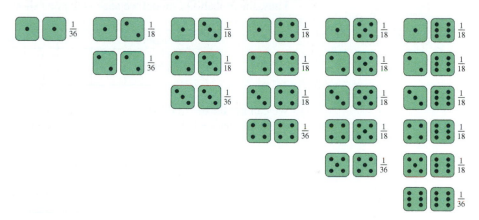

Figure 5

EXAMPLE 3   **Weighted Dice**

To impress your friends with your die-rolling skills, you have surreptitiously weighted your die in such a way that 6 is three times as likely to come up as any one of the other numbers. (All the other outcomes are equally likely.) Obtain a probability distribution for a roll of the die, and use it to calculate the probability of an even number coming up.

Solution Let us label our unknowns (there appear to be two of them):

x = probability of rolling a 6

y = probability of rolling any one of the other numbers.

We are first told that "6 is three times as likely to come up as any one of the other numbers." If we rephrase this in terms of our unknown probabilities, we get "the probability of rolling a 6 is three times the probability of rolling any one of the other numbers." In symbols,

$x = 3y.$

We must also use a piece of information that has not been given to us but that we know must be true: The sum of the probabilities of all the outcomes is 1:

$$x + y + y + y + y + y = 1$$

or

$$x + 5y = 1.$$

We now have two linear equations in two unknowns, and we solve for x and y. Substituting the value of x in the first equation ($x = 3y$) in the second ($x + 5y = 1$) gives

$$8y = 1$$

or

$$y = \frac{1}{8}.$$

To get x, we substitute the value of y back into either equation and find

$$x = \frac{3}{8}.$$

Thus, the probability model we seek is the one shown in the following table:

Outcome	1	2	3	4	5	6
Probability	$\frac{1}{8}$	$\frac{1}{8}$	$\frac{1}{8}$	$\frac{1}{8}$	$\frac{1}{8}$	$\frac{3}{8}$

We can use the distribution to calculate the probability of an even number coming up by adding the probabilities of the favorable outcomes:

$$P(\{2, 4, 6\}) = \frac{1}{8} + \frac{1}{8} + \frac{3}{8} = \frac{5}{8}$$

Thus, there is a $5/8 = .625$ chance that an even number will come up.

➡ **Before we go on ...** We should check that the probability distribution in Example 3 satisfies the requirements: 6 is indeed three times as likely to come up as any other number. Also, the probabilities that we calculated do add up to 1:

$$\frac{1}{8} + \frac{1}{8} + \frac{1}{8} + \frac{1}{8} + \frac{1}{8} + \frac{3}{8} = 1. \blacksquare$$

Probability of Unions, Intersections, and Complements

So far, all we know about computing the probability of an event E is that $P(E)$ is the sum of the probabilities of the individual outcomes in E. Suppose, though, that we do not know the probabilities of the individual outcomes in E but we do know that $E = A \cup B$, where we happen to know $P(A)$ and $P(B)$. How do we compute the probability of $A \cup B$? We might be tempted to say that $P(A \cup B)$ is $P(A) + P(B)$, but let us look at an example using the probability distribution in Quick Example 5 at the beginning of this section:

Outcome	1	2	3	4	5	6
Probability	.3	.3	0	.1	.2	.1

For A, let us take the event $\{2, 4, 5\}$, and for B, let us take $\{2, 4, 6\}$. $A \cup B$ is then the event $\{2, 4, 5, 6\}$. We know that we can find the probabilities $P(A)$, $P(B)$, and $P(A \cup B)$ by adding the probabilities of all the outcomes in these events, so

$$P(A) = P(\{2, 4, 5\}) = .3 + .1 + .2 = .6$$
$$P(B) = P(\{2, 4, 6\}) = .3 + .1 + .1 = .5$$
$$P(A \cup B) = P(\{2, 4, 5, 6\}) = .3 + .1 + .2 + .1 = .7.$$

Our first guess was wrong: $P(A \cup B) \neq P(A) + P(B)$. Notice, however, that the outcomes in $A \cap B$ are counted twice in computing $P(A) + P(B)$ but only once in computing $P(A \cup B)$:

$$P(A) + P(B) = P(\{2, 4, 5\}) + P(\{2, 4, 6\}) \qquad A \cap B = \{2, 4\}$$
$$= (.3 + .1 + .2) + (.3 + .1 + .1) \qquad P(A \cap B) \text{ counted twice}$$
$$= 1.1,$$

whereas

$$P(A \cup B) = P(\{2, 4, 5, 6\}) = .3 + .1 + .2 + .1 \qquad P(A \cap B) \text{ counted once}$$
$$= .7.$$

Thus, if we take $P(A) + P(B)$ and then subtract the surplus $P(A \cap B)$, we get $P(A \cup B)$. In symbols,

$$P(A \cup B) = P(A) + P(B) - P(A \cap B)$$
$$.7 = .6 + .5 - .4$$

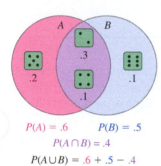

$P(A) = .6 \qquad P(B) = .5$

$P(A \cap B) = .4$

$P(A \cup B) = .6 + .5 - .4$

Figure 6

(see Figure 6). We call this formula the **addition principle**. One more thing: Notice that our original guess $P(A \cup B) = P(A) + P(B)$ would have worked if we had chosen A and B with no outcomes in common; that is, if $A \cap B = \varnothing$. When $A \cap B = \varnothing$, recall that we say that A and B are mutually exclusive.

Visualizing the Addition Principle

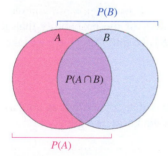

$P(A \cup B) = P(A) + P(B) - P(A \cap B)$

Mutually Exclusive Events

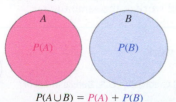

$P(A \cup B) = P(A) + P(B)$

Addition Principle

If A and B are any two events, then

$$P(A \cup B) = P(A) + P(B) - P(A \cap B).$$

Addition Principle for Mutually Exclusive Events

If $A \cap B = \varnothing$, we say that A and B are **mutually exclusive**, and we have

$$P(A \cup B) = P(A) + P(B). \qquad \text{Because } P(A \cap B) = 0$$

This holds true also for more than two events: If $A_1, A_2, \ldots, A_n$ are mutually exclusive events (that is, the intersection of every pair of them is empty), then

$$P(A_1 \cup A_2 \cup \cdots \cup A_n) = P(A_1) + P(A_2) + \cdots + P(A_n).$$

Addition principle for many mutually exclusive events

> **Quick Examples**
>
> **14.** There is a 10% chance of rain (R) tomorrow, a 20% chance of high winds (W), and a 5% chance of both. The probability of either rain or high winds (or both) is
>
> $$P(R \cup W) = P(R) + P(W) - P(R \cap W)$$
> $$= .10 + .20 - .05 = .25.$$

15. The probability that you will be in Cairo at 6:00 am tomorrow (C) is .3, while the probability that you will be in Alexandria at 6:00 am tomorrow (A) is .2. Thus, the probability that you will be in either Cairo or Alexandria at 6:00 am tomorrow is

$$P(C \cup A) = P(C) + P(A) \qquad \text{\textcolor{red}{A and C are mutually exclusive.}}$$
$$= .3 + .2 = .5.$$

16. When a pair of fair dice is rolled, the probability of the numbers that face up adding to 7 is 6/36, the probability of their adding to 8 is 5/36, and the probability of their adding to 9 is 4/36. Thus, the probability of the numbers adding to 7, 8, or 9 is

$$P(\{7\} \cup \{8\} \cup \{9\}) = P(7) + P(8) + P(9) \qquad \text{\textcolor{blue}{The events are mutually exclusive.}}$$
$$= \frac{6}{36} + \frac{5}{36} + \frac{4}{36} = \frac{15}{36} = \frac{5}{12}.$$

* The sum of the numbers that face up cannot equal two different numbers at the same time.

EXAMPLE 4   School and Work

A survey[27] conducted by the Bureau of Labor Statistics found that 68% of the high school graduating class of 2010 went on to college the following year, while 42% of the class was working. Furthermore, 92% were either in college or working (or both).

a. What percentage went on to college and work at the same time?

b. What percentage went on to college but not work?

Solution We can think of the experiment of choosing a member of the high school graduating class of 2010 at random. The sample space is the set of all these graduates.

a. We are given information about two events:

A: A graduate went on to college; $P(A) = .68$.

B: A graduate went on to work; $P(B) = .42$.

We are also told that $P(A \cup B) = .92$. We are asked for the probability that a graduate went on to both college and work, $P(A \cap B)$. To find $P(A \cap B)$, we take advantage of the fact that the formula

$$P(A \cup B) = P(A) + P(B) - P(A \cap B)$$

can be used to calculate any one of the four quantities that appear in it as long as we know the other three. Substituting the quantities we know, we get

$$.92 = .68 + .42 - P(A \cap B),$$

so

$$P(A \cap B) = .68 + .42 - .92 = .18.$$

Thus, 18% of the graduates went on to college and work at the same time.

b. We are asked for the probability of a new event:

C: A graduate went on to college but not to work.

You can use the formula
$P(A \cup B) = P(A) + P(B) - P(A \cap B)$
to calculate any of the four quantities in the formula if you know the other three.

---

[27] Source: "College Enrollment and Work Activity of High School Graduates," U.S. Bureau of Labor Statistics (www.bls.gov/news.release/hsgec.htm).

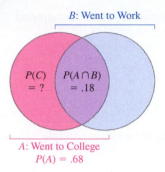

B: Went to Work

$P(C)$ = ? $P(A \cap B)$ = .18

A: Went to College
$P(A)$ = .68

Figure 7

C is the part of A outside of $A \cap B$, so $C \cup (A \cap B) = A$, and C and $A \cap B$ are mutually exclusive. (See Figure 7.)

Thus, applying the addition principle, we have

$$P(C) + P(A \cap B) = P(A).$$

From part (a) we know that $P(A \cap B) = .18$, so

$$P(C) + .18 = .68$$

giving

$$P(C) = .50.$$

In other words, 50% of the graduates went on to college but not to work.

We can use the addition principle to deduce other useful properties of probability distributions.

> ## More Principles of Probability Distributions
>
> The following rules hold for any sample space S and any event A:
>
> $$P(S) = 1$$ The probability of *something* happening is 1.
>
> $$P(\varnothing) = 0$$ The probability of *nothing* happening is 0.
>
> $$P(A') = 1 - P(A).$$ The probability of A *not* happening is 1 minus the probability of A.
>
> **Note** We can also write the third equation as
>
> $$P(A) = 1 - P(A')$$
>
> or
>
> $$P(A) + P(A') = 1. \blacksquare$$

**Visualizing the Rule
for Complements**

Sample Space S

A A'

$P(A) + P(A') = 1$

> ### Quick Examples
>
> **17.** There is a 10% chance of rain (R) tomorrow. Therefore, the probability that it will *not* rain is
>
> $$P(R') = 1 - P(R) = 1 - .10 = .90.$$
>
> **18.** The probability that Eric Ewing will score at least two goals is .6. Therefore, the probability that he will score at most one goal is $1 - .6 = .4$.

Q : *Can you persuade me that all of these principles are true?*

A : Let us take them one at a time. We know that $S = \{s_1, s_2, \ldots, s_n\}$ is the set of all outcomes, so

$$P(S) = P(\{s_1, s_2, \ldots, s_n\})$$
$$= P(s_1) + P(s_2) + \cdots + P(s_n)$$ We add the probabilities of the outcomes to obtain the probability of an event.
$$= 1.$$ By the definition of a probability distribution

Now, note that $S \cap \varnothing = \varnothing$, so that S and $\varnothing$ are mutually exclusive. Applying the addition principle gives

$$P(S) = P(S \cup \varnothing) = P(S) + P(\varnothing).$$

Subtracting $P(S)$ from both sides gives $0 = P(\varnothing)$.

If A is any event in S, then we can write

$$S = A \cup A',$$

where A and A' are mutually exclusive. (Why?) Thus, by the addition principle,

$$P(S) = P(A) + P(A').$$

Because $P(S) = 1$, we get

$$1 = P(A) + P(A')$$

or $P(A') = 1 - P(A)$.

EXAMPLE 5   Subprime Mortgages during the Housing Bubble

A home loan is either current, 30–59 days past due, 60–89 days past due, 90 or more days past due, in foreclosure, or repossessed by the lender. In November 2008 the probability that a randomly selected subprime home mortgage in California was not current was .51. The probability that a mortgage was not current, but neither in foreclosure nor repossessed, was .28.[28] Calculate the probabilities of the following events:

a. A California home mortgage was current.

b. A California home mortgage was in foreclosure or repossessed.

Solution

a. Let us write C for the event that a randomly selected subprime home mortgage in California was current. The event that the home mortgage was *not* current is its complement C', and we are given that $P(C') = .51$. We have

$$P(C) + P(C') = 1$$
$$P(C) + .51 = 1,$$

so $P(C) = 1 - .51 = .49$.

b. Take

F: A mortgage was in foreclosure or repossessed.

N: A mortgage was neither current, in foreclosure, nor repossessed.

We are given $P(N) = .28$. Further, the events F and N are mutually exclusive with union C', the set of all noncurrent mortgages. Hence,

$$P(C') = P(F) + P(N)$$
$$.51 = P(F) + .28,$$

giving

$$P(F) = .51 - .28 = .23.$$

Thus, there was a 23% chance that a subprime home mortgage was either in foreclosure or repossessed.

---

[28] Source: Federal Reserve Bank of New York (www.newyorkfed.org/regional/subprime.html).

EXAMPLE 6　Apple Sales

The following table shows sales, in millions of items, of Macs, iPhones, and iPads sold by **Apple** in 2012, 2013, and 2014:[29]

	Macs (A)	iPhones (B)	iPads (C)	Total
2012 (U)	18	125	58	201
2013 (V)	16	150	71	238
2014 (W)	19	169	68	256
Total	53	444	197	695

If one of these 695 million items sold is selected at random, find the probabilities of the following events:

a. It is a Mac.

b. It was sold in 2013.

c. It is a Mac sold in 2013.

d. Either it is a Mac or it was sold in 2013.

e. It is not a Mac.

Solution　Before answering the questions, first notice that the sample space S is the set of all items represented in the table, so S has a total of 695 million outcomes.

a. When we say that an item is being selected at random, we mean that all the outcomes are equally likely. If A is the event that the selected item is a Mac, then

$$P(A) = \frac{n(A)}{n(S)} = \frac{53}{695} \approx .076.$$

The event A is represented by the blue shaded region in the table:

	Macs (A)	iPhones (B)	iPads (C)	Total
2012 (U)	18	125	58	201
2013 (V)	16	150	71	238
2014 (W)	19	169	68	256
Total	53	444	197	695

b. If V is the event that the selected item was sold in 2013, then

$$P(V) = \frac{n(V)}{n(S)} = \frac{238}{695} \approx .342.$$

In the table, V is represented as shown:

	Macs (A)	iPhones (B)	iPads (C)	Total
2012 (U)	18	125	58	201
2013 (V)	16	150	71	238
2014 (W)	19	169	68	256
Total	53	444	197	695

---

[29] Figures are rounded. Source: Apple quarterly press releases, www.investor.apple.com.

c. The event that the selected item is a Mac sold in 2013 is the event $A \cap V$:

$$P(A \cap V) = \frac{n(A \cap V)}{n(S)} = \frac{16}{695} \approx .023.$$

In the table, $A \cap V$ is represented by the overlap of the regions representing A and V:

	Macs (*A*)	iPhones (*B*)	iPads (*C*)	Total
2012 (*U*)	18	125	58	201
2013 (*V*)	16	150	71	238
2014 (*W*)	19	169	68	256
Total	53	444	197	695

d. The event that the selected item either is a Mac or was sold in 2013 is the event $A \cup V$ and is represented by the entire blue shaded area in the table:

	Macs (*A*)	iPhones (*B*)	iPads (*C*)	Total
2012 (*U*)	18	125	58	201
2013 (*V*)	16	150	71	238
2014 (*W*)	19	169	68	256
Total	53	444	197	695

We can compute its probability in two ways:

1. Directly from the table:

$$P(A \cup V) = \frac{n(A \cup V)}{n(S)} = \frac{53 + 238 - 16}{695} \approx .396$$

2. Using the addition principle:

$$P(A \cup V) = P(A) + P(V) - P(A \cap V)$$
$$\approx .076 + .342 - .023 = .395$$ Slightly less accurate, as we rounded the three intermediate answers $P(A)$, $P(V)$, and $P(A \cap V)$.

e. The event that the selected item is not a Mac is the event A'. Its probability may be computed by using the formula for the probability of the complement:

$$P(A') = 1 - P(A) \approx 1 - .076 = .924$$

FAQs

Distinguishing Probability from Relative Frequency

Q: *Relative frequency and modeled probability using equally likely outcomes have essentially the same formula: (Number of favorable outcomes)/(Total number of outcomes). How do I know whether a given probability is one or the other?*

A: Ask yourself this: Has the probability been arrived at experimentally, by performing a number of trials and counting the number of times the event occurred? If so, the probability is estimated; that is, it is the relative frequency. If, on the other hand, the probability was computed by analyzing the experiment under consideration rather than by performing actual trials of the experiment, it is a probability model.

> **Q**: Out of every 100 homes, 67 have broadband Internet service. Thus, the probability that a house has broadband Internet service is .67. Is this probability estimated (relative frequency) or theoretical (a probability model)?
>
> **A**: That depends on how the ratio 67 out of 100 was arrived at. If it is based on a poll of *all* homes, then the probability is theoretical. If it is based on a survey of only a *sample* of homes, it is estimated (see the Q&A following Example 1).

7.3 EXERCISES

▼ more advanced   ◆ challenging
T indicates exercises that should be solved using technology

1. Complete the following probability distribution table, and then calculate the stated probabilities. [**HINT**: See Quick Example 5.]

Outcome	a	b	c	d	e
Probability	.1	.05	.6	.05	

   a. $P(\{a, c, e\})$
   b. $P(E \cup F)$, where $E = \{a, c, e\}$ and $F = \{b, c, e\}$
   c. $P(E')$, where E is as in part (b)
   d. $P(E \cap F)$, where E and F are as in part (b)

2. Repeat Exercise 1 using the following table. [**HINT**: See Quick Example 5.]

Outcome	a	b	c	d	e
Probability	.1		.65	.1	.05

In Exercises 3–8, calculate the (modeled) probability $P(E)$ using the given information, assuming that all outcomes are equally likely. [**HINT**: See Quick Examples 11–13.]

3. $n(S) = 20, n(E) = 5$ 4. $n(S) = 8, n(E) = 4$

5. $n(S) = 10, n(E) = 10$ 6. $n(S) = 10, n(E) = 0$

7. $S = \{a, b, c, d\}, E = \{a, b, d\}$

8. $S = \{1, 3, 5, 7, 9\}, E = \{3, 7\}$

In Exercises 9–18 an experiment is given together with an event. Find the (modeled) probability of each event, assuming that the coins and dice are distinguishable and fair and that what are observed are the faces or numbers uppermost. (Compare with Exercises 1–10 in Section 7.1.)

9. Two coins are tossed; the result is at most one tail.

10. Two coins are tossed; the result is one or more heads.

11. Three coins are tossed; the result is at most one head.

12. Three coins are tossed; the result is more tails than heads.

13. Two dice are rolled; the numbers add to 5.

14. Two dice are rolled; the numbers add to 9.

15. Two dice are rolled; the numbers add to 1.

16. Two dice are rolled; one of the numbers is even, and the other is odd.

17. Two dice are rolled; both numbers are prime.[30]

18. Two dice are rolled; neither number is prime.

19. If two indistinguishable dice are rolled, what is the probability of the event $\{(4, 4), (2, 3)\}$? What is the corresponding event for a pair of distinguishable dice? [**HINT**: See Example 2.]

20. If two indistinguishable dice are rolled, what is the probability of the event $\{(5, 5), (2, 5), (3, 5)\}$? What is the corresponding event for a pair of distinguishable dice? [**HINT**: See Example 2.]

21. A die is weighted in such a way that each of 2, 4, and 6 is twice as likely to come up as each of 1, 3, and 5. Find the probability distribution. What is the probability of rolling less than 4? [**HINT**: See Example 3.]

22. Another die is weighted in such a way that each of 1 and 2 is three times as likely to come up as each of the other numbers. Find the probability distribution. What is the probability of rolling an even number?

23. A tetrahedral die has four faces, numbered 1–4. If the die is weighted in such a way that each number is twice as likely to land facing down as the next number (1 twice as likely as 2, 2 twice as likely as 3, and 3 twice as likely as 4), what is the probability distribution for the face landing down?

24. A dodecahedral die has 12 faces, numbered 1–12. If the die is weighted in such a way that 2 is twice as likely to land facing up as 1, 3 is three times as likely to land facing up as 1, and so on, what is the probability distribution for the face landing up?

In Exercises 25–40, use the given information to find the indicated probability. [**HINT**: See Quick Examples 14–16.]

25. $P(A) = .1, P(B) = .6, P(A \cap B) = .05$. Find $P(A \cup B)$.

26. $P(A) = .3, P(B) = .4, P(A \cap B) = .02$. Find $P(A \cup B)$.

27. $A \cap B = \varnothing, P(A) = .3, P(A \cup B) = .4$. Find $P(B)$.

28. $A \cap B = \varnothing, P(B) = .8, P(A \cup B) = .8$. Find $P(A)$.

---

[30] A positive integer is prime if it is neither 1 nor a product of smaller integers.

29. $A \cap B = \varnothing$, $P(A) = .3$, $P(B) = .4$. Find $P(A \cup B)$.

30. $A \cap B = \varnothing$, $P(A) = .2$, $P(B) = .3$. Find $P(A \cup B)$.

31. $P(A \cup B) = .9$, $P(B) = .6$, $P(A \cap B) = .1$. Find $P(A)$.

32. $P(A \cup B) = 1.0$, $P(A) = .6$, $P(A \cap B) = .1$. Find $P(B)$.

33. $P(A) = .75$. Find $P(A')$.   **34.** $P(A) = .22$. Find $P(A')$.

35. A, B, and C are mutually exclusive. $P(A) = .3$, $P(B) = .4$, $P(C) = .3$. Find $P(A \cup B \cup C)$.

36. A, B, and C are mutually exclusive. $P(A) = .2$, $P(B) = .6$, $P(C) = .1$. Find $P(A \cup B \cup C)$.

37. A and B are mutually exclusive. $P(A) = .3$, $P(B) = .4$. Find $P((A \cup B)')$.

38. A and B are mutually exclusive. $P(A) = .4$, $P(B) = .4$. Find $P((A \cup B)')$.

39. $A \cup B = S$ and $A \cap B = \varnothing$. Find $P(A) + P(B)$.

40. $P(A \cup B) = .3$ and $P(A \cap B) = .1$. Find $P(A) + P(B)$.

In Exercises 41–46, determine whether the information shown is consistent with a probability distribution. If not, say why.

41. $P(A) = .2$; $P(B) = .1$; $P(A \cup B) = .4$

42. $P(A) = .2$; $P(B) = .4$; $P(A \cup B) = .2$

43. $P(A) = .2$; $P(B) = .4$; $P(A \cap B) = .2$

44. $P(A) = .2$; $P(B) = .4$; $P(A \cap B) = .3$

45. $P(A) = 0.1$; $P(B) = 0$; $P(A \cup B) = 0$

46. $P(A) = .1$; $P(B) = 0$; $P(A \cap B) = 0$

Applications

47. *Subprime Mortgages during the 2000–2008 Housing Bubble* (Compare Exercise 27 in Section 7.2.) The following chart shows the (approximate) total number of subprime home mortgages in Texas in November 2008, broken down into four categories:[31]

Mortgage Status	Current	Past Due	In Foreclosure	Repossessed	Total
Number	136,330	53,310	8,750	5,090	203,480

(The four categories are mutually exclusive; for instance, "Past Due" refers to a mortgage whose payment status is past due but is not in foreclosure, and "In Foreclosure" refers to a mortgage that is in the process of being foreclosed but not yet repossessed.)

a. Find the probability that a randomly selected subprime mortgage in Texas during November 2008 was neither in foreclosure nor repossessed. [**HINT:** See Example 1.]

b. What is the probability that a randomly selected subprime mortgage in Texas during November 2008 was not current?

48. *Subprime Mortgages during the 2000–2008 Housing Bubble* (Compare Exercise 28 in Section 7.2.) The following chart shows the (approximate) total number of subprime home mortgages in Florida in November 2008, broken down into four categories:[32]

Mortgage Status	Current	Past Due	In Foreclosure	Repossessed	Total
Number	130,400	73,260	72,380	17,000	293,040

(The four categories are mutually exclusive; for instance, "Past Due" refers to a mortgage whose payment status is past due but is not in foreclosure, and "In Foreclosure" refers to a mortgage that is in the process of being foreclosed but not yet repossessed.)

a. Find the probability that a randomly selected subprime mortgage in Florida during November 2008 was either in foreclosure or repossessed. [**HINT:** See Example 1.]

b. What is the probability that a randomly selected subprime mortgage in Florida during November 2008 was not repossessed?

49. *Ethnic Diversity* The following pie chart shows the ethnic makeup of California schools in the 2006–2007 academic year.[33]

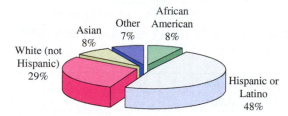

Write down the probability distribution showing the probability that a randomly selected California student in 2006–2007 belonged to one of the ethnic groups named. What is the probability that a student is neither white nor Asian?

50. *Ethnic Diversity* (Compare Exercise 49.) The following pie chart shows the ethnic makeup of California schools in the 1981–1982 academic year.[34]

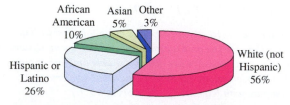

Write down the probability distribution showing the probability that a randomly selected California student in

[31] Data are rounded to the nearest 10 units. Source: Federal Reserve Bank of New York (www.newyorkfed.org/regional/subprime.html).

[32] *Ibid.*

[33] Source: CBEDS data collection, Educational Demographics, October 2006 (www.cde.ca.gov).

[34] *Ibid.*

1981–1982 belonged to one of the ethnic groups named. What is the probability that a student is neither Hispanic, Latino, nor African American?

51. ▼ *Internet Investments in the 1990s* The following excerpt is from an article in *The New York Times* in July 1999:[35]

    While statistics are not available for web entrepreneurs who fail, the venture capitalists that finance such Internet start-up companies have a rule of thumb. For every 10 ventures that receive financing—and there are plenty that do not—2 will be stock market successes, which means spectacular profits for early investors; 3 will be sold to other concerns, which translates into more modest profits; and the rest will fail.

    a. What is a sample space for the scenario?
    b. Write down the associated probability distribution.
    c. What is the probability that a start-up venture that receives financing will realize profits for early investors?

52. ▼ *Internet Investments in the 1990s* The following excerpt is from an article in *The New York Times* in July 1999:[36]

    Right now, the market for Web stocks is sizzling. Of the 126 initial public offerings of Internet stocks priced this year, 73 are trading above the price they closed on their first day of trading. . . . Still, 53 of the offerings have failed to live up to their fabulous first-day billings, and 17 [of these] are below the initial offering price.

    Assume that, on the first day of trading, all stocks closed higher than their initial offering price.
    a. What is a sample space for the scenario?
    b. Write down the associated probability distribution. (Round your answers to two decimal places.)
    c. What is the probability that an Internet stock purchased during the period reported ended either below its initial offering price or above the price it closed on its first day of trading? [**HINT:** See Example 3.]

53. ▼ *Market Share: Light Vehicles* In 2003, 25% of all light vehicles sold (SUVs, pickups, passenger cars, and minivans) in the United States were SUVs, and 15% were pickups. Moreover, a randomly chosen vehicle sold that year was five times as likely to be a passenger car as a minivan.[37] Find the associated probability distribution.

54. ▼ *Market Share: Light Vehicles* In 2000, 15% of all light vehicles (SUVs, pickups, passenger cars, and minivans)

sold in the United States were pickups, and 55% were passenger cars. Moreover, a randomly chosen vehicle sold that year was twice as likely to be an SUV as a minivan.[38] Find the associated probability distribution.

Gambling In Exercises 55–62 are detailed some of the nefarious dicing practices of the Win Some/Lose Some Casino. *In each case, find the probabilities of all the possible outcomes and also the probability that an odd number or an odd sum faces up.* [**HINT:** See Example 3.]

55. Some of the dice are specially designed so that 1 and 6 never come up and all the other outcomes are equally likely.

56. Other dice are specially designed so that 1 comes up half the time, 6 never comes up, and all the other outcomes are equally likely.

57. Some of the dice are cleverly weighted so that each of 2, 3, 4, and 5 is twice as likely to come up as 1 is, and 1 and 6 are equally likely.

58. Other dice are weighted so that each of 2, 3, 4, and 5 is half as likely to come up as 1 is, and 1 and 6 are equally likely.

59. ▼ Some pairs of dice are magnetized so that each pair of mismatching numbers is twice as likely to come up as each pair of matching numbers.

60. ▼ Other pairs of dice are so strongly magnetized that mismatching numbers never come up.

61. ▼ Some dice are constructed in such a way that deuce (2) is five times as likely to come up as 4 and three times as likely to come up as each of 1, 3, 5, and 6.

62. ▼ Other dice are constructed in such a way that deuce is six times as likely to come up as 4 and four times as likely to come up as each of 1, 3, 5, and 6.

63. *Astrology* The astrology software package *Turbo Kismet*[39] works by first generating random number sequences and then interpreting them numerologically. When I ran it yesterday, it informed me that there was a 1/3 probability that I would meet a tall, dark stranger this month, a 2/3 probability that I would travel this month, and a 1/6 probability that I would meet a tall, dark stranger and also travel this month. What is the probability that I will either meet a tall, dark stranger or travel this month? [**HINT:** See Quick Example 14.]

64. *Astrology* Another astrology software package, *Java Kismet*, is designed to help day traders choose stocks based on the position of the planets and constellations. When I ran it yesterday, it informed me that there was a .5 probability that Amazon.com will go up this afternoon, a .2 probability that Yahoo.com will go up this afternoon, and a .2 chance that

[35] Article: "Not All Hit It Rich in the Internet Gold Rush," *New York Times*, July 20, 1999, p. A1.

[36] Article: Ibid. Source for data: Comm-Scan/*New York Times*, July 20, 1999, p. A1.

[37] Source: Environmental Protection Agency/*New York Times*, June 28, 2003.

[38] *Ibid.*

[39] The name and concept were borrowed from a hilarious (as yet unpublished) novel by the science-fiction writer William Orr, who also happened to be a faculty member at Hofstra University.

both will go up this afternoon. What is the probability that either Amazon.com or Yahoo.com will go up this afternoon? [**HINT:** See Quick Example 14.]

65. *Polls* According to a *New York Times*/CBS poll released in March 2005, 61% of those polled ranked jobs or health care as the top domestic priority.[40] What is the probability that a randomly selected person polled did not rank either as the top domestic priority? [**HINT:** See Example 5.]

66. *Polls* According to *The New York Times*/CBS poll of March 2005 referred to in Exercise 65, 72% of those polled ranked neither Iraq nor North Korea as the top foreign policy issue.[41] What is the probability that a randomly selected person polled ranked either Iraq or North Korea as the top foreign policy issue? [**HINT:** See Example 5.]

67. *Electric Car Sales* In 2014 the probability that a randomly chosen electric car sold in the United States was manufactured by **Tesla** was .30, while the probability that it was manufactured by **Nissan** was .35.[42] What is the probability that a randomly chosen electric car was manufactured by neither company?

68. *Hybrid Auto Sales* In 2010 the probability that a randomly chosen hybrid vehicle sold in the United States was manufactured by **Ford** was .12, while the probability that it was manufactured by **Nissan** was .02.[43] What is the probability that a randomly chosen hybrid vehicle was manufactured by neither company?

Student Admissions Exercises 69–84 are based on the following table, which shows the profile, by the math section of the SAT Reasoning Test, of admitted students at UCLA for the Fall 2014 semester:[44]

SAT Reasoning Test—Math Section

	700–800	600–699	500–599	400–499	200–399	Total
Admitted	8,398	3,517	1,410	358	9	13,692
Not Admitted	16,599	18,363	13,119	6,714	1,652	56,447
Total Applicants	24,997	21,880	14,529	7,072	1,661	70,139

Determine the probabilities of the following events. (Round your answers to the nearest .01.) [**HINT:** Example 6.]

69. An applicant was admitted.

70. An applicant had a Math SAT below 400.

71. An applicant had a Math SAT below 400 and was admitted.

72. An applicant had a Math SAT of 700 or above and was admitted.

73. An applicant was not admitted.

74. An applicant did not have a Math SAT below 400.

75. An applicant had a Math SAT in the range 500–599 or was admitted.

76. An applicant had a Math SAT of 700 or above or was admitted.

77. An applicant neither was admitted nor had a Math SAT in the range 500–599.

78. An applicant neither had a Math SAT of 700 or above nor was admitted.

79. ▼ An applicant who had a Math SAT below 400 was admitted.

80. ▼ An applicant who had a Math SAT of 700 or above was admitted.

81. ▼ An admitted student had a Math SAT of 700 or above.

82. ▼ An admitted student had a Math SAT below 400.

83. ▼ A rejected applicant had a Math SAT below 600.

84. ▼ A rejected applicant had a Math SAT of at least 600.

85. ▼ *Social Security* According to *The New York Times*/CBS poll of March 2005 referred to in Exercise 65, 79% agreed that it should be the government's responsibility to provide a decent standard of living for the elderly, and 43% agreed that it would be a good idea to invest part of their Social Security taxes on their own. What is the smallest percentage of people who could have agreed with both statements? What is the largest percentage of people who could have agreed with both statements?

86. ▼ *Social Security* According to *The New York Times*/CBS poll of March 2005 referred to in Exercise 65, 49% agreed that Social Security taxes should be raised if necessary to keep the system afloat, and 43% agreed that it would be a good idea to invest part of their Social Security taxes on their own. What is the largest percentage of people who could have agreed with at least one of these statements? What is the smallest percentage of people who could have agreed with at least one of these statements?

87. ▼ *Greek Life* The TΦΦ Sorority has a tough pledging program: It requires its pledges to master the Greek alphabet forward, backward, and "sideways." During the last pledge period, two thirds of the pledges failed to learn it backward, and three quarters of them failed to learn it sideways; 5 of the 12 pledges failed to master it either backward or sideways. Because admission into the sisterhood requires both backward and sideways mastery, what fraction of the pledges were disqualified on this basis?

88. ▼ *Swords and Sorcery* Lance the Wizard has been informed that tomorrow there will be a 50% chance of encountering the evil Myrmidons and a 20% chance of meeting up with the dreadful Balrog. Moreover, Hugo the Elf has predicted that

[40] Source: *New York Times*, March 3, 2005, p. A20.

[41] *Ibid.*

[42] Probabilities are approximate. Source for sales data: www.wikipedia.com.

[43] *Ibid.*

[44] Source: University of California (www.admissions.ucla.edu/Prospect/Adm_fr/Frosh_Prof14.htm).

there is a 10% chance of encountering both tomorrow. What is the probability that Lance will be lucky tomorrow and encounter neither the Myrmidons nor the Balrog?

89. ▼ *Public Health* A study shows that 80% of the population was vaccinated against the Venusian flu but 2% of the vaccinated population got the flu anyway. If 10% of the total population got this flu, what percent of the population either got the vaccine or got the disease?

90. ▼ *Public Health* A study shows that 75% of the population was vaccinated against the Martian ague but 4% of this group got this disease anyway. If 10% of the total population got this disease, what is the probability that a randomly selected person neither was vaccinated nor contracted Martian ague?

Communication and Reasoning Exercises

91. Design an experiment based on rolling a fair die for which there are exactly three outcomes with the same probabilities.

92. Design an experiment based on rolling a fair die for which there are at least three outcomes with different probabilities.

93. ▼ Tony has had a losing streak at the casino: The chances of winning the game he is playing are 40%, but he has lost five times in a row. Tony argues that, because he should have won two times, the game must obviously be rigged. Comment on his reasoning.

94. ▼ Maria is on a winning streak at the casino. She has already won four times in a row and concludes that her chances of winning a fifth time are good. Comment on her reasoning.

95. Complete the following sentence. The probability of the union of two events is the sum of the probabilities of the two events if _____.

96. A friend of yours asserted at lunch today that, according to the weather forecast for tomorrow, there is a 52% chance of rain and a 60% chance of snow. "But that's impossible!" you blurted out, "the percentages add up to more than 100%." Explain why you were wrong.

97. ▼ A certain experiment is performed a large number of times, and the event E has relative frequency equal to zero. This means that it should have modeled probability zero—right? [**HINT:** See the definition of a probability model.]

98. ▼ (Refer to Exercise 97.) How can the modeled probability of winning the lottery be nonzero if you have never won it despite having played 600 times? [**HINT:** See the definition of a probability model.]

99. ▼ Explain how the addition principle for mutually exclusive events follows from the general addition principle.

100. ▼ Explain how the property $P(A') = 1 - P(A)$ follows directly from the properties of a probability distribution.

101. ◆ It is said that lightning never strikes twice in the same spot. Assuming this to be the case, what should be the modeled probability that lightning will strike a given spot during a thunderstorm? Explain. [**HINT:** See the definition of a probability model.]

102. ◆ A certain event has modeled probability equal to zero. This means it will never occur—right? [**HINT:** See the definition of a probability model.]

103. ◆ Find a formula for the probability of the union of three (not necessarily mutually exclusive) events A, B, and C.

104. ◆ Four events A, B, C, and D have the following property: If any two events have an outcome in common, that outcome is common to all four events. Find a formula for the probability of their union.

7.4 Probability and Counting Techniques

Counting Techniques Return

We saw in Section 7.3 that, when all outcomes in a sample space are equally likely, we can use the following formula to model the probability of each event.

Modeling Probability: Equally Likely Outcomes

In an experiment in which all outcomes are equally likely, the probability of an event E is given by

$$P(E) = \frac{\text{Number of favorable outcomes}}{\text{Total number of outcomes}} = \frac{n(E)}{n(S)}.$$

This formula is simple, but calculating $n(E)$ and $n(S)$ may not be. In this section we look at some examples in which we need to use the counting techniques discussed in Chapter 6.

S is the set of *all* outcomes that can occur, and has nothing to do with having green marbles.

EXAMPLE 1 Marbles

A bag contains four red marbles and two green ones. Upon seeing the bag, Suzan (who has compulsive marble-grabbing tendencies) sticks her hand in and grabs three at random. Find the probability that she will get both green marbles.

Solution According to the formula, we need to know these numbers:

- The number of elements in the sample space *S*
- The number of elements in the event *E*.

First of all, what is the sample space? The sample space is the set of all possible outcomes, and each outcome consists of a set of three marbles (in Suzan's hand). So the set of outcomes is the set of all sets of three marbles chosen from a total of six marbles (four red and two green). Thus,

$$n(S) = C(6, 3) = 20.$$

Now what about *E*? This is the event that Suzan gets both green marbles. We must *rephrase this as a subset of S* in order to deal with it: "*E* is the collection of sets of three marbles such that one is red and two are green." Thus, $n(E)$ is the *number* of such sets, which we determine using a decision algorithm.

Step 1 Choose a red marble: $C(4, 1) = 4$ possible outcomes.

Step 2 Choose the two green marbles: $C(2, 2) = 1$ possible outcome.

We get $n(E) = 4 \times 1 = 4$. Now,

$$P(E) = \frac{n(E)}{n(S)} = \frac{4}{20} = \frac{1}{5}.$$

Thus, there is a one in five chance of Suzan's getting both the green marbles.

EXAMPLE 2 Investment Lottery

After a down day on the stock market, you decide to ignore your broker's cautious advice and purchase three stocks at random from the six most active stocks listed on the New York Stock Exchange at the end of the day's trading.[45]

Symbol	Company	Price ($)	% Change
BAC	Bank of America	16.78	−1.81
KEY	KeyCorp	12.42	−7.17
GE	General Electric	28.92	−1.43
PFE	Pfizer	33.82	−2.73
VRX	Valeant Pharmaceuticals	93.77	−15.90
NYCB	New York Community Bancorp	16.52	−2.02

Find the probabilities of the following events:

a. You purchase BAC and KEY.

b. At most two of the stocks you purchase declined in value by more than 2.5%.

[45] Most active stocks on October 30, 2015. Source: www.nasdaq.com.

Solution First, the sample space is the set of all collections of three stocks chosen from the six. Thus,

$$n(S) = C(6, 3) = 20.$$

a. The event E of interest is the event that you purchase BAC and KEY. Thus, E is the set of all groups of three stocks that include BAC and KEY. Because there is only one more stock left to choose,

$$n(E) = C(4, 1) = 4.$$

We now have

$$P(E) = \frac{n(E)}{n(S)} = \frac{4}{20} = \frac{1}{5} = .2.$$

b. Let F be the event that at most two of the stocks you purchase declined in value by more than 2.5%. Thus, F is the set of all groups of three stocks of which at most two declined in value by more than 2.5%. To calculate $n(F)$, we use the following decision algorithm.

Alternative 1: None of the stocks declined in value by more than 2.5%.
    Step 1 Choose three stocks that did not decline in value by more than 2.5%: $C(3, 3) = 1$ possibility.

Alternative 2: One of the stocks declined in value by more than 2.5%.
    Step 1 Choose one stock that declined in value by more than 2.5%: $C(3, 1) = 3$ possibilities.
    Step 2 Choose two stocks that did not decline in value by more than 2.5%: $C(3, 2) = 3$ possibilities.
    This gives $3 \times 3 = 9$ possibilities for this alternative.

Alternative 3: Two of the stocks declined in value by more than 2.5%.
    Step 1 Choose two stocks that declined in value by more than 2.5%: $C(3, 2) = 3$ possibilities.
    Step 2 Choose one stock that did not decline in value by more than 2.5%: $C(3, 1) = 3$ possibilities.
    This gives $3 \times 3 = 9$ possibilities for this alternative.

So we have a total of $1 + 9 + 9 = 19$ possible outcomes. Thus,

$$n(F) = 19$$

and

$$P(F) = \frac{n(F)}{n(S)} = \frac{19}{20} = .95.$$

At most two of the stocks declined in value by more than 2.5%.

Complementary Events

At least three of the stocks declined in value by more than 2.5%.

➡ **Before we go on . . .** When we are counting the number of outcomes in an event, the calculation is sometimes easier if we look at the *complement* of that event. In the case of part (b) of Example 2, the complement of the event F is

    F': At least three of the stocks you purchase declined in value by more than 2.5%.

Because there are only three stocks in your portfolio, this is the same as the event that all three stocks you purchase declined in value by more than 2.5%. The decision algorithm for $n(F')$ is far simpler:

Step 1 Choose three stocks that declined in value by more than 2.5%: $C(3, 3) = 1$ possibility.

So $n(F') = 1$, giving

$$n(F) = n(S) - n(F') = 20 - 1 = 19,$$

as we calculated above. ∎

EXAMPLE 3 **Poker Hands**

You are dealt 5 cards from a well-shuffled standard deck of 52. Find the probability that you have a full house. (Recall that a full house consists of 3 cards of one denomination and 2 of another.)

Solution The sample space S is the set of all possible 5-card hands dealt from a deck of 52. Thus,

$$n(S) = C(52, 5) = 2{,}598{,}960.$$

If the deck is thoroughly shuffled, then each of these 5-card hands is equally likely. Now consider the event E, the set of all possible 5-card hands that constitute a full house. To calculate $n(E)$, we use a decision algorithm, which we show in the following compact form:

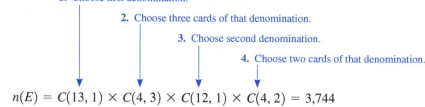

1. Choose first denomination.
2. Choose three cards of that denomination.
3. Choose second denomination.
4. Choose two cards of that denomination.

$$n(E) = C(13, 1) \times C(4, 3) \times C(12, 1) \times C(4, 2) = 3{,}744$$

Thus,

$$P(E) = \frac{n(E)}{n(S)} = \frac{3{,}744}{2{,}598{,}960} \approx .00144.$$

In other words, there is an approximately 0.144% chance that you will be dealt a full house.

EXAMPLE 4 **More Poker Hands**

You are playing poker, and you have been dealt the following hand:

J♠, J♦, J♥, 2♣, 10♠.

You decide to exchange the last two cards. The exchange works as follows: The two cards are discarded (not replaced in the deck), and you are dealt two new cards.

a. Find the probability that you end up with a full house.

b. Find the probability that you end up with four jacks.

c. What is the probability that you end up with either a full house or four jacks?

Solution

a. To get a full house, you must be dealt two of a kind. The sample space S is the set of all pairs of cards selected from what remains of the original deck of 52. You were dealt 5 cards originally, so there are $52 - 5 = 47$ cards left in the deck. Thus, $n(S) = C(47, 2) = 1{,}081$. The event E is the set of all pairs of cards that constitute two of a kind. Note that you cannot get two jacks because only one is left in the deck. Also, only three 2s and three 10s are left in the deck. We have

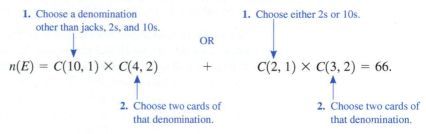

Thus,

$$P(E) = \frac{n(E)}{n(S)} = \frac{66}{1{,}081} \approx .0611.$$

b. We have the same sample space as in part (a). Let F be the set of all pairs of cards that include the missing jack of clubs. So

    1. Choose the jack of clubs.

        2. Choose one card from the remaining 46.

$$n(F) = C(1, 1) \times C(46, 1) = 46.$$

Thus,

$$P(F) = \frac{n(F)}{n(S)} = \frac{46}{1{,}081} \approx .0426.$$

c. We are asked to calculate the probability of the event $E \cup F$. From the addition principle we have

$$P(E \cup F) = P(E) + P(F) - P(E \cap F).$$

Because $E \cap F$ means "E and F," $E \cap F$ is the event that the pair of cards you are dealt are two of a kind and include the jack of clubs. But this is impossible because only one jack is left. Thus, $E \cap F = \varnothing$, so $P(E \cap F) = 0$. This gives us

$$P(E \cup F) = P(E) + P(F) \approx .0611 + .0426 = .1037.$$

In other words, there is slightly better than a 1 in 10 chance that you will wind up with either a full house or four of a kind, given the original hand.

➡ **Before we go on . . .** A more accurate answer to part (c) of Example 4 is $(66 + 46)/1{,}081 \approx .1036$. We lost some accuracy in rounding the answers to parts (a) and (b). ∎

EXAMPLE 5   **Committees**

The University Senate bylaws at Hofstra University state the following:[46]

The Student Affairs Committee shall consist of one elected faculty senator, one faculty senator-at-large, one elected student senator, five student senators-at-

---

[46] As of 2011. Source: Hofstra University Senate Bylaws.

large (including one from the graduate school), two delegates from the Student Government Association, the President of the Student Government Association or his/her designate, and the President of the Graduate Student Organization. It shall be chaired by the elected student senator on the Committee and it shall be advised by the Dean of Students or his/her designate.

You are an undergraduate student, and even though you are not an elected student senator, you would very much like to serve on the Student Affairs Committee. The senators-at-large as well as the Student Government delegates are chosen by means of a random drawing from a list of candidates. There are already 13 undergraduate candidates for the position of senator-at-large and 6 candidates for Student Government delegates, and you have been offered a position on the Student Government Association by the president (who happens to be a good friend of yours), should you wish to join it. (This would make you ineligible for a senator-at-large position.) What should you do?

Solution You have two options. Option 1 is to include your name on the list of candidates for the senator-at-large position. Option 2 is to join the Student Government Association (SGA) and add your name to its list of candidates. Let us look at the two options separately.

Option 1: Add your name to the senator-at-large list.
This will result in a list of 14 undergraduates for 4 undergraduate positions. The sample space is the set of all possible outcomes of the random drawing. Each outcome consists of a set of 4 lucky students chosen from 14. Thus,

$$n(S) = C(14, 4) = 1{,}001.$$

We are interested in the probability that you are among the chosen four. Thus, E is the set of sets of four that include you:

1. Choose yourself.

2. Choose three from the remaining 13.

$$n(E) = C(1, 1) \times C(13, 3) = 286.$$

So

$$P(E) = \frac{n(E)}{n(S)} = \frac{286}{1{,}001} = \frac{2}{7} \approx .2857.$$

Option 2: Join the SGA and add your name to its list.
This results in a list of seven candidates from which two are selected. For this case, the sample space consists of all sets of two chosen from seven, so

$$n(S) = C(7, 2) = 21,$$

and

1. Choose yourself.

2. Choose one from the remaining six.

$$n(E) = C(1, 1) \times C(6, 1) = 6.$$

Thus,

$$P(E) = \frac{n(E)}{n(S)} = \frac{6}{21} = \frac{2}{7} \approx .2857.$$

In other words, the probability of being selected is exactly the same for Option 1 as it is for Option 2! Thus, you can choose either option, and you will have slightly less than a 29% chance of being selected.

7.4 EXERCISES

▼ more advanced   ◆ challenging
T indicates exercises that should be solved using technology

Recall from Example 1 that whenever Suzan sees a bag of marbles, she grabs a handful at random. In Exercises 1–10, she has seen a bag containing four red marbles, three green ones, two white ones, and one purple one. She grabs five of them. Find the probabilities of the following events, expressing each as a fraction in lowest terms. [**HINT:** See Example 1.]

1. She has all the red ones.

2. She has none of the red ones.

3. She has at least one white one.

4. She has at least one green one.

5. She has two red ones and one of each of the other colors.

6. She has two green ones and one of each of the other colors.

7. She has at most one green one.

8. She has no more than one white one.

9. She does not have all the red ones.

10. She does not have all the green ones.

Dogs of the Dow *The "Dogs of the Dow" are the stocks listed on the Dow with the highest dividend yield. Exercises 11–16 are based on the following table, which shows the top ten stocks of the "Dogs of the Dow" list for 2015, based on their performance the preceding year.*[47] [**HINT:** See Example 2.]

Symbol	Company	Price	Yield
T	AT&T	33.59	5.48%
VZ	Verizon	46.78	4.70%
CVX	Chevron	112.18	3.82%
MCD	McDonald's	93.70	3.63%
PFE	Pfizer	31.15	3.60%
GE	General Electric	25.27	3.48%
MRK	Merck	56.79	3.17%
CAT	Caterpillar	91.53	3.06%
XOM	ExxonMobil	92.45	2.99%
KO	Coca-Cola	42.22	2.89%

11. If you selected two of these stocks at random, what is the probability that both the stocks in your selection had yields of 3.75% or more?

12. If you selected three of these stocks at random, what is the probability that all three of the stocks in your selection had yields of 3.75% or more?

13. If you selected four of these stocks at random, what is the probability that your selection included the company with the highest yield and excluded the company with the lowest yield?

14. If you selected four of these stocks at random, what is the probability that your selection included KO and VZ but excluded PFE and GE?

15. ▼ If your portfolio included 100 shares of PFE and you then purchased 100 shares each of any two companies on the list at random, find the probability that you ended up with a total of 200 shares of PFE.

16. ▼ If your portfolio included 100 shares of PFE and you then purchased 100 shares each of any three companies on the list at random, find the probability that you ended up with a total of 200 shares of PFE.

17. *Tests* A test has three parts. Part A consists of eight true-false questions, Part B consists of five multiple-choice questions with five choices each, and Part C requires you to match five questions with five different answers one-to-one. Assuming that you make random guesses in filling out your answer sheet, what is the probability that you will earn 100% on the test? (Leave your answer as a formula.)

18. *Tests* A test has three parts. Part A consists of four true-false questions, Part B consists of four multiple-choice questions with five choices each, and Part C requires you to match six questions with six different answers one-to-one. Assuming that you make random choices in filling out your answer sheet, what is the probability that you will earn 100% on the test? (Leave your answer as a formula.)

Poker *In Exercises 19–24 you are asked to calculate the probability of being dealt various poker hands. (Recall that a poker player is dealt 5 cards at random from a standard deck of 52.) Express each of your answers as a decimal rounded to four decimal places unless otherwise stated.* [**HINT:** See Example 3.]

19. **Two of a kind:** Two cards with the same denomination and three cards with other denominations (different from each other and that of the pair). Example: K♣, K♥, 2♠, 4♦, J♠

20. **Three of a kind:** Three cards with the same denomination and two cards with other denominations (different from each other and that of the three). Example: Q♣, Q♥, Q♠, 4♦, J♠

---
[47] Source: www.dogsofthedow.com.

21. **Two pairs:** Two cards with one denomination, two with another, and one with a third. Example: 3♣, 3♥, Q♠, Q♥, 10♠

22. **Straight flush:** Five cards of the same suit with consecutive denominations but not a royal flush. (A royal flush consists of the 10, J, Q, K, and A of one suit.) Round the answer to one significant digit. Examples: A♣, 2♣, 3♣, 4♣, 5♣, or 9♦, 10♦, J♦, Q♦, K♦, or A♥, 2♥, 3♥, 4♥, 5♥, but *not* 10♦, J♦, Q♦, K♦, A♦

23. **Flush:** Five cards of the same suit but not a straight flush or royal flush. Example: A♣, 5♣, 7♣, 8♣, K♣

24. **Straight:** Five cards with consecutive denominations but not all of the same suit. Examples: 9♦, 10♦, J♣, Q♥, K♦, and 10♥, J♦, Q♦, K♦, A♦

25. ***The Monkey at the Typewriter*** Suppose that a monkey is seated at a computer keyboard and randomly strikes the 26 letter keys and the space bar. Find the probability that its first 39 characters (including spaces) will be "to be or not to be that is the question". (Leave your answer as a formula.)

26. ***The Cat on the Piano*** A standard piano keyboard has 88 different keys. Find the probability that a cat, jumping on 4 keys in sequence and at random (possibly with repetition), will strike the first four notes of Beethoven's Fifth Symphony. (Leave your answer as a formula.)

27. (***Based on a question from the GMAT***) Tyler and Gebriella are among seven contestants from whom four semifinalists are to be selected at random. Find the probability that neither Tyler nor Gebriella is selected.

28. (***Based on a question from the GMAT***) Tyler and Gebriella are among seven contestants from whom four semifinalists are to be selected at random. Find the probability that Tyler but not Gebriella is selected.

29. ▼ ***Lotteries*** The *Sorry State Lottery* requires you to select five different numbers from 0 through 49. (Order is not important.) You are a Big Winner if the five numbers you select agree with those in the drawing, and you are a Small-Fry Winner if four of your five numbers agree with those in the drawing. What is the probability of being a Big Winner? What is the probability of being a Small-Fry Winner? What is the probability that you are either a Big Winner or a Small-Fry winner?

30. ▼ ***Lotteries*** The *Sad State Lottery* requires you to select a sequence of three different numbers from 0 through 49. (Order is important.) You are a Winner if your sequence agrees with that in the drawing, and you are a Booby Prize Winner if your selection of numbers is correct but in the wrong order. What is the probability of being a Winner? What is the probability of being a Booby Prize Winner? What is the probability that you are either a Winner or a Booby Prize Winner?

31. ▼ ***Transfers*** Your company is considering offering 400 employees the opportunity to transfer to its new headquarters in Ottawa, and, as personnel manager, you decide that it would be fairest if the transfer offers are decided by means of a lottery. Assuming that your company currently employs 100 managers, 100 factory workers, and 500 miscellaneous staff, find the following probabilities, leaving the answers as formulas:

   a. All the managers will be offered the opportunity.
   b. You will be offered the opportunity.

32. ▼ ***Transfers*** (Refer back to Exercise 31.) After thinking about your proposed method of selecting employees for the opportunity to move to Ottawa, you decide that it might be a better idea to select 50 managers, 50 factory workers, and 300 miscellaneous staff, all chosen at random. Find the probability that you will be offered the opportunity. (Leave your answer as a formula.)

33. ▼ ***Lotteries*** In a New York State daily lottery game, a sequence of three digits (not necessarily different) in the range 0–9 are selected at random. Find the probability that all three are different.

34. ▼ ***Lotteries*** Refer back to Exercise 33. Find the probability that two of the three digits are the same.

35. ▼ ***Elimination Tournaments*** In an elimination tournament the teams are arranged in opponent pairs for the first round, and the winner of each round goes on to the next until the champion emerges. What is the probability that North Carolina will beat Central Connecticut but lose to Virginia in the following (fictitious) soccer tournament? (Assume that all outcomes are equally likely.)

36. ▼ ***Elimination Tournaments*** In a (fictitious) soccer tournament involving the four teams San Diego State, De Paul, Colgate, and Hofstra, find the probability that Hofstra will play Colgate in the finals and win. (Assume that all outcomes are equally likely and that the teams not listed in the first round slots are placed at random.)

Elimination Tournaments *The following diagram illustrates a 16-team tournament bracket, in which the 16 participating teams are arranged on the left under Round 1 and the winners of each round are added as the tournament progresses. The top* *team in each game is considered the "home" team, so the top-to-bottom order matters.*

Round 1   Round 2   Round 3   Round 4
           (Quarter final) (Semifinal)   (Final)

Champion

To seed *a tournament means to select which teams to play each other in the first round according to their preliminary ranking. For instance, in professional tennis and NCAA basketball the seeding is set up in the following order based on the preliminary rankings: 1 versus 16, 8 versus 9, 5 versus 12, 4 versus 13, 6 versus 11, 3 versus 14, 7 versus 10, and 2 versus 15.[48] Exercises 37–40 are based on various types of elimination tournaments. (Leave each answer as a formula.)*

37. In a randomly chosen seeding of a 16-team tournament, what is the probability that the top-ranked team plays the bottom-ranked team, the second-ranked team plays the second-lowest ranked team, and so on? [**HINT:** See Exercise 65 in Section 6.4.]

38. In a randomly chosen seeding of an 8-team tournament, what is the probability that each team plays a team with adjacent ranking? [**HINT:** See Exercise 66 in Section 6.4.]

39. ▼ In 2014, after the NCAA basketball 64-team tournament had already been seeded, **Quicken Loans**, backed by investor Warren Buffett, offered a billion dollar prize for picking all the winners.[49] An *upset* occurs when a team beats a higher-ranked team.

a. If you picked the winners completely at random, what is the probability that you would win the prize?

b. If you picked the winners completely at random, what is the probability that your choice would give 15 upsets in the first four rounds? [**HINT:** See Exercise 67 in Section 6.4.]

40. ▼ Refer to Exercise 39. In the 2013 NCAA playoffs there were 10 upsets in the first round, 4 in the second round, and 3 in each of the third and fourth rounds.[50] If you had picked the winners completely at random, what is the probability that your choice would have given the same number of upsets in each of the first four rounds as above?

41. ▼ *Sports* The following table shows the results of the Big Eight Conference for the 1988 college football season:[51]

Team	Won	Lost
Nebraska (NU)	7	0
Oklahoma (OU)	6	1
Oklahoma State (OSU)	5	2
Colorado (CU)	4	3
Iowa State (ISU)	3	4
Missouri (MU)	2	5
Kansas (KU)	1	6
Kansas State (KSU)	0	7

An arrangement such as the one above in which each team has a different number of wins (from 0 to 7) is called a *perfect progression.* Assuming that the "Won" score for each team is chosen at random in the range 0–7, find the probability that the results form a perfect progression.[52] (Leave your answer as a formula.)

42. ▼ *Sports* Refer back to Exercise 41. Find the probability of a perfect progression with Nebraska scoring seven wins and zero losses. (Leave your answer as a formula.)

43. ▼ *Graph Searching* A graph consists of a collection of **nodes** (the dots in the figure) connected by **edges** (line segments from one node to another). A **move on a graph** is a move from one node to another along a single edge. Find the probability of going from Start to Finish in a sequence

[48] Source: www.wikipedia.com, www.ncaa.com

[49] No one won the prize, and the offer was scrapped the following year as a result of a series of lawsuits and countersuits by Yahoo, SCA Promotions (a sweepstakes company), and Berkshire-Hathaway. Sources: http://abcnews.go.com/Sports/warren-buffet-backs-billion-dollar-march-madness-challenge/story?id=21615743, http://money.cnn.com/2015/03/12/news/buffett-ncaa-bracket-bet/index.html.

[50] Source: *Washington Post*, March 16, 2014 (www.washingtonpost.com).

[51] Source: On the probability of a perfect progression, *The American Statistician,* August 1991, vol. 45, no. 3, p. 214.

[52] Even if all the teams are equally likely to win each game, the chances of a perfect progression actually coming up are a little more difficult to estimate, because the number of wins by one team directly affects the number of wins by the others. For instance, it is impossible for all eight teams to show a score of seven wins and zero losses at the end of the season—someone must lose! It is, however, not too hard to come up with a counting argument to estimate the total number of win-loss scores actually possible.

of two random moves in the graph shown. (All directions are equally likely.)

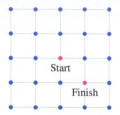

44. ▼ *Graph Searching* Refer back to Exercise 43. Find the probability of going from Start to one of the Finish nodes in a sequence of two random moves in the following figure. (All directions are equally likely.)

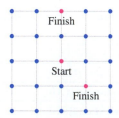

45. ◆ *Product Design* Your company has patented an electronic digital padlock that a user can program with his or her own four-digit code. (Each digit can be 0 through 9, and repetitions are allowed.) The padlock is designed to open either if the correct code is keyed in or—and this is helpful for forgetful people—if exactly one of the digits is incorrect. What is the probability that a randomly chosen sequence of four digits will open a programmed padlock?

46. ◆ *Product Design* Assume that you already know the first digit of the combination for the lock described in Exercise 45. Find the probability that a random guess of the remaining three digits will open the lock. [**HINT:** See Example 5.]

47. ◆ *Committees* An investigatory committee in the Kingdom of Utopia consists of a chief investigator (a Royal Party member), an assistant investigator (a Birthday Party member), two at-large investigators (either party), and five ordinary members (either party). Royal Party member Larry Sifford is hoping to avoid serving on the committee unless he is the Chief Investigator and Otis Taylor, a Birthday Party member, is the Assistant Investigator. The committee is to be selected at random from a pool of 12 candidates (including Larry Sifford and Otis Taylor), half of whom are Royal Party and half of whom are Birthday Party.
   a. How many different committees are possible?
   [**HINT:** See Example 5.]
   b. How many committees are possible in which Larry's hopes are fulfilled? (This includes the possibility that he's not on the committee at all.)

   c. What is the probability that he'll be happy with a randomly selected committee?

48. ◆ *Committees* A committee is to consist of a chair, three hagglers, and four do-nothings. The committee is formed by choosing randomly from a pool of 10 people and assigning them to the various "jobs."
   a. How many different committees are possible?
   [**HINT:** See Example 5.]
   b. Norman is eager to be the chair of the committee. What is the probability that he will get his wish?
   c. Norman's girlfriend Norma is less ambitious and would be happy to hold any position on the committee provided that Norman is also selected as a committee member. What is the probability that she will get her wish and serve on the committee?
   d. Norma does not get along with Oona (who is also in the pool of prospective members) and would be most unhappy if Oona were to chair the committee. Find the probability that all Norma's wishes will be fulfilled: She and Norman are on the committee, and it is not chaired by Oona.

Communication and Reasoning Exercises

49. What is wrong with the following argument? A bag contains two blue marbles and two red ones; two are drawn at random. Because there are four possibilities—(red, red), (blue, blue), (red, blue) and (blue, red)—the probability that both are red is 1/4.

50. What is wrong with the following argument? When we roll two indistinguishable dice, the number of possible outcomes (unordered groups of two not necessarily distinct numbers) is 21 and the number of outcomes in which both numbers are the same is 6. Hence, the probability of throwing a double is 6/21 = 2/7.

51. ▼ Suzan grabs two marbles out of a bag of five red marbles and four green ones. She could do so in two ways: She could take them out one at a time so that there is a first and a second marble, or she could grab two at once so that there is no order. Does the method she uses to grab the marbles affect the probability that she gets two red marbles?

52. ▼ If Suzan grabs two marbles, one at a time, out of a bag of five red marbles and four green ones, find an event with a probability that depends on the order in which the two marbles are drawn.

53. Create an interesting application whose solution requires finding a probability using combinations.

54. Create an interesting application whose solution requires finding a probability using permutations.

7.5 | Conditional Probability and Independence

Conditional Probability

Cyber Video Games, Inc., ran a television ad in advance of the release of its latest game, Ultimate Hockey. As Cyber Video's director of marketing, you would like to assess the ad's effectiveness, so you ask your market research team to survey video game players. The results of its survey of 800 video game players are summarized in the following table:

	Saw Ad	Did Not See Ad	Total
Purchased Game	20	70	90
Did Not Purchase Game	80	630	710
Total	100	700	800

At first glance, it looks as though potential customers are being *put off* by the ad: Only 20 people who saw the ad purchased the game, whereas 70 people purchased the game without seeing the ad at all. But let us analyze the figures a little more carefully.

First, let's restrict attention to those players who saw the ad (first column of data: "Saw Ad") and compute the estimated probability that a player *who saw the ad* purchased Ultimate Hockey.

	Saw Ad
Purchased Game	20
Did Not Purchase Game	80
Total	100

To compute this probability, we calculate

Probability that someone who saw the ad purchased the game

$$= \frac{\text{Number of people who saw the ad and bought the game}}{\text{Total number of people who saw the ad}} = \frac{20}{100} = .2.$$

In other words, 20% of game players who saw the ad went ahead and purchased the game. Let us compare this with the corresponding probability for those players who did *not* see the ad (second column of data "Did Not See Ad"):

	Did Not See Ad
Purchased Game	70
Did Not Purchase Game	630
Total	700

Probability that someone who did not see the ad purchased the game

$$= \frac{\text{Number of people who did not see the ad and purchased the game}}{\text{Total number of people who did not see the ad}} = \frac{70}{700} = .1.$$

Thus, only 10% of game players who did not see the ad purchased the game, whereas 20% of those who *did* see the ad purchased the game. Thus, it appears that the ad *was* highly persuasive: Seeing the ad appears to have made one twice as likely to purchase the game as not seeing the ad.

Here's some terminology. In this example there were two related events of importance:

> *A*: A video game player purchased Ultimate Hockey.

> *B*: A video game player saw the ad.

The first probability we computed was the estimated probability that a video game player purchased Ultimate Hockey *given that* he or she saw the ad. We call the latter probability the (estimated) **probability of *A*, given *B***, and we write it as $P(A|B)$. We call $P(A|B)$ a **conditional probability**—it is the probability of *A* under the condition that *B* occurred. Put another way, it is the probability of *A* occurring if the sample space is reduced to just those outcomes in *B*:

$$P(\text{Purchased game } \textit{given that } \text{saw the ad}) = P(A|B) = .2.$$

The second probability we computed was the estimated probability that a video game player purchased Ultimate Hockey *given that* he or she did not see the ad, or the **probability of *A*, given *B'***:

$$P(\text{Purchased game } \textit{given that } \text{did not see the ad}) = P(A|B') = .1.$$

Calculating Conditional Probabilities

How do we calculate conditional probabilities? In the example above, we used the ratio

$$P(A|B) = \frac{\text{Number of people who saw the ad and bought the game}}{\text{Total number of people who saw the ad}}.$$

The numerator is the frequency of $A \cap B$, and the denominator is the frequency of *B*:

$$P(A|B) = \frac{fr(A \cap B)}{fr(B)}.$$

Now, we can write this formula in another way:

$$P(A|B) = \frac{fr(A \cap B)}{fr(B)} = \frac{fr(A \cap B)/N}{fr(B)/N} = \frac{P(A \cap B)}{P(B)}.$$

We therefore have the following definition, which applies to general probability distributions.

Visualizing Conditional Probability

In the figure, $P(A|B)$ is represented by the fraction of *B* that is covered by *A*.

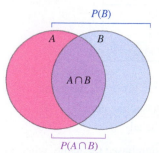

$$P(A|B) = \frac{P(A \cap B)}{P(B)}$$

Conditional Probability

If *A* and *B* are events with $P(B) \neq 0$, then the **(conditional) probability of *A* given *B*** is

$$P(A|B) = \frac{P(A \cap B)}{P(B)}.$$

Quick Examples

1. If there is a 50% chance of rain (*R*) and a 10% chance of both rain and lightning (*L*), then the probability of lightning, given that it rains, is

$$P(L|R) = \frac{P(L \cap R)}{P(R)} = \frac{.10}{.50} = .20.$$

Here are two more ways to express the result:

- If it rains, the probability of lightning is .20.
- Assuming that it rains, there is a 20% chance of lightning.

2. Referring to the Cyber Video data at the beginning of this section, the probability that a video game player did not purchase the game (A'), given that she did not see the ad (B'), is

$$P(A'|B') = \frac{P(A' \cap B')}{P(B')} = \frac{630/800}{700/800} = \frac{9}{10} = .9.$$

Q: *Returning to the video game sales survey, how do we compute the ordinary probability of A, not "given" anything?*

A: We look at the event A that a randomly chosen game player purchased Ultimate Hockey *regardless of whether or not he or she saw the ad*. In the "Purchased Game" row we see that a total of 90 people purchased the game out of a total of 800 surveyed. Thus, the (estimated) probability of A is

$$P(A) = \frac{fr(A)}{N} = \frac{90}{800} \approx .11.$$

We sometimes refer to $P(A)$ as the **unconditional** probability of A to distinguish it from conditional probabilities such as $P(A|B)$ and $P(A|B')$.

Now let's see some more examples involving conditional probabilities.

EXAMPLE 1   **Dice**

If you roll a fair die twice and observe the numbers that face up, find the probability that the sum of the numbers is 8, given that the first number is 3.

Solution We begin by recalling that the sample space when we roll a fair die twice is the set $S = \{(1, 1), (1, 2), \ldots, (6, 6)\}$ containing the 36 different equally likely outcomes.

The two events under consideration are

A: The sum of the numbers is 8.

B: The first number is 3.

We also need

$A \cap B$: The sum of the numbers is 8 and the first number is 3.

But this can happen in only one way: $A \cap B = \{(3, 5)\}$. From the formula, then,

$$P(A|B) = \frac{P(A \cap B)}{P(B)} = \frac{1/36}{6/36} = \frac{1}{6}.$$

➡ **Before we go on ...** There is another way to think about Example 1. When we say that the first number is 3, we are restricting the sample space to the six outcomes $(3, 1), (3, 2), \ldots, (3, 6)$, all still equally likely. Of these six, only one has a sum of 8, so the probability of the sum being 8, given that the first number is 3, is 1/6. ■

Notes

1. Remember that, in the expression $P(A|B)$, A is the event whose probability you want, given that you know the event B has occurred.

2. From the formula, notice that $P(A|B)$ is not defined if $P(B) = 0$. Could $P(A|B)$ make any sense if the event B were impossible? ■

EXAMPLE 2 **School and Work**

A survey[53] of the high school graduating class of 2010, conducted by the Bureau of Labor Statistics, found that, if a graduate went on to college, there was a 40% chance that he or she would work at the same time. On the other hand, there was a 68% chance that a randomly selected graduate would go on to college. What is the probability that a graduate went to college and work at the same time?

Solution To understand what the question asks and what information is given, it is helpful to rephrase everything using the standard wording "*the probability that ___* " and "*the probability that ___ given that ___.*" Now we have "The probability that a graduate worked, given that the graduate went on to college, equals .40. (See Figure 8.) The probability that a graduate went on to college is .68." The events in question are as follows:

W: A high school graduate went on to work.

C: A high school graduate went on to college.

From our rephrasing of the question we can write

$$P(W|C) = .40. \qquad P(C) = .68. \qquad \text{Find } P(W \cap C).$$

The definition

$$P(W|C) = \frac{P(W \cap C)}{P(C)}$$

can be used to find $P(W \cap C)$:

$$P(W \cap C) = P(W|C)\,P(C)$$
$$= (.40)(.68) \approx .27.$$

Thus, there is a 27% chance that a member of the high school graduating class of 2010 went on to college and work at the same time.

If a graduate went on to college, there was a 40% chance that he or she would work.

Rephrase by filling in the blanks:

The probability that_____ given that_____ equals____.

The probability that a graduate worked, given that the graduate went on to college, equals .40.

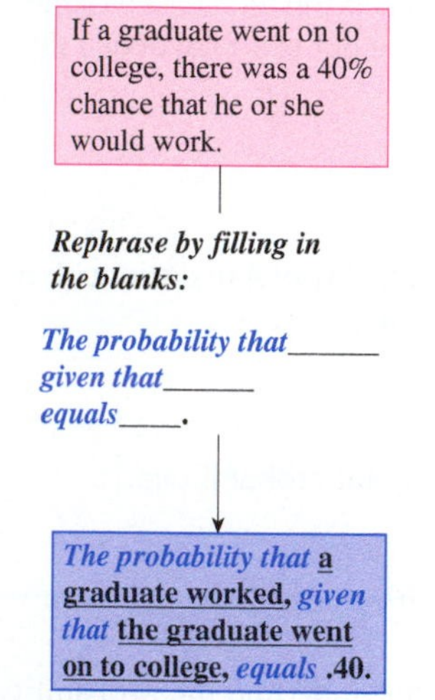

$P(\text{Worked} \mid \text{Went to college}) = .40$

Figure 8

The Multiplication Principle and Trees

In Example 2 we saw that the formula

$$P(A|B) = \frac{P(A \cap B)}{P(B)}$$

---

[53] Source: "College Enrollment and Work Activity of High School Graduates," U.S. Bureau of Labor Statistics (www.bls.gov/news.release/hsgec.htm).

can be used to calculate $P(A \cap B)$ if we rewrite the formula in the following form, known as the **multiplication principle for conditional probability**.

Multiplication Principle for Conditional Probability

If A and B are events, then

$$P(A \cap B) = P(A|B)\,P(B).$$

Quick Example

3. If there is a 50% chance of rain (R) and a 20% chance of lightning (L) if it rains, then the probability of both rain and lightning is

$$P(R \cap L) = P(L|R)\,P(R) = (.20)(.50) = .10.$$

The multiplication principle is often used in conjunction with **tree diagrams**. Let's return to *Cyber Video Games, Inc.*, and its television ad campaign. Its marketing survey was concerned with the following events:

A: A video game player purchased Ultimate Hockey.

B: A video game player saw the ad.

We can illustrate the various possibilities by means of the two-stage "tree" shown in Figure 9.

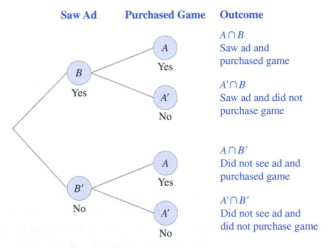

Figure 9

Consider the outcome $A \cap B$. To get there from the starting position on the left, we must first travel up to the B node. (In other words, B must occur.) Then we must travel up the branch from the B node to the A node. We are now going to associate a probability with each branch of the tree: the probability of traveling along that branch *given that we have gotten to its beginning node*. For instance, the probability of traveling up the branch from the starting position to the B node is $P(B) = 100/800 = .125$. (See the data in the survey.) The probability of going up the branch from the B node to the A node is the probability that A occurs, given that B has occurred. In other words, it is the *conditional* probability $P(A|B) = .2$.

(We calculated this probability at the beginning of the section.) The probability of the outcome $A \cap B$ can then be computed by using the multiplication principle:

$$P(A \cap B) = P(B) P(A \mid B) = (.125)(.2) = .025.$$

In other words, *to obtain the probability of the outcome $A \cap B$, we multiply the probabilities on the branches leading to that outcome* (Figure 10).

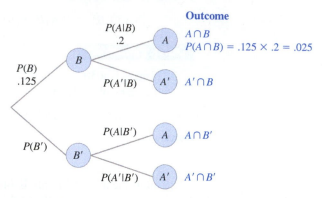

Figure 10

The same argument holds for the remaining three outcomes, and we can use the table given at the beginning of this section to calculate all the conditional probabilities shown in Figure 11.

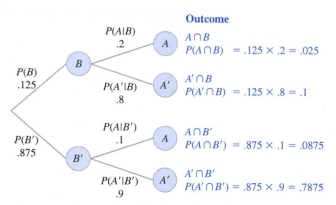

Figure 11

Note The sum of the probabilities on the branches leaving any node is always 1. (Why?) This observation often speeds things up because after we have labeled one branch (or all but one if a node has more than two branches leaving it), we can easily label the remaining one. ∎

EXAMPLE 3   Unfair Coins

An experiment consists of tossing two coins. The first coin is fair, while the second coin is twice as likely to land with heads facing up as it is with tails facing up. Draw a tree diagram to illustrate all the possible outcomes, and use the multiplication principle to compute the probabilities of all the outcomes.

Solution A quick calculation shows that the probability distribution for the second coin is $P(\text{H}) = 2/3$ and $P(\text{T}) = 1/3$. (How did we get that?) Figure 12 shows the tree diagram and the calculations of the probabilities of the outcomes.

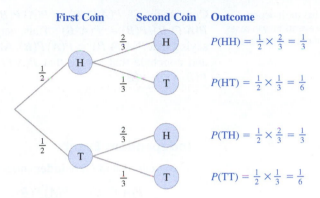

Figure 12

Independence

Let us go back once again to *Cyber Video Games, Inc.*, and its ad campaign. How did we assess the ad's effectiveness? We considered the following events:

> *A*: A video game player purchased Ultimate Hockey.

> *B*: A video game player saw the ad.

We used the survey data to calculate $P(A)$, the probability that a video game player purchased Ultimate Hockey, and $P(A|B)$, the probability that a video game player *who saw the ad* purchased Ultimate Hockey. When these probabilities are compared, one of three things can happen.

Case 1 $P(A|B) > P(A)$
This is what the survey data actually showed: A video game player was more likely to purchase Ultimate Hockey if he or she saw the ad. This indicates that the ad is effective; seeing the ad had a positive effect on a player's decision to purchase the game.

Case 2 $P(A|B) < P(A)$
If this had happened, then a video game player would have been *less* likely to purchase Ultimate Hockey if he or she saw the ad. This would have indicated that the ad had backfired; it had, for some reason, put potential customers off. In this case, just as in the first case, the event *B* would have had an effect—a negative one—on the event *A*.

Case 3 $P(A|B) = P(A)$
In this case, seeing the ad would have had absolutely no effect on a potential customer's buying Ultimate Hockey. Put another way, the probability of *A* occurring *does not depend* on whether *B* occurred or not. We say in a case like this that the events *A* and *B* are **independent**.

In general, we say that two events *A* and *B* are independent if $P(A|B) = P(A)$. When this happens, we have

$$P(A) = P(A|B) = \frac{P(A \cap B)}{P(B)},$$

so

$$P(A \cap B) = P(A) P(B).$$

*We shall discuss the independence of two events only in cases in which their probabilities are both nonzero.

Conversely, if $P(A \cap B) = P(A)P(B)$, then, assuming that $P(B) \neq 0$,* $P(A) = P(A \cap B)/P(B) = P(A|B)$. Thus, saying that $P(A) = P(A|B)$ is the same as saying that $P(A \cap B) = P(A)P(B)$. Also, we can switch A and B in this last formula and conclude that saying that $P(A \cap B) = P(A)P(B)$ is the same as saying that $P(B|A) = P(B)$.

Independent Events

The events A and B are **independent** if

$$P(A \cap B) = P(A)P(B).$$

Equivalent formulas (assuming that neither A nor B is impossible) are

$$P(A|B) = P(A)$$

and

$$P(B|A) = P(B).$$

If two events A and B are not independent, then they are **dependent**.

* See Exercises 111 and 112.

Note It can be verified* that if A and B are independent, then A' and B' are also independent, as are A and B', and A' and B. ∎

The property $P(A \cap B) = P(A)P(B)$ can be extended to three or more independent events. If, for example, A, B, and C are three mutually independent events (that is, each one of them is independent of each of the other two and of their intersection), then, among other things,

$$P(A \cap B \cap C) = P(A)P(B)P(C).$$

Quick Examples

4. If A and B are independent, and if A has a probability of .2 and B has a probability of .3, then $A \cap B$ has a probability of $(.2)(.3) = .06$.

5. Let us assume that the phase of the moon has no effect on whether or not my newspaper is delivered. The probability of a full moon (M) on a randomly selected day is about .034, and the probability that my newspaper will be delivered (D) on the random day is .2. Therefore, the probability that it is a full moon and my paper is delivered is

$$P(M \cap D) = P(M)P(D) = (.034)(.2) = .0068.$$

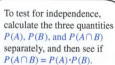
To test for independence, calculate the three quantities $P(A)$, $P(B)$, and $P(A \cap B)$ separately, and then see if $P(A \cap B) = P(A) \cdot P(B)$.

Testing for Independence

To check whether two events A and B are independent, we compute $P(A)$, $P(B)$, and $P(A \cap B)$. If $P(A \cap B) = P(A)P(B)$, the events are independent; otherwise, they are dependent. Sometimes it is obvious that two events, by their nature, are independent, so a test is not necessary. For example, the event that a die you roll comes up 1 is clearly independent of whether or not a coin you toss comes up heads.

Quick Examples

6. Roll two distinguishable dice (one red, one green), and observe the numbers that face up.

$$A: \text{The red die is even; } P(A) = \frac{18}{36} = \frac{1}{2}.$$

$$B: \text{The dice have the same parity}^*; P(B) = \frac{18}{36} = \frac{1}{2}.$$

$$A \cap B: \text{Both dice are even; } P(A \cap B) = \frac{9}{36} = \frac{1}{4}.$$

$P(A \cap B) = P(A)P(B)$, so A and B are independent.

7. Roll two distinguishable dice, and observe the numbers that face up.

$$A: \text{The sum of the numbers is 6; } P(A) = \frac{5}{36}.$$

$$B: \text{Both numbers are odd; } P(B) = \frac{9}{36} = \frac{1}{4}.$$

$$A \cap B: \text{The sum is 6, and both are odd; } P(A \cap B) = \frac{3}{36} = \frac{1}{12}.$$

$P(A \cap B) \neq P(A)P(B)$, so A and B are dependent.

* Two numbers have the **same parity** if both are even or both are odd. Otherwise, they have **opposite parity**.

EXAMPLE 4   Weather Prediction

According to the weather service, there is a 50% chance of rain in New York and a 30% chance of rain in Honolulu. Assuming that New York's weather is independent of Honolulu's, find the probability that it will rain in at least one of these cities.

Solution We take A to be the event that it will rain in New York and B to be the event that it will rain in Honolulu. We are asked to find the probability of $A \cup B$, the event that it will rain in at least one of the two cities. We use the addition principle:

$$P(A \cup B) = P(A) + P(B) - P(A \cap B).$$

We know that $P(A) = .50$ and $P(B) = .30$. But what about $P(A \cap B)$? Because the events A and B are independent, we can compute

$$P(A \cap B) = P(A)P(B)$$
$$= (.50)(.30) = .15.$$

Thus,

$$P(A \cup B) = P(A) + P(B) - P(A \cap B)$$
$$= .50 + .30 - .15$$
$$= .65.$$

So there is a 65% chance that it will rain either in New York or in Honolulu (or in both).

EXAMPLE 5   **Roulette**

You are playing roulette and have decided to leave all 10 of your $1 chips on black for five consecutive rounds, hoping for a sequence of five blacks, which, according to the rules, will leave you with $320. There is a 50% chance of black coming up on each spin, ignoring the complicating factor of zero or double zero. What is the probability that you will be successful?

Solution Because the roulette wheel has no memory, each spin is independent of the others. Thus, if A_1 is the event that black comes up the first time, A_2 the event that it comes up the second time, and so on, then

$$P(A_1 \cap A_2 \cap A_3 \cap A_4 \cap A_5) = P(A_1)\,P(A_2)\,P(A_3)\,P(A_4)\,P(A_5) = \left(\frac{1}{2}\right)^5 = \frac{1}{32}.$$

The next example is a version of a well-known brain teaser that forces one to think carefully about conditional probability.

EXAMPLE 6   **Legal Argument**

A man was arrested for attempting to smuggle a bomb on board an airplane. During the subsequent trial, his lawyer claimed that, by means of a simple argument, she would prove beyond a shadow of a doubt that her client not only was innocent of any crime, but was in fact contributing to the safety of the other passengers on the flight. This was her eloquent argument: "Your Honor, first of all, my client had absolutely no intention of setting off the bomb. As the record clearly shows, the detonator was unarmed when he was apprehended. In addition—and your Honor is certainly aware of this—there is a small but definite possibility that there will be a bomb on any given flight. On the other hand, the chances of there being *two* bombs on a flight are so remote as to be negligible. There is in fact no record of this having *ever* occurred. Thus, because my client had already brought one bomb on board (with no intention of setting it off) and because we have seen that the chances of there being a second bomb on board were vanishingly remote, it follows that the flight was far safer as a result of his action! I rest my case." This argument was so elegant in its simplicity that the judge acquitted the defendant. Where is the flaw in the argument? (Think about this for a while before reading the solution.)

Solution The lawyer has cleverly confused the phrases "two bombs on board" and "a second bomb on board." To pinpoint the flaw, let us take B to be the event that there is one bomb on board a given flight, and let A be the event that there are two independent bombs on board. Let us assume for argument's sake that $P(B) = 1/1,000,000 = .000001$. Then the probability of the event A is

$$(.000001)(.000001) = .000000000001.$$

This *is* vanishingly small, as the lawyer contended. It was at this point that the lawyer used a clever maneuver: She assumed in concluding her argument that the probability of having two bombs on board was the same as the probability of having a *second* bomb on board. But to say that there is a *second* bomb on board is to imply that there already is one bomb on board. This is therefore a *conditional* event: the event that there are two bombs on board, *given that there is already one bomb on board*. Thus, the probability that there is a second bomb on board is the probability that there are two bombs on board, given that there is already one bomb on board, which is

$$P(A \mid B) = \frac{P(A \cap B)}{P(B)} = \frac{.000000000001}{.000001} = .000001.$$

In other words, it is the same as the probability of there being a single bomb on board to begin with! Thus the man's carrying the bomb onto the plane did not improve the flight's safety at all.[*]

✱ If we want to be picky, there was a *slight* decrease in the probability of a second bomb because there was one less seat for a potential second bomb bearer to occupy. In terms of our analysis, this is saying that the event of one passenger with a bomb and the event of a second passenger with a bomb are not completely independent.

FAQs

Probability of what given what?

Q : *How do I tell whether a statement in an application is talking about conditional probability or unconditional probability? And if it is talking about conditional probability, how do I determine what to use as A and B in $P(A \mid B)$?*

A : Look carefully at the wording of the statement. If there is some kind of qualification or restriction to a smaller set than the entire sample space, then it is probably talking about conditional probability, as in the following examples:

60% of veterans vote Republican, while 40% of the entire voting population vote Republican.

Here the sample space can be taken to be the entire voting population.
Reworded (see Example 2): *The probability of voting Republican (R) is 60% given that the person is a veteran (V); $P(R \mid V) = .60$, whereas the probability of voting Republican is .40: $P(R) = .40$.*

The likelihood of being injured if in an accident is 80% for a driver not wearing a seatbelt but it is 50% for all drivers.

Here, the sample space can be taken to be the set of drivers involved in an accident—these are the only drivers discussed.
Reworded: *The probability of a driver being injured (I) is .80 given that the driver is not wearing a seatbelt (B); $P(I \mid B) = .80$ whereas, for all drivers, the probability of being injured is .50: $P(I) = .50$.*

7.5 EXERCISES

▼ more advanced   ◆ challenging
[T] indicates exercises that should be solved using technology

In Exercises 1–10, compute the indicated quantity.

1. $P(B) = .5$, $P(A \cap B) = .2$. Find $P(A \mid B)$.

2. $P(B) = .6$, $P(A \cap B) = .3$. Find $P(A \mid B)$.

3. $P(A \mid B) = .2$, $P(B) = .4$. Find $P(A \cap B)$.

4. $P(A \mid B) = .1$, $P(B) = .5$. Find $P(A \cap B)$.

5. $P(A \mid B) = .4$, $P(A \cap B) = .3$. Find $P(B)$.

6. $P(A \mid B) = .4$, $P(A \cap B) = .1$. Find $P(B)$.

7. $P(A) = .5$, $P(B) = .4$. A and B are independent. Find $P(A \cap B)$.

8. $P(A) = .2$, $P(B) = .2$. A and B are independent. Find $P(A \cap B)$.

9. $P(A) = .5$, $P(B) = .4$. A and B are independent. Find $P(A \mid B)$.

10. $P(A) = .3$, $P(B) = .6$. A and B are independent. Find $P(B \mid A)$.

In Exercises 11–16, fill in the blanks using the named events.
[**HINT:** See Example 2 and the FAQ at the end of the section.]

11. 10% of all Anchovians detest anchovies (*D*), whereas 30% of all married Anchovians (*M*) detest them. $P(\_\_\_) = \_\_\_$; $P(\_\_\_ \mid \_\_\_) = \_\_\_$

12. 95% of all music composers can read music (M), whereas 99% of all classical music composers (C) can read music. $P(\_\_) = \_\_$; $P(\_\_|\_\_) = \_\_$

13. 30% of all lawyers who lost clients (L) were antitrust lawyers (A), whereas 10% of all antitrust lawyers lost clients. $P(\_\_|\_\_) = \_\_$; $P(\_\_|\_\_) = \_\_$

14. 2% of all items bought on my auction site (B) were works of art (A), whereas only 1% of all works of art on the site were bought. $P(\_\_|\_\_) = \_\_$; $P(\_\_|\_\_) = \_\_$

15. 55% of those who go out in the midday sun (M) are Englishmen (E), whereas only 5% of those who do not go out in the midday sun are Englishmen. $P(\_\_|\_\_) = \_\_$; $P(\_\_|\_\_) = \_\_$

16. 80% of those who have a Mac now (M) will purchase a Mac next time (X), whereas 20% of those who do not have a Mac now will purchase a Mac next time. $P(\_\_|\_\_) = \_\_$; $P(\_\_|\_\_) = \_\_$

In Exercises 17–22, find the conditional probability of the indicated event when two fair dice (one red and one green) are rolled. [**HINT:** *See Example 1.*]

17. The sum is 5, given that the green one is not a 1.

18. The sum is 6, given that the green one is either 4 or 3.

19. The red one is 5, given that the sum is 6.

20. The red one is 4, given that the green one is 4.

21. The sum is 5, given that the dice have opposite parity.

22. The sum is 6, given that the dice have opposite parity.

Exercises 23–28 require the use of counting techniques from Chapter 6. A bag contains three red marbles, two green ones, one fluorescent pink one, two yellow ones, and two orange ones. Suzan grabs four at random. Find the probability of the indicated event.

23. She gets all the red ones, given that she gets the fluorescent pink one.

24. She gets all the red ones, given that she does not get the fluorescent pink one.

25. She gets none of the red ones, given that she gets the fluorescent pink one.

26. She gets one of each color other than fluorescent pink, given that she gets the fluorescent pink one.

27. She gets one of each color other than fluorescent pink, given that she gets at least one red one.

28. She gets at least two red ones, given that she gets at least one green one.

In Exercises 29–32, supply the missing quantities. [**HINT:** *See Example 3 and the discussion preceding it.*]

29.

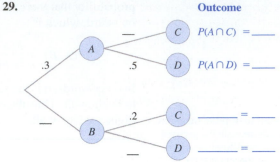

30.

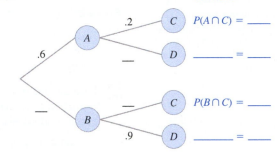

31.

32.

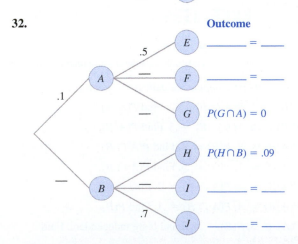

In Exercises 33–36, say whether the given pair of events are independent, mutually exclusive, or neither.

33. *A*: Your new skateboard design is a success.

B: Your new skateboard design is a failure.

34. *A*: Your new skateboard design is a success.

B: There is life in the Andromeda galaxy.

35. *A*: Your new skateboard design is a success.

B: Your competitor's new skateboard design is a failure.

36. *A*: Your first coin flip results in heads.

B: Your second coin flip results in heads.

In Exercises 37–42, two dice (one red and one green) are rolled, and the numbers that face up are observed. Test the given pair of events for independence. [**HINT:** See Quick Examples 6 and 7.]

37. *A*: The red die is 1, 2, or 3; *B*: The green die is even.

38. *A*: The red die is 1; *B*: The sum is even.

39. *A*: Exactly one die is 1; *B*: The sum is even.

40. *A*: Neither die is 1 or 6; *B*: The sum is even.

41. *A*: Neither die is 1; *B*: Exactly one die is 2.

42. *A*: Both dice are 1; *B*: Neither die is 2.

43. If a coin is tossed 11 times, find the probability of the sequence H, T, T, H, H, H, T, H, H, T, T. [**HINT:** See Example 5.]

44. If a die is rolled four times, find the probability of the sequence 4, 3, 2, 1. [**HINT:** See Example 5.]

Applications

45. *Personal Bankruptcy* In 2004 the probability that a person in the United States would declare personal bankruptcy was .006. The probability that a person in the United States would declare personal bankruptcy and had recently experienced a "big three" event (loss of job, medical problem, or divorce or separation) was .005.[54] What was the probability that a person had recently experienced one of the "big three" events, given that she had declared personal bankruptcy? (Round your answer to one decimal place.)

46. *Personal Bankruptcy* In 2004 the probability that a person in the United States would declare personal bankruptcy was .006. The probability that a person in the United States would declare personal bankruptcy and had recently overspent credit cards was .0004.[55] What was the probability that a person had recently overspent credit cards given that he had declared personal bankruptcy?

47. *Existing Home Sales* During the year ending April 30, 2015, there were approximately 5.0 million sales of existing homes in the United States, of which 1.2 million were sold in the West. During April 2015 there were a total of 450,000 existing homes sold in the United States, of which 110,000 were sold in the West.[56]

a. Find the probability that a home sale in the year ending April 30, 2015, took place in the West, given that the home was sold during April of that year.

b. Find the probability that a home sale in the year ending April 30, 2015, took place in April of that year, given that it took place in the West.

48. *Existing Home Sales* Refer to the data given in Exercise 47.

a. Find the probability that a home sale in the year ending April 30, 2015, took place outside the West, given that the home was sold during April of that year.

b. Find the probability that a home sale in the year ending April 30, 2015, took place in April of that year, given that it took place outside the West.

49. *Social Security* According to a *New York Times*/CBS poll released in March 2005, 79% of respondents agreed that it should be the government's responsibility to provide a decent standard of living for the elderly, and 43% agreed that it would be a good idea to invest part of their Social Security taxes on their own.[57] If agreement with one of these propositions is independent of agreement with the other, what is the probability that a person agreed with both propositions? (Round your answer to two decimal places.) [**HINT:** See Quick Examples 4 and 5.]

50. *Social Security* According to *The New York Times*/CBS poll of March 2005 referred to in Exercise 49, 49% of respondents agreed that Social Security taxes should be raised if necessary to keep the system afloat, and 43% agreed that it would be a good idea to invest part of their Social Security taxes on their own.[58] If agreement with one of these propositions is independent of agreement with the other, what is the probability that a person agreed with both propositions? (Round your answer to two decimal places.) [**HINT:** See Quick Examples 4 and 5.]

51. *Marketing* A market survey shows that 40% of the population used Brand X laundry detergent last year, 5% of the population gave up doing its laundry last year, and 4% of the population used Brand X and then gave up doing laundry last year. Are the events of using Brand X and giving up doing laundry independent? Is a user of Brand X detergent more or less likely to give up doing laundry than a randomly chosen person?

52. *Marketing* A market survey shows that 60% of the population used Brand Z computers last year, 5% of the population

---

[54] Probabilities are approximate. Source: *New York Times*, March 13, 2005, p, WK3.

[55] The .0004 figure is an estimate by the authors. Source: *Ibid.*

[56] Source: National Association of Realtors (www.realtor.org).

[57] Source: *New York Times*, March 3, 2005, p. A20.

[58] *Ibid.*

quit their jobs last year, and 3% of the population used Brand Z computers and then quit their jobs. Are the events of using Brand Z computers and quitting one's job independent? Is a user of Brand Z computers more or less likely to quit a job than a randomly chosen person?

53. **Road Safety** In 1999 the probability that a randomly selected vehicle would be involved in a deadly tire-related accident was approximately 3×10^{-6}, whereas the probability that a tire-related accident would prove deadly was .02.[59] What was the probability that a vehicle would be involved in a tire-related accident?

54. **Road Safety** In 1998 the probability that a randomly selected vehicle would be involved in a deadly tire-related accident was approximately 2.8×10^{-6}, while the probability that a tire-related accident would prove deadly was .016.[60] What was the probability that a vehicle would be involved in a tire-related accident?

Publishing *Exercises 55–62 are based on the following table, which shows the results of a survey of 100 authors by a publishing company:*

	New Authors	**Established Authors**	**Total**
Successful	5	25	30
Unsuccessful	15	55	70
Total	20	80	100

Compute the following conditional probabilities:

55. An author is established, given that she is successful.

56. An author is successful, given that he is established.

57. An author is unsuccessful, given that he is a new author.

58. An author is a new author, given that she is unsuccessful.

59. An author is unsuccessful, given that she is established.

60. An author is established, given that he is unsuccessful.

61. An unsuccessful author is established.

62. An established author is successful.

In Exercises 63–68, draw an appropriate tree diagram, and use the multiplication principle to calculate the probabilities of all the outcomes. [**HINT:** See Example 3.]

63. **Sales** Each day, there is a 40% chance that you will sell an automobile. You know that 30% of all the automobiles you sell are two-door models and the rest are four-door models.

64. **Product Reliability** You purchase Brand X memory chips one quarter of the time and Brand Y memory chips the rest of the time. Brand X memory chips have a 1% failure rate, while Brand Y memory chips have a 3% failure rate.

65. **Car Rentals** Your auto rental company rents out 30 small cars, 24 luxury sedans, and 46 slightly damaged "budget" vehicles. The small cars break down 14% of the time, the luxury sedans break down 8% of the time, and the "budget" cars break down 40% of the time.

66. **Travel** It appears that there is only a one in five chance that you will be able to take your spring vacation to the Greek Islands. If you are lucky enough to go, you will visit either Corfu (20% chance) or Rhodes. On Rhodes there is a 20% chance of meeting a tall, dark stranger, while on Corfu there is no such chance.

67. **Weather Prediction** There is a 50% chance of rain today and a 50% chance of rain tomorrow. Assuming that the event that it rains today is independent of the event that it rains tomorrow, draw a tree diagram showing the probabilities of all outcomes. What is the probability that there will be no rain today or tomorrow?

68. **Weather Prediction** There is a 20% chance of snow today and a 20% chance of snow tomorrow. Assuming that the event that it snows today is independent of the event that it snows tomorrow, draw a tree diagram showing the probabilities of all outcomes. What is the probability that it will snow by the end of tomorrow?

Education and Employment *Exercises 69–78 are based on the following table, which shows U.S. employment figures for 2014, broken down by educational attainment.[61] All numbers are in millions and represent civilians aged 25 years and over. Those classed as "not in labor force" were not employed nor actively seeking employment. Round all answers to two decimal places.*

	Employed	**Unemployed**	**Not in Labor Force**	**Total**
Less Than High School Diploma	9.9	1.0	13.2	24.1
High School Diploma Only	33.9	2.2	25.9	62.0
Some College or Associate's Degree	35.3	2.0	18.4	55.7
Bachelor's Degree or Higher	48.8	1.6	16.9	67.3
Total	127.9	6.8	74.4	209.1

69. Find the probability that a person was employed, given that the person had a bachelor's degree or higher.

70. Find the probability that a person was employed, given that the person had attained less than a high school diploma.

71. Find the probability that a person had a bachelor's degree or higher, given that the person was employed.

---

[59] The original data reported three tire-related deaths per million vehicles. Source: *New York Times* analysis of National Traffic Safety Administration crash data/Polk Company vehicle registration data/ *New York Times*, Nov. 22, 2000, p. C5.

[60] The original data reported 2.8 tire-related deaths per million vehicles. Source: *Ibid.*

[61] Source: Bureau of Labor Statistics (www.bls.gov).

72. Find the probability that a person had attained less than a high school diploma, given that the person was employed.

73. ▼ Find the probability that a person who had not completed a bachelor's degree or higher was not in the labor force.

74. ▼ Find the probability that a person who had completed at least a high school diploma was not in the labor force.

75. ▼ Find the probability that a person who had completed a bachelor's degree or higher and was in the labor force was employed.

76. ▼ Find the probability that a person who had completed less than a high school diploma and was in the labor force was employed.

77. ▼ Your friend claims that an unemployed person is more likely to have a high school diploma only than an employed person. Respond to this claim by citing actual probabilities.

78. ▼ Your friend claims that a person not in the labor force is more likely to have less than a high school diploma than an employed person. Respond to this claim by citing actual probabilities.

79. ▼ *Air Bag Safety* According to a 2000 study conducted by the Harvard School of Public Health, a child seated in the front seat who was wearing a seatbelt was 31% more likely to be killed in an accident if the car had an air bag that deployed than if it did not.[62] Let the sample space S be the set of all accidents involving a child seated in the front seat wearing a seatbelt. Let K be the event that the child was killed, and let D be the event that the air bag deployed. Fill in the missing terms and quantities: $P(\_\_\_|\_\_\_) = \_\_\_ \times P(\_\_\_|\_\_\_)$. [**HINT:** When we say, "A is 31% more likely than B," we mean that the probability of A is 1.31 times the probability of B.]

80. ▼ *Air Bag Safety* According to the study cited in Exercise 79, a child seated in the front seat not wearing a seatbelt was 84% more likely to be killed in an accident if the car had an air bag that deployed than if it did not.[63] Let the sample space S be the set of all accidents involving a child seated in the front seat not wearing a seatbelt. Fill in the missing terms and quantities: $P(\_\_\_|\_\_\_) = \_\_\_ \times P(\_\_\_|\_\_\_)$. [**HINT:** When we say, "A is 84% more likely than B," we mean that the probability of A is 1.84 times the probability of B.]

81. ▼ *Productivity* A company wishes to enhance productivity by running a one-week training course for its employees. Let T be the event that an employee participated in the course, and let I be the event that an employee's productivity improved the week after the course was run.
  a. Assuming that the course has a positive effect on productivity, how are $P(I|T)$ and $P(I)$ related?
  b. If T and I are independent, what can one conclude about the training course?

82. ▼ *Productivity* Consider the events T and I in Exercise 81.
  a. Assuming that everyone who improved took the course but that not everyone took the course, how are $P(T|I)$ and $P(T)$ related?
  b. If half the employees who improved took the course and half the employees took the course, are T and I independent?

83. ▼ *Internet Use in 2000* The following pie chart shows the percentage of the population that used the Internet in 2000, broken down further by family income and based on a survey taken in August 2000:[64]

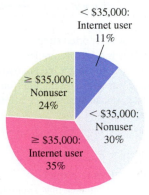

  a. Determine the probability that a randomly chosen person was an Internet user, given that his or her family income was at least $35,000.
  b. Based on the data, was a person more likely to be an Internet user if his or her family income was less than $35,000 or $35,000 or more? (Support your answer by citing the relevant conditional probabilities.)

84. ▼ *Internet Use in 2001* Repeat Exercise 83 using the following pie chart, which shows the results of a similar survey taken in September 2001:[65]

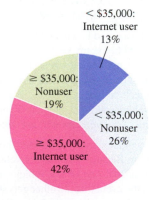

[62] The study was conducted by Dr. Segul-Gomez at the Harvard School of Public Health. Source: *New York Times*, December 1, 2000, p. F1.

[63] *Ibid.*

[64] Source: *Falling Through the Net: Toward Digital Inclusion, A Report on Americans' Access to Technology Tools*, U.S. Department of Commerce, October 2000. Available at www.ntia.doc.gov/ntiahome/fttn00/contents00.html.

[65] Source: *A Nation Online: How Americans Are Expanding Their Use of the Internet*, U.S. Department of Commerce, February 2002. Available at www.ntia.doc.gov/ntiahome/dn/index.html.

Auto Theft *Exercises 85–90 are based on the following table, which shows the probability that an owner of the given model would report his or her vehicle stolen in a 1-year period:*[66]

Model	Jeep Wrangler	Suzuki Sidekick (two-door)	Toyota Land Cruiser	Geo Tracker (two-door)	Acura Integra (two-door)
Probability	.0170	.0154	.0143	.0142	.0123
Model	Mitsubishi Montero	Acura Integra (four-door)	BMW 3-series (two-door)	Lexus GS300	Honda Accord (two-door)
Probability	.0108	.0103	.0077	.0074	.0070

In an experiment in which a vehicle is selected, consider the following events:

    R: The vehicle was reported stolen.
    J: The vehicle was a Jeep Wrangler.
    A2: The vehicle was an Acura Integra (two-door).
    A4: The vehicle was an Acura Integra (four-door).
    A: The vehicle was an Acura Integra (either two-door or four-door).

85. ▼ Fill in the blanks: $P(\_\_\_|\_\_\_) = .0170$.

86. ▼ Fill in the blanks: $P(\_\_\_|A4) = \_\_\_$.

87. ▼ Which of the following is true?
    (A) There is a 1.43% chance that a vehicle reported stolen was a Toyota Land Cruiser.
    (B) Of all the vehicles reported stolen, 1.43% of them were Toyota Land Cruisers.
    (C) Given that a vehicle was reported stolen, there is a .0143 probability that it was a Toyota Land Cruiser.
    (D) Given that a vehicle was a Toyota Land Cruiser, there was a 1.43% chance that it was reported stolen.

88. ▼ Which of the following is true?
    (A) $P(R|A) = .0123 + .0103 = .0226$
    (B) $P(R'|A2) = 1 - .0123 = .9877$
    (C) $P(A2|A) = .0123/(.0123 + .0103) \approx .544$
    (D) $P(R|A2') = 1 - .0123 = .9877$

89. ▼ It is now January, and I own a BMW 3-series and a Lexus GS300. Because I house my vehicles in different places, the event that one of my vehicles gets stolen does not depend on the event that the other gets stolen. Compute each probability to six decimal places.
    a. Both my vehicles will get stolen this year.
    b. At least one of my vehicles will get stolen this year.

90. ▼ It is now December, and I own a Mitsubishi Montero and a Jeep Wrangler. Because I house my vehicles in different places, the event that one of my vehicles gets stolen does not depend on the event that the other gets stolen. I have just returned from a 1-year trip to the Swiss Alps.
    a. What is the probability that my Montero, but not my Wrangler, has been stolen?

b. Which is more likely: the event that my Montero was stolen or the event that *only* my Montero was stolen?

91. ▼ *Drug Tests* If 90% of the athletes who test positive for steroids in fact use them, and 10% of all athletes use steroids and test positive, what percentage of athletes test positive?

92. ▼ *Fitness Tests* If 80% of candidates for the soccer team pass the fitness test, and only 20% of all athletes are soccer team candidates who pass the test, what percentage of the athletes are candidates for the soccer team?

93. ▼ *Food Safety* According to a University of Maryland study of 200 samples of ground meats,[67] the probability that a sample was contaminated by *Salmonella* was .20. The probability that a *Salmonella*-contaminated sample was contaminated by a strain resistant to at least three antibiotics was .53. What was the probability that a ground meat sample was contaminated by a strain of *Salmonella* resistant to at least three antibiotics?

94. ▼ *Food Safety* According to the study mentioned in Exercise 93, the probability that a ground meat sample was contaminated by *Salmonella* was .20. The probability that a *Salmonella*-contaminated sample was contaminated by a strain resistant to at least one antibiotic was .84. What was the probability that a ground meat sample was contaminated by a strain of *Salmonella* resistant to at least one antibiotic?

95. ◆ *Food Safety* According to the study mentioned in Exercise 93, the probability that one of the samples was contaminated by *Salmonella* was .20. The probability that a *Salmonella*-contaminated sample was contaminated by a strain resistant to at least one antibiotic was .84, and the probability that a *Salmonella*-contaminated sample was contaminated by a strain resistant to at least three antibiotics was .53. Find the probability that a ground meat sample that was contaminated by an antibiotic-resistant strain was contaminated by a strain resistant to at least three antibiotics.

96. ◆ *Food Safety* According to the study mentioned in Exercise 93, the probability that a ground meat sample was contaminated by a strain of *Salmonella* resistant to at least three antibiotics was .11. The probability that someone infected with any strain of *Salmonella* will become seriously ill

is .10. What is the probability that someone eating a randomly chosen ground meat sample will not become seriously ill with a strain of *Salmonella* resistant to at least three antibiotics?

97. ▼ *Ultimate Hockey* *Cyber Video Games, Inc.*, the makers of Ultimate Hockey (see the discussion at the beginning of this section), switched to a cheaper ad agency whose TV ad had no effect whatsoever on sales according to a survey. Unfortunately, a student intern accidentally erased some of the survey data in the table below:

	Saw Ad	Did Not See Ad	Total
Purchased Game	20	40	60
Did Not Purchase Game	180	?	?
Total	200	?	?

Calculate the missing quantities.

98. ▼ *Ultimate Hockey* (See Exercise 97.) *Cyber Video Games, Inc.*, the makers of Ultimate Hockey, switched to a more expensive ad agency whose TV ad *still* had no effect whatsoever on sales according to a survey. Unfortunately, the same student intern again erased some of the survey data in the table below:

	Saw Ad	Did Not See Ad	Total
Purchased Game	20	60	80
Did Not Purchase Game	?	180	?
Total	?	240	?

Calculate the missing quantities.

Communication and Reasoning Exercises

99. The probability of misspelling "Waner" is p and is equal to the probability of misspelling "Costenoble." If the event of misspelling one author's name is independent of the event of misspelling the other, what percentage of people can be expected to misspell both names?

100. The probability of spelling "Waner" correctly is p, and the probability of spelling "Costenoble" correctly is q. If the event of misspelling one author's name is independent of the event of correctly spelling the other, what is the probability of misspelling both names?

101. Name three events, each independent of the others, when a fair coin is tossed four times.

102. Name three pairs of independent events when a pair of distinguishable and fair dice is rolled and the numbers that face up are observed.

103. You wish to ascertain the probability of an event E, but you happen to know that the event F has occurred. Is the probability you are seeking $P(E)$ or $P(E|F)$? Give the reason for your answer.

104. Your television advertising campaign seems to have been very persuasive: 10,000 people who saw the ad purchased your product, while only 2,000 people purchased the product without seeing the ad. Explain how additional data could show that your ad campaign was, in fact, unpersuasive.

105. You are having trouble persuading your friend Iliana that conditional probability is different from unconditional probability. She just said, "Look here, Saul, the probability of throwing a double-six is 1/36, and that's that! That probability is not affected by anything, including the 'given' that the sum is larger than 7." How do you persuade her otherwise?

106. Your other friend Giuseppe is spreading rumors that the conditional probability $P(E|F)$ is always bigger than $P(E)$. Is he right? (If he is, explain why. If he is not, give an example to prove him wrong.)

107. ▼ If $A \subseteq B$ and $P(B) \neq 0$, why is $P(A|B) = \dfrac{P(A)}{P(B)}$?

108. ▼ If $B \subseteq A$ and $P(B) \neq 0$, why is $P(A|B) = 1$?

109. ▼ Your best friend thinks that it is impossible for two mutually exclusive events with nonzero probabilities to be independent. Establish whether or not he is correct.

110. ▼ Another of your friends thinks that two mutually exclusive events with nonzero probabilities can never be dependent. Establish whether or not she is correct.

111. ◆ Show that if A and B are independent, then so are A' and B' (assuming that none of these events has zero probability). [**HINT:** $A' \cap B'$ is the complement of $A \cup B$.]

112. ◆ Show that if A and B are independent, then so are A and B' (assuming that none of these events has zero probability). [**HINT:** $P(B'|A) + P(B|A) = 1$.]

7.6 Bayes' Theorem and Applications

Motivating Bayes' Theorem

Should schools test their athletes for drug use? A problem with drug testing is that there are always false positive results, so one can never be certain that an athlete who tests positive is in fact using drugs. Here is a typical scenario.

EXAMPLE 1   Steroids Testing

Gamma Chemicals advertises its anabolic steroid detection test as being 95% effective at detecting steroid use, meaning that the test will show a positive result on 95% of all anabolic steroid users. The company also states that its test has a false positive rate of 6%. This means that the probability of a nonuser testing positive is .06. Estimating that about 10% of its athletes are using anabolic steroids, *Enormous State University* begins testing its football players. The quarterback, Hugo V. Huge, tests positive and is promptly dropped from the team. Hugo claims that he is not using anabolic steroids. How confident can we be that he is not telling the truth?

Solution There are two events of interest here: the event T that a person tests positive and the event A that the person tested uses anabolic steroids. Here are the probabilities we are given:

$$P(T|A) = .95$$
$$P(T|A') = .06$$
$$P(A) = .10.$$

Figure 13

We are asked to find $P(A|T)$, the probability that someone who tests positive is using anabolic steroids. We can use a tree diagram to calculate $P(A|T)$. The trick to setting up the tree diagram is to use as the first branching the events with *unconditional* probabilities we know. Because the only unconditional probability we are given is $P(A)$, we use A and A' as our first branching (Figure 13). For the second branching, we use the outcomes of the drug test: positive (T) or negative (T'). The probabilities on these branches are conditional probabilities because they depend on whether or not an athlete uses steroids. (See Figure 14.) (We fill in the probabilities that are not supplied by remembering that the sum of the probabilities on the branches leaving any node must be 1.)

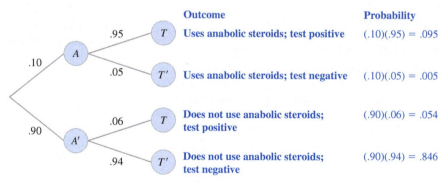

Figure 14

We can now calculate the probability we are asked to find:

$$P(A|T) = \frac{P(A \cap T)}{P(T)} = \frac{P(\text{Uses anabolic steroids and tests positive})}{P(\text{Tests positive})}$$

$$= \frac{P(\text{Using } A \text{ and } T \text{ branches})}{\text{Sum of } P(\text{Using branches ending in } T)}.$$

From the tree diagram, we see that $P(A \cap T) = .095$. To calculate $P(T)$, the probability of testing positive, notice that there are two outcomes on the tree diagram that reflect a positive test result. The probabilities of these events are .095 and .054. Because these two events are mutually exclusive (an athlete either uses steroids or does not, but not both), the probability of a test being positive (ignoring whether or not steroids are used) is the sum of these probabilities, .149. Thus,

$$P(A|T) = \frac{.095}{.095 + .054} = \frac{.095}{.149} \approx .64.$$

Thus, there is a 64% chance that a randomly selected athlete who tests positive, such as Hugo, is using steroids. In other words, we can be 64% confident that Hugo is lying.

⟶ **Before we go on...** Note that the correct answer in Example 1 is 64%, *not* the 94% we might suspect from the test's false positive rating. In fact, we can't answer the question asked without knowing the percentage of athletes who actually use steroids. For instance, if *no* athletes at all use steroids, then Hugo must be telling the truth, so the test result has no significance whatsoever. On the other hand, if *all* athletes use steroids, then Hugo is definitely lying, regardless of the outcome of the test.

False positive rates are determined by testing a large number of samples known not to contain drugs and computing estimated probabilities. False negative rates are computed similarly by testing samples known to contain drugs. However, the accuracy of the tests depends also on the skill of those administering them. False positives were a significant problem when drug testing started to become common, with estimates of false positive rates for common immunoassay tests ranging from 10% to 30% on the high end,[*] but the accuracy has improved since then. Because of the possibility of false positive results, positive immunoassay tests need to be confirmed by the more expensive and much more reliable gas chromatograph/mass spectrometry (GC/MS) test. See also the National Collegiate Athletic Association's (NCAA) Drug-Testing Program handbook, available at www.ncaa.org/health-and-safety/policy/drug-testing. The section on Institutional Drug Testing addresses the problem of false positives. ∎

[*] *Drug Testing in the Workplace*, ACLU Briefing Paper, 1996.

Bayes' Theorem Formula

The calculation we used to answer the question in Example 1 can be recast as a formula known as **Bayes' theorem**. Figure 15 shows a general form of the tree we used in Example 1.

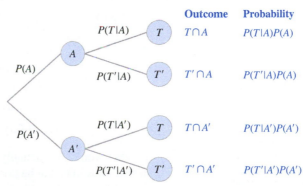

		Outcome	Probability

Figure 15

We calculated

$$P(A|T) = \frac{P(A \cap T)}{P(T)}$$

as follows. We first calculated the numerator $P(A \cap T)$ using the multiplication principle:

$$P(A \cap T) = P(T|A)\,P(A).$$

We then calculated the denominator $P(T)$ by using the addition principle for mutually exclusive events together with the multiplication principle:

$$P(T) = P(A \cap T) + P(A' \cap T)$$
$$= P(T|A)P(A) + P(T|A')P(A').$$

Substituting gives

$$P(A|T) = \frac{P(T|A)P(A)}{P(T|A)P(A) + P(T|A')P(A')}.$$

This is the short form of Bayes' theorem.

Bayes' Theorem (Short Form)

If A and T are events, then

Bayes' Formula

$$P(A|T) = \frac{P(T|A)P(A)}{P(T|A)P(A) + P(T|A')P(A')}.$$

Using a Tree

$$P(A|T) = \frac{P(\text{Using } A \text{ and } T \text{ branches})}{\text{Sum of } P(\text{Using branches ending in } T)}$$

Quick Examples

1. Let us calculate the probability that an athlete from Example 1 who tests positive is actually using steroids if only 5% of ESU athletes are using steroids. Thus,

$$P(T|A) = .95$$
$$P(T|A') = .06$$
$$P(A) = .05$$
$$P(A') = .95,$$

so

$$P(A|T) = \frac{P(T|A)P(A)}{P(T|A)P(A) + P(T|A')P(A')}$$
$$= \frac{(.95)(.05)}{(.95)(.05) + (.06)(.95)} \approx .45.$$

In other words, it is actually more likely that such an athlete does *not* use steroids than that he does.[*]

[*] Without knowing the results of the test, we would have said that there was a probability of $P(A) = .05$ that the athlete is using steroids. The positive test result raises the probability to $P(A|T) = .45$, but the test gives too many false positives for us to be any more than 45% certain that the athlete is actually using steroids.

Remembering the Formula

Although the formula looks complicated at first sight, it is not hard to remember if you notice the pattern. Or you could re-derive it yourself by thinking of the tree diagram.

The next example illustrates that we can use either a tree diagram or the Bayes' theorem formula.

EXAMPLE 2   **Lie Detectors**

The *Sherlock Lie Detector Company* manufactures the latest in lie detectors, and the *Count-Your-Pennies* (CYP) store chain is eager to use them to screen its employees for theft. Sherlock's advertising claims that the test misses a lie only once in every 100 instances. On the other hand, an analysis by a consumer group reveals 20% of people who are telling the truth fail the test anyway.* The local police department estimates that 1 out of every 200 employees has engaged in theft. When the CYP store first screened its employees, the test indicated that Prudence V. Good was lying when she claimed that she had never stolen from CYP. What is the probability that she was lying and had in fact stolen from the store?

* The reason for this is that many people show physical signs of distress when asked accusatory questions. Many people are nervous around police officers even if they have done nothing wrong.

Solution   We are asked for the probability that Prudence Good was lying, and in the preceding sentence we are told that the lie detector test showed her to be lying. So we are looking for a conditional probability: the probability that she is lying, given that the lie detector test is positive. Now we can start to give names to the events:

        L: A subject is lying.
        T: The test is positive (indicated that the subject was lying).

We are looking for $P(L|T)$. We know that 1 out of every 200 employees engages in theft. Let us assume that no employee admits to theft while taking a lie detector test, so the probability $P(L)$ that a test subject is lying is $1/200$. We also know the false negative and false positive rates $P(T'|L)$ and $P(T|L')$.

Using a Tree Diagram   Figure 16 shows the tree diagram.

Figure 16

We see that

$$P(L|T) = \frac{P(\text{Using } L \text{ and } T \text{ branches})}{\text{Sum of } P(\text{Using branches ending in } T)}$$

$$= \frac{.00495}{.00495 + .199} \approx .024.$$

This means that there was only a 2.4% chance that Prudence Good was lying and had stolen from the store!

Using Bayes' Theorem   We have

        $P(L) = .005$
        $P(T|L') = .2$
        $P(T'|L) = .01,$

from which we obtain

        $P(T|L) = .99,$

so

$$P(L|T) = \frac{P(T|L)\,P(L)}{P(T|L)\,P(L) + P(T|L')\,P(L')} = \frac{(.99)(.005)}{(.99)(.005) + (.2)(.995)} \approx .024.$$

Expanded Form of Bayes' Theorem

We have seen the "short form" of Bayes' theorem. What is the "long form"? To motivate an expanded form of Bayes' theorem, look again at the formula we've been using:

$$P(A|T) = \frac{P(T|A)\,P(A)}{P(T|A)\,P(A) + P(T|A')\,P(A')}.$$

The events A and A' form a **partition** of the sample space S; that is, their union is the whole of S, and their intersection is empty (Figure 17).

The expanded form of Bayes' theorem applies to a partition of S into three or more events, as shown in Figure 18.

By saying that the events A_1, A_2, and A_3 form a partition of S, we mean that their union is the whole of S and the intersection of any two of them is empty, as in the figure. When we have a partition into three events as shown, the formula gives us $P(A_1|T)$ in terms of $P(T|A_1)$, $P(T|A_2)$, $P(T|A_3)$, $P(A_1)$, $P(A_2)$, and $P(A_3)$.

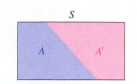

A and A' form a partition of S.

Figure 17

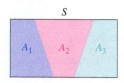

A_1, A_2, and A_3 form a partition of S.

Figure 18

Bayes' Theorem (Expanded Form)

If the events A_1, A_2, and A_3 form a partition of the sample space S, then

$$P(A_1|T) = \frac{P(T|A_1)\,P(A_1)}{P(T|A_1)\,P(A_1) + P(T|A_2)\,P(A_2) + P(T|A_3)\,P(A_3)}$$

As for why this is true and what happens when we have a partition into *four or more* events, we will wait for the exercises. In practice, as was the case with a partition into two events, we can often compute $P(A_1|T)$ by constructing a tree diagram.

EXAMPLE 3 School and Work

A survey[68] conducted by the Bureau of Labor Statistics found that approximately 27% of the high school graduating class of 2010 went on to a 2-year college, 41% went on to a 4-year college, and the remaining 32% did not go on to college. Of those who went on to a 2-year college, 52% worked at the same time, 32% of those going on to a 4-year college worked, and 78% of those who did not go on to college worked. What percentage of those working had not gone on to college?

Solution We can interpret these percentages as probabilities if we consider the experiment of choosing a member of the high school graduating class of 2010 at random. The events we are interested in are these:

R_1: A graduate went on to a 2-year college.

R_2: A graduate went on to a 4-year college.

R_3: A graduate did not go to college.

A: A graduate went on to work.

---

[68] Source: "College Enrollment and Work Activity of High School Graduates," U.S. Bureau of Labor Statistics (www.bls.gov/news.release/hsgec.htm).

The three events R_1, R_2, and R_3 partition the sample space of all graduates into three events. We are given the following probabilities:

$$P(R_1) = .27 \qquad P(R_2) = .41 \qquad P(R_3) = .32$$
$$P(A|R_1) = .52 \qquad P(A|R_2) = .32 \qquad P(A|R_3) = .78.$$

We are asked to find the probability that a graduate who went on to work did not go to college, so we are looking for $P(R_3|A)$. Bayes' formula for these events is

$$P(R_3|A) = \frac{P(A|R_3)\,P(R_3)}{P(A|R_1)\,P(R_1) + P(A|R_2)\,P(R_2) + P(A|R_3)\,P(R_3)}$$

$$= \frac{(.78)(.32)}{(.52)(.27) + (.32)(.41) + (.78)(.32)} \approx .48.$$

Thus, we conclude that 48% of all those working had not gone on to college.

➡ **Before we go on...** We could also solve Example 3 using a tree diagram. As before, the first branching corresponds to the events with unconditional probabilities that we know: R_1, R_2, and R_3. You should complete the tree and check that you obtain the same result as above. ∎

7.6   EXERCISES

▼ more advanced   ◆ challenging
Ⓣ indicates exercises that should be solved using technology

In Exercises 1–8, use Bayes' theorem or a tree diagram to calculate the indicated probability. Round all answers to four decimal places. [**HINT:** See Quick Example 1 and Example 3.]

1. $P(A|B) = .8$, $P(B) = .2$, $P(A|B') = .3$. Find $P(B|A)$.

2. $P(A|B) = .6$, $P(B) = .3$, $P(A|B') = .5$. Find $P(B|A)$.

3. $P(X|Y) = .8$, $P(Y') = .3$, $P(X|Y') = .5$. Find $P(Y|X)$.

4. $P(X|Y) = .6$, $P(Y') = .4$, $P(X|Y') = .3$. Find $P(Y|X)$.

5. Y_1, Y_2, Y_3 form a partition of S. $P(X|Y_1) = .4$, $P(X|Y_2) = .5$, $P(X|Y_3) = .6$, $P(Y_1) = .8$, $P(Y_2) = .1$. Find $P(Y_1|X)$.

6. Y_1, Y_2, Y_3 form a partition of S. $P(X|Y_1) = .2$, $P(X|Y_2) = .3$, $P(X|Y_3) = .6$, $P(Y_1) = .3$, $P(Y_2) = .4$. Find $P(Y_1|X)$.

7. Y_1, Y_2, Y_3 form a partition of S. $P(X|Y_1) = .4$, $P(X|Y_2) = .5$, $P(X|Y_3) = .6$, $P(Y_1) = .8$, $P(Y_2) = .1$. Find $P(Y_2|X)$.

8. Y_1, Y_2, Y_3 form a partition of S. $P(X|Y_1) = .2$, $P(X|Y_2) = .3$, $P(X|Y_3) = .6$, $P(Y_1) = .3$, $P(Y_2) = .4$. Find $P(Y_2|X)$.

Applications

9. *Music Downloading* According to a study on the effect of music downloading on spending on music, 11% of all Internet users had decreased their spending on music.[69] We estimate that 40% of all music fans used the Internet at the time of the study.[70] If 20% of non–Internet users had decreased their spending on music, what percentage of those who had decreased their spending on music were Internet users? [**HINT:** See Examples 1 and 2.]

10. *Music Downloading* According to the study cited in Exercise 9, 36% of experienced file-sharers with broadband access had decreased their spending on music. Let us estimate that 3% of all music fans were experienced file-sharers with broadband access at the time of the study.[71] If 20% of the other music fans had decreased their spending on music, what percentage of those who had decreased their spending on music were experienced file-sharers with broadband access? [**HINT:** See Examples 1 and 2.]

11. *Weather* It snows in Greenland an average of once every 25 days, and when it does, glaciers have a 20% chance of growing. When it does not snow in Greenland, glaciers have only a 4% chance of growing. What is the probability that it is snowing in Greenland when glaciers are growing?

---

[69] Regardless of whether they used the Internet to download music. Source: *New York Times*, May 6, 2002, p. C6.

[70] According to the U.S. Department of Commerce, 51% of all U.S. households had computers in 2001.

[71] Around 15% of all online households had broadband access in 2001, according to a *New York Times* article (Dec. 24, 2001, p. C1).

12. **Weather** It rains in Spain an average of once every 10 days, and when it does, hurricanes have a 2% chance of happening in Hartford. When it does not rain in Spain, hurricanes have a 1% chance of happening in Hartford. What is the probability that it rains in Spain when hurricanes happen in Hartford?

13. **University Admissions** In fall 2014, 34% of applicants with a Math SAT of 700 or more were admitted by the University of California, Los Angeles (UCLA), while 12% with a Math SAT of less than 700 were admitted. Further, 36% of all applicants had a Math SAT score of 700 or more.[72] What percentage of admitted applicants had a Math SAT of 700 or more? (Round your answer to the nearest percentage point.)

14. **University Admissions** In fall 2014, 71% of rejected applicants to UCLA had a Math SAT of less than 700, while 39% of accepted applicants to UCLA had a Math SAT of less than 700. Further, 80% of all applicants were rejected.[73] What percentage of applicants with a Math SAT of less than 700 were rejected? (Round your answer to the nearest percentage point.)

15. **Side-Impact Hazard** In 2004, 45.4% of all light vehicles were cars, and the rest were light trucks or SUVs. The probability that a severe side-impact crash would prove deadly to a driver depended on the type of vehicle he or she was driving at the time, as shown in the table:[74]

Car	1.0
Light Truck or SUV	.3

What is the probability that the victim of a deadly side-impact accident was driving a car?

16. **Side-Impact Hazard** In 2004, 27.3% of all light vehicles were light trucks, and the rest were cars or SUVs. The probability that a severe side-impact crash would prove deadly to a driver depended on the type of vehicle he or she was driving at the time, as shown in the table:[75]

Light Truck	.2
Car or SUV	.7

What is the probability that the victim of a deadly side-impact accident was driving a car or SUV?

17. **Athletic Fitness Tests** Any athlete who fails the *Enormous State University*'s women's soccer fitness test is automatically dropped from the team. Last year, Mona Header failed

the test but claimed that this was due to the early hour. (The fitness test is traditionally given at 5 am on a Sunday morning.) In fact, a study by the ESU Physical Education Department suggested that 50% of athletes fit enough to play on the team would fail the soccer test, although no unfit athlete could possibly pass the test. It also estimated that 45% of the athletes who take the test are fit enough to play soccer. Assuming that these estimates are correct, what is the probability that Mona was justifiably dropped?

18. **Academic Testing** Professor Frank Nabarro insists that all senior physics majors take his notorious physics aptitude test. The test is so tough that anyone *not* going on to a career in physics has no hope of passing, whereas 60% of the seniors who do go on to a career in physics still fail the test. Further, 75% of all senior physics majors in fact go on to a career in physics. Assuming that you fail the test, what is the probability that you will not go on to a career in physics?

19. **Side-Impact Hazard** (Compare Exercise 15.) In 2004, 27.3% of all light vehicles were light trucks, 27.3% were SUVs, and 45.4% were cars. The probability that a severe side-impact crash would prove deadly to a driver depended on the type of vehicle he or she was driving at the time, as shown in the table:[76]

Light Truck	.210
SUV	.371
Car	1.000

What is the probability that the victim of a deadly side-impact accident was driving an SUV? [**HINT:** See Example 3.]

20. **Side-Impact Hazard** In 1986, 23.9% of all light vehicles were light trucks, 5.0% were SUVs, and 71.1% were cars. Refer to Exercise 19 for the probabilities that a severe side-impact crash would prove deadly. What is the probability that the victim of a deadly side-impact accident was driving a car? [**HINT:** See Example 3.]

21. **University Admissions** In fall 2008, UCLA admitted 22% of its California resident applicants, 28% of its applicants from other U.S. states, and 22% of its international student applicants. Of all its applicants, 84% were California residents, 10% were from other U.S. states, and 6% were international students.[77] What percentage of all admitted students were California residents? (Round your answer to the nearest 1%.)

22. **University Admissions** In fall 2002, UCLA admitted 26% of its California resident applicants, 18% of its applicants from other U.S. states, and 13% of its international student applicants. Of all its applicants, 86% were California residents, 11% were from other U.S. states, and 3% were international

---

[72] Percentages are rounded. Source: University of California (www.admissions.ucla.edu/Prospect/Adm_fr/Frosh_Prof14.htm).

[73] *Ibid.*

[74] A "serious" side-impact accident is defined as one in which the driver of a car would be killed. Source: National Highway Traffic Safety Administration/*New York Times*, May 30, 2004, p. BU 9.

[75] *Ibid.*

[76] *Ibid.*

[77] Source: University of California (www.admissions.ucla.edu/Prospect/Adm_fr/Frosh_Prof08.htm).

students.[78] What percentage of all admitted students were California residents? (Round your answer to the nearest 1%.)

23. **Internet Use (Historical)** In 2000, 86% of all Caucasians in the United States, 77% of all African-Americans, 77% of all Hispanics, and 85% of residents not classified into one of these groups used the Internet for email.[79] At that time, the U.S. population was 69% Caucasian, 12% African-American, and 13% Hispanic. What percentage of U.S. residents who used the Internet for email were Hispanic?

24. **Internet Use (Historical)** In 2000, 59% of all Caucasians in the United States, 57% of all African-Americans, 58% of all Hispanics, and 54% of residents not classified into one of these groups used the Internet to search for information.[80] At that time, the U.S. population was 69% Caucasian, 12% African-American, and 13% Hispanic. What percentage of U.S. residents who used the Internet for information search were African-American?

25. ▼ **Market Surveys** A *New York Times* survey of homeowners in the 1990s showed that 86% of those with swimming pools were married couples, and the other 14% were single.[81] It also showed that 15% of all homeowners had pools.
    a. Assuming that 90% of all homeowners without pools are married couples, what percentage of homes owned by married couples have pools?
    b. Would it have been more profitable for pool manufacturers to go after single homeowners or married homeowners? Explain.

26. ▼ **Crime and Preschool.** A *New York Times* survey[82] of needy and disabled youths showed that 51% of those who had no preschool education were arrested or charged with a crime by the time they were 19, whereas only 31% who had preschool education wound up in this category. The survey did not specify what percentage of the youths in the survey had preschool education, so let us take a guess at that and estimate that 20% of them had attended preschool.
    a. What percentage of the youths arrested or charged with a crime had no preschool education?
    b. What would this figure be if 80% of the youths had attended preschool? Would youths who had preschool education be more likely to be arrested or charged with a crime than those who did not? Support your answer by quoting probabilities.

27. ▼ **Grade Complaints** Two of the mathematics professors at *Enormous State University* are Professor A (known for easy grading) and Professor F (known for tough grading). Last semester, roughly three quarters of Professor F's class consisted of former students of Professor A; these students apparently felt encouraged by their (utterly undeserved) high grades. (Professor F's own former students had fled in droves to Professor A's class to try to shore up their grade-point averages.) At the end of the semester, as might have been predicted, all of Professor A's former students wound up with a C– or lower. The rest of the students in the class—former students of Professor F who had decided to "stick it out"—fared better, and two thirds of them earned higher than a C–. After discovering what had befallen them, all the students who earned C– or lower got together and decided to send a delegation to the department chair to complain that their grade-point averages had been ruined by this callous and heartless beast! The contingent was to consist of 10 representatives selected at random from among them. How many of the 10 would you estimate to have been former students of Professor A?

28. ▼ **Weather Prediction** A local TV station employs Desmorelda, "Mistress of the Zodiac," as its weather forecaster. Now, when it rains, Sagittarius is in the shadow of Jupiter one-third of the time, and it rains on 4 out of every 50 days. Sagittarius falls in Jupiter's shadow on only 1 in every 5 rainless days. The powers that be at the station notice a disturbing pattern to Desmorelda's weather predictions. It seems that she always predicts that it will rain when Sagittarius is in the shadow of Jupiter. What percentage of the time is she correct? Should they replace her?

29. ▼ **Employment in the 1980s** In a 1987 survey of married couples with earnings, 95% of all husbands were employed. Of all employed husbands, 71% of their wives were also employed.[83] Noting that either the husband or wife in a couple with earnings had to be employed, find the probability that the husband of an employed woman was also employed.

30. ▼ **Employment in the 1980s** Repeat Exercise 29 in the event that 50% of all husbands were employed.

31. ▼ **Juvenile Delinquency** According to a study at the Oregon Social Learning Center, boys who had been arrested by age 14 were 17.9 times more likely to become chronic offenders than those who had not.[84] Use these data to estimate the percentage of chronic offenders who had been arrested by age 14 in a city where 0.1% of all boys have been arrested by age 14. [**HINT:** Use Bayes' formula rather than a tree.]

[78] Source: UCLA website, May 2002 (www.admissions.ucla.edu/Prospect/Adm_fr/Frosh_Prof.htm).

[79] Source: NTIA and ESA, U.S. Department of Commerce, using August 2000 U.S. Bureau of The Census Current Population Survey Supplement.

[80] *Ibid.*

[81] Source: "All about Swimming Pools," *New York Times*, September 13, 1992.

[82] Source: "Governors Develop Plan to Help Preschool Children," *New York Times*, August 2, 1992.

[83] Source: *Statistical Abstract of the United States*, 111th Ed., 1991, U.S. Dept. of Commerce/U.S. Bureau of Labor Statistics. Figures rounded to the nearest 1%.

[84] Based on a study by Marion S. Forgatch of 319 boys from high-crime neighborhoods in Eugene, Oregon. Source: W. Wayt Gibbs, "Seeking the Criminal Element," *Scientific American*, March 1995, pp. 101–107.

32. ▼ *Crime* According to the same study at the Oregon Social Learning Center, chronic offenders were 14.3 times more likely to commit violent offenses than people who were not chronic offenders.[85] In a neighborhood where 2 in every 1,000 residents is a chronic offender, estimate the probability that a violent offender is also a chronic offender. [**HINT:** Use Bayes' formula rather than a tree.]

33. ▼ *Benefits of Exercise* According to a study in *The New England Journal of Medicine*,[86] 202 of a sample of 5,990 middle-aged men had developed diabetes. It also found that men who were very active (burning about 3,500 calories daily) were half as likely to develop diabetes compared with men who were sedentary. Assume that one third of all middle-aged men are very active, and the rest are classified as sedentary. What is the probability that a middle-aged man with diabetes is very active?

34. ▼ *Benefits of Exercise* Repeat Exercise 33 assuming that only 1 in 10 middle-aged men is very active.

35. ◆ *Air Bag Safety* According to a 2000 study conducted by the Harvard School of Public Health, a child seated in the front seat who was wearing a seatbelt was 31% more likely to be killed in an accident if the car had an air bag that deployed than if it did not.[87] Air bags deployed in 25% of all accidents. For a child seated in the front seat wearing a seatbelt, what is the probability that the air bag deployed in an accident in which the child was killed? (Round your answer to two decimal places.) [**HINT:** When we say that "A is 31% more likely than B," we mean that the probability of A is 1.31 times the probability of B.]

36. ◆ *Air Bag Safety* According to the study cited in Exercise 35, a child seated in the front seat who was not wearing a seatbelt was 84% more likely to be killed in an accident if the car had an air bag that deployed than if it did not.[88] Air bags deployed in 25% of all accidents. For a child seated in the front seat not wearing a seatbelt, what is the probability that the air bag deployed in an accident in which the child was killed? (Round your answer to two decimal places.) [**HINT:** When we say that "A is 84% more likely than B," we mean that the probability of A is 1.84 times the probability of B.]

---

[85] See footnote for Exercise 31.

[86] As cited in an article in *New York Times* on July 18, 1991.

[87] The study was conducted by Dr. Segui-Gomez at the Harvard School of Public Health. Source: *New York Times*, December 1, 2000, p. F1.

[88] *Ibid.*

Communication and Reasoning Exercises

37. Your friend claims that the probability of *A* given *B* is the same as the probability of *B* given *A*. How would you convince him that he is wrong?

38. Complete the following sentence. To use Bayes' formula to compute $P(E|F)$, you need to be given _____.

39. ▼ Give an example in which a steroids test gives a false positive only 1% of the time, yet if an athlete tests positive, the chance that he or she has used steroids is under 10%.

40. ▼ Give an example in which a steroids test gives a false positive 30% of the time, yet if an athlete tests positive, the chance that he or she has used steroids is over 90%.

41. ▼ Use a tree to derive the expanded form of Bayes' theorem for a partition of the sample space S into three events R_1, R_2, and R_3.

42. ▼ Write down an expanded form of Bayes' theorem that applies to a partition of the sample space S into four events R_1, R_2, R_3, and R_4.

43. ◆ *Politics* The following letter appeared in *The New York Times*:[89]

> To the Editor:
>
> It stretches credulity when William Safire contends (column, Jan. 11) that 90 percent of those who agreed with his Jan. 8 column, in which he called the First Lady, Hillary Rodham Clinton, "a congenital liar," were men and 90 percent of those who disagreed were women.
>
> Assuming these percentages hold for Democrats as well as Republicans, only 10 percent of Democratic men disagreed with him. Is Mr. Safire suggesting that 90 percent of Democratic men supported him? How naive does he take his readers to be?
>
> A. D.
> New York, Jan. 12, 1996

Comment on the letter writer's reasoning.

44. ◆ *Politics* Refer back to Exercise 43. If the letter writer's conclusion was correct, what percentage of all Democrats would have agreed with Safire's column?

---

[89] The original letter appeared in *New York Times*, January 16, 1996, p. A16. We have edited the first phrase of the second paragraph slightly for clarity; the original sentence read: "Assuming the response was equally divided between Democrats and Republicans, . . . "

<table><tr><td>**7.7**</td><td>**Markov Systems**</td></tr></table>

Markov Systems, States, and Transition Probabilities

Many real-life situations can be modeled by processes that pass from state to state with given probabilities. A simple example of such a **Markov system** is the fluctuation of a gambler's fortune as he or she continues to bet. Other examples come from the study of trends in the commercial world and the study of neural networks and artificial intelligence. The mathematics of Markov systems is an interesting combination of probability and matrix arithmetic.

Here is a basic example we shall use many times: A market analyst for *Gamble Detergents* is interested in whether consumers prefer powdered laundry detergents or liquid detergents. Two market surveys taken 1 year apart revealed that 20% of powdered detergent users had switched to liquid 1 year later, while the rest were still using powder. Only 10% of liquid detergent users had switched to powder 1 year later, with the rest still using liquid.

We analyze this example as follows: Every year, a consumer may be in one of two possible **states**: He may be a powdered detergent user or a liquid detergent user. Let us number these states: A consumer is in state 1 if he uses powdered detergent and in state 2 if he uses liquid. There is a basic **time step** of 1 year. If a consumer happens to be in state 1 during a given year, then there is a probability of 20% = .2 (the chance that a randomly chosen powder user will switch to liquid) that he will be in state 2 the next year. We write

$$p_{12} = .2$$

to indicate that the probability of going *from* state 1 *to* state 2 in one time step is .2. The other 80% of the powder users are using powder the next year. We write

$$p_{11} = .8$$

to indicate that the probability of *staying* in state 1 from one year to the next is .8.***** What if a consumer is in state 2? Then the probability of going to state 1 is given as 10% = .1, so the probability of remaining in state 2 is .9. Thus,

$$p_{21} = .1$$

and

$$p_{22} = .9.$$

We can picture this system as in Figure 19, which shows the **state transition diagram** for this example. The numbers p_{ij}, which appear as labels on the arrows, are the **transition probabilities**.

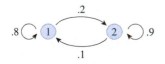

Figure 19

Markov System, States, and Transition Probabilities

A **Markov system**† (or **Markov process** or **Markov chain**) is a system that can be in one of several specified **states**. There is specified a certain **time step**, and at each step, the system will randomly change states or remain where it is. The probability of going from state i to state j is a fixed number p_{ij}, called the **transition probability**.

Quick Example

1. The Markov system depicted in Figure 19 has two states: state 1 and state 2. The transition probabilities are as follows:

$$p_{11} = \text{Probability of going from state 1 to state 1} = .8$$
$$p_{12} = \text{Probability of going from state 1 to state 2} = .2$$
$$p_{21} = \text{Probability of going from state 2 to state 1} = .1$$
$$p_{22} = \text{Probability of going from state 2 to state 2} = .9.$$

Notice that, because the system must go somewhere at each time step, the transition probabilities originating at a particular state always add up to 1. For example, in the transition diagram above, when we add the probabilities originating at state 1, we get $.8 + .2 = 1$.

The transition probabilities may be conveniently arranged in a matrix.

Transition Matrix

The **transition matrix** associated with a given Markov system is the matrix P whose ijth entry is the transition probability p_{ij}, the transition probability of going *from* state i *to* state j. In other words, the entry in position ij is the *label on the arrow going from state i to state j* in a state transition diagram.

Thus, the transition matrix for a system with two states would be set up as follows:

$$\textbf{From} \quad \begin{array}{c} \\ \mathbf{1} \\ \mathbf{2} \end{array} \overset{\displaystyle \textbf{To}}{\begin{array}{cc} \mathbf{1} & \mathbf{2} \\ \begin{bmatrix} p_{11} & p_{12} \\ p_{21} & p_{22} \end{bmatrix} \end{array}}. \qquad \begin{array}{l} \text{Arrows originating in state 1} \\ \text{Arrows originating in state 2} \end{array}$$

Quick Example

2. In the system pictured in Figure 19, the transition matrix is

$$P = \begin{bmatrix} .8 & .2 \\ .1 & .9 \end{bmatrix}.$$

Note Notice that because the sum of the transition probabilities that originate at any state is 1, *the sum of the entries in any row of a transition matrix is* 1. ∎

Distribution Vectors and Powers of the Transition Matrix

EXAMPLE 1   Laundry Detergent Switching

Consider the Markov system found by *Gamble Detergents* at the beginning of this section. Suppose that 70% of consumers are now using powdered detergent, while the other 30% are using liquid.

a. What will be the distribution 1 year from now? (That is, what percentage will be using powdered and what percentage liquid detergent?)

b. Assuming that the probabilities remain the same, what will be the distribution 2 years from now? 3 years from now?

Solution

a. First, let us think of the statement that 70% of consumers are using powdered detergent as telling us a probability: The probability that a randomly chosen consumer uses powdered detergent is .7. Similarly, the probability that a randomly chosen consumer uses liquid detergent is .3. We want to find the corresponding probabilities 1 year from now. To do this, consider the tree diagram in Figure 20.

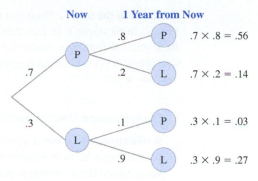

Figure 20

The first branching shows the probabilities now, and the second branching shows the (conditional) transition probabilities. So if we want to know the probability that a consumer is using powdered detergent 1 year from now, it will be

Probability of using powder after 1 year $= .7 \times .8 + .3 \times .1 = .59$.

On the other hand, we have:

Probability of using liquid after 1 year $= .7 \times .2 + .3 \times .9 = .41$.

Now, here's the crucial point: *These are exactly the same calculations as in the matrix product*

$$[.7 \quad .3]\begin{bmatrix} .8 & .2 \\ .1 & .9 \end{bmatrix} = [.59 \quad .41].$$

↑ Initial distribution   ↑ Transition matrix   ↑ Distribution after one step

Thus, to get the distribution of detergent users after 1 year, all we have to do is multiply the **initial distribution vector** $[.7 \quad .3]$ by the transition matrix P. The result is $[.59 \quad .41]$, the **distribution vector after one step**.

b. Now what about the distribution after 2 years? If we assume that the same fraction of consumers switch or stay put in the second year as in the first, we can simply repeat the calculation we did above, using the new distribution vector:

$$[.59 \quad .41]\begin{bmatrix} .8 & .2 \\ .1 & .9 \end{bmatrix} = [.513 \quad .487].$$

↑ Distribution after one step   ↑ Transition matrix   ↑ Distribution after two steps

Thus, after 2 years we can expect 51.3% of consumers to be using powdered detergent and 48.7% to be using liquid detergent. Similarly, after 3 years we have

$$[.513 \quad .487]\begin{bmatrix} .8 & .2 \\ .1 & .9 \end{bmatrix} = [.4591 \quad .5409].$$

So after 3 years, 45.91% of consumers will be using powdered detergent, and 54.09% will be using liquid. Slowly but surely, liquid detergent seems to be winning.

➡ **Before we go on . . .** Note that the sum of the entries is 1 in each of the distribution vectors in Example 1. In fact, these vectors are giving the probability distributions for each year of finding a randomly chosen consumer using either powdered or liquid detergent. A vector having nonnegative entries adding up to 1 is called a **probability vector.** ∎

Distribution Vector after *m* Steps

A **distribution vector** is a probability vector giving the probability distribution for finding a Markov system in its various possible states. If v is a distribution vector, then the distribution vector one step later will be vP. The distribution m steps later will be

$$\text{Distribution after } m \text{ steps} = v \cdot P \cdot P \cdot \ldots \cdot P \ (m \text{ times}) = vP^m.$$

Quick Example

3. If $P = \begin{bmatrix} 0 & 1 \\ .5 & .5 \end{bmatrix}$ and $v = \begin{bmatrix} .2 & .8 \end{bmatrix}$, then we can calculate the following distribution vectors:

$$vP = \begin{bmatrix} .2 & .8 \end{bmatrix} \begin{bmatrix} 0 & 1 \\ .5 & .5 \end{bmatrix} = \begin{bmatrix} .4 & .6 \end{bmatrix} \qquad \text{Distribution after one step}$$

$$vP^2 = (vP)P = \begin{bmatrix} .4 & .6 \end{bmatrix} \begin{bmatrix} 0 & 1 \\ .5 & .5 \end{bmatrix} = \begin{bmatrix} .3 & .7 \end{bmatrix} \qquad \text{Distribution after two steps}$$

$$vP^3 = (vP^2)P = \begin{bmatrix} .3 & .7 \end{bmatrix} \begin{bmatrix} 0 & 1 \\ .5 & .5 \end{bmatrix} = \begin{bmatrix} .35 & .65 \end{bmatrix}. \qquad \text{Distribution after three steps}$$

What about the matrix P^m that appears above? Multiplying a distribution vector v times P^m gives us the distribution m steps later, so we can think of P^m as the m-step transition matrix. More explicitly, consider the following example.

EXAMPLE 2   Powers of the Transition Matrix

Continuing the example of detergent switching, suppose that a consumer is now using powdered detergent. What are the probabilities that the consumer will be using powdered or liquid detergent 2 years from now? What if the consumer is now using liquid detergent?

Solution To record the fact that we know that the consumer is using powdered detergent, we can take as our initial distribution vector $v = \begin{bmatrix} 1 & 0 \end{bmatrix}$. To find the distribution 2 years from now, we compute vP^2. To make a point, we do the calculation slightly differently:

$$vP^2 = \begin{bmatrix} 1 & 0 \end{bmatrix} \begin{bmatrix} .8 & .2 \\ .1 & .9 \end{bmatrix} \begin{bmatrix} .8 & .2 \\ .1 & .9 \end{bmatrix}$$

$$= \begin{bmatrix} 1 & 0 \end{bmatrix} \begin{bmatrix} .66 & .34 \\ .17 & .83 \end{bmatrix}$$

$$= \begin{bmatrix} .66 & .34 \end{bmatrix}.$$

So the probability that our consumer is using powdered detergent 2 years from now is .66, while the probability of using liquid detergent is .34. The point to notice is that these are the entries in the first row of P^2. Similarly, if we consider a consumer now using liquid detergent, we should take the initial distribution vector to be $v = \begin{bmatrix} 0 & 1 \end{bmatrix}$ and compute

$$vP^2 = \begin{bmatrix} 0 & 1 \end{bmatrix} \begin{bmatrix} .66 & .34 \\ .17 & .83 \end{bmatrix} = \begin{bmatrix} .17 & .83 \end{bmatrix}.$$

Thus, the bottom row gives the probabilities that a consumer, now using liquid detergent, will be using either powdered or liquid detergent 2 years from now.

In other words, the ijth entry of P^2 gives the probability that a consumer, starting in state i, will be in state j after two time steps.

What is true in Example 2 for two time steps is true for any number of time steps, which gives us the following.

Powers of the Transition Matrix

P^m ($m = 1, 2, 3, \ldots$) is the **m-step transition matrix**. The ijth entry in P^m is the probability of a transition from state i to state j in m steps.

Quick Example

4. If $P = \begin{bmatrix} 0 & 1 \\ .5 & .5 \end{bmatrix}$, then

$$P^2 = P \cdot P = \begin{bmatrix} .5 & .5 \\ .25 & .75 \end{bmatrix} \qquad \text{Two-step transition matrix}$$

$$P^3 = P \cdot P^2 = \begin{bmatrix} .25 & .75 \\ .375 & .625 \end{bmatrix}. \qquad \text{Three-step transition matrix}$$

The probability of going from state 1 to state 2 in two steps = (1, 2)-entry of P^2 = .5.

The probability of going from state 1 to state 2 in three steps = (1, 2)-entry of P^3 = .75.

Steady-State Distribution Vector

What happens if we follow our laundry detergent–using consumers for many years?

EXAMPLE 3   Long-Term Behavior

Suppose that 70% of consumers are now using powdered detergent while the other 30% are using liquid. Assuming that the transition matrix remains valid the whole time, what will be the distribution 1, 2, 3, . . . , and 50 years later?

Solution Of course, to do this many matrix multiplications, we're best off using technology. We already did the first three calculations in an earlier example.

Distribution after 1 year: $\quad \begin{bmatrix} .7 & .3 \end{bmatrix} \begin{bmatrix} .8 & .2 \\ .1 & .9 \end{bmatrix} = \begin{bmatrix} .59 & .41 \end{bmatrix}$

Distribution after 2 years: $\quad \begin{bmatrix} .59 & .41 \end{bmatrix} \begin{bmatrix} .8 & .2 \\ .1 & .9 \end{bmatrix} = \begin{bmatrix} .513 & .487 \end{bmatrix}$

Distribution after 3 years: $\quad \begin{bmatrix} .513 & .487 \end{bmatrix} \begin{bmatrix} .8 & .2 \\ .1 & .9 \end{bmatrix} = \begin{bmatrix} .4591 & .5409 \end{bmatrix}$

$\vdots$

Distribution after 48 years: $\quad \begin{bmatrix} .33333335 & .66666665 \end{bmatrix}$

Distribution after 49 years: $\quad \begin{bmatrix} .33333334 & .66666666 \end{bmatrix}$

Distribution after 50 years: $\quad \begin{bmatrix} .33333334 & .66666666 \end{bmatrix}$

Thus, the distribution after 50 years is approximately $\begin{bmatrix} .33333334 & .66666666 \end{bmatrix}$.

Something interesting seems to be happening in Example 3. The distribution seems to be getting closer and closer to

$$\begin{bmatrix} .333333\ldots & .666666\ldots \end{bmatrix} = \begin{bmatrix} \frac{1}{3} & \frac{2}{3} \end{bmatrix}.$$

Let's call this distribution vector v_∞. Notice two things about v_∞:

- v_∞ is a probability vector.
- If we calculate $v_\infty P$, we find

$$v_\infty P = \begin{bmatrix} \frac{1}{3} & \frac{2}{3} \end{bmatrix} \begin{bmatrix} .8 & .2 \\ .1 & .9 \end{bmatrix} = \begin{bmatrix} \frac{1}{3} & \frac{2}{3} \end{bmatrix} = v_\infty.$$

In other words,

$$v_\infty P = v_\infty.$$

We call a probability vector v with the property that $vP = v$ a **steady-state (probability) vector**.

Q: *Where does the name steady-state vector come from?*

A: If $vP = v$, then v is a distribution that will not change from time step to time step. In the example above, because $\begin{bmatrix} 1/3 & 2/3 \end{bmatrix}$ is a steady-state vector, if 1/3 of consumers use powdered detergent and 2/3 use liquid detergent one year, then the proportions will be the same the next year. Individual consumers may still switch from year to year, but as many will switch from powder to liquid as will switch from liquid to powder, so the number using each will remain constant.

But how do we find a steady-state vector?

EXAMPLE 4 Calculating the Steady-State Vector

Calculate the steady-state probability vector for the transition matrix in the preceding examples:

$$P = \begin{bmatrix} .8 & .2 \\ .1 & .9 \end{bmatrix}.$$

Solution We are asked to find

$$v_\infty = [x \quad y].$$

This vector must satisfy the equation

$$v_\infty P = v_\infty$$

or

$$[x \quad y]\begin{bmatrix} .8 & .2 \\ .1 & .9 \end{bmatrix} = [x \quad y].$$

Doing the matrix multiplication gives

$$[.8x + .1y \quad .2x + .9y] = [x \quad y].$$

Equating corresponding entries gives

$$.8x + .1y = x$$
$$.2x + .9y = y$$

or

$$-.2x + .1y = 0$$
$$.2x - .1y = 0.$$

Now, these equations are really the same equation. (Do you see that?) There is one more thing we know, though: Because $[x \quad y]$ is a probability vector, its entries must add up to 1. This gives one more equation:

$$x + y = 1.$$

Taking this equation together with one of the two equations above gives us the following system:

$$x + \quad y = 1$$
$$-.2x + .1y = 0.$$

We now solve this system using any of the techniques we learned for solving systems of linear equations. We find that the solution is $x = 1/3$, and $y = 2/3$, so the steady-state vector is

$$v_\infty = [x \quad y] = \left[\tfrac{1}{3} \quad \tfrac{2}{3}\right],$$

as suggested in Example 3.

Using Technology

Technology can be used to compute the steady-state vector in Example 4.

TI-83/84 Plus
Define [A] as the coefficient matrix of the system of equations being solved and [B] as the column matrix of the right-hand sides.
Then compute [A]$^{-1}$[B]
[More details in the Technology Guide.]

Spreadsheet
Use the MMULT and MINVERSE commands to solve the necessary system of equations.
[More details in the Technology Guide.]

WM Website
www.WanerMath.com
→ Online Utilities
→ Pivot and Gauss-Jordan Tool

You can use the Pivot and Gauss-Jordan Tool to solve the system of equations that gives you the steady-state vector.

The method we just used works for any size transition matrix and can be summarized as follows.

Calculating the Steady-State Distribution Vector

To calculate the steady-state probability vector for a Markov system with transition matrix P, we solve the system of equations given by

$$x + y + z + \cdots = 1$$
$$[x \quad y \quad z \quad \cdots]P = [x \quad y \quad z \quad \cdots],$$

where we use as many unknowns as there are states in the Markov system. The steady-state probability vector is then

$$v_\infty = [x \quad y \quad z \quad \cdots].$$

Q : *Is there always a steady-state distribution vector?*

A : Yes, although the explanation why is more involved than we can give here.

Q : *In Example 3 we started with a distribution vector v and found that vP^m got closer and closer to v_∞ as m got larger. Does that always happen?*

A : It does if the Markov system is **regular**, as we define below, but may not for other kinds of systems. Again, we shall not prove this fact here.

Regular Markov Systems

A **regular** Markov system is one for which some power of its transition matrix P has no zero entries. If a Markov system is regular, then

1. It has a unique steady-state probability vector v_∞, and

2. If v is any probability vector whatsoever, then vP^m approaches v_∞ as m gets large. We say that the **long-term behavior** of the system is to have distribution (close to) v_∞.

Interpreting the Steady-State Vector

In a regular Markov system, the entries in the steady-state probability vector give the long-term probabilities that the system will be in the corresponding states, or the fractions of time one can expect to find the Markov system in the corresponding states.

Quick Examples

5. The system with transition matrix $P = \begin{bmatrix} .8 & .2 \\ .1 & .9 \end{bmatrix}$ is regular because $P(= P^1)$ has no zero entries.

6. The system with transition matrix $P = \begin{bmatrix} 0 & 1 \\ .5 & .5 \end{bmatrix}$ is regular because $P^2 = \begin{bmatrix} .5 & .5 \\ .25 & .75 \end{bmatrix}$ has no zero entries.

7. The system with transition matrix $P = \begin{bmatrix} 0 & 1 \\ 1 & 0 \end{bmatrix}$ is *not* regular: $P^2 = \begin{bmatrix} 1 & 0 \\ 0 & 1 \end{bmatrix}$ and $P^3 = P$ again, so the powers of P alternate between these two matrices. Thus, every power of P has zero entries. Although this system has a steady-state vector, namely, $[.5 \quad .5]$, if we take $v = [1 \quad 0]$, then $vP = [0 \quad 1]$ and $vP^2 = v$, so the distribution vectors vP^m just alternate between these two vectors, not approaching v_∞.

We finish with one more example.

EXAMPLE 5   **Gambler's Ruin**

A timid gambler, armed with her annual bonus of $20, decides to play roulette using the following scheme. At each spin of the wheel, she places $10 on red. If red comes up, she wins an additional $10; if black comes up, she loses her $10. For the sake of simplicity, assume that she has a probability of 1/2 of winning. (In the real game, the probability is slightly lower—a fact that many gamblers forget.) She keeps playing until she has either reached $30 or lost it all. In either case she then packs up and leaves. Model this situation as a Markov system, and find the associated transition matrix. What can we say about the long-term behavior of this system?

Solution We must first decide on the states of the system. A good choice is the gambler's financial state, the amount of money she has at any stage of the game. According to her rules, she can be broke or can have $10, $20, or $30. Thus, there are four states: $1 = \$0, 2 = \$10, 3 = \$20$, and $4 = \$30$. Because she bets $10 each time, she moves down $10 if she loses (with probability 1/2) and up $10 if she wins (with probability also 1/2) until she reaches one of the extremes. The transition diagram is shown in Figure 21.

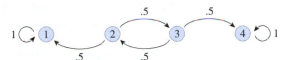

Figure 21

Note that once the system enters state 1 or state 4, it does not leave; with probability 1, it stays in the same state.[*] We call such states **absorbing states**. We can now write down the transition matrix:

$$P = \begin{bmatrix} 1 & 0 & 0 & 0 \\ .5 & 0 & .5 & 0 \\ 0 & .5 & 0 & .5 \\ 0 & 0 & 0 & 1 \end{bmatrix}.$$

(Notice all the 0 entries, corresponding to possible transitions that we did not draw in the transition diagram because they have 0 probability of occurring. We usually leave out such arrows.) Is this system regular? Take a look at P^2:

$$P^2 = \begin{bmatrix} 1 & 0 & 0 & 0 \\ .5 & .25 & 0 & .25 \\ .25 & 0 & .25 & .5 \\ 0 & 0 & 0 & 1 \end{bmatrix}.$$

Notice that the first and last rows haven't changed. After two steps, there is still no chance of leaving state 1 or state 4. In fact, no matter how many powers we take, no matter how many steps we look at, there will still be no way to leave either of those states, and the first and last rows will still have plenty of zeros. This system is not regular.

Nonetheless, we can try to find a steady-state probability vector. If we do this (and you should set up the system of linear equations and solve it), we find that there are infinitely many steady-state probability vectors, namely, all vectors of the form

*These states are like "Roach Motels" ("Roaches check in, but they don't check out")!

$[x \quad 0 \quad 0 \quad 1-x]$ for $0 \le x \le 1$. (You can check directly that these are all steady-state vectors.) As with a regular system, if we start with any distribution, the system will tend toward one of these steady-state vectors. In other words, eventually the gambler will either lose all her money or leave the table with $30.

But which outcome is more likely and with what probability? One way to approach this question is to try computing the distribution after many steps. The distribution that represents the gambler starting with $20 is $v = [0 \quad 0 \quad 1 \quad 0]$. Using technology, it's easy to compute vP^n for some large values of n:

$$vP^{10} \approx [.333008 \quad 0 \quad .000977 \quad .666016]$$
$$vP^{50} \approx [.333333 \quad 0 \quad 0 \quad .666667].$$

So it looks like the probability that she will leave the table with $30 is approximately 2/3, while the probability that she loses it all is 1/3.

What if she started with only $10? Then our initial distribution would be $v = [0 \quad 1 \quad 0 \quad 0]$, and

$$vP^{10} \approx [.666016 \quad .000977 \quad 0 \quad .333008]$$
$$vP^{50} \approx [.666667 \quad 0 \quad 0 \quad .333333].$$

So this time, the probability of her losing everything is about 2/3, while the probability of her leaving with $30 is 1/3. There is a way of calculating these probabilities exactly using matrix arithmetic; however, it would take us too far afield to describe it here.

➡ **Before we go on . . .** Another interesting question is, How long will it take the gambler in Example 5 to get to $30 or lose it all? This is called the *time to absorption* and can also be calculated by using matrix arithmetic. ■

7.7 EXERCISES

▼ more advanced   ◆ challenging
Ⓣ indicates exercises that should be solved using technology

In Exercises 1–10, write down the transition matrix associated with each state transition diagram.

1.

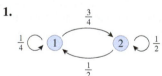

2.

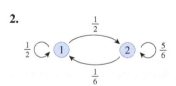

3.

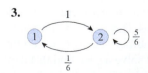

4.

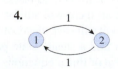

5.

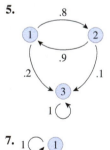

6.

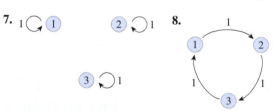

7.

8.

9.

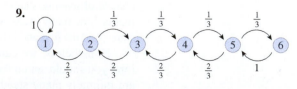

10.

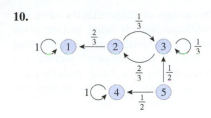

In Exercises 11–24, you are given a transition matrix P and initial distribution vector v. Find **(a)** the two-step transition matrix and **(b)** the distribution vectors after one, two, and three steps. [**HINT:** See Quick Examples 3 and 4.]

11. $P = \begin{bmatrix} .5 & .5 \\ 0 & 1 \end{bmatrix}, v = \begin{bmatrix} 1 & 0 \end{bmatrix}$

12. $P = \begin{bmatrix} 1 & 0 \\ .5 & .5 \end{bmatrix}, v = \begin{bmatrix} 0 & 1 \end{bmatrix}$

13. $P = \begin{bmatrix} .2 & .8 \\ .4 & .6 \end{bmatrix}, v = \begin{bmatrix} .5 & .5 \end{bmatrix}$

14. $P = \begin{bmatrix} \frac{1}{3} & \frac{2}{3} \\ \frac{1}{2} & \frac{1}{2} \end{bmatrix}, v = \begin{bmatrix} \frac{1}{4} & \frac{3}{4} \end{bmatrix}$

15. $P = \begin{bmatrix} \frac{1}{2} & \frac{1}{2} \\ 1 & 0 \end{bmatrix}, v = \begin{bmatrix} \frac{2}{3} & \frac{1}{3} \end{bmatrix}$

16. $P = \begin{bmatrix} 0 & 1 \\ \frac{1}{4} & \frac{3}{4} \end{bmatrix}, v = \begin{bmatrix} \frac{1}{5} & \frac{4}{5} \end{bmatrix}$

17. $P = \begin{bmatrix} \frac{3}{4} & \frac{1}{4} \\ \frac{3}{4} & \frac{1}{4} \end{bmatrix}, v = \begin{bmatrix} \frac{1}{2} & \frac{1}{2} \end{bmatrix}$

18. $P = \begin{bmatrix} \frac{2}{3} & \frac{1}{3} \\ \frac{2}{3} & \frac{1}{3} \end{bmatrix}, v = \begin{bmatrix} \frac{1}{7} & \frac{6}{7} \end{bmatrix}$

19. $P = \begin{bmatrix} .5 & .5 & 0 \\ 0 & 1 & 0 \\ 0 & .5 & .5 \end{bmatrix}, v = \begin{bmatrix} 1 & 0 & 0 \end{bmatrix}$

20. $P = \begin{bmatrix} .5 & 0 & .5 \\ 1 & 0 & 0 \\ 0 & .5 & .5 \end{bmatrix}, v = \begin{bmatrix} 0 & 1 & 0 \end{bmatrix}$

21. $P = \begin{bmatrix} 0 & 1 & 0 \\ \frac{1}{3} & \frac{1}{3} & \frac{1}{3} \\ 1 & 0 & 0 \end{bmatrix}, v = \begin{bmatrix} \frac{1}{2} & 0 & \frac{1}{2} \end{bmatrix}$

22. $P = \begin{bmatrix} \frac{1}{2} & \frac{1}{2} & 0 \\ \frac{1}{2} & \frac{1}{2} & 0 \\ \frac{1}{2} & 0 & \frac{1}{2} \end{bmatrix}, v = \begin{bmatrix} 0 & 0 & 1 \end{bmatrix}$

23. $P = \begin{bmatrix} .1 & .9 & 0 \\ 0 & 1 & 0 \\ 0 & .2 & .8 \end{bmatrix}, v = \begin{bmatrix} .5 & 0 & .5 \end{bmatrix}$

24. $P = \begin{bmatrix} .1 & .1 & .8 \\ .5 & 0 & .5 \\ .5 & 0 & .5 \end{bmatrix}, v = \begin{bmatrix} 0 & 1 & 0 \end{bmatrix}$

In Exercises 25–36, you are given a transition matrix P. Find the steady-state distribution vector. [**HINT:** See Example 4.]

25. $P = \begin{bmatrix} \frac{1}{2} & \frac{1}{2} \\ 1 & 0 \end{bmatrix}$

26. $P = \begin{bmatrix} 0 & 1 \\ \frac{1}{4} & \frac{3}{4} \end{bmatrix}$

27. $P = \begin{bmatrix} \frac{1}{3} & \frac{2}{3} \\ \frac{1}{2} & \frac{1}{2} \end{bmatrix}$

28. $P = \begin{bmatrix} .2 & .8 \\ .4 & .6 \end{bmatrix}$

29. $P = \begin{bmatrix} .1 & .9 \\ .6 & .4 \end{bmatrix}$

30. $P = \begin{bmatrix} .2 & .8 \\ .7 & .3 \end{bmatrix}$

31. $P = \begin{bmatrix} .5 & 0 & .5 \\ 1 & 0 & 0 \\ 0 & .5 & .5 \end{bmatrix}$

32. $P = \begin{bmatrix} 0 & .5 & .5 \\ .5 & .5 & 0 \\ 1 & 0 & 0 \end{bmatrix}$

33. $P = \begin{bmatrix} 0 & 1 & 0 \\ \frac{1}{3} & \frac{1}{3} & \frac{1}{3} \\ 1 & 0 & 0 \end{bmatrix}$

34. $P = \begin{bmatrix} \frac{1}{2} & \frac{1}{2} & 0 \\ \frac{1}{2} & \frac{1}{2} & 0 \\ \frac{1}{2} & 0 & \frac{1}{2} \end{bmatrix}$

35. $P = \begin{bmatrix} .1 & .9 & 0 \\ 0 & 1 & 0 \\ 0 & .2 & .8 \end{bmatrix}$

36. $P = \begin{bmatrix} .1 & .1 & .8 \\ .5 & 0 & .5 \\ .5 & 0 & .5 \end{bmatrix}$

Applications

37. *Marketing* A market survey shows that half the owners of *Sorey State Boogie Boards* became disenchanted with the product and switched to *C&T Super Professional Boards* the next surf season, while the other half remained loyal to Sorey State. On the other hand, three quarters of the C&T Boogie Board users remained loyal to C&T, while the rest switched to Sorey State. Set these data up as a Markov transition matrix, and calculate the probability that a Sorey State Board user will be using the same brand two seasons later. [**HINT:** See Example 1.]

38. *Major Switching* At *Suburban Community College*, 10% of all business majors switched to another major the next semester, while the remaining 90% continued as business majors. Of all non–business majors, 20% switched to a business major the following semester, while the rest did not. Set up these data as a Markov transition matrix, and calculate the probability that a business major will no longer be a business major in two semesters' time. [**HINT:** See Example 1.]

39. *Pest Control* In an experiment to test the effectiveness of the latest roach trap, the "Roach Resort," 50 roaches were placed in the vicinity of the trap and left there for an hour. At the end of the hour, it was observed that 30 of the roaches had "checked in," while the rest were still scurrying around. (Remember that "once a roach checks in, it never checks out.")

a. Set up the transition matrix P for the system with decimal entries, and calculate P^2 and P^3.

b. If a roach begins outside the "Resort," what is the probability of its "checking in" by the end of 1 hour? 2 hours? 3 hours?

c. What do you expect to be the long-term impact on the number of roaches? [**HINT:** See Example 5.]

40. *Employment* You have worked for the Department of Administrative Affairs (DAA) for 27 years, and you still have little or no idea exactly what your job entails. To make your life a little more interesting, you have decided on the following course of action. Every Friday afternoon, you will use your desktop computer to generate a random digit from 0 to 9 (inclusive). If the digit is a zero, you will immediately quit your job, never to return. Otherwise, you will return to work the following Monday.

  a. Use the states (1) employed by the DAA and (2) not employed by the DAA to set up a transition probability matrix P with decimal entries, and calculate P^2 and P^3.

  b. What is the probability that you will still be employed by the DAA after each of the next 3 weeks?

  c. What are your long-term prospects for employment at the DAA? [**HINT:** See Example 5.]

41. *Risk Analysis* An auto insurance company classifies each motorist as "high risk" if the motorist has had at least one moving violation during the past calendar year and "low risk" if the motorist has had no violations during the past calendar year. According to the company's data, a high-risk motorist has a 50% chance of remaining in the high-risk category the next year and a 50% chance of moving to the low-risk category. A low-risk motorist has a 10% chance of moving to the high-risk category the next year and a 90% chance of remaining in the low-risk category. In the long term, what percentage of motorists fall in each category?

42. *Debt Analysis* A credit card company classifies its cardholders as falling into one of two credit ratings: "good" and "poor." On the basis of its rating criteria, the company finds that a cardholder with a good credit rating has an 80% chance of remaining in that category the following year and a 20% chance of dropping into the poor category. A cardholder with a poor credit rating has a 40% chance of moving into the good rating the following year and a 60% chance of remaining in the poor category. In the long term, what percentage of cardholders fall in each category?

43. *Textbook Adoptions* College instructors who adopt this book are (we hope!) twice as likely to continue to use the book the following semester as they are to drop it, whereas nonusers are nine times as likely to remain nonusers the following year as they are to adopt this book.

  a. Determine the probability that a nonuser will be a user in 2 years.

  b. In the long term, what proportion of college instructors will be users of this book?

44. *Confidence Level* Tommy the Dunker's performance on the basketball court is influenced by his state of mind: If he scores, he is twice as likely to score on the next shot as he is to miss, whereas if he misses a shot, he is three times as likely to miss the next shot as he is to score.

  a. If Tommy has missed a shot, what is the probability that he will score two shots later?

  b. In the long term, what percentage of shots are successful?

45. *Debt Analysis* As the manager of a large retailing outlet, you have classified all credit customers as falling into one of the following categories: Paid Up, Outstanding 0–90 Days, Bad Debts. Based on an audit of your company's records, you have come up with the following table, which gives the probabilities that a single credit customer will move from one category to the next in the period of 1 month.

		To		
		Paid Up	**0–90 Days**	**Bad Debts**
	Paid Up	.5	.5	0
From	**0–90 Days**	.5	.3	.2
	Bad Debts	0	.5	.5

How do you expect the company's credit customers to be distributed in the long term?

46. *Debt Analysis* Repeat Exercise 45 using the following table:

		To		
		Paid Up	**0–90 Days**	**Bad Debts**
	Paid Up	.8	.2	0
From	**0–90 Days**	.5	.3	.2
	Bad Debts	0	.5	.5

47. ▼ *Income Brackets* The following diagram shows the movement of U.S. households among three income groups—affluent, middle class, and poor—over the 11-year period 1980–1991.[90]

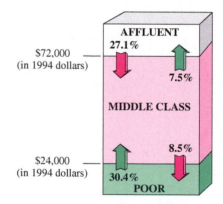

  a. Use the transitions shown in the diagram to construct a transition matrix (assuming zero probabilities for the transitions between affluent and poor).

---

[90] The figures were based on household after-tax income. The study was conducted by G. J. Duncan of Northwestern University and T. Smeeding of Syracuse University and was based on annual surveys of the personal finances of 5,000 households since the late 1960s. (The surveys were conducted by the University of Michigan.) Source: *New York Times*, June 4, 1995, p. E4.

b. Assuming that the trend shown were to continue, what percentage of households classified as affluent in 1980 were predicted to become poor in 2002? (Give your answer to the nearest 0.1%.)

c. ⊤ According to the model, what percentage of all U.S. households will be in each income bracket in the long term? (Give your answer to the nearest 0.1%.)

48. ▼ *Income Brackets* The following diagram shows the movement of U.S. households among three income groups—affluent, middle class, and poor—over the 12-year period 1967–1979.[91]

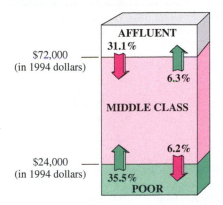

a. Use the transitions shown in the diagram to construct a transition matrix (assuming zero probabilities for the transitions between affluent and poor).

b. Assuming that the trend shown had continued, what percentage of households classified as affluent in 1967 would have been poor in 1991? (Give your answer to the nearest 0.1%.)

c. ⊤ According to the model, what percentage of all U.S. households will be in each income bracket in the long term? (Give your answer to the nearest 0.1%.)

49. ⊤ ▼ *Income Distribution* A University of Michigan study shows the following one-generation transition probabilities among four major income groups.[92]

Eldest Son's Income

		Bottom 10%	10–50%	50–90%	Top 10%
	Bottom 10%	.30	.52	.17	.01
Father's Income	10–50%	.10	.48	.38	.04
	50–90%	.04	.38	.48	.10
	Top 10%	.01	.17	.52	.30

In the long term, what percentage of male earners would you expect to find in each category? Why are the long-range figures not necessarily 10% in the lowest 10% income bracket, 40% in the 10–50% range, 40% in the 50–90% range, and 10% in the top 10% range? (Use technology to compute an approximation of the steady-state transition matrix.)

50. ⊤ ▼ *Income Distribution* Repeat Exercise 49, using the following data:

Eldest Son's Income

		Bottom 10%	10–50%	50–90%	Top 10%
	Bottom 10%	.50	.32	.17	.01
Father's Income	10–50%	.10	.48	.38	.04
	50–90%	.04	.38	.48	.10
	Top 10%	.01	.17	.32	.50

Market Share: Cell Phones *Three of the largest cellular phone companies in 2004 were* **Verizon**, **Cingular**, *and* **AT&T Wireless**. *Exercises 51 and 52 are based on the following figure, which shows percentages of subscribers who switched from one company to another during the third quarter of 2003:*[93]

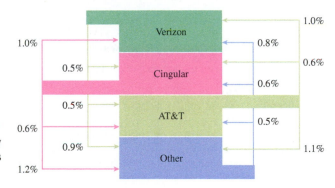

51. a. ⊤ ▼ Use the diagram to set up an associated transition matrix.

b. At the end of the third quarter of 2003, the market shares were Verizon: 29.7%, Cingular: 19.3%, AT&T: 18.1%, and Other: 32.9%. Use your Markov system to estimate the percentage shares at the *beginning* of the third quarter of 2003.

c. Using the information from part (b), estimate the market shares at the end of 2005. Which company is predicted to have gained the most in market share?

---

[91] *Ibid.*

[92] Source: Gary Solon, University of Michigan/*New York Times*, May 18, 1992, p. D5. We have adjusted some of the figures so that the probabilities add to 1; they did not do so in the original table because of rounding.

[93] Published market shares and "churn rate" (percentage drops for each company) were used to estimate the individual transition percentages. "Other" consists of Sprint, Nextel, and T-Mobile. No other cellular companies are included in this analysis. Source: The Yankee Group/*New York Times*, January 21, 2004, p. C1.

52. a. [T] ▼ Use the diagram to set up an associated transition matrix.

b. At the end of the third quarter of 2003, the market shares were Verizon: 29.7%, Cingular: 19.3%, AT&T: 18.1%, Other: 32.9%. Use your Markov system to estimate the percentage shares 1 year earlier.

c. Using the information from part (b), estimate the market shares at the end of 2010. Which company is predicted to have lost the most in market share?

53. ▼ *Dissipation* Consider the following five-state model of one-dimensional dissipation without drift:

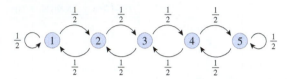

Find the steady-state distribution vector.

54. ▼ *Dissipation with Drift* Consider the following five-state model of one-dimensional dissipation with drift:

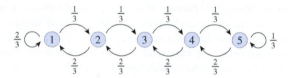

Find the steady-state distribution vector.

Communication and Reasoning Exercises

55. Describe an interesting situation that can be modeled by the transition matrix

$$P = \begin{bmatrix} .2 & .8 & 0 \\ 0 & 1 & 0 \\ .4 & .6 & 0 \end{bmatrix}.$$

56. Describe an interesting situation that can be modeled by the transition matrix

$$P = \begin{bmatrix} .8 & .1 & .1 \\ 1 & 0 & 0 \\ .3 & .3 & .4 \end{bmatrix}.$$

57. ▼ Describe some drawbacks to using Markov processes to model the behavior of the stock market with states (1) bull market and (2) bear market.

58. ▼ Can the repeated toss of a fair coin be modeled by a Markov process? If so, describe a model. If not, explain the reason.

59. ▼ Explain: If Q is a matrix whose rows are steady-state distribution vectors for P then $QP = Q$.

60. ▼ Construct a four-state Markov system so that both $[.5 \quad .5 \quad 0 \quad 0]$ and $[0 \quad 0 \quad .5 \quad .5]$ are steady-state vectors. [**HINT:** Try one in which no arrows link the first two states to the last two.]

61. ▼ Refer to the following state transition diagram, and explain in words (without doing any calculation) why the steady-state vector has a zero in position 1.

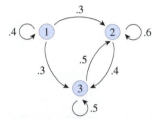

62. ▼ Without doing any calculation, find the steady-state distribution of the following system and explain the reasoning behind your claim.

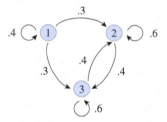

63. ◆ Construct a regular state transition diagram that possesses the steady-state vector $[.3 \quad .3 \quad .4]$.

64. ◆ Construct a regular state transition diagram that possesses the steady-state vector $[.6 \quad .3 \quad 0 \quad .1]$.

65. ◆ Show that if a Markov system has two distinct steady-state distributions v and w, then $\dfrac{v + w}{2}$ is another steady-state distribution.

66. ◆ If higher and higher powers of P approach a fixed matrix Q, explain why the rows of Q must be steady-state distributions vectors.

CHAPTER 7 REVIEW

KEY CONCEPTS

 www.WanerMath.com
Go to the Website to find a
comprehensive and interactive
Web-based summary of Chapter 7.

7.1 Sample Spaces and Events
Experiment, outcome, sample space [p. 480]
Event [p. 482]
The complement of an event [p. 485]
Unions of events [p. 486]
Intersections of events [p. 486]
Mutually exclusive events [p. 489]

7.2 Relative Frequency
Relative frequency or estimated probability [p. 494]
Relative frequency distribution [p. 494]
Properties of relative frequency distribution [p. 497]

7.3 Probability and Probability Models
Probability distribution, probability: $0 \leq P(s_i) \leq 1$ and
$P(s_1) + \cdots + P(s_n) = 1$ [p. 502]
If $P(E) = 0$, we call E an impossible event [p. 503]
Probability models [p. 503]
Probability models for equally likely outcomes:
$P(E) = n(E)/n(S)$ [p. 505]
Addition principle: $P(A \cup B) = P(A) + P(B) - P(A \cap B)$
[p. 509]
If A and B are mutually exclusive, then
$P(A \cup B) = P(A) + P(B)$ [p. 509]
If S is the sample space, then $P(S) = 1$, $P(\varnothing) = 0$, and
$P(A') = 1 - P(A)$ [p. 511]

7.4 Probability and Counting Techniques
Use counting techniques from Chapter 6 to calculate
probability [p. 520]

7.5 Conditional Probability and Independence
Conditional probability: $P(A|B) = P(A \cap B)/P(B)$ [p. 530]

Multiplication principle for conditional probability:
$P(A \cap B) = P(A|B)P(B)$ [p. 533]
Independent events: $P(A \cap B) = P(A)P(B)$ [p. 536]

7.6 Bayes' Theorem and Applications
Bayes' theorem (short form):
$$P(A|T) = \frac{P(T|A)P(A)}{P(T|A)P(A) + P(T|A')P(A')} \quad \text{[p. 548]}$$
Bayes' theorem (partition of sample space into three events):
$$P(A_1|T) = \frac{P(T|A_1)P(A_1)}{P(T|A_1)P(A_1) + P(T|A_2)P(A_2) + P(T|A_3)P(A_3)}$$
[p. 550]

7.7 Markov Systems
Markov system, Markov process, states, transition
probabilities [p. 555]
Transition matrix associated with a given Markov system
[p. 556]
A vector having nonnegative entries adding up to 1 is called a
probability vector [p. 558]
A distribution vector is a probability vector giving the probability distribution for finding a Markov system in its various
possible states [p. 558]
If v is a distribution vector, then the distribution vector one
step later will be vP. The distribution after m steps will
be vP^m [p. 558]
P^m is the m-step transition matrix. The ijth entry in P^m is the
probability of a transition from state i to state j in m steps
[p. 559]
A steady-state (probability) vector is a probability vector v
such that $vP = v$ [p. 560]
Calculation of the steady-state distribution vector [p. 561]
A regular Markov system, long-term behavior of a regular
Markov system [p. 562]
An absorbing state is one for which the probability of staying
in the state is 1 (and the probability of leaving it for any
other state is 0) [p. 563]

REVIEW EXERCISES

*In Exercises 1–6, say how many elements are in the sample
space S, list the elements of the given event E, and compute the
probability of E.*

1. Three coins are tossed; the result is one or more tails.

2. Four coins are tossed; the result is fewer heads than tails.

3. Two distinguishable dice are rolled; the numbers facing up
   add to 7.

4. Three distinguishable dice are rolled; the numbers facing
   up add to 5.

5. A die is weighted so that each of 2, 3, 4, and 5 is half as
   likely to come up as either 1 or 6; however, 2 comes up.

6. Two indistinguishable dice are rolled; the numbers facing
   up add to 7.

In Exercises 7–10, calculate the relative frequency P(E).

7. Two coins are tossed 50 times, and two heads come up
   12 times. E is the event that at least one tail comes up.

8. Ten stocks are selected at random from a portfolio. Seven of
   them have increased in value since their purchase, and the
   rest have decreased. Eight of them are Internet stocks, and
   two of those have decreased in value. E is the event that a
   stock either has increased in value or is an Internet stock.

9. You have read 150 of the 400 novels in your home, but your sister Roslyn has read 200, of which only 50 are novels you have read as well. *E* is the event that a novel has been read by neither you nor your sister.

10. You roll two dice 10 times. Both dice show the same number 3 times, and on 2 rolls, exactly one number is odd. *E* is the event that the sum of the numbers is even.

In Exercises 11–14, calculate the probability P(E).

11. There are 32 students in categories A and B combined. Some students are in both, 24 are in A, and 24 are in B. *E* is the event that a randomly selected student (among the 32) is in both categories.

12. You roll two dice, one red and one green. Losing combinations are doubles (both dice showing the same number) and outcomes in which the green die shows an odd number and the red die shows an even number. The other combinations are winning ones. *E* is the event that you roll a winning combination.

13. The *jPlay* portable music/photo/video player and bottle opener comes in three models: A, B, and C, each with five colors to choose from, and there are equal numbers of each combination. *E* is the event that a randomly selected *jPlay* is either orange (one of the available colors), a Model A, or both.

14. The *Heavy Weather Service* predicts that for tomorrow there is a 50% chance of tornadoes, a 20% chance of a monsoon, and a 10% chance of both. What is the probability that we will be lucky tomorrow and encounter neither tornadoes nor a monsoon?

A bag contains four red marbles, two green ones, one transparent one, three yellow ones, and two orange ones. You select five at random. In Exercises 15–20, compute the probability of the given event.

15. You have selected all the red ones.

16. You have selected all the green ones.

17. All are different colors.

18. At least one is not red.

19. At least two are yellow.

20. None are yellow and at most one is red.

In Exercises 21–26, find the probability of being dealt the given type of 5-card hand from a standard deck of 52 cards. (None of these is a recognized poker hand.) Express your answer in terms of combinations.

21. **Kings and Queens:** Each of the five cards is either a king or a queen.

22. **Five Pictures:** Each card is a picture card (J, Q, K).

23. **Fives and Queens:** Three fives, the queen of spades, and one other queen.

24. **Prime Full House:** A full house (three cards of one denomination, two of another) with the face value of each card a prime number (ace = 1, J = 11, Q = 12, K = 13).

25. **Full House of Commons:** A full house (three cards of one denomination, two of another) with no royal cards (that is, no J, Q, K, or ace).

26. **Black Two Pair:** Five black cards (spades or clubs), two with one denomination, two with another, and one with a third.

Two dice, one green and one yellow, are rolled. In Exercises 27–32, find the conditional probability, and say whether the indicated pair of events is independent.

27. The sum is 5, given that the green one is not 1 and the yellow one is 1.

28. The sum is 6, given that the green one is either 1 or 3 and the yellow one is 1.

29. The yellow one is 4, given that the green one is 4.

30. The yellow one is 5, given that the sum is 6.

31. The dice have the same parity, given that both of them are odd.

32. The sum is 7, given that the dice do not have the same parity.

A poll shows that half the consumers who use Brand A switched to Brand B the following year, while the other half stayed with Brand A. Three quarters of the Brand B users stayed with Brand B the following year, while the rest switched to Brand A. Use this information to answer Exercises 33–36.

33. Give the associated Markov transition matrix, with state 1 representing using Brand A and state 2 representing using Brand B.

34. Compute the associated two- and three-step transition matrices. What is the probability that a Brand A user will be using Brand B 3 years later?

35. If two thirds of consumers are presently using Brand A and one third are using Brand B, how are these consumers distributed in 3 years' time?

36. In the long term, what fraction of the time will a user spend using each of the two brands?

Applications: OHaganBooks.com
[Try the game at www.OHaganBooks.com]

Inventory OHaganBooks.com currently operates three warehouses: one in Washington, one in California, and the new one in Texas. Book inventories are shown in the following table, which should be used for Exercises 37–42.

	Sci Fi	Horror	Romance	Other	Total
Washington	10,000	12,000	12,000	30,000	64,000
California	8,000	12,000	6,000	16,000	42,000
Texas	15,000	15,000	20,000	44,000	94,000
Total	33,000	39,000	38,000	90,000	200,000

A book is selected at random. Compute the probability of the given event.

37. That it is either a sci-fi book or stored in Texas (or both)

38. That it is a sci-fi book stored in Texas

39. That it is a sci-fi book, given that it is stored in Texas

40. That it is stored in Texas, given that it is a sci-fi book

41. That it is stored in Texas, given that it is not a sci-fi book

42. That it is not stored in Texas, given that it is a sci-fi book

Marketing To gauge the effectiveness of the OHaganBooks .com site, you recently commissioned a survey of online shoppers. According to the results, 2% of online shoppers visited OHaganBooks.com during a 1-week period, while 5% of them visited at least one of OHaganBooks.com's two main competitors: JungleBooks.com *and* FarmerBooks.com. *Use this information to answer Exercises 43–50.*

43. What percentage of online shoppers never visited OHaganBooks.com?

44. What percentage of online shoppers never visited either of OHaganBooks.com's main competitors?

45. Assuming that visiting OHaganBooks.com was independent of visiting a competitor, what percentage of online shoppers visited either OHaganBooks.com or a competitor?

46. Assuming that visiting OHaganBooks.com was independent of visiting a competitor, what percentage of online shoppers visited OHaganBooks.com but not a competitor?

47. Under the assumption of Exercise 45, what is the probability that an online shopper will visit none of the three sites during a week?

48. If no one who visited OHaganBooks.com ever visited any of the competitors, what is the probability that an online shopper will visit none of the three sites during a week?

49. Actually, the assumption in Exercise 45 is not what was found by the survey, because an online shopper visiting a competitor was in fact more likely to visit OHaganBooks .com than a randomly selected online shopper. Let H be the event that an online shopper visits OHaganBooks.com, and let C be the event that the shopper visits a competitor. Which is greater: $P(H \cap C)$ or $P(H)P(C)$? Why?

50. What the survey found is that 25% of online shoppers who visited a competitor also visited OHaganBooks.com. Given this information, what percentage of online shoppers visited OHaganBooks.com and neither of its competitors?

51. *Sales* According to statistics gathered by OHaganBooks .com, 2% of online shoppers visited the OHaganBooks.com website during the course of a week, and 8% of those purchased books from the company. Further, 0.5% of online shoppers who did not visit the OHaganBooks.com website during the course of a week nonetheless purchased books from the company (through mail-order catalogs and other

sites such as LemmaZorn.com). What is the probability that an online shopper who purchased books from OHaganBooks .com during a given week visited the site?

52. *Sales* Repeat Exercise 51 in the event that 1% of online shoppers visited OHaganBooks.com during the week.

53. *University Admissions* In the year when Billy-Sean O'Hagan applied to *Suburban State U*, 56% of in-state applicants were admitted, while only 15% of out-of-state applicants were admitted. Further, 72% of all applicants were in-state. What percentage of admitted applicants were in-state? (Round your answer to the nearest percentage point.)

54. *University Admissions* Billy-Sean O'Hagan had also applied to *Gigantic State U*, at which time 75% of all applicants were from the United States and 22% of those applicants were admitted. Also, 14% of the applicants from foreign countries were admitted. What percentage of admitted applicants were from the United States? (Round your answer to the nearest percentage point.)

*Competition As was mentioned earlier, OHaganBooks.com has two main competitors—*JungleBooks.com *and* FarmerBooks .com*—and no other competitors of any significance. Exercises 55–58 are based on the following table, which shows the movement of customers during July.[94] (Thus, for instance, the first row tells us that 80% of OHaganBooks.com's customers remained loyal, 10% of them went to JungleBooks.com, and the remaining 10% went to FarmerBooks.com.)*

		To		
		OHaganBooks	**JungleBooks**	**FarmerBooks**
	OHaganBooks	80%	10%	10%
From	**JungleBooks**	40%	60%	0%
	FarmerBooks	20%	0%	80%

At the beginning of July, OHaganBooks.com had an estimated market share of one fifth of all customers, while its two competitors had two fifths each.

55. Estimate the market shares each company had at the end of July.

56. Assuming that the July trends continue in August, predict the market shares of each company at the end of August.

57. Name one or more important factors that the Markov model does not take into account.

58. Assuming that the July trend were to continue indefinitely, predict the market share enjoyed by each of the three e-commerce sites.

---

[94] By a "customer" of one of the three e-commerce sites, we mean someone who purchases more at that site than at any of the two competitors' sites.

CASE STUDY

* This problem caused quite a stir in late 1991 when it was discussed in Marilyn vos Savant's column in *Parade* magazine. Vos Savant gave the answer that you should switch. She received about 10,000 letters in response, most of them disagreeing with her. Several of those disagreeing with her were mathematicians.

The Monty Hall Problem

Here is a famous "paradox" that even mathematicians find counterintuitive. On the game show *Let's Make a Deal*, you are shown three doors, A, B, and C, and behind one of them is the Big Prize. After you select one of them—say, door A—to make things more interesting the host (Monty Hall), who knows what is behind each door, opens one of the other doors—say, door B—to reveal that the Big Prize is not there. He then offers you the opportunity to change your selection to the remaining door, door C. Should you switch or stick with your original guess? Does it make any difference?

Most people would say that the Big Prize is equally likely to be behind door A or door C, so there is no reason to switch.* In fact, this is wrong: The prize is more likely to be behind door C! There are several ways of seeing why this is so. Here is how you might work it out using Bayes' theorem.

Let A be the event that the Big Prize is behind door A, let B be the event that it is behind door B, and let C be the event that it is behind door C. Let F be the event that Monty has opened door B and revealed that the prize is not there. You wish to find $P(C|F)$ using Bayes' theorem. To use that formula you need to find $P(F|A)$ and $P(A)$ and similarly for B and C. Now, $P(A) = P(B) = P(C) = 1/3$ because at the outset, the prize is equally likely to be behind any of the doors. $P(F|A)$ is the probability that Monty will open door B if the prize is actually behind door A, and this is $1/2$ because we assume that he will choose either B or C randomly in this case. On the other hand, $P(F|B) = 0$, because he will never open the door that hides the prize. Also, $P(F|C) = 1$ because if the prize is behind door C, he must open door B to keep from revealing that the prize is behind door C. Therefore,

$$P(C|F) = \frac{P(F|C)P(C)}{P(F|A)P(A) + P(F|B)P(B) + P(F|C)P(C)}$$

$$= \frac{1 \cdot \frac{1}{3}}{\frac{1}{2} \cdot \frac{1}{3} + 0 \cdot \frac{1}{3} + 1 \cdot \frac{1}{3}} = \frac{2}{3}.$$

You conclude from this that you *should* switch to door C because it is more likely than door A to be hiding the prize.

Here is a more elementary way you might work it out. Consider the tree diagram of possibilities shown in Figure 22. The top two branches of the tree give the cases in

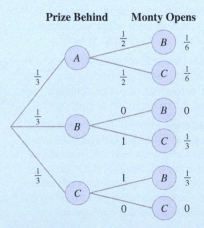

Prize Behind   Monty Opens

Figure 22

which the prize is behind door A, and there is a total probability of 1/3 for that case. The remaining two branches with nonzero probabilities give the cases in which the prize is behind the door that you did not choose, and there is a total probability of 2/3 for that case. Again, you conclude that you should switch your choice of doors because the one you did not choose is twice as likely as door A to be hiding the Big Prize.

EXERCISES

1. The answer you came up with, to switch to the other door, depends on the strategy Monty Hall uses in picking the door to open. Suppose that he actually picks one of doors B and C at random so that there is a chance that he will reveal the Big Prize. If he opens door B and it happens that the prize is not there, should you switch or not?

2. What if you know that Monty's strategy is always to open door B if possible (i.e., it does not hide the Big Prize) after you choose A?
   a. If he opens door B, should you switch?
   b. If he opens door C, should you switch?

3. Repeat the analysis of the original game, but suppose that the game uses four doors instead of three (and still only one prize).

4. Repeat the analysis of the original game, but suppose that the game uses 1,000 doors instead of 3 (and still only one prize).

Section 7.2

The TI-83/84 Plus has a random number generator that we can use to simulate experiments. For the following example, recall that a fair coin has probability $1/2$ of coming up heads and $1/2$ of coming up tails.

Example Use a simulated experiment to check the following.

a. The estimated probability of heads coming up in a toss of a fair coin approaches $1/2$ as the number of trials gets large.

b. The estimated probability of heads coming up in two consecutive tosses of a fair coin approaches $1/4$ as the number of trials gets large.[95]

Solution

a. Let us use 1 to represent heads and 0 to represent tails. We need to generate a list of **random binary digits** (0 or 1). One way to do this—a method that works for most forms of technology—is to generate a random number between 0 and 1 and then round it to the nearest whole number, which will be either 0 or 1.

We generate random numbers on the TI-83/84 Plus using the "rand" function. To round the number X to the nearest whole number on the TI-83/84 Plus, follow MATH → NUM, select "round," and enter round(X,0). This instruction rounds X to zero decimal places—that is, to the nearest whole number. Since we wish to round a random number, we need to enter

round(rand,0)

To obtain rand, follow MATH → PRB.

The result will be either 0 or 1. Each time you press ENTER, you will now get another 0 or 1. The TI-83/84 Plus can also generate a random integer directly (without the need for rounding) through the instruction

randInt(0,1)

To obtain randInt, follow MATH → PRB.

In general, the command randInt(*m*, *n*) generates a random integer in the range $[m, n]$. The

following sequence of 100 random binary digits was produced by using technology.[96]

0	1	0	0	1	1	0	1	0	0
0	1	0	0	0	0	0	0	1	0
1	1	0	0	0	1	0	0	1	1
1	1	1	0	1	0	0	0	1	0
1	1	1	1	1	1	1	0	0	1
1	0	1	1	1	0	0	1	1	0
0	1	0	1	1	1	0	1	1	1
1	0	0	0	0	0	0	1	1	1
1	1	1	1	0	0	1	1	1	0
1	1	1	0	1	1	0	1	0	0

If we use only the first row of data (corresponding to the first ten tosses), we find

$$P(\text{H}) = \frac{fr(1)}{N} = \frac{4}{10} = .4.$$

Using the first two rows ($N = 20$) gives

$$P(\text{H}) = \frac{fr(1)}{N} = \frac{6}{20} = .3.$$

Using all ten rows ($N = 100$) gives

$$P(\text{H}) = \frac{fr(1)}{N} = \frac{54}{100} = .54.$$

This is somewhat closer to the theoretical probability of $1/2$ and supports our intuitive notion that the larger the number of trials, the more closely the estimated probability should approximate the theoretical value.[97]

b. We need to generate pairs of random binary digits and then check whether they are both 1s. Although the TI-83/84 Plus will generate a pair of random digits if you enter round(rand(2),0), it would be a lot more convenient if the calculator could tell you right away whether both digits are 1s (corresponding to two consecutive heads in a coin toss). Here is a simple way of accomplishing this. Notice that if we *add* the two random binary digits, we obtain either 0, 1, or 2, telling us the number of heads that result

---

[95] Since the set of outcomes of a pair of coin tosses is {HH, HT, TH, TT}, we expect HH to come up once in every four trials, on average.

[96] The instruction randInt(0,1,100)→L_1 will generate a list of 100 random 0s and 1s and store it in L_1, where it can be summed with Sum(L_1) (under 2ND LIST →MATH).

[97] Do not expect this to happen every time. Compare, for example, $P(\text{H})$ for the first five rows and for all ten rows.

from the two consecutive throws. Therefore, all we need to do is add the pairs of random digits and then count the number of times 2 comes up. A formula we can use is

```
randInt(0,1)+randInt(0,1)
```

What would be even *more* convenient would be if the result of the calculation were either 0 or 1, with 1 signifying success (two consecutive heads) and 0 signifying failure. Then we could simply add up all the results to obtain the number of times two heads occurred. To do this, we first divide the result of the previous calculation above by 2 (obtaining 0, .5, or 1, where now 1 signifies success) and then round *down* to an integer, using a function called "int":

```
int(0.5*(randInt(0,1)+randInt(0,1)))
```

Following is the result of 100 such pairs of coin tosses, with 1 signifying success (two heads) and 0 signifying failure (all other outcomes). The last column records the number of successes in each row and the total number at the end.

1	1	0	0	0	0	0	0	0	0	2
0	1	0	0	0	0	0	1	0	1	3
0	1	0	0	1	1	0	0	0	1	4
0	0	0	0	0	0	0	0	1	0	1
0	1	0	0	1	0	0	1	0	0	3
1	0	1	0	0	0	0	0	0	0	2
0	0	0	0	0	0	0	0	0	1	1
0	1	1	1	1	0	0	0	0	1	5
1	1	0	1	0	0	1	1	0	0	5
0	0	0	0	0	0	0	0	1	0	1
										27

Now, as in part (a), we can compute estimated probabilities, with D standing for the outcome "two heads":

First 10 trials: $P(D) = \dfrac{fr(1)}{N} = \dfrac{2}{10} = .2$

First 20 trials: $P(D) = \dfrac{fr(1)}{N} = \dfrac{5}{20} = .25$

First 50 trials: $P(D) = \dfrac{fr(1)}{N} = \dfrac{13}{50} = .26$

100 trials: $P(D) = \dfrac{fr(1)}{N} = \dfrac{27}{100} = .27.$

Q: *What is happening with the data? The probabilities seem to be getting less accurate as N increases!*

A: Quite by chance, exactly 5 of the first 20 trials resulted in success, which matches the theoretical probability. The figure below shows an Excel plot of estimated probability versus N (for N a multiple of 10). Notice that, as N increases, the graph seems to meander within smaller distances of .25.

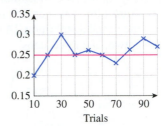

Trials

Q: *The previous techniques work fine for simulating coin tosses. What about rolls of a fair die, in which we want outcomes between 1 and 6?*

A: We can simulate a roll of a die by generating a random integer in the range 1 through 6. The following formula accomplishes this:

```
1 + int(5.99999*rand).
```

(We used 5.99999 instead of 6 to avoid the outcome 7.)

Section 7.7

Example 1 (page 556) Consider the Markov system found by *Gamble Detergents* at the beginning of this section. Suppose that 70% of consumers are now using powdered detergent while the other 30% are using liquid. What will be the distribution 1 year from now? 2 years from now? 3 years from now?

Solution

In Chapter 4 we saw how to set up and multiply matrices. For this example we can use the matrix editor to define `[A]` as the initial distribution and `[B]` as the transition matrix. (Remember that the only names we can use are `[A]` through `[J]`.)

Entering [A] (obtained by pressing MATRIX 1 ENTER) will show you the initial distribution. To obtain the distribution after one step, press X MATRIX 2 ENTER, which has the effect of multiplying the previous answer by the transition matrix [B]. Now just press ENTER repeatedly to continue multiplying by the transition matrix and obtain the distribution after any number of steps. The screenshot shows the initial distribution [A] and the distributions after one, two, and three steps.

```
[A]
        [[.7 .3]]
Ans*[B]
        [[.59 .41]]
      [[.513 .487]]
    [[.4591 .5409]]
```

Example 4 (page 560)
Calculate the steady-state probability vector for the transition matrix in the preceding examples.

Solution

Finding the steady-state probability vector comes down to solving a system of equations. As discussed in Chapters 3 and 4, there are several ways to use a calculator to help. The most straightforward is to use matrix inversion to solve the matrix form of the system. In this case, as in the text, the system of equations we need to solve is

$$x + \quad y = 1$$
$$-.2x + .1y = 0.$$

We write this as the matrix equation $AX = B$ with

$$A = \begin{bmatrix} 1 & 1 \\ -.2 & .1 \end{bmatrix} \qquad B = \begin{bmatrix} 1 \\ 0 \end{bmatrix}.$$

To find $X = A^{-1}B$ using the TI-83/84 Plus, we first use the matrix editor to enter these matrices as [A] and [B], then compute $[A]^{-1}[B]$ on the Home screen.

```
[A]⁻¹[B]
    [[.3333333333]
     [.6666666667]]
```

To convert the entries to fractions, we can follow this by the command.

▸Frac   MATH ENTER ENTER

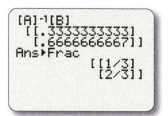

```
[A]⁻¹[B]
    [[.3333333333]
     [.6666666667]]
Ans▸Frac
              [[1/3]
               [2/3]]
```

Spreadsheet | Technology Guide

Section 7.2

Spreadsheets have random number generators that we can use to simulate experiments. For the following example, recall that a fair coin has probability 1/2 of coming up heads and 1/2 of coming up tails.

Example Use a simulated experiment to check the following.

a. The estimated probability of heads coming up in a toss of a fair coin approaches 1/2 as the number of trials gets large.

b. The estimated probability of heads coming up in two consecutive tosses of a fair coin approaches 1/4 as the number of trials gets large.[98]

Solution

a. Let us use 1 to represent heads and 0 to represent tails. We need to generate a list of **random binary digits**

---

[98] Because the set of outcomes of a pair of coin tosses is {HH, HT, TH,TT}, we expect HH to come up once in every four trials, on average.

(0 or 1). One way to do this—a method that works for most forms of technology—is to generate a random number between 0 and 1 and then round it to the nearest whole number, which will be either 0 or 1.

In a spreadsheet, the formula RAND() gives a random number between 0 and 1.[99] The function ROUND(X,0) rounds X to zero decimal places—that is, to the nearest integer. Therefore, to obtain a random binary digit in any cell, just enter the following formula:

=ROUND(RAND(),0)

Spreadsheets can also generate a random integer directly (without the need for rounding) through the formula

=RANDBETWEEN(0,1)

To obtain a whole array of random numbers, just drag this formula into the cells you wish to use.

b. We need to generate pairs of random binary digits and then check whether they are both 1s. It would be convenient if the spreadsheet could tell you right away whether both digits are 1s (corresponding to two consecutive heads in a coin toss). Here is a simple way of accomplishing this. Notice that if we *add* two random binary digits, we obtain either 0, 1, or 2, telling us the number of heads that result from the two consecutive throws. Therefore, all we need to do is add pairs of random digits and then count the number of times 2 comes up. A formula we can use is

=RANDBETWEEN(0,1)+RANDBETWEEN(0,1)

What would be even *more* convenient would be if the result of the calculation were either 0 or 1, with 1 signifying success (two consecutive heads) and 0 signifying failure. Then we could simply add up all the results to obtain the number of times two heads occurred. To do this, we first divide the result of the calculation above by 2 (obtaining 0, .5, or 1, where now 1 signifies success) and then round *down* to an integer using a function called "int":

=INT(0.5*(RANDBETWEEN(0,1)+
RANDBETWEEN(0,1)))

Following is the result of 100 such pairs of coin tosses, with 1 signifying success (two heads) and 0 signifying failure (all other outcomes). The last column records the number of successes in each row and the total number at the end.

1	1	0	0	0	0	0	0	0	0	2
0	1	0	0	0	0	0	1	0	1	3
0	1	0	0	1	1	0	0	0	1	4
0	0	0	0	0	0	0	0	1	0	1
0	1	0	0	1	0	0	1	0	0	3
1	0	1	0	0	0	0	0	0	0	2
0	0	0	0	0	0	0	0	0	1	1
0	1	1	1	1	0	0	0	0	1	5
1	1	0	1	0	0	1	1	0	0	5
0	0	0	0	0	0	0	0	1	0	1
										27

Now, as in part (a), we can compute estimated probabilities, with D standing for the outcome "two heads":

First 10 trials: $P(D) = \dfrac{fr(1)}{N} = \dfrac{2}{10} = .2$

First 20 trials: $P(D) = \dfrac{fr(1)}{N} = \dfrac{5}{20} = .25$

First 50 trials: $P(D) = \dfrac{fr(1)}{N} = \dfrac{13}{50} = .26$

100 trials: $P(D) = \dfrac{fr(1)}{N} = \dfrac{27}{100} = .27.$

Q: *What is happening with the data? The probabilities seem to be getting less accurate as N increases!*

A: Quite by chance, exactly 5 of the first 20 trials resulted in success, which matches the theoretical probability. The figure below shows an Excel plot of estimated probability versus N (for N a multiple of 10). Notice that, as N increases, the graph seems to meander within smaller distances of .25.

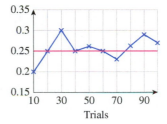

Q: *The previous techniques work fine for simulating coin tosses. What about rolls of a fair die, in which we want outcomes between 1 and 6?*

A: We can simulate a roll of a die by generating a random integer in the range 1 through 6. The following formula accomplishes this:

=1 + INT(5.99999*RAND())

(We used 5.99999 instead of 6 to avoid the outcome 7.)

---

[99] The parentheses after RAND are necessary even though the function takes no arguments.

Section 7.7

Example 1 (page 556) Consider the Markov system found by *Gamble Detergents* at the beginning of this section. Suppose that 70% of consumers are now using powdered detergent while the other 30% are using liquid. What will be the distribution 1 year from now? 2 years from now? 3 years from now?

Solution

In your spreadsheet, enter the initial distribution vector in cells A1 and B1 and the transition matrix to the right of that, as shown.

	A	B	C	D	E
1	0.7	0.3		0.8	0.2
2				0.1	0.9

To calculate the distribution after one step, use the array formula

$$\texttt{=MMULT(A1:B1,\$D\$1:\$E\$2)}$$

The absolute cell references (dollar signs) ensure that the formula always refers to the same transition matrix, even if we copy it into other cells. To use the array formula, select cells A2 and B2, where the distribution vector will go, enter this formula, and then press Control+Shift+Enter.[100]

	A	B	C	D	E
1	0.7	0.3		0.8	0.2
2	=MMULT(A1:B1,$D$1:$E$2)			0.1	0.9

The result is the following, with the distribution after one step highlighted.

	A	B	C	D	E
1	0.7	0.3		0.8	0.2
2	0.59	0.41		0.1	0.9

To calculate the distribution after two steps, select cells A2 and B2, and drag the fill handle down to copy the formula to cells A3 and B3. Note that the formula now takes the vector in A2:B2 and multiplies it by the transition matrix to get the vector in A3:B3. To calculate several more steps, drag down as far as desired.

	A	B	C	D	E
1	0.7	0.3		0.8	0.2
2	0.59	0.41		0.1	0.9
3					
4					

---

[100] On a Macintosh you can also use Command+Enter.

	A	B	C	D	E
1	0.7	0.3		0.8	0.2
2	0.59	0.41		0.1	0.9
3	0.513	0.487			
4	0.4591	0.5409			

Example 4 (page 560) Calculate the steady-state probability vector for the transition matrix in the preceding examples.

Solution

Finding the steady-state probability vector comes down to solving a system of equations. As was discussed in Chapters 3 and 4, there are several ways to use Excel to help. The most straightforward is to use matrix inversion to solve the matrix form of the system. In this case, as in the text, the system of equations we need to solve is

$$x + \quad y = 1$$
$$-.2x + .1y = 0.$$

We write this as the matrix equation $AX = B$ with

$$A = \begin{bmatrix} 1 & 1 \\ -.2 & .1 \end{bmatrix} \qquad B = \begin{bmatrix} 1 \\ 0 \end{bmatrix}.$$

We enter A in cells A1:B2, B in cells D1:D2, and the formula for $X = A^{-1}B$ in a convenient location, say, B4:B5.

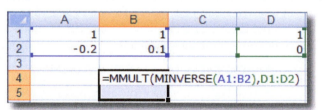

	A	B	C	D
1	1	1		1
2	-0.2	0.1		0
3				
4		=MMULT(MINVERSE(A1:B2),D1:D2)		
5				

When we press Control+Shift+Enter, we see the result.

	A	B	C	D
1	1	1		1
2	-0.2	0.1		0
3				
4		0.3333333		
5		0.6666667		

If we want to see the answer in fraction rather than decimal form, we format the cells as fractions.

	A	B	C	D
1	1	1		1
2	-0.2	0.1		0
3				
4		1/3		
5		2/3		

8

RANDOM VARIABLES AND STATISTICS

CASE STUDY

Spotting Tax Fraud with Benford's Law

You are a tax fraud specialist working for the Internal Revenue Service (IRS), and you have just been handed a portion of the tax return from *Colossal Conglomerate*. The IRS suspects that the portion you were handed may be fraudulent and would like your opinion.

Is there any mathematical test, you wonder, that can point to a suspicious tax return based on nothing more than the numbers entered?

Paula Borchardt/AGE Fotostock

 **www.WanerMath.com**

At the Website, in addition to the resources listed in the Preface, you will find:

• Histogram, Bernoulli trials, and normal distribution utilities

The following optional extra sections:

• Sampling Distributions and the Central Limit Theorem
• Confidence Intervals
• Calculus and Statistics

Introduction

Statistics is the branch of mathematics concerned with organizing, analyzing, and interpreting numerical data. For example, given the current annual incomes of 1,000 lawyers selected at random, you might wish to answer some questions: If I become a lawyer, what income am I likely to earn? Do lawyers' salaries vary widely? If so, how widely?

To answer questions like these, it helps to begin by organizing the data in the form of tables or graphs. This is the topic of the first section of the chapter. The second section describes an important class of examples that are applicable to a wide range of situations, from tossing a coin to product testing.

Once the data are organized, the next step is to apply mathematical tools for analyzing the data and answering questions like those posed above. Numbers such as the **mean** and the **standard deviation** can be computed to reveal interesting facts about the data. These numbers can then be used to make predictions about future events.

The chapter ends with a section on one of the most important distributions in statistics, the **normal distribution**. This distribution describes many sets of data and also plays an important role in the underlying mathematical theory.

8.1 Random Variables and Distributions

Random Variables

In many experiments we can assign numerical values to the outcomes. For instance, if we roll a die, each outcome has a value from 1 through 6. If you select a lawyer and ascertain his or her annual income, the outcome is again a number. We call a rule that assigns a number to each outcome of an experiment a **random variable**.

> **Random Variable**
>
> A **random variable** X is a rule that assigns a number, or **value**, to each outcome in the sample space of an experiment.*
>
> #### Quick Examples
>
> 1. Roll a die; X = The number facing up.
> 2. Select a mutual fund; X = The number of companies in the fund portfolio.
> 3. Select a computer; X = The number of gigabytes of memory it has.
> 4. Survey a group of 20 college students; X = The mean SAT.

* In the language of functions (Chapter 1), a random variable is a *real-valued function* whose domain is the sample space.

Visualizing a Random Variable

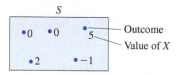

Discrete and Continuous Random Variables
A **discrete** random variable can take on only specific, isolated numerical values, such as the outcome of a roll of a die or the number of dollars in a randomly chosen bank account. A **continuous** random variable, on the other hand, can take on any values within a continuum or an interval, such as the temperature in Central Park or the height of an athlete in centimeters. Discrete random variables that can take on only finitely many values (such as the outcome of a roll of a die) are called **finite** random variables.

Quick Examples

	Random Variable	Values	Type
5.	Select a mutual fund; $X =$ The number of companies in the fund portfolio.	$\{1, 2, 3, \ldots\}$	Discrete infinite
6.	Take five shots at the goal during a soccer match; $X =$ The number of times you score.	$\{0, 1, 2, 3, 4, 5\}$	Finite
7.	Measure the length of an object; $X =$ Its length in centimeters.	Any positive real number	Continuous
8.	Roll a die until you get a 6; $X =$ The number of times you roll the die.	$\{1, 2, 3, \ldots\}$	Discrete infinite
9.	Bet a whole number of dollars in a race where the betting limit is $100; $X =$ The amount you bet.	$\{0, 1, \ldots, 100\}$	Finite
10.	Bet a whole number of dollars in a race where there is no betting limit; $X =$ The amount you bet.	$\{0, 1, \ldots, 100, 101, \ldots\}$	Discrete infinite

Notes

1. In Chapter 7 the only sample spaces that we considered in detail were finite sample spaces. However, in general, sample spaces can be infinite, as in many of the experiments mentioned above.
2. There are some borderline situations. For instance, if X is the salary of a factory worker, then X is, strictly speaking, discrete. However, the values of X are so numerous and close together that in some applications it makes sense to model X as a continuous random variable. ■

For the moment we shall consider only finite random variables.

EXAMPLE 1　Finite Random Variable

Let X be the number of heads that come up when a coin is tossed three times. List the value of X for each possible outcome. What are the possible values of X?

Solution　First, we describe X as a random variable:

X is the rule that assigns to each outcome the number of heads that come up.

We take as the outcomes of this experiment all possible sequences of three heads and tails. Then, for instance, if the outcome is HTH, the value of X is 2. An easy way to list the values of X for all the outcomes is by means of a table:

Outcome	HHH	HHT	HTH	HTT	THH	THT	TTH	TTT
Value of X	3	2	2	1	2	1	1	0

From the table we also see that the possible values of X are 0, 1, 2, and 3.

2 Heads ($X = 2$)

W **Website**
www.WanerMath.com
Go to the Chapter 8 Topic Summary to find an interactive simulation based on Example 1.

➡ **Before we go on ...** Remember that X is just a rule we decide on. In Example 1 we could have taken X to be a different rule, such as the number of tails or perhaps the number of heads minus the number of tails. These different rules are examples of different random variables associated with the same experiment. ■

EXAMPLE 2   Stock Prices

You have purchased $10,000 worth of stock in a biotech company whose newest arthritis drug is awaiting approval by the Food and Drug Administration (FDA). If the drug is approved this month, the value of the stock will double by the end of the month. If the drug is rejected this month, the stock's value will decline by 80%. If no decision is reached this month, its value will decline by 10%. Let X be the value of your investment at the end of this month. List the value of X for each possible outcome.

Solution There are three possible outcomes: the drug is approved this month, it is rejected this month, and no decision is reached. Once again, we express the random variable as a rule:

The random variable X is the rule that assigns to each outcome the value, in dollars, of your investment at the end of this month.

We can now tabulate the values of X as follows:

Outcome	Approved this month	Rejected this month	No decision
Value of X	20,000	2,000	9,000

Probability Distribution of a Finite Random Variable

Given a random variable X, it is natural to look at certain *events*—for instance, the event that $X = 2$. By this, we mean the event consisting of all outcomes that have an assigned X-value of 2. Looking once again at the chart in Example 1, with X being the number of heads that face up when a coin is tossed three times, we find the following events:

The event that $X = 0$ is {TTT}.
The event that $X = 1$ is {HTT, THT, TTH}.
The event that $X = 2$ is {HHT, HTH, THH}.
The event that $X = 3$ is {HHH}.
The event that $X = 4$ is $\varnothing$.   There are no outcomes with four heads.

Each of these events has a certain probability. For instance, the probability of the event that $X = 2$ is 3/8 because the event in question consists of three of the eight possible (equally likely) outcomes. We shall abbreviate this by writing

$$P(X = 2) = \frac{3}{8}.$$ The probability that $X = 2$ is 3/8.

Similarly,

$$P(X = 4) = 0.$$ The probability that $X = 4$ is 0.

When X is a finite random variable, the collection of the probabilities of X equaling each of its possible values is called the **probability distribution** of X. Because the probabilities in a probability distribution can be estimated or theoretical, we shall discuss both *estimated probability distributions* (or *relative frequency distributions*) and *theoretical (modeled) probability distributions* of random variables. (See the next two examples.)

Probability Distribution of a Finite Random Variable

If X is a finite random variable with values $n_1, n_2, \ldots$, then its **probability distribution** lists the probabilities that $X = n_1, X = n_2, \ldots$. The sum of these probabilities is always 1.

Visualizing the Probability Distribution of a Random Variable

If each outcome in S is equally likely, we get the probability distribution shown for the random variable X.

S

$\bullet$ 0	$\bullet$ 0	$\bullet$ 5
$\bullet$ 2	$\bullet$ -1	

Probability Distribution of X

x	-1	0	2	5
$P(X = x)$	$\dfrac{1}{5} = .2$	$\dfrac{2}{5} = .4$	$\dfrac{1}{5} = .2$	$\dfrac{1}{5} = .2$

Here, $P(X = x)$ means "the probability that the random variable X has the specific value x."

Quick Example

11. Roll a fair die; $X =$ the number facing up. Then the probability that any specific value of X occurs is $\frac{1}{6}$. So the probability distribution of X is the following (notice that the probabilities add up to 1):

x	1	2	3	4	5	6
$P(X = x)$	$\dfrac{1}{6}$	$\dfrac{1}{6}$	$\dfrac{1}{6}$	$\dfrac{1}{6}$	$\dfrac{1}{6}$	$\dfrac{1}{6}$

Using this probability distribution, we can calculate the probabilities of certain events; for instance,

$$P(X < 3) = \frac{1}{3} \qquad \text{The event that } X < 3 \text{ is the event } \{1, 2\}.$$

$$P(1 < X < 5) = \frac{1}{2} \qquad \text{The event that } 1 < X < 5 \text{ is the event } \{2, 3, 4\}.$$

Note The distinction between X (uppercase) and x (lowercase) in the tables above is important; X stands for the random variable in question, whereas x stands for a specific *value* of X (so x is always a number). Thus, if, say, $x = 2$, then $P(X = x)$ means $P(X = 2)$, the probability that X is 2. Similarly, if Y is a random variable, then $P(Y = y)$ is the probability that Y has the specific value y. ■

$P(X = x)$

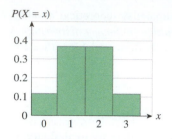

Figure 1

EXAMPLE 3 Probability Distribution

Let X be the number of heads that face up in three tosses of a coin. Give the probability distribution of X. What is the probability of throwing at least two heads?

Solution X is the random variable of Example 1, so its values are 0, 1, 2, and 3. The probability distribution of X is given in the following table:

x	0	1	2	3
$P(X = x)$	$\dfrac{1}{8}$	$\dfrac{3}{8}$	$\dfrac{3}{8}$	$\dfrac{1}{8}$

Notice that the probabilities add to 1, as we might expect. From the distribution the probability of throwing at least two heads is

$$P(X \ge 2) = P(\{2, 3\}) = \frac{3}{8} + \frac{1}{8} = \frac{1}{2}.$$

We can use a bar graph to visualize a probability distribution. Figure 1 shows the bar graph for the probability distribution we obtained. Such a graph is sometimes called a **histogram**.

➡ **Before we go on ...** The probabilities in the table in Example 3 are *modeled* probabilities. To obtain a similar table of relative frequencies, we would have to repeatedly toss a coin three times and calculate the fraction of times we got 0, 1, 2, and 3 heads. ■

Note The table of probabilities in Example 3 looks like the probability distribution associated with an experiment, as we studied in Section 7.3. In fact, the probability distribution of a random variable is not really new. Consider the following experiment: Toss three coins, and count the number of heads. The associated probability distribution (per Section 7.3) would be this:

Outcome	0	1	2	3
Probability	$\dfrac{1}{8}$	$\dfrac{3}{8}$	$\dfrac{3}{8}$	$\dfrac{1}{8}$

The difference is that in this chapter we are thinking of 0, 1, 2, and 3 not as the outcomes of the experiment, but as values of the random variable X. ■

EXAMPLE 4 Relative Frequency Distribution

The following table shows the (fictitious) income brackets of a sample of 1,000 lawyers in their first year out of law school:

Income Bracket	$20,000–$29,999	$30,000–$39,999	$40,000–$49,999	$50,000–$59,999	$60,000–$69,999	$70,000–$79,999	$80,000–$89,999
Number	20	80	230	400	170	70	30

Think of the experiment of choosing a first-year lawyer at random (all being equally likely), and assign to each lawyer the number X that is the midpoint of his or her income bracket. Find the relative frequency distribution of X.

Solution Statisticians refer to the income brackets as **measurement classes**. Because the first measurement class contains incomes that are at least $20,000 but less than

* One might argue that the midpoint should be (20,000 + 29,999)/2 = 24,999.50, but we round this to 25,000. So, technically we are using "rounded" midpoints of the measurement classes.

Using Technology

Technology can be used to automate the calculations in Example 4. Here is an outline.

TI-83/84 Plus
STAT EDIT values of x in L_1 and frequencies in L_2.
Home screen: $L_2/sum(L_2) \rightarrow L_3$
[More details in the Technology Guide.]

Spreadsheet
Headings x, Fr, and $P(X = x)$ in A1–C1
x-values and frequencies in columns A2–B8
=B2/SUM(B:B) in C2
Copy down column C.
[More details in the Technology Guide.]

W Website
www.WanerMath.com
→ Online Utilities
→ Histogram Utility

Enter the x-values and frequencies as shown:

```
25000, 20
35000, 80
45000, 230
55000, 400
65000, 170
75000, 70
85000, 30
```

Make sure "Show probability distribution" is checked, and press "Results". The relative frequency distribution will appear at the bottom of the page.

$30,000, its midpoint is $25,000.* Similarly, the second measurement class has midpoint $35,000, and so on. We can rewrite the table with the midpoints as follows:

x	25,000	35,000	45,000	55,000	65,000	75,000	85,000
Frequency	20	80	230	400	170	70	30

We have used the term *frequency* rather than *number*, although it means the same thing. This table is called a **frequency table**. It is *almost* the relative frequency distribution for X except that we must replace frequencies by relative frequencies. (We did this in calculating relative frequencies in Chapter 7.) We start with the lowest measurement class. Because 20 of the 1,000 lawyers fall in this group, we have

$$P(X = 25,000) = \frac{20}{1,000} = .02.$$

We can calculate the remaining relative frequencies similarly to obtain the following distribution:

x	25,000	35,000	45,000	55,000	65,000	75,000	85,000
$P(X = x)$	.02	.08	.23	.40	.17	.07	.03

Note again the distinction between X and x: X stands for the random variable in question, whereas x stands for a specific value (25,000, 35,000, . . . , or 85,000) of X.

EXAMPLE 5   Probability Distribution: Greenhouse Gases

The following table shows per capita emissions of greenhouse gases for the 30 countries with the highest per capita carbon dioxide emissions. (Emissions are rounded to the nearest 5 metric tons.)[1]

Country	Per Capita Emissions (metric tons)	Country	Per Capita Emissions (metric tons)
Qatar	45	Canada	15
Trinidad and Tobago	35	Russian Federation	15
Kuwait	30	Turkmenistan	10
Brunei Darussalam	25	Korea, Republic of	10
Aruba	25	Czech Republic	10
Oman	20	Finland	10
Luxembourg	20	Netherlands	10
United Arab Emirates	20	Equatorial Guinea	10
Saudi Arabia	20	Japan	10
Bahrain	20	Israel	10
United States	15	Norway	10
Kazakhstan	15	Souh Africa	10
Australia	15	Belgium	10
New Caledonia	15	Germany	10
Estonia	15	Poland	10

---

[1] Figures are based on 2011 data. Source: United Nations Millennium Development Goals Indicators, based on information from the U.S. Department of Energy's Carbon Dioxide Information Analysis Center (CDIAC) (http://mdgs.un.org).

Consider the experiment in which a country is selected at random from this list, and let X be the per capita carbon dioxide emissions for that country. Find the probability distribution of X, and graph it with a histogram. Use the probability distribution to compute $P(X \geq 20)$ (the probability that X is 20 or more), and interpret the result.

Solution The values of X are the possible emissions figures, which we can take to be 0, 5, 10, 15, ... , 45. In the table below, we first compute the frequency of each value of X by counting the number of countries that produce that per capita level of greenhouse gases. For instance, there are seven countries that have $X = 15$. Then we divide each frequency by the sample size $N = 30$ to obtain the probabilities.[*]

* Even though we are using the term "frequency," we are really calculating *modeled* probability based on the assumption of equally likely outcomes. In this context, the frequencies are the number of favorable outcomes for each value of X. (See the Q&A discussion at the end of Section 7.3.)

x	0	5	10	15	20	25	30	35	40	45
Frequency	0	0	13	7	5	2	1	1	0	1
$P(X = x)$	0	0	$\dfrac{13}{30}$	$\dfrac{7}{30}$	$\dfrac{5}{30}$	$\dfrac{2}{30}$	$\dfrac{1}{30}$	$\dfrac{1}{30}$	0	$\dfrac{1}{30}$

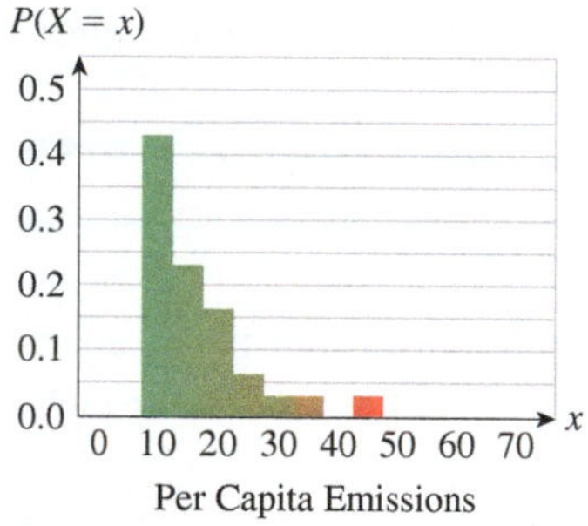

Figure 2 shows the resulting histogram.

Finally, we compute $P(X \geq 20)$, the probability of the event that X has a value of 20 or more, which is the sum of the probabilities $P(X = 20)$, $P(X = 25)$, and so on. From the table we obtain

$$P(X \geq 20) = \frac{5}{30} + \frac{2}{30} + \frac{1}{30} + \frac{1}{30} + 0 + \frac{1}{30} = \frac{10}{30} \approx .33.$$

Thus, there is an approximately 33% chance that a country randomly selected from the given list produces 20 or more metric tons per capita of carbon dioxide.

$P(X = x)$

Per Capita Emissions

Figure 2

Recognizing What to Use as a Random Variable and Deciding on Its Values

Q: *In an application, how, exactly, do I decide what to use as a random variable X?*

A: Be as systematic as possible: First, decide what the experiment is and what its sample space is. Then, on the basis of what is asked for in the application, complete the following sentence: "X assigns ___ to each outcome." For instance, "X assigns the number of flavors to each packet of gummy bears selected" or "X assigns the average faculty salary to each college selected."

Q: *Once I have decided what X should be, how do I decide what values to assign it?*

A: Ask yourself: What are the conceivable values I could get for X? Then choose a collection of values that includes all of these. For instance, if X is the number of heads obtained when a coin is tossed five times, then the possible values of X are 0, 1, 2, 3, 4, and 5. If X is the average faculty salary in dollars, rounded to the nearest \$5,000, then possible values of X could be 20,000, 25,000, 30,000, and so on, up to the highest salary in your data.

8.1 EXERCISES

▼ more advanced   ◆ challenging
T indicates exercises that should be solved using technology

In Exercises 1–10, classify the random variable X as finite, discrete infinite, or continuous, and indicate the values that X can take. [**HINT:** See Quick Examples 5–10.]

1. Roll two dice; $X =$ the sum of the numbers facing up.

2. Open a 500-page book on a random page; $X =$ the page number.

3. Select a stock at random; $X =$ your profit, to the nearest dollar, if you purchase one share and sell it one year later.

4. Select an electric utility company at random; $X =$ the exact amount of electricity, in gigawatt hours, it supplies in a year.

5. Look at the second hand of your watch; X is the time it reads in seconds.

6. Watch a soccer game; $X =$ the total number of goals scored.

7. Watch a soccer game; $X =$ the total number of goals scored, up to a maximum of 10.

8. Your class is given a mathematics exam worth 100 points; X is the average score, rounded to the nearest whole number.

9. According to quantum mechanics, the energy of an electron in a hydrogen atom can assume only the values $k/1, k/4, k/9, k/16, \ldots$ for a certain constant value k. $X =$ the energy of an electron in a hydrogen atom.

10. According to classical mechanics, the energy of an electron in a hydrogen atom can assume any positive value. $X =$ the energy of an electron in a hydrogen atom.

In Exercises 11–18, (a) say what an appropriate sample space is; (b) complete the following sentence: "X is the rule that assigns to each . . . "; and (c) list the values of X for all the outcomes. [**HINT:** See Example 1.]

11. X is the number of tails that come up when a coin is tossed twice.

12. X is the largest number of consecutive times heads comes up in a row when a coin is tossed three times.

13. X is the sum of the numbers that face up when two dice are rolled.

14. X is the value of the larger number when two dice are rolled.

15. X is the number of red marbles that Tonya has in her hand after she selects four marbles from a bag containing four red marbles and two green ones and then notes how many there are of each color.

16. X is the number of green marbles that Stej has in his hand after he selects four marbles from a bag containing three red marbles and two green ones and then notes how many there are of each color.

17. The mathematics final exam scores for the students in your study group are 89%, 85%, 95%, 63%, 92%, and 80%.

18. The capacities of the hard drives of your dormitory suite mates' computers are 1,000 GB, 1,500 GB, 2,000 GB, 2,500 GB, 3,000 GB, and 3,500 GB.

19. The random variable X has this probability distribution table:

x	2	4	6	8	10
$P(X = x)$	.1	.2	—	—	.1

   a. Assuming that $P(X = 8) = P(X = 6)$, find each of the missing values. [**HINT:** See Quick Example 11.]

   b. Calculate $P(X \geq 6)$ and $P(2 < X < 8)$.

20. The random variable X has the probability distribution table shown below:

x	−2	−1	0	1	2
$P(X = x)$	—	—	.4	.1	.1

   a. Calculate $P(X \geq 0)$ and $P(X < 0)$. [**HINT:** See Quick Example 11.]

   b. Assuming that $P(X = -2) = P(X = -1)$, find each of the missing values.

In Exercises 21–28, give the probability distribution for the indicated random variable, draw the corresponding histogram, and calculate the indicated probability. [**HINT:** See Example 3.]

21. A fair die is rolled, and X is the number facing up. Calculate $P(X < 5)$.

22. A fair die is rolled, and X is the square of the number facing up. Calculate $P(X > 9)$.

23. Three fair coins are tossed, and X is the square of the number of heads showing. Calculate $P(1 \leq X \leq 9)$.

24. Three fair coins are tossed, and X is the number of heads minus the number of tails. Calculate $P(-3 \leq X \leq -1)$.

25. A red die and a green die are rolled, and X is the sum of the numbers facing up. Calculate $P(X \neq 7)$.

26. A red die and a green die are rolled, and

$$X = \begin{cases} 0 & \text{if the numbers are the same} \\ 1 & \text{if the numbers are different.} \end{cases}$$

Calculate $P(X > 1)$.

27. ▼ A red die and a green die are rolled, and X is the larger of the two numbers facing up. Calculate $P(X \leq 3)$.

28. ▼ A red die and a green die are rolled, and X is the smaller of the two numbers facing up. Calculate $P(X \geq 4)$.

Applications

29. *2010 Income Distribution up to $100,000* The following table shows the distribution of household incomes in 2010 for a sample of 1,000 households in the United States with incomes up to $100,000:[2]

Income Bracket ($)	0–19,999	20,000–39,999	40,000–59,999	60,000–79,999	80,000–99,999
Households	240	290	180	170	120

a. Let X be the (rounded) midpoint of a bracket in which a household falls. Find the relative frequency distribution of X, and graph its histogram. [**HINT:** See Example 4.]

b. Shade the area of your histogram corresponding to the probability that a randomly selected U.S. household in the sample has a value of X above 50,000. What is this probability?

30. *2003 Income Distribution up to $100,000* Repeat Exercise 29, using the following data from a sample of 1,000 households in the United States in 2003:[3]

Income Bracket ($)	0–19,999	20,000–39,999	40,000–59,999	60,000–79,999	80,000–99,999
Households	270	280	200	150	100

31. *Population Age in Mexico* The following chart shows the ages of 250 randomly selected residents of Mexico:[4]

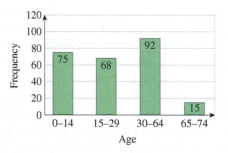

What is the associated random variable? Represent the data as a relative frequency distribution using the (rounded) midpoints of the given measurement classes. [**HINT:** See Example 4.]

32. *Population Age in the United States* Repeat Exercise 31, using the following chart, which shows the ages of 250 randomly selected residents of the United States:[5]

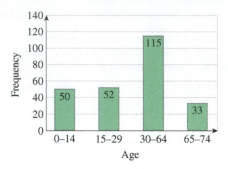

33. *Sport Utility Vehicles—Tow Ratings* The following table shows tow ratings (in pounds) for some popular sports utility vehicles in 2000:[6]

Vehicle	Tow Rating
Mercedes Grand Marquis V8	2,000
Jeep Wrangler I6	2,000
Ford Explorer V6	3,000
Dodge Dakota V6	4,000
Mitsubishi Montero V6	5,000
Ford Explorer V8	6,000
Dodge Durango V8	6,000
Dodge Ram 1500 V8	8,000
Ford Expedition V8	8,000
Hummer 2-Door Hardtop	8,000

Let X be the tow rating of a randomly chosen popular SUV from the list above.

a. What are the values of X?

b. Compute the frequency and probability distributions of X. [**HINT:** See Example 5.]

c. What is the probability that an SUV (from the list above) is rated to tow no more than 5,000 pounds?

34. *Housing Prices Going into the Real Estate Bubble* The following table shows the average percentage increase in the price of a house from 1980 to 2001 in nine regions of the United States:[7]

---

[2] Based on actual income distribution in 2010. Source: U.S. Census Bureau, Current Population Survey, 2010 American Community Survey (www.census.gov).

[3] Based on actual income distribution in 2003 (not adjusted for inflation). Source: U.S. Census Bureau, Current Population Survey, 2004 Annual Social and Economic Supplement (www.census.gov).

[4] Based on population distribution in 2010. Source: Instituto Nacional de Estadística y Geografía (www.inegi.org.mx).

[5] *Ibid.* (The data for the United States was also provided by the Instituto Nacional de Estadística y Geografía.)

[6] Tow ratings are for 2000 models and vary considerably within each model. Figures cited are rounded. For more detailed information, consult www.rvsafety.com/towrate2k.htm.

[7] Percentages are rounded to the nearest 25%. Source: Third Quarter 2001 House Price Index, released November 30, 2001, by the Office of Federal Housing Enterprise Oversight; available online at www.ofheo .gov/house/3q01hpi.pdf.

Region	Percent Increase
New England	300
Pacific	225
Middle Atlantic	225
South Atlantic	150
Mountain	150
West North Central	125
West South Central	75
East North Central	150
East South Central	125

Let X be the percentage increase in the price of a house in a randomly selected region.
a. What are the values of X?
b. Compute the frequency and probability distribution of X. [**HINT:** See Example 5.]
c. What is the probability that, in a randomly selected region, the percentage increase in the cost of a house exceeded 200%?

35. *Stock Market Gyrations* The following chart shows the day-by-day change, rounded to the nearest 100 points, in the Dow Jones Industrial Average during 20 successive business days around the start of the financial crisis in October 2008:[8]

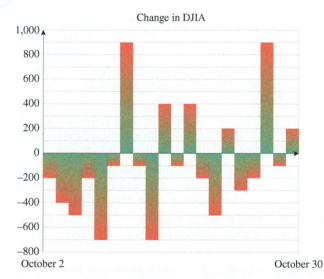

Change in DJIA

October 2   October 30

Let X be the (rounded) change in the Dow on a randomly selected day.
a. What are the values of X?
b. Compute the frequency and probability distribution of X. [**HINT:** See Example 5.]
c. What is the probability that, on a randomly selected day, the Dow decreased by more than 250 points?

36. *Stock Market Gyrations* Repeat Exercise 35 using the following chart for November–December 2008:[9]

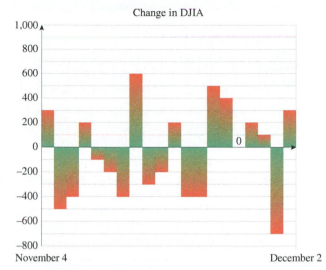

Change in DJIA

November 4   December 2

37. *Grade-Point Averages* The grade-point averages of the students in your mathematics class are

$$3.2, 3.5, 4.0, 2.9, 2.0, 3.3, 3.5, 2.9, 2.5, 2.0,$$
$$2.1, 3.2, 3.6, 2.8, 2.5, 1.9, 2.0, 2.2, 3.9, 4.0.$$

Use these raw data to construct a frequency table with the measurement classes 1.1–2.0, 2.1–3.0, 3.1–4.0, and find the probability distribution using the (rounded) midpoint values as the values of X. [**HINT:** See Example 5.]

38. *Test Scores* Your scores for the 20 surprise math quizzes last semester were (out of 10)

$$4.5, 9.5, 10.0, 3.5, 8.0, 9.5, 7.5, 6.5, 7.0, 8.0,$$
$$8.0, 8.5, 7.5, 7.0, 8.0, 9.0, 10.0, 8.5, 7.5, 8.0.$$

Use these raw data to construct a frequency table with the brackets 2.1–4.0, 4.1–6.0, 6.1–8.0, 8.1–10.0, and find the probability distribution using the (rounded) midpoint values as the values of X. [**HINT:** See Example 5.]

39. ▼ *Car Purchases* To persuade his parents to contribute to his new car fund, Carmine has spent the last week surveying the ages of 2,000 cars on campus. His findings are reflected in the following frequency table:

Age of Car (years)	0	1	2	3	4	5	6	7	8	9	10
Number of Cars	140	350	450	650	200	120	50	10	5	15	10

Carmine's jalopy is 6 years old. He would like to make the following claim to his parents: "x percent of students have cars newer than mine." Use a relative frequency distribution to find x.

[8] Source: http://finance.google.com.

[9] *Ibid.*

40. ▼ *Car Purchases* Carmine's parents, not convinced of his need for a new car, produced the following statistics showing the ages of cars owned by students on the dean's list:

Age of Car (years)	0	1	2	3	4	5	6	7	8	9	10
Number of Cars	0	2	5	5	10	10	15	20	20	20	40

They then claimed that if he kept his 6-year-old car for another year, his chances of getting on the dean's list would be increased by x percent. Use a relative frequency distribution to find x.

Highway Safety Exercises 41–50 are based on the following table, which shows crashworthiness ratings for several categories of motor vehicles.[10] In all of these exercises, take X as the crash-test rating of a small car, Y as the crash-test rating for a small SUV, and so on, as shown in the table.

	Number Tested	Overall Frontal Crash Test Rating			
		3 (Good)	2 (Acceptable)	1 (Marginal)	0 (Poor)
Small Cars, X	16	1	11	2	2
Small SUVs, Y	10	1	4	4	1
Medium SUVs, Z	15	3	5	3	4
Passenger Vans, U	13	3	0	3	7
Midsize Cars, V	15	3	5	0	7
Large Cars, W	19	9	5	3	2

41. Compute the relative frequency distribution for X.

42. Compute the relative frequency distribution for Y.

43. Compute $P(X \geq 2)$, and interpret the result.

44. Compute $P(Y \leq 1)$, and interpret the result.

45. Compare $P(Y \geq 2)$ and $P(Z \geq 2)$. What does the result suggest about SUVs?

46. Compare $P(V \geq 2)$ and $P(Z \geq 2)$. What does the result suggest?

47. ▼ Which of the six categories shown has the *lowest* probability of a Good rating?

48. ▼ Which of the six categories shown has the *highest* probability of a Poor rating?

49. ▼ You choose, at random, a small car and a small SUV. What is the probability that both will be rated at least 2?

---

[10] Ratings are by the Insurance Institute for Highway Safety. Sources: Oak Ridge National Laboratory: "An Analysis of the Impact of Sport Utility Vehicles in the United States," Stacy C. Davis, Lorena F. Truett (August 2000)/Insurance Institute for Highway Safety (www-cta.ornl.gov/Publications/Final SUV report.pdf, www.highwaysafety.org/vehicle_ratings).

50. ▼ You choose, at random, a small car and a midsize car. What is the probability that both will be rated at most 1?

Exercises 51 and 52 assume familiarity with counting arguments and probability (see Section 7.4).

51. ▼ *Camping Kent's Tents* has four red tents and three green tents in stock. Karin selects four of them at random. Let X be the number of red tents she selects. Give the probability distribution, and find $P(X \geq 2)$.

52. ▼ *Camping Kent's Tents* has five green knapsacks and four yellow ones in stock. Curt selects four of them at random. Let X be the number of green knapsacks he selects. Give the probability distribution, and find $P(X \leq 2)$.

53. ◆ *Testing Your Calculator* Use your calculator or computer to generate a sequence of 100 random digits in the range 0–9, and test the random number generator for uniformness by drawing the distribution histogram.

54. ◆ *Testing Your Dice* Repeat Exercise 53, but this time, use a die to generate a sequence of 50 random numbers in the range 1–6.

Communication and Reasoning Exercises

55. Are all infinite random variables necessarily continuous? Explain.

56. Are all continuous random variables necessarily infinite? Explain.

57. If you are unable to compute the (theoretical) probability distribution for a random variable X, how can you estimate the distribution?

58. What do you expect to happen to the probabilities in a probability distribution as you make the measurement classes smaller?

59. ▼ Give an example of a real-life situation that can be modeled by a random variable with a probability distribution whose histogram is highest on the left.

60. ▼ Give an example of a real-life situation that can be modeled by a random variable with a probability distribution whose histogram is highest on the right.

61. ▼ How wide should the bars in a histogram be so that the area of each bar equals the probability of the corresponding range of values of X?

62. ▼ How wide should the bars in a histogram be so that the probability $P(a \leq X \leq b)$ equals the area of the corresponding portion of the histogram?

63. ▼ Give at least one scenario in which you might prefer to model the number of pages in a randomly selected book using a continuous random variable rather than a discrete random variable.

64. ▼ Give at least one scenario in which you might prefer to model a temperature using a discrete random variable rather than a continuous random variable.

8.2 Bernoulli Trials and Binomial Random Variables

Your electronics production plant produces video game joysticks. Unfortunately, quality control at the plant leaves much to be desired, and 10% of the joysticks the plant produces are defective. A large corporation has expressed interest in adopting your product for its new game console, and today an inspection team will be visiting to test video game joysticks as they come off the assembly line. If the team tests five joysticks, what is the probability that none will be defective? What is the probability that more than one will be defective?

In this scenario we are interested in the following, which is an example of a particular type of finite random variable called a **binomial random variable**: Think of the experiment as a sequence of five "trials" (in each trial the inspection team chooses one joystick at random and tests it) each with two possible outcomes: "success" (a defective joystick) and "failure" (a nondefective one).[*] If we now take X to be the number of successes (defective joysticks) the inspection team finds, we can recast the questions above as follows: Find $P(X = 0)$ and $P(X > 1)$.

[*] These are customary names for the two possible outcomes, and they often do not indicate actual success or failure at anything. "Success" is the label we give the outcome of interest—in this case, finding a defective joystick.

[†] Jakob Bernoulli (1654–1705) was one of the pioneers of probability theory.

Bernoulli Trial

A **Bernoulli[†] trial** is an experiment that has two possible outcomes, called **success** and **failure**. If the probability of success is p, then the probability of failure is $q = 1 - p$.

Visualizing a Bernoulli Trial

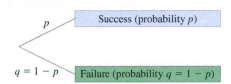

Tossing a coin three times is an example of a **sequence of independent Bernoulli trials**: a sequence of Bernoulli trials in which the outcomes in any one trial are independent (in the sense of Chapter 7) of those in any other trial and in which the probability of success is the same for all the trials.

Quick Examples

1. Roll a die, and take success to be the event that you roll a 6. Then $p = 1/6$ and $q = 5/6$. Rolling the die 10 times is then an example of a sequence of 10 independent Bernoulli trials.

2. Provide a property with flood insurance for 20 years, and take success to be the event that the property is flooded during a particular year. Observing whether or not the property is flooded each year for 20 years is then an example of 20 independent Bernoulli trials (assuming that the occurrence of flooding one year is independent of whether there was flooding in earlier years).

3. You know that 60% of all bond funds will depreciate in value next year. Take success to be the event that a randomly chosen fund depreciates next year. Then $p = .6$ and $q = .4$. Choosing five funds at random for your portfolio from a very large number of possible funds is, approximately,[§] an example of five independent Bernolli trials.

[§] Choosing a "loser" (a fund that will depreciate next year) slightly depletes the pool of "losers" and hence slightly decreases the probability of choosing another one. However, the fact that the pool of funds is very large means that this decrease is extremely small. Hence, p is very nearly constant.

4. Suppose that E is an event in an experiment with sample space S. Then we can think of the experiment as a Bernoulli trial with two outcomes; success if E occurs and failure if E' occurs. The probability of success is then

$$p = P(E), \qquad \text{Success is the occurrence of } E.$$

and the probability of failure is

$$q = P(E') = 1 - P(E) = 1 - p. \qquad \text{Failure is the occurrence of } E'.$$

Repeating the experiment 30 times, say, is then an example of 30 independent Bernoulli trials.

Note Quick Example 4 tells us that Bernoulli trials are not very special kinds of experiments; in fact, we are performing a Bernoulli trial every time we repeat *any* experiment and observe whether a specific event E occurs. Thinking of an experiment in this way amounts, mathematically, to thinking of $\{E, E'\}$ as our sample space (E = success, E' = failure). ∎

Binomial Random Variable

A **binomial random variable** is one that counts the number of successes in a sequence of independent Bernoulli trials, where the number of trials is fixed.

Visualizing a Binomial Random Variable

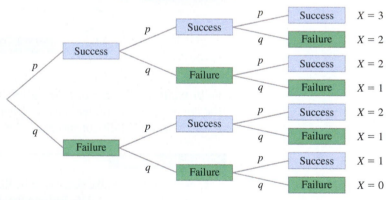

$X =$ Number of Successes

Quick Examples

5. Roll a die 10 times; X is the number of times you roll a 6.

6. Provide a property with flood insurance for 20 years; X is the number of years, during the 20-year period, during which the property is flooded (assuming that the occurrence of flooding in any year is independent of whether there was flooding in earlier years).

7. You know that 60% of all bond funds will depreciate in value next year, and you randomly select four from a very large number of possible choices; X is the number of bond funds you hold that will depreciate next year. (X is approximately binomial; see the margin note for Quick Example 3.)

EXAMPLE 1   **Probability Distribution of a Binomial Random Variable**

Suppose that we have a possibly unfair coin with the probability of heads p and the probability of tails $q = 1 - p$.

a. Let X be the number of heads you get in a sequence of five tosses. Find $P(X = 2)$.

b. Let X be the number of heads you get in a sequence of n tosses. Find $P(X = x)$.

Solution

a. We are looking for the probability of getting exactly two heads in a sequence of five tosses. Let's start with a simpler question: What is the probability that we will get the sequence HHTTT?

   The probability that the first toss will come up heads is p.

   The probability that the second toss will come up heads is also p.

   The probability that the third toss will come up tails is q.

   The probability that the fourth toss will come up tails is q.

   The probability that the fifth toss will come up tails is q.

   The probability that the first toss will be heads *and* the second will be heads *and* the third will be tails *and* the fourth will be tails *and* the fifth will be tails equals the probability of the *intersection* of these five events. Because these are independent events, the probability of the intersection is the product of the probabilities, which is

$$p \times p \times q \times q \times q = p^2 q^3.$$

   Now HHTTT is only one of several outcomes with two heads and three tails. Two others are HTHTT and TTTHH. How many such outcomes are there altogether? This is the number of "words" with two H's and three T's, and we know from Chapter 6 that the answer is $C(5, 2) = 10$.

   Each of the 10 outcomes with two H's and three T's has the same probability: $p^2 q^3$. (Why?) Thus, the probability of getting one of these 10 outcomes is the probability of the union of all these (mutually exclusive) events, and we saw in Chapter 7 that this is just the sum of the probabilities. In other words, the probability we are after is

$$P(X = 2) = p^2 q^3 + p^2 q^3 + \cdots + p^2 q^3 \qquad \text{\color{blue}{$C(5, 2)$ times}}$$
$$= C(5, 2) p^2 q^3.$$

The structure of this formula is as follows:

$$P(X = 2) = C(5, 2) p^2 q^3$$

Number of heads   Number of tails

Number of tosses   Probability of tails
Number of heads   Probability of heads

b. What we did using the numbers 5 and 2 in part (a) works as well in general. For the general case, with n tosses and x heads, replace 5 with n and replace 2 with x to get

$$P(X = x) = C(n, x) p^x q^{n-x}.$$

(Note that the exponent of q is the number of tails, which is $n - x$.)

The calculation in Example 1 applies to any binomial random variable, so we can say the following.

Probability Distribution of Binomial Random Variables

If X is the number of successes in a sequence of n independent Bernoulli trials, then

$$P(X = x) = C(n, x)p^x q^{n-x},$$

where

n = Number of trials

p = Probability of success

q = Probability of failure $= 1 - p$.

Quick Example

8. If you roll a fair die five times, the probability of throwing exactly two 6s is

$$P(X = 2) = C(5, 2)\left(\frac{1}{6}\right)^2 \left(\frac{5}{6}\right)^3 = 10 \times \frac{1}{36} \times \frac{125}{216} \approx .1608.$$

Here, we used $n = 5$ and $p = 1/6$, the probability of rolling a 6 on one roll of the die.

EXAMPLE 2 Aging

By 2030 the probability that a randomly chosen resident in the United States will be 65 years old or older is projected[11] to be .2.

a. What is the probability that, in a randomly selected sample of six U.S. residents, exactly four of them will be 65 or older?

b. If X is the number of people aged 65 or older in a sample of six, construct the probability distribution of X and plot its histogram.

c. Compute $P(X \le 2)$.

d. Compute $P(X \ge 2)$.

Solution

a. The experiment is a sequence of Bernoulli trials; in each trial we select a person and ascertain his or her age. If we take "success" to mean selection of a person aged 65 or older, then the probability distribution is

$$P(X = x) = C(n, x)p^x q^{n-x},$$

where

n = Number of trials = 6

p = Probability of success = .2

q = Probability of failure = .8.

---

[11] Source: U.S. Census Bureau, Decennial Census, Population Estimates and Projections (www.agingstats.gov/agingstatsdotnet/Main_Site/Data/2012_Documents/Population.aspx).

So

$$P(X = 4) = C(6, 4)(.2)^4(.8)^2$$
$$= 15 \times .0016 \times .64 = .01536.$$

b. We have already computed $P(X = 4)$. Here are all the calculations:

$$P(X = 0) = C(6, 0)(.2)^0(.8)^6$$
$$= 1 \times 1 \times .262144 = .262144$$

$$P(X = 1) = C(6, 1)(.2)^1(.8)^5$$
$$= 6 \times .2 \times .32768 = .393216$$

$$P(X = 2) = C(6, 2)(.2)^2(.8)^4$$
$$= 15 \times .04 \times .4096 = .24576$$

$$P(X = 3) = C(6, 3)(.2)^3(.8)^3$$
$$= 20 \times .008 \times .512 = .08192$$

$$P(X = 4) = C(6, 4)(.2)^4(.8)^2$$
$$= 15 \times .0016 \times .64 = .01536$$

$$P(X = 5) = C(6, 5)(.2)^5(.8)^1$$
$$= 6 \times .00032 \times .8 = .001536$$

$$P(X = 6) = C(6, 6)(.2)^6(.8)^0$$
$$= 1 \times .000064 \times 1 = .000064.$$

The probability distribution is therefore as follows:

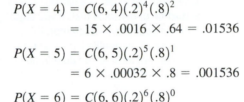

$P(X = x)$

Figure 3

x	0	1	2	3	4	5	6
$P(X = x)$	.262144	.393216	.24576	.08192	.01536	.001536	.000064

Figure 3 shows its histogram.

c. $P(X \le 2)$—the probability that the number of people selected who are at least 65 years old is either 0, 1, or 2—is the probability of the union of these events and is thus the sum of the three probabilities:

$$P(X \le 2) = P(X = 0) + P(X = 1) + P(X = 2)$$
$$= .262144 + .393216 + .24576$$
$$= .90112.$$

d. To compute $P(X \ge 2)$, we *could* compute the sum

$$P(X \ge 2) = P(X = 2) + P(X = 3) + P(X = 4) + P(X = 5) + P(X = 6),$$

but it is far easier to compute the probability of the complement of the event,

$$P(X < 2) = P(X = 0) + P(X = 1)$$
$$= .262144 + .393216 = .65536$$

and then subtract the answer from 1:

$$P(X \ge 2) = 1 - P(X < 2)$$
$$= 1 - .65536 = .34464.$$

FAQs

Terminology and Recognizing When to Use the Binomial Distribution

Q : *What is the difference between Bernoulli trials and a binomial random variable?*

A : A Bernoulli trial is a type of experiment, whereas a binomial random variable is the resulting kind of random variable. More precisely, if your experiment consists of performing a sequence of n Bernoulli trials (think of throwing a dart n times at random points on a dartboard hoping to hit the bull's-eye), then the random variable X that counts the number of successes (the number of times you actually hit the bull's-eye) is a binomial random variable.

Q : *How do I recognize when a situation gives a binomial random variable?*

A : Make sure that the experiment consists of a sequence of independent Bernoulli trials, that is, a sequence of a fixed number of trials of an experiment that has two outcomes, where the outcome of each trial does not depend on the outcomes in previous trials and where the probability of success is the same for all the trials. For instance, repeatedly throwing a dart at a dartboard hoping to hit the bull's-eye does not constitute a sequence of Bernoulli trials if you adjust your aim each time depending on the outcome of your previous attempt. This dart-throwing experiment can be modeled by a sequence of Bernoulli trials if you make no adjustments after each attempt and your aim does not improve (or deteriorate) with time.

8.2 EXERCISES

▼ more advanced ◆ challenging
T indicates exercises that should be solved using technology

In Exercises 1–10, you are performing five independent Bernoulli trials with $p = .1$ and $q = .9$. Calculate the probability of the stated outcome. Check your answer using technology. [**HINT:** See Quick Example 8.]

1. Two successes
2. Three successes
3. No successes
4. No failures
5. All successes
6. All failures
7. At most two successes
8. At least four successes
9. At least three successes
10. At most three successes

In Exercises 11–18, X is a binomial variable with $n = 6$ and $p = .4$. Compute the given probability. Check your answer using technology. [**HINT:** See Example 2.]

11. $P(X = 3)$
12. $P(X = 4)$
13. $P(X \leq 2)$
14. $P(X \leq 1)$
15. $P(X \geq 5)$
16. $P(X \geq 4)$
17. $P(1 \leq X \leq 3)$
18. $P(3 \leq X \leq 5)$

In Exercises 19 and 20, graph the histogram of the given binomial distribution. Check your answer using technology.

19. $n = 5, p = \dfrac{1}{4}, q = \dfrac{3}{4}$ **20.** $n = 5, p = \dfrac{1}{3}, q = \dfrac{2}{3}$

In Exercises 21 and 22, graph the histogram of the given binomial distribution, and compute the given quantity, indicating the corresponding region on the graph.

21. $n = 4, p = \dfrac{1}{3}, q = \dfrac{2}{3}; P(X \leq 2)$

22. $n = 4, p = \dfrac{1}{4}, q = \dfrac{3}{4}; P(X \leq 1)$

Applications

23. *Internet Addiction* The probability that a randomly chosen person in the Netherlands connects to the Internet immediately upon waking[12] is approximately .25. What is the probability that, in a randomly selected sample of five people, two connect to the Internet immediately upon waking? [**HINT:** See Example 2.]

24. *Alien Retirement* The probability that a randomly chosen citizen-entity of Cygnus is of pension age[13] is approximately .8. What is the probability that, in a randomly selected sample of four citizen-entities, all of them are of pension age? [**HINT:** See Example 2.]

---

[12] Source: *Webwereld* November 17, 2008 (http://webwereld.nl/article/view/id/53599).

[13] The retirement age in Cygnus is 12,000 bootlags, which is equivalent to approximately 20 minutes Earth time.

25. *1990s Internet Stock Boom* According to a July 1999 article in *The New York Times*,[14] venture capitalists had this "rule of thumb": The probability that an Internet start-up company will be a "stock market success" resulting in "spectacular profits for early investors" is .2. If you were a venture capitalist who invested in 10 Internet start-up companies, what was the probability that at least 1 of them would be a stock market success? (Round your answer to four decimal places.)

26. *1990s Internet Stock Boom* According to the article cited in Exercise 25, 13.5% of Internet stocks that entered the market in 1999 ended up trading below their initial offering prices. If you were an investor who purchased five Internet stocks at their initial offering prices, what was the probability that at least four of them would end up trading at or above their initial offering price? (Round your answer to four decimal places.)

27. *Job Training* *(from the GRE Exam in Economics)* In a large on-the-job training program, half of the participants are female and half are male. In a random sample of three participants, what is the probability that an investigator will draw at least one male?

28. *Job Training* *(based on a question from the GRE Exam in Economics)* In a large on-the-job training program, half of the participants are female and half are male. In a random sample of five participants, what is the probability that an investigator will draw at least two males?

29. *Manufacturing* Your manufacturing plant produces air bags, and it is known that 10% of them are defective. Five air bags are tested.
  a. Find the probability that three of them are defective.
  b. Find the probability that at least two of them are defective.

30. *Manufacturing* Compute the probability distribution of the binomial variable described in Exercise 29, and use it to compute the probability that if five air bags are tested, at least one will be defective and at least one will not.

31. *Teenage Pastimes* According to a study,[15] the probability that a randomly selected teenager watched a rented video at least once during a week was .71. What is the probability that at least 8 teenagers in a group of 10 watched a rented movie at least once last week?

32. *Other Teenage Pastimes* According to the study cited in Exercise 31, the probability that a randomly selected teenager studied at least once during a week was only .52. What is the probability that less than half of the students in your study group of 10 have studied in the last week?

33. *Subprime Mortgages during the Housing Bubble* In November 2008[16] the probability that a randomly selected subprime home mortgage in Florida was in foreclosure was .24. Choose 10 subprime home mortgages at random.
  a. What is the probability that exactly 5 of them were in foreclosure?
  b. **T** Use technology to generate the probability distribution for the associated binomial random variable.
  c. Fill in the blank: If 10 subprime home mortgages were chosen at random, the number of them most likely to have been in foreclosure was _____.

34. *Subprime Mortgages during the Housing Bubble* In November 2008[17] the probability that a randomly selected subprime home mortgage in Texas was current in its payments was .67. Choose 10 subprime home mortgages at random.
  a. What is the probability that exactly 4 of them were current?
  b. **T** Use technology to generate the probability distribution for the associated binomial random variable.
  c. Fill in the blank: If 10 subprime home mortgages were chosen at random, the number of them most likely to have been current was _____.

35. ▼ *Triple Redundancy* To ensure reliable performance of vital computer systems, aerospace engineers sometimes employ the technique of "triple redundancy," in which three identical computers are installed in a space vehicle. If one of the three computers gives results different from the other two, it is assumed to be malfunctioning and is ignored. This technique will work as long as no more than one computer malfunctions. Assuming that an onboard computer is 99% reliable (that is, the probability of its failing is .01), what is the probability that at least two of the three computers will malfunction?

36. ▼ *IQ Scores* Mensa is a club for people who have high IQ scores. To qualify, your IQ must be at least 132, putting you in the top 2% of the general population. If a group of 10 people are chosen at random, what is the probability that at least 2 of them qualify for Mensa?

37. T ▼ *Standardized Tests* Assume that on a standardized test of 100 questions, a person has a probability of 80% of answering any particular question correctly. Find the probability of correctly answering between 75 and 85 questions, inclusive. (Assume independence, and round your answer to four decimal places.)

38. T ▼ *Standardized Tests* Assume that on a standardized test of 100 questions, a person has a probability of 80% of answering any particular question correctly. Find the probability of correctly answering at least 90 questions. (Assume independence, and round your answer to four decimal places.)

39. T ▼ *Product Testing* It is known that 43% of all the ZeroFat hamburger patties produced by your factory actually contain

[14] "Not All Hit It Rich in the Internet Gold Rush," *New York Times*, July 20, 1999, p. A1.

[15] Sources: Rand Youth Poll/Teen-Age Research Unlimited/*New York Times*, March 14, 1998, p. D1.

[16] Source: Federal Reserve Bank of New York (www.newyorkfed.org/regional/subprime.html).

[17] *Ibid.*

more than 10 grams of fat. Compute the probability distribution for $n = 50$ Bernoulli trials.

a. What is the most likely value for the number of burgers in a sample of 50 that contain more than 10 grams of fat?

b. Complete the following sentence: There is an approximately 71% chance that a batch of 50 ZeroFat patties contains ____ or more patties with more than 10 grams of fat.

c. Compare the graphs of the distributions for $n = 50$ trials and $n = 20$ trials. What do you notice?

40. **T** ▼ *Product Testing* It is known that 65% of all the ZeroCal hamburger patties produced by your factory actually contain more than 1,000 calories. Compute the probability distribution for $n = 50$ Bernoulli trials.

a. What is the most likely value for the number of burgers in a sample of 50 that contain more than 1,000 calories?

b. Complete the following sentence: There is an approximately 73% chance that a batch of 50 ZeroCal patties contains ____ or more patties with more than 1,000 calories.

c. Compare the graphs of the distributions for $n = 50$ trials and $n = 20$ trials. What do you notice?

41. **T** ▼ *Quality Control* A manufacturer of light bulbs chooses bulbs at random from its assembly line for testing. If the probability of a bulb's being bad is .01, how many bulbs does the manufacturer need to test before the probability of finding at least one bad one rises to more than .5? (You may have to use trial and error to solve this.)

42. **T** ▼ *Quality Control* A manufacturer of light bulbs chooses bulbs at random from its assembly line for testing. If the probability of a bulb's being bad is .01, how many bulbs does the manufacturer need to test before the probability of finding at least two bad ones rises to more than .5? (You may have to use trial and error to solve this.)

43. ▼ *Highway Safety* According to a study,[18] a male driver in the United States will average 562 accidents per 100 million miles. Regard an n-mile trip as a sequence of n Bernoulli trials with "success" corresponding to having an accident during a particular mile. What is the probability that a male driver will have an accident in a 1-mile trip?

44. ▼ *Highway Safety:* According to the study cited in Exercise 43, a female driver in the United States will average 611 accidents per 100 million miles. Regard an n-mile trip as a sequence of n Bernoulli trials with "success" corresponding to having an accident during a particular mile. What is the probability that a female driver will have an accident in a 1-mile trip?

45. ◆ *Mad Cow Disease* In March 2004 the U.S. Department of Agriculture announced plans to test approximately 243,000 slaughtered cows per year for mad cow disease

(bovine spongiform encephalopathy).[19] When announcing the plan, the Agriculture Department stated that "by the laws of probability, that many tests should detect mad cow disease even if it is present in only 5 cows out of the 45 million in the nation."[20] Test the Department's claim by computing the probability that, if only 5 out of 45 million cows had mad cow disease, at least 1 cow would test positive in a year (assuming that the testing was done randomly).

46. ◆ *Mad Cow Disease* According to the article cited in Exercise 45, only 223,000 of the cows being tested for bovine spongiform encephalopathy were to be "downer cows," that is, cows unable to walk to their slaughter. Assuming that just one downer cow in 500,000 is infected on average, use a binomial distribution to find the probability that two or more cows would test positive in a year. Your associate claims that "by the laws of probability, that many tests should detect at least two cases of mad cow disease even if it is present in only two cows out of a million downers." Comment on that claim.

Communication and Reasoning Exercises

47. A soccer player is more likely to score on his second shot if he was successful on his first. Can we model a succession of shots a player takes as a sequence of Bernoulli trials? Explain.

48. A soccer player takes repeated shots on goal. What assumption must we make if we want to model a succession of shots by a player as a sequence of Bernoulli trials?

49. Your friend just told you that "misfortunes always occur in threes." If life is just a sequence of Bernoulli trials, is this possible? Explain.

50. Suppose an experiment consists of repeatedly (every week) checking whether your graphing calculator battery has died. Is this a sequence of Bernoulli trials? Explain.

51. In an experiment with sample space S, a certain event E has a probability p of occurring. What has this scenario to do with Bernoulli trials?

52. An experiment consists of removing a gummy bear from a bag originally containing 10 and then eating it. Regard eating a lime-flavored bear as success. Repeating the experiment five times is a sequence of Bernoulli trials—right?

53. ▼ Why is the following not a binomial random variable? Select, without replacement, five marbles from a bag containing six red marbles and two blue ones, and let X be the number of red marbles you have selected.

54. ▼ By contrast with Exercise 53, why can the following be modeled by a binomial random variable? Select, without replacement, 5 electronic components from a batch of 10,000 in which 1,000 are defective, and let X be the number of defective components you select.

---

[18] Data are based on a report by the National Highway Traffic Safety Administration released in January, 1996. Source for data: U.S. Department of Transportation/*New York Times*, April 9, 1999, p. F1.

[19] Source: *New York Times*, March 17, 2004, p. A19.

[20] As stated in the *New York Times* article.

8.3   Measures of Central Tendency

Mean, Median, and Mode of a Set of Data

One day you decide to measure the popularity rating of your statistics instructor, Mr. Pelogrande. Ideally, you should poll all of Mr. Pelogrande's students, which is what statisticians would refer to as the **population**. However, it would be difficult to poll all the members of the population in question. (Mr. Pelogrande teaches more than 400 students.) Instead, you decide to survey 10 of his students, chosen at random, and ask them to rate Mr. Pelogrande on a scale of 0–100. The survey results in the following set of data:

$$60, 50, 55, 0, 100, 90, 40, 20, 40, 70.$$

Such a collection of data is called a **sample**, because the 10 people polled represent only a (small) sample of Mr. Pelogrande's students. We should think of the individual scores $60, 50, 55, \ldots$ as values of a random variable: Choose one of Mr. Pelogrande's students at random, and let X be the rating the student gives to Mr. Pelogrande.

How do we distill a single measurement, or **statistic**, from this sample that would describe Mr. Pelogrande's popularity? Perhaps the most commonly used statistic is the **average**, or **mean**, which is computed by adding the scores and dividing the sum by the number of scores in the sample:

$$\text{Sample mean} = \frac{60 + 50 + 55 + 0 + 100 + 90 + 40 + 20 + 40 + 70}{10}$$

$$= \frac{525}{10} = 52.5.$$

We might then conclude, on the basis of the sample, that Mr. Pelogrande's average popularity rating is about 52.5. The usual notation for the sample mean is $\bar{x}$, and the formula that we use to compute it is

$$\bar{x} = \frac{x_1 + x_2 + \cdots + x_n}{n},$$

where $x_1, x_2, \ldots, x_n$ are the values of X in the sample.

A convenient way of writing the sum that appears in the numerator is to use **summation** or **sigma notation**. We write the sum $x_1 + x_2 + \cdots + x_n$ as

$$\sum_{i=1}^{n} x_i.$$

$\sum\limits_{i=1}^{n}$ by itself stands for "the sum, from $i = 1$ to n."

$\sum\limits_{i=1}^{n} x_i$ stands for "the sum of the x_i, from $i = 1$ to n."

We think of i as taking on the values $1, 2, \ldots, n$ in turn, making x_i equal $x_1, x_2, \ldots, x_n$ in turn, and we then add up these values.

Visualizing the Mean

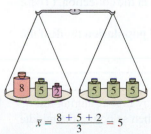

$$\bar{x} = \frac{8 + 5 + 2}{3} = 5$$

Sample and Mean

A **sample** is a sequence of values (or scores) of a random variable X. (The process of collecting such a sequence is sometimes called **sampling** X.) The **sample mean** is the average of the values, or **scores**, in the sample. To compute the sample mean, we use the following formula:

$$\bar{x} = \frac{x_1 + x_2 + \cdots + x_n}{n} = \frac{\sum_{i=1}^{n} x_i}{n}$$

* In Section 1.4 we simply wrote $\sum x$ for the sum of all the x_i, but here we will use the subscripts to make it easier to interpret formulas in this and the next section.

† When we talk about *populations*, the understanding is that the underlying experiment consists of selecting a member of a given population and ascertaining the value of X.

or simply

$$\bar{x} = \frac{\sum_i x_i}{n}. \qquad \text{$\sum_i$ stands for "sum over all i."*}$$

Here, n is the **sample size** (number of scores), and $x_1, x_2, \ldots, x_n$ are the individual values.

If the sample $x_1, x_2, \ldots, x_n$ consists of all the values of X from the entire population† (for instance, the ratings given Mr. Pelogrande by *all* of his students), we refer to the mean as the **population mean**, and write it as μ (Greek "mu") instead of $\bar{x}$.

Quick Examples

1. The mean of the sample 1, 2, 3, 4, 5 is $\bar{x} = 3$.

2. The mean of the sample $-1, 0, 2$ is $\bar{x} = \dfrac{-1 + 0 + 2}{3} = \dfrac{1}{3}$.

3. The mean of the population $-3, -3, 0, 0, 1$ is

$$\mu = \frac{-3 - 3 + 0 + 0 + 1}{5} = -1.$$

Note: Sample Mean versus Population Mean Determining a population mean can be difficult or even impossible. For instance, computing the mean household income for the United States would entail recording the income of every single household in the United States. Instead of attempting to do this, we usually use sample means instead. The larger the sample used, the more accurately we expect the sample mean to approximate the population mean. Estimating how accurately a sample mean based on a given sample size approximates the population mean is possible, but we will not go into that in this book. ■

The mean $\bar{x}$ is an attempt to describe where the "center" of the sample is. It is therefore called a **measure of central tendency**. There are two other common measures of central tendency: the "middle score," or **median**, and the "most frequent score," or **mode**. These are defined as follows.

Visualizing the Median and Mode

Median = Middle score = 4

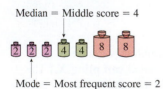

Mode = Most frequent score = 2

Median and Mode

The **sample median** m is the middle score (in the case of an odd-size sample), or average of the two middle scores (in the case of an even-size sample) when the scores in a sample are arranged in ascending order.

A **sample mode** is a score that appears most often in the collection. (There may be more than one mode in a sample.)

As before, we refer to the **population median** and **population mode** if the sample consists of the data from the entire population.

Quick Examples

4. The sample median of 2, -3, -1, 4, 2 is found by first arranging the scores in ascending order: $-3, -1, 2, 2, 4$ and then selecting the middle

(third) score: $m = 2$. The sample mode is also 2 because this is the score that appears most often.

5. The sample 2, 5, 6, −1, 0, 6 has median $m = (2 + 5)/2 = 3.5$ and mode 6.

The mean tends to give more weight to scores that are farther away from the center than does the median. For example, if you take the largest score in a collection of more than two numbers and make it larger, the mean will increase but the median will remain the same. For this reason the median is often preferred for collections that contain a wide range of scores. The mode can sometimes lie far from the center and is therefore used less often as an indication of where the "center" of a sample lies.

EXAMPLE 1   Teenage Spending in the 1990s

A 10-year survey of spending patterns of U.S. teenagers in the 1990s yielded the following figures (in billions of dollars spent in a year):[21] 90, 90, 85, 80, 80, 80, 80, 85, 90, 100. Compute and interpret the mean, median, and mode, and illustrate the data on a graph.

Solution The *mean* is given by

$$\bar{x} = \frac{\sum_i x_i}{n}$$

$$= \frac{90 + 90 + 85 + 80 + 80 + 80 + 80 + 85 + 90 + 100}{10} = \frac{860}{10} = 86.$$

Thus, spending by teenagers averaged $86 billion per year.

For the *median* we arrange the sample data in ascending order:

$$80, 80, 80, 80, 85, 85, 90, 90, 90, 100.$$

We then take the average of the two middle scores:

$$m = \frac{85 + 85}{2} = 85.$$

This means that in half the years in question, teenagers spent $85 billion or less, and in half they spent $85 billion or more.

For the *mode* we choose the score (or scores) that occurs most frequently: 80. Thus, teenagers spent $80 billion per year more often than any other amount.

The frequency histogram in Figure 4 illustrates these three measures.

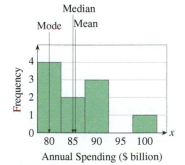

Figure 4

➡ **Before we go on . . .** There is a nice geometric interpretation of the difference between the median and mode: The median line shown in Figure 4 divides the total area of the histogram into two equal pieces, whereas the mean line passes through its "center of gravity"; if you placed the histogram on a knife-edge along the mean line, it would balance. ∎

Expected Value of a Finite Random Variable

Now, instead of looking at a sample of values of a given random variable, let us look at the probability distribution of the random variable itself and see if we can predict

---

[21] Spending figures are rounded, and cover the years 1988 through 1997. Source: Rand Youth Poll/Teen-Age Research Unlimited/*New York Times*, March 14, 1998, p. D1.

the sample mean without actually taking a sample. This prediction is what we call the *expected value* of the random variable.

EXAMPLE 2   Expected Value of a Random Variable

Suppose you roll a fair die a large number of times. What do you expect to be the average of the numbers that face up?

Solution   Suppose we take a sample of n rolls of the die (where n is large). Because the probability of rolling a 1 is $1/6$, we would expect that we would roll a 1 one sixth of the time, or $n/6$ times. Similarly, each other number should also appear $n/6$ times. The frequency table should then look like this:

x	1	2	3	4	5	6
Number of Times x Is Rolled (Frequency)	$\dfrac{n}{6}$	$\dfrac{n}{6}$	$\dfrac{n}{6}$	$\dfrac{n}{6}$	$\dfrac{n}{6}$	$\dfrac{n}{6}$

Note that we would not really expect the scores to be evenly distributed in practice, although for very large values of n we would expect the frequencies to vary only by a small percentage. To calculate the sample mean, we would add up all the scores and divide by the sample size. Now, the table tells us that there are $n/6$ ones, $n/6$ twos, $n/6$ threes, and so on, up to $n/6$ sixes. Adding these all up gives

$$\sum_i x_i = \frac{n}{6} \cdot 1 + \frac{n}{6} \cdot 2 + \frac{n}{6} \cdot 3 + \frac{n}{6} \cdot 4 + \frac{n}{6} \cdot 5 + \frac{n}{6} \cdot 6.$$

(Notice that we can obtain this number by multiplying the frequencies by the values of X and then adding.) Thus, the mean is

$$\bar{x} = \frac{\sum_i x_i}{n}$$

$$= \frac{\frac{n}{6} \cdot 1 + \frac{n}{6} \cdot 2 + \frac{n}{6} \cdot 3 + \frac{n}{6} \cdot 4 + \frac{n}{6} \cdot 5 + \frac{n}{6} \cdot 6}{n}$$

$$= \frac{1}{6} \cdot 1 + \frac{1}{6} \cdot 2 + \frac{1}{6} \cdot 3 + \frac{1}{6} \cdot 4 + \frac{1}{6} \cdot 5 + \frac{1}{6} \cdot 6 \qquad \text{Divide top and bottom by } n.$$

$$= 3.5.$$

This is the average value we expect to get after a large number of rolls or, in short, the **expected value** of a roll of the die. More precisely, we say that this is the expected value of the random variable X whose value is the number we get by rolling a die. Notice that n, the number of rolls, does not appear in the expected value. In fact, we could redo the calculation more simply by dividing the frequencies in the table by n *before* adding. Doing this replaces the frequencies with the *probabilities*, $1/6$. That is, it *replaces the frequency distribution with the probability distribution.*

x	1	2	3	4	5	6
$P(X = x)$	$\dfrac{1}{6}$	$\dfrac{1}{6}$	$\dfrac{1}{6}$	$\dfrac{1}{6}$	$\dfrac{1}{6}$	$\dfrac{1}{6}$

The expected value of X is then the sum of the products $x \cdot P(X = x)$. This is how we shall compute it from now on.

To obtain the expected value, multiply the values of X by their probabilities, and then add the results.

Expected Value of a Finite Random Variable

If X is a finite random variable that takes on the values $x_1, x_2, \ldots, x_n$, then the **expected value** of X, written $E(X)$ or μ, is

$$\mu = E(X) = x_1 \cdot P(X = x_1) + x_2 \cdot P(X = x_2) + \cdots + x_n \cdot P(X = x_n)$$

$$= \sum_i x_i \cdot P(X = x_i).$$

In Words

To compute the expected value from the probability distribution of X, we multiply the values of X by their probabilities and add up the results.

Interpretation

We interpret the expected value of X as a *prediction* of the mean of a large random sample of measurements of X; in other words, it is what we "expect" the mean of a large number of scores to be. (The larger the sample, the more accurate this prediction will tend to be.)

Quick Example

6. If X has the distribution shown,

x	−1	0	4	5
$P(X = x)$	.3	.5	.1	.1

then

$$\mu = E(X) = -1(.3) + 0(.5) + 4(.1) + 5(.1) = -.3 + 0 + .4 + .5 = .6.$$

EXAMPLE 3 **Sports Injuries**

According to historical data, the number of injuries that a member of the *Enormous State University* women's soccer team will sustain during a typical season is given by the following probability distribution table:

Injuries	0	1	2	3	4	5	6
Probability	.20	.20	.22	.20	.15	.01	.02

If X denotes the number of injuries sustained by a player during one season, compute $E(X)$, and interpret the result.

Solution We can compute the expected value using the following tabular approach: Take the probability distribution table, add another row in which we compute the product $xP(X = x)$, and then add these products together.

x	0	1	2	3	4	5	6	
$P(X = x)$	.20	.20	.22	.20	.15	.01	.02	**Total**
$xP(X = x)$	0	.20	.44	.60	.60	.05	.12	2.01

The total of the entries in the bottom row is the expected value. Thus,

$$E(X) = 2.01.$$

We interpret the result as follows: If many soccer players are observed for a season, we predict that the average number of injuries each will sustain is about two.

EXAMPLE 4   Roulette

A roulette wheel (of the kind used in the United States) has the numbers 1 through 36, 0 and 00. A bet on a single number pays 35 to 1. This means that if you place a $1 bet on a single number and win (your number comes up), you get your $1 back plus $35 (that is, you gain $35). If your number does not come up, you lose the $1 you bet. What is the expected gain from a $1 bet on a single number?

Solution   The probability of winning is $1/38$, so the probability of losing is $37/38$. Let X be the gain from a $1 bet. X has two possible values: $X = -1$ if you lose and $X = 35$ if you win. $P(X = -1) = 37/38$ and $P(X = 35) = 1/38$. This probability distribution and the calculation of the expected value are given in the following table:

x	-1	35	
$P(X = x)$	$\dfrac{37}{38}$	$\dfrac{1}{38}$	Total
$xP(X = x)$	$-\dfrac{37}{38}$	$\dfrac{35}{38}$	$-\dfrac{2}{38}$

So we expect to average a small loss of $2/38 \approx \$0.0526$ on each spin of the wheel.

➡ **Before we go on . . .**   Of course, you cannot actually lose the expected $0.0526 on one spin of the roulette wheel in Example 4. However, if you play many times, this is what you expect your *average* loss per bet to be. For example, if you played 100 times, you could expect to lose about $100 \times 0.0526 = \$5.26$. ∎

A betting game in which the expected value is zero is called a **fair game**. For example, if you and I flip a coin, and I give you $1 each time it comes up heads but you give me $1 each time it comes up tails, then the game is fair. Over the long run, we expect to come out even. On the other hand, a game like roulette, in which the expected value is not zero, is **biased**. Most casino games are slightly biased in favor of the house.[*] Thus, most gamblers will lose only a small amount, and many gamblers will actually win something (and return to play some more). However, when the earnings are averaged over the huge numbers of people playing, the house is guaranteed to come out ahead. This is how casinos make (lots of) money.

[*] Only rarely are games not biased in favor of the house. However, blackjack played without continuous shuffle machines can be beaten by card counting.

Expected Value of a Binomial Random Variable

Suppose you guess all the answers to the questions on a multiple-choice test. What score can you expect to get? This scenario is an example of a sequence of Bernoulli trials (see the preceding section), and the number of correct guesses is therefore a binomial random variable whose expected value we wish to know. There is a simple formula for the expected value of a binomial random variable.

Expected Value of a Binomial Random Variable

If X is the binomial random variable associated with n independent Bernoulli trials, each with probability p of success, then the expected value of X is

$$\mu = E(X) = np.$$

Quick Examples

7. If X is the number of successes in 20 Bernoulli trials with $p = .7$, then the expected number of successes is $\mu = E(X) = (20)(.7) = 14$.

8. If an event F in some experiment has $P(F) = .25$, the experiment is repeated 100 times, and X is the number of times F occurs, then $E(X) = (100)(.25) = 25$ is the number of times we expect F to occur.

Where does this formula come from? We *could* use the formula for expected value and compute the sum

$$E(X) = 0C(n, 0)p^0q^n + 1C(n, 1)p^1q^{n-1} + 2C(n, 2)p^2q^{n-2} + \cdots + nC(n, n)p^nq^0$$

directly (using the binomial theorem), but this is one of the many places in mathematics where a less direct approach is much easier. X is the number of successes in a sequence of n Bernoulli trials, each with probability p of success. Thus, p is the fraction of time we expect a success, so out of n trials we expect np successes. Because X counts successes, we expect the value of X to be np. (With a little more effort, this can be made into a formal proof that the sum above equals np.)

EXAMPLE 5   Guessing on an Exam

An exam has 50 multiple-choice questions, each having four choices. If a student randomly guesses on each question, how many correct answers can he or she expect to get?

Solution Each guess is a Bernoulli trial with probability of success 1 in 4, so $p = .25$. Thus, for a sequence of $n = 50$ trials,

$$\mu = E(X) = np = (50)(.25) = 12.5.$$

Thus, the student can expect to get about 12.5 correct answers.

Q: *Wait a minute. How can a student get a fraction of a correct answer?*

A: Remember that the expected value is the average number of correct answers a student will get if he or she guesses on a large number of such tests. Or we can say that if many students use this strategy of guessing, they will average about 12.5 correct answers each.

Estimating the Expected Value from a Sample

It is not always possible to know the probability distribution of a random variable. For instance, if we take X to be the income of a randomly selected lawyer, we could not be expected to know the probability distribution of X. However, we can still obtain a good *estimate* of the expected value of X (the average income of all lawyers) by using the relative frequency distribution based on a large random sample.

EXAMPLE 6   Estimating an Expected Value

The following table shows the (fictitious) incomes of a random sample of 1,000 lawyers in the United States in their first year out of law school.

Income Bracket	$20,000–$29,999	$30,000–$39,999	$40,000–$49,999	$50,000–$59,999	$60,000–$69,999	$70,000–$79,999	$80,000–$89,999
Number	20	80	230	400	170	70	30

Estimate the average of the incomes of all lawyers in their first year out of law school.

Solution We first interpret the question in terms of a random variable. Let X be the income of a lawyer selected at random from among all currently practicing first-year lawyers in the United States. We are given a sample of 1,000 values of X, and we are asked to find the expected value of X. First, we use the midpoints of the income brackets to set up a relative frequency distribution for X:

x	25,000	35,000	45,000	55,000	65,000	75,000	85,000
$P(X = x)$	.02	.08	.23	.40	.17	.07	.03

Our estimate for $E(X)$ is then

$$E(X) = \sum_i x_i \cdot P(X = x_i)$$

$$= (25,000)(.02) + (35,000)(.08) + (45,000)(.23) + (55,000)(.40)$$
$$+ (65,000)(.17) + (75,000)(.07) + (85,000)(.03) = \$54,500.$$

Thus, $E(X)$ is approximately \$54,500. That is, the average income of all currently practicing first-year lawyers in the United States is approximately \$54,500.

FAQs

Recognizing When to Compute the Mean and When to Compute the Expected Value

Q : *When am I supposed to compute the mean (add the values of X and divide by n) and when am I supposed to use the expected value formula?*

A : The formula for the mean (adding and dividing by the number of observations) is used to compute the mean of a sequence of random scores, or sampled values of X. If, on the other hand, you are given the probability distribution for X (even if it is only an estimated probability distribution), then you need to use the expected value formula.

8.3 EXERCISES

▼ more advanced   ◆ challenging
T indicates exercises that should be solved using technology

Compute the mean, median, and mode of the data samples in Exercises 1–8. [**HINT:** See Quick Examples 1–5.]

1. $-1, 5, 5, 7, 14$

2. $2, 6, 6, 7, -1$

3. $2, 5, 6, 7, -1, -1$

4. $3, 1, 6, -3, 0, 5$

5. $\frac{1}{2}, \frac{3}{2}, -4, \frac{5}{4}$

6. $-\frac{3}{2}, \frac{3}{8}, -1, \frac{5}{2}$

7. $2.5, -5.4, 4.1, -0.1, -0.1$

8. $4.2, -3.2, 0, 1.7, 0$

9. ▼ Give a sample of six scores with mean 1 and with median $\neq$ mean. (Arrange the scores in ascending order.)

10. ▼ Give a sample of five scores with mean 100 and median 1. (Arrange the scores in ascending order.)

In Exercises 11–16, calculate the expected value of X for the given probability distribution. [**HINT:** See Quick Example 6.]

11.

x	0	1	2	3
$P(X = x)$	.5	.2	.2	.1

12.

x	1	2	3	4
$P(X = x)$	.1	.2	.5	.2

13.

x	10	20	30	40
$P(X = x)$	$\frac{15}{50}$	$\frac{20}{50}$	$\frac{10}{50}$	$\frac{5}{50}$

14.

x	2	4	6	8
$P(X = x)$	$\frac{1}{20}$	$\frac{15}{20}$	$\frac{2}{20}$	$\frac{2}{20}$

15.

x	-5	-1	0	2	5	10
$P(X = x)$	.2	.3	.2	.1	.2	0

16.

x	-20	-10	0	10	20	30
$P(X = x)$	.2	.4	.2	.1	0	.1

In Exercises 17–28, calculate the expected value of the given random variable X. [Exercises 23, 24, 27, and 28 assume familiarity with counting arguments and probability (see Section 7.4).] [**HINT:** See Quick Example 6.]

17. X is the number that faces up when a fair die is rolled.

18. X is a number selected at random from the set $\{1, 2, 3, 4\}$.

19. X is the number of tails that come up when a coin is tossed twice.

20. X is the number of tails that come up when a coin is tossed three times.

21. ▼ X is the higher number when two dice are rolled.

22. ▼ X is the lower number when two dice are rolled.

23. ▼ X is the number of red marbles that Suzan has in her hand after she selects four marbles from a bag containing four red marbles and two green ones.

24. ▼ X is the number of green marbles that Suzan has in her hand after she selects four marbles from a bag containing three red marbles and two green ones.

25. ▼ Twenty darts are thrown at a dartboard. The probability of hitting a bull's-eye is .1. Let X be the number of bull's-eyes hit.

26. ▼ Thirty darts are thrown at a dartboard. The probability of hitting a bull's-eye is $\frac{1}{5}$. Let X be the number of bull's-eyes hit.

27. T ▼ Select 5 cards without replacement from a standard deck of 52, and let X be the number of queens you draw.

28. T ▼ Select 5 cards without replacement from a standard deck of 52, and let X be the number of red cards you draw.

Applications

29. *Stock Market Gyrations* Following is a sample of the day-by-day change, rounded to the nearest 100 points, in the Dow Jones Industrial Average during 10 successive business days around the start of the financial crisis in October 2008:[22]

$-400, -500, -200, -700, -100, 900, -100, -700, 400, -100$

Compute the mean and median of the given sample. Fill in the blank: There were as many days with a change in the Dow above _____ points as there were with changes below that. [**HINT:** See Quick Examples 1–5.]

30. *Stock Market Gyrations* Following is a sample of the day-by-day change, rounded to the nearest 100 points, in the Dow Jones Industrial Average during 10 successive business days around the start of the financial crisis in October 2008:[23]

$-100, 400, -200, -500, 200, -300, -200, 900, -100, 200$

Compute the mean and median of the given sample. Fill in the blank: There were as many days with a change in the Dow above _____ points as there were with changes below that. [**HINT:** See Quick Examples 1–5.]

31. *Gold* The following figures show the price of gold per ounce, in dollars, for the 10-business-day period Dec. 7–Dec. 18, 2015:[24]

$$1,076, 1,072, 1,081, 1,071, 1,072,$$
$$1,068, 1,062, 1,075, 1,049, 1,062$$

Find the sample mean, median, and mode(s). What do your answers tell you about the price of gold?

---

[22] Source: http://finance.google.com.

[23] *Ibid.*

[24] Prices rounded to the nearest $1. Source: www.kitco.com/gold .londonfix.html.

32. Silver The following figures show the price of silver per ounce, in dollars, for the 10-business-day period Dec. 7– Dec. 18, 2015:[25]

14.5, 14.2, 14.3, 14.2, 14.0, 13.7, 13.7, 13.7, 14.1, 13.8

Find the sample mean, median, and mode(s). What do your answers tell you about the price of silver?

33. Supermarkets A survey of 52 U.S. supermarkets yielded the following relative frequency table, where X is the number of checkout lanes at a randomly chosen supermarket:[26]

x	1	2	3	4	5	6	7	8	9	10
$P(X = x)$	.01	.04	.04	.08	.10	.15	.25	.20	.08	.05

  a. Compute $\mu = E(X)$, and interpret the result. [**HINT:** See Example 3.]
  b. Which is larger: $P(X < \mu)$ or $P(X > \mu)$? Interpret the result.

34. Video Arcades Your company, *Sonic Video, Inc.*, has conducted research that shows the following probability distribution, where X is the number of video arcades in a randomly chosen city with more than 500,000 inhabitants:

x	0	1	2	3	4	5	6	7	8	9
$P(X = x)$	.07	.09	.35	.25	.15	.03	.02	.02	.01	.01

  a. Compute $\mu = E(X)$, and interpret the result. [**HINT:** See Example 3.]
  b. Which is larger: $P(X < \mu)$ or $P(X > \mu)$? Interpret the result.

35. School Enrollment The following table shows the approximate numbers of school goers in the United States (residents who attended some educational institution) in 1998, broken down by age group:[27]

Age	3–6.9	7–12.9	13–16.9	17–22.9	23–26.9	27–42.9
Population (millions)	12	24	15	14	2	5

Use the rounded midpoints of the given measurement classes to compute the probability distribution of the age X of a school goer. (Round probabilities to two decimal places.) Hence, compute the expected value of X. What information does the expected value give about residents enrolled in schools?

36. School Enrollment Repeat Exercise 35, using the following data from 1980:[28]

Age	3–6.9	7–12.9	13–16.9	17–22.9	23–26.9	27–42.9
Population (millions)	8	20	11	13	1	3

37. Population Age in Mexico The following chart shows the ages of 250 randomly selected residents of Mexico:[29]

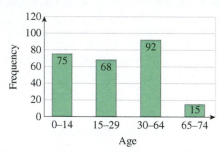

Use the relative frequency distribution based on the (rounded) midpoints of the given measurement classes to obtain an estimate of the average age of a resident in Mexico. (Round the answer to one decimal place.) [**HINT:** See Example 6.]

38. Population Age in the United States Repeat Exercise 37, using the following chart, which shows the ages of 250 randomly selected residents of the United States:[30]

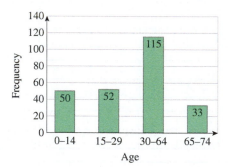

39. 2010 Income Distribution up to $100,000 The following table shows the distribution of household incomes in 2010[31] for a sample of 1,000 households in the United States with incomes up to $100,000:

Income Bracket ($)	0– 19,999	20,000– 39,999	40,000– 59,999	60,000– 79,999	80,000– 99,999
Households	240	290	180	170	120

---

[25] Prices rounded to the nearest $0.10. Source: www.kitco.com/gold .londonfix.html.

[26] Sources: J. T. McClave, P. G. Benson, T. Sincich, *Statistics for Business and Economics,* 7th Ed. (Prentice Hall, 1998), p. 177; W. Chow et al., "A model for predicting a supermarket's annual sales per square foot," Graduate School of Management, Rutgers University.

[27] Data are approximate. Source: Statistical Abstract of the United States: 2000.

[28] *Ibid.*

[29] Based on population distribution in 2010. Source: Instituto Nacional de Estadística y Geografía (www.inegi.org.mx).

[30] *Ibid.*

[31] Based on actual income distribution in 2010. Source: U.S. Census Bureau, Current Population Survey, 2010 American Community Survey (www.census.gov).

Use this information to estimate, to the nearest $1,000, the average household income for such households. [**HINT:** See Example 6.]

40. *2003 Income Distribution up to $100,000* Repeat Exercise 39, using the following data[32] from a sample of 1,000 households in the United States in 2003:

Income Bracket ($)	0–19,999	20,000–39,999	40,000–59,999	60,000–79,999	80,000–99,999
Households	270	280	200	150	100

Highway Safety *Exercises 41–44 are based on the following table, which shows crashworthiness ratings for several categories of motor vehicles.[33] In all of these exercises, take X as the crash-test rating of a small car, Y as the crash-test rating for a small SUV, and so on as shown in the table.*

	Number Tested	Overall Frontal Crash Test Rating			
		3 (Good)	2 (Acceptable)	1 (Marginal)	0 (Poor)
Small Cars X	16	1	11	2	2
Small SUVs Y	10	1	4	4	1
Medium SUVs Z	15	3	5	3	4
Passenger Vans U	13	3	0	3	7
Midsize Cars V	15	3	5	0	7
Large Cars W	19	9	5	3	2

41. ▼ Compute the probability distributions and expected values of X and Y. On the basis of the results, which of the two types of vehicles performed better in frontal crashes?

42. ▼ Compute the probability distributions and expected values of Z and V. On the basis of the results, which of the two types of vehicles performed better in frontal crashes?

43. [T] ▼ On the basis of expected values, which of the following categories performed best in crash tests: small cars, midsize cars, or large cars?

44. [T] ▼ On the basis of expected values, which of the following categories performed best in crash tests: small SUVs, medium SUVs, or passenger vans?

45. ▼ *Roulette* A roulette wheel has the numbers 1 through 36, 0, and 00. Half of the numbers from 1 through 36 are red,

and a bet on red pays even money (that is, if you bet $1 and win, you will get back your $1 plus another $1). How much do you expect to win with a $1 bet on red? [**HINT:** See Example 4.]

46. ▼ *Roulette* A roulette wheel has the numbers 1 through 36, 0, and 00. A bet on two numbers pays 17 to 1 (that is, if you bet $1 and one of the two numbers you bet comes up, you get back your $1 plus another $17). How much do you expect to win with a $1 bet on two numbers? [**HINT:** See Example 4.]

47. *Teenage Pastimes* According to a study,[34] the probability that a randomly selected teenager shopped at a mall at least once during a week was .63. How many teenagers in a randomly selected group of 40 would you expect to shop at a mall during the next week? [**HINT:** See Example 5.]

48. *Other Teenage Pastimes* According to the study referred to in Exercise 47, the probability that a randomly selected teenager played a computer game at least once during a week was .48. How many teenagers in a randomly selected group of 30 would you expect to play a computer game during the next 7 days? [**HINT:** See Example 5.]

49. ▼ *Manufacturing* Your manufacturing plant produces air bags, and it is known that 10% of them are defective. A random collection of 20 air bags is tested.
  a. How many of them would you expect to be defective?
  b. In how large a sample would you expect to find 12 defective air bags?

50. ▼ *Spiders* Your pet tarantula, Spider, has a .12 probability of biting an acquaintance who comes into contact with him. Next week, you will be entertaining 20 friends (all of whom will come into contact with Spider).
  a. How many guests should you expect Spider to bite?
  b. At your last party, Spider bit 6 of your guests. Assuming that Spider bit the expected number of guests, how many guests did you have?

Exercises 51 and 52 assume familiarity with counting arguments and probability (see Section 7.4).

51. ▼ *Camping* Kent's Tents has four red tents and three green tents in stock. Karin selects four of them at random. Let X be the number of red tents she selects. Give the probability distribution of X and find the expected number of red tents selected.

52. ▼ *Camping* Kent's Tents has five green knapsacks and four yellow ones in stock. Curt selects four of them at random. Let X be the number of green knapsacks he selects. Give the probability distribution of X, and find the expected number of green knapsacks selected.

---

[32] Based on actual income distribution in 2003 (not adjusted for inflation). Source: U.S. Census Bureau, Current Population Survey, 2004 Annual Social and Economic Supplement (www.census.gov).

[33] Ratings are by the Insurance Institute for Highway Safety. Sources: Oak Ridge National Laboratory: "An Analysis of the Impact of Sport Utility Vehicles in the United States," Stacy C. Davis, Lorena F. Truett, August 2000, Insurance Institute for Highway Safety (www-cta.ornl.gov/Publications/Final SUV report.pdf, www.highwaysafety.org/vehicle_ratings).

[34] Source: Rand Youth Poll/Teen-Age Research Unlimited/*New York Times*, March 14, 1998, p. D1.

Elimination Tournaments In an elimination tournament the teams are arranged in opponent pairs for the first round, and the winner of each round goes on to the next round until the champion emerges. The following diagram illustrates a 16-team tournament bracket, in which the 16 participating teams are arranged on the left under Round 1 and the winners of each round are added as the tournament progresses. The top team in each game is considered the "home" team, so the top-to-bottom order matters.

To seed *a tourment means to select which teams to play each other in the first round according to their preliminary ranking. For instance, in professional tennis and NCAA basketball the seeding is set up in the following order based on the preliminary rankings: 1 versus 16, 8 versus 9, 5 versus 12, 4 versus 13, 6 versus 11, 3 versus 14, 7 versus 10, and 2 versus 15.*[35] *Exercises 53–56 are based on various types of elimination tournaments.*

53. Someone offers you the following bet: If a randomly chosen seeding of a 16-team tournament results in the top-ranked team playing the bottom-ranked team, the second-ranked team playing the second-lowest ranked team, and so on, you win $1 million; otherwise, you lose $1. What are your expected winnings on this bet? [**HINT:** See Exercise 37 in Section 7.4.]

54. Someone offers you the following bet: If a randomly chosen seeding of an 8-team tournament results in each team playing a team with adjacent ranking, you win $100; otherwise, you lose $1. What are your expected winnings on this bet? [**HINT:** See Exercise 38 in Section 7.4.]

55. ▼ In 2014, after the NCAA basketball 64-team tournament had already been seeded, **Quicken Loans**, backed by investor

Warren Buffett, offered a billion-dollar prize for picking all the winners.[36] If 50,000,000 people entered the contest by picking winners at random, how much money did Quicken Loans expect to have to pay out? [**HINT:** See Exercise 39 in Section 7.4.]

56. ▼ Refer to Exercise 55. An *upset* occurs when a team beats a higher-ranked team. Suppose that, on average, there are 20 upsets in the first four rounds of the NCAA tournament each year. Assuming that the higher-ranked team always has the same probability of winning and that the outcomes of the games are independent, what is the probability that the higher-ranked team will win?

57. [T] ▼ *Stock Portfolios* You are required to choose between two stock portfolios: *FastForward Funds* and *SolidState Securities*. Stock analysts have constructed the following probability distributions for next year's rate of return for the two funds.

FastForward Funds

Rate of Return	−0.4	−0.3	−0.2	−0.1	0	0.1	0.2	0.3	0.4
Probability	.015	.025	.043	.132	.289	.323	.111	.043	.019

SolidState Securities

Rate of Return	−0.4	−0.3	−0.2	−0.1	0	0.1	0.2	0.3	0.4
Probability	.012	.023	.050	.131	.207	.330	.188	.043	.016

Which of the two funds gives the higher expected rate of return?

58. [T] ▼ *Risk Management* Before making your final decision whether to invest in *FastForward Funds* or *SolidState Securities* (see Exercise 57), you consult your colleague in the risk management department of your company. She informs you that, in the event of a stock market crash, the following probability distributions for next year's rate of return would apply:

FastForward Funds

Rate of Return	−0.8	−0.7	−0.6	−0.5	−0.4	−0.2	−0.1	0	0.1
Probability	.028	.033	.043	.233	.176	.230	.111	.044	.102

SolidState Securities

Rate of Return	−0.8	−0.7	−0.6	−0.5	−0.4	−0.2	−0.1	0	0.1
Probability	.033	.036	.038	.167	.176	.230	.211	.074	.035

Which of the two funds offers the lower risk in case of a market crash?

---

[35] Source: www.wikipedia.com, www.ncaa.com.

[36] No one won the prize, and the offer was scrapped the following year as a result of a series of lawsuits and countersuits by Yahoo, SCA Promotions (a sweepstakes company), and Berkshire-Hathaway. Sources: http://abcnews.go.com/Sports/warren-buffet-backs-billion-dollar-march-madness-challenge/story?id=21615743, http://money.cnn.com/2015/03/12/news/buffett-ncaa-bracket-bet/index.html.

59. ◆ *Insurance Schemes* The *Acme Insurance Company* is launching a drive to generate greater profits, and it decides to insure racetrack drivers against wrecking their cars. The company's research shows that, on average, a racetrack driver races 4 times a year and has a 1 in 10 chance of wrecking a vehicle, worth an average of $100,000, in every race. The annual premium is $5,000, and Acme automatically drops any driver who is involved in an accident (after paying for a new car) but does not refund the premium. How much profit (or loss) can the company expect to earn from a typical driver in a year? [**HINT:** Use a tree diagram to compute the probabilities of the various outcomes.]

60. ◆ *Insurance* The *Blue Sky Flight Insurance Company* insures passengers against air disasters, charging a prospective passenger $20 for coverage on a single plane ride. In the event of a fatal air disaster, it pays out $100,000 to the named beneficiary. In the event of a nonfatal disaster, it pays out an average of $25,000 for hospital expenses. Given that the probability of a plane's crashing on a single trip[37] is .00000087, and assuming that a passenger involved in a plane crash has a .9 chance of being killed, determine the profit (or loss) per passenger that the insurance company expects to make on each trip. [**HINT:** Use a tree to compute the probabilities of the various outcomes.]

Communication and Reasoning Exercises

61. In a certain set of five scores, there are as many values above the mean as below it. It follows that
(A) The median and mean are equal.
(B) The mean and mode are equal.
(C) The mode and median are equal.
(D) The mean, mode, and median are all equal.

62. In a certain set of scores, the median occurs more often than any other score. It follows that
(A) The median and mean are equal.
(B) The mean and mode are equal.
(C) The mode and median are equal.
(D) The mean, mode, and median are all equal.

63. Your friend Charlesworth claims that the median of a collection of data is always close to the mean. Is he correct? If so, say why. If not, give an example to prove him wrong.

64. Your other friend Imogen asserts that Charlesworth is wrong and that it is the mode and the median that are always close to each other. Is she correct? If so, say why. If not, give an example to prove her wrong.

65. Must the expected number of times you hit a bull's-eye after 50 attempts always be a whole number? Explain.

66. Your statistics instructor tells you that the expected score of the upcoming midterm test is 75%. That means that 75% is the most likely score to occur—right?

67. ▼ Your grade in a recent midterm was 80%, but the class average was 83%. Most people in the class scored better than you—right?

68. ▼ Your grade in a recent midterm was 80%, but the class median was 100%. Your score was lower than the average score—right?

69. ▼ Slim tells you that the population mean is just the mean of a suitably large sample. Is he correct? Explain.

70. ▼ Explain how you can use a sample to estimate an expected value.

71. ▼ Following is an excerpt from a full-page ad by **MoveOn .org** in the *New York Times* criticizing President G.W. Bush:[38]

On Tax Cuts:

George Bush: ". . . Americans will keep, this year, an average of almost $1,000 more of their own money."

The Truth: Nearly half of all taxpayers get less than $100. And 31% of all taxpayers get nothing at all.

The statements referred to as "The Truth" contradict the statement attributed to President Bush—right? Explain.

72. ▼ Following is an excerpt from a five-page ad by WeissneggerForGov.org in *The Martian Enquirer* criticizing Supreme Martian Administrator, Gov. Red Davis:

On Worker Accommodation:

Gov. Red Davis: "The median size of Government worker habitats in Valles Marineris is at least 400 square feet."

Weissnegger: "The average size of a Government worker habitat in Valles Marineris is a mere 150 square feet."

The statements attributed to Weissnegger do not contradict the statement attributed to Gov. Davis—right? Explain.

73. ▼ Sonia has just told you that the expected household income in the United States is the same as the population mean of all U.S. household incomes. Clarify her statement by describing an experiment and an associated random variable X so that the expected household income is the expected value of X.

74. ▼ If X is a random variable, what is the difference between a sample mean of measurements of X and the expected value of X? Illustrate by means of an example.

---

[37] This was the probability of a passenger plane crashing per departure in 1990. (Source: National Transportation Safety Board)

[38] Source: Full-page ad in the *New York Times*, September 17, 2003, p. A25.

Variance and Standard Deviation of a Set of Scores

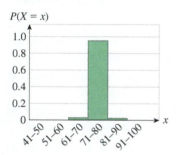

$P(X = x)$

Figure 5(a)

$P(X = x)$

Figure 5(b)

Your grade on a recent midterm was 68%; the class average was 72%. How do you stand in comparison with the rest of the class? If the grades were widely scattered, then your grade may be close to the mean, and a fair number of people may have done a lot worse than you (Figure 5(a)). If, on the other hand, almost all the grades were within a few points of the average, then your grade may not be much higher than the lowest grade in the class (Figure 5(b)).

This scenario suggests that it would be useful to have a way of measuring not only the central tendency of a set of scores (mean, median, or mode) but also the amount of "scatter" or "dispersion" of the data.

If the scores in our set are $x_1, x_2, \ldots, x_n$ and their mean is $\bar{x}$ (or μ in the case of a population mean), we are really interested in the distribution of the differences $x_i - \bar{x}$. We could compute the *average* of these differences, but this average will always be 0. (Why?) It is really the *sizes* of these differences that interest us, so we might try computing the average of the absolute values of the differences. This idea is reasonable, but it leads to technical difficulties that are avoided by a slightly different approach: The statistic we use is based on the average of the *squares* of the differences, as explained in the following definitions.

Population Variance and Standard Deviation

If the values $x_1, x_2, \ldots, x_n$ are all the measurements of X in the entire population, then the **population variance** is given by

$$\sigma^2 = \frac{(x_1 - \mu)^2 + (x_2 - \mu)^2 + \cdots + (x_n - \mu)^2}{n} = \frac{\sum_{i=1}^{n} (x_i - \mu)^2}{n}.$$

(Remember that μ is the symbol we use for the *population* mean.) The **population standard deviation** is the square root of the population variance:

$$\sigma = \sqrt{\sigma^2}.$$

Sample Variance and Standard Deviation

The **sample variance** of a sample $x_1, x_2, \ldots, x_n$ of n values of X is given by

$$s^2 = \frac{(x_1 - \bar{x})^2 + (x_2 - \bar{x})^2 + \cdots + (x_n - \bar{x})^2}{n - 1} = \frac{\sum_{i=1}^{n} (x_i - \bar{x})^2}{n - 1}.$$

The **sample standard deviation** is the square root of the sample variance:

$$s = \sqrt{s^2}.$$

Visualizing Small and Large Variance

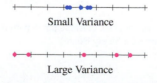

Small Variance

Large Variance

> **Quick Examples**
>
> 1. The sample variance of the scores 1, 2, 3, 4, 5 is the sum of the squares of the differences between the scores and the mean $\bar{x} = 3$, divided by $n - 1 = 4$:
>
> $$s^2 = \frac{(1-3)^2 + (2-3)^2 + (3-3)^2 + (4-3)^2 + (5-3)^2}{4}$$
>
> $$= \frac{10}{4} = 2.5,$$
>
> so
>
> $$s = \sqrt{2.5} \approx 1.58.$$
>
> 2. The population variance of the scores 1, 2, 3, 4, 5 is the sum of the squares of the differences between the scores and the mean $\mu = 3$, divided by $n = 5$:
>
> $$\sigma^2 = \frac{(1-3)^2 + (2-3)^2 + (3-3)^2 + (4-3)^2 + (5-3)^2}{5}$$
>
> $$= \frac{10}{5} = 2,$$
>
> so
>
> $$\sigma = \sqrt{2} \approx 1.41.$$

Q: *The population variance is the average of the squares of the differences between the values and the mean. But why do we divide by n − 1 instead of n when calculating the sample variance?*

A: In real-life applications we would like the variance we calculate from a sample to approximate the variance of the whole population. In statistical terms, we would like the expected value of the sample variance s^2 to be the same as the population variance σ^2. The sample variance s^2 as we have defined it is the "unbiased estimator" of the population variance σ^2 that accomplishes this task; if, instead, we divided by n in the formula for s^2, we would, on average, tend to underestimate the population variance. (See the online text on Sampling Distributions at the Website for further discussion of unbiased estimators.) Note that as the sample size gets larger and larger, the discrepancy between the formulas for s^2 and σ^2 becomes negligible; dividing by n gives almost the same answer as dividing by $n - 1$. It is traditional, nonetheless, to use the sample variance in preference to the population variance when working with samples, and we do that here. In practice we should not try to draw conclusions about the entire population from samples so small that the difference between the two formulas matters. As one book puts it, "If the difference between n and $n - 1$ ever matters to you, then you are probably up to no good anyway—e.g., trying to substantiate a questionable hypothesis with marginal data."[*]

[*] W. H. Press, S. A. Teukolsky, E. T. Vetterling, and B. P. Flannery, *Numerical Recipes: The Art of Scientific Computing,* Cambridge University Press, 2007.

Here's a simple example of calculating standard deviation.

EXAMPLE 1 Income

Following is a sample of the incomes (in thousands of dollars) of eight U.S. residents selected at random:[39]

$$50, 40, 60, 20, 90, 10, 30, 20.$$

Compute the sample mean and standard deviation, rounded to one decimal place. What percentage of the scores fall within one standard deviation of the mean? What percentage fall within two standard deviations of the mean?

Solution The sample mean is

$$\bar{x} = \frac{\sum_i x_i}{n} = \frac{50 + 40 + 60 + 20 + 90 + 10 + 30 + 20}{8} = \frac{320}{8} = 40.$$

The sample variance is

$$
\begin{aligned}
s^2 &= \frac{\sum_i (x_i - \bar{x})^2}{n - 1} \\
&= \frac{1}{7}[(50 - 40)^2 + (40 - 40)^2 + (60 - 40)^2 + (20 - 40)^2 \\
&\quad + (90 - 40)^2 + (10 - 40)^2 + (30 - 40)^2 + (20 - 40)^2] \\
&= \frac{1}{7}(100 + 0 + 400 + 400 + 2{,}500 + 900 + 100 + 400) \\
&= \frac{4{,}800}{7}.
\end{aligned}
$$

Thus, the sample standard deviation is

$$s = \sqrt{\frac{4{,}800}{7}} \approx 26.2. \qquad \textcolor{blue}{\text{Rounded to one decimal place}}$$

To ask which scores fall "within one standard deviation of the mean" is to ask which scores fall in the interval $[\bar{x} - s, \bar{x} + s]$, or about $[40 - 26.2, 40 + 26.2] = [13.8, 66.2]$. Six out of the eight scores fall in this interval, so the percentage of scores that fall within one standard deviation of the mean is $6/8 = .75$, or 75%.

For two standard deviations, the interval in question is $[\bar{x} - 2s, \bar{x} + 2s] \approx [40 - 52.4, 40 + 52.4] = [-12.4, 92.4]$, which includes all of the scores. In other words, 100% of the scores fall within two standard deviations of the mean.

Q: *In Example 1, 75% of the scores fell within one standard deviation of the mean, and all of them fell within two standard deviations of the mean. Is this typical?*

A: Actually, the percentage of scores within a number of standard deviations of the mean depends a great deal on the way the scores are distributed. There are two useful methods for *estimating* the percentage of scores that fall within any number of standard deviations of the mean. The first method applies to any set of data and is due to P.L. Chebyshev (1821–1894), while the second applies to "nice" sets of data and is based on the "normal distribution," which we shall discuss in Section 8.5.

---

[39] The sample is roughly statistically representative of the actual income distribution in the United States in 2010 for incomes up to $100,000. (See Exercise 35.) Source for income distribution: U.S. Census Bureau (www.census.gov).

Chebyshev's Rule

For any set of data the following statements are true:

At least 3/4 of the scores fall within two standard deviations of the mean (within the interval $[\bar{x} - 2s, \bar{x} + 2s]$ for samples or $[\mu - 2\sigma, \mu + 2\sigma]$ for populations).

At least 8/9 of the scores fall within three standard deviations of the mean (within the interval $[\bar{x} - 3s, \bar{x} + 3s]$ for samples or $[\mu - 3\sigma, \mu + 3\sigma]$ for populations).

At least 15/16 of the scores fall within four standard deviations of the mean (within the interval $[\bar{x} - 4s, \bar{x} + 4s]$ for samples or $[\mu - 4\sigma, \mu + 4\sigma]$ for populations).

$\vdots$

At least $1 - 1/k^2$ of the scores fall within k standard deviations of the mean (within the interval $[\bar{x} - ks, \bar{x} + ks]$ for samples or $[\mu - k\sigma, \mu + k\sigma]$ for populations).

Visualizing Chebyshev's Rule

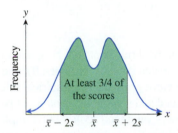
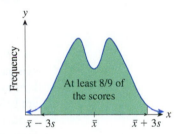

Empirical Rule*

*Unlike Chebyshev's rule, which is a precise theorem, the empirical rule is a rule of thumb that is intentionally vague about what exactly is meant by a "bell-shaped distribution" and "approximately such-and-such %." (As a result, the rule is often stated differently in different textbooks.) We will see in Section 8.5 that if the distribution is a *normal* one, the empirical rule translates to a precise statement.

For a set of data whose frequency distribution is bell-shaped and symmetric (see Figure 6), the following is true:

Approximately 68% of the scores fall within one standard deviation of the mean (within the interval $[\bar{x} - s, \bar{x} + s]$ for samples or $[\mu - \sigma, \mu + \sigma]$ for populations).

Approximately 95% of the scores fall within two standard deviations of the mean (within the interval $[\bar{x} - 2s, \bar{x} + 2s]$ for samples or $[\mu - 2\sigma, \mu + 2\sigma]$ for populations).

Approximately 99.7% of the scores fall within three standard deviations of the mean (within the interval $[\bar{x} - 3s, \bar{x} + 3s]$ for samples or $[\mu - 3\sigma, \mu + 3\sigma]$ for populations).

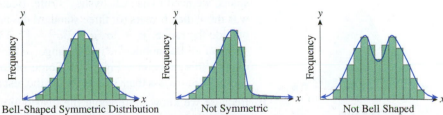

Figure 6

Visualizing the Empirical Rule

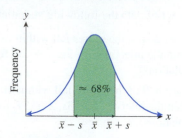

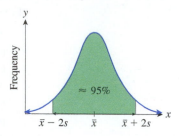

Quick Examples

3. If the mean of a sample is 20 with standard deviation $s = 2$, then at least 15/16, or 93.75%, of the scores lie within four standard deviations of the mean—that is, in the interval $[12, 28]$.

4. If the mean of a sample with a bell-shaped symmetric distribution is 20 with standard deviation $s = 2$, then approximately 95% of the scores lie in the interval $[16, 24]$.

The empirical rule could not be applied in Example 1. The distribution there is not symmetric (sketch it to see for yourself), and the fact that there were only eight scores limits the accuracy further. The empirical rule is, however, accurate in distributions that are bell shaped and symmetric, even if not perfectly so. Chebyshev's rule, on the other hand, is always valid (and applies in Example 1 in particular) but tends to be "overcautious" and in practice underestimates how much of a distribution lies in a given interval.

EXAMPLE 2 **Automobile Life**

The average life span of a Batmobile is 9 years, with a standard deviation of 2 years. My own Batmobile lasted less than 3 years before being condemned to the bat-junkyard.

a. Without any further knowledge about the distribution of Batmobile life spans, what can one say about the percentage of Batmobiles that last less than 3 years?

b. Refine the answer in part (a), assuming that the distribution of Batmobile life spans is bell shaped and symmetric.

Solution

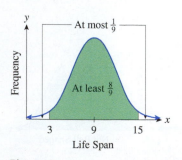

Figure 7

a. If we are given no further information about the distribution of Batmobile life spans, we need to use Chebyshev's rule. Because the life span of my Batmobile was more than 6 years (or three standard deviations) shorter than the mean, it lies outside the range $[\mu - 3\sigma, \mu + 3\sigma] = [3, 15]$. Because *at least* 8/9 of the life spans of all Batmobiles lie in this range, *at most* 1/9, or 11%, of the life spans lie outside this range (see Figure 7). Some of these, like the life span of my own Batmobile, are less than 3 years, while the rest are more than $\mu + 3\sigma = 15$ years.

b. Because we know more about the distribution now than we did in part (a), we can use the empirical rule and obtain sharper results. The empirical rule predicts that approximately 99.7% of the life spans of Batmobiles lie in the range

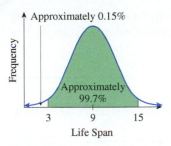

Figure 8

$[\mu - 3\sigma, \mu + 3\sigma] = [3, 15]$. Thus, approximately $1 - 99.7\% = 0.3\%$ of them lie outside that range. Because the distribution is symmetric, however, more can be said: Half of that 0.3%, or 0.15% of Batmobiles, will last longer than 15 years, while the other 0.15% are, like my own ill-fated Batmobile, doomed to a life span of less than 3 years (see Figure 8).

Variance and Standard Deviation of a Finite Random Variable

Recall that the expected value of a random variable X is a prediction of the average of a large sample of values of X. Can we similarly predict the variance of a large sample? Suppose we have a sample $x_1, x_2, \ldots, x_n$. If n is large, the sample and population variances are essentially the same, so we concentrate on the population variance, which is the average of the numbers $(x_i - \bar{x})^2$. This average can be predicted by using $E([X - \mu]^2)$, the expected value of $(X - \mu)^2$. In general, we make the following definition.

Variance and Standard Deviation of a Finite Random Variable

If X is a finite random variable taking on values $x_1, x_2, \ldots, x_n$, then the **variance** of X is

$$\sigma^2 = E([X - \mu]^2)$$
$$= (x_1 - \mu)^2 P(X = x_1) + (x_2 - \mu)^2 P(X = x_2) + \cdots + (x_n - \mu)^2 P(X = x_n)$$
$$= \sum_i (x_i - \mu)^2 P(X = x_i).$$

The **standard deviation** of X is then the square root of the variance:

$$\sigma = \sqrt{\sigma^2}.$$

To compute the variance from the probability distribution of X, first compute the expected value μ, and then compute the expected value of $(X - \mu)^2$.

Quick Example

5. The following distribution has expected value $\mu = E(X) = 2$:

x	-1	2	3	10
$P(X = x)$	.3	.5	.1	.1

The variance of X is

$$\sigma^2 = (x_1 - \mu)^2 P(X = x_1) + (x_2 - \mu)^2 P(X = x_2) + \cdots + (x_n - \mu)^2 P(X = x_n)$$
$$= (-1 - 2)^2(.3) + (2 - 2)^2(.5) + (3 - 2)^2(.1) + (10 - 2)^2(.1) = 9.2.$$

The standard deviation of X is

$$\sigma = \sqrt{9.2} \approx 3.03.$$

Note We can interpret the variance of X as the number we expect to get for the variance of a large sample of values of X, and similarly for the standard deviation. ∎

We can calculate the variance and standard deviation of a random variable using a tabular approach just as when we calculated the expected value in Example 3 in Section 8.3.

EXAMPLE 3   Variance of a Random Variable

Compute the variance and standard deviation for the following probability distribution:

x	10	20	30	40	50	60
$P(X = x)$	.2	.2	.3	.1	.1	.1

Solution We first compute the expected value, μ, in the usual way:

x	10	20	30	40	50	60	
$P(X = x)$	.2	.2	.3	.1	.1	.1	
$xP(X = x)$	2	4	9	4	5	6	$\mu = 30$

Next, we add an extra three rows:

- a row for the differences $(x - \mu)$, which we get by subtracting μ from the values of X
- a row for the squares $(x - \mu)^2$, which we obtain by squaring the values immediately above
- a row for the products $(x - \mu)^2 P(X = x)$, which we obtain by multiplying the values in the second and the fifth rows.

x	10	20	30	40	50	60	
$P(X = x)$	.2	.2	.3	.1	.1	.1	
$xP(X = x)$	2	4	9	4	5	6	$\mu = 30$
$x - \mu$	-20	-10	0	10	20	30	
$(x - \mu)^2$	400	100	0	100	400	900	
$(x - \mu)^2 P(X = x)$	80	20	0	10	40	90	$\sigma^2 = 240$

The sum of the values in the last row is the variance. The standard deviation is then the square root of the variance:

$$\sigma = \sqrt{240} \approx 15.49.$$

Note Chebyshev's rule and the empirical rule apply to random variables just as they apply to samples and populations, as we illustrate in the following example. ∎

EXAMPLE 4   Internet Commerce

Your newly launched company, *CyberPromo, Inc.*, sells computer games on the Internet.

a. Statistical research indicates that the life span of an Internet marketing company such as yours is symmetrically distributed with an expected value of 30 months and standard deviation of 4 months. Complete the following sentence:

There is (at least/at most/approximately)_____ a _____ percent chance that CyberPromo will still be around for more than 3 years.

b. How would the answer to part (a) be affected if the distribution of life spans was not known to be symmetric?

Solution

a. Do we use Chebyshev's rule or the empirical rule? Because the empirical rule requires that the distribution be both symmetric and bell shaped—not just symmetric—we cannot conclude that it applies here, so we are forced to use Chebyshev's rule instead.

Let X be the life span of an Internet commerce site. The expected value of X is 30 months, and the hoped-for life span of CyberPromo, Inc., is 36 months, which is 6 months, or $6/4 = 1.5$ standard deviations, above the mean. Chebyshev's rule tells us that X is within $k = 1.5$ standard deviations of the mean at least $1 - 1/k^2$ of the time; that is,

$$P(24 \leq X \leq 36) \geq 1 - \frac{1}{k^2} = 1 - \frac{1}{1.5^2} \approx .56.$$

In other words, at least 56% of all Internet marketing companies have life spans in the range of 24 to 36 months. Thus, *at most* 44% have life spans outside this range. Because the distribution is symmetric, at most 22% have life spans longer than 36 months. Thus, we can complete the sentence as follows:

There is <u>at most</u> a <u>22</u> percent chance that CyberPromo will still be around for more than 3 years.

b. If the given distribution was not known to be symmetric, how would this affect the answer? We saw above that regardless of whether the distribution is symmetric or not, at most 44% have life spans outside the range 24 to 36 months. Because the distribution is not symmetric, we cannot conclude that at most half of the 44% have life spans longer than 36 months, and all we can say is that *no more than 44% can possibly have life spans longer than 36 years.* In other words:

There is <u>at most</u> a <u>44</u> percent chance that CyberPromo will still be around for more than 3 years.

Variance and Standard Deviation of a Binomial Random Variable

We saw that there is an easy formula for the expected value of a binomial random variable: $\mu = np$, where n is the number of trials and p is the probability of success. Similarly, there is a simple formula for the variance and standard deviation.

Variance and Standard Deviation of a Binomial Random Variable

If X is a binomial random variable associated with n independent Bernoulli trials, each with probability p of success, then the variance and standard deviation of X are given by

$$\sigma^2 = npq \quad \text{and} \quad \sigma = \sqrt{npq}$$

where $q = 1 - p$ is the probability of failure.

> **Quick Example**
>
> **6.** If X is the number of successes in 20 Bernoulli trials with $p = .7$, then the standard deviation is $\sigma = \sqrt{npq} = \sqrt{(20)(.7)(.3)} \approx 2.05$.

For values of p near $1/2$ and large values of n, a binomial distribution is bell shaped and (nearly) symmetric; hence, the empirical rule applies. One rule of thumb is that we can use the empirical rule when both $np \geq 10$ and $nq \geq 10$.*

* Remember that the empirical rule gives only an *estimate* of probabilities. In Section 8.5 we give a more accurate approximation that takes into account the fact that the binomial distribution is not continuous.

EXAMPLE 5   Internet Commerce

You have calculated that there is a 40% chance that a hit on your web page results in a fee paid to your company, *CyberPromo, Inc.* Your web page receives 25 hits per day. Let X be the number of hits that result in payment of the fee ("successful hits").

a. What are the expected value and standard deviation of X?

b. Complete the following: On approximately 95 out of 100 days, I will get between ___ and ___ successful hits.

Solution

a. The random variable X is binomial with $n = 25$ and $p = .4$. To compute μ and σ, we use the formulas

$$\mu = np = (25)(.4) = 10 \text{ successful hits}$$
$$\sigma = \sqrt{npq} = \sqrt{(25)(.4)(.6)} \approx 2.45 \text{ hits.}$$

b. Because $np = 10 \geq 10$ and $nq = (25)(.6) = 15 \geq 10$, we can use the empirical rule, which tells us that there is an approximately 95% probability that the number of successful hits is within two standard deviations of the mean—that is, in the interval

$$[\mu - 2\sigma, \mu + 2\sigma] = [10 - 2(2.45), 10 + 2(2.45)] = [5.1, 14.9].$$

Thus, on approximately 95 out of 100 days, I will get between 5.1 and 14.9 successful hits.

FAQs

Recognizing When to Use the Empirical Rule or Chebyshev's Rule

Q : *How do I decide whether to use Chebyshev's rule or the empirical rule?*

A : Check to see whether the probability distribution you are considering is both symmetric and bell shaped. If so, you can use the empirical rule. If not, then you must use Chebyshev's rule. Thus, for instance, if the distribution is symmetric but not known to be bell shaped, you must use Chebyshev's rule.

8.4   EXERCISES

▼ more advanced   ◆ challenging

[T] indicates exercises that should be solved using technology

In Exercises 1–8, compute the (sample) variance and standard deviation of the given data sample. (You calculated the means in the Section 8.3 exercises. Round all answers to two decimal places.) [**HINT:** See Quick Examples 1 and 2.]

1. $-1, 5, 5, 7, 14$ **2.** $2, 6, 6, 7, -1$

3. $2, 5, 6, 7, -1, -1$ **4.** $3, 1, 6, -3, 0, 5$

5. $\dfrac{1}{2}, \dfrac{3}{2}, -4, \dfrac{5}{4}$ **6.** $-\dfrac{3}{2}, \dfrac{3}{8}, -1, \dfrac{5}{2}$

7. $2.5, -5.4, 4.1, -0.1, -0.1$ **8.** $4.2, -3.2, 0, 1.7, 0$

In Exercises 9–14, calculate the standard deviation of X for each probability distribution. (You calculated the expected values in the Section 8.3 exercises. Round all answers to two decimal places.) [**HINT:** See Quick Example 5.]

9.

x	0	1	2	3
$P(X = x)$	.5	.2	.2	.1

10.

x	1	2	3	4
$P(X = x)$	.1	.2	.5	.2

11.

x	10	20	30	40
$P(X = x)$	$\dfrac{3}{10}$	$\dfrac{2}{5}$	$\dfrac{1}{5}$	$\dfrac{1}{10}$

12.

x	2	4	6	8
$P(X = x)$	$\dfrac{1}{20}$	$\dfrac{15}{20}$	$\dfrac{2}{20}$	$\dfrac{2}{20}$

13.

x	-5	-1	0	2	5	10
$P(X = x)$	.2	.3	.2	.1	.2	0

14.

x	-20	-10	0	10	20	30
$P(X = x)$	.2	.4	.2	.1	0	.1

In Exercises 15–24, calculate the expected value, the variance, and the standard deviation of the given random variable X. (You calculated the expected values in the Section 8.3 exercises. Round all answers to two decimal places.)

15. X is the number that faces up when a fair die is rolled.

16. X is the number selected at random from the set $\{1, 2, 3, 4\}$.

17. X is the number of tails that come up when a coin is tossed twice.

18. X is the number of tails that come up when a coin is tossed three times.

19. ▼ X is the higher number when two dice are rolled.

20. ▼ X is the lower number when two dice are rolled.

21. ▼ X is the number of red marbles that Suzan has in her hand after she selects four marbles from a bag containing four red marbles and two green ones.

22. ▼ X is the number of green marbles that Suzan has in her hand after she selects four marbles from a bag containing three red marbles and two green ones.

23. ▼ Twenty darts are thrown at a dartboard. The probability of hitting a bull's-eye is .1. Let X be the number of bull's-eyes hit.

24. ▼ Thirty darts are thrown at a dartboard. The probability of hitting a bull's-eye is $\frac{1}{5}$. Let X be the number of bull's-eyes hit.

Applications

25. *Popularity Ratings* In your bid to be elected class representative, you have your election committee survey five randomly chosen students in your class and ask them to rank you on a scale of 0–10. Your rankings are 3, 2, 0, 9, 1.
   a. Find the sample mean and standard deviation. (Round your answers to two decimal places.) [**HINT:** See Example 1 and Quick Examples 1 and 2.]
   b. Assuming that the sample mean and standard deviation are indicative of the class as a whole, in what range does the empirical rule predict that approximately 68% of the class will rank you? What other assumptions must we make to use the rule?

26. *Popularity Ratings* Your candidacy for elected class representative is being opposed by Slick Sally. Your election committee has surveyed six of the students in your class and had them rank Sally on a scale of 0–10. The rankings were 2, 8, 7, 10, 5, 8.
   a. Find the sample mean and standard deviation. (Round your answers to two decimal places.) [**HINT:** See Example 1 and Quick Examples 1 and 2.]
   b. Assuming that the sample mean and standard deviation are indicative of the class as a whole, in what range does the empirical rule predict that approximately 95% of the class will rank Sally? What other assumptions must we make to use the rule?

27. *Unemployment* Following is a sample of unemployment rates (in percentage points) in the United States sampled from the period 1990–2004:[40]

$$4.2, 4.7, 5.4, 5.8, 4.9.$$

   a. Compute the mean and standard deviation of the given sample. (Round your answers to one decimal place.)
   b. Assuming that the distribution of unemployment rates in the population is symmetric and bell shaped, 95% of the time, the unemployment rate is between _____ and _____ percent.

---

[40] Sources for data: Bureau of Labor Statistics (BLS) (www.bls.gov).

28. *Unemployment* Following is a sample of unemployment rates among Hispanics (in percentage points) in the United States sampled from the period 1990–2004:[41]

$$7.7, 7.5, 9.3, 6.9, 8.6$$

a. Compute the mean and standard deviation of the given sample. (Round your answers to one decimal place.)

b. Assuming that the distribution of unemployment rates in the population of interest is symmetric and bell shaped, 68% of the time, the unemployment rate is between _____ and _____ percent.

29. *Stock Market Gyrations* Following is a sample of the day-by-day change, rounded to the nearest 100 points, in the Dow Jones Industrial Average during 10 successive business days around the start of the financial crisis in October 2008:[42]

$$-400, -500, -200, -700, -100, 900, -100, -700, 400, -100.$$

a. Compute the mean and standard deviation of the given sample. (Round your answers to the nearest whole number.)

b. Assuming that the distribution of day-by-day changes of the Dow during financial crises is symmetric and bell shaped, then the Dow falls by more than _____ points 16% of the time. What is the percentage of times in the sample that the Dow actually fell by more than that amount? [**HINT:** See Example 2(b).]

30. *Stock Market Gyrations* Following is a sample of the day-by-day change, rounded to the nearest 100 points, in the Dow Jones Industrial Average during 10 successive business days around the start of the financial crisis in October 2008:[43]

$$-100, 400, -200, -500, 200, -300, -200, 900, -100, 200.$$

a. Compute the mean and standard deviation of the given sample. (Round your answers to the nearest whole number.)

b. Assuming that the distribution of day-by-day changes of the Dow during financial crises is symmetric and bell shaped, then the Dow rises by more than _____ points 2.5% of the time. What is the percentage of times in the sample that the Dow actually rose by more than that amount? [**HINT:** See Example 2(b).]

31. **T** *Sport Utility Vehicles* Following are highway driving gas mileages of a selection of medium-sized sport utility vehicles (SUVs):[44]

$$17, 18, 17, 18, 21, 16, 21, 18, 16, 14, 15, 22, 17, 19, 17, 18.$$

a. Find the sample standard deviation (rounded to two decimal places).

b. In what gas mileage range does Chebyshev's inequality predict that at least 8/9 (approximately 89%) of the selection will fall?

c. What is the actual percentage of SUV models of the sample that fall in the range predicted in part (b)? Which gives the more accurate prediction of this percentage: Chebyshev's rule or the empirical rule?

32. **T** *Sport Utility Vehicles* Following are the city driving gas mileages of a selection of sport utility vehicles (SUVs):[45]

$$14, 15, 14, 15, 13, 16, 12, 14, 19, 18, 16, 16, 12, 15, 15, 13.$$

a. Find the sample standard deviation (rounded to two decimal places).

b. In what gas mileage range does Chebyshev's inequality predict that at least 75% of the selection will fall?

c. What is the actual percentage of SUV models of the sample that fall in the range predicted in part (b)? Which gives the more accurate prediction of this percentage: Chebyshev's rule or the empirical rule?

33. *Shopping Malls* A survey of all the shopping malls in your region yields the following probability distribution, where X is the number of movie theater screens in a selected mall:

Number of Movie Screens	0	1	2	3	4
Probability	.4	.1	.2	.2	.1

Compute the expected value μ and the standard deviation σ of X. (Round answers to two decimal places.) What percentage of malls have a number of movie theater screens within two standard deviations of μ?

34. *Pastimes* A survey of all the students in your school yields the following probability distribution, where X is the number of movies that a selected student has seen in the past week:

Number of Movies	0	1	2	3	4
Probability	.5	.1	.2	.1	.1

Compute the expected value μ and the standard deviation σ of X. (Round answers to two decimal places.) For what percentage of students is X within two standard deviations of μ?

35. **T** *2010 Income Distribution up to $100,000* The following table shows the distribution of household incomes in 2010[46] for a sample of 1,000 households in the United States with incomes up to $100,000:

Income ($1,000)	10	30	50	70	90
Households	240	290	180	170	120

[41] Sources for data: Bureau of Labor Statistics (BLS) (www.bls.gov).

[42] Source: http://finance.google.com.

[43] *Ibid.*

[44] Figures are the low-end of ranges for 1999 models tested. Source: Oak Ridge National Laboratory: "An Analysis of the Impact of Sport Utility Vehicles in the United States," Stacy C. Davis, Lorena F. Truett (August 2000)/Insurance Institute for Highway Safety (http://cta.ornl.gov/cta/ Publications/pdf/ORNL_TM_2000_147.pdf).

[45] *Ibid.*

[46] Based on actual income distribution in 2010. Source: U.S. Census Bureau, Current Population Survey, 2010 American Community Survey (www.census.gov).

Compute the expected value μ and the standard deviation σ of the associated random variable X. If we define a "lower-income" family as one whose income is more than one standard deviation below the mean and a "higher-income" family as one whose income is at least one standard deviation above the mean, what is the income gap between higher- and lower-income families in the United States? (Round your answers to the nearest $1,000.)

36. ▥ *2003 Income Distribution up to $100,000* Repeat Exercise 35, using the following data from a sample of 1,000 households in the United States in 2003:[47]

Income ($1,000)	10	30	50	70	90
Households	270	280	200	150	100

37. *Hispanic Employment: Male* The following table shows the approximate number of males of Hispanic origin employed in the United States in 2005, broken down by age group:[48]

Age	15–24.9	25–54.9	55–64.9
Employment (thousands)	16,000	13,000	1,600

    a. Use the rounded midpoints of the given measurement classes to compute the expected value and the standard deviation of the age X of a male Hispanic worker in the United States. (Round all probabilities and intermediate calculations to two decimal places.)

    b. Into what age interval does the empirical rule predict that 68% of all male Hispanic workers will fall? (Round answers to the nearest year.)

38. *Hispanic Employment: Female* Repeat Exercise 37, using the corresponding data for females of Hispanic origin:[49]

Age	15–24.9	25–54.9	55–64.9
Employment (thousands)	1,200	5,000	600

39. *Commerce* You have been told that the average life span of an Internet-based company is 2 years, with a standard deviation of 0.15 years. Further, the associated distribution is highly skewed (not symmetric). Your Internet company is now 2.6 years old. What percentage of all Internet-based companies have enjoyed a life span at least as long as yours? Your answer should contain one of the following phrases: *At least; At most; Approximately.* [**HINT**: See Example 2.]

40. *Commerce* You have been told that the average life span of a car-compounding service is 3 years, with a standard deviation of 0.2 years. Further, the associated distribution is symmetric but not bell shaped. Your car-compounding service is exactly 2.6 years old. What fraction of car-compounding services last at most as long as yours? Your answer should contain one of the following phrases: *At least; At most; Approximately.* [**HINT**: See Example 2.]

41. *Batmobiles* The average life span of a Batmobile is 9 years, with a standard deviation of 2 years.[50] Further, the probability distribution of the life spans of Batmobiles is symmetric but is not known to be bell shaped.

Because my old Batmobile has been sold as bat-scrap, I have decided to purchase a new one. According to the above information, there is

(A) at least   **(B)** at most   **(C)** approximately

a _____ percent chance that my new Batmobile will last 13 years or more.

42. *Spiderman Coupés* The average life span of a Spiderman Coupé is 8 years, with a standard deviation of 2 years. Further, the probability distribution of the life spans of Spiderman Coupés is not known to be bell shaped or symmetric. I have just purchased a brand-new Spiderman Coupé. According to the above information, there is

(A) at least   **(B)** at most   **(C)** approximately

a _____ percent chance that my new Spiderman Coupé will last for less than 4 years.

43. *Teenage Pastimes* According to a study,[51] the probability that a randomly selected teenager shopped at a mall at least once during a week was .63. Let X be the number of teenagers in a randomly selected group of 40 that will shop at a mall during the next week.

[47]Based on actual income distribution in 2003 (not adjusted for inflation). Source: U.S. Census Bureau, Current Population Survey, 2004 Annual Social and Economic Supplement (www.census.gov).

[48]Figures are rounded. Bounds for the age groups for the first and third categories were adjusted for computational convenience. Source: Bureau of Labor Statistics (ftp://ftp.bls.gov/pub/suppl/empsit.cpseed15.txt).

[49] *Ibid.*

[50] See Example 2.

[51] Source: Rand Youth Poll/Teen-Age Research Unlimited/*New York Times*, March 14, 1998, p. D1.

a. Compute the expected value and standard deviation of *X*. (Round answers to two decimal places.) [**HINT:** See Example 5.]

b. Fill in the missing quantity: There is an approximately 2.5% chance that ___ or more teenagers in the group will shop at a mall during the next week.

44. *Other Teenage Pastimes* According to the study referred to in Exercise 43, the probability that a randomly selected teenager played a computer game at least once during a week was .48. Let *X* be the number of teenagers in a randomly selected group of 30 who will play a computer game during the next 7 days.

a. Compute the expected value and standard deviation of *X*. (Round answers to two decimal places.) [**HINT:** See Example 5.]

b. Fill in the missing quantity: There is an approximately 16% chance that ___ or more teenagers in the group will play a computer game during the next 7 days.

45. ▼ *Teenage Marketing* In 2000, 22% of all teenagers in the United States had checking accounts.[52] Your bank, *TeenChex, Inc.,* is interested in targeting teenagers who do not already have a checking account.

a. If TeenChex selects a random sample of 1,000 teenagers, what number of teenagers *without* checking accounts can it expect to find? What is the standard deviation of this number? (Round the standard deviation to one decimal place.)

b. Fill in the missing quantities: There is an approximately 95% chance that between ___ and ___ teenagers in the sample will not have checking accounts. (Round answers to the nearest whole number.)

46. ▼ *Teenage Marketing* In 2000, 18% of all teenagers in the United States owned stocks or bonds.[53] Your brokerage company, *TeenStox, Inc.,* is interested in targeting teenagers who do not already own stocks or bonds.

a. If TeenStox selects a random sample of 2,000 teenagers, what number of teenagers who *do not* own stocks or bonds can it expect to find? What is the standard deviation of this number? (Round the standard deviation to one decimal place.)

b. Fill in the missing quantities: There is an approximately 99.7% chance that between ___ and ___ teenagers in the sample will not own stocks or bonds. (Round answers to the nearest whole number.)

47. ⊤ ▼ *Supermarkets* A survey of supermarkets in the United States yielded the following relative frequency table, where *X* is the number of checkout lanes at a randomly chosen supermarket:[54]

x	1	2	3	4	5	6	7	8	9	10
$P(X = x)$	.01	.04	.04	.08	.10	.15	.25	.20	.08	.05

a. Compute the mean, variance, and standard deviation (accurate to one decimal place).

b. As financial planning manager at *Express Lane Mart,* you wish to install a number of checkout lanes that is in the range of at least 75% of all supermarkets. What is this range, according to Chebyshev's inequality? What is the *least* number of checkout lanes you should install so as to fall within this range?

48. ⊤ ▼ *Video Arcades* Your company, *Sonic Video, Inc.,* has conducted research that shows the following probability distribution, where *X* is the number of video arcades in a randomly chosen city with more than 500,000 inhabitants:

x	0	1	2	3	4	5	6	7	8	9
$P(X = x)$	.07	.09	.35	.25	.15	.03	.02	.02	.01	.01

a. Compute the mean, variance, and standard deviation (accurate to one decimal place).

b. As CEO of Sonic Video, you wish to install a chain of video arcades in Sleepy City, U.S.A. The city council regulations require that the number of arcades be within the range shared by at least 75% of all cities. What is this range? What is the *largest* number of video arcades you should install so as to comply with this regulation?

Distribution of Wealth *If we model after-tax household income by a normal distribution, then the figures of a 1995 study imply the information in the following table, which should be used for Exercises 49–60.[55] Assume that the distribution of incomes in each country is bell shaped and symmetric.*

Country	United States	Canada	Switzerland	Germany	Sweden
Mean Household Income	$38,000	$35,000	$39,000	$34,000	$32,000
Standard Deviation	$21,000	$17,000	$16,000	$14,000	$11,000

49. If we define a "poor" household as one whose after-tax income is at least 1.3 standard deviations below the mean, what is the household income of a poor family in the United States?

---

[52] Source: Teen-Age Research Unlimited, January 25, 2001 (www.teenresearch.com).

[53] *Ibid.*

[54] Source: J. T. McClave, P. G. Benson, T. Sincich, *Statistics for Business and Economics*, 7th Ed. (Prentice Hall, 1998) p. 177; W. Chow et al., "A model for predicting a supermarket's annual sales per square foot," Graduate School of Management, Rutgers University.

[55] The data are rounded to the nearest $1,000 and based on a report published by the Luxembourg Income Study. The report shows after-tax income, including government benefits (such as food stamps) of households with children. Our figures were obtained from the published data by assuming a normal distribution of incomes. All data were based on constant 1991 U.S. dollars and converted foreign currencies (adjusted for differences in buying power). Source: Luxembourg Income Study/*New York Times*, August 14, 1995, p. A9.

50. If we define a "poor" household as one whose after-tax income is at least 1.3 standard deviations below the mean, what is the household income of a poor family in Switzerland?

51. If we define a "rich" household as one whose after-tax income is at least 1.3 standard deviations above the mean, what is the household income of a rich family in the United States?

52. If we define a "rich" household as one whose after-tax income is at least 1.3 standard deviations above the mean, what is the household income of a rich family in Sweden?

53. ▼ Refer to Exercise 49. Which of the five countries listed has the poorest households (i.e., the lowest cutoff for considering a household poor)?

54. ▼ Refer to Exercise 52. Which of the five countries listed has the wealthiest households (i.e., the highest cutoff for considering a household rich)?

55. ▼ Which of the five countries listed has the largest gap between rich and poor?

56. ▼ Which of the five countries listed has the smallest gap between rich and poor?

57. What percentage of U.S. families earned an after-tax income of $17,000 or less?

58. What percentage of U.S. families earned an after-tax income of $80,000 or more?

59. What was the after-tax income range of approximately 99.7% of all Germans?

60. What was the after-tax income range of approximately 99.7% of all Swedes?

T *Aging Exercises 61–68 are based on the following list, which shows the percentage of aging population (residents of age 65 and older) in each of the 50 states in 1990, 2000, and 2010:*[56]

1990

4, 9, 10, 10, 10, 10, 10, 11, 11, 11, 11, 11, 11,
11, 11, 12, 12, 12, 12, 12, 12, 13, 13, 13, 13, 13,
13, 13, 13, 13, 13, 13, 13, 13, 13, 14, 14, 14, 14,
14, 14, 14, 14, 15, 15, 15, 15, 15, 15, 18

2000

6, 9, 10, 10, 10, 11, 11, 11, 11, 11, 11, 11, 12,
12, 12, 12, 12, 12, 12, 12, 12, 12, 13, 13, 13,
13, 13, 13, 13, 13, 13, 13, 13, 13, 14, 14,
14, 14, 14, 14, 14, 15, 15, 15, 15, 16, 18

2010

8, 9, 10, 11, 11, 11, 12, 12, 12, 12, 12, 12, 12,
13, 13, 13, 13, 13, 13, 13, 13, 13, 14, 14, 14, 14,
14, 14, 14, 14, 14, 14, 14, 14, 14, 14, 14, 14,
14, 14, 14, 15, 15, 15, 15, 15, 16, 16, 17

61. **T** Compute the population mean and standard deviation for the 2000 and 2010 data.

62. **T** Compute the population mean and standard deviation for the 1990 and 2000 data.

63. The answer to Exercise 61 suggests that the 2010 population was:
(A) older on average and more diverse with respect to age
(B) older on average and less diverse with respect to age
(C) younger on average and more diverse with respect to age
(D) younger on average and less diverse with respect to age than the 2000 population.

64. The answer to Exercise 62 suggests that the 1990 population was:
(A) older on average and more diverse with respect to age
(B) older on average and less diverse with respect to age
(C) younger on average and more diverse with respect to age
(D) younger on average and less diverse with respect to age than the 2000 population.

65. **T** Compare the actual percentage of states whose aging population in 2010 was within one standard deviation of the mean to the percentage predicted by the empirical rule. Comment on your answer.

66. **T** Compare the actual percentage of states whose aging population in 1990 was within one standard deviation of the mean to the percentage predicted by the empirical rule. Comment on your answer.

67. **T** What was the actual percentage of states whose aging population in 2010 was within two standard deviations of the mean? Does Chebyshev's rule apply to the data? Explain.

68. **T** What was the actual percentage of states whose aging population in 2000 was within two standard deviations of the mean? Does Chebyshev's rule apply to the data? Explain.

Electric Grid Stress In early 2000 a Federal Energy Regulatory Commission order (FERC Order 888, first issued in 1996 and advocated by various energy companies, including **Enron***) went into effect, mandating that owners of power transmission lines make them available on the open market. The subsequent levels of stress on the electric grid are believed by many to have led to the Northeast blackout of August 14, 2003. Exercises 69–72 deal with the stress on the national electric grid before and after Order 888 went into effect.*

69. ▼ The following chart shows the approximate standard deviation of the power grid frequency, in 1/1,000 cycles per second, taken over 6-month periods. (0.9 is the average standard deviation.)[57]

[56] Percentages are rounded and listed in ascending order. Source: U.S. Census Bureau, Census 2000 Summary File 1 (www.census.gov/prod/2001pubs/c2kbr01-10.pdf), Age and Sex Composition: 2010 (www.census.gov/prod/cen2010/briefs/c2010br-03.pdf).

[57] Source: Robert Blohm, energy consultant and adviser to the North American Electric Reliability Council/*New York Times*, August 20, 2003, p. A16.

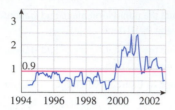

Which of the following statements are true? (More than one may be true.)

(A) The power grid frequency was at or below the mean until late 1999.

(B) The power grid frequency was more stable in mid-1999 than in 1995.

(C) The power grid frequency was more stable in mid-2002 than in mid-1999.

(D) The greatest fluctuations in the power grid frequency occurred in 2000–2001.

(E) The power grid frequency was more stable around January 1995 than around January 1999.

70. ▼ The following chart shows the approximate monthly means of the power grid frequency, in $1/1,000$ cycles per second. 0.0 represents the desired frequency of exactly 60 cycles per second.[58]

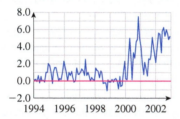

Which of the following statements are true? (More than one may be true.)

(A) Both the mean and the standard deviation show an upward trend from 2000 on.

(B) The mean, but not the standard deviation, shows an upward trend from 2000 on.

(C) The demand for electric power peaked in the second half of 2001.

(D) The standard deviation was larger in the second half of 2002 than in the second half of 1999.

(E) The mean of the monthly means in 2000 was lower than that for 2002, but the standard deviation of the monthly means was higher.

71. ▼ The following chart shows the approximate monthly means of the power grid frequency, in $1/1,000$ cycles per second. 0.0 represents the desired frequency of exactly

60 cycles per second. (Note that this is the same data as in Exercise 70 but over a different period of time.)[59]

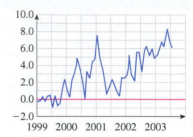

Which of the following statements are true? (More than one may be true.)

(A) Both the mean and the standard deviation show an upward trend from 2002 on.

(B) The mean, but not the standard deviation, shows an upward trend from 2002 on.

(C) The standard deviation, but not the mean, shows an upward trend from 2002 on.

(D) The standard deviation was greater in 2003 than in 2001.

(E) The mean of the monthly means in 2002 was higher than that for 2000, but the standard deviation of the monthly means was lower.

72. ▼ The following chart shows the number of transmission loading relief procedures (procedures undertaken to relieve excessive transmission loads by shifting power to other lines) per month.[60]

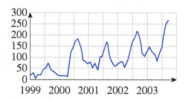

Which of the following statements are true? (More than one may be true.)

(A) Both the annual mean and the standard deviation show a significant upward trend from 2000 on.

(B) The annual mean, but not the standard deviation, shows a significant upward trend from 2000 on.

(C) The annual standard deviation, but not the mean, shows a significant upward trend from 2000 on.

(D) The standard deviation in the second half of 2000 is significantly greater than that in the first half.

(E) The standard deviation in 2000 is significantly greater than that in 2001.

[58] Source: Robert Blohm, energy consultant and adviser to the North American Electric Reliability Council/*New York Times*, August 20, 2003, p. A16.

[59] Source: Eric J. Lerner, "What's wrong with the electric grid?" *The Industrial Physicist,* Oct./Nov. 2003, American Institute of Physics (www.aip.org/tip/INPHFA/vol-9/iss-5/p8.pdf).
[60] *Ibid.*

Communication and Reasoning Exercises

73. Which is greater for a given set of data: the sample standard deviation or the population standard deviation? Explain.

74. Suppose you take larger and larger samples of a given population. Would you expect the sample and population standard deviations to get closer or farther apart? Explain.

75. In one Finite Math class, the average grade was 75, and the standard deviation of the grades was 5. In another Finite Math class, the average grade was 65, and the standard deviation of the grades was 20. What conclusions can you draw about the distributions of the grades in each class?

76. You are a manager in a precision manufacturing firm, and you must evaluate the performance of two employees. You do so by examining the quality of the parts they produce. One particular item should be 50.0 ± 0.3 mm long to be usable. The first employee produces parts that are an average of 50.1 mm long with a standard deviation of 0.15 mm. The second employee produces parts that are an average of 50.0 mm long with a standard deviation of 0.4 mm. Which employee do you rate higher? Why? (Assume that the empirical rule applies.)

77. ▼ If a finite random variable has an expected value of 10 and a standard deviation of 0, what must its probability distribution be?

78. ▼ If the values of X in a population consist of an equal number of 1s and -1s, what is its standard deviation?

79. ◆ Find an algebraic formula for the population standard deviation of a population $\{x, y\}$ of two scores $(x \le y)$.

80. ◆ Find an algebraic formula for the sample standard deviation of a sample $\{x, y\}$ of two scores $(x \le y)$.

8.5 Normal Distributions

Continuous Random Variables

Figure 9 shows the probability distributions for the number of successes in sequences of 10 and 15 independent Bernoulli trials, each with probability of success $p = .5$.

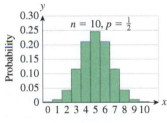

Figure 9(a)

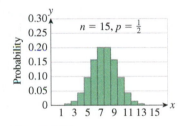

Figure 9(b)

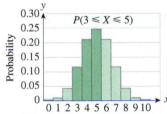

Figure 10

Because each column is 1 unit wide, its area is numerically equal to its height. Thus, the area of each rectangle can be interpreted as a probability. For example, in Figure 9(a) the area of the rectangle over $X = 3$ represents $P(X = 3)$. If we want to find $P(3 \le X \le 5)$, we can add up the areas of the three rectangles over 3, 4, and 5, shown shaded in Figure 10. Notice that if we add up the areas of *all* the rectangles in Figure 9(a), the total is 1 because $P(0 \le X \le 10) = 1$. We can summarize these observations.

Properties of the Probability Distribution Histogram

In a probability distribution histogram in which each column is 1 unit wide:

- The total area enclosed by the histogram is 1 square unit.

- $P(a \le X \le b)$ is the area enclosed by the rectangles lying between and including $X = a$ and $X = b$.

This discussion is motivation for considering another kind of random variable, one whose probability distribution is specified not by a bar graph, as above, but by the graph of a function.

Continuous Random Variable; Probability Density Function

A **continuous random variable** X may take on any real value whatsoever. The probabilities $P(a \leq X \leq b)$ are defined by means of a **probability density function**, a function whose graph lies above the x-axis with the total area between the graph and the x-axis being 1. The probability $P(a \leq X \leq b)$ is defined to be the area enclosed by the curve, the x-axis, and the lines $x = a$ and $x = b$ (see Figure 11).

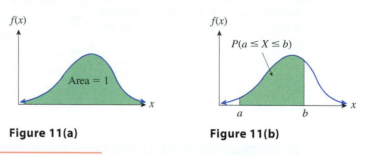

Figure 11(a)   **Figure 11(b)**

Notes

1. In Chapter 7 we defined probability distributions only for *finite* sample spaces. Because continuous random variables have infinite sample spaces, we need the definition above to give meaning to $P(a \leq X \leq b)$ if X is a continuous random variable.

2. If $a = b$, then $P(X = a) = P(a \leq X \leq a)$ is the area under the curve between the lines $x = a$ and $x = a$—no area at all! Thus, when X is a continuous random variable, $P(X = a) = 0$ for every value of a.

3. Whether we take the region in Figure 11(b) to include the boundary or not does not affect the area. The probability $P(a < X < b)$ is defined as the area strictly between the vertical lines $x = a$ and $x = b$ but is, of course, the same as $P(a \leq X \leq b)$, because the boundary contributes nothing to the area. When we are calculating probabilities associated with a continuous random variable,

$$P(a \leq X \leq b) = P(a < X \leq b) = P(a \leq X < b) = P(a < X < b). \quad ■$$

Normal Density Functions

Among all the possible probability density functions, there is an important class of functions called **normal density functions**, or **normal distributions**. The graph of a normal density function is bell shaped and symmetric, as the following figure shows. The formula for a normal density function is rather complicated looking:

$$f(x) = \frac{1}{\sigma\sqrt{2\pi}} e^{-(x-\mu)^2/(2\sigma^2)}.$$

The quantity μ is called the **mean** and can be any real number. The quantity σ is called the **standard deviation** and can be any positive real number. The number $e = 2.7182\ldots$ is a useful constant that shows up many places in mathematics, much as the constant π does. Finally, the constant $1/(\sigma\sqrt{2\pi})$ that appears in front is there to make the total area come out to be 1. We rarely use the actual formula in computations; instead, we use tables or technology.

Normal Density Function; Normal Distribution

A **normal density function**, or **normal distribution**, is a function of the form

$$f(x) = \frac{1}{\sigma\sqrt{2\pi}} e^{-(x-\mu)^2/(2\sigma^2)},$$

where μ is the mean and σ is the standard deviation. Its graph is bell shaped and symmetric, and has the following form:

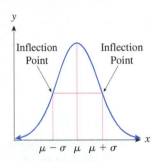

The "inflection points" are the points where the curve changes from bending in one direction to bending in another.*

Figure 12 shows the graphs of several normal density functions. The third of these has mean 0 and standard deviation 1, and is called the **standard normal distribution**. We use Z rather than X to refer to the standard normal variable.

Using Technology

The graphs in Figure 12 can be drawn on a TI-83/84 Plus or the Website grapher.

TI-83/84 Plus
Figure 12(a):
`Y₁=normalpdf(x,2,1)`
Figure 12(b):
`Y₁=normalpdf(x,0,2)`
Figure 12(c):
`Y₁=normalpdf(x)`

WW Website
www.WanerMath.com

→ Online Utilities

→ Function Evaluator and Grapher

Figure 12(a):
`normalpdf(x,2,1)`
Figure 12(b):
`normalpdf(x,0,2)`
Figure 12(c):
`normalpdf(x)`

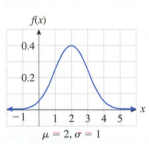

$\mu = 2, \sigma = 1$

Figure 12(a)

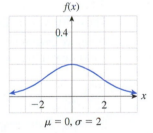

$\mu = 0, \sigma = 2$

Figure 12(b)

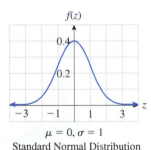

$\mu = 0, \sigma = 1$
Standard Normal Distribution

Figure 12(c)

Calculating Probabilities for the Standard Normal Distribution

The standard normal distribution has $\mu = 0$ and $\sigma = 1$. The corresponding variable is called the **standard normal variable**, which we always denote by Z. Recall that to calculate the probability $P(a \leq Z \leq b)$, we need to find the area under the distribution curve between the vertical lines $z = a$ and $z = b$. We can use the table in the Appendix to look up these areas, or we can use technology. Here is an example.

EXAMPLE 1 Standard Normal Distribution

Let Z be the standard normal variable. Calculate the following probabilities:

a. $P(0 \leq Z \leq 2.4)$ **b.** $P(0 \leq Z \leq 2.43)$

c. $P(-1.37 \leq Z \leq 2.43)$ **d.** $P(1.37 \leq Z \leq 2.43)$

Figure 13

Solution

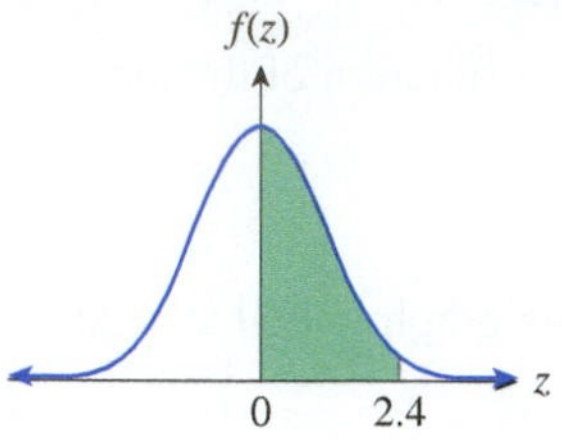

a. We are asking for the shaded area under the standard normal curve shown in Figure 13. We can find this area, correct to four decimal places, by looking at the table in the Appendix, which lists the area under the standard normal curve from $Z = 0$ to $Z = b$ for any value of b between 0 and 3.09. To use the table, write 2.4 as 2.40, and read the entry in the row labeled 2.4 and the column labeled 0.00 $(2.4 + 0.00 = 2.40)$. Here is the relevant portion of the table:

Z	0.00	0.01	0.02	0.03
2.3	.4893	.4896	.4898	.4901
→ 2.4	.4918	.4920	.4922	.4925
2.5	.4938	.4940	.4941	.4943

Thus, $P(0 \leq Z \leq 2.40) = .4918$.

b. The area we require can be read from the same portion of the table shown above. Write 2.43 as $2.4 + 0.03$, and read the entry in the row labeled 2.4 and the column labeled 0.03:

Z	0.00	0.01	0.02	0.03
2.3	.4893	.4896	.4898	.4901
→ 2.4	.4918	.4920	.4922	.4925
2.5	.4938	.4940	.4941	.4943

Thus, $P(0 \leq Z \leq 2.43) = .4925$.

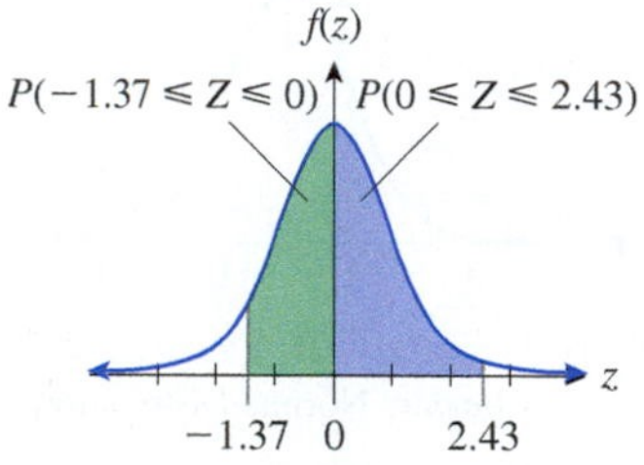

Figure 14

c. Here, we cannot use the table directly because the range $-1.37 \leq Z \leq 2.43$ does not start at 0. But we can break the area up into two smaller areas that start or end at 0:

$$P(-1.37 \leq Z \leq 2.43) = P(-1.37 \leq Z \leq 0) + P(0 \leq Z \leq 2.43).$$

In terms of the graph, we are splitting the desired area into two smaller areas (Figure 14). We already calculated the area of the right-hand piece in part (b):

$$P(0 \leq Z \leq 2.43) = .4925.$$

For the left-hand piece, the symmetry of the normal curve tells us that

$$P(-1.37 \leq Z \leq 0) = P(0 \leq Z \leq 1.37).$$

This we can find on the table. Look at the row labeled 1.3 and the column labeled 0.07, and read

$$P(-1.37 \leq Z \leq 0) = P(0 \leq Z \leq 1.37) = .4147.$$

Thus,

$$P(-1.37 \leq Z \leq 2.43) = P(-1.37 \leq Z \leq 0) + P(0 \leq Z \leq 2.43)$$
$$= .4147 + .4925$$
$$= .9072.$$

Using Technology

Technology can be used to calculate the probabilities in Example 1. For instance, the calculation for part (c) is as follows:

TI-83/84 Plus
`normalcdf(-1.37,2.43)`
(`normalcdf` is in 2ND VARS .)
[More details in the Technology Guide.]

Spreadsheet
`=NORMSDIST(2.43)`
`-NORMSDIST(-1.37)`
[More details in the Technology Guide.]

Website
www.WanerMath.com
→ Online Utilities
→ Normal Distribution
   Utility
Set up as shown, and press "Calculate Probability".

d. The range $1.37 \leq Z \leq 2.43$ does not contain 0, so we cannot use the technique of part (c). Instead, the corresponding area can be computed as the *difference* of two areas:

$$P(1.37 \leq Z \leq 2.43) = P(0 \leq Z \leq 2.43) - P(0 \leq Z \leq 1.37)$$
$$= .4925 - .4147$$
$$= .0778.$$

Calculating Probabilities for Any Normal Distribution

Although we have tables to compute the area under the *standard* normal curve, there are no readily available tables for nonstandard distributions. For example, if $\mu = 2$ and $\sigma = 3$, then how would we calculate $P(0.5 \leq X \leq 3.2)$? The following conversion formula provides a method for doing so.

Standardizing a Normal Distribution

If X has a normal distribution with mean μ and standard deviation σ, and if Z is the standard normal variable, then

$$P(a \leq X \leq b) = P\left(\frac{a - \mu}{\sigma} \leq Z \leq \frac{b - \mu}{\sigma}\right).$$

Quick Example

1. If $\mu = 2$ and $\sigma = 3$, then

$$P(0.5 \leq X \leq 3.2) = P\left(\frac{0.5 - 2}{3} \leq Z \leq \frac{3.2 - 2}{3}\right)$$
$$= P(-0.5 \leq Z \leq 0.4) = .1915 + .1554 = .3469.$$

To completely justify the above formula requires more mathematics than we shall discuss here. However, here is the main idea: If X is normal with mean μ and standard deviation σ, then $X - \mu$ is normal with mean 0 and standard deviation still σ, while $(X - \mu)/\sigma$ is normal with mean 0 and standard deviation 1. In other words, $(X - \mu)/\sigma = Z$. Therefore,

$$P(a \leq X \leq b) = P\left(\frac{a - \mu}{\sigma} \leq \frac{X - \mu}{\sigma} \leq \frac{b - \mu}{\sigma}\right) = P\left(\frac{a - \mu}{\sigma} \leq Z \leq \frac{b - \mu}{\sigma}\right).$$

EXAMPLE 2   Quality Control

Pressure gauges manufactured by *Precision Corp.* must be checked for accuracy before being placed on the market. To test a pressure gauge, a worker uses it to measure the pressure of a sample of compressed air known to be at a pressure of exactly 50 pounds per square inch. If the gauge reading is off by more than 1% (0.5 pounds), it is rejected. Assuming that the reading of a pressure gauge under these circumstances is a normal random variable with mean 50 and standard deviation 0.4, find the percentage of gauges rejected.

Using Technology

Technology can be used to calculate the probability $P(49.5 \leq X \leq 50.5)$ in Example 2:

TI-83/84 Plus

```
normalcdf(49.5,50.5,50,
0.4)
```
(`normalcdf` is in [2ND] [VARS].)
[More details in the Technology Guide.]

Spreadsheet

```
=NORMDIST(50.5,50,0.4,1)
-NORMDIST(49.5,50,0.4,1)
```
[More details in the Technology Guide.]

 Website

www.WanerMath.com

→ Online Utilities

→ Normal Distribution Utility

Set up as shown, and press "Calculate Probability".

Solution If X is the reading of the gauge, then X has a normal distribution with $\mu = 50$ and $\sigma = 0.4$. We are asking for $P(X < 49.5 \text{ or } X > 50.5) = 1 - P(49.5 \leq X \leq 50.5)$. We calculate

$$P(49.5 \leq X \leq 50.5) = P\left(\frac{49.5 - 50}{0.4} \leq Z \leq \frac{50.5 - 50}{0.4}\right) \quad \text{Standardize}$$

$$= P(-1.25 \leq Z \leq 1.25)$$

$$= 2 \cdot P(0 \leq Z \leq 1.25)$$

$$= 2(.3944) = .7888.$$

So $\quad P(X < 49.5 \text{ or } X > 50.5) = 1 - P(49.5 \leq X \leq 50.5)$

$$= 1 - .7888 = .2112.$$

In other words, about 21% of the gauges will be rejected.

In many applications we need to know the probability that a value of a normal random variable will lie within one standard deviation of the mean, or within two standard deviations, or within some number of standard deviations. To compute these probabilities, we first notice that, if X has a normal distribution with mean μ and standard deviation σ, then

$$P(\mu - k\sigma \leq X \leq \mu + k\sigma) = P(-k \leq Z \leq k)$$

by the standardizing formula. We can compute these probabilities for various values of k using the table in the Appendix, and we obtain the following results.

Probability of a Normal Distribution Being within k Standard Deviations of Its Mean

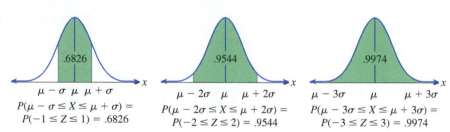

Now you can see where the empirical rule in Section 8.4 comes from! Notice also that the probabilities above are a good deal larger than the lower bounds given by Chebyshev's rule. Chebyshev's rule must work for distributions that are skew or any shape whatsoever.

EXAMPLE 3 Loans

The values of mortgage loans made by a certain bank one year were normally distributed with a mean of $120,000 and a standard deviation of $40,000.

a. What is the probability that a randomly selected mortgage loan was in the range of $40,000–$200,000?

b. You would like to state in your annual report that 50% of all mortgage loans were in a certain range with the mean in the center. What is that range?

Solution

a. We are asking for the probability that a loan was within two standard deviations ($80,000) of the mean. By the calculation done previously, this probability is .9544.

b. We look for the k such that

$$P(120,000 - k \cdot 40,000 \leq X \leq 120,000 + k \cdot 40,000) = .5$$

Because

$$P(120,000 - k \cdot 40,000 \leq X \leq 120,000 + k \cdot 40,000) = P(-k \leq Z \leq k),$$

we look in the Appendix to see for which k we have

$$P(0 \leq Z \leq k) = .25$$

so that $P(-k \leq Z \leq k) = .5$. That is, we look *inside* the table to see where 0.25 is, and find the corresponding k. We find

$$P(0 \leq Z \leq 0.67) = .2486$$

and

$$P(0 \leq Z \leq 0.68) = .2517.$$

Therefore, the k that we want is about halfway between 0.67 and 0.68—call it 0.675. This tells us that 50% of all mortgage loans were in the range

$$120,000 - 0.675 \cdot 40,000 = \$93,000$$

to

$$120,000 + 0.675 \cdot 40,000 = \$147,000.$$

Normal Approximation to a Binomial Distribution

You might have noticed that the histograms of some of the binomial distributions we have drawn (for example, those in Figure 9) have a very rough bell shape. In fact, in many cases it is possible to draw a normal curve that closely approximates a given binomial distribution.

Normal Approximation to a Binomial Distribution

If X is the number of successes in a sequence of n independent Bernoulli trials, with probability p of success in each trial, and if the range of values of X within three standard deviations of the mean lies entirely within the range 0 to n (the possible values of X), then

$$P(a \leq X \leq b) \approx P(a - 0.5 \leq Y \leq b + 0.5),$$

where Y has a normal distribution with the same mean and standard deviation as X; that is, $\mu = np$ and $\sigma = \sqrt{npq}$, where $q = 1 - p$.

Notes

1. The condition that $0 \leq \mu - 3\sigma < \mu + 3\sigma \leq n$ is satisfied if n is sufficiently large and p is not too close to 0 or 1; it ensures that most of the normal curve lies in the range 0 to n.

2. In the formula $P(a \le X \le b) \approx P(a - 0.5 \le Y \le b + 0.5)$, we assume that a and b are integers. The use of $a - 0.5$ and $b + 0.5$ is called the **continuity correction**. To see that it is necessary, think about what would happen if you wanted to approximate, say, $P(X = 2) = P(2 \le X \le 2)$. Should the answer be 0? ∎

Figures 15 and 16 show two binomial distributions with their normal approximations superimposed and illustrate how closely the normal approximation fits the binomial distribution.

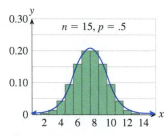

Figure 15

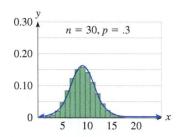

Figure 16

EXAMPLE 4   **Coin Tosses**

a. If you flip a fair coin 100 times, what is the probability of getting more than 55 heads or fewer than 45 heads?

b. What number of heads (out of 100) would make you suspect that the coin is not fair?

Solution

a. We are asking for

$$P(X < 45 \text{ or } X > 55) = 1 - P(45 \le X \le 55).$$

We *could* compute this by calculating

$$1 - [C(100, 45)(.5)^{45}(.5)^{55} + C(100, 46)(.5)^{46}(.5)^{54} + \cdots + C(100, 55)(.5)^{55}(.5)^{45}],$$

but we can much more easily *approximate* it by looking at a normal distribution with mean $\mu = 50$ and standard deviation $\sigma = \sqrt{(100)(.5)(.5)} = 5$. (Notice that three standard deviations above and below the mean is the range 35 to 65, which is well within the range of possible values for X, which is 0 to 100, so the approximation should be a good one.) Let Y have this normal distribution. Then

$$P(45 \le X \le 55) \approx P(44.5 \le Y \le 55.5)$$
$$= P(-1.1 \le Z \le 1.1)$$
$$= .7286.$$

Therefore,

$$P(X < 45 \text{ or } X > 55) \approx 1 - .7286 = .2714.$$

b. This is a deep question that touches on the concept of **statistical significance**: What evidence is strong enough to overturn a reasonable assumption (the assumption that the coin is fair)? Statisticians have developed sophisticated ways of answering this question, but we can look at one simple test now. Suppose we tossed a coin 100 times and got 66 heads. If the coin were fair, then $P(X > 65) \approx P(Y > 65.5) = P(Z > 3.1) \approx .001$. This is small enough to raise a reasonable doubt that the coin is fair. However, we should not be too surprised if we threw 56 heads because we can calculate $P(X > 55) \approx .1357$, which is not such a small probability. As we said, the actual tests of statistical significance are more sophisticated than this, but we shall not go into them.

8.5 EXERCISES

▼ more advanced   ◆ challenging
T indicates exercises that should be solved using technology

Note: Answers for Section 8.5 were computed by using the four-digit table in the Appendix and may differ slightly from the more accurate answers generated by using technology.

In Exercises 1–8, Z is the standard normal variable. Find the indicated probabilities. [**HINT:** See Example 1.]

1. $P(0 \leq Z \leq 0.5)$

2. $P(0 \leq Z \leq 1.5)$

3. $P(-0.71 \leq Z \leq 0.71)$

4. $P(-1.71 \leq Z \leq 1.71)$

5. $P(-0.71 \leq Z \leq 1.34)$

6. $P(-1.71 \leq Z \leq 0.23)$

7. $P(0.5 \leq Z \leq 1.5)$

8. $P(0.71 \leq Z \leq 1.82)$

In Exercises 9–14, X has a normal distribution with the given mean and standard deviation. Find the indicated probabilities. [**HINT:** See Quick Example 1.]

9. $\mu = 50, \sigma = 10$, find $P(35 \leq X \leq 65)$

10. $\mu = 40, \sigma = 20$, find $P(35 \leq X \leq 45)$

11. $\mu = 50, \sigma = 10$, find $P(30 \leq X \leq 62)$

12. $\mu = 40, \sigma = 20$, find $P(30 \leq X \leq 53)$

13. $\mu = 100, \sigma = 15$, find $P(110 \leq X \leq 130)$

14. $\mu = 100, \sigma = 15$, find $P(70 \leq X \leq 80)$

15. ▼ Find the probability that a normal variable takes on values within 0.5 standard deviations of its mean.

16. ▼ Find the probability that a normal variable takes on values within 1.5 standard deviations of its mean.

17. ▼ Find the probability that a normal variable takes on values more than $\frac{2}{3}$ standard deviations away from its mean.

18. ▼ Find the probability that a normal variable takes on values more than $\frac{5}{3}$ standard deviations away from its mean.

19. ▼ Suppose X is a normal random variable with mean $\mu = 100$ and standard deviation $\sigma = 10$. Find b such that $P(100 \leq X \leq b) = .3$. [**HINT:** See Example 3.]

20. ▼ Suppose X is a normal random variable with mean $\mu = 10$ and standard deviation $\sigma = 5$. Find b such that $P(10 \leq X \leq b) = .4$. [**HINT:** See Example 3.]

21. ▼ Suppose X is a normal random variable with mean $\mu = 100$ and standard deviation $\sigma = 10$. Find a such that $P(X \geq a) = .04$.

22. ▼ Suppose X is a normal random variable with mean $\mu = 10$ and standard deviation $\sigma = 5$. Find a such that $P(X \geq a) = .03$.

23. If you roll a die 100 times, what is the approximate probability that you will roll between 10 and 15 ones, inclusive? (Round your answer to two decimal places.) [**HINT:** See Example 4.]

24. If you roll a die 100 times, what is the approximate probability that you will roll between 15 and 20 ones, inclusive? (Round your answer to two decimal places.) [**HINT:** See Example 4.]

25. If you roll a die 200 times, what is the approximate probability that you will roll fewer than 25 ones? (Round your answer to two decimal places.)

26. If you roll a die 200 times, what is the approximate probability that you will roll more than 40 ones? (Round your answer to two decimal places.)

Applications

27. *SAT Scores* SAT test scores are normally distributed with a mean of 500 and a standard deviation of 100. Find the probability that a randomly chosen test-taker will score between 450 and 550. [**HINT:** See Example 3.]

28. *SAT Scores* SAT test scores are normally distributed with a mean of 500 and a standard deviation of 100. Find the probability that a randomly chosen test-taker will score 650 or higher. [**HINT:** See Example 3.]

29. *LSAT Scores* LSAT test scores are normally distributed with a mean of 151 and a standard deviation of 7. Find the probability that a randomly chosen test-taker will score between 137 and 158.

30. *LSAT Scores* LSAT test scores are normally distributed with a mean of 151 and a standard deviation of 7. Find the probability that a randomly chosen test-taker will score 144 or lower.

31. *IQ Scores* IQ scores (as measured by the Stanford-Binet intelligence test) are normally distributed with a mean of 100 and a standard deviation of 16. What percentage of the population has an IQ score between 110 and 140? (Round your answer to the nearest percentage point.)

32. *IQ Scores* Refer to Exercise 31. What percentage of the population has an IQ score between 80 and 90? (Round your answer to the nearest percentage point.)

33. *IQ Scores* Refer to Exercise 31. Find the approximate number of people in the United States (assuming a total population of 323,000,000) with an IQ of 120 or higher.

34. *IQ Scores* Refer to Exercise 31. Find the approximate number of people in the United States (assuming a total population of 323,000,000) with an IQ of 140 or higher.

35. *SAT Scores* SAT test scores are normally distributed with a mean of 500 and a standard deviation of 100. What score would place you in the top 5% of test-takers? [**HINT:** See Example 3.]

36. *LSAT Scores* LSAT test scores are normally distributed with a mean of 151 and a standard deviation of 7. What score would place you in the top 2% of test-takers? [**HINT:** See Example 3.]

37. *Baseball* The mean batting average in major league baseball is about 0.250. If batting averages are normally distributed, the standard deviation in the averages is 0.03, and there are 250 batters, what is the expected number of batters with an average of at least 0.400?

38. *Baseball* The mean batting average in major league baseball is about 0.250. If batting averages are normally distributed, the standard deviation in the averages is 0.05, and

there are 250 batters, what is the expected number of batters with an average of at least 0.400?[61]

39. *Marketing* Your pickle company rates its pickles on a scale of spiciness from 1 to 10. Market research shows that customer preferences for spiciness are normally distributed, with a mean of 7.5 and a standard deviation of 1. Assuming that you sell 100,000 jars of pickles, how many jars with a spiciness of 9 or above do you expect to sell?

40. *Marketing* Your hot sauce company rates its sauce on a scale of spiciness of 1 to 20. Market research shows that customer preferences for spiciness are normally distributed, with a mean of 12 and a standard deviation of 2.5. Assuming that you sell 300,000 bottles of sauce, how many bottles with a spiciness below 9 do you expect to sell?

Distribution of Income *If we model after-tax household income with a normal distribution, then the figures of a 1995 study imply the information in the following table, which should be used for Exercises 41–46.*[62] *Assume that the distribution of incomes in each country was normal, and round all percentages to the nearest whole number.*

Country	United States	Canada	Switzerland	Germany	Sweden
Mean Household Income ($)	38,000	35,000	39,000	34,000	32,000
Standard Deviation	21,000	17,000	16,000	14,000	11,000

41. What percentage of U.S. households had an income of $50,000 or more?

42. What percentage of German households had an income of $50,000 or more?

43. What percentage of Swiss households were either very wealthy (income at least $100,000) or very poor (income at most $12,000)?

---

[61] The last time that a batter ended the year with an average above 0.400 was in 1941. The batter was Ted Williams of the Boston Red Sox, and his average was 0.406. Over the years, as pitching and batting have improved, the standard deviation in batting averages has declined from around 0.05 when professional baseball began to around 0.03 by the end of the twentieth century. For a very interesting discussion of statistics in baseball and in evolution, see Stephen Jay Gould, *Full House: The Spread of Excellence from Plato to Darwin*, Random House, 1997.

[62] The data are rounded to the nearest $1,000 and are based on a report published by the Luxembourg Income Study. The report shows after-tax income, including government benefits (such as food stamps), of households with children. Our figures were obtained from the published data by assuming a normal distribution of incomes. All data were based on constant 1991 U.S. dollars and converted foreign currencies (adjusted for differences in buying power). Source: Luxembourg Income Study/*New York Times*, August 14, 1995, p. A9.

44. What percentage of Swedish households were either very wealthy (income at least $100,000) or very poor (income at most $12,000)?

45. Which country had a higher proportion of very poor families (income $12,000 or less): the United States or Canada?

46. Which country had a higher proportion of very poor families (income $12,000 or less): Canada or Switzerland?

47. ▼ *Comparing IQ Tests* IQ scores as measured by both the Stanford-Binet intelligence test and the Wechsler intelligence test have a mean of 100. The standard deviation for the Stanford-Binet test is 16, while that for the Wechsler test is 15. For which test do a smaller percentage of test-takers score less than 80? Why?

48. ▼ *Comparing IQ Tests* Referring to Exercise 47, for which test do a larger percentage of test-takers score more than 120?

49. ▼ *Product Repairs* The new copier your business bought lists a mean time between failures of 6 months, with a standard deviation of 1 month. One month after a repair, it breaks down again. Is this surprising? (Assume that the times between failures are normally distributed.)

50. ▼ *Product Repairs* The new computer your business bought lists a mean time between failures of 1 year, with a standard deviation of 2 months. Ten months after a repair, it breaks down again. Is this surprising? (Assume that the times between failures are normally distributed.)

Software Testing *Exercises 51–56 are based on the following information, gathered from student testing of a statistical software package called MODSTAT.*[63] *Students were asked to complete certain tasks using the software, without any instructions. The results were as follows. (Assume that the time for each task is normally distributed.)*

Task	Mean Time (minutes)	Standard Deviation
Task 1: Descriptive Analysis of Data	11.4	5.0
Task 2: Standardizing Scores	11.9	9.0
Task 3: Poisson Probability Table	7.3	3.9
Task 4: Areas under Normal Curve	9.1	5.5

51. Find the probability that a student will take at least 10 minutes to complete Task 1.

52. Find the probability that a student will take at least 10 minutes to complete Task 3.

53. ▼ Assuming that the time it takes a student to complete each task is independent of the others, find the probability that a student will take at least 10 minutes to complete each of Tasks 1 and 2.

54. ▼ Assuming that the time it takes a student to complete each task is independent of the others, find the probability that a student will take at least 10 minutes to complete each of Tasks 3 and 4.

55. ◆ It can be shown that if X and Y are independent normal random variables with means μ_X and μ_Y, and standard deviations σ_X and σ_Y respectively, then their sum $X + Y$ is also normally distributed and has mean $\mu = \mu_X + \mu_Y$ and standard deviation $\sigma = \sqrt{\sigma_X^2 + \sigma_Y^2}$. Assuming that the time it takes a student to complete each task is independent of the others, find the probability that a student will take at least 20 minutes to complete both Tasks 1 and 2.

56. ◆ Referring to Exercise 55, compute the probability that a student will take at least 20 minutes to complete both Tasks 3 and 4.

57. *Internet Access* In 2010, 71% of all households in the United States had Internet access.[64] Find the probability that, in a small town with 1,200 households, at least 840 had Internet access in 2010. [**HINT:** See Example 4.]

58. *Television Ratings* According to data from the **Nielsen Company**, there is a 1.8% chance that any television that is turned on during the time of the evening newscasts will be tuned to **ABC**'s evening news show.[65] Your company wishes to advertise on a local station carrying ABC that serves a community with 5,000 households that regularly watch TV during this time slot. Find the approximate probability that at least 100 households will be tuned in to the show. [**HINT:** See Example 4.]

59. *Aviation* The probability of a plane crashing on a single trip in 2010 was .00000276.[66] Find the approximate probability that in 10 million flights, there will be fewer than 35 crashes.

60. *Aviation* The probability of a plane crashing on a single trip in 1990 was .00000087. Find the approximate probability that in 100 million flights, there will be more than 110 crashes.

---

[63] Data are rounded to one decimal place. Source: *Student Evaluations of MODSTAT,* by Joseph M. Nowakowski, Muskingum College, New Concord, OH, 1997.

[64] Source: *Digital Nation: Expanding Internet Usage,* National Telecommunications and Information Administration, February 2011 (www.ntia.doc.gov/data).

[65] As of November 2011. Source: The Nielsen Company via the TVbytheNumbers website (http://tvbythenumbers.zap2it.com/category/ratings/evening-news).

[66] Figures are for scheduled commercial flights. Source for this exercise and the following three: National Transportation Safety Board (www.ntsb.gov).

61. ▼ *Insurance* Your company issues flight insurance. You charge $3, and in the event of a plane crash, you will pay out $1 million to the victim or his or her family. In 2010 the probability of a plane crashing on a single trip was .00000276. If 10 people per flight buy insurance from you, what was your approximate probability of losing money over the course of 10 million flights in 2010? [**HINT:** First determine how many crashes there must be for you to lose money.]

62. ▼ *Insurance* Refer back to Exercise 61. What is your approximate probability of losing money over the course of 100 million flights?

63. ◆ *Polls* In a certain political poll, each person polled has a 90% probability of telling his or her real preference. Suppose that 55% of the population really prefer candidate Goode and 45% prefer candidate Slick. First find the probability that a person polled will say that he or she prefers Goode. Then find the approximate probability that, if 1,000 people are polled, more than 52% will say that they prefer Goode.

64. ◆ *Polls* In a certain political poll, each person polled has a 90% probability of telling his or her real preference. Suppose that 1,000 people are polled and 51% say that they prefer candidate Goode, while 49% say that they prefer candidate Slick. Find the approximate probability that Goode could do at least this well if, in fact, only 49% prefer Goode.

65. ◆ *IQ Scores* Mensa is a club for people with high IQs. To qualify, you must be in the top 2% of the population. One way of qualifying is by having an IQ of at least 148, as measured by the Cattell intelligence test. Assuming that scores on this test are normally distributed with a mean of 100, what is the standard deviation? [**HINT:** Use the table in the Appendix "backward."]

66. ◆ *SAT Scores* Another way to qualify for Mensa (see Exercise 65) is to score at least 1,250 on the SAT [combined Critical Reading (Verbal, before March 2005) and Math scores], which puts you in the top 2%. Assuming that SAT scores are normally distributed with a mean of 1,000, what is the standard deviation? (See the hint for Exercise 65.)

Communication and Reasoning Exercises

67. Under what assumptions are the estimates in the empirical rule exact?

68. If X is a continuous random variable, what values can the quantity $P(X = a)$ have?

69. Which is larger for a continuous random variable: $P(X \le a)$ or $P(X < a)$?

70. Which of the following is greater: $P(X \le b)$ or $P(a \le X \le b)$?

71. ▼ A uniform continuous distribution is one with a probability density curve that is a horizontal line. If X takes on values between the numbers a and b with a uniform distribution, find the height of its probability density curve.

72. ▼ Which would you expect to have the greater variance: the standard normal distribution or the uniform distribution taking values between -1 and 1? Explain.

73. ◆ Which would you expect to have a density curve that is higher at the mean: the standard normal distribution or a normal distribution with standard deviation 0.5? Explain.

74. ◆ Suppose students must perform two tasks: Task 1 and Task 2. Which of the following would you expect to have a smaller standard deviation?
    (A) The time it takes a student to perform both tasks if the time it takes to complete Task 2 is independent of the time it takes to complete Task 1.
    (B) The time it takes a student to perform both tasks if students will perform similarly in both tasks.
    Explain.

KEY CONCEPTS

W www.WanerMath.com
Go to the Website to find a comprehensive and interactive Web-based summary of Chapter 8.

8.1 Random Variables and Distributions
Random variable; discrete vs. continuous random variable [p. 580]
Probability distribution of a finite random variable [p. 583]
Using measurement classes [p. 584]

8.2 Bernoulli Trials and Binomial Random Variables
Bernoulli trial [p. 591]
Binomial random variable [p. 592]
Probability distribution of binomial random variable:
$P(X = x) = C(n, x)p^x q^{n-x}$ [p. 594]

8.3 Measures of Central Tendency
Sample, sample mean; population, population mean [p. 599]

Median, mode [p. 600]
Expected value of a random variable:
$\mu = E(X) = \sum_i x_i \cdot P(X = x_i)$ [p. 603]
Expected value of a binomial random variable: $\mu = E(X) = np$ [p. 605]

8.4 Measures of Dispersion
Population variance:
$$\sigma^2 = \frac{\sum\limits_{i=1}^{n}(x_i - \mu)^2}{n}$$
Population standard deviation:
$\sigma = \sqrt{\sigma^2}$ [p. 612]
Sample variance:
$$s^2 = \frac{\sum\limits_{i=1}^{n}(x_i - \bar{x})^2}{n - 1}$$
Sample standard deviation: $s = \sqrt{s^2}$ [p. 612]
Chebyshev's rule [p. 615]
Empirical rule [p. 615]

Variance of a random variable:
$\sigma^2 = \sum_i (x_i - \mu)^2 P(X = x_i)$ [p. 617]
Standard deviation of X: $\sigma = \sqrt{\sigma^2}$ [p. 617]
Variance and standard deviation of a binomial random variable:
$\sigma^2 = npq, \sigma = \sqrt{npq}$ [p. 619]

8.5 Normal Distributions
Probability density function [p. 628]
Normal density function; normal distribution; standard normal distribution [p. 629]
Calculating probabilities based on the standard normal distribution [p. 629]
Standardizing a normal distribution [p. 631]
Calculating probabilities based on non-standard normal distributions [p. 631]
Normal approximation to a binomial distribution [p. 633]

REVIEW EXERCISES

In Exercises 1–6, find the probability distribution for the given random variable and draw a histogram.

1. A couple has two children; $X = $ the number of boys. (Assume an equal likelihood of a child being a boy or a girl.)

2. A couple has three children; $X = $ the number of girls. (Assume an equal likelihood of a child being a boy or a girl.)

3. A four-sided die (with sides numbered 1 through 4) is rolled twice in succession; $X = $ the sum of the two numbers.

4. 48.2% of Xbox players are in their teens, 38.6% are in their twenties, 11.6% are in their thirties, and the rest are in their forties; $X = $ age of an Xbox player. (Use the midpoints of the measurement classes.)

5. From a bin that contains 20 defective joysticks and 30 good ones, 3 are chosen at random; $X = $ the number of defective joysticks chosen. (Round all probabilities to four decimal places.)

6. Two dice are weighted so that each number 2, 3, 4, and 5 is half as likely to face up as each of 1 and 6; $X = $ the number of 1s that face up when both are thrown.

7. Use any method to calculate the sample mean, median, and standard deviation of the following sample of scores: $-1, 2, 0, 3, 6$.

8. Use any method to calculate the sample mean, median, and standard deviation of the following sample of scores: 4, 4, 5, 6, 6.

9. Give an example of a sample of four scores with mean 1 and median 0. (Arrange them in ascending order.)

10. Give an example of a sample of six scores with sample standard deviation 0 and mean 2.

11. Give an example of a population of six scores with mean 0 and population standard deviation 1.

12. Give an example of a sample of five scores with mean 0 and sample standard deviation 1.

A die is constructed in such a way that rolling a 6 is twice as likely as rolling each other number. That die is rolled four times. Let X be the number of times a 6 is rolled. Evaluate the probabilities in Exercises 13–20.

13. $P(X = 1)$
14. $P(X = 3)$

15. The probability that 6 comes up at most twice

16. The probability that 6 comes up at most once

17. The probability that X is more than 3

18. The probability that X is at least 2

19. $P(1 \leq X \leq 3)$
20. $P(X \leq 3)$

21. A couple has three children; X = the number of girls. (Assume an equal likelihood of a child being a boy or a girl.) Find the expected value and standard deviation of X, and complete the following sentence with the smallest possible whole number: All values of X lie within ___ standard deviations of the expected value.

22. A couple has four children; X = the number of boys. (Assume only a 25% chance of a child being a boy.) Find the expected value and standard deviation of X, and complete the following sentence with the smallest possible whole number: All values of X lie within ___ standard deviations of the expected value.

23. A random variable X has the following frequency distribution:

x	-3	-2	-1	0	1	2	3
$fr(X = x)$	1	2	3	4	3	2	1

Find the probability distribution, expected value, and standard deviation of X, and complete the following sentence: 87.5% (or 14/16) of the time, X is within ___ (round to one decimal place) standard deviations of the expected value.

24. A random variable X has the following frequency distribution:

x	-4	-2	0	2	4	6
$fr(X = x)$	3	3	4	5	3	2

Find the probability distribution, expected value, and standard deviation of X, and complete the following sentence: ___ percent of the values of X lie within one standard deviation of the expected value.

25. A random variable X has expected value $\mu = 100$ and standard deviation $\sigma = 16$. Use Chebyshev's rule to find an interval in which X is guaranteed to lie with a probability of at least 90%.

26. A random variable X has a symmetric distribution and an expected value $\mu = 200$ and standard deviation $\sigma = 5$. Use Chebyshev's rule to find a value that X is guaranteed to exceed with a probability of at most 10%.

27. A random variable X has a bell-shaped symmetric distribution, with expected value $\mu = 200$ and standard deviation $\sigma = 20$. The empirical rule tells us that X has a value greater than ___ approximately 0.15% of the time.

28. A random variable X has a bell-shaped symmetric distribution, with expected value $\mu = 100$ and standard deviation $\sigma = 30$. Use the empirical rule to give an interval in which X lies approximately 95% of the time.

In Exercises 29–34 the mean and standard deviation of a normal variable X are given. Find the indicated probability.

29. X is the standard normal variable Z; $P(0 \le X \le 1.5)$.

30. X is the standard normal variable Z; $P(X \le -1.5)$.

31. X is the standard normal variable Z; $P(|X| \ge 2.1)$.

32. $\mu = 100, \sigma = 16; P(80 \le X \le 120)$

33. $\mu = 0, \sigma = 2; P(X \le -1)$

34. $\mu = -1, \sigma = 0.5; P(X \ge 1)$

Applications: OHaganBooks.com
[Try the game at www.OHaganBooks.com]

Marketing   As a promotional gimmick, OHaganBooks.com has been selling copies of Encyclopædia Galactica *at an extremely low price that is changed each week at random in a nationally televised drawing. Exercises 35–40 are based on the following table, which gives the frequency with which each price will be chosen and summarizes the anticipated sales:*

Price	$5.50	$10	$12	$15
Frequency (weeks)	1	2	3	4
Weekly Sales	6,200	3,500	3,000	1,000

35. What is the expected value of the price of *Encyclopædia Galactica*?

36. What are the expected weekly sales of *Encyclopædia Galactica*?

37. What is the expected weekly revenue from sales of *Encyclopædia Galactica*? (Revenue = Price per copy sold × Number of copies sold.)

38. OHaganBooks.com originally paid *Duffin House* $20 per copy for the *Encyclopædia Galactica*. What is the expected weekly loss from sales of the encyclopædia? (Loss = Loss per copy sold × Number of copies sold.)

39. True or false? If X and Y are two random variables, then $E(XY) = E(X)E(Y)$. (The expected value of the product of two random variables is the product of the expected values.) Support your claim by referring to the answers of Exercises 35–37.

40. True or false? If X and Y are two random variables, then $E(X/Y) = E(X)/E(Y)$ (the expected value of the ratio of two random variables is the ratio of the expected values). Support your claim by referring to the answers of Exercises 36 and 38.

41. *Online Sales*   The following table shows the number of online orders at OHaganBooks.com per million residents in 100 U.S. cities during 1 month:

Orders (per million residents)	1–2.9	3–4.9	5–6.9	7–8.9	9–10.9
Number of Cities	25	35	15	15	10

a. Let X be the number of orders per million residents in a randomly chosen U.S. city. (Use rounded midpoints of the given measurement classes.) Construct the probability distribution for X, and hence compute the expected value μ of X and standard deviation σ. (Round answers to four decimal places.)

b. What range of orders per million residents does the empirical rule predict from approximately 68% of all cities? Would you judge that the empirical rule applies? Why?

c. The actual percentage of cities from which you obtain between 3 and 8 orders per million residents is (choose the correct answer that gives the most specific information):

(A) Between 50% and 65%
(B) At least 65%
(C) At least 50%
(D) 57.5%

42. *Pollen* Marjory Duffin is planning a joint sales meeting with OHaganBooks.com in Atlanta at the end of March but is extremely allergic to pollen, so she went online to find pollen counts for the period. The following table shows the results of her search:

Pollen Count	0–1.9	2–3.9	4–5.9	6–7.9	8–9.9	10–11.9
Number of Days	3	5	7	2	1	2

a. Let X be the pollen count on a given day. (Use rounded midpoints of the given measurement classes.) Construct the probability distribution for X, and hence compute the expected value μ of X and standard deviation σ. (Round answers to four decimal places.)

b. What range of pollen counts does the empirical rule predict on approximately 95% of the days? Would you judge that the empirical rule applies? Why?

c. The actual percentage of days on which the pollen count is between 2 and 7 is (choose the correct answer that gives the most specific information):

(A) Between 50% and 60%
(B) At least 60%
(C) At most 70%
(D) Between 60% and 70%

Mac vs. Windows On average, 5% of all hits by Mac OS users and 10% of all hits by Windows users result in orders for books at OHaganBooks.com. Due to online promotional efforts, the site traffic is approximately 10 hits per hour by Mac OS users, and 20 hits per hour by Windows users. Compute the probabilities in Exercises 43–48. (Round all answers to three decimal places.)

43. What is the probability that exactly three Windows users will order books in the next hour?

44. What is the probability that at most three Windows users will order books in the next hour?

45. What is the probability that exactly one Mac OS user and three Windows users will order books in the next hour?

46. What assumption must you make to justify your calculation in Exercise 45?

47. How many orders for books can OHaganBooks.com expect in the next hour from Mac OS users?

48. How many orders for books can OHaganBooks.com expect in the next hour from Windows users?

Online Cosmetics OHaganBooks.com has launched a subsidiary, GnuYou.com, which sells beauty products online. Most products sold by GnuYou.com are skin creams and hair products. Exercises 49–52 are based on the following table, which shows monthly revenues earned through sales of these products. (Assume a normal distribution. Round all answers to three decimal places.)

Product	Skin Creams	Hair Products
Mean Monthly Revenue ($)	38,000	34,000
Standard Deviation ($)	21,000	14,000

49. What is the probability that GnuYou.com will sell *at least* $50,000 worth of skin cream next month?

50. What is the probability that GnuYou.com will sell *at most* $50,000 worth of hair products next month?

51. What is the probability that GnuYou.com will sell less than $12,000 of skin creams next month?

52. What is the probability that GnuYou.com will sell less than $12,000 of hair products next month?

53. *Intelligence* Billy-Sean O'Hagan, now a senior at *Suburban State University*, has done exceptionally well and has just joined Mensa, a club for people with high IQs. Within Mensa is a group called the Three Sigma Club because its members' IQ scores are at least three standard deviations higher than the U.S. mean. Assuming a U.S. population of 323,000,000 and assuming that IQ scores are normally distributed, how many people in the United States are qualified for the Three Sigma Club? (Round your answer to the nearest 1,000 people.)

54. *Intelligence* To join Mensa (not necessarily the Three Sigma Club), one needs an IQ of at least 132, corresponding to the top 2% of the population. Assuming that scores on this test are normally distributed with a mean of 100, what is the standard deviation? (Round your answer to the nearest whole number.)

55. *Intelligence* On the basis of information given in Exercises 53 and 54, what score must Billy-Sean have to get into the Three Sigma Club? (Assume that IQ scores are normally distributed with a mean of 100, and use the rounded standard deviation.)

56. *Intelligence* Mensa allows the results of various standardized tests to be used to gain membership. Suppose that there was such a test with a mean of 500 on which one needed to score at least 600 to be in the top 2% of the population, hence eligible to join Mensa. What would Billy-Sean need to score on this test to get into the Three Sigma Club?

CASE STUDY

Spotting Tax Fraud with Benford's Law[67]

You are a tax fraud specialist working for the Internal Revenue Service (IRS), and you have just been handed a portion of the tax return from *Colossal Conglomerate*. The IRS suspects that the portion you were handed may be fraudulent, and would like your opinion. Is there any mathematical test, you wonder, that can point to a suspicious tax return based on nothing more than the numbers entered?

You decide, on an impulse, to make a list of the first digits of all the numbers entered in the portion of the Colossal Conglomerate tax return (there are 625 of them). You reason that, if the tax return is an honest one, the first digits of the numbers should be uniformly distributed. More precisely, if the experiment consists of selecting a number at random from the tax return, and the random variable X is defined to be the first digit of the selected number, then X should have the following probability distribution:

x	1	2	3	4	5	6	7	8	9
$P(X = x)$	$\frac{1}{9}$	$\frac{1}{9}$	$\frac{1}{9}$	$\frac{1}{9}$	$\frac{1}{9}$	$\frac{1}{9}$	$\frac{1}{9}$	$\frac{1}{9}$	$\frac{1}{9}$

You then do a quick calculation based on this probability distribution and find an expected value of $E(X) = 5$. Next, you turn to the Colossal Conglomerate tax return data and calculate the relative frequency (estimated probability) of the actual numbers in the tax return. You find the following results:

Colossal Conglomerate Return

y	1	2	3	4	5	6	7	8	9
$P(Y = y)$	.29	.1	.04	.15	.31	.08	.01	.01	.01

It certainly does look suspicious! For one thing, the digits 1 and 5 seem to occur a lot more often than any of the other digits and roughly three times what you predicted. Moreover, when you compute the expected value, you obtain $E(Y) = 3.48$, considerably lower than the value of 5 you predicted. "Gotcha!" you exclaim.

You are about to file a report recommending a detailed audit of Colossal Conglomerate when you recall an article you once read about first digits in lists of numbers. The article dealt with a remarkable discovery in 1938 by Dr. Frank Benford, a physicist at General Electric. What Dr. Benford noticed was that the pages of logarithm tables that listed numbers starting with the digits 1 and 2 tended to be more soiled and dog-eared than the pages that listed numbers starting with higher digits—say, 8. For some reason, numbers that start with low digits seemed more prevalent than numbers that start with high digits. He subsequently analyzed more than 20,000 sets of numbers, such as tables of baseball statistics, listings of widths of rivers, half-lives of radioactive elements, street addresses, numbers in magazine articles. The result was always the same: Inexplicably, numbers that start with low digits tended to appear more frequently than those that start with high ones, numbers beginning with

---

[67] The discussion is based on the article "Following Benford's Law, or Looking Out for No. 1" by Malcolm W. Browne, *New York Times*, August 4, 1998, p. F4. The use of Benford's Law in detecting tax evasion is discussed in a Ph.D. dissertation by Dr. Mark J. Nigrini (Southern Methodist University, Dallas).

the digit 1 being most prevalent of all.[68] Moreover, the expected value of the first digit was not the expected 5, but 3.44.

Because the first digits in Colossal Conglomerate's return have an expected value of 3.48, very close to Benford's value, it might appear that your suspicion was groundless after all. (Back to the drawing board . . .)

Out of curiosity, you decide to investigate Benford's discovery more carefully. What you find is that Benford did more than simply observe a strange phenomenon in lists of numbers. He went further and derived the following formula for the probability distribution of first digits in lists of numbers:

$$P(X = x) = \log(1 + 1/x) \quad (x = 1, 2, \ldots, 9).$$

You compute these probabilities and find the following distribution. (The probabilities are all rounded and thus do not add to exactly 1.)

x	1	2	3	4	5	6	7	8	9
$P(X = x)$	.30	.18	.12	.10	.08	.07	.06	.05	.05

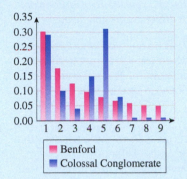

Figure 17

You then enter these data along with the Colossal Conglomerate tax return data in your spreadsheet program and obtain the graph shown in Figure 17.

The graph shows something awfully suspicious happening with the digit 5. The percentage of numbers in the Colossal Conglomerate return that begin with 5 far exceeds Benford's prediction that approximately 8% of all numbers should begin with 5.

Now it seems fairly clear that you are justified in recommending Colossal Conglomerate for an audit after all.

Q: *Because no given set of data can reasonably be expected to satisfy Benford's law exactly, how can I be certain that the Colossal Conglomerate data are not simply due to chance?*

A: You can never be 100% certain. It is certainly conceivable that the tax figures just happen to result in the "abnormal" distribution in the Colossal Conglomerate tax return. However—and this is the subject of what is known as inferential statistics— there is a method for deciding whether you can be, say, "95% certain" that the anomaly reflected in the data is not due to chance. To check, you must first compute a statistic that determines how far a given set of data deviates from satisfying a theoretical prediction (Benford's law, in this case). This statistic is called a **sum-of-squares error** and is given by the following formula (reminiscent of the variance):

$$\text{SSE} = n\left[\frac{[P(y_1) - P(x_1)]^2}{P(x_1)} + \frac{[P(y_2) - P(x_2)]^2}{P(x_2)} + \cdots + \frac{[P(y_9) - P(x_9)]^2}{P(x_9)}\right].$$

Here, n is the sample size: 625 in the case of Colossal Conglomerate. The quantities $P(x_i)$ are the theoretically predicted probabilities according to Benford's Law, and the $P(y_i)$ are the probabilities in the Colossal Conglomerate return. Notice that if the Colossal Conglomerate return probabilities had exactly matched the theoretically predicted probabilities, then SSE would have been zero. Notice also the effect of multiplying by the sample size n: The larger the sample, the more likely that the

---

[68] This does not apply to all lists of numbers. For instance, a list of randomly chosen numbers between 100 and 999 will have first digits uniformly distributed between 1 and 9.

discrepancy between the $P(x_i)$ and the $P(y_i)$ is not due to chance. Substituting the numbers gives[69]

$$SSE \approx 625\left[\frac{[.29 - .30]^2}{.30} + \frac{[.1 - .18]^2}{.18} + \cdots + \frac{[.01 - .05]^2}{.05}\right]$$

$$\approx 552.$$

Q : *The value of SSE does seem quite large. But how can I use this figure in my report? I would like to say something impressive, such as "Based on the portion of the Colossal Conglomerate tax return analyzed, one can be 95% certain that the figures are anomalous."*

A : The error SSE is used by statisticians to answer exactly such a question. What they would do is compare this figure to the largest SSE we would have expected to get by chance in 95 out of 100 selections of data that *do* satisfy Benford's law. This "biggest error" is computed using a "chi-squared" distribution and can be found in Excel by entering

```
=CHIINV(0.05,8)
```

Here, the 0.05 is $1 - 0.95$, encoding the "95% certainty," and the 8 is called the "number of degrees of freedom" = number of outcomes (9) minus 1.

You now find, using Excel, that the chi-squared figure is 15.5, meaning that the largest SSE that you could have expected purely by chance is 15.5. Because Colossal Conglomerate's error is much larger at 552, you can now justifiably say in your report that there is a 95% certainty that the figures are anomalous.[70]

EXERCISES

Which of the following lists of data would you expect to follow Benford's law? If the answer is "no," give a reason.

1. Distances between cities in France, measured in kilometers

2. Distances between cities in France, measured in miles

3. The grades (0–100) in your math instructor's grade book

4. The Dow Jones averages for the past 100 years

5. Verbal SAT scores of college-bound high school seniors

6. Life spans of companies

T *Use technology to determine whether the given distribution of first digits fails, with 95% certainty, to follow Benford's law.*

7. *Good Neighbor, Inc.*'s tax return ($n = 1,000$)

y	1	2	3	4	5	6	7	8	9
$P(Y = y)$	.31	.16	.13	.11	.07	.07	.05	.06	.04

8. *Honest Growth Funds* stockholder report ($n = 400$)

y	1	2	3	4	5	6	7	8	9
$P(Y = y)$	.28	.16	.1	.11	.07	.09	.05	.07	.07

---

[69] If you use more accurate values for the probabilities in Benford's distribution, the value is approximately 560.

[70] What this actually means is that, if you were to do a similar analysis on a large number of tax returns and you designated as "not conforming to Benford's law" all of those whose value of SSE was larger than 15.5, you would be justified in 95% of the cases.

Section 8.1

Example 3 (page 584) Let X be the number of heads that face up in three tosses of a coin. We obtained the following probability distribution of X in the text:

x	0	1	2	3
$P(X = x)$	$\frac{1}{8}$	$\frac{3}{8}$	$\frac{3}{8}$	$\frac{1}{8}$

Use technology to obtain the corresponding histogram.

Solution

1. In the TI-83/84 Plus, you can enter a list of probabilities as follows: Press STAT, choose EDIT, and then press ENTER. Clear columns L_1 and L_2 if they are not already cleared. (Select the heading of a column and press CLEAR ENTER to clear it.) Enter the values of X in the column under L_1 (pressing ENTER after each entry), and enter the frequencies in the column under L_2.

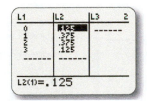

2. To graph the data as in Figure 1, first set the WINDOW to $0 \le X \le 4$, $0 \le Y \le 0.5$, and Xscl = 1 (the width of the bars). Then turn STAT PLOT on ([2nd] Y=), and configure it by selecting the histogram icon, setting Xlist = L_1 and Freq = L_2. Then hit GRAPH.

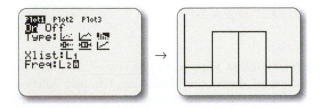

Example 4 (page 584) We obtained the following frequency table in the text:

x	25,000	35,000	45,000	55,000	65,000	75,000	85,000
Frequency	20	80	230	400	170	70	30

Find the probability distribution of X.

Solution

We need to divide each frequency by the sum. Although the computations in this example (dividing the seven frequencies by 1,000) are simple to do by hand, they could become tedious in general, so technology is helpful.

1. On the TI-83/84 Plus, press STAT, select EDIT, enter the values of X in the L_1 list, and enter the frequencies in the L_2 list as in Example 3 (below left).

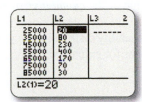

2. Then, on the Home screen, enter

$$L_2/1000 \to L_3$$ L_2 is 2ND 2, L_3 is 2ND 3.

or, better yet,

$$L_2/\text{sum}(L_2) \to L_3$$ Sum is found in 2ND STAT, under MATH.

3. After pressing ENTER, you can now go back to the STAT EDIT screen, and you will find the probabilities displayed in L_3 as shown above on the right.

Section 8.2

Example 2(b) (page 594) By 2030 the probability that a randomly chosen resident in the United States will be 65 years old or older is projected to be .2. If X is the number of people aged 65 or older in a sample of 6, construct the probability distribution of X.

Solution

In the "Y=" screen, you can enter the binomial distribution formula

$$Y_1 = 6 \text{ nCr } X*0.2^X*0.8^{(6-X)}$$

directly (to get nCr, press MATH and select PRB), and hit TABLE. You can then replicate the table in the text by choosing $X = 0, 1, \ldots, 6$ (use the TBLSET screen to set "Indpnt" to "Ask" if you have not already done so).

The TI-83/84 Plus also has a built-in binomial distribution function that you can use in place of the explicit formula:

$$Y_1 = \texttt{binompdf(6,0.2,X)} \quad \text{In } \boxed{\text{2ND}} \, \boxed{\text{VARS}}$$

The TI-83/84 Plus function `binomcdf` (directly following `binompdf`) gives the value of the *cumulative* distribution function, $P(0 \le X \le x)$.

To graph the resulting probability distribution on your calculator, follow the instructions for graphing a histogram in Section 8.1.

Section 8.3

Example 3 (page 603) According to historical data, the number of injuries that a member of the *Enormous State University* women's soccer team will sustain during a typical season is given by the following probability distribution table:

Injuries	0	1	2	3	4	5	6
Probability	.20	.20	.22	.20	.15	.01	.02

If X denotes the number of injuries sustained by a player during one season, compute $E(X)$.

Solution

To obtain the expected value of a probability distribution on the TI-83/84 Plus, press $\boxed{\text{STAT}}$, select EDIT, and then press $\boxed{\text{ENTER}}$, and enter the values of X in the L_1 list and the probabilities in the column in the L_2 list. Then, on the Home screen, you can obtain the expected value as

$$\texttt{sum(L}_1\texttt{*L}_2\texttt{)} \qquad \begin{array}{l}\text{L_1 is } \boxed{\text{2ND}}\,\boxed{1}\; \text{L_2 is } \boxed{\text{2ND}}\,\boxed{2} \\ \text{Sum is found in } \boxed{\text{2ND}}\,\boxed{\text{STAT}}\text{, under MATH.}\end{array}$$

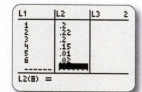 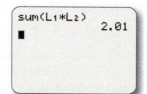

Section 8.4

Example 3 (page 618) Compute the variance and standard deviation for the following probability distribution:

x	10	20	30	40	50	60
$P(X = x)$	.2	.2	.3	.1	.1	.1

Solution

1. As in Example 3 in Section 8.3, begin by entering the probability distribution of X into columns L_1 and L_2 in the LIST screen (press $\boxed{\text{STAT}}$ and select EDIT). (See below left.)

2. Then, on the Home screen, enter

$$\texttt{sum(L}_1\texttt{*L}_2\texttt{)} \rightarrow \text{M} \qquad \begin{array}{l}\text{Stores the value of } \mu \text{ as M} \\ \text{Sum is found in } \boxed{\text{2ND}}\,\boxed{\text{STAT}}, \\ \text{under MATH.}\end{array}$$

3. To obtain the variance, enter

$$\texttt{sum((L}_1\texttt{-M)}\texttt{\^{}2*L}_2\texttt{)} \qquad \begin{array}{l}\text{Computation of} \\ \Sigma(x - \mu)^2 \, P(X = x)\end{array}$$

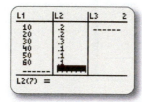 

Section 8.5

Example 1(b), (c) (page 629) Let Z be the standard normal variable. Calculate the following probabilities:

b. $P(0 \le Z \le 2.43)$

c. $P(-1.37 \le Z \le 2.43)$

Solution

On the TI-83/84 Plus, press $\boxed{\text{2ND}}\,\boxed{\text{VARS}}$ to obtain the selection of distribution functions. The first function, `normalpdf`, gives the values of the normal density function (whose graph is the normal curve). The second, `normalcdf`, gives $P(a \le Z \le b)$. For example, to compute $P(0 \le Z \le 2.43)$, enter

$$\texttt{normalcdf(0,2.43)}$$

To compute $P(-1.37 \leq Z \leq 2.43)$, enter

$$\texttt{normalcdf(-1.37,2.43)}$$

Example 2 (page 631) Pressure gauges manufactured by *Precision Corp.* must be checked for accuracy before being placed on the market. To test a pressure gauge, a worker uses it to measure the pressure of a sample of compressed air known to be at a pressure of exactly 50 pounds per square inch. If the gauge reading is off by more than 1% (0.5 pounds), it is rejected. Assuming that the reading of a pressure gauge under these circumstances is a normal random variable with mean 50 and standard deviation 0.4, find the percentage of gauges rejected.

Solution

As seen in the text, we need to compute $1 - P(49.5 \leq X \leq 50.5)$ with $\mu = 50$ and $\sigma = 0.4$. On the TI-83/84 Plus, the built-in $\texttt{normalcdf}$ function permits us to compute $P(a \leq X \leq b)$ for nonstandard normal distributions as well. The format is

$$\texttt{normalcdf(a,b,}\mu\texttt{,}\sigma\texttt{)} \qquad P(a \leq X \leq b)$$

For example, we can compute $P(49.5 \leq X \leq 50.5)$ by entering

$$\texttt{normalcdf(49.5,50.5,50,0.4)}$$

Then subtract it from 1 to obtain the answer:

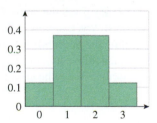

Spreadsheet Technology Guide

Section 8.1

Example 3 (page 584) Let X be the number of heads that face up in three tosses of a coin. We obtained the following probability distribution of X in the text:

x	0	1	2	3
$P(X = x)$	$\frac{1}{8}$	$\frac{3}{8}$	$\frac{3}{8}$	$\frac{1}{8}$

Use technology to obtain the corresponding histogram.

Solution

1. In your spreadsheet, enter the values of X in one column and the probabilities in another.

	A	B
1	x	P(X=x)
2	0	0.125
3	1	0.375
4	2	0.375
5	3	0.125

2. Next, select *only* the column of probabilities (B2–B5), and then insert a column chart. The procedure for doing this depends heavily on the spreadsheet and platform you are using.

Example 4 (page 584) We obtained the following frequency table in the text:

x	25,000	35,000	45,000	55,000	65,000	75,000	85,000
Frequency	20	80	230	400	170	70	30

Find the probability distribution of X.

Solution

We need to divide each frequency by the sum. Although the computations in this example (dividing the seven frequencies by 1,000) are simple to do by hand, they could become tedious in general, so technology is helpful. Spreadsheets manipulate lists with ease. Set up your spreadsheet as shown.

	A	B	C
1	x	Fr	P(X=x)
2	25000	20	=B2/SUM(B:B)
3	35000	80	
4	45000	230	
5	55000	400	
6	65000	170	
7	75000	70	
8	85000	30	

↓

	A	B	C
1	x	Fr	P(X=x)
2	25000	20	0.02
3	35000	80	0.08
4	45000	230	0.23
5	55000	400	0.4
6	65000	170	0.17
7	75000	70	0.07
8	85000	30	0.03

The formula `SUM(B:B)` gives the sum of all the numerical entries in column B. You can now change the frequencies to see the effect on the probabilities. You can also add new values and frequencies to the list if you copy the formula in column C farther down the column.

Section 8.2

Example 2(b) (page 594) By 2030 the probability that a randomly chosen resident in the United States will be 65 years old or older is projected to be .2. If X is the number of people aged 65 or older in a sample of 6, construct the probability distribution of X.

Solution

You can generate the binomial distribution as follows in your spreadsheet:

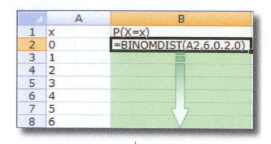

	A	B
1	x	P(X=x)
2	0	=BINOMDIST(A2,6,0.2,0)
3	1	
4	2	
5	3	
6	4	
7	5	
8	6	

↓

	A	B
1	x	P(X=x)
2	0	0.262144
3	1	0.393216
4	2	0.24576
5	3	0.08192
6	4	0.01536
7	5	0.001536
8	6	6.4E-05

The values of X are shown in column A, and the probabilities are computed in column B. The arguments of the `BINOMDIST` function are as follows:

`BINOMDIST`$(x, n, p,$ Cumulative $(0 = $ no, $1 = $ yes$)$).

Setting the last argument to 0 (as shown) gives $P(X = x)$. Setting it to 1 gives $P(X \le x)$.

To graph the resulting probability distribution using your spreadsheet, insert a bar chart as in Section 8.1.

Section 8.3

Example 3 (page 603) According to historical data, the number of injuries that a member of the *Enormous State University* women's soccer team will sustain during a typical season is given by the following probability distribution table:

Injuries	0	1	2	3	4	5	6
Probability	.20	.20	.22	.20	.15	.01	.02

If X denotes the number of injuries sustained by a player during one season, compute $E(X)$.

Solution

As the method we used suggests, the calculation of the expected value from the probability distribution is particularly easy to do using a spreadsheet program such as Excel.

The following worksheet shows one way to do it. (The first two columns contain the probability distribution of X; the quantities $xP(X = x)$ are summed in cell C9.)

	A	B	C
1	x	P(X=x)	x*P(X=x)
2	0	0.2	=A2*B2
3	1	0.2	
4	2	0.22	
5	3	0.2	
6	4	0.15	
7	5	0.01	
8	6	0.02	
9			=SUM(C2:C8)

↓

	A	B	C
1	x	P(X=x)	x*P(X=x)
2	0	0.2	0
3	1	0.2	0.2
4	2	0.22	0.44
5	3	0.2	0.6
6	4	0.15	0.6
7	5	0.01	0.05
8	6	0.02	0.12
9			2.01

An alternative is to use the SUMPRODUCT function: Once we enter the first two columns above, the formula

    =SUMPRODUCT(A2:A8,B2:B8)

computes the sum of the products of corresponding entries in the columns, giving us the expected value.

Section 8.4

Example 3 (page 618)   Compute the variance and standard deviation for the following probability distribution:

x	10	20	30	40	50	60
$P(X = x)$	.2	.2	.3	.1	.1	.1

Solution

As in Example 3 in Section 8.3, begin by entering the probability distribution into columns A and B, and then proceed as shown:

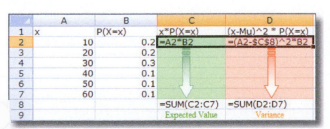

The variance then appears in cell D8:

	A	B	C	D
1	x	P(X=x)	x*P(X=x)	(x-Mu)^2 * P(X=x)
2	10	0.2	2	80
3	20	0.2	4	20
4	30	0.3	9	0
5	40	0.1	4	10
6	50	0.1	5	40
7	60	0.1	6	90
8			30	240
9			Expected Value	Variance

Section 8.5

Example 1(b), (c) (page 629)   Let Z be the standard normal variable. Calculate the following probabilities:

b. $P(0 \le Z \le 2.43)$

c. $P(-1.37 \le Z \le 2.43)$

Solution

In spreadsheets the function NORMSDIST (<u>Norm</u>al <u>S</u>tandard <u>Dist</u>ribution) gives the area shown on the left in the figure below. (Tables such as the one in the Appendix give the area shown on the right.)

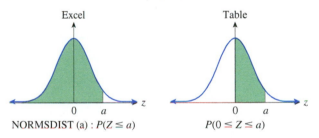

To compute a general area, $P(a \le Z \le b)$ in your spreadsheet, subtract the cumulative area to a from that to b:

    =NORMSDIST(b)-NORMSDIST(a) $P(a \le Z \le b)$

In particular, to compute $P(0 \le Z \le 2.43)$, use

    =NORMSDIST(2.43)-NORMSDIST(0)

and to compute $P(-1.37 \le Z \le 2.43)$, use

    =NORMSDIST(2.43)-NORMSDIST(-1.37)

	A	B	C
1	=NORMSDIST(2.43)-NORMSDIST(0)		
2	=NORMSDIST(2.43)-NORMSDIST(-1.37)		

↓

	A
1	0.49245059
2	0.90710714

Example 2 (page 631) Pressure gauges manufactured by *Precision Corp.* must be checked for accuracy before being placed on the market. To test a pressure gauge, a worker uses it to measure the pressure of a sample of compressed air known to be at a pressure of exactly 50 pounds per square inch. If the gauge reading is off by more than 1% (0.5 pounds), it is rejected. Assuming that the reading of a pressure gauge under these circumstances is a normal random variable with mean 50 and standard deviation 0.4, find the percentage of gauges rejected.

Solution

In spreadsheets we use the function NORMDIST instead of NORMSDIST. Its format is similar to NORMSDIST but includes extra arguments as shown:

$$=\text{NORMDIST}(a,\mu,\sigma,1) \quad P(X \le a)$$

(The last argument, set to 1, tells the spreadsheet that we want the cumulative distribution.) To compute $P(a \le X \le b)$, we enter the following in any vacant cell:

$$=\text{NORMDIST}(b,\mu,\sigma,1)$$
$$-\text{NORMDIST}(a,\mu,\sigma,1) \quad P(a \le X \le b)$$

For example, we can compute $P(49.5 \le X \le 50.5)$ by entering

$$=\text{NORMDIST}(50.5,50,0.4,1)$$
$$-\text{NORMDIST}(49.5,50,0.4,1)$$

We then subtract it from 1 to obtain the answer:

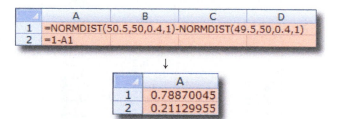

	A	B	C	D
1	=NORMDIST(50.5,50,0.4,1)-NORMDIST(49.5,50,0.4,1)			
2	=1-A1			

↓

	A
1	0.78870045
2	0.21129955

APPENDIX A | LOGIC

Introduction

Logic is the underpinning of all reasoned argument. The ancient Greeks recognized its role in mathematics and philosophy, and studied it extensively. Aristotle, in his *Organon*, wrote the first systematic treatise on logic. His work had a heavy influence on philosophy, science, and religion through the Middle Ages.

But Aristotle's logic was expressed in ordinary language and so was subject to the ambiguities of ordinary language. Philosophers came to want to express logic more formally and symbolically, more like the way that mathematics is written. (Leibniz, in the 17th century, was probably the first to envision and call for such a formalism.) It was with the publication in 1847 of G. Boole's *The Mathematical Analysis of Logic* and A. DeMorgan's *Formal Logic* that **symbolic logic** came into being, and logic became recognized as part of mathematics. Since Boole and DeMorgan, logic and mathematics have been inextricably intertwined. Logic is part of mathematics, but at the same time it is the language of mathematics.

The study of symbolic logic is usually broken into several parts. The first and most fundamental is the **propositional logic**. Built on top of this is the **predicate logic**, which is the language of mathematics. In this appendix we give an introduction to propositional logic.

A.1 Statements and Logical Operators

Propositional logic is the study of *propositions*. A **statement**, or **proposition**, is any declarative sentence which is either true (T) or false (F). We refer to T or F as the **truth value** of the statement.

EXAMPLE 1 Statements

a. "$2 + 2 = 4$" is a statement because it can be either true or false.* Because it happens to be a true statement, its truth value is T.

b. "$1 = 0$" is also a statement, but its truth value is F.

c. "It will rain tomorrow" is a statement. To determine its truth value, we shall have to wait for tomorrow.

d. "Solve the following equation for x" is not a statement, because it cannot be assigned any truth value whatsoever. It is an imperative, or command, rather than a declarative sentence.

e. "The number 5" is not a statement, because it is not even a complete sentence.

f. "This statement is false" gets us into a bind: If it were true, then, because it is declaring itself to be false, it must be false. On the other hand, if it were false, then its declaring itself false is a lie, so it is true! In other words, if it is true, then it is

* Is "$2 + 2 = 4$" a sentence? Read it aloud: "Two plus two equals four" is a perfectly respectable English sentence.

651

false, and if it is false, then it is true, and we go around in circles. We get out of this bind by saying that because the sentence cannot be either true or false, we refuse to call it a statement. An equivalent pseudo-statement is: "I am lying," so this sentence is known as **the liar's paradox**.

Note Sentences that refer to themselves, or *self-referential sentences*, as illustrated in Example 1(f), are not permitted to be statements. This eliminates the liar's paradox and several similar problems. ■

We shall use letters like p, q, r, and so on to stand for statements. Thus, for example, we might decide that p should stand for the statement "The moon is round." We write

p: "The moon is round"   *p is the statement that the moon is round.*

to express this.

We can form new statements from old ones in several different ways. For example, starting with p: "I am an Anchovian," we can form the **negation** of p: "It is not the case that I am an Anchovian" or simply "I am not an Anchovian."

Negation of a Statement

If p is a statement, then its **negation** is the statement "not p" and is denoted by $\sim p$. We mean by this that, if p is true, then $\sim p$ is false, and vice versa.

Quick Examples

1. If p: "$2 + 2 = 4$," then $\sim p$: "It is not the case that $2 + 2 = 4$," or, more simply, $\sim p$: "$2 + 2 \neq 4$."
2. If q: "$1 = 0$," then $\sim q$: "$1 \neq 0$."
3. If r: "Diamonds are a pearl's best friend," then $\sim r$: "Diamonds are not a pearl's best friend."
4. If s: "All politicians are crooks," then $\sim s$: "Not all politicians are crooks."
5. **Double Negation:** If p is any statement, then the negation of $\sim p$ is $\sim(\sim p)$: "not (not p)," or, in other words, p. Thus, $\sim(\sim p)$ has the same meaning as p.

Notes

1. Notice in Quick Example 1 that $\sim p$ is false, because p is true. However, in Quick Example 2, $\sim q$ is true, because q is false. A statement of the form $\sim q$ can very well be true; it is a common mistake to think that it must be false.
2. Saying that not all politicians are crooks is not the same as saying that no politicians are crooks but is the same as saying that some (meaning one or more) politicians are not crooks.
3. The symbol $\sim$ is our first example of a **logical operator**.
4. When we say in Quick Example 5 that $\sim(\sim p)$ has the same meaning as p, we mean that they are *logically equivalent*—a notion we will make precise below. ■

Here is another way we can form a new statement from old ones. Starting with p: "I am wise," and q: "I am strong," we can form the statement "I am wise and I am strong." We denote this new statement by $p \wedge q$, read "p and q." In order for $p \wedge q$ to be true, *both p and q* must be true. Thus, for example, if I am wise but not strong, then $p \wedge q$ is false. The symbol $\wedge$ is another logical operator. The statement $p \wedge q$ is called the **conjunction** of p and q.

Conjunction

The **conjunction** of p and q is the statement $p \wedge q$, which we read "p and q." It can also be said in a number of different ways, such as "p even though q." The statement $p \wedge q$ is true when both p and q are true, and $p \wedge q$ is false otherwise.

Quick Examples

6. If p: "This galaxy will ultimately disappear into a black hole" and q: "$2 + 2 = 4$," then $p \wedge q$ is the statement "Not only will this galaxy ultimately disappear into a black hole, but $2 + 2 = 4$!"

7. If p: "$2 + 2 = 4$" and q: "$1 = 0$," then $p \wedge q$: "$2 + 2 = 4$ and $1 = 0$." Its truth value is F because q is F.

8. With p and q as in Quick Example 6, the statement $p \wedge (\sim q)$ says: "This galaxy will ultimately disappear into a black hole and $2 + 2 \neq 4$," or, more colorfully, as "Contrary to your hopes, this galaxy is doomed to disappear into a black hole; moreover, two plus two is decidedly *not* equal to four!"

Notes

1. We sometimes use the word "but" as an emphatic form of "and." For instance, if p: "It is hot," and q: "It is not humid," then we can read $p \wedge q$ as "It is hot but not humid." There are always many ways of saying essentially the same thing in a natural language; one of the purposes of symbolic logic is to strip away the verbiage and record the underlying logical structure of a statement.

2. A **compound statement** is a statement formed from simpler statements via the use of logical operators. Examples are $\sim p$, $(\sim p) \wedge (q \wedge r)$, and $p \wedge (\sim p)$. A statement that cannot be expressed as a compound statement is called an **atomic statement.*** For example, "I am clever" is an atomic statement. In a compound statement such as $(\sim p) \wedge (q \wedge r)$, we refer to p, q, and r as the **variables** of the statement. Thus, for example, $\sim p$ is a compound statement in the single variable p. ∎

Before discussing other logical operators, we pause for a moment to talk about **truth tables**, which give a convenient way to analyze compound statements.

★ "Atomic" comes from the Greek for "not divisible." Atoms were originally thought to be the indivisible components of matter. Although the march of science proved that wrong, the name stuck.

Truth Table

The **truth table** for a compound statement shows, for each combination of possible truth values of its variables, the corresponding truth value of the statement.

Quick Examples

9. The truth table for negation, that is, for $\sim p$, is as follows:

p	$\sim p$
T	F
F	T

Each row shows a possible truth value for p and the corresponding value of $\sim p$.

10. The truth table for conjunction, that is, for $p \wedge q$, is as follows:

p	q	$p \wedge q$
T	T	T
T	F	F
F	T	F
F	F	F

Each row shows a possible combination of truth values of p and q and the corresponding value of $p \wedge q$.

EXAMPLE 2   Construction of Truth Tables

Construct truth tables for the following compound statements:

a. $\sim(p \wedge q)$ **b.** $(\sim p) \wedge q$

Solution

a. Whenever we encounter a complex statement, we work from the inside out, just as we might do if we had to evaluate an algebraic expression such as $-(a + b)$. Thus, we start with the p and q columns, then construct the $p \wedge q$ column, and finally, construct the $\sim(p \wedge q)$ column.

p	q	$p \wedge q$	$\sim(p \wedge q)$
T	T	T	F
T	F	F	T
F	T	F	T
F	F	F	T

Notice how we get the $\sim(p \wedge q)$ column from the $p \wedge q$ column: We reverse all the truth values.

b. Because there are two variables, p and q, we again start with the p and q columns. We then evaluate $\sim p$ and finally take the conjunction of the result with q.

p	q	$\sim p$	$(\sim p) \wedge q$
T	T	F	F
T	F	F	F
F	T	T	T
F	F	T	F

Because we are "and-ing" $\sim p$ with q, we look at the values in the $\sim p$ and q columns and combine these according to the instructions for "and." Thus, for example, in the first row we have $F \wedge T = F$, and in the third row we have $T \wedge T = T$.

Here is a third logical operator. Starting with p: "You are over 18" and q: "You are accompanied by an adult," we can form the statement "You are over 18 or are accompanied by an adult," which we write symbolically as $p \vee q$, read "p or q." Now in English the word "or" has several possible meanings, so we have to agree on which one we want here. Mathematicians have settled on the **inclusive or:** $p \vee q$ means p is true or q is true *or both are true*.* With p and q as above, $p \vee q$ stands for "You are over 18 or are accompanied by an adult, or both." We shall sometimes include the phrase "or both" for emphasis, but even if we leave it off, we still interpret "or" as inclusive.

* There is also the **exclusive or:** "*p* or *q* but not both*.*" This can be expressed as $(p \vee q) \wedge \sim(p \wedge q)$. Do you see why?

Disjunction

The **disjunction** of p and q is the statement $p \vee q$, which we read "p or q." Its truth value is defined by the following truth table:

p	q	$p \vee q$
T	T	T
T	F	T
F	T	T
F	F	F

This is the **inclusive or**, so $p \vee q$ is true when p is true or q is true *or both* are true.

Quick Examples

11. Let p: "The butler did it," and let q: "The cook did it." Then $p \vee q$: "Either the butler or the cook did it."

12. Let p: "The butler did it," and let q: "The cook did it," and let r: "The lawyer did it." Then $(p \vee q) \wedge (\sim r)$: "Either the butler or the cook did it, but not the lawyer."

Note The only way for $p \vee q$ to be false is for *both* p and q to be false. For this reason we can say that $p \vee q$ also means "p and q are not both false." ∎

To introduce our next logical operator, we ask you to consider the following statement: "If you earn an A in logic, then I'll buy you a new car." It seems to be made up out of two simpler statements:

p: "You earn an A in logic," and
q: "I will buy you a new car."

The original statement says: *If p is true, then q is true*, or, more simply, **if** p, **then** q. We can also phrase this as p **implies** q, and we write the statement symbolically as $p \rightarrow q$.

Now let us suppose for the sake of argument that the original statement: "If you earn an A in logic, then I'll buy you a new car," is true. This does *not* mean that you *will* earn an A in logic. All it says is that *if* you do so, then I will buy you that car. If we think of this as a promise, the only way that it can be broken is if you *do* earn an A and I do *not* buy you a new car. With this in mind, we define the logical statement $p \to q$ as follows.

Conditional

The **conditional** $p \to q$, which we read "if p, then q" or "p implies q," is defined by the following truth table:

p	q	$p \to q$
T	T	T
T	F	F
F	T	T
F	F	T

The arrow "$\to$" is the **conditional** operator, and in $p \to q$ the statement p is called the **antecedent** or **hypothesis**, and q is called the **consequent**, or **conclusion**. A statement of the form $p \to q$ is also called an **implication**.

Quick Examples

13. "If $1 + 1 = 2$ then the sun rises in the east" has the form $p \to q$ where p: "$1 + 1 = 2$" is true and q: "the sun rises in the east" is also true. Therefore, the statement is true.

14. "If the moon is made of green cheese, then I am Arnold Schwarzenegger" has the form $p \to q$ where p is false. From the truth table, we see that $p \to q$ is therefore true, regardless of whether or not I am Arnold Schwarzenegger.

15. "If $1 + 1 = 2$ then $0 = 1$" has the form $p \to q$ where this time p is true but q is false. Therefore, by the truth table, the given statement is false.

Notes

1. The only way that $p \to q$ can be false is if p is true and q is false. This is the case of the "broken promise" in the car example above.

2. If you look at the last two rows of the truth table, you see that we say that "$p \to q$" is true when p is false, *no matter what the truth value of q*. Think again about the promise: If you don't get that A, then whether or not I buy you a new car, I have not broken my promise. It may seem strange at first to say that $F \to T$ is T and $F \to F$ is also T, but, as they did in choosing to say that "or" is always inclusive, mathematicians agreed that the truth table above gives the most useful definition of the conditional. ∎

It is usually misleading to think of $p \to q$ as meaning that one of them causes the other. For instance, take p: "no one likes algebra," and q: "there are seven days in a week." Then $p \to q$ is true whereas neither causes the other. Here is a list of some English phrases that *do* have the same meaning as $p \to q$.

Some Phrasings of the Conditional

We interpret each of the following as equivalent to the conditional $p \to q$.

If p, then q.	p implies q.
q follows from p.	Not p unless q.
q if p.	p only if q.
Whenever p, q.	q whenever p.
p is sufficient for q.	q is necessary for p.
p is a sufficient condition for q.	q is a necessary condition for p.

Quick Example

16. "If it's Tuesday, this must be Belgium" can be rephrased in several ways as follows:

    "Its being Tuesday implies that this is Belgium."
    "This is Belgium if it's Tuesday."
    "It's Tuesday only if this is Belgium."
    "It can't be Tuesday unless this is Belgium."
    "Its being Tuesday is sufficient for this to be Belgium."
    "That this is Belgium is a necessary condition for its being Tuesday."

Notice the difference between "if" and "only if." We say that "p only if q" means $p \to q$ because, assuming that $p \to q$ is true, p can be true only if q is also. In other words, the only line of the truth table that has $p \to q$ true and p true also has q true. The phrasing "p is a sufficient condition for q" says that it suffices to know that p is true to be able to conclude that q is true. For example, it is sufficient that you get an A in logic for me to buy you a new car. Other things might induce me to buy you the car, but an A in logic would suffice. The phrasing "q is necessary for p" says that for p to be true, q must be true (just as we said for "p only if q").

Q: *Does the commutative law hold for the conditional? In other words, is $p \to q$ the same as $q \to p$?*

A: *No, as we can see in the following truth table:*

p	q	$p \to q$	$q \to p$
T	T	T	T
T	F	F	T
F	T	T	F
F	F	T	T

Not the same

Converse and Contrapositive

The statement $q \to p$ is called the **converse** of the statement $p \to q$. A conditional and its converse are *not* the same.

The statement $\sim q \to \sim p$ is the **contrapositive** of the statement $p \to q$. A conditional and its contrapositive are logically equivalent in the sense we define below: They have the same truth value for all possible values of p and q.

EXAMPLE 3 Converse and Contrapositive

Give the converse and contrapositive of the statement "If you earn an A in logic, then I'll buy you a new car."

Solution This statement has the form $p \to q$, where p: "you earn an A" and q: "I'll buy you a new car." The converse is $q \to p$. In words, this is "If I buy you a new car then you earned an A in logic."

The contrapositive is $(\sim q) \to (\sim p)$. In words, this is "If I don't buy you a new car, then you didn't earn an A in logic."

Assuming that the original statement is true, notice that the converse is not necessarily true. There is nothing in the original promise that prevents me from buying you a new car if you do not earn the A. On the other hand, the contrapositive is true. If I don't buy you a new car, it must be that you didn't earn an A; otherwise I would be breaking my promise.

It sometimes happens that we do want both a conditional and its converse to be true. The conjunction of a conditional and its converse is called a **biconditional**.

Biconditional

The **biconditional**, written $p \leftrightarrow q$, is defined to be the statement $(p \to q) \land (q \to p)$. Its truth table is the following:

p	q	$p \leftrightarrow q$
T	T	T
T	F	F
F	T	F
F	F	T

Phrasings of the Biconditional
We interpret each of the following as equivalent to $p \leftrightarrow q$:

p if and only if q.
p is necessary and sufficient for q.
p is equivalent to q.

Quick Example

17. "I teach math if and only if I am paid a large sum of money" can be rephrased in several ways as follows:

"I am paid a large sum of money if and only if I teach math."
"My teaching math is necessary and sufficient for me to be paid a large sum of money."
"For me to teach math, it is necessary and sufficient that I be paid a large sum of money."

A.2 # Logical Equivalence

We mentioned above that we say that two statements are **logically equivalent** if, for all possible truth values of the variables involved, the two statements always have the same truth values. If s and t are equivalent, we write $s \equiv t$. This is *not* another logical statement. It is simply the claim that the two statements s and t are logically equivalent. Here are some examples.

EXAMPLE 4   Logical Equivalence

Use truth tables to show the following:

a. $p \equiv \sim(\sim p)$. This is called **double negation**.

b. $\sim(p \wedge q) \equiv (\sim p) \vee (\sim q)$. This is one of **DeMorgan's laws**.

Solution

a. To demonstrate the logical equivalence of these two statements, we construct a truth table with columns for both p and $\sim(\sim p)$:

Same

p	$\sim p$	$\sim(\sim p)$
T	F	T
F	T	F

Because the p and $\sim(\sim p)$ columns contain the same truth values in all rows, the two statements are logically equivalent.

b. We construct a truth table showing both $\sim(p \wedge q)$ and $(\sim p) \vee (\sim q)$:

Same

p	q	$p \wedge q$	$\sim(p \wedge q)$	$\sim p$	$\sim q$	$(\sim p) \vee (\sim q)$
T	T	T	F	F	F	F
T	F	F	T	F	T	T
F	T	F	T	T	F	T
F	F	F	T	T	T	T

Because the $\sim(p \wedge q)$ column and $(\sim p) \vee (\sim q)$ column agree, the two statements are equivalent.

➡ **Before we go on . . .** The statement $\sim(p \wedge q)$ can be read as "It is not the case that both p and q are true" or "p and q are not both true." We have just shown that this is equivalent to "Either p is false or q is false." ■

Here are the two equivalences known as DeMorgan's laws.

DeMorgan's Laws

If p and q are statements, then

$$\sim(p \wedge q) \equiv (\sim p) \vee (\sim q)$$
$$\sim(p \vee q) \equiv (\sim p) \wedge (\sim q)$$

Quick Example

18. Let p: "The President is a Democrat," and let q: "The President is a Republican." Then the following two statements say the same thing:

$\sim(p \wedge q)$: "The President is not both a Democrat and a Republican."

$(\sim p) \vee (\sim q)$: "Either the President is not a Democrat, or is not a Republican (or is neither)."

Here is a list of some important logical equivalences, some of which we have already encountered. All of them can be verified by using truth tables as in Example 4. (The verifications of some of these are in the exercise set.)

Important Logical Equivalences

$\sim(\sim p) \equiv p$	The double negative law
$p \wedge q \equiv q \wedge p$	The commutative law for conjunction
$p \vee q \equiv q \vee p$	The commutative law for disjunction
$(p \wedge q) \wedge r \equiv p \wedge (q \wedge r)$	The associative law for conjunction
$(p \vee q) \vee r \equiv p \vee (q \vee r)$	The associative law for disjunction
$\sim(p \vee q) \equiv (\sim p) \wedge (\sim q)$	DeMorgan's laws
$\sim(p \wedge q) \equiv (\sim p) \vee (\sim q)$	
$p \wedge (q \vee r) \equiv (p \wedge q) \vee (p \wedge r)$	The distributive laws
$p \vee (q \wedge r) \equiv (p \vee q) \wedge (p \vee r)$	
$p \wedge p \equiv p$	The absorption laws
$p \vee p \equiv p$	
$p \rightarrow q \equiv (\sim q) \rightarrow (\sim p)$	The contrapositive law

Note that these logical equivalences apply to *any* statement. The ps, qs, and rs can stand for atomic statements or compound statements, as we see in the next example.

EXAMPLE 5 Applying Logical Equivalences

a. Apply DeMorgan's law (once) to the statement $\sim([p \wedge (\sim q)] \wedge r)$.
b. Apply the distributive law to the statement $(\sim p) \wedge [q \vee (\sim r)]$.
c. Consider: "You will get an A if either you are clever and the sun shines or you are clever and it rains." Rephrase the condition more simply using the distributive law.

Solution

a. We can analyze the given statement from the outside in. It is first of all a negation, but further, it is the negation $\sim(A \wedge B)$, where A is the compound statement $[p \wedge (\sim q)]$ and B is r:

$$\sim(\overbrace{A}^{} \wedge B)$$
$$\sim([p \wedge (\sim q)] \wedge r)$$

Now one of DeMorgan's laws is

$$\sim(A \wedge B) \equiv (\sim A) \vee (\sim B).$$

Applying this equivalence gives

$$\sim([p \wedge (\sim q)] \wedge r) \equiv (\sim[p \wedge (\sim q)]) \vee (\sim r).$$

b. The given statement has the form $A \wedge [B \vee C]$, where $A = (\sim p)$, $B = q$, and $C = (\sim r)$. So we apply the distributive law $A \wedge [B \vee C] \equiv [A \wedge B] \vee [A \wedge C]$:

$$(\sim p) \wedge [q \vee (\sim r)] \equiv [(\sim p) \wedge q] \vee [(\sim p) \wedge (\sim r)].$$

(We need not stop here: The second expression on the right is just begging for an application of DeMorgan's law . . .)

c. The condition is "either you are clever and the sun shines or you are clever and it rains." Let's analyze this symbolically: Let p: "You are clever," q: "The sun shines," and r: "It rains." The condition is then $(p \wedge q) \vee (p \wedge r)$. We can "factor out" the p using one of the distributive laws in reverse, getting

$$(p \wedge q) \vee (p \wedge r) \equiv p \wedge (q \vee r).$$

We are taking advantage of the fact that the logical equivalences we listed can be read from right to left as well as from left to right. Putting $p \wedge (q \vee r)$ back into English, we can rephrase the sentence as "You will get an A if you are clever and either the sun shines or it rains."

➡ **Before we go on . . .** In Example 5(a) we could, if we wanted, apply DeMorgan's law again, this time to the statement $\sim[p \wedge (\sim q)]$ that is part of the answer. Doing so gives

$$\sim[p \wedge (\sim q)] \equiv (\sim p) \vee \sim(\sim q) \equiv (\sim p) \vee q.$$

Notice that we've also used the double negative law. Therefore, the original expression can be simplified as follows:

$$\sim([p \wedge (\sim q)] \wedge r) \equiv (\sim[p \wedge (\sim q)]) \vee (\sim r) \equiv ((\sim p) \vee q) \vee (\sim r),$$

which we can write as

$$(\sim p) \vee q \vee (\sim r)$$

because the associative law tells us that it does not matter which two expressions we "or" first. ∎

A.3 Tautologies, Contradictions, and Arguments

Tautologies and Contradictions

A compound statement is a **tautology** if its truth value is always T, regardless of the truth values of its variables. It is a **contradiction** if its truth value is always F, regardless of the truth values of its variables.

19. $p \vee (\sim p)$ has truth table

p	$\sim p$	$p \vee (\sim p)$
T	F	T
F	T	T

all T's

and is therefore a tautology.

20. $p \wedge (\sim p)$ has truth table

p	$\sim p$	$p \wedge (\sim p)$
T	F	F
F	T	F

and is therefore a contradiction.

When a statement is a tautology, we also say that the statement is **tautological**. In common usage this sometimes means simply that the statement is self-evident. In logic it means something stronger: that the statement is always true under all circumstances. In contrast, a contradiction, or **contradictory** statement, is *never* true under any circumstances.

Some of the most important tautologies are the **tautological implications**, tautologies that have the form of implications. We look at two of them: direct reasoning and indirect reasoning.

Modus Ponens or Direct Reasoning

The following tautology is called *modus ponens* or **direct reasoning**:

$$[(p \to q) \wedge p] \to q.$$

In Words

If an implication and its antecedent (p) are both true, then so is its consequent (q).

21. *If my loving math implies that I will pass this course, and if I do love math, then I will pass this course.*

Note You can check that the statement $[(p \to q) \wedge p] \to q$ is a tautology by constructing its truth table. ∎

Tautological implications are useful mainly because they allow us to check the validity of **arguments**.

Argument

An **argument** is a list of statements called **premises** followed by a statement called the **conclusion**. If the premises are $P_1, P_2, \ldots, P_n$ and the conclusion is C, then we say that the argument is **valid** if the statement $(P_1 \wedge P_2 \wedge \ldots \wedge P_n) \to C$ is a tautology. In other words, an argument is valid if the truth of all its premises logically implies the truth of its conclusion.

Quick Examples

22. The following is a valid argument:

$$p \to q$$
$$\underline{p}$$
$$\therefore \quad q$$

(This is the traditional way of writing an argument: We list the premises above a line and then put the conclusion below; the symbol "$\therefore$" stands for the word "therefore.") This argument is valid because the statement $[(p \to q) \wedge p] \to q$ is a tautology, namely, *modus ponens*.

23. The following is an invalid argument:

$$p \to q$$
$$\underline{q}$$
$$\therefore \quad p$$

The argument is invalid because the statement $[(p \to q) \wedge q] \to p$ is not a tautology. In fact, if p is F and q is T, then the whole statement is F.

The argument in Quick Example 23 is known as the *fallacy of affirming the consequent*. It is a common invalid argument and not always obviously flawed at first sight, so it is often exploited by advertisers. For example, consider the following claim: All Olympic athletes drink Boors, so you should too. The suggestion is that, if you drink Boors, you will be an Olympic athlete:

If you are an Olympic Athlete you drink Boors.   Premise (Let's pretend this is true.)

You drink Boors.   Premise (True)

$\therefore$ You are an Olympic Athlete.   Conclusion (May be false!)

This is an error that Boors hopes you will make!

There is, however, a correct argument in which we *deny* the consequent.

Modus Tollens or Indirect Reasoning

The following tautology is called *modus tollens* or **indirect reasoning:**

$$[(p \to q) \wedge (\sim q)] \to (\sim p)$$

In Words

If an implication is true but its consequent (q) is false, then its antecedent (p) is false.

In Argument Form

$$p \to q$$
$$\frac{\sim q}{\therefore \ \sim p}$$

Quick Example

24. *If my loving math implies that I will pass this course, and if I do not pass the course, then it must be the case that I do not love math.*
In argument form:

> If I love math, then I will pass this course.
>
> I will not pass the course.
> ────────────────────────────
> Therefore, I do not love math.

Note This argument is not as direct as *modus ponens*; it contains a little twist: "If I loved math, I would pass this course. However, I will not pass this course. Therefore, it must be that I don't love math (else I *would* pass this course)." Hence the name "indirect reasoning."

Note that, again, there is a similar but fallacious argument to avoid, for instance: "If I were an Olympic athlete then I would drink Boors ($p \to q$). However, I am not an Olympic athlete ($\sim p$). Therefore, I won't drink Boors ($\sim q$)." This is a mistake Boors certainly hopes you do *not* make! ■

There are other interesting tautologies that we can use to justify arguments. We mention one more and refer the interested reader to the Website for more examples and further study.

Website
www.WanerMath.com

For an extensive list of tautologies go online and follow:

Chapter L Logic

→ List of Tautologies and Tautological Implications

Disjunctive Syllogism or "One or the Other"

The following tautologies are both known as the **disjunctive syllogism** or **one-or-the-other:**

$$[(p \vee q) \wedge (\sim p)] \to q \qquad [(p \vee q) \wedge (\sim q)] \to p$$

In Words

If one or the other of two statements is true, but one is known to be false, then the other must be true.

In Argument Form

$$p \vee q \qquad\qquad p \vee q$$
$$\frac{\sim p}{\therefore \ q} \qquad\qquad \frac{\sim q}{\therefore \ p}$$

> **Quick Example**
>
> **25.** *The butler or the cook did it. The butler didn't do it. Therefore, the cook did it.*
> In argument form:
>
> > The butler or the cook did it.
> >
> > The butler did not do it.
> > _____
> > Therefore, the cook did it.

A   EXERCISES

*Which of Exercises 1–10 are statements? Comment on the truth values of all the statements you encounter. If a sentence fails to be a statement, explain why. [**HINT:** See Example 1.]*

1. All swans are white.   **2.** The fat cat sat on the mat.

3. Look in thy glass and tell whose face thou viewest.[1]

4. My glass shall not persuade me I am old.[2]

5. There is no largest number.

6. 1,000,000,000 is the largest number.

7. Intelligent life abounds in the universe.

8. There may or may not be a largest number.

9. This is exercise number 9.   **10.** This sentence no verb.[3]

*Let p: "Our mayor is trustworthy," q: "Our mayor is a good speller," and r = "Our mayor is a patriot." Express each of the statements in Exercises 11–16 in logical form: [**HINT:** See Quick Examples 1–8, 11, and 12.]*

11. Although our mayor is not trustworthy he is a good speller.

12. Either our mayor is trustworthy or he is a good speller.

13. Our mayor is a trustworthy patriot who spells well.

14. While our mayor is both trustworthy and patriotic, he is not a good speller.

15. It may or may not be the case that our mayor is trustworthy.

16. Our mayor is either not trustworthy or not a patriot, yet he is an excellent speller.

Let p: "Willis is a good teacher," q: "Carla is a good teacher," r: "Willis' students hate math," and s: "Carla's students hate math." Express the statements in Exercises 17–24 in words.

17. $p \wedge (\sim r)$ **18.** $(\sim p) \wedge (\sim q)$

19. $q \vee (\sim q)$ **20.** $((\sim p) \wedge (\sim s)) \vee q$

21. $r \wedge (\sim r)$ **22.** $(\sim s) \vee (\sim r)$

23. $\sim(q \vee s)$ **24.** $\sim(p \wedge r)$

Assume that it is true that "Polly sings well," it is false that "Quentin writes well," and it is true that "Rita is good at math." Determine the truth of each of the statements in Exercises 25–32.

25. Polly sings well and Quentin writes well.

26. Polly sings well or Quentin writes well.

27. Polly sings poorly and Quentin writes well.

28. Polly sings poorly or Quentin writes poorly.

29. Either Polly sings well and Quentin writes poorly, or Rita is good at math.

30. Either Polly sings well and Quentin writes poorly, or Rita is not good at math.

31. Either Polly sings well or Quentin writes well, or Rita is good at math.

32. Either Polly sings well and Quentin writes well, or Rita is bad at math.

*Find the truth value of each of the statements in Exercises 33–48. [**HINT:** See Quick Examples 13–15.]*

33. "If $1 = 1$, then $2 = 2$."   **34.** "If $1 = 1$, then $2 = 3$."

35. "If $1 \neq 0$, then $2 \neq 2$."   **36.** "If $1 = 0$, then $1 = 1$."

37. "A sufficient condition for 1 to equal 2 is $1 = 3$."

38. "$1 = 1$ is a sufficient condition for 1 to equal 0."

39. "$1 = 0$ is a necessary condition for 1 to equal 1."

40. "$1 = 1$ is a necessary condition for 1 to equal 2."

41. "If I pay homage to the great Den, then the sun will rise in the east."

42. "If I fail to pay homage to the great Den, then the sun will still rise in the east."

_____

[1] William Shakespeare, Sonnet 3.

[2] *Ibid.*, Sonnet 22.

[3] From *Metamagical Themas: Questing for the Essence of Mind and Pattern* by Douglas R. Hofstadter (Bantam Books, New York 1986).

43. "In order for the sun to rise in the east, it is necessary that it sets in the west."

44. "In order for the sun to rise in the east, it is sufficient that it sets in the west."

45. "The sun rises in the west only if it sets in the west."

46. "The sun rises in the east only if it sets in the east."

47. "In order for the sun to rise in the east, it is necessary and sufficient that it sets in the west."

48. "In order for the sun to rise in the west, it is necessary and sufficient that it sets in the east."

Construct the truth tables for the statements in Exercises 49–62. [**HINT:** See Example 2.]

49. $p \land (\sim q)$

50. $p \lor (\sim q)$

51. $\sim(\sim p) \lor p$

52. $p \land (\sim p)$

53. $(\sim p) \land (\sim q)$

54. $(\sim p) \lor (\sim q)$

55. $(p \land q) \land r$

56. $p \land (q \land r)$

57. $p \land (q \lor r)$

58. $(p \land q) \lor (p \land r)$

59. $p \to (q \lor p)$

60. $(p \lor q) \to \sim p$

61. $p \leftrightarrow (p \lor q)$

62. $(p \land q) \leftrightarrow \sim p$

Use truth tables to verify the logical equivalences given in Exercises 63–72.

63. $p \land p \equiv p$

64. $p \lor p \equiv p$

65. $p \lor q \equiv q \lor p$
(Commutative law for disjunction)

66. $p \land q \equiv q \land p$
(Commutative law for conjunction)

67. $\sim(p \lor q) \equiv (\sim p) \land (\sim q)$

68. $\sim(p \land (\sim q)) \equiv (\sim p) \lor q$

69. $(p \land q) \land r \equiv p \land (q \land r)$
(Associative law for conjunction)

70. $(p \lor q) \lor r \equiv p \lor (q \lor r)$
(Associative law for disjunction)

71. $p \to q \equiv (\sim q) \to (\sim p)$ **72.** $\sim(p \to q) \equiv p \land (\sim q)$

In Exercises 73–78, use truth tables to check whether the given statement is a tautology, a contradiction, or neither. [**HINT:** See Quick Examples 19 and 20.]

73. $p \land (\sim p)$

74. $p \land p$

75. $p \land \sim(p \lor q)$

76. $p \lor \sim(p \lor q)$

77. $p \lor \sim(p \land q)$

78. $q \lor \sim(p \land (\sim p))$

Apply the stated logical equivalence to the given statement in Exercises 79–84. [**HINT:** See Example 5(a), (b).]

79. $p \lor (\sim p)$; the commutative law

80. $p \land (\sim q)$; the commutative law

81. $\sim(p \land (\sim q))$; DeMorgan's law

82. $\sim(q \lor (\sim q))$; DeMorgan's law

83. $p \lor ((\sim p) \land q)$; the distributive law

84. $(\sim q) \land ((\sim p) \lor q)$; the distributive law

In Exercises 85–88, use the given logical equivalence to rewrite the given sentence. [**HINT:** See Example 5(c).]

85. It is not true that both I am Julius Caesar and you are a fool. DeMorgan's law.

86. It is not true that either I am Julius Caesar or you are a fool. DeMorgan's law.

87. Either it is raining and I have forgotten my umbrella, or it is raining and I have forgotten my hat. The distributive law.

88. I forgot my hat or my umbrella, and I forgot my hat or my glasses. The distributive law.

Give the contrapositive and converse of each of the statements in Exercises 89 and 90, phrasing your answers in words.

89. "If I think, then I am."

90. "If these birds are of a feather, then they flock together."

Exercises 91 and 92 are multiple choice. Indicate which statement is equivalent to the given statement, and say why that statement is equivalent to the given one.

91. "In order for you to worship Den, it is necessary for you to sacrifice beasts of burden."
   (A) "If you are not sacrificing beasts of burden, then you are not worshiping Den."
   (B) "If you are sacrificing beasts of burden, then you are worshiping Den."
   (C) "If you are not worshiping Den, then you are not sacrificing beasts of burden."

92. "In order to read the Tarot, it is necessary for you to consult the Oracle."
   (A) "In order to consult the Oracle, it is necessary to read the Tarot."
   (B) "In order not to consult the Oracle, it is necessary not to read the Tarot."
   (C) "In order not to read the Tarot, it is necessary not to read the Oracle."

In Exercises 93–102, write the given argument in symbolic form (use the underlined letters to represent the statements containing them), then decide whether it is valid or not. If it is valid, name the validating tautology. [**HINT:** See Quick Examples 19–25.]

93. If I am <u>h</u>ungry I am also <u>t</u>hirsty. I am hungry. Therefore, I am thirsty.

94. If I am not <u>h</u>ungry, then I certainly am not <u>t</u>hirsty either. I am not thirsty, and so I cannot be hungry.

95. For me to bring my umbrella, it's sufficient that it rain. It is not raining. Therefore, I will not bring my umbrella.

96. For me to bring my umbrella, it's necessary that it rain. But it is not raining. Therefore, I will not bring my umbrella.

97. For me to pass math, it is sufficient that I have a good teacher. I will not pass math. Therefore, I have a bad teacher.

98. For me to pass math, it is necessary that I have a good teacher. I will pass math. Therefore, I have a good teacher.

99. I will either pass math or I have a bad teacher. I have a good teacher. Therefore, I will pass math.

100. Either roses are not red or violets are not blue. But roses are red. Therefore, violets are not blue.

101. I am either smart or athletic, and I am athletic. So I must not be smart.

102. The president is either wise or strong. She is strong. Therefore, she is not wise.

In Exercises 103–108, use the stated tautology to complete the argument.

103. If John is a swan, it is necessary that he is green. John is indeed a swan. Therefore, _____. (*Modus ponens.*)

104. If Jill had been born in Texas, then she would be able to ride horses. But Jill cannot ride horses. Therefore, ____. (*Modus tollens.*)

105. If John is a swan, it is necessary that he is green. But John is not green. Therefore, _____. (*Modus tollens.*)

106. If Jill had been born in Texas, then she would be able to ride horses. Jill was born in Texas. Therefore, ___ (*Modus ponens.*)

107. Peter is either a scholar or a gentleman. He is not, however, a scholar. Therefore, ___. (Disjunctive syllogism.)

108. Pam is either a plumber or an electrician. She is not, however, an electrician. Therefore, ___ (Disjunctive syllogism.)

Communication and Reasoning Exercises

109. If two statements are logically equivalent, what can be said about their truth tables?

110. If a proposition is neither a tautology nor a contradiction, what can be said about its truth table?

111. If A and B are two compound statements such that $A \vee B$ is a contradiction, what can you say about A and B?

112. If A and B are two compound statements such that $A \wedge B$ is a tautology, what can you say about A and B?

113. Give an example of an instance where $p \rightarrow q$ means that q causes p.

114. Complete the following. If $p \rightarrow q$, then its converse, ___ , is the statement that ___ and (is/is not) logically equivalent to $p \rightarrow q$.

115. Give an instance of a true biconditional $p \leftrightarrow q$ where neither one of p or q causes the other.

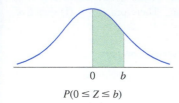

$P(0 \leq Z \leq b)$

Z	0.00	0.01	0.02	0.03
2.3	.4893	.4896	.4898	.4901
→ 2.4	.4918	.4920	.4922	.4925
2.5	.4938	.4940	.4941	.4943

The table below gives the probabilities $P(0 \leq Z \leq b)$, where Z is a standard normal variable. For example, to find $P(0 \leq Z \leq 2.43)$, write 2.43 as 2.4 + 0.03, and read the entry in the row labeled 2.4 and the column labeled 0.03. From the portion of the table shown at left, you will see that $P(0 \leq Z \leq 2.43) = .4925$.

Z	0.00	0.01	0.02	0.03	0.04	0.05	0.06	0.07	0.08	0.09
0.0	.0000	.0040	.0080	.0120	.0160	.0199	.0239	.0279	.0319	.0359
0.1	.0398	.0438	.0478	.0517	.0557	.0596	.0636	.0675	.0714	.0753
0.2	.0793	.0832	.0871	.0910	.0948	.0987	.1026	.1064	.1103	.1141
0.3	.1179	.1217	.1255	.1293	.1331	.1368	.1406	.1443	.1480	.1517
0.4	.1554	.1591	.1628	.1664	.1700	.1736	.1772	.1808	.1844	.1879
0.5	.1915	.1950	.1985	.2019	.2054	.2088	.2123	.2157	.2190	.2224
0.6	.2257	.2291	.2324	.2357	.2389	.2422	.2454	.2486	.2517	.2549
0.7	.2580	.2611	.2642	.2673	.2704	.2734	.2764	.2794	.2823	.2852
0.8	.2881	.2910	.2939	.2967	.2995	.3023	.3051	.3078	.3106	.3133
0.9	.3159	.3186	.3212	.3238	.3264	.3289	.3315	.3340	.3365	.3389
1.0	.3413	.3438	.3461	.3485	.3508	.3531	.3554	.3577	.3599	.3621
1.1	.3643	.3665	.3686	.3708	.3729	.3749	.3770	.3790	.3810	.3830
1.2	.3849	.3869	.3888	.3907	.3925	.3944	.3962	.3980	.3997	.4015
1.3	.4032	.4049	.4066	.4082	.4099	.4115	.4131	.4147	.4162	.4177
1.4	.4192	.4207	.4222	.4236	.4251	.4265	.4279	.4292	.4306	.4319
1.5	.4332	.4345	.4357	.4370	.4382	.4394	.4406	.4418	.4429	.4441
1.6	.4452	.4463	.4474	.4484	.4495	.4505	.4515	.4525	.4535	.4545
1.7	.4554	.4564	.4573	.4582	.4591	.4599	.4608	.4616	.4625	.4633
1.8	.4641	.4649	.4656	.4664	.4671	.4678	.4686	.4693	.4699	.4706
1.9	.4713	.4719	.4726	.4732	.4738	.4744	.4750	.4756	.4761	.4767
2.0	.4772	.4778	.4783	.4788	.4793	.4798	.4803	.4808	.4812	.4817
2.1	.4821	.4826	.4830	.4834	.4838	.4842	.4846	.4850	.4854	.4857
2.2	.4861	.4864	.4868	.4871	.4875	.4878	.4881	.4884	.4887	.4890
2.3	.4893	.4896	.4898	.4901	.4904	.4906	.4909	.4911	.4913	.4916
2.4	.4918	.4920	.4922	.4925	.4927	.4929	.4931	.4932	.4934	.4936
2.5	.4938	.4940	.4941	.4943	.4945	.4946	.4948	.4949	.4951	.4952
2.6	.4953	.4955	.4956	.4957	.4959	.4960	.4961	.4962	.4963	.4964
2.7	.4965	.4966	.4967	.4968	.4969	.4970	.4971	.4972	.4973	.4974
2.8	.4974	.4975	.4976	.4977	.4977	.4978	.4979	.4979	.4980	.4981
2.9	.4981	.4982	.4982	.4983	.4984	.4984	.4985	.4985	.4986	.4986
3.0	.4987	.4987	.4987	.4988	.4988	.4989	.4989	.4989	.4990	.4990

Answers to Selected Exercises

Chapter 0

Section 0.1

1. -48 **3.** $2/3$ **5.** -1 **7.** 9 **9.** 1 **11.** 33 **13.** 14
15. $5/18$ **17.** 13.31 **19.** 6 **21.** $43/16$ **23.** 0
25. `3*(2-5)` **27.** `3/(2-5)` **29.** `(3-1)/(8+6)`
31. `3-(4+7)/8` **33.** `2/(3+x)-x*y^2`
35. `3.1x^3-4x^(-2)-60/(x^2-1)`
37. `(2/3)/5` **39.** `3^(4-5)*6`
41. `3*(1+4/100)^(-3)` **43.** `3^(2*x-1)+4^x-1`
45. `2^(2x^2-x+1)`
47. `4*e^(-2*x)/(2-3e^(-2*x))` or
`(4*e^(-2*x))/(2-3e^(-2*x))`
49. `3(1-(-1/2)^2)^2+1`

Section 0.2

1. 27 **3.** -36 **5.** $4/9$ **7.** $-1/8$ **9.** 16 **11.** 2 **13.** 32
15. 2 **17.** x^5 **19.** $-\dfrac{y}{x}$ **21.** $\dfrac{1}{x}$ **23.** $x^3 y$ **25.** $\dfrac{z^4}{y^3}$ **27.** $\dfrac{x^6}{y^6}$
29. $\dfrac{x^4 y^6}{z^4}$ **31.** $\dfrac{3}{x^4}$ **33.** $\dfrac{3}{4x^{2/3}}$ **35.** $1 - 0.3x^2 - \dfrac{6}{5x}$ **37.** 2
39. $1/2$ **41.** $4/3$ **43.** $2/5$ **45.** 7 **47.** 5 **49.** -2.668
51. $3/2$ **53.** 2 **55.** 2 **57.** ab **59.** $x + 9$ **61.** $x\sqrt[3]{a^3 + b^3}$
63. $\dfrac{2y}{\sqrt{x}}$ **65.** $3^{1/2}$ **67.** $x^{3/2}$ **69.** $(xy^2)^{1/3}$ **71.** $x^{3/2}$
73. $\dfrac{3}{5}x^{-2}$ **75.** $\dfrac{3}{2}x^{-1.2} - \dfrac{1}{3}x^{-2.1}$ **77.** $\dfrac{2}{3}x - \dfrac{1}{2}x^{0.1} + \dfrac{4}{3}x^{-1.1}$
79. $\dfrac{3}{4}x^{1/2} - \dfrac{5}{3}x^{-1/2} + \dfrac{4}{3}x^{-3/2}$ **81.** $\dfrac{3}{4}x^{2/5} - \dfrac{7}{2}x^{-3/2}$
83. $(x^2 + 1)^{-3} - \dfrac{3}{4}(x^2 + 1)^{-1/3}$ **85.** $\sqrt[3]{2^2}$ **87.** $\sqrt[3]{x^4}$
89. $\sqrt[5]{\sqrt{x}\sqrt[3]{y}}$ **91.** $-\dfrac{3}{2\sqrt[4]{x}}$ **93.** $\dfrac{0.2}{\sqrt[3]{x^2}} + \dfrac{3\sqrt{x}}{7}$
95. $\dfrac{3}{4\sqrt{(1-x)^5}}$ **97.** 64 **99.** $\sqrt{3}$ **101.** $1/x$ **103.** xy
105. $\left(\dfrac{y}{x}\right)^{1/3}$ **107.** ± 4 **109.** $\pm 2/3$ **111.** $-1, -1/3$
113. -2 **115.** 16 **117.** ± 1 **119.** $33/8$

Section 0.3

1. $4x^2 + 6x$ **3.** $2xy - y^2$ **5.** $x^2 - 2x - 3$
7. $2y^2 + 13y + 15$ **9.** $4x^2 - 12x + 9$
11. $x^2 + 2 + 1/x^2$ **13.** $4x^2 - 9$ **15.** $y^2 - 1/y^2$
17. $2x^3 + 6x^2 + 2x - 4$ **19.** $x^4 - 4x^3 + 6x^2 - 4x + 1$
21. $y^5 + 4y^4 + 4y^3 - y$ **23.** $(x + 1)(2x + 5)$
25. $(x^2 + 1)^5 (x + 3)^3 (x^2 + x + 4)$
27. $-x^3 (x^3 + 1)\sqrt{x + 1}$ **29.** $(x + 2)\sqrt{(x + 1)^3}$

31. a. $x(2 + 3x)$ **b.** $x = 0, -2/3$
33. a. $2x^2(3x - 1)$ **b.** $x = 0, 1/3$
35. a. $(x - 1)(x - 7)$ **b.** $x = 1, 7$
37. a. $(x - 3)(x + 4)$ **b.** $x = 3, -4$
39. a. $(2x + 1)(x - 2)$ **b.** $x = -1/2, 2$
41. a. $(2x + 3)(3x + 2)$ **b.** $x = -3/2, -2/3$
43. a. $(3x - 2)(4x + 3)$ **b.** $x = 2/3, -3/4$
45. a. $(x + 2y)^2$ **b.** $x = -2y$
47. a. $(x^2 - 1)(x^2 - 4)$ **b.** $x = \pm 1, \pm 2$

Section 0.4

1. $\dfrac{2x^2 - 7x - 4}{x^2 - 1}$ **3.** $\dfrac{3x^2 - 2x + 5}{x^2 - 1}$ **5.** $\dfrac{x^2 - x + 1}{x + 1}$
7. $\dfrac{x^2 - 1}{x}$ **9.** $\dfrac{2x - 3}{x^2 y}$ **11.** $\dfrac{(x + 1)^2}{(x + 2)^4}$ **13.** $\dfrac{-1}{\sqrt{(x^2 + 1)^3}}$
15. $\dfrac{-(2x + y)}{x^2(x + y)^2}$

Section 0.5

1. -1 **3.** 5 **5.** $13/4$ **7.** $43/7$ **9.** -1 **11.** $(c - b)/a$
13. $x = -4, 1/2$ **15.** No solutions **17.** $\pm\sqrt{\dfrac{5}{2}}$ **19.** -1
21. $-1, 3$ **23.** $\dfrac{1 \pm \sqrt{5}}{2}$ **25.** 1 **27.** $\pm 1, \pm 3$
29. $\pm\sqrt{\dfrac{-1 + \sqrt{5}}{2}}$ **31.** $-1, -2, -3$ **33.** -3 **35.** 1
7. -2 **39.** $1, \pm\sqrt{5}$ **41.** $\pm 1, \pm\dfrac{1}{\sqrt{2}}$ **43.** $-2, -1, 2, 3$

Section 0.6

1. $0, 3$ **3.** $\pm\sqrt{2}$ **5.** $-1, -5/2$ **7.** -3 **9.** $0, -1$
11. $x = -1$ ($x = -2$ is not a solution.)
13. $-2, -3/2, -1$ **15.** -1 **17.** $\pm\sqrt[4]{2}$ **19.** ± 1
21. ± 3 **23.** $2/3$ **25.** $-4, -1/4$

Section 0.7

1. $P(0, 2), Q(4, -2), R(-2, 3), S(-3.5, -1.5), T(-2.5, 0), U(2, 2.5)$

3. 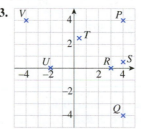 **5.**

7. 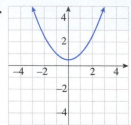   **9.**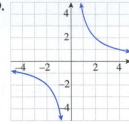

11. a. Not defined **b.** Not defined **c.** Yes, $f(-10) = 0$
13. a. -7 **b.** -3 **c.** 1 **d.** $4y - 3$ **e.** $4(a + b) - 3$
15. a. 3 **b.** 6 **c.** 2 **d.** 6 **e.** $a^2 + 2a + 3$
f. $(x + h)^2 + 2(x + h) + 3$ **17. a.** 2 **b.** 0 **c.** 65/4
d. $x^2 + 1/x$ **e.** $(s + h)^2 + 1/(s + h)$
f. $(s + h)^2 + 1/(s + h) - (s^2 + 1/s)$

11. 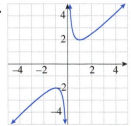   **13.** $\sqrt{2}$
15. $\sqrt{a^2 + b^2}$
17. 1/2
19. Circle with center $(0, 0)$ and radius 3

19. 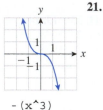   **21.**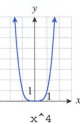

$-(x\text{\textasciicircum}3)$ $x\text{\textasciicircum}4$

23.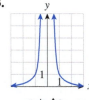

$1/x\text{\textasciicircum}2$

25. a. (A) **b.** (D) **c.** (E) **d.** (F) **e.** (C) **f.** (B)
27. $0.1*x\text{\textasciicircum}2-4*x+5$

x	0	1	2	3
$f(x)$	5	1.1	-2.6	-6.1
x	4	5	6	7
$f(x)$	-9.4	-12.5	-15.4	-18.1
x	8	9	10	
$f(x)$	-20.6	-22.9	-25	

29. $(x\text{\textasciicircum}2-1)/(x\text{\textasciicircum}2+1)$

x	0.5	1.5	2.5	3.5
$h(x)$	-0.6000	0.3846	0.7241	0.8491
x	4.5	5.5	6.5	7.5
$h(x)$	0.9059	0.9360	0.9538	0.9651
x	8.5	9.5	10.5	
$h(x)$	0.9727	0.9781	0.9820	

31. a. -1 **b.** 2 **c.** 2   **33. a.** 1 **b.** 0 **c.** 1

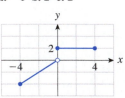   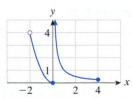

$x*(x<0)+2*(x>=0)$ $(x\text{\textasciicircum}2)*(x<=0)+(1/x)*$
$(0<x)$

Section 0.8

1.

Exponential Form	$10^2 = 100$	$4^3 = 64$	$4^4 = 256$
Logarithmic Form	$\log_{10} 100 = 2$	$\log_4 64 = 3$	$\log_4 256 = 4$

Exponential Form	$0.45^0 = 1$	$8^{1/2} = 2\sqrt{2}$	$4^{-3} = \dfrac{1}{64}$
Logarithmic Form	$\log_{0.45} 1 = 0$	$\log_8 2\sqrt{2} = \dfrac{1}{2}$	$\log_4\left(\dfrac{1}{64}\right) = -3$

3.

Exponential Form	$0.3^2 = 0.09$	$\left(\dfrac{1}{2}\right)^0 = 1$	$10^{-3} = 0.001$
Logarithmic Form	$\log_{0.3} 0.09 = 2$	$\log_{1/2} 1 = 0$	$\log_{10} 0.001 = -3$

Exponential Form	$9^{-2} = \dfrac{1}{81}$	$2^{10} = 1,024$	$64^{-1/3} = \dfrac{1}{4}$
Logarithmic Form	$\log_9 \dfrac{1}{81} = -2$	$\log_2 1,024 = 10$	$\log_{64} \dfrac{1}{4} = -\dfrac{1}{3}$

5. 2 **7.** -2 **9.** 5 **11.** 1 **13.** -2 **15.** 1/2 **17.** 12
19. 1/10 **21.** 3/8 **23.** $x^4 y^5$ **25.** $\dfrac{x^2 y^3}{z^4}$ **27.** $\dfrac{2^x}{x^2}$
29. $b + c$ **31.** $a + b + c$ **33.** $-c$ **35.** $a - b$
37. $2a - c$ **39.** $4a$ **41.** $b - 2$ **43.** $1 - a$ **45.** $c/2$
47. 2 **49.** 2 **51.** 4 **53.** 0.4210 **55.** 1.3972

Chapter 1

Section 1.1

1. a. 2 **b.** -0.5 **3. a.** -2.5 **b.** 8 **c.** -8 **5. a.** 20 **b.** 30
c. 30 **d.** 20 **e.** 0 **f.** 20   **7. a.** 0 **b.** -3 **c.** 3 **d.** 3
9. a. Yes; $f(4) = 63/16$ **b.** Not defined **c.** Yes; -2

35. a. 0 **b.** 2 **c.** 3 **d.** 3

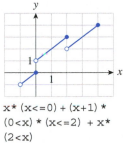

x* (x<=0) + (x+1) *
(0<x) * (x<=2) + x*
(2<x)

37. a. $h(2x + h)$ **b.** $2x + h$ **39. a.** $-h(2x + h)$ **b.** $-(2x + h)$
41. a. $p(2) = 2.95$; Pemex produced 2.95 million barrels of crude oil per day in 2010. $p(3) = 2.94$; Pemex produced 2.94 million barrels of crude oil per day in 2011. $p(6) = 2.79$; Pemex produced 2.79 million barrels of crude oil per day in 2014. **b.** $p(4) - p(2) = 2.91 - 2.95 = -0.04$; Crude oil production by Pemex decreased by 0.04 million barrels/day from 2010 ($t = 2$) to 2012 ($t = 4$).
43. a. Graph of p:

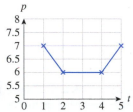

$p(4.5) \approx 6.5$. Interpretation: The popularity of Twitter midway through 2012 was about 6.5%. **b.** (D) **45.** $f(7) \approx 1,000$. Interpretation: Approximately 1,000,000 homes were started in 2007. $f(14) \approx 600$: Approximately 600,000 homes were started in 2014. **b.** $f(9.5) \approx 450$. Interpretation: 450,000 homes were started in the year beginning July 2009.
47. $f(7 - 3) \approx 1,600$, $f(7) - f(3) \approx -500$. Interpretation: 1,600,000 homes were started in 2004; there were 500,000 fewer housing starts in 2007 than in 2003. **49.** $t = 0$. Interpretation: The greatest 5-year increase in the number of housing starts occurred in 2000–2005. **51. a.** $n(2) \approx 400$, $n(4) \approx 400$, $n(4.5) \approx 350$. Interpretation: Abercrombie & Fitch's net income was $400 million in 2006, $400 million in 2008, and $350 million in the year ending June 2009.
b. $t \approx 8$. Interpretation: Between Dec. 2007 and Dec. 2012, Abercrombie & Fitch's net income was increasing most rapidly in Dec. 2012. **c.** $t \approx 5$. Interpretation: Between Dec. 2007 and Dec. 2012, Abercrombie & Fitch's net income was decreasing most rapidly in Dec. 2009. **53. a.** $[0, 8]$. $t \geq 0$ is not an appropriate domain because it would predict federal funding of NASA beyond 1966, whereas the model is based only on data up to 1966. **b.** $p(5) \approx 2.4$. In 1963, 2.4% of the U.S. federal budget was allocated to NASA. **c.** $t = 5$. The percentage of the budget allocated to NASA was increasing most rapidly in 1963. **55. a.** 100*(1-12200/t^4.48)

b. Graph:

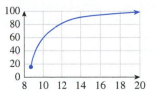

c. Table:

t	9	10	11	12	13	14
$p(t)$	35.2	59.6	73.6	82.2	87.5	91.1
t	15	16	17	18	19	20
$p(t)$	93.4	95.1	96.3	97.1	97.7	98.2

d. 82.2% **e.** 14 months **57. a.** $v(10) \approx 58$, $v(16) = 200$, $v(28) = 3,800$. Processor speeds were about 58 MHz in 1990, 200 MHz in 1996, and 3,800 MHz in 2008.
b. (8*(1.22)^x)*(x<16)+(400*x-6200)*
(x>=16)*(x<25)+3800*(x>=25)
c. Graph:

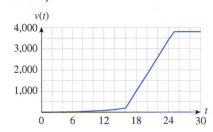

Table:

t	0	2	4	6	8	10
$v(t)$	8.0	12	18	26	39	58
t	12	14	16	18	20	22
$v(t)$	87	130	200	1,000	1,800	2,600
t	24	26	28	30		
$v(t)$	3,400	3,800	3,800	3,800		

d. 2003
59. a.

$$T(x) = \begin{cases} 0.10x & \text{if } 0 < x \leq 9{,}225 \\ 922.50 + 0.15(x - 9{,}225) & \text{if } 9{,}225 < x \leq 37{,}450 \\ 5{,}156.25 + 0.25(x - 37{,}450) & \text{if } 37{,}450 < x \leq 90{,}750 \\ 18{,}481.25 + 0.28(x - 90{,}750) & \text{if } 90{,}750 < x \leq 189{,}300 \\ 46{,}075.25 + 0.33(x - 189{,}300) & \text{if } 189{,}300 < x \leq 411{,}500 \\ 119{,}401.25 + 0.35(x - 411{,}500) & \text{if } 411{,}500 < x \leq 413{,}200 \\ 119{,}996.25 + 0.396(x - 413{,}200) & \text{if } 413{,}200 < x \end{cases}$$

b. $7,043.75 **61.** t; m
63. $y(x) = 4x^2 - 2$ (or $f(x) = 4x^2 - 2$)
65. False. A graph usually gives infinitely many values of the function, while a numerical table will give only a finite number of values.

67. False. In a numerically specified function, only certain values of the function are specified, so we cannot know its value on every real number in $[0, 10]$, whereas an algebraically specified function would give values for every real number in $[0, 10]$. **69.** False: Functions with infinitely many points in their domain (such as $f(x) = x^2$) cannot be specified numerically. **71.** As the text reminds us, to evaluate f of a quantity (such as $x + h$) replace x everywhere by the *whole quantity* $x + h$, getting $f(x + h) = (x + h)^2 - 1$.
73. They are different portions of the graph of the associated equation $y = f(x)$. **75.** The graph of g is the same as the graph of f but shifted 5 units to the right.

Section 1.2

1. a. $s(x) = x^2 + x$ **b.** Domain: $(-\infty, +\infty)$ **c.** 6
3. a. $p(x) = (x - 1)\sqrt{x + 10}$ **b.** Domain: $[-10, 0)$ **c.** -14
5. a. $q(x) = \frac{\sqrt{10 - x}}{x - 1}$ **b.** Domain: $0 \leq x \leq 10; x \neq 1$
c. Undefined **7. a.** $m(x) = 5(x^2 + 1)$ **b.** Domain: $(-\infty, +\infty)$
c. 10 **9.** $N(t) = 200 + 10t$ (N = number of music files, t = time in days) **11.** $A(x) = x^2/2$ **13.** $C(x) = 12x$

15. $h(n) = \begin{cases} 4 & \text{if } 1 \leq n \leq 5 \\ 0 & \text{if } n > 5 \end{cases}$ **17.** $C(x) = 1{,}500x + 1{,}000$

per day **a.** $5,500 **b.** $1,500 **c.** $1,500 **d.** Variable cost = $1,500x$; Fixed cost = $1,000; Marginal cost = $1,500 per piano
e. Graph:

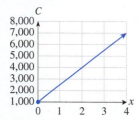

19. a. $C(x) = 0.4x + 70$, $R(x) = 0.5x$, $P(x) = 0.1x - 70$
b. $P(500) = -20$; a loss o $20 **c.** 700 copies
21. $R(x) = 100x$, $P(x) = -2{,}000 + 90x - 0.2x^2$; at least 24 jerseys **23.** $P(x) = -1.7 - 0.02x + 0.0001x^2$; approximately 264 thousand sq. ft. **25.** $P(x) = 100x - 5{,}132$, with domain $[0, 405]$. For profit, $x \geq 52$ **27.** 5,000 units
29. $FC/(SP - VC)$ **31.** $P(x) = 579.7x - 20{,}000$, with domain $x \geq 0$; $x = 34.50$ g per day for breakeven
33. a. Graph:

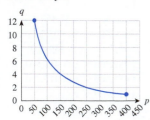

b. Demand decreases by 3.8 million units per year. **c.** (D)
35. a. 1,027 million units **b.** 1,607 million units
c. $R(p) = 0.17p^3 - 63p^2 + 5{,}900p$ million dollars per year; $113 billion per year **d.** Increases;

Graph:

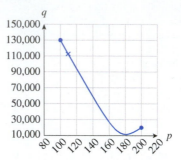

37. $240 per skateboard **39. a.** $110 per phone **b.** Shortage of 25 million phones **41. a.** Equilibrium price: $200; equilibrium demand: 2.8 million units
b. Graph:

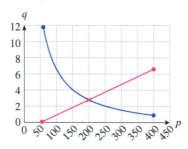

c. A shortage of around 9.2 million e-readers **43. a.** $12,000
b. $N(q) = 2{,}000 + 100q^2 - 500q$; this is the cost of removing q lb of PCPs per day after the subsidy is taken into account.
c. $N(20) = 32{,}000$; the net cost of removing 20 lb of PCPs per day is $32,000. **45. a.** (B) **b.** $37 billion **47. a.** (C)
b. $20.80 per shirt if the team buys 70 shirts
Graph:

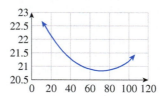

49. a. (A), (D); Graph:

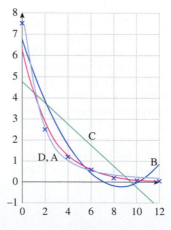

b. Model (A); Approximately $0.0021   **51.** A linear model (A) is the best choice; a plot of the given points gives a straight line.   **53.** Model (D) is the best choice; Model (A) would predict increasing demand with increasing price, Model (B) would correspond to a curve that becomes less steep as p increases, and Model (C) would give a concave-up parabola.   **55.** $A(t) = 5,000(1 + 0.0005/12)^{12t}$; $5,018
57. 2033   **59.** 31.0 g, 9.25 g, 2.76 g   **61.** 20,000 years
63. a. 1,000 years: 65%, 2,000 years: 42%, 3,000 years: 27%
b. 1,600 years   **65.** 30   **67.** Curve fitting. The model is based on fitting a curve to a given set of observed data.   **69.** The cost of downloading a movie was $4 in January and is decreasing by 20¢ per month.   **71.** Variable; marginal.   **73.** Yes, as long as the supply is going up at a faster rate, as illustrated by the following graph:

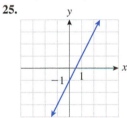

Equilibrium price

75. Extrapolate both models and choose the one that gives the most reasonable predictions.   **77.** They are ≥ 0.
79. Books per person

Section 1.3

1. 11; $m = 3$ **3.** -4; $m = -1$ **5.** 7; $m = 3/2$
7. $f(x) = -x/2 - 2$ **9.** $f(0) = -5$, $f(x) = -x - 5$
11. f is linear: $f(x) = 4x + 6$ **13.** g is linear: $g(x) = 2x - 1$
15. $-3/2$ **17.** $1/6$ **19.** Undefined   **21.** 0   **23.** $-4/3$

25.

27.

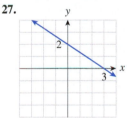

29.

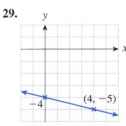

31.

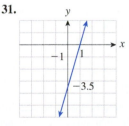

33.

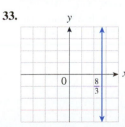

35.

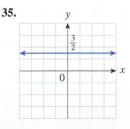

37.

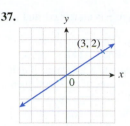

39. 2   **41.** 2   **43.** -2 **45.** Undefined   **47.** 1.5   **49.** -0.09
51. 1/2   **53.** $(d - b)/(c - a)$ **55.** Undefined   **57.** $-b/a$
59. a. 1 **b.** 1/2 **c.** 0 **d.** 3 **e.** $-1/3$ **f.** -1 **g.** Undefined

h. $-1/4$ **i.** -2 **61.** $y = 3x$ **63.** $y = \dfrac{1}{4}x - 1$

65. $y = 10x - 203.5$ **67.** $y = -5x + 6$
69. $y = -3x + 2.25$ **71.** $y = -x + 12$

73. $y = 2x + 4$ **75.** $y = \dfrac{q}{p}x$ **77.** $y = q$ **79.** $y = -\dfrac{q}{p}x$

81. Fixed cost = $8,000, marginal cost = $25 per bicycle   **83.** $C = 205x + 20$; $205 per iPhone; $8,220
85. $q = -40p + 2,000$ **87. a.** $q = -5.8p + 2,953$;
1,416 million phones **b.** $1; 5.8 million
89. a. $q = -4,500p + 41,500$ **b.** Rides per day per $1 increase in the fare; ridership decreases by 4,500 rides per day for every $1 increase in the fare. **c.** 14,500 rides per day
91. a. $y = 40t + 290$ million pounds of pasta **b.** 890 million pounds   **93. a.** $N = -0.29t + 0.92$ **b.** Billions of dollars per year; Amazon's net income decreased at a rate of $0.29 billion per year. **c.** $0.05 billion   **95. a.** 2.5 ft/sec **b.** 20 ft along the track **c.** after 6 sec   **97. a.** 130 mph **b.** $s = 130t - 1,300$
99. a. $L = 42.5n + 500$ **b.** Pages per edition; *Applied Calculus* is growing at a rate of 42.5 pages per edition. **c.** 24th edition
101. $F = 1.8C + 32$; 86°F; 72°F; 14°F; 7°F
103. a. $J = 0.54S - 86$ **b.** $57 million **c.** Millions of dollars of JetBlue Airways net income per million dollars of Southwest Airlines net income; JetBlue Airways earned an additional net income of $0.54 per $1 additional net income earned by Southwest Airlines.   **105.** $I(N) = 0.05N + 50,000$; $N = $1,000,000; marginal income is $m = 5$¢ per dollar of net profit
107. Increasing at 400 MHz per year
109. a. $y = 31.1t + 78$ **b.** $y = 112.5t - 1,550$

c. $y = \begin{cases} 31.1t + 78 & \text{if } 0 \le t < 20 \\ 112.5t - 1,550 & \text{if } 20 \le t \le 40 \end{cases}$ or

$y = \begin{cases} 31.1t + 78 & \text{if } 0 \le t \le 20 \\ 112.5t - 1,550 & \text{if } 20 < t \le 40 \end{cases}$

d. $2,275,000, in good agreement with the actual value shown in the graph.

111. $N = \begin{cases} 0.22t + 3 & \text{if } 0 \le t \le 5 \\ -0.15t + 4.85 & \text{if } 5 < t \le 9 \end{cases}$

3.8 million jobs
113. Compute the corresponding successive changes Δx in x and Δy in y, and compute the ratios $\Delta y/\Delta x$. If the answer is always the same number, then the values in the table come from a linear function.

115. $f(x) = -\dfrac{a}{b}x + \dfrac{c}{b}$. If $b = 0$, then $\dfrac{a}{b}$ is undefined, and y cannot be specified as a function of x. (The graph of the resulting equation would be a vertical line.)　**117.** slope, 3.
119. If m is positive, then y will increase as x increases; if m is negative, then y will decrease as x increases; if m is zero, then y will not change as x changes.　**121.** The slope increases, because an increase in the y-coordinate of the second point increases Δy while leaving Δx fixed.　**123.** Bootlags per zonar; bootlags　**125.** It must increase by 10 units each day, including the third.　**127.** (B)　**129.** It is linear with slope $m + n$.　**131.** Answers may vary. For example, $f(x) = x^{1/3}, g(x) = x^{2/3}$　**133.** Increasing the number of items from the break-even number results in a profit: Because the slope of the revenue graph is larger than the slope of the cost graph, it is higher than the cost graph to the right of the point of intersection, and hence corresponds to a profit.

Section 1.4

1. 6　**3.** 86　**5. a.** 0.5 (better fit) **b.** 0.75
7. a. 27.42 **b.** 27.16 (better fit)
9. $y = 1.5x - 0.6667$
Graph:

11. $y = 0.7x + 0.85$
Graph:

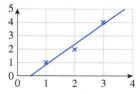
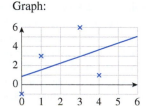

13. a. $r = 0.9959$ (best, not perfect) **b.** $r = 0.9538$
c. $r = 0.3273$ (worst)

15.

x	y	xy	x^2
0	800	0	0
2	1,600	3,200	4
4	2,300	9,200	16
6	4,700	12,400	20

$y = 375x + 816.7$; 3,066.7 million
17. $q = -0.7p + 3.2$; 750 million smartphones
19. $y = 0.135x + 0.15$; 6.9 million jobs
21. a. $I = -0.007S + 0.78$
Graph:

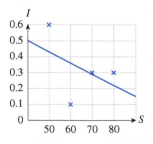

b. Amazon lost $7 million in net income per billion dollars earned in net sales.　**c.** $40 billion　**d.** The graph shows a poor fit, so the linear model does not seem reasonable.
23. a. $L = 45.8n + 507$
Graph:

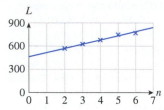

b. *Applied Calculus* is growing at a rate of 45.8 pages per edition.　**25. a.** $y = 1.62x - 23.87$
Graph:

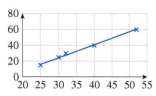

b. Each acre of cultivated land produces about 1.62 tons of soybeans.　**27. a.** $y = -11.85x + 797.71$; $r \approx -0.414$
b. Continental's net income is not correlated with the price of oil.　**c.** The points are nowhere near the regression line, confirming the conclusion in part (b).
Graph:

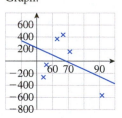

29. a. $y = 1.00x - 102$
Graph:

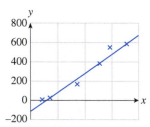

b. There is about one additional doctorate in engineering per additional doctorate in the natural sciences.　**c.** $r \approx 0.976$; a strong correlation.　**d.** Yes; the graph suggests a linear relationship; the data points are close to the regression line and show no obvious pattern (such as a curve).

31. a. $y = 28.9t + 37.0$
Graph:

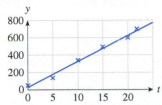

$r \approx 0.992$ **b.** The number of natural science doctorates
has been increasing at a rate of about 28.9 per year.
c. More-or-less constant rate; the slopes of successive pairs
of points do not show an increasing or decreasing trend as we
go from left to right.   **d.** Yes; if r had been equal to 1, then the
points would lie exactly on the regression line, which would
indicate that the number of doctorates is growing at a constant
rate.   **33. a.** More-or-less constant rate; Exercise 29 suggests
a roughly linear relationship between the number of natural
science doctorates and the number of engineering doctorates,
and Exercise 31 suggests that the number of natural science
doctorates has been increasing at a more-or less constant
rate. Therefore, the number of engineering doctorates is also
increasing at a more-or-less constant rate.   **b.** No; $r = 1$ in
Exercise 29 would indicate an exactly linear relationship
between the number of natural science doctorates and the
number of engineering doctorates, so the conclusion would be
the same.   **c.** No; $r = 1$ in Exercise 31 would indicate that the
number of natural science doctorates has been increasing at a
constant rate, so the conclusion would be the same.
35. a. $p = 0.13t + 0.22$; $r \approx 0.97$
Graph:

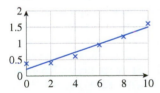

b. Yes; the first and last points lie above the regression line,
while the central points lie below it, suggesting a curve.
c.

◇	A	B	C	D
1	t	p (observed)	p (predicted)	Residual
2	0	0.38	0.22	0.16
3	2	0.4	0.48	-0.08
4	4	0.6	0.74	-0.14
5	6	0.95	1	-0.05
6	8	1.2	1.26	-0.06
7	10	1.6	1.52	0.08

Notice that the residuals are positive at first, then become
negative, and then become positive, confirming the impression
from the graph.   **37.** The line that passes through (a, b)
and (c, d) gives a sum-of-squares error SSE $= 0$, which is
the smallest value possible.   **39.** The regression line is the
line passing through the given points.   **41.** 0   **43.** No. The

regression line through $(-1, 1)$, $(0, 0)$, and $(1, 1)$ passes through
none of these points.   **45.** (Answers may vary.) The data in
Exercise 35 give $r \approx 0.97$, yet the plotted points suggest a
curve, not a straight line.

Chapter 1 Review

1. a. 1 **b.** -2 **c.** 0 **d.** -1 **3. a.** 1 **b.** 0 **c.** 0 **d.** -1
5.   **7.**

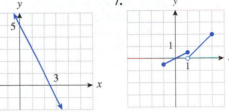

9. Absolute value   **11.** Linear   **13.** Quadratic
15. $y = -3x + 11$ **17.** $y = 1.25x - 4.25$
19. $y = (1/2)x + 3/2$ **21.** $y = 4x - 12$

23. $y = -\dfrac{x}{4} + 1$ **25.** $y = -0.214x + 1.14$, $r \approx -0.33$

27. a. Exponential
Graph:

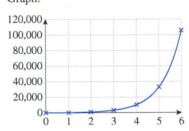

b. The ratios (rounded to 1 decimal place) are:

V(1)/V(0)	V(2)/V(1)	V(3)/V(2)	V(4)/V(3)	V(5)/V(4)	V(6)/V(5)
3	3.3	3.3	3.2	3.2	3.2

They are close to 3.2.   **c.** About 343,700 visits per day
29. a. 2.3; 3.5; 6   **b.** For website traffic of up to 50,000 visits
per day, the number of crashes is increasing by 0.03 per
additional thousand visits.   **c.** 140,000   **31. a.** (A)
b. (A) Leveling off (B) Rising (C) Rising; begins to fall after
7 months (D) Rising   **33. a.** The number of visits would
increase by 30 per day.   **b.** No; it would increase at a slower
and slower rate and then begin to decrease.   **c.** Probably
not. This model predicts that website popularity will start to
decrease as advertising increases beyond $8,500 per month and
then drop toward zero.   **35. a.** $v = 0.05c + 1,800$
b. 2,150 new visits per day   **c.** $14,000 per month
37. $d = 0.95w + 8$; 86 kg   **39. a.** Cost: $C = 5.5x + 500$;
Revenue: $R = 9.5x$; Profit $P = 4x - 500$ **b.** More than
125 albums per week   **c.** More than 200 albums per week
41. a. $q = -80p + 1,060$ **b.** 100 albums per week
c. $9.50, for a weekly profit of $700
43. a. $q = -74p + 1,015.5$ **b.** 239 albums per week

Chapter 2

Section 2.1

1. $INT = \$120$, $FV = \$2,120$
3. $INT = \$160$, $FV = \$4,160$
5. $INT = \$505$, $FV = \$20,705$
7. $INT = \$250$, $FV = \$10,250$
9. $INT = \$60$, $FV = \$12,060$ **11.** $PV = \$9,090.91$
13. $PV = \$966.18$ **15.** $PV = \$14,932.80$ **17.** \$5,200
19. \$997.61 **21.** 5% **23.** \$170; \$3,400 **25.** \$8,000
27. Wells Fargo; \$1,225.00 **29.** In 2 years **31.** Weekly rate
of 1.25%; annual rate of 65% **33.** 10% **35.** 86.957%
37. 4.00% **39.** 4.91% if you had sold in November 2010
41. No. Simple interest increase is linear. The graph is visibly
not linear in that time period. Further, the slopes of the lines
through the successive pairs of marked points are quite different. **43.** 9.2% **45.** 3,260,000 **47.** $P = 500 + 46t$ thousand
($t =$ time in years since 1950)

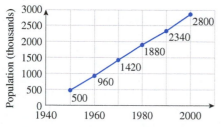

49. About 0.2503% **51.** 3.775% **53.** Graph (A) is the only
possible choice, because the equation $FV = PV(1 + rt) =$
$PV + PVrt$ gives the future value as a linear function of time.
55. 0.05% **57.** $FV = PV(1 + rt) = PV(1 + (12i)(n/12)) =$
$PV(1 + in)$ **59.** Wrong. In simple interest growth the change
each year is a fixed percentage of the *starting* value, not the
preceding year's value. (Also see Exercise 60.) **61.** Simple
interest is always calculated on a constant amount, PV. If interest is paid into your account, then the amount on which interest
is calculated does not remain constant.

Section 2.2

1. \$10,304.24 **3.** \$12,709.44 **5.** \$13,439.16 **7.** \$11,327.08
9. \$19,154.30 **11.** \$613.91 **13.** \$810.65 **15.** \$1,227.74
17. 5.09% **19.** 10.47% **21.** 10.52% **23.** \$268.99
25. \$728.91 **27.** \$2,927.15 **29.** \$21,161.79
31. \$163,414.56 **33.** \$174,110 **35.** \$750.00 **37.** \$9,000
39. \$7,462.65 **41.** 0.43% **43.** \$55,526.45 per year
45. \$27,171.92 **47.** \$111,678.96 **49.** \$1,039.21 **51.** The
one earning 11.9% compounded monthly **53.** Yes. The
investment would have grown to about \$503,096 million.
55. 160 reals **57.** 656 bolivianos **59.** 12 bolivars **61.** The
Nicaragua investment is better: It is worth about 1.03 units of
currency (in constant units) per unit invested as opposed to
about 1.027 units for Mexico. **63.** 53.81% **65.** 65.99% if
you had sold in November 2010 **67.** No. Compound interest
increase is exponential, and exponential curves either increase

continually (in the case of appreciation) or decrease continually (in the case of depreciation). The graph of the stock price
has both increases and decreases during the given period,
so the curve cannot model compound interest change.
69. 31 years; about \$26,100 **71.** 2.3 years
73. a. \$1,510.31 **b.** \$54,701.29 **c.** 23.51% **75.** The function
$y = P(1 + r/m)^{mx}$ is not a linear function of x, but an exponential function. Thus, its graph is not a straight line.
77. Wrong. Its growth is exponential and can be modeled by
$0.01(1.10)^t$. **79.** The graphs are the same because the formulas give the same function of x; a compound interest investment behaves as though it were being compounded once a year
at the effective rate. **81.** The effective rate exceeds the nominal rate when the interest is compounded more than once a
year because then interest is being paid on interest accumulated during each year, resulting in a larger effective rate.
Conversely, if the interest is compounded less often than
once a year, the effective rate is less than the nominal rate.
83. Compare their future values in constant dollars. The investment with the larger future value is the better investment.
85. The graphs are approaching a particular curve as m gets
larger, approximately the curve given by the largest two values
of m.

Section 2.3

1. \$15,528.23 **3.** \$171,793.82 **5.** \$23,763.28 **7.** \$147.05
9. \$491.12 **11.** \$105.38 **13.** \$90,155.46 **15.** \$69,610.99
17. \$95,647.68 **19.** \$554.60 **21.** \$1,366.41 **23.** \$524.14
25. \$248.85 **27.** \$1,984.65 **29.** \$494.87 **31.** \$5,615.31
33. \$79,573.29 **35.** \$923,373.42 **37.** \$50,000.46. This is
more than the original value of the loan. In a 200-year mortgage, the fraction of the initial payments going toward reducing the principal is so small that the rounding upward of the
payment makes it appear that, for many years at the start of the
mortgage, more is owed than the original amount borrowed.
39. \$999.61 **41.** \$998.47 **43.** \$917.45 **45.** \$584,686.94
47. \$348,312.44 **49.** \$2,677.02 **51.** Stock fund: \$206.33;
Bond fund: \$825.32 **53.** \$318,794.79 **55.** \$7,451.49
57. \$973.54 **59.** \$278.92

61.

Age	Male	Female
30	\$276.62	\$235.24
50	\$927.11	\$756.08
70	\$9,469.40	\$5,020.06

63. \$131.28 **65.** \$144,321.81 **67.** Wait until December, and
pay \$34.41 less per month. **69.** \$96,454.02 **71.** \$153.07
73. November: \$34,991.58; December: \$34,965.95
75. \$74.95 **77.** You should take the loan from Solid Savings
& Loan: It will have payments of \$248.85 per month. The
payments on the other loan would be more than \$300 per
month. **79.** Original monthly payments were \$824.79. The
new monthly payments will be \$613.46. You will save
\$36,488.88 in interest.

81. Answers using correctly rounded intermediate results:

Year	Interest	Payment on Principal
1	$3,934.98	$1,798.98
2	$3,785.69	$1,948.27
3	$3,623.97	$2,109.99
4	$3,448.84	$2,285.12
5	$3,259.19	$2,474.77
6	$3,053.77	$2,680.19
7	$2,831.32	$2,902.64
8	$2,590.39	$3,143.57
9	$2,329.48	$3,404.48
10	$2,046.91	$3,687.05
11	$1,740.88	$3,993.08
12	$1,409.47	$4,324.49
13	$1,050.54	$4,683.42
14	$661.81	$5,072.15
15	$240.84	$5,491.80

83. 10.81%   **85.** 13 years   **87.** 4.5 years   **89.** 24 years
91. He is wrong because his estimate ignores the interest that will be earned by your annuity—both while it is increasing and while it is decreasing. Your payments will be considerably smaller (depending on the interest earned).   **93.** Wrong; the split investment earns more. For instance, after 10 years it earns $31,056.46 + $32,775.87 = $63,832.33, which is more than the $63,803.03 earned by the single investment.   **95.** He is not correct. For instance, the payments on a $100,000 10-year mortgage at 12% are $1,434.71, while for a 20-year mortgage at the same rate, they are $1,101.09, which is a lot more than half the 10-year mortgage payment.   **97.** 3.617%

99. $PV = FV(1 + i)^{-n} = PMT\dfrac{(1 + i)^n - 1}{i}(1 + i)^{-n} = PMT\dfrac{1 - (1 + i)^{-n}}{i}$

Chapter 2 Review

1. $7,425.00   **3.** $7,604.88   **5.** $6,757.41   **7.** $4,848.48
9. $4,733.80   **11.** $5,331.37   **13.** $177.58   **15.** $112.54
17. $187.57   **19.** 14.0 years   **21.** 10.8 years
23. 7.0 years   **25.** $9,584.17   **27.** 5.346%   **29.** 168.85%
31. 85.28% if she sold in February 2010.   **33.** No.
Simple interest increase is linear. We can compare slopes between successive points to see whether the slope remained roughly constant: From December 2002 to August 2004 the slope was $(16.31 - 3.28)/(20/12) = 7.818$, while from August 2004 to March 2005 the slope was $(33.95 - 16.31)/(7/12) = 30.24$. These slopes are quite different.   **35.** 2013

Year	2010	2011	2012	2013	2014
Revenue	$180,000	$216,000	$259,200	$311,040	$373,248

37. At least 52,515 shares   **39.** $3,234.94   **41.** $231,844
43. 7.75%   **45.** $420,275   **47.** $140,778   **49.** $1,453.06
51. $2,239.90 per month   **53.** $53,055.66   **55.** 5.99%

Chapter 3

Section 3.1

1. Particular solutions (answers may vary): $(-1, -3)$, $(0, -1)$, $(1, 1)$; general solution parameterized by x: $(x, 2x - 1)$; x arbitrary; general solution parameterized by y: $(\frac{1}{2}(y + 1), y)$; y arbitrary   **3.** Particular solutions (answers may vary): $(-2, 2)$, $(0, 1/2)$, $(2, -1)$; general solution parameterized by x: $(x, -\frac{3}{4}x + \frac{1}{2})$; x arbitrary; general solution parameterized by y: $(\frac{1}{3}(-4y + 2), y)$; y arbitrary   **5.** Particular solutions (answers may vary): $(-5/4, -1)$, $(-5/4, 0)$, $(-5/4, 1)$; general solution parameterized by y: $(-5/4, y)$; y arbitrary
7. $(2, 2)$ **9.** $(3, 1)$ **11.** $(6, 6)$ **13.** $(5/3, -4/3)$ **15.** $(0, -2)$
17. $(x, (1 - 2x)/3)$ or $((1 - 3y)/2, y)$ **19.** No solution
21. $(5, 0)$ **23.** $(0.3, -1.1)$ **25.** $(116.6, -69.7)$
27. $(3.3, 1.8)$ **29.** $(3.4, 1.9)$ **31.** $x - 2y = 0$
33. $x - 1.10y = 0$ **35.** $x - 3y = 0$; $x + y = 12$
37. $x - 4y = 0$; $x - y = 15$ **39.** 200 qt of vanilla, 100 qt of mocha   **41.** 2 servings of Mixed Cereal, 1 serving of Mango Tropical Fruit   **43. a.** 4 servings of beans, 5 slices of bread **b.** No. One of the variables in the solution of the system has a negative value.   **45.** Mix 12 servings of Designer Whey, 2 servings of Muscle Milk for a cost of $9.20.   **47.** 65 g
49. 200 TWTR, 300 MSFT   **51.** 200 TD, 400 CNA   **53.** 242 in favor, 193 against   **55.** 5 soccer games, 7 football games
57. 7   **59.** $1.50 each   **61.** 55 widgets   **63.** Demand: $q = -4p + 47$; supply: $q = 4p - 29$; equilibrium price: $9.50   **65.** 33 pairs of dirty socks, 11 T-shirts   **67.** $1,200
69. The three lines in a plane must intersect in a single point for there to be a unique solution. This can happen in two ways: (1) The three lines intersect in a single point, or (2) two of the lines are the same, and the third line intersects it in a single point.   **71.** Yes. Even if two lines have negative slope, they will still intersect if the slopes differ.   **73.** You cannot round both of them up, since there will not be sufficient eggs and cream. Rounding both answers down will ensure that you will not run out of ingredients. It may be possible to round one answer down and the other up, and this should be tried.
75. (B)   **77.** (B)   **79.** Answers will vary.   **81.** It is very likely. Two randomly chosen straight lines are unlikely to be parallel.

Section 3.2

1. $(3, 1)$ **3.** $(6, 6)$ **5.** $(\frac{1}{2}(1 - 3y), y)$; y arbitrary
7. No solution   **9.** $(1/4, 3/4)$ **11.** No solution
13. $(10/3, 1/3)$ **15.** $(4, 4, 4)$ **17.** $(-1, -3, 1/2)$
19. (z, z, z); z arbitrary   **21.** No solution   **23.** $(-1, 1, 1)$
25. $(1, z - 2, z)$; z arbitrary   **27.** $(4 + y, y, -1)$; y arbitrary
29. $(4 - y/3 + z/3, y, z)$; y arbitrary, z arbitrary
31. $(-17, 20, -2)$ **33.** $(-3/2, 0, 1/2, 0)$
35. $(-3z, 1 - 2z, z, 0)$; z arbitrary   **37.** $(7/5 - 17z/5 + 8w/5, 1/5 - 6z/5 - 6w/5, z, w)$; z, w arbitrary
39. $(1, 2, 3, 4, 5)$ **41.** $(-2, -2 + z - u, z, u, 0)$;

z, u arbitrary   **43.** $(16, 12/7, -162/7, -88/7)$
45. $(-8/15, 7/15, 7/15, 7/15, 7/15)$ **47.** $(1.0, 1.4, 0.2)$
49. $(-5.5, -0.9, -7.4, -6.6)$ **51.** A pivot is an entry in a
matrix that is selected to "clear a column"; that is, use the row
operations of a certain type to obtain zeros everywhere above
and below it. "Pivoting" is the procedure of clearing a column
using a designated pivot.   **53.** $2R_1 + 5R_4$, or $6R_1 + 15R_4$
(which is less desirable)   **55.** It will include a row of zeros.
57. The claim is wrong. If there are more equations than
unknowns, there can be a unique solution as well as row(s) of
zeros in the reduced matrix, as in Example 6.   **59.** Two
61. The number of pivots must equal the number of variables,
since no variable will be used as a parameter.   **63.** A simple
example is $x = 1; y - z = 1; x + y - z = 2.$ **65.** It has to be
the zero solution (each unknown is equal to zero): Putting each
unknown equal to zero causes each equation to be satisfied
because the right-hand sides are zero. Thus, the zero solution is
in fact a solution. Because the solution is unique, this solution is
the *only* solution.   **67.** No: As was pointed out in Exercise 65,
every homogeneous system has at least one solution (namely,
the zero solution) and hence cannot be inconsistent.

Section 3.3

1. 100 batches of vanilla, 50 batches of mocha, 100 batches of
strawberry   **3.** 3 sections of Finite Math, 2 sections of Applied
Calculus, 1 section of Computer Methods   **5.** $32 million for
regional music, $18 million for pop/rock music, $8 million for
tropical music   **7.** 6 Airbus A330-300s, 4 Boeing 767-300ERs,
8 Boeing Dreamliner 787-9s   **9.** 22 tons from Cheesy Cream,
56 tons from Super Smooth & Sons, 22 tons from Bagel's Best
Friend   **11.** 10 evil sorcerers, 50 trolls, 500 orcs   **13.** $600 to
each of the MPBF and the SCN and $1,200 to the Jets
15. United Continental: 120; American: 40; Southwest: 50
17. $5,000 in SHPIX, $2,000 in RYURX, $2,000 in RYCWX
19. 100 shares of WSR, 50 shares of HCC, 50 shares of SNDK
21. Microsoft: 88 million, Time Warner: 79 million, Yahoo:
75 million, Google: 42 million   **23.** The third equation is
$x + y + z + w = 100.$ General solution: $x = 50.5 - 0.5w,$
$y = 33.5 - 0.3w, z = 16 - 0.2w, w$ arbitrary. State Farm is
most affected by other companies.   **25. a.** Brooklyn to Long
Island: 500 books; Brooklyn to Manhattan: 500 books; Queens
to Long Island: 1,000 books; Queens to Manhattan: 1,000
books. **b.** Brooklyn to Long Island: none; Brooklyn to
Manhattan: 1,000 books; Queens to Long Island: 1,500 books;
Queens to Manhattan: 500 books, giving a total cost of $8,000
27. a. The associated system of equations has infinitely many
solutions. **b.** No; the associated system of equations still has
infinitely many solutions. **c.** Yes; North America to Australia:
440,000, North America to South Africa: 190,000, Europe to
Australia: 950,000, Europe to South Africa: 950,000.
29. a. $x + y = 14,000; z + w = 95,000; x + z = 63,550,$
$y + w = 45,450.$ The system does not have a unique solution,
indicating that the given data are insufficient to obtain the
missing data. **b.** $(x, y, z, w) = (5,600, 8,400, 57,950, 37,050)$
31. a. No; The general solution is: Eastward Blvd.: $S + 200;$

Northwest La.: $S + 50;$ Southwest La.: S, where S is arbitrary.
Thus, it would suffice to know the traffic along Southwest La.
b. Yes, as it leads to the solution Eastward Blvd.: 260;
Northwest La.: 110; Southwest La.: 60.   **c.** 50 vehicles per day
33. a. With $x =$ traffic on middle section of Bree, $y =$ traffic
on middle section of Jeppe, $z =$ traffic on middle section of
Simmons, $w =$ traffic on middle section of Harrison, the general solution is $x = w - 100, y = w, z = w, w$ arbitrary.
b. No; there are infinitely many possible values for y.
c. 500 cars per minute   **d.** 100 cars per minute   **e.** No; a large
number of cars can circulate around the middle block without
affecting the numbers shown.   **35. a.** No; the corresponding
system of equations is underdetermined. The net flow of traffic
along any of the three stretches of Broadway would suffice.
b. West   **37.** $10 billion   **39.** $x =$ water, $y =$ gray matter,
$z =$ tumor   **41.** $x =$ water, $y =$ bone, $z =$ tumor, $u =$ air
43. tumor   **45.** 200 Democrats, 20 Republicans, 13 of other
parties   **47.** Yes; $20m in Company X, $5m in Company Y,
$10m in Company Z, $30m in Company W   **49.** It is not realistic to expect to use exactly all of the ingredients. Solutions of the
associated system may involve negative numbers or not exist.
Only solutions with nonnegative values for all the unknowns
correspond to being able to use up all of the ingredients.
51. Yes; $x = 100$ **53.** Yes; $0.3x - 0.7y + 0.3z = 0$ is one
form of the equation.   **55.** No; represented by an inequality
rather than an equation.   **57.** Answers will vary.

Chapter 3 Review

1. One solution

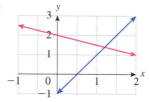

3. Infinitely many solutions

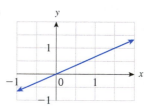

5. One solution

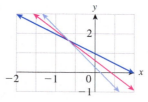

7. $(6/5, 7/5)$ **9.** $(3y/2, y); y$ arbitrary   **11.** $(-0.7, 1.7)$
13. $(-1, -1, -1)$ **15.** $(z - 2, 4(z - 1), z); z$ arbitrary
17. No solution   **19.** $-40°$ **21.** It is impossible; setting
$F = 1.8C$ leads to an inconsistent system of equations.
23. $x + y + z + w = 10$; linear   **25.** $w = 0$; linear

27. $-1.3y + z = 0$ or $1.3y - z = 0$; linear
29. 550 packages from Duffin House, 350 from Higgins Press
31. 1,200 packages from Duffin House, 200 from Higgins Press   **33.** \$40   **35.** 7 of each   **37.** 5,000 hits per day at OHaganBooks.com, 1,250 at JungleBooks.com, 3,750 at FarmerBooks.com   **39.** 100 shares of HAL, 20 shares of POM, 80 shares of WELL   **41.** Billy-Sean is forced to take exactly the following combination: Liberal Arts: 52 credits, Sciences: 12 credits, Fine Arts: 12 credits, Mathematics: 48 credits.   **43. a.** $x = 100$, $y = 100 + w$, $z = 300 - w$, w arbitrary   **b.** 100 book orders per day   **c.** 300 book orders per day   **d.** $x = 100$, $y = 400$, $z = 0$, $w = 300$ **e.** 100 book orders per day   **45.** Yes; New York to OHaganBooks.com: 450 packages, New York to FantasyBooks.com: 50 packages, Illinois to OHaganBooks.com: 150 packages, Illinois to FantasyBooks.com: 150 packages

Chapter 4

Section 4.1

1. 1×4; 0   **3.** 4×1; 5/2   **5.** $p \times q$; e_{22} **7.** 2×2; 3
9. $1 \times n$; d_r **11.** $x = 1$, $y = 2$, $z = 3$, $w = 4$

13. $\begin{bmatrix} 0.25 & -2 \\ 1 & 0.5 \\ -2 & 5 \end{bmatrix}$ **15.** $\begin{bmatrix} -0.75 & -1 \\ 0 & -0.5 \\ -1 & 6 \end{bmatrix}$

17. $\begin{bmatrix} -1 & -1 \\ 1 & -1 \\ -1 & 5 \end{bmatrix}$ **19.** $\begin{bmatrix} 0 & 2 & -2 \\ -2 & 0 & 4 \end{bmatrix}$

21. $\begin{bmatrix} 4 & -1 & -1 \\ 5 & 1 & 0 \end{bmatrix}$ **23.** $\begin{bmatrix} -2 + x & 0 & 1 + w \\ -5 + z & 3 + r & 2 \end{bmatrix}$

25. $\begin{bmatrix} -1 & -2 & 1 \\ -5 & 5 & -3 \end{bmatrix}$ **27.** $\begin{bmatrix} 9 & 15 \\ 0 & -3 \\ -3 & 3 \end{bmatrix}$

29. $\begin{bmatrix} -8.5 & -22.35 & -24.4 \\ 54.2 & 20 & 42.2 \end{bmatrix}$

31. $\begin{bmatrix} 1.54 & 8.58 \\ 5.94 & 0 \\ 6.16 & 7.26 \end{bmatrix}$ **33.** $\begin{bmatrix} 7.38 & 76.96 \\ 20.33 & 0 \\ 29.12 & 39.92 \end{bmatrix}$

35. $\begin{bmatrix} -19.85 & 115.82 \\ -50.935 & 46 \\ -57.24 & 94.62 \end{bmatrix}$

37. 2013: $[17.2 \quad 153.4 \quad 74.2]$;   2014: $[19.6 \quad 192.6 \quad 63.4]$

39. Sales $= \begin{bmatrix} 700 & 1,300 & 2,000 \\ 400 & 300 & 500 \end{bmatrix}$;

Inventory $-$ Sales $= \begin{bmatrix} 300 & 700 & 3,000 \\ 600 & 4,700 & 1,500 \end{bmatrix}$

41. Profit $=$ Revenue $-$ Cost $= \begin{bmatrix} 8,000 & 7,200 & 8,800 \\ 5,600 & 5,760 & 7,040 \\ 2,800 & 3,500 & 4,000 \end{bmatrix}$

43. 2000 distribution $= A = [53.6 \quad 64.4 \quad 100.2 \quad 63.2]$;
2010 distribution $= B = [55.3 \quad 66.9 \quad 114.6 \quad 71.9]$;
Net change 2000 to 2010 $= D = B - A =$
$[1.7 \quad 2.5 \quad 14.4 \quad 8.7]$; 2020 Distribution $= B + D =$
$[57.0 \quad 69.4 \quad 129.0 \quad 80.6]$
45. Total foreclosures $=$ Foreclosures in California $+$ Foreclosures in Florida $+$ Foreclosures in Texas $=$
$[55,900 \quad 51,900 \quad 54,100 \quad 56,200 \quad 59,400] +$
$[19,600 \quad 19,200 \quad 23,800 \quad 22,400 \quad 23,600] +$
$[8,800 \quad 9,100 \quad 9,300 \quad 10,600 \quad 10,100] =$
$[84,300 \quad 80,200 \quad 87,200 \quad 89,200 \quad 93,100]$
47. Difference $=$ Foreclosures in California $-$
Foreclosures in Florida $= [55,900 \quad 51,900 \quad 54,100$
$56,200 \quad 59,400] - [19,600 \quad 19,200 \quad 23,800 \quad 22,400$
$23,600] = [36,300 \quad 32,700 \quad 30,300 \quad 33,800 \quad 35,800]$
The difference was greatest in April.

49. a. Use $= \begin{matrix} & \text{Proc} \quad \text{Mem} \quad \text{Tubes} \\ \text{Pom II} \\ \text{Pom Classic} \end{matrix} \begin{bmatrix} 2 & 16 & 20 \\ 1 & 4 & 40 \end{bmatrix}$;

Inventory $= \begin{bmatrix} 500 & 5,000 & 10,000 \\ 200 & 2,000 & 20,000 \end{bmatrix}$;

Inventory $- 100 \times$ Use $= \begin{bmatrix} 300 & 3,400 & 8,000 \\ 100 & 1,600 & 16,000 \end{bmatrix}$

b. After 4 months

51. a. $A = \begin{bmatrix} 440 & 190 \\ 950 & 950 \\ 1,790 & 200 \end{bmatrix}$, $D = \begin{bmatrix} -20 & 40 \\ 50 & 50 \\ 0 & 100 \end{bmatrix}$

2008 Tourism $= A + D = \begin{bmatrix} 420 & 230 \\ 1,000 & 1,000 \\ 1,790 & 300 \end{bmatrix}$

b. $\frac{1}{2}(A + B)$; $\begin{bmatrix} 430 & 210 \\ 975 & 975 \\ 1,790 & 250 \end{bmatrix}$

53. No; for two matrices to be equal, they must have the same dimensions.   **55.** The ijth entry of the sum $A + B$ is obtained by adding the ijth entries of A and B.
57. It would have zeros down the main diagonal:

$A = \begin{bmatrix} 0 & \# & \# & \# & \# \\ \# & 0 & \# & \# & \# \\ \# & \# & 0 & \# & \# \\ \# & \# & \# & 0 & \# \\ \# & \# & \# & \# & 0 \end{bmatrix}$

The symbols # indicate arbitrary numbers.
59. $(A^T)_{ij} = A_{ji}$ **61.** Answers will vary.

a. $\begin{bmatrix} 0 & -4 \\ 4 & 0 \end{bmatrix}$ **b.** $\begin{bmatrix} 0 & -4 & 5 \\ 4 & 0 & 1 \\ -5 & -1 & 0 \end{bmatrix}$

63. The associativity of matrix addition is a consequence of the associativity of addition of numbers, since we add matrices by adding the corresponding entries (which are real numbers).
65. Answers will vary.

Section 4.2

1. $[13]$ **3.** $[5/6]$ **5.** $[-2y + z]$ **7.** Undefined
9. $[3 \quad 0 \quad -6 \quad -2]$ **11.** $[-6 \quad 37 \quad 7]$
13. $\begin{bmatrix} -4 & -7 & -1 \\ 9 & 17 & 0 \end{bmatrix}$ **15.** $\begin{bmatrix} 0 & 1 \\ 0 & 0 \end{bmatrix}$ **17.** $\begin{bmatrix} 1 & -1 \\ 1 & -1 \end{bmatrix}$

19. $\begin{bmatrix} 0 & 0 \\ 0 & 0 \end{bmatrix}$ **21.** Undefined   **23.** $\begin{bmatrix} 1 & -5 & 3 \\ 0 & 0 & 9 \\ 0 & 4 & 1 \end{bmatrix}$

25. $\begin{bmatrix} 3 \\ -4 \\ 0 \\ 3 \end{bmatrix}$ **27.** $\begin{bmatrix} 0.23 & 5.36 & -21.65 \\ -13.18 & -5.82 & -16.62 \\ -11.21 & -9.9 & 0.99 \\ -2.1 & 2.34 & 2.46 \end{bmatrix}$

29. $A^2 = \begin{bmatrix} 0 & 0 & 1 & 2 \\ 0 & 0 & 0 & 1 \\ 0 & 0 & 0 & 0 \\ 0 & 0 & 0 & 0 \end{bmatrix}$; $A^3 = \begin{bmatrix} 0 & 0 & 0 & 1 \\ 0 & 0 & 0 & 0 \\ 0 & 0 & 0 & 0 \\ 0 & 0 & 0 & 0 \end{bmatrix}$;

$A^4 = \begin{bmatrix} 0 & 0 & 0 & 0 \\ 0 & 0 & 0 & 0 \\ 0 & 0 & 0 & 0 \\ 0 & 0 & 0 & 0 \end{bmatrix}$; ...; $A^{100} = \begin{bmatrix} 0 & 0 & 0 & 0 \\ 0 & 0 & 0 & 0 \\ 0 & 0 & 0 & 0 \\ 0 & 0 & 0 & 0 \end{bmatrix}$;

31. $\begin{bmatrix} 4 & -1 \\ -1 & -7 \end{bmatrix}$ **33.** $\begin{bmatrix} 4 & -1 \\ -12 & 2 \end{bmatrix}$

35. $\begin{bmatrix} -2 & 1 & -2 \\ 10 & -2 & 2 \\ -10 & 2 & -2 \end{bmatrix}$

37. $\begin{bmatrix} -2 + x - z & 2 - r & -6 + w \\ 10 + 2z & -2 + 2r & 10 \\ -10 - 2z & 2 - 2r & -10 \end{bmatrix}$

39. a.–d. $P^2 = P^4 = P^8 = P^{1,000} = \begin{bmatrix} 0.2 & 0.8 \\ 0.2 & 0.8 \end{bmatrix}$

41. a. $P^2 = \begin{bmatrix} 0.01 & 0.99 \\ 0 & 1 \end{bmatrix}$

b. $P^4 = \begin{bmatrix} 0.0001 & 0.9999 \\ 0 & 1 \end{bmatrix}$

c. and d. $P^8 \approx P^{1,000} \approx \begin{bmatrix} 0 & 1 \\ 0 & 1 \end{bmatrix}$

43. a.–d. $P^2 = P^4 = P^8 = P^{1,000} = \begin{bmatrix} 0.3 & 0.3 & 0.4 \\ 0.3 & 0.3 & 0.4 \\ 0.3 & 0.3 & 0.4 \end{bmatrix}$

The rows of P are the same, and the entries in each row add up to 1. If P is any square matrix with identical rows such that the entries in each row add up to 1, then $P \cdot P = P$.

45. $2x - y + 4z = 3$; $-4x + \dfrac{3}{4}y + \dfrac{1}{3}z = -1$; $-3x = 0$

47. $x - y + w = -1$; $x + y + 2z + 4w = 2$

49. $\begin{bmatrix} 1 & -1 \\ 2 & -1 \end{bmatrix}\begin{bmatrix} x \\ y \end{bmatrix} = \begin{bmatrix} 4 \\ 0 \end{bmatrix}$

51. $\begin{bmatrix} 1 & 1 & -1 \\ 2 & 1 & 1 \\ \frac{3}{4} & 0 & \frac{1}{2} \end{bmatrix}\begin{bmatrix} x \\ y \\ z \end{bmatrix} = \begin{bmatrix} 8 \\ 4 \\ 1 \end{bmatrix}$

53. Revenue = Price × Quantity =
$[15 \quad 10 \quad 12]\begin{bmatrix} 50 \\ 40 \\ 30 \end{bmatrix} = [1{,}510]$

55. $92.5 million
57. Revenue = Quantity × Price =
$\begin{bmatrix} 700 & 1{,}300 & 2{,}000 \\ 400 & 300 & 500 \end{bmatrix}\begin{bmatrix} 30 \\ 10 \\ 15 \end{bmatrix} = \begin{bmatrix} 64{,}000 \\ 22{,}500 \end{bmatrix}$

59. $5,100 billion (or $5.1 trillion)
61. $D = N(F - M)$, where N is the income per person, and F and M are, respectively, the female and male populations in 2020; $400 billion
63. $[23.1 \quad 5.2]$, which represents the amount, in pounds, by which per capita consumption of ice cream exceeded that of yogurt in 1983 and 2013.
65. Number of foreclosures filings handled by firm = Percentage handled by firm × Total number =
$[8{,}460 \quad 8{,}860 \quad 9{,}140]$
67. The number of foreclosures in California and Florida combined in each of the months shown.
69. $[1 \quad -1 \quad 1]\begin{bmatrix} 54{,}100 & 56{,}200 & 59{,}400 \\ 23{,}800 & 22{,}400 & 23{,}600 \\ 9{,}300 & 10{,}600 & 10{,}100 \end{bmatrix}\begin{bmatrix} 1 \\ 1 \\ 1 \end{bmatrix}$
$= [129{,}900]$

71. $\begin{bmatrix} 2 & 16 & 20 \\ 1 & 4 & 40 \end{bmatrix}\begin{bmatrix} 100 & 150 \\ 50 & 40 \\ 10 & 15 \end{bmatrix} = \begin{bmatrix} 1{,}200 & 1{,}240 \\ 700 & 910 \end{bmatrix}$

73. $AB = \begin{bmatrix} 29.6 \\ 85.5 \\ 97.5 \end{bmatrix}$, $AC = \begin{bmatrix} 22 & 7.6 \\ 47.5 & 38 \\ 89.5 & 8 \end{bmatrix}$

The entries of AB give the number of people from each of the three regions who settle in Australia or South Africa, while the entries in AC break those figures down further into settlers in South Africa and settlers in Australia.
75. $[54.6 \quad 66.0 \quad 111.8 \quad 70.6]$
77. Answers will vary. One example:
$A = [1 \quad 2]$, $B = \begin{bmatrix} 1 & 2 & 3 \\ 4 & 5 & 6 \end{bmatrix}$.
Another example: $A = [1]$, $B = [1 \quad 2]$
79. We find that the addition and multiplication of 1×1 matrices is identical to the addition and multiplication of numbers.
81. The claim is correct. Every matrix equation represents the equality of two matrices. When two matrices are equal, each of their corresponding entries must be equal. Equating the corresponding entries gives a system of equations.

83. Here is a possible scenario: Costs of items A, B, and C in 2013 = $[10 \quad 20 \quad 30]$, Percentage increases in these costs in 2014 = $[0.5 \quad 0.1 \quad 0.20]$, Actual increases in costs = $[10 \times 0.5 \quad 20 \times 0.1 \quad 30 \times 0.20]$.
85. It produces a matrix whose ij entry is the product of the ij entries of the two matrices.

Section 4.3

1. Yes   **3.** Yes   **5.** No

7. $\begin{bmatrix} -1 & 1 \\ 2 & -1 \end{bmatrix}$ **9.** $\begin{bmatrix} 0 & 1 \\ 1 & 0 \end{bmatrix}$ **11.** $\begin{bmatrix} 1 & -1 \\ -1 & 2 \end{bmatrix}$

13. Singular   **15.** $\begin{bmatrix} 1 & -1 & 0 \\ 0 & 1 & -1 \\ 0 & 0 & 1 \end{bmatrix}$

17. $\begin{bmatrix} 1 & -1 & 1 \\ \frac{1}{2} & 0 & -\frac{1}{2} \\ -\frac{1}{2} & 1 & -\frac{1}{2} \end{bmatrix}$ **19.** $\begin{bmatrix} 1 & \frac{1}{3} & -\frac{1}{3} \\ 1 & -\frac{2}{3} & -\frac{1}{3} \\ -1 & \frac{1}{3} & \frac{2}{3} \end{bmatrix}$

21. Singular   **23.** $\begin{bmatrix} 0 & 1 & -2 & 1 \\ 0 & 1 & -1 & 0 \\ 1 & -1 & 2 & -1 \\ 0 & 1 & -1 & 1 \end{bmatrix}$

25. $\begin{bmatrix} 1 & -2 & 1 & 0 \\ 0 & 1 & -2 & 1 \\ 0 & 0 & 1 & -2 \\ 0 & 0 & 0 & 1 \end{bmatrix}$ **27.** $-2; \begin{bmatrix} \frac{1}{2} & \frac{1}{2} \\ \frac{1}{2} & -\frac{1}{2} \end{bmatrix}$

29. $-2; \begin{bmatrix} -2 & 1 \\ \frac{3}{2} & -\frac{1}{2} \end{bmatrix}$ **31.** $1/36; \begin{bmatrix} 6 & 6 \\ 0 & 6 \end{bmatrix}$

33. 0; Singular

35. $\begin{bmatrix} 0.38 & 0.45 \\ 0.49 & -0.41 \end{bmatrix}$ **37.** $\begin{bmatrix} 0.00 & -0.99 \\ 0.81 & 2.87 \end{bmatrix}$

39. Singular

41. $\begin{bmatrix} 91.35 & -8.65 & 0 & -71.30 \\ -0.07 & -0.07 & 0 & 2.49 \\ 2.60 & 2.60 & -4.35 & 1.37 \\ 2.69 & 2.69 & 0 & -2.10 \end{bmatrix}$

43. $(5/2, 3/2)$ **45.** $(6, -4)$ **47.** $(6, 6, 6)$
49. a. $(10, -5, -3)$ **b.** $(6, 1, 5)$ **c.** $(0, 0, 0)$
51. a. 10/3 servings of beans, 5/6 slices of bread

b. $\begin{bmatrix} -\frac{1}{2} & \frac{1}{6} \\ \frac{7}{8} & -\frac{5}{24} \end{bmatrix} \begin{bmatrix} A \\ B \end{bmatrix} = \begin{bmatrix} -\frac{A}{2} + \frac{B}{6} \\ \frac{7A}{8} - \frac{5B}{24} \end{bmatrix}$; that is,

$-A/2 + B/6$ servings of beans, $7A/8 - 5B/24$ slices of bread   **53. a.** 100 batches of vanilla, 50 batches of mocha, 100 batches of strawberry **b.** 100 batches of vanilla, no mocha, 200 batches of strawberry

c. $\begin{bmatrix} 1 & -\frac{1}{3} & -\frac{1}{3} \\ -1 & 0 & 1 \\ 0 & \frac{2}{3} & -\frac{1}{3} \end{bmatrix} \begin{bmatrix} A \\ B \\ C \end{bmatrix}$, or $A - B/3 - C/3$ batches of

vanilla, $-A + C$ batches of mocha, $2B/3 - C/3$ batches of strawberry

55. \$5,000 in MYY, \$2,000 in SH, \$2,000 in REW
57. 100 shares of WSR, 50 shares of HCC, 50 shares of SNDK
59. $[54.1 \quad 65.7 \quad 112.9 \quad 70.3]$
61. a. $(-0.7071, 3.5355)$ **b.** R^2, R^3 **c.** R^{-1}
63. $[37 \quad 81 \quad 40 \quad 80 \quad 15 \quad 45 \quad 40 \quad 96 \quad 29 \quad 59 \quad 4 \quad 8]$
65. CORRECT ANSWER   **67.** (A)   **69.** The inverse does not exist; the matrix is singular. (If two rows of a matrix are the same, then row reducing it will lead to a row of zeros, so it cannot be reduced to the identity.)
71. Calculation (See the Student Solutions Manual.)
73. When one or more of the d_i are zero; if that is the case, then the matrix $[D \,|\, I]$ easily reduces to a matrix that has a row of zeros on the left-hand portion, so D is singular. Conversely, if none of the d_i are zero, then $[D \,|\, I]$ easily reduces to a matrix of the form $[I \,|\, E]$, showing that D is invertible.
75. $(AB)(B^{-1}A^{-1}) = A(BB^{-1})A^{-1} = AIA^{-1} = AA^{-1} = I$
77. If A has an inverse, then every system of equations $AX = B$ has a unique solution, namely, $X = A^{-1}B$. But if A reduces to a matrix with a row of zeros, then such a system has either infinitely many solutions or no solution at all.

Section 4.4

1. -1 **3.** -0.25 **5.** $[0 \quad 0 \quad 1 \quad 0]$; $e = 2.25$
7. $[1 \quad 0 \quad 0]^T$ or $[0 \quad 1 \quad 0]^T$; $e = 1/4$

9. $\begin{matrix} & p & r \\ a & \begin{bmatrix} 1 & 10 \\ b & 2 & -4 \end{bmatrix} \end{matrix}$ **11.** $3[-1]$ **13.** $q[0]$

15. Strictly determined. The row player's optimal strategy is a; the column player's optimal strategy is q; Value: 1
17. Not strictly determined.   **19.** Not strictly determined.
21. $R = \begin{bmatrix} \frac{1}{4} & \frac{3}{4} \end{bmatrix}, C = \begin{bmatrix} \frac{3}{4} & \frac{1}{4} \end{bmatrix}^T, e = -1/4$
23. $R = \begin{bmatrix} \frac{3}{4} & \frac{1}{4} \end{bmatrix}, C = \begin{bmatrix} \frac{3}{4} & \frac{1}{4} \end{bmatrix}^T, e = -5/4$
25. Row player: you; Column player: your friend;

$$\begin{matrix} & H & T \\ H & \begin{bmatrix} -1 & 1 \\ T & 1 & -1 \end{bmatrix} \end{matrix}$$

27. F = France; S = Sweden; N = Norway

Your Opponent Defends

		F	S	N
You Invade	F	-1	1	1
	S	1	-1	1
	N	1	1	-1

29. Row player: you; Column player: your opponent; B = Brakpan, N = Nigel, S = Springs;

Your Opponent

		B	N	S
You	B	0	0	$1{,}000$
	N	0	0	$1{,}000$
	S	$-1{,}000$	$-1{,}000$	0

31. P = PleasantTap; T = Thunder Rumble; S = Strike the Gold, N = None;

		Winner			
		P	T	S	N
You Bet	P	25	−10	−10	−10
	T	−10	35	−10	−10
	S	−10	−10	40	−10

33. You can expect to lose 39 customers.
35. Option II: Move to the suburbs. **37. a.** About 66%
b. Yes; spend the whole night studying game theory; 75%
c. Game theory; 57.5% **39. a.** Lay off 10 workers; cost: $40,000 **b.** 0 inches of snow, costing $300,000 **c.** Lay off 15 workers. **41. a.** CE should charge $1,000, and GCS should charge $900; 15% gain in market share for CE. **b.** CE should charge $1,200. (The more CE can charge for the same market, the better!) **43.** Pablo vs. Noto; evenly matched **45.** Both commanders should use the northern route; 2 days.
47. Confess

49. a.

		F	O
	F	24	21
	O	25	24

b. Both candidates should visit Ohio, leaving Romney with a 24% chance of winning the election. **51.** Allocate 1/7 of the budget to WISH and the rest (6/7) to WASH. Softex will lose approximately $2,860. **53.** Like a saddle point in a payoff matrix, the center of a saddle is a low point (minimum height) in one direction and a high point (maximum) in a perpendicular direction. **55.** Although there is a saddle point in the (2, 4) position, you would be wrong to use saddle points (based on the minimax criterion) to reach the conclusion that row strategy 2 is best. One reason is that the entries in the matrix do not represent payoffs, since high numbers of employees in an area do not necessarily represent benefit to the row player. Another reason for this is that there is no opponent deciding what your job will be in such a way as to force you into the least populated job. **57.** If you strictly alternate the two strategies, the column player will know which pure strategy you will play on each move and can choose a pure strategy accordingly. For example, consider the game

$$\begin{array}{c} \\ A \\ B \end{array} \begin{array}{cc} a & b \\ \begin{bmatrix} 1 & 0 \\ 0 & 1 \end{bmatrix} \end{array}.$$

By the analysis of Example 3 (or the symmetry of the game), the best strategy for the row player is $[0.5 \quad 0.5]$, and the best strategy for the column player is $[0.5 \quad 0.5]^T$. This gives an expected value of 0.5 for the game. However, suppose that the row player alternates A and B strictly and that the column player catches on to this. Then, whenever the row player plays A the column player will play b, and whenever the row player plays B, the column player will play a. This gives a payoff of 0 each time, worse for the row player than the expected value of 0.5.

Section 4.5

1. a. 0.8 **b.** 0.2 **c.** 0.05 **3.** $\begin{bmatrix} 0.2 & 0.1 \\ 0.5 & 0 \end{bmatrix}$
5. $[52{,}000 \quad 40{,}000]^T$ **7.** $[50{,}000 \quad 50{,}000]^T$
9. $[2{,}560 \quad 2{,}800 \quad 4{,}000]^T$
11. $[27{,}000 \quad 28{,}000 \quad 17{,}000]^T$
13. Increase of 100 units in each sector.
15. Increase of $[1.5 \quad 0.2 \quad 0.1]^T$; the ith column of $(I - A)^{-1}$ gives the change in production necessary to meet an increase in external demand of one unit for the product of Sector i.

17. $A = \begin{bmatrix} 0.2 & 0.4 & 0.5 \\ 0 & 0.8 & 0 \\ 0 & 0.2 & 0.5 \end{bmatrix}$

19. Main DR: $80,000, Bits & Bytes: $38,000 **21.** Equipment Sector production approximately $86,000 million, Components Sector production approximately $140,000 million
23. a. 0.006 **b.** textiles; clothing and footwear

25.

From	To		
	Primary	**Secondary**	**Tertiary**
Primary	53.1	330	0
Secondary	82.6	2,530	768
Tertiary	41.3	1,320	1,440
Total Output	590	11,000	9,600

(Entries are in billions of pesos.)
27. 210 billion pesos of raw materials, 7,600 billion pesos of manufactured goods, 6,800 billion pesos of services.
29. Production in the primary sector would rise by around 2,208 billion pesos, production in the secondary sector would rise by around 303 billion pesos, and production in the tertiary sector would drop by around 952 billion pesos.

31. Entries of $\begin{bmatrix} 1{,}177 \\ 1{,}517 \\ 1{,}033 \\ 1{,}421 \end{bmatrix}$ (in millions of dollars)

33. a. $0.78 **b.** Other food products **35.** It would mean that all of the sectors require neither their own product nor the product of any other sector. **37.** It would mean that all of the output of that sector was used internally in the economy; none of the output was available for export, and no importing was necessary. **39.** If an entry in the matrix $(I - A)^{-1}$ is zero, then an increase in demand for one sector (the column sector) has no effect on the production of another sector (the row sector). **41.** Usually, to produce one unit of one sector requires less than one unit of input from another. We would expect then that an increase in demand of one unit for one sector would require a smaller increase in production in another sector.

Chapter 4 Review

1. Undefined   **3.** $\begin{bmatrix} 1 & 8 \\ 5 & 11 \\ 6 & 13 \end{bmatrix}$ **5.** $\begin{bmatrix} 1 & 3 \\ 2 & 3 \\ 3 & 3 \end{bmatrix}$ **7.** $\begin{bmatrix} 1 & -2 \\ 0 & 1 \end{bmatrix}$

9. $\begin{bmatrix} 2 & 4 \\ 1 & 12 \end{bmatrix}$ **11.** $\begin{bmatrix} 1 & 1 \\ 0 & 1 \end{bmatrix}$ **13.** $\begin{bmatrix} 1 & -\frac{1}{2} & -\frac{5}{2} \\ 0 & \frac{1}{4} & -\frac{1}{4} \\ 0 & 0 & 1 \end{bmatrix}$

15. Singular

17. $\begin{bmatrix} 1 & 2 \\ 3 & 4 \end{bmatrix}\begin{bmatrix} x \\ y \end{bmatrix} = \begin{bmatrix} 0 \\ 2 \end{bmatrix}$; $\begin{bmatrix} x \\ y \end{bmatrix} = \begin{bmatrix} 2 \\ -1 \end{bmatrix}$

19. $\begin{bmatrix} 1 & 1 & 1 \\ 1 & 2 & 1 \\ 1 & 1 & 2 \end{bmatrix}\begin{bmatrix} x \\ y \\ z \end{bmatrix} = \begin{bmatrix} 2 \\ 3 \\ 1 \end{bmatrix}$; $\begin{bmatrix} x \\ y \\ z \end{bmatrix} = \begin{bmatrix} 2 \\ 1 \\ -1 \end{bmatrix}$

21. $R = \begin{bmatrix} 1 & 0 & 0 \end{bmatrix}$, $C = \begin{bmatrix} 0 & 1 & 0 & 0 \end{bmatrix}^T$, $e = 1$

23. $R = \begin{bmatrix} 0 & 0.8 & 0.2 \end{bmatrix}$, $C = \begin{bmatrix} 0.2 & 0 & 0.8 \end{bmatrix}^T$, $e = -0.2$

25. $\begin{bmatrix} 1,100 \\ 700 \end{bmatrix}$ **27.** $\begin{bmatrix} 48,125 \\ 22,500 \\ 10,000 \end{bmatrix}$

29. Inventory $-$ Sales $+$ Purchases $=$

$\begin{bmatrix} 2,500 & 4,000 & 3,000 \\ 1,500 & 3,000 & 1,000 \end{bmatrix} - \begin{bmatrix} 300 & 500 & 100 \\ 100 & 600 & 200 \end{bmatrix} +$

$\begin{bmatrix} 400 & 400 & 300 \\ 200 & 400 & 300 \end{bmatrix} = \begin{bmatrix} 2,600 & 3,900 & 3,200 \\ 1,600 & 2,800 & 1,100 \end{bmatrix}$

31. $N =$

$\begin{bmatrix} 2,600 & 3,900 & 3,200 \\ 1,600 & 2,800 & 1,100 \end{bmatrix} - x\begin{bmatrix} 280 & 550 & 100 \\ 50 & 500 & 120 \end{bmatrix} +$

$x\begin{bmatrix} 400 & 400 & 300 \\ 200 & 400 & 300 \end{bmatrix} = \begin{bmatrix} 2,600 & 3,900 & 3,200 \\ 1,600 & 2,800 & 1,100 \end{bmatrix} +$

$x\begin{bmatrix} 120 & -150 & 200 \\ 150 & -100 & 180 \end{bmatrix}$; 28 months from now (July 1)

33. Revenue $=$ Quantity $\times$ Price $=$

$\begin{bmatrix} 280 & 550 & 100 \\ 50 & 500 & 120 \end{bmatrix}\begin{bmatrix} 5 \\ 6 \\ 5.5 \end{bmatrix} = \begin{bmatrix} 5,250 \\ 3,910 \end{bmatrix}$ Texas, Nevada

35. July 1: 1,000 shares, August 1: 2,000 shares, September 1: 2,000 shares
37. Loss $=$ Number of shares $\times$ (Purchase price $-$ Dividends $-$ Selling price) $= \begin{bmatrix} 1,000 & 2,000 & 2,000 \end{bmatrix}$

$\left(\begin{bmatrix} 20 \\ 10 \\ 5 \end{bmatrix} - \begin{bmatrix} 0.10 \\ 0.10 \\ 0 \end{bmatrix} - \begin{bmatrix} 3 \\ 1 \\ 1 \end{bmatrix}\right) = \begin{bmatrix} 42,700 \end{bmatrix}$

39. $\begin{bmatrix} 2,000 & 4,000 & 4,000 \end{bmatrix}\begin{bmatrix} 0.8 & 0.1 & 0.1 \\ 0.4 & 0.6 & 0 \\ 0.2 & 0 & 0.8 \end{bmatrix} =$

$\begin{bmatrix} 4,000 & 2,600 & 3,400 \end{bmatrix}$ **41.** The matrix shows that no JungleBooks.com customers switched directly to FarmerBooks.com, so the only way to get to FarmerBooks.com is via

OHaganBooks.com.   **43.** Go with the "3 for 1" promotion and gain 20,000 customers from JungleBooks.com.
45. JungleBooks.com will go with "3 for 2," and OHaganBooks.com will go with Finite Math, resulting in a gain of 15,000 customers to OHaganBooks.com.   **47.** Choose between the "3 for 1" and Finite Math promotions with probabilities 60% and 40%, respectively. You would expect to gain 12,000 customers from JungleBooks.com (if you played this game many times).

49. $A = \begin{bmatrix} 0.1 & 0.5 \\ 0.01 & 0.05 \end{bmatrix}$

51. $1,190 worth of paper, $1,802 worth of books

Chapter 5

Section 5.1

1.
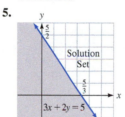
Unbounded

3.
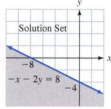
Unbounded

5.
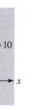
Unbounded

7.
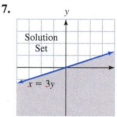
Unbounded

9.
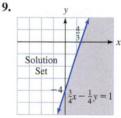
Unbounded

11.
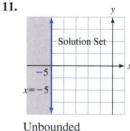
Unbounded

13.
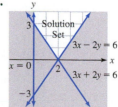
Unbounded;
corner point: (2, 0)

15.
Unbounded;
corner points:
(2, 0), (0, 3)

17.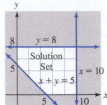

Bounded; corner points:
$(5, 0)$, $(10, 0)$, $(10, 8)$,
$(0, 8)$, $(0, 5)$

19.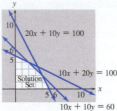

Bounded; corner points:
$(0, 0)$, $(5, 0)$, $(0, 5)$,
$(2, 4)$, $(4, 2)$

21.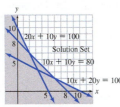

Unbounded; corner points:
$(0, 10)$, $(10, 0)$, $(2, 6)$,
$(6, 2)$

23.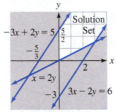

Unbounded;
corner points: $(0, 0)$,
$(0, 5/2)$, $(3, 3/2)$

25.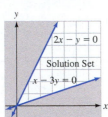

Unbounded;
corner point: $(0, 0)$

27.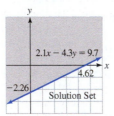

29.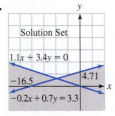

Corner point:
$(-7.74, 2.50)$

31.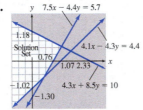

Corner points:
$(0.36, -0.68)$, $(1.12, 0.61)$

33. $x =$ Number of quarts of Creamy Vanilla,
$y =$ Number of quarts of Continental Mocha

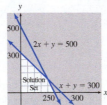

Corner points: $(0, 0)$, $(250, 0)$, $(0, 300)$, $(200, 100)$

35. $x =$ Number of ounces of chicken,
$y =$ Number of ounces of grain

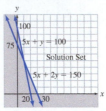

Corner points: $(30, 0)$, $(10, 50)$, $(0, 100)$

37. $x =$ Number of servings of Mixed Cereal for Baby,
$y =$ Number of servings of Mango Tropical Fruit Dessert

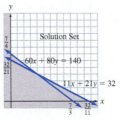

Corner points: $(0, 7/4)$, $(1, 1)$, $(32/11, 0)$

39. $x =$ Number of dollars in MHI,
$y =$ Number of dollars in BKK

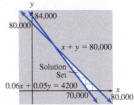

Corner points: $(70,000, 0)$, $(80,000, 0)$, $(20,000, 60,000)$

41. $x =$ Number of shares of TD,
$y =$ Number of shares of CNA

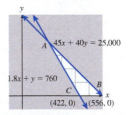

Corner points: $(200, 400)$, $(556, 0)$, $(422, 0)$

43. $x =$ Number of full-page ads in *Sports Illustrated*,
$y =$ Number of full-page ads in *GQ*

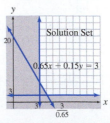

Corner points (rounded): $(3, 7)$, $(4, 3)$

45. An example is $x \geq 0$, $y \geq 0$, $x + y \geq 1$.   **47.** The given triangle can be described as the solution set of the system $x \geq 0$, $y \geq 0$, $x + 2y \leq 2$.   **49.** Answers may vary. One limitation is that the method is suitable only for situations with two unknown quantities. Accuracy is also limited when graphing. **51.** (C)   **53.** (B)   **55.** There are no feasible solutions; that is, it is impossible to satisfy all the constraints.   **57.** Answers will vary.

Section 5.2

1. $p = 6$, $x = 3$, $y = 3$ **3.** $c = 4$, $x = 2$, $y = 2$
5. $p = 24$, $x = 7$, $y = 3$ **7.** $p = 16$, $x = 4$, $y = 2$
9. $c = 1.8$, $x = 6$, $y = 2$ **11.** Max: $p = 16$, $x = 4$, $y = 6$.
Min: $p = 2$, $x = 2$, $y = 0$ **13.** $p = 16$, $x = 8$, $y = 0$
15. No optimal solution; objective function unbounded
17. No optimal solution; objective function unbounded
19. $c = 28$; $(x, y) = (14, 0)$ and $(6, 4)$ and the line connecting them   **21.** $c = 3$, $x = 3$, $y = 2$ **23.** No solution; feasible region empty   **25.** 200 quarts of Creamy Vanilla and 100 quarts of Continental Mocha   **27.** 100 ounces of grain and no chicken   **29.** 1 serving of cereal and 1 serving of dessert   **31.** Purchase 60 compact fluorescent light bulbs and 960 square feet of insulation for a saving of $312 per year in energy costs.   **33.** Mix 6 servings of Xtend and 7 servings of Gainz for a cost of $13.70.   **35. a.** Yes; use 15 servings of Gainz and 4 servings of Strongevity. **b.** No; the value of the objective function can be made arbitrarily large. (You can use as much Strongevity as you like without violating your trainer's specifications.)   **37.** Make 200 Dracula Salamis and 400 Frankenstein Sausages for a profit of $1,400.
39. 30 spots on *Big Bang Theory* and 30 spots on *American Dad*   **41.** No shares of OCR and 300 shares of RCKY.
43. 422.2 shares of TD and no shares of CNA; minimum risk index $\approx 1,267$ **45.** 10 hours in diplomacy and 40 hours in battle   **47. a.** Use 7 sleep spells and 3 shock spells. **b.** Use 6 sleep spells and 6 shock spells. **c.** No optimal solution; net expenditure of aural energy can be an arbitrarily large negative number.   **49.** 100 hours per week for new customers and 60 hours per week for old customers.   **51.** (A)
53. Every point along the line connecting them is also an optimal solution.   **55. a.** Not necessary; optimal solutions exist. **b.** Necessary **c.** Necessary **d.** Not necessary; no optimal solution exists.   **57.** Answers may vary. Maximize $p = x + y$ subject to $x + y \leq 10$; $x + y \geq 11$, $x \geq 0$, $y \geq 0$.
59. Answers may vary.   **61.** A simple example is the following: Maximize profit $p = 2x + y$ subject to $x \geq 0$, $y \geq 0$. Then p can be made as large as we like by choosing large values of x and/or y. Thus, there is no optimal solution to the problem.   **63.** Mathematically, this means that there are infinitely many possible solutions: one for each point along the line joining the two corner points in question. In practice, select those points with integer solutions (since x and y must be whole numbers in this problem) that are in the feasible region and close to this line, and choose the one that gives the largest profit.   **65.** Proof

Section 5.3

1. $p = 8$; $x = 4$, $y = 0$ **3.** $p = 4$; $x = 4$, $y = 0$ **5.** $p = 80$; $x = 10$, $y = 0$, $z = 10$ **7.** $p = 53$; $x = 5$, $y = 0$, $z = 3$
9. $z = 14,500$; $x_1 = 0$, $x_2 = 500/3$, $x_3 = 5,000/3$
11. $p = 6$; $x = 2$, $y = 1$, $z = 0$, $w = 3$ **13.** $p = 7$; $x = 1$, $y = 0$, $z = 2$, $w = 0$, $v = 4$ (or: $x = 1$, $y = 0$, $z = 2$, $w = 1$, $v = 3$)   **15.** $p = 21$; $x = 0$, $y = 2.27$, $z = 5.73$
17. $p = 4.52$; $x = 1$, $y = 0$, $z = 0.67$, $w = 1.52$ (or: $x = 1.67$, $y = 0.67$, $z = 0$, $w = 1.52$)   **19.** $p = 7.7$; $x = 1.1$, $y = 0$, $z = 2.2$, $w = 0$, $v = 4.4$ **21.** You should purchase 500 calculus texts, no history texts, and no marketing texts. The maximum profit is $5,000 per semester.   **23.** The company can make a maximum profit of $650 by making 100 gallons of PineOrange, 200 gallons of PineKiwi, and 150 gallons of OrangeKiwi.   **25.** The department should offer no Ancient History, 30 sections of Medieval History, and 15 sections of Modern History, for a profit of $1,050,000. There will be 500 students without classes, but all time slots and professors are used.   **27.** Plant 80 acres of tomatoes, and leave the other 20 acres unplanted. This will give you a profit of $160,000.
29. It can make a profit of $10,000 by selling 1,000 servings of granola, 500 servings of nutty granola, and no nuttiest granola. It is left with 2,000 ounces of almonds.   **31.** Achlúk can inflict a maximum of 70,000 units of damage using an arsenal of 5,000 axes, no maces, and 5,000 spears.   **33.** Allocate 5 million gallons to process A and 45 million gallons to process C. Another solution: Allocate 10 million gallons to process B and 40 million gallons to process C.   **35.** Use 20 servings of Gainz and none of the others for 120 grams of BCAAs.   **37.** She is wrong; you should buy 1,000 shares of OCR and no others.
39. Allocate $2,250,000 to automobile loans, $500,000 to signature loans, and $2,250,000 to any combination of furniture loans and other secured loans.   **41.** Invest $75,000 in Universal, none in the rest. Another optimal solution is: Invest $18,750 in Universal, and $75,000 in EMI.   **43.** Tucson to Honolulu: 290 boards; Tucson to Venice Beach: 330 boards; Toronto to Honolulu: 0 boards; Toronto to Venice Beach: 200 boards, giving 820 boards shipped.   **45.** Fly 10 people from Chicago to Los Angeles, 5 people from Chicago to New York, and 10 people from Denver to New York.   **47.** Yes; the given problem can be stated as: Maximize $p = 3x - 2y$ subject to $-x + y - z \leq 0$, $x - y - z \leq 6$, $x \geq 0$, $y \geq 0$, $z \geq 0$ **49.** The graphical method applies only to LP problems in two unknowns, whereas the simplex method can be used to solve LP problems with any number of unknowns.   **51.** She is correct. There are only two constraints, so there can be only two active variables, giving two or fewer nonzero values for the unknowns at each stage.   **53.** A basic solution to a system of linear equations is a solution in which all the nonpivotal variables are taken to be zero; that is, all variables whose values are arbitrary are assigned the value zero. To obtain a basic solution for a given system of linear equations, one can row-reduce the associated augmented matrix, write down the general solution, and then set all the parameters (variables with "arbitrary" values) equal to zero.   **55.** No. Let us assume for

the sake of simplicity that all the pivots are 1s. (They certainly be changed to 1s without affecting the value of any of the variables.) Because the entry at the bottom of the pivot column is negative, the bottom row gets replaced by itself plus a positive multiple of the pivot row. The value of the objective function (bottom right entry) is thus replaced by itself plus a positive multiple of the nonnegative rightmost entry of the pivot row. Therefore, it cannot decrease.

Section 5.4

1. $p = 20/3$; $x = 4/3$, $y = 16/3$ **3.** $p = 850/3$; $x = 50/3$, $y = 25/3$ **5.** $p = 750$; $x = 0$, $y = 150$, $z = 0$
7. $p = 450$; $x = 10$, $y = 10$, $z = 10$ **9.** $p = 260/3$; $x = 50/3$, $y = 0$, $z = 70/3$, $w = 0$ **11.** $c = 80$; $x = 20/3$, $y = 20/3$ **13.** $c = 100$; $x = 0$, $y = 100$, $z = 0$
15. $c = 111$; $x = 1$, $y = 1$, $z = 1$ **17.** $c = 200$; $x = 200$, $y = 0$, $z = 0$, $w = 0$ **19.** $p = 136.75$; $x = 0$, $y = 25.25$, $z = 0$, $w = 15.25$ **21.** $c = 66.67$; $x = 0$, $y = 66.67$, $z = 0$
23. $c = -250$; $x = 0$, $y = 500$, $z = 500$, $w = 1,500$
25. Plant 100 acres of tomatoes and no other crops. This will give you a profit of $200,000. (You will be using all 100 acres of your farm.)   **27.** 10 mailings to the East Coast, none to the Midwest, and 10 to the West Coast. Cost: $900. Another solution resulting in the same cost is no mailings to the East Coast, 15 to the Midwest, and none to the West Coast.   **29.** 10,000 quarts of orange juice and 2,000 quarts of orange concentrate
31. Sell 24,000 regional music albums, 10,000 pop/rock music albums, and 6,000 tropical music albums per day for a maximum revenue of $196,000.   **33.** Achlúk can inflict a maximum of 1,000 units of damage using an arsenal of no axes, 100 maces, and 50 spears.   **35.** Use 200 axes, no maces, and 100 spears for a minimum cost of 1,800 gold pieces.
37. One serving of cereal, one serving of juice, and no dessert!
39. 15 bundles from Nadir, 5 from Sonny, and none from Blunt. Cost: $70,000. Another solution resulting in the same cost is 10 bundles from Nadir, none from Sonny, and 10 from Blunt.   **41.** Use no Xtend, 10 servings of Gainz, and 20 servings of Strongevity for a total cost of $37.   **43. a.** Build 1 convention-style hotel, 4 vacation-style hotels, and 2 small motels. The total cost will amount to $188 million. **b.** Because 20% of this is $37.6 million, you will still be covered by the subsidy.   **45.** Tucson to Honolulu: 500 boards per week; Tucson to Venice Beach: 120 boards per week; Toronto to Honolulu: 0 boards per week; Toronto to Venice Beach: 410 boards per week. Minimum weekly cost is $9,700.   **47.** $2,500 from Congressional Integrity Bank, $0 from Citizens' Trust, $7,500 from Checks R Us.   **49.** Fly 5 people from Chicago to Los Angeles, 15 from Chicago to New York, 5 from Denver to Los Angeles, and none from Denver to New York at a total cost of $4,000.
51. Hire no more cardiologists, 12 rehabilitation specialists, and 5 infectious disease specialists.   **53.** The solution $x = 0$, $y = 0, \ldots$, represented by the initial tableau may not be feasible. In Phase I we use pivoting to arrive at a basic solution that is feasible.   **55.** The basic solution corresponding to the initial tableau has all the unknowns equal to zero, and this is not a fea-

sible solution because it does not satisfy the given inequality.
57. (C)   **59.** Answers may vary. Examples are Exercises 1 and 2.   **61.** Answers may vary. A simple example is: Maximize $p = x + y$ subject to $x + y \le 10$, $x + y \ge 20$, $x \ge 0$, $y \ge 0$.

Section 5.5

1. Minimize $c = 6s + 2t$ subject to $s - t \ge 2$, $2s + t \ge 1$, $s \ge 0$, $t \ge 0$.   **3.** Maximize $p = 100x + 50y$ subject to $x + 2y \le 2$, $x + y \le 1$, $x \le 3$, $x \ge 0$, $y \ge 0$.   **5.** Minimize $c = 3s + 4t + 5u + 6v$ subject to $s + u + v \ge 1$, $s + t + v \ge 1$, $s + t + u \ge 1$, $t + u + v \ge 1$, $s \ge 0$, $t \ge 0$, $u \ge 0$, $v \ge 0$.   **7.** Maximize $p = 1{,}000x + 2{,}000y + 500z$ subject to $5x + z \le 1$, $-x + z \le 3$, $y \le 1$, $x - y \le 0$, $x \ge 0$, $y \ge 0$, $z \ge 0$.   **9.** $c = 4$; $s = 2$, $t = 2$ **11.** $c = 80$; $s = 20/3$, $t = 20/3$ **13.** $c = 1.8$; $s = 6$, $t = 2$ **15.** $c = 25$; $s = 5$, $t = 15$ **17.** $c = 30$; $s = 30$, $t = 0$, $u = 0$
19. $c = 100$; $s = 0$, $t = 100$, $u = 0$ **21.** $c = 30$; $s = 10$, $t = 10$, $u = 10$ **23.** $R = \begin{bmatrix} 3/5 & 2/5 \end{bmatrix}$, $C = \begin{bmatrix} 2/5 & 3/5 & 0 \end{bmatrix}^T$, $e = 1/5$ **25.** $R = \begin{bmatrix} 1/4 & 0 & 3/4 \end{bmatrix}$, $C = \begin{bmatrix} 1/2 & 0 & 1/2 \end{bmatrix}^T$, $e = 1/2$ **27.** $R = \begin{bmatrix} 0 & 3/11 & 3/11 & 5/11 \end{bmatrix}$, $C = \begin{bmatrix} 8/11 & 0 & 2/11 & 1/11 \end{bmatrix}^T$, $e = 9/11$ **29.** 4 ounces each of fish and cornmeal for a total cost of 40¢ per can; 5/12¢ per gram of protein, 5/12¢ per gram of fat.   **31.** 100 ounces of grain and no chicken for a total cost of $1; 1/2¢ per gram of protein, 0¢ per gram of fat.   **33.** One serving of cereal, one serving of juice, and no dessert! for a total cost of 37¢; 1/6¢ per calorie and 17/120¢ per % U.S. RDA of vitamin C.
35. 10 mailings to the East Coast, none to the Midwest, and 10 to the West Coast. Cost: $900; 20¢ per Democrat and 40¢ per Republican. OR 15 mailings to the Midwest and no mailing to the coasts. Cost: $900; 20¢ per Democrat and 40¢ per Republican.   **37.** Place 13 ads in *Sports Illustrated* and 8 in *GQ*. Cost: $34,000.   **39.** Gillian should use use 12 sleep spells and 12 shock spells, costing 840 therms of energy.   **41.** T. N. Spend should spend about 73% of the days in Littleville, 27% in Metropolis, and skip Urbantown. T. L. Down should spend about 91% of the days in Littleville, 9% in Metropolis, and skip Urbantown. The expected outcome is that T. L. Down will lose about 227 votes per day of campaigning.   **43.** Each player should show one finger with probability 1/2, two fingers with probability 1/3, and three fingers with probability 1/6. The expected outcome is that player A will win 2/3 point per round, on average.   **45.** Write moves as (x, y), where x represents the number of regiments sent to the first location and y represents the number sent to the second location. Colonel Blotto should play $(0, 4)$ with probability 4/9, $(2, 2)$ with probability 1/9, and $(4, 0)$ with probability 4/9. Captain Kije has several optimal strategies, one of which is to play $(0, 3)$ with probability 1/30, $(1, 2)$ with probability 8/15, $(2, 1)$ with probability 16/45, and $(3, 0)$ with probability 7/90. The expected outcome is that Colonel Blotto will win 14/9 points on average.   **47.** Two variables and three constraints: In the matrix formulation of an LP problem, the number of rows is one more than the number of constraints, and the number of columns is one more than the number of variables. Thus, the

matrix form of the primal problem has three rows and four columns, so its transpose has four rows and three columns, translating to three constraints and two variables.   **49.** The dual of a standard minimization problem satisfying the nonnegative objective condition is a standard maximization problem, which can be solved by using the standard simplex algorithm, thus avoiding the need to do Phase I.   **51.** Answers will vary. An example is: Minimize $c = x - y$ subject to $x - y \geq 100$, $x + y \geq 200$, $x \geq 0$, $y \geq 0$. This problem can be solved by using the techniques in Section 5.4.   **53.** The dual problem is a nonstandard maximization problem, because the right-hand sides of its constraints are the entries in the bottom row of the matrix representation of the primal problem, and at least one of those entries is negative   **55.** If the given problem is a standard minimization problem satisfying the nonnegative objective condition, its dual is a standard maximization problem and so can be solved by using a single-phase simplex method. Otherwise, dualizing may not save any labor, since the dual will not be a standard maximization problem.

Chapter 5 Review

1.

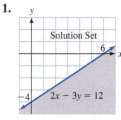

Unbounded

3.

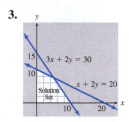

Bounded; corner points: $(0, 0)$, $(0, 10)$, $(5, 15/2)$, $(10, 0)$

5. $p = 21$; $x = 9$, $y = 3$ **7.** $c = 22$; $x = 8$, $y = 6$
9. $p = 45$; $x = 0$, $y = 15$, $z = 15$ **11.** $p = 220$; $x = 20$, $y = 20$, $z = 60$ **13.** $c = 30$; $x = 30$, $y = 0$, $z = 0$
15. No solution; feasible region unbounded
17. $c = 50$; $x = 20$, $y = 10$, $z = 0$, $w = 20$, OR $x = 30$, $y = 0$, $z = 0$, $w = 20$ **19.** $c = 60$; $x = 24$, $y = 12$ OR $x = 0$, $y = 60$ **21.** $c = 20$; $x = 0$, $y = 20$
23. $R = \begin{bmatrix} 1/2 & 1/2 & 0 \end{bmatrix}$, $C = \begin{bmatrix} 0 & 1/3 & 2/3 \end{bmatrix}^T$, $e = 0$
25. $R = \begin{bmatrix} 1/27 & 7/9 & 5/27 \end{bmatrix}$, $C = \begin{bmatrix} 8/27 & 5/27 & 14/27 \end{bmatrix}^T$, $e = 8/27$ **27.** (A)   **29.** 35   **31.** 400 packages from each for a minimum cost of $52,000   **33. a.** (B), (D) **b.** 450 packages from Duffin House and 375 from Higgins Press for a minimum cost of $52,500   **35.** 220 shares of EEE and 20 shares of RRR. The minimum total risk index is 500.   **37.** 240 Sprinkles, 120 Storms, and no Hurricanes   **39.** Order 600 packages from Higgins and none from the others for a total cost of $90,000.   **41. a.** Let x = Number of science credits, y = Number of fine arts credits, z = Number of liberal arts credits, and w = Number of math credits. Minimize $C = 300x + 300y + 200z + 200w$ subject to: $x + y + z + w \geq 120$; $x - y \geq 0$; $-2x + w \leq 0$; $-y + 3z - 3w \leq 0$; $x \geq 0$, $y \geq 0$, $z \geq 0$, $w \geq 0$
b. Billy-Sean should take the following combination:

Sciences—24 credits, Fine Arts—no credits, Liberal Arts—48 credits, Mathematics—48 credits for a total cost of $26,400.   **43.** Smallest cost is $20,000; New York to OHaganBooks.com: 600 packages, New York to FantasyBooks.com: 0 packages, Illinois to OHaganbooks.com: 0 packages, Illinois to FantasyBooks.com: 200 packages.
45. FantasyBooks.com should choose between "2 for 1" and "3 for 2" with probabilities 20% and 80%, respectively. OHaganBooks.com should choose between "3 for 1" and "Finite Math" with probabilities 60% and 40%, respectively. OHaganBooks.com expects to gain 12,000 customers from FantasyBooks.com.

Chapter 6

Section 6.1

1. $F = \{\text{spring, summer, fall, winter}\}$
3. $I = \{1, 2, 3, 4, 5, 6\}$ **5.** $A = \{1, 2, 3\}$
7. $B = \{2, 4, 6, 8\}$ **9. a.** $S = \{(\text{H, H}), (\text{H, T}), (\text{T, H}), (\text{T, T})\}$ **b.** $S = \{(\text{H, H}), (\text{H, T}), (\text{T, T})\}$
11. $S = \{(1, 5), (2, 4), (3, 3), (4, 2), (5, 1)\}$
13. $S = \{(1, 5), (2, 4), (3, 3)\}$ **15.** $S = \varnothing$

17.

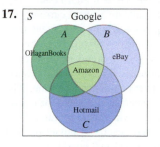

19.

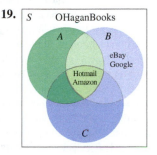

21. A **23.** A **25.** {June, Janet, Jill, Justin, Jeffrey, Jello, Sally, Solly, Molly, Jolly}   **27.** {Jello}   **29.** $\varnothing$
31. {Jello}   **33.** {Janet, Justin, Jello, Sally, Solly, Molly, Jolly}   **35.** {(small, triangle), (small, square), (medium, triangle), (medium, square), (large, triangle), (large, square)}   **37.** {(small, blue), (small, green), (medium, blue), (medium, green), (large, blue), (large, green)}

39.

	A	B	C
1		Triangle	Square
2	Blue	Blue Triangle	Blue Square
3	Green	Green Triangle	Green Square

41.

	A	B	C
1		Blue	Green
2	Small	Small Blue	Small Green
3	Medium	Medium Blue	Medium Green
4	Large	Large Blue	Large Green

43. $B \times A = \{$1H, 1T, 2H, 2T, 3H, 3T, 4H, 4T, 5H, 5T, 6H, 6T$\}$

45. $A \times A \times A = \{$HHH, HHT, HTH, HTT, THH, THT, TTH, TTT$\}$

47. $\{(1, 1), (1, 3), (1, 5), (3, 1), (3, 3), (3, 5), (5, 1), (5, 3), (5, 5)\}$ **49.** $\varnothing$ **51.** $\{(1, 1), (1, 3), (1, 5), (3, 1), (3, 3), (3, 5), (5, 1), (5, 3), (5, 5), (2, 2), (2, 4), (2, 6), (4, 2), (4, 4), (4, 6), (6, 2), (6, 4), (6, 6)\}$ **53–59.** Answers will vary.

61. $A \cap B = \{$Acme, Crafts$\}$ **63.** $B \cup C = \{$Acme, Brothers, Crafts, Dion, Effigy, Global, Hilbert$\}$

65. $A' \cap C = \{$Dion, Hilbert$\}$ **67.** $A \cap B' \cap C' = \varnothing$

69. $\{2003, 2004, 2005, 2006\} \times \{$Sailboats, Motor Boats, Yachts$\}$

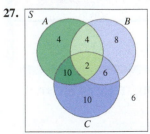

	A	B	C	D
1		Sailboats	Motor Boats	Yachts
2	2003	(2003 Sailboats)	(2003 Motor Boats)	(2003 Yachts)
3	2004	(2004 Sailboats)	(2004 Motor Boats)	(2004 Yachts)
4	2005	(2005 Sailboats)	(2005 Motor Boats)	(2005 Yachts)
5	2006	(2006 Sailboats)	(2006 Motor Boats)	(2006 Yachts)

71. $I \cup J$ **73.** (B) **75.** Answers may vary. Let $A = \{1\}$, $B = \{2\}$, and $C = \{1, 2\}$. Then $(A \cap B) \cup C = \{1, 2\}$, but $A \cap (B \cup C) = \{1\}$. In general, $A \cap (B \cup C)$ must be a subset of A, but $(A \cap B) \cup C$ need not be; also, $(A \cap B) \cup C$ must contain C as a subset, but $A \cap (B \cup C)$ need not.

77. A universal set is a set containing all "things" currently under consideration. When discussing sets of positive integers, the universe might be the set of all positive integers, or the set of all integers (positive, negative, and 0), or any other set containing the set of all positive integers. **79.** Answers will vary. A is the set of suppliers who deliver components on time, B is the set of suppliers whose components are known to be of high quality, and C is the set of suppliers who do not promptly replace defective components. **81.** Let $A = \{$movies that are violent$\}$, $B = \{$movies that are shorter than 2 hours$\}$, $C = \{$movies that have a tragic ending$\}$, and $D = \{$movies that have an unexpected ending$\}$. The given sentence can be rewritten as "She prefers movies in $A' \cap B \cap (C \cup D)'$." It can also be rewritten as "She prefers movies in $A' \cap B \cap C' \cap D'$."

83. Removing the comma would cause the statement to be ambiguous, as it could then correspond to either WWII $\cup$ (Comix $\cap$ Aliens') or to (WWII $\cup$ Comix) $\cap$ Aliens'. (See Exercise 76.)

Section 6.2

1. 9 **3.** 7 **5.** 4 **7.** $n(A \cup B) = 7$, $n(A) + n(B) - n(A \cap B) = 4 + 5 - 2 = 7$

9. 4 **11.** 18 **13.** 72 **15.** 60 **17.** 20 **19.** 6 **21.** 9

23. 4 **25.** $n((A \cap B)') = 9$ $n(A') + n(B') - n((A \cup B)') = 6 + 7 - 4 = 9$

27. 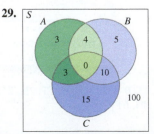 **29.**

31. 115.4 million **33.** 2 **35.** 9 million **37.** $C \cap N$ is the set of authors who are both successful and new. $C \cup N$ is the set of authors who are either successful or new (or both). $n(C) = 30$; $n(N) = 20$; $n(C \cap N) = 5$; $n(C \cup N) = 45$; $45 = 30 + 20 - 5$ **39.** $C \cap N'$ is the set of authors who are successful but not new. $n(C \cap N') = 25$ **41.** 31.25%; 83.33%

43. $M \cap C$; $n(M \cap C) = 110$ thousand units

45. $B \cap T'$; $n(B \cap T') = 290$ thousand units

47. $B \cap (W \cup M)$; $n(B \cap (W \cup M)) = 230$ thousand units

49. $X \cap M'$; $n(X \cap M') = 16$ **51.** 35; the number of industries that either were not in the health-care sector or were unchanged in value (or both) **53.** 1/10; the fraction of industries that decreased that were from the information technology sector **55. a.** 931 **b.** 382

57. a. 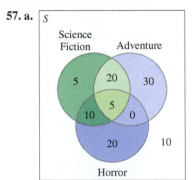 **b.** 37.5%

59. 17 **61.** $n(A) < n(B)$ **63.** The number of elements in the Cartesian product of two finite sets is the product of the number of elements in the two sets. **65.** Answers will vary.

67. When $A \cap B \neq \varnothing$ **69.** When $B \subseteq A$

71. $n(A \cup B \cup C) = n(A) + n(B) + n(C) - n(A \cap B) - n(B \cap C) - n(A \cap C) + n(A \cap B \cap C)$

Section 6.3

1. 10 **3.** 30 **5.** 6 outcomes **7.** 15 outcomes

9. 13 outcomes **11.** 25 outcomes **13.** 4 **15.** 93

17. 16 **19.** 30 **21.** 13 **23.** 18 **25.** 25,600 **27.** 3,381

29. a. 288 **b.** 288 **31.** 256 **33.** 10 **35.** 32,760

37. 4 **39. a.** 8,000,000 **b.** 30,000 **c.** 4,251,528

41. a. $2 \times 10^9 = 2$ billion possible card numbers

b. $10^9 \times 9 = 9$ billion possible card numbers

43. a. $4^3 = 64$ **b.** 4^n **c.** $4^{2.1 \times 10^{10}}$

45. a. $16^6 = 16,777,216$ **b.** $16^3 = 4,096$ **c.** $16^2 = 256$ **d.** 766

47. $(10 \times 9 \times 8 \times 7 \times 6 \times 5 \times 4) \times (8 \times 7 \times 6 \times 5) = 1,016,064,000$ possible casts **49. a.** $26^3 \times 10^3 = 17,576,000$ **b.** $26^2 \times 23 \times 10^3 = 15,548,000$

c. $15,548,000 - 3 \times 10^3 = 15,545,000$ **51. a.** 4 **b.** 4

c. There would be an infinite number of routes. **53. a.** 72

b. 36 **55.** 96 **57. a.** 36 **b.** 37 **59.** Step 1: Choose a day of the week on which Jan. 1 will fall: seven choices. Step 2: Decide whether or not it is a leap year; two choices. Total: $7 \times 2 = 14$ possible calendars. **61.** 1,900 **63.** Step 1: Choose a position in the left-right direction; m choices. Step 2: Choose a position in the front-back direction; n choices. Step 3: Choose a position in the up-down direction; r choices. Hence, there are

$m \cdot n \cdot r$ possible outcomes.   **65.** 4   **67.** Cartesian product
69. The decision algorithm produces every pair of shirts twice, first in one order and then in the other.   **71.** Think of placing the five squares in a row of five empty slots. Step 1: Choose a slot for the blue square; five choices. Step 2: Choose a slot for the green square; four choices. Step 3: Choose the remaining three slots for the yellow squares; one choice. Hence, there are 20 possible five-square sequences.

Section 6.4

1. 720   **3.** 56   **5.** 360   **7.** 15   **9.** 3   **11.** 45   **13.** 20
15. 4,950   **17.** 360   **19.** 35   **21.** 120   **23.** 120   **25.** 20
27. 60   **29.** 210   **31.** 7   **33.** 35   **35.** 24   **37.** 126
39. 196   **41.** 105   **43.** $\dfrac{C(30, 5) \times 5^{25}}{6^{30}} \approx 0.192$

45. $\dfrac{C(30, 15) \times 3^{15} \times 3^{15}}{6^{30}} \approx 0.144$

47. $C(11, 1)C(10, 4)C(6, 4)C(2, 2)$
49. $C(11, 2)C(9, 1)C(8, 1)C(7, 3)C(4, 1) \times C(3, 1)C(2, 1)C(1, 1)$
51. $C(10, 2)C(8, 4)C(4, 1)C(3, 1)C(2, 1)C(1, 1)$
53. 24   **55.** $C(13, 2)C(4, 2)C(4, 2) \times 44 = 123{,}552$
57. $13 \times C(4, 2)C(12, 3) \times 4 \times 4 \times 4 = 1{,}098{,}240$
59. $10 \times 4^5 - 10 \times 4 = 10{,}200$ **61. a.** 252 **b.** 20 **c.** 26
63. a. 300 **b.** 3 **c.** 1 in 100 or .01   **65. a.** 16! **b.** $8! \times 2^8$
67. $C(60, 15) \times 8$ **69. a.** 210 **b.** 91 **c.** No   **71. a.** 23!
b. 18! **c.** $19 \times 18!$ **73.** (A)   **75.** (D)   **77. a.** 9,880
b. 1,560 **c.** 11,480   **79. a.** $C(20, 2) = 190$ **b.** $C(n, 2)$
81. The multiplication principle; it can be used to solve all problems that use the formulas for permutations.   **83.** (A), (D)
85. A permutation. Changing the order in a list of driving instructions can result in a different outcome; for instance, "1. Turn left. 2. Drive one mile." and "1. Drive one mile. 2. Turn left." will take you to different locations.   **87.** Urge your friend not to focus on formulas but instead to learn to formulate decision algorithms and use the principles of counting.   **89.** It is ambiguous on the following point: Are the three students to play different characters, or are they to play a group of three, such as "three guards"? This should be made clear in the exercise.

Chapter 6 Review

1. $N = \{-3, -2, -1\}$ **3.** $S = \{(1, 2), (1, 3), (1, 4),$
$(1, 5), (1, 6), (2, 1), (2, 3), (2, 4), (2, 5), (2, 6), (3, 1),$
$(3, 2), (3, 4), (3, 5), (3, 6), (4, 1), (4, 2), (4, 3), (4, 5),$
$(4, 6), (5, 1), (5, 2), (5, 3), (5, 4), (5, 6), (6, 1), (6, 2),$
$(6, 3), (6, 4), (6, 5)\}$ **5.** $A \cup B' = \{a, b, d\}$,
$A \times B' = \{(a, a), (a, d), (b, a), (b, d)\}$ **7.** $A \times B$
9. $(E' \cap Q)'$ or $E \cup Q'$ **11.** $n(A \cup B) =$
$n(A) + n(B) - n(A \cap B), n(C') = n(S) - n(C); 100$
13. $n(A \times B) = n(A)n(B), n(A \cup B) = n(A) +$
$n(B) - n(A \cap B), n(A') = n(S) - n(A); 21$
15. $C(12, 1)C(4, 2)C(11, 3)C(4, 1)C(4, 1)C(4, 1)$
17. $C(4, 1)C(10, 1)$ **19.** 6   **21.** $C(4, 4)C(8, 1) = 8$
23. $C(3, 2)C(9, 3) + C(3, 3)C(9, 2) = 288$

25. The set of books that are either sci fi or stored in Texas (or both); $n(S \cup T) = 112{,}000$ **27.** The set of books that are either stored in California or not sci fi; $n(C \cup S') = 175{,}000$
29. The romance books that are also horror books or stored in Texas; $n(R \cap (T \cup H)) = 20{,}000$ **31.** 1,000
33. FarmerBooks.com; 1,800   **35.** FarmerBooks.com; 5,400   **37.** $26 \times 26 \times 26 = 17{,}576$
39. $26 \times 25 \times 9 \times 10 = 58{,}500$ **41.** 60,000   **43.** 19,600

Chapter 7

Section 7.1

1. $S = \{HH, HT, TH, TT\}$; $E = \{HH, HT, TH\}$
3. $S = \{HHH, HHT, HTH, HTT, THH, THT, TTH, TTT\}$;
$E = \{HTT, THT, TTH, TTT\}$

5. $S = \left\{ \begin{array}{l} (1, 1), (1, 2), (1, 3), (1, 4), (1, 5), (1, 6), \\ (2, 1), (2, 2), (2, 3), (2, 4), (2, 5), (2, 6), \\ (3, 1), (3, 2), (3, 3), (3, 4), (3, 5), (3, 6), \\ (4, 1), (4, 2), (4, 3), (4, 4), (4, 5), (4, 6), \\ (5, 1), (5, 2), (5, 3), (5, 4), (5, 5), (5, 6), \\ (6, 1), (6, 2), (6, 3), (6, 4), (6, 5), (6, 6) \end{array} \right\}$;

$E = \{(1, 4), (2, 3), (3, 2), (4, 1)\}$

7. $S = \left\{ \begin{array}{l} (1, 1), (1, 2), (1, 3), (1, 4), (1, 5), (1, 6), \\ (2, 2), (2, 3), (2, 4), (2, 5), (2, 6), \\ (3, 3), (3, 4), (3, 5), (3, 6), \\ (4, 4), (4, 5), (4, 6), \\ (5, 5), (5, 6), \\ (6, 6) \end{array} \right\}$;

$E = \{(1, 3), (2, 2)\}$
9. S as in Exercise 7; $E = \{(2, 2), (2, 3), (2, 5), (3, 3),$
$(3, 5), (5, 5)\}$ **11.** $S = \{m, o, z, a, r, t\}$: $E = \{o, a\}$
13. $S = \{(s, o), (s, r), (s, e), (o, s), (o, r), (o, e), (r, s),$
$(r, o), (r, e), (e, s), (e, o), (e, r)\}$; $E = \{(o, s), (o, r),$
$(o, e), (e, s), (e, o), (e, r)\}$ **15.** $S = \{01, 02, 03, 04, 10, 12,$
$13, 14, 20, 21, 23, 24, 30, 31, 32, 34, 40, 41, 42, 43\}$;
$E = \{10, 20, 21, 30, 31, 32, 40, 41, 42, 43\}$
17. $S = \{\text{domestic car, imported car, van, antique car, antique truck}\}$; $E = \{\text{van, antique truck}\}$ **19. a.** all sets of four gummy candies chosen from the packet of 12. **b.** all sets of four gummy candies in which two are strawberry and two are black currant.
21. a. all lists of 15 people chosen from 20. **b.** all lists of 15 people chosen from 20, in which Hillary Clinton occupies the eleventh position.   **23.** $A \cap B; n(A \cap B) = 1$
25. $B'; n(B') = 33$ **27.** $B' \cap D'; n(B' \cap D') = 2$
29. $C \cup B; n(C \cup B) = 12$ **31.** $W \cap I$ **33.** $E \cup I'$
35. $I \cup (W \cap E')$ **37.** $(I \cup W) \cap E'$ **39.** 56; 4
41. $C(4, 1)C(2, 1)C(2, 1) = 16$
43. $E = \{\text{Pacific, Mountain, West South Central, South Atlantic}\}$ **45.** $E \cup F$ is the event that you choose a region that saw an increase in housing prices of 6% or more or is on the east coast. $E \cup F = \{\text{Pacific, Mountain, West South Central, New England, Middle Atlantic, South Atlantic}\}$. $E \cap F$ is the

event that you choose a region that saw an inecrease in housing prices of 6% or more and is on the east coast. $E \cap F =$ {South Atlantic}. **47. a.** Not mutually exclusive **b.** Mutually exclusive **49.** $S \cap N$ is the event that an author is successful and new. $S \cup N$ is the event that an author is either successful or new; $n(S \cap N) = 5$; $n(S \cup N) = 45$ **51.** N and E **53.** $S \cap N'$ is the event that an author is successful but not a new author. $n(S \cap N') = 25$ **55.** 31.25%; 83.33% **57.** $X \cap M'$; $n(X \cap M') = 16$ **59.** The event that an industry either was not in the health care sector or was unchanged in value (or both); 35 **61.** F and M, F and T, M and T, X and Y, Y and Z, X and Z, T and Z **63. a.** $E' \cap H$ **b.** $E \cup H$ **c.** $(E \cup G)' = E' \cap G'$ **65. a.** {9} **b.** {6} **67. a.** The dog's fight drive is weakest. **b.** The dog's fight and flight drives are either both strongest or both weakest. **c.** Either the dog's fight drive is strongest, or its flight drive is strongest. **69.** $C(6, 4) = 15$; $C(1, 1)C(5, 3) = 10$ **71. a.** $n(S) = P(7, 3) = 210$ **b.** $E \cap F$ is the event that Celera wins and Electoral College is in second or third place. In other words, it is the set of all lists of three horses in which Celera is first and Electoral College is second or third. $n(E \cap F) = 10$ **73.** subset of the sample space **75.** E and F do not both occur **77.** (B) **79.** True; consider the following experiment: Select an element of the set S at random. **81.** Answers may vary. Cast a die, and record the remainder when the number facing up is divided by 2. **83.** Yes. For instance, $E = \{(2, 5), (5, 1)\}$ and $F = \{(4, 3)\}$ are two such events.

Section 7.2

1. .4 **3.** .8 **5.** .6

7.

Outcome	HH	HT	TH	TT
Rel. Frequency	.275	.2375	.3	.1875

9. .575 **11.** The second coin *seems* slightly biased in favor of heads, as heads comes up approximately 58% of the time. On the other hand, it is conceivable that the coin is fair and that heads came up 58% of the time purely by chance. Deciding which conclusion is more reasonable requires some knowledge of inferential statistics. **13.** Yes **15.** No; relative frequencies cannot be negative. **17.** Yes **19.** Missing value: .3 **a.** .6 **b.** .4 **21.** Answers will vary. **23.** Answers will vary. **25. a.** .54 **b.** .16 **c.** .98

27. a.

Mortgage Status	Current	Past Due	In Foreclosure	Repossessed
Rel. Frequency	.67	.26	.045	.025

b. .33

29. a.

Age	0–14	15–29	30–64	>64
Rel. Frequency	.30	.27	.37	.06

b. .36

31. a.

Test Rating	3	2	1	0
Rel. Frequency	.1	.4	.4	.1

b. .5 **33.** Dialup: 1,256, cable modem: 412, DSL: 302, Other: 30

35.

Outcome	Surge	Plunge	Steady
Rel. Frequency	.2	.3	.5

37. .25 **39.** .2 **41.** .7 **43.** 5/6 **45.** 5/16

47.

Outcome	U	C	R
Rel. Frequency	.2	.64	.16

49.

Conventional	No pesticide	Single pesticide	Multiple pesticide
Probability	.27	.13	.60
Organic	No pesticide	Single pesticide	Multiple pesticide
Probability	.77	.13	.10

51. $P(\text{false negative}) = 10/400 = .025$, $P(\text{false positive}) = 10/200 = .05$ **53.** Answers will vary. **55.** The fraction of times E occurs **57.** 101; $fr(E)$ can be any number between 0 and 100 inclusive, so the possible answers are $0/100 = 0$, $1/100 = .01$, $2/100 = .02, \ldots, 99/100 = .99$, $100/100 = 1$. **59.** Wrong. For a pair of fair dice, the probability of a pair of matching numbers is 1/6, as Ruth says. However, it is quite possible, although not very likely, that if you cast a pair of fair dice 20 times, you will never obtain a matching pair. (In fact, there is approximately a 2.6% chance that this will happen.) In general, a nontrivial claim about probability can never be absolutely validated or refuted experimentally. All we can say is that the evidence suggests that the dice are not fair. **61.** For a (large) number of days, record the temperature prediction for the next day, and then check the actual high temperature the next day. Record whether the prediction was accurate (within, say, 2°F of the actual temperature). The fraction of times the prediction was accurate is the relative frequency.

Section 7.3

1. $P(e) = .2$ **a.** .9 **b.** .95 **c.** .1 **d.** .8 **3.** $P(E) = 1/4$ **5.** $P(E) = 1$ **7.** $P(E) = 3/4$ **9.** $P(E) = 3/4$ **11.** $P(E) = 1/2$ **13.** $P(E) = 1/9$ **15.** $P(E) = 0$ **17.** $P(E) = 1/4$ **19.** $1/12$; $\{(4, 4), (2, 3), (3, 2)\}$

21.

Outcome	1	2	3	4	5	6
Probability	$\frac{1}{9}$	$\frac{2}{9}$	$\frac{1}{9}$	$\frac{2}{9}$	$\frac{1}{9}$	$\frac{2}{9}$

$P(\{1, 2, 3\}) = 4/9$

23.

Outcome	1	2	3	4
Probability	$\frac{8}{15}$	$\frac{4}{15}$	$\frac{2}{15}$	$\frac{1}{15}$

25. .65   **27.** .1   **29.** .7   **31.** .4   **33.** .25   **35.** 1.0   **37.** .3
39. 1.0   **41.** No; $P(A \cup B)$ should be $\leq P(A) + P(B)$.
43. Yes   **45.** No; $P(A \cup B)$ should be $\geq P(A)$.   **47. a.** .93
b. .33
49.

Outcome	Hispanic or Latino	White (not Hispanic)	African American	Asian	Other
Probability	0.48	0.29	0.08	0.08	0.07

P(neither White nor Asian) = .63
51. a. $S = \{$stock market success, sold to other concern, fail$\}$

b.

Outcome	Stock Market Success	Sold to Other Concern	Fail
Probability	.2	.3	.5

c. .5

53.

Outcome	SUV	Pickup	Passenger Car	Minivan
Probability	.25	.15	.50	.10

55. $P(1) = 0$, $P(6) = 0$; $P(2) = P(3) = P(4) = P(5) = 1/4 = .25$; $P(\text{odd}) = .5$ **57.** $P(1) = P(6) = 1/10$; $P(2) = P(3) = P(4) = P(5) = 1/5$, $P(\text{odd}) = 1/2$
59. $P(1, 1) = P(2, 2) = \ldots = P(6, 6) = 1/66$; $P(1, 2) = \ldots = P(6, 5) = 1/33$, $P(\text{odd sum}) = 6/11$
61. $P(2) = 15/38$; $P(4) = 3/38$, $P(1) = P(3) = P(5) = P(6) = 5/38$, $P(\text{odd}) = 15/38$ **63.** 5/6
65. .39   **67.** .35   **69.** .20   **71.** .00   **73.** .80   **75.** .38
77. .62   **79.** .01   **81.** .61   **83.** .38   **85.** 22%; 43%   **87.** All
of them   **89.** 88.4%   **91.** Here is one possible experiment:
Roll a die, and observe which of the following outcomes
occurs. Outcome A: 1 or 2 facing up; $P(A) = 1/3$, Outcome B:
3 or 4 facing up; $P(B) = 1/3$, Outcome C: 5 or 6 facing up;
$P(C) = 1/3$.   **93.** He is wrong. It is possible to have a run of
losses of any length. Tony may have grounds to *suspect* that
the game is rigged, but he has no proof.   **95.** they are mutually
exclusive.   **97.** Wrong. For example, the modeled probability of
winning a state lottery is small but nonzero. However, the vast
majority of people who play the lottery every day of their lives
never win no matter how frequently they play, so the relative
frequency is zero for these people.   **99.** When $A \cap B = \varnothing$
we have $P(A \cap B) = P(\varnothing) = 0$, so $P(A \cup B) = P(A) + P(B) - P(A \cap B) = P(A) + P(B) - 0 = P(A) + P(B)$.   **101.** Zero. According to the assumption,
no matter how many thunderstorms occur, lightning cannot
strike a given spot more than once, so, after n trials the relative
frequency will never exceed $1/n$ and so will approach zero as
the number of trials gets large. Since the modeled probability
models the limit of relative frequencies as the number of trials
gets large, it must therefore be zero.   **103.** $P(A \cup B \cup C) = P(A) + P(B) + P(C) - P(A \cap B) - P(A \cap C) - P(B \cap C) + P(A \cap B \cap C)$

Section 7.4

1. 1/42   **3.** 7/9   **5.** 1/7   **7.** 1/2   **9.** 41/42   **11.** 1/15
13. 4/15   **15.** 1/5   **17.** $1/(2^8 \times 5^5 \times 5!)$ **19.** .4226
21. .0475   **23.** .0020   **25.** $1/27^{39}$ **27.** 1/7
29. Probability of being a Big Winner $= 1/2,118,760 \approx$ .000000472. Probability of being a Small-Fry Winner $= 225/2,118,760 \approx .000106194$. Probability of being either a
Big Winner or a Small-Fry Winner $= 226/2,118,760 \approx$ .000106666.   **31. a.** $C(600, 300)/C(700, 400)$
b. $C(699, 399)/C(700, 400)$ or $400/700$ **33.** $P(10, 3)/10^3 = 18/25 = .72$ **35.** 1/8   **37.** $(8! \times 2^8)/16!$ **39. a.** $1/2^{63}$
b. $[C(60, 15) \times 8]/2^{63}$ **41.** $8!/8^8$ **43.** 1/8   **45.** 37/10,000
47. a. 90,720   **b.** 25,200   **c.** $25,200/90,720 \approx .28$
49. The four outcomes listed are not equally likely; for
example, (red, blue) can occur in four ways. The methods
of this section yield a probability for (red, red) of
$C(2, 2)/C(4, 2) = 1/6$.   **51.** No. If we do not pay attention to
order, the probability is $C(5, 2)/C(9, 2) = 10/36 = 5/18$.
If we do pay attention to order, the probability is
$P(5, 2)/P(9, 2) = 20/72 = 5/18$ again. The difference
between permutations and combinations cancels when we
compute the probability.   **53.** Answers will vary.

Section 7.5

1. .4   **3.** .08   **5.** .75   **7.** .2   **9.** .5   **11.** $P(D) = .10$;
$P(D|M) = .30$ **13.** $P(A|L) = .30$; $P(L|A) = .10$
15. $P(E|M) = .55$; $P(E|M') = .05$ **17.** 1/10   **19.** 1/5
21. 2/9   **23.** 1/84   **25.** 5/21   **27.** 24/175
29.

31.

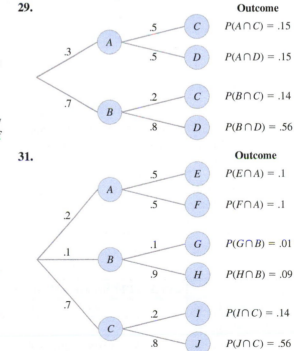

33. Mutually exclusive   **35.** Neither

37. $\frac{1}{2} \cdot \frac{1}{2} = \frac{1}{4}$ Independent **39.** $\frac{5}{18} \cdot \frac{1}{2} \neq \frac{1}{9}$ Dependent

41. $\frac{25}{36} \cdot \frac{5}{18} \neq \frac{2}{9}$ Dependent **43.** $(1/2)^{11} = 1/2{,}048$

45. .8 **47. a.** .24 **b.** .092 **49.** .34 **51.** Not independent; $P(\text{giving up} \mid \text{used Brand X}) = .1$ is larger than $P(\text{giving up}) = .05$. **53.** .00015 **55.** 5/6 **57.** 3/4
59. 11/16 **61.** 11/14

63.

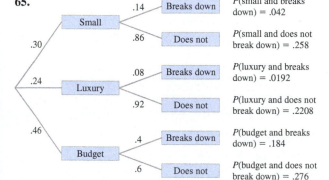

65.

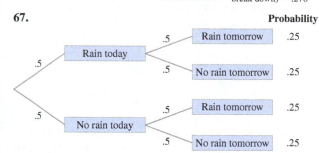

67.

Probability

Rain today .5 → Rain tomorrow .25
Rain today .5 → No rain tomorrow .25
No rain today .5 → Rain tomorrow .25
No rain today .5 → No rain tomorrow .25

69. .73 **71.** .38 **73.** .41 **75.** .97 **77.** The claim is correct. The probability that an unemployed person has a high school diploma only is .32, while the corresponding figure for an employed person is .27. **79.** $P(K \mid D) = 1.31 P(K \mid D')$
81. a. $P(I \mid T) > P(I)$ **b.** It was ineffective. **83. a.** .59
b. \$35,000 or more: $P(\text{Internet user} \mid < \$35{,}000) \approx$
$.27 < P(\text{Internet user} \mid \geq \$35{,}000) \approx .59$. **85.** $P(R \mid J)$
87. (D) **89. a.** .000057 **b.** .015043 **91.** 11% **93.** 106
95. .631
97.

	Saw Ad	Did Not See Ad	Total
Purchased Game	20	40	60
Did Not Purchase Game	180	360	540
Total	200	400	600

99. $100p^2\%$ **101.** Answers will vary. Here is a simple one:
E: The first toss is a head; F: The second toss is a head; G: The

third toss is a head. **103.** The probability you seek is $P(E \mid F)$, or should be. If, for example, you were going to place a wager on whether E occurs or not, it is crucial to know that the sample space has been reduced to F. (You know that F did occur.) If you base your wager on $P(E)$ rather than $P(E \mid F)$ you will misjudge your likelihood of winning. **105.** You might explain that the conditional probability of E is not the *a priori* probability of E, but it is the probability of E in a hypothetical world in which the outcomes are restricted to be what is given. In the example she is citing, yes, the probability of throwing a double-six is $1/36$ in the absence of any other knowledge. However, by the "conditional probability" of throwing a double-six given that the sum is larger than 7, we might mean the probability of a double-six in a situation in which the dice have already been thrown, but all we know is that the sum is greater than 7. Because there are only 15 ways in which that can happen, the conditional probability is $1/15$. For a more extreme case, consider the conditional probability of throwing a double-six given that the sum is 12. **107.** If $A \subseteq B$ then $A \cap B = A$, so $P(A \cap B) = P(A)$ and $P(A \mid B) = P(A \cap B)/P(B) = P(A)/P(B)$. **109.** Your friend is correct. If A and B are mutually exclusive, then $P(A \cap B) = 0$. On the other hand, if A and B are independent, then $P(A \cap B) = P(A)P(B)$. Thus, $P(A)P(B) = 0$. If a product is 0, then one of the factors must be 0, so either $P(A) = 0$ or $P(B) = 0$. Thus, it cannot be true that A and B are mutually exclusive, have nonzero probabilities, and are independent all at the same time.
111. $P(A' \cap B') = 1 - P(A \cup B) =$
$1 - [P(A) + P(B) - P(A \cap B)] =$
$1 - [P(A) + P(B) - P(A)P(B)] =$
$(1 - P(A))(1 - P(B)) = P(A')P(B')$

Section 7.6

1. .4 **3.** .7887 **5.** .7442 **7.** .1163 **9.** 26.8% **11.** .1724
13. 61% **15.** .73 **17.** .71 **19.** .165 **21.** 82% **23.** 12%
25. a. 14.43%; **b.** 19.81% of single homeowners have pools. Thus, they should go after the single homeowners. **27.** 9
29. .9310 **31.** 1.76% **33.** .20 **35.** .30 **37.** Show him an example such as Example 1 of this section, where $P(T \mid A) = .95$ but $P(A \mid T) \approx .64$. **39.** Suppose that the steroid test gives 10% false negatives and that only 0.1% of the tested population uses steroids. Then the probability that an athlete uses steroids, given that he or she has tested positive, is

$$\frac{(.9)(.001)}{(.9)(.001) + (.01)(.999)} \approx .083.$$ **41.** Draw a tree in which

the first branching shows which of R_1, R_2, or R_3 occurred and the second branching shows which of T or T' then occurred. There are three final outcomes in which T occurs:
$P(R_1 \cap T) = P(T \mid R_1)P(R_1)$, $P(R_2 \cap T) = P(T \mid R_2)P(R_2)$, and $P(R_3 \cap T) = P(T \mid R_3)P(R_3)$. In only one of these, the

first, does R_1 occur. Thus, $P(R_1 \mid T) = \dfrac{P(R_1 \cap T)}{P(T)} =$

$$\frac{P(T \mid R_1)P(R_1)}{P(T \mid R_1)P(R_1) + P(T \mid R_2)P(R_2) + P(T \mid R_3)P(R_3)}$$

43. The reasoning is flawed. Let A be the event that a Democrat agrees with Safire's column, and let F and M be the events that a Democrat reader is female and male, respectively. Then A. D. makes the following argument: $P(M|A) = .9$, $P(F|A') = .9$. Therefore, $P(A|M) = .9$. According to Bayes' theorem, we cannot conclude anything about $P(A|M)$ unless we know $P(A)$, the percentage of all Democrats who agreed with Safire's column. This was not given.

Section 7.7

1. $\begin{bmatrix} \frac{1}{4} & \frac{3}{4} \\ \frac{1}{2} & \frac{1}{2} \end{bmatrix}$ **3.** $\begin{bmatrix} 0 & 1 \\ \frac{1}{6} & \frac{5}{6} \end{bmatrix}$

5. $\begin{bmatrix} 0 & .8 & .2 \\ .9 & 0 & .1 \\ 0 & 0 & 1 \end{bmatrix}$ **7.** $\begin{bmatrix} 1 & 0 & 0 \\ 0 & 1 & 0 \\ 0 & 0 & 1 \end{bmatrix}$

9. $\begin{bmatrix} 1 & 0 & 0 & 0 & 0 & 0 \\ \frac{2}{3} & 0 & \frac{1}{3} & 0 & 0 & 0 \\ 0 & \frac{2}{3} & 0 & \frac{1}{3} & 0 & 0 \\ 0 & 0 & \frac{2}{3} & 0 & \frac{1}{3} & 0 \\ 0 & 0 & 0 & \frac{2}{3} & 0 & \frac{1}{3} \\ 0 & 0 & 0 & 0 & 1 & 0 \end{bmatrix}$

11. a. $\begin{bmatrix} .25 & .75 \\ 0 & 1 \end{bmatrix}$ **b.** distribution after one step: $[.5 \quad .5]$;
after two steps: $[.25 \quad .75]$; after three steps: $[.125 \quad .875]$

13. a. $\begin{bmatrix} .36 & .64 \\ .32 & .68 \end{bmatrix}$ **b.** distribution after one step: $[.3 \quad .7]$;
after two steps: $[.34 \quad .66]$; after three steps: $[.332 \quad .668]$

15. a. $\begin{bmatrix} \frac{3}{4} & \frac{1}{4} \\ \frac{1}{2} & \frac{1}{2} \end{bmatrix}$ **b.** distribution after one step: $[\frac{2}{3} \quad \frac{1}{3}]$;
after two steps: $[\frac{2}{3} \quad \frac{1}{3}]$; after three steps: $[\frac{2}{3} \quad \frac{1}{3}]$

17. a. $\begin{bmatrix} \frac{3}{4} & \frac{1}{4} \\ \frac{3}{4} & \frac{1}{4} \end{bmatrix}$ **b.** distribution after one step:
$[\frac{3}{4} \quad \frac{1}{4}]$; after two steps: $[\frac{3}{4} \quad \frac{1}{4}]$; after three steps: $[\frac{3}{4} \quad \frac{1}{4}]$

19. a. $\begin{bmatrix} .25 & .75 & 0 \\ 0 & 1 & 0 \\ 0 & .75 & .25 \end{bmatrix}$ **b.** distribution after one step:
$[.5 \quad .5 \quad 0]$; after two steps: $[.25 \quad .75 \quad 0]$; after three steps: $[.125 \quad .875 \quad 0]$

21. a. $\begin{bmatrix} \frac{1}{3} & \frac{1}{3} & \frac{1}{3} \\ \frac{4}{9} & \frac{4}{9} & \frac{1}{9} \\ 0 & 1 & 0 \end{bmatrix}$ **b.** distribution after one step: $[\frac{1}{2} \quad \frac{1}{2} \quad 0]$;
after two steps: $[\frac{1}{6} \quad \frac{2}{3} \quad \frac{1}{6}]$; after three steps: $[\frac{7}{18} \quad \frac{7}{18} \quad \frac{2}{9}]$

23. a. $\begin{bmatrix} .01 & .99 & 0 \\ 0 & 1 & 0 \\ 0 & .36 & .64 \end{bmatrix}$ **b.** distribution after one step:
$[.05 \quad .55 \quad .4]$; after two steps: $[.005 \quad .675 \quad .32]$; after three steps: $[.0005 \quad .7435 \quad .256]$ **25.** $[\frac{2}{3} \quad \frac{1}{3}]$

27. $[\frac{3}{7} \quad \frac{4}{7}]$ **29.** $[\frac{2}{5} \quad \frac{3}{5}]$ **31.** $[\frac{2}{5} \quad \frac{1}{5} \quad \frac{2}{5}]$
33. $[\frac{1}{3} \quad \frac{1}{2} \quad \frac{1}{6}]$ **35.** $[0 \quad 1 \quad 0]$

37. 1 = Sorey State, 2 = C&T;
$P = \begin{bmatrix} \frac{1}{2} & \frac{1}{2} \\ \frac{1}{4} & \frac{3}{4} \end{bmatrix}$; $\frac{3}{8} = .375$ **39. a.** 1 = not checked in;
2 = checked in;
$P = \begin{bmatrix} .4 & .6 \\ 0 & 1 \end{bmatrix}$, $P^2 = \begin{bmatrix} .16 & .84 \\ 0 & 1 \end{bmatrix}$,
$P^3 = \begin{bmatrix} .064 & .936 \\ 0 & 1 \end{bmatrix}$ **b.** 1 hour: .6; 2 hours: .84; 3 hours: .936
c. Eventually, all the roaches will have checked in.
41. 16.67% fall into the high-risk category, and 83.33% into the low-risk category. **43. a.** $47/300 \approx .156667$ **b.** 3/13
45. 41.67% of the customers will be in the Paid Up category, 41.67% in the 0–90 days category, and 16.67% in the bad debt category.

47. a. $P = \begin{bmatrix} .729 & .271 & 0 \\ .075 & .84 & .085 \\ 0 & .304 & .696 \end{bmatrix}$ **b.** 2.3% **c.** Affluent:

17.8%; middle class: 64.3%; poor: 18.0%
49. Long-term income distribution (top to bottom): $[8.43\%, \ 41.57\%, \ 41.57\%, \ 8.43\%]$. The reason they are not the "expected" 10%, 40%, 40%, and 10% is that the movement between the income groups changes their relative sizes.

51. a. $P = \begin{bmatrix} .981 & .005 & .005 & .009 \\ .01 & .972 & .006 & .012 \\ .01 & .006 & .973 & .011 \\ .008 & .006 & .005 & .981 \end{bmatrix}$

b. Verizon: 29.6%, Cingular: 19.4%, AT&T: 18.2%, Other: 32.8% **c.** Verizon: 30.3%, Cingular: 18.6%, AT&T: 17.6%, Other: 33.5%. The biggest gainers are Verizon and Other, each gaining 0.6%. **53.** $[\frac{1}{5} \quad \frac{1}{5} \quad \frac{1}{5} \quad \frac{1}{5} \quad \frac{1}{5}]$ **55.** Answers will vary. **57.** There are two assumptions made by Markov systems that may not be true about the stock market: the assumption that the transition probabilities do not change over time and the assumption that the transition probability depends only on the current state. **59.** If q is a row of Q, then by assumption, $qP = q$. Thus, when we multiply the rows of Q by P, nothing changes, and $QP = Q$. **61.** At each step, only 0.4 of the population in state 1 remains there, and nothing enters from any other state. Thus, when the first entry in the steady-state distribution vector is multiplied by 0.4 it must remain unchanged. The only number for which this is true is 0.
63. An example is

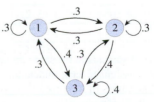

65. If $vP = v$ and $wP = w$, then $\frac{1}{2}(v + w)P = \frac{1}{2}vP + \frac{1}{2}wP = \frac{1}{2}v + \frac{1}{2}w = \frac{1}{2}(v + w)$. Further, if the entries of v and w add up to 1, then so do the entries of $(v + w)/2$.

Chapter 7 Review

1. $n(S) = 8$, $E = \{$HHT, HTH, HTT, THH, THT, TTH, TTT$\}$, $P(E) = 7/8$ **3.** $n(S) = 36$; $E = \{(1, 6), (2, 5), (3, 4), (4, 3), (5, 2), (6, 1)\}$; $P(E) = 1/6$ **5.** $n(S) = 6$; $E = \{2\}$; $P(E) = 1/8$ **7.** .76   **9.** .25   **11.** .5   **13.** 7/15   **15.** 8/792
17. 48/792   **19.** 288/792   **21.** $C(8, 5)/C(52, 5)$
23. $C(4, 3)C(1, 1)C(3, 1)/C(52, 5)$
25. $C(9, 1)C(8, 1)C(4, 3)C(4, 2)/C(52, 5)$ **27.** 1/5; dependent   **29.** 1/6; independent   **31.** 1; dependent

33. $P = \begin{bmatrix} \frac{1}{2} & \frac{1}{2} \\ \frac{1}{4} & \frac{3}{4} \end{bmatrix}$ **35.** Brand A: 65/192 ≈ .339,

Brand B: 127/192 ≈ .661   **37.** 14/25   **39.** 15/94
41. 79/167   **43.** 98%   **45.** 6.9%   **47.** .931   **49.** $P(H \cap C)$, since $P(H \mid C) > P(H)$ gives $P(H \cap C) > P(H)P(C)$
51. .246   **53.** 91%   **55.** 40% for OHaganBooks.com, 26% for JungleBooks.com, and 34% for FarmerBooks.com
57. Here are three: (1) It is possible for someone to be a customer at two different enterprises; (2) some customers may stop using all three of the companies; and (3) new customers can enter the field.

Chapter 8

Section 8.1

1. Finite; $\{2, 3, \ldots, 12\}$ **3.** Discrete infinite; $\{0, 1, -1, 2, -2, \ldots\}$ (Negative profits indicate loss.)
5. Continuous; X can assume any value between 0 and 60 (including 0).   **7.** Finite; $\{0, 1, 2, \ldots, 10\}$ **9.** Discrete infinite; $\{k/1, k/4, k/9, k/16, \ldots\}$ **11. a.** $S = \{$HH, HT, TH, TT$\}$
b. X is the rule that assigns to each outcome the number of tails.

c.
Outcome	HH	HT	TH	TT
Value of X	0	1	1	2

13. a. $S = \{(1, 1), (1, 2), \ldots, (1, 6), (2, 1), (2, 2), \ldots, (6, 6)\}$ **b.** X is the rule that assigns to each outcome the sum of the two numbers.

c.
Outcome	(1, 1)	(1, 2)	(1, 3)	...	(6, 6)
Value of X	2	3	4	...	12

15. a. $S = \{(4, 0), (3, 1), (2, 2)\}$ (listed in order (red, green))
b. X is the rule that assigns to each outcome the number of red marbles.

c.
Outcome	(4, 0)	(3, 1)	(2, 2)
Value of X	4	3	2

17. a. $S = $ the set of students in the study group.
b. X is the rule that assigns to each student his or her final exam score.   **c.** The values of X, in the order given, are 89%, 85%, 95%, 63%, 92%, 80%.
19. a. $P(X = 8) = P(X = 6) = .3$ **b.** .7; .5

21.
x	1	2	3	4	5	6
$P(X = x)$	$\frac{1}{6}$	$\frac{1}{6}$	$\frac{1}{6}$	$\frac{1}{6}$	$\frac{1}{6}$	$\frac{1}{6}$

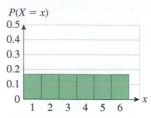

$P(X < 5) = 2/3$

23.
x	0	1	4	9
$P(X = x)$	$\frac{1}{8}$	$\frac{3}{8}$	$\frac{3}{8}$	$\frac{1}{8}$

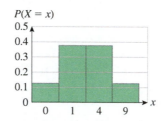

$P(1 \le X \le 9) = 7/8$

25.
x	2	3	4	5	6	7	8	9	10	11	12
$P(X = x)$	$\frac{1}{36}$	$\frac{2}{36}$	$\frac{3}{36}$	$\frac{4}{36}$	$\frac{5}{36}$	$\frac{6}{36}$	$\frac{5}{36}$	$\frac{4}{36}$	$\frac{3}{36}$	$\frac{2}{36}$	$\frac{1}{36}$

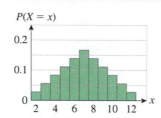

$P(X \ne 7) = 5/6$

27.
x	1	2	3	4	5	6
$P(X = x)$	$\frac{1}{36}$	$\frac{3}{36}$	$\frac{5}{36}$	$\frac{7}{36}$	$\frac{9}{36}$	$\frac{11}{36}$

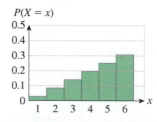

$P(X \le 3) = 1/4$

29. a.
x	10,000	30,000	50,000	70,000	90,000
$P(X = x)$	.24	.29	.18	.17	.12

b. .29; histogram:

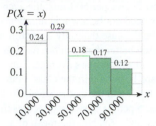

$P(X = x)$

31. The random variable is X = age of a resident in Mexico.

x	7	22	47	70
$P(X = x)$	.300	.272	.368	.060

33. a. 2,000, 3,000, 4,000, 5,000, 6,000, 7,000, 8,000 (7,000 is optional)
b.

x	2,000	3,000	4,000	5,000	6,000	7,000	8,000
Freq.	2	1	1	1	2	0	3
$P(X = x)$	.2	.1	.1	.1	.2	0	.3

c. $P(X \le 5,000) = .5$ **35. a.** $-700, -600, -500, -400,$ $-300, -200, -100, 0, 100, 200, 300, 400, 500, 600, 700,$ $800, 900$ ($-600, 0, 100, 300, 500, 600, 700,$ and 800 are optional.)
b.

x	-700	-600	-500	-400	-300	-200	-100	0
Freq.	2	0	2	1	1	4	4	0
$P(X = x)$	.1	0	.1	.05	.05	.2	.2	0

x	100	200	300	400	500	600	700	800	900
Freq.	0	2	0	2	0	0	0	0	2
$P(X = x)$	0	.1	0	.1	0	0	0	0	.1

c. .3

37.

Class	1.1–2.0	2.1–3.0	3.1–4.0
Freq.	4	7	9

x	1.5	2.5	3.5
$P(X = x)$	.20	.35	.45

39. 95.5%

41.

x	3	2	1	0
$P(X = x)$	.0625	.6875	.125	.125

43. .75; the relative frequency that a randomly selected small car is rated Good or Acceptable is .75.   **45.** $P(Y \ge 2) = .50$, $P(Z \ge 2) \approx .53$, suggesting that medium SUVs are safer than small SUVs in frontal crashes   **47.** Small cars   **49.** .375

51.

x	1	2	3	4
$P(X = x)$	$\frac{4}{35}$	$\frac{18}{35}$	$\frac{12}{35}$	$\frac{1}{35}$

$P(X \ge 2) = 31/35 \approx .886$
53. Answers will vary.   **55.** No; for instance, if X is the number of times you must toss a coin until heads comes up, then X is infinite but not continuous.   **57.** By measuring the values of X for a large number of outcomes and then using the estimated probability (relative frequency)   **59.** Answers will vary. Here is an example: Let X be the number of days a diligent student waits before beginning to study for an exam scheduled in 10 days' time.   **61.** The bars should be 1 unit wide so that their height is numerically equal to their area.   **63.** Answers may vary. If we are interested in exact page counts, then the number of possible values is very large, and the values are (relatively speaking) close together, so using a continuous random variable might be advantageous. In general, the finer and more numerous the measurement classes, the more likely it becomes that a continuous random variable could be advantageous.

Section 8.2

1. .0729   **3.** .59049   **5.** .00001   **7.** .99144   **9.** .00856   **11.** .27648   **13.** .54432   **15.** .04096   **17.** .77414

19. $P(X = x)$

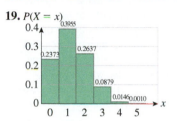

21. $P(X = x)$

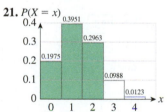

$P(X \le 2) \approx .8889$
23. .2637   **25.** .8926   **27.** .875   **29. a.** .0081 **b.** .08146   **31.** .41   **33. a.** .0509 **b.** Probability distribution (entries rounded to four decimal places):

x	0	1	2	3	4
$P(X = x)$	.0643	.2030	.2885	.2429	.1343
	5	6	7	8	9	10
	.0509	.0134	.0024	.0003	.0000	.0000

c. 2   **35.** .000298   **37.** .8321   **39. a.** 21 **b.** 20 **c.** The graph for $n = 50$ trials is more widely distributed than the graph for $n = 20$.   **41.** 69 trials   **43.** $.562 \times 10^{-5}$ **45.** .0266; because there is only a 2.66% chance of detecting the disease in a given

year, the government's claim seems dubious.   **47.** No; in a sequence of Bernoulli trials, the occurrence of one success does not affect the probability of success on the next attempt.
49. No; if life is a sequence of Bernoulli trials, then the occurrence of one misfortune ("success") does not affect the probability of a misfortune on the next trial. Hence, misfortunes may very well not "occur in threes."   **51.** Think of performing the experiment as a Bernoulli trial with "success" being the occurrence of E. Performing the experiment n times independently in succession would then be a sequence of n Bernoulli trials.
53. The probability of selecting a red marble changes after each selection, as the number of marbles left in the bag decreases. This violates the requirement that, in a sequence of Bernoulli trials, the probability of "success" does not change.

Section 8.3

1. $\bar{x} = 6$, median $= 5$, mode $= 5$ **3.** $\bar{x} = 3$, median $= 3.5$, mode $= -1$ **5.** $\bar{x} = -0.1875$, median $= 0.875$, every value is a mode   **7.** $\bar{x} = 0.2$, median $= -0.1$, mode $= -0.1$
9. Answers may vary. Two examples are: 0, 0, 0, 0, 0, 6 and 0, 0, 0, 1, 2, 3   **11.** 0.9   **13.** 21   **15.** -0.1 **17.** 3.5   **19.** 1
21. 4.472   **23.** 2.667   **25.** 2   **27.** 0.385   **29.** $\bar{x} = -150$, $m = -150$; -150 **31.** $\bar{x} = 1,068.8$, median $= 1,071.5$; modes $= 1,062$ and $1,072$; Over the 10-business-day period sampled, the price of gold averaged $1,068.8 per ounce. It was above $1,071.5 as many times as it was below that price, and stood at $1,062 and $1,072 per ounce more often than at any other price.   **33. a.** 6.5; there were, on average, 6.5 checkout lanes in each supermarket that was surveyed.
b. $P(X < \mu) = .42$; $P(X > \mu) = .58$ and is thus larger. Most supermarkets have more than the average number of checkout lanes.

35.

x	5	10	15	20	25	35
$P(X = x)$	.17	.33	.21	.19	.03	.07

$E(X) = 14.3$;

the average age of a school goer in 1998 was 14.3.
37. 29.6   **39.** $43,000

41.

x	3	2	1	0
$P(X = x)$	.0625	.6875	.125	.125

$E(X) = 1.6875$

y	3	2	1	0
$P(Y = y)$	.1	.4	.4	.1

$E(Y) = 1.5$;

small cars
43. Large cars   **45.** Expect to lose 5.3¢.   **47.** 25.2 students   **49. a.** Two defective air bags   **b.** 120 air bags

51.

x	1	2	3	4
$P(X = x)$	$\frac{4}{35}$	$\frac{18}{35}$	$\frac{12}{35}$	$\frac{1}{35}$

$E(X) = 16/7 \approx 2.2857$ tents

53. A loss of about 51¢   **55.** About half a cent   **57.** FastForward: 3.97%; SolidState: 5.51%; SolidState gives the higher expected return.   **59.** A loss of $29,390   **61.** (A)   **63.** He is wrong; for example, the collection 0, 0, 300 has mean 100 and

median 0.   **65.** No; the expected number is the average number of times you will hit the bull's-eye per 50 shots; the average of a set of whole numbers need not be a whole number.
67. Not necessarily; it might be the case that only a small fraction of people in the class scored better than you but received exceptionally high scores that raised the class average. Suppose, for instance, that there are 10 people in the class. Four received 100%, you received 80%, and the rest received 70%. Then the class average is 83%, 5 people have lower scores than you, but only 4 have higher scores.   **69.** No; the mean of a very large sample is only an *estimate* of the population mean. The means of larger and larger samples *approach* the population mean as the sample size increases.   **71.** Wrong; the statement attributed to President Bush asserts that the mean tax saving would be $1,000, whereas the statements referred to as "The Truth" suggest that the *median* tax saving would be close to $100 (and that the 31st percentile would be zero).
73. Select a U.S. household at random, and let X be the income of that household. The expected value of X is then the population mean of all U.S. household incomes.

Section 8.4

1. $s^2 = 29$; $s = 5.39$ **3.** $s^2 = 12.4$; $s = 3.52$ **5.** $s^2 = 6.64$; $s = 2.58$ **7.** $s^2 = 13.01$; $s = 3.61$ **9.** 1.04   **11.** 9.43
13. 3.27   **15.** Expected value $= 3.5$, variance ≈ 2.92, standard deviation ≈ 1.71 **17.** Expected value $= 1$, variance $= 0.5$, standard deviation ≈ 0.71 **19.** Expected value ≈ 4.47, variance ≈ 1.97, standard deviation ≈ 1.40 **21.** Expected value ≈ 2.67, variance ≈ 0.36, standard deviation ≈ 0.60
23. Expected value $= 2$, variance $= 1.8$, standard deviation ≈ 1.34 **25. a.** $\bar{x} = 3$, $s \approx 3.54$ **b.** $[0, 6.54]$ We must assume that the population distribution is bell shaped and symmetric.
27. a. $\bar{x} = 5.0$, $s \approx 0.6$ **b.** 3.8, 6.2   **29. a.** $\bar{x} = -150$, $s \approx 495$ **b.** 645, 20%   **31. a.** 2.18   **b.** $[11.22, 24.28]$ **c.** 100%; empirical rule   **33.** $\mu = 1.5$, $\sigma = 1.43$; 100%   **35.** $\mu = 43$, $\sigma \approx 26.6$; $53,000   **37. a.** $\mu \approx 30.2$ years old, $\sigma \approx 11.78$ years
b. 18–42   **39.** At most 6.25%   **41.** At most; 12.5
43. a. $\mu = 25.2$, $\sigma = 3.05$ **b.** 31   **45. a.** $\mu = 780$, $\sigma \approx 13.1$
b. 754, 806   **47. a.** $\mu = 6.5$, $\sigma^2 = 4.0$, $\sigma = 2.0$ **b.** $[2.5, 10.5]$; three checkout lanes   **49.** $10,700 or less   **51.** $65,300 or more
53. U.S.   **55.** U.S.   **57.** 16%   **59.** 0–$76,000   **61.** 2000 data: $\mu = 12.56$, $\sigma \approx 1.8885$; 2010 data: $\mu = 13.30$, $\sigma \approx 1.6643$
63. (B)   **65.** 72%; the empirical rule predicts 68%. The associated probability distribution is roughly bell shaped but not symmetric.   **67.** 94%; Chebyshev's rule is valid, since it predicts that *at least* 75% of the scores are in this range.   **69.** (B), (D)
71. (B), (E)   **73.** The sample standard deviation is bigger; the formula for sample standard deviation involves division by the smaller term $n - 1$ instead of n, which makes the resulting number larger.   **75.** The grades in the first class were clustered fairly close to 75. By Chebyshev's inequality, at least 88% of the class had grades in the range 60–90. On the other hand, the grades in the second class were widely dispersed. The second class had a much wider spread of ability than did the first class.   **77.** The variable must take on only the value 10, with probability 1.   **79.** $(y - x)/2$

Section 8.5

1. .1915 **3.** .5222 **5.** .6710 **7.** .2417 **9.** .8664 **11.** .8621
13. .2286 **15.** .3830 **17.** .5028 **19.** 108.4 **21.** 117.5
23. .35 **25.** .05 **27.** .3830 **29.** .8185 **31.** 26%
33. 34,100,000 **35.** 665 **37.** 0 **39.** About 6,680 **41.** 28%
43. 5% **45.** U.S. **47.** Wechsler; as this test has a smaller
standard deviation, a greater percentage of scores fall within
20 points of the mean. **49.** This is surprising, because the
time between failures was more than five standard deviations
away from the mean, which happens with an extremely small
probability. **51.** .6103 **53.** .6103 × .5832 ≈ .3559
55. .6255 **57.** .7881 **59.** .9049 **61.** .2912 **63.** Probability
that a person will say Goode = .54. Probability that Goode
polls more than 52% ≈ .8925. **65.** 23.4 **67.** When the
distribution is normal **69.** Neither. They are equal.
71. $1/(b - a)$ **73.** A normal distribution with standard deviation 0.5, because it is narrower near the mean but must enclose
the same amount of area as the standard curve, so it must be
higher.

Chapter 8 Review

1.

x	0	1	2
$P(X = x)$	$\frac{1}{4}$	$\frac{1}{2}$	$\frac{1}{4}$

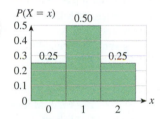

3.

x	2	3	4	5	6	7	8
$P(X = x)$	$\frac{1}{16}$	$\frac{2}{16}$	$\frac{3}{16}$	$\frac{4}{16}$	$\frac{3}{16}$	$\frac{2}{16}$	$\frac{1}{16}$

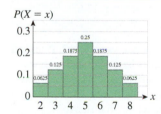

5.

x	0	1	2	3
$P(X = x)$	.2071	.4439	.2908	.0582

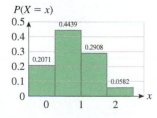

7. $\bar{x} = 2$, $m = 2$, $s \approx 2.7386$ **9.** Two examples are: 0, 0, 0, 4
and $-1, -1, 1, 5$ **11.** An example is $-1, -1, -1, 1, 1, 1$
13. .4165 **15.** .9267 **17.** .0067 **19.** .7330 **21.** $\mu = 1.5$,
$\sigma = 0.8660$; 2
23.

x	-3	-2	-1	0	1	2	3
$P(X = x)$	$\frac{1}{16}$	$\frac{2}{16}$	$\frac{3}{16}$	$\frac{4}{16}$	$\frac{3}{16}$	$\frac{2}{16}$	$\frac{1}{16}$

$\mu = 0$, $\sigma = 1.5811$; within 1.3 standard deviations of the
mean **25.** [49.4, 150.6] **27.** 260 **29.** .4332 **31.** .0358
33. .3085 **35.** \$12.15 **37.** \$27,210 **39.** False; let $X =$ price
and $Y =$ weekly sales. Then weekly revenue $= XY$. However,
27,210 ≠ 12.15 × 2,620. In other words, $E(XY) \neq E(X)E(Y)$.
41. a.

x	2	4	6	8	10
$P(X = x)$	.25	.35	.15	.15	.10

$\mu = 5$, $\sigma = 2.5690$ **b.** Between 2.431 and 7.569 orders per
million residents; the empirical rule does not apply because the
distribution is not symmetric. **c.** (A) **43.** .190 **45.** .060
47. 0.5 **49.** .284 **51.** .108 **53.** Using normal distribution
table: 420,000 people; more accurate answer, using technology: 436,000 people **55.** 148

Appendix A

1. False statement **3.** Not a statement, because it is not a
declarative sentence **5.** True statement **7.** True (we hope!)
statement **9.** Not a statement, because it is self-referential
11. $(\sim p) \wedge q$ **13.** $(p \wedge r) \wedge q$ or just $p \wedge q \wedge r$
15. $p \vee (\sim p)$ **17.** Willis is a good teacher and his students do
not hate math. **19.** Either Carla is a good teacher or she is
not. **21.** Willis' students both hate and do not hate math.
23. It is not true that either Carla is a good teacher or her students hate math. **25.** F **27.** F **29.** T **31.** T **33.** T **35.** F
37. T **39.** F **41.** T **43.** T **45.** T **47.** T

49.

p	q	$\sim q$	$p \wedge \sim q$
T	T	F	F
T	F	T	T
F	T	F	F
F	F	T	F

51.

p	$\sim p$	$\sim(\sim p)$	$\sim(\sim p) \vee p$
T	F	T	T
F	T	F	F

53.

p	q	$\sim p$	$\sim q$	$(\sim p) \wedge (\sim q)$
T	T	F	F	F
T	F	F	T	F
F	T	T	F	F
F	F	T	T	T

55.

p	q	r	$p \wedge q$	$(p \wedge q) \wedge r$
T	T	T	T	T
T	T	F	T	F
T	F	T	F	F
T	F	F	F	F
F	T	T	F	F
F	T	F	F	F
F	F	T	F	F
F	F	F	F	F

57.

p	q	r	$q \vee r$	$p \wedge (q \vee r)$
T	T	T	T	T
T	T	F	T	T
T	F	T	T	T
T	F	F	F	F
F	T	T	T	F
F	T	F	T	F
F	F	T	T	F
F	F	F	F	F

59.

p	q	$q \vee p$	$p \to (q \vee p)$
T	T	T	T
T	F	T	T
F	T	T	T
F	F	F	T

61.

p	q	$p \vee q$	$p \leftrightarrow (p \vee q)$
T	T	T	T
T	F	T	T
F	T	T	F
F	F	F	T

63.

p	$p \wedge p$
T	T
F	F

Same

65.

p	q	$p \vee q$	$q \vee p$
T	T	T	T
T	F	T	T
F	T	T	T
F	F	F	F

Same

67.

p	q	$p \vee q$	$\sim(p \vee q)$	$\sim p$	$\sim q$	$(\sim p) \wedge (\sim q)$
T	T	T	F	F	F	F
T	F	T	F	F	T	F
F	T	T	F	T	F	F
F	F	F	T	T	T	T

Same

69.

p	q	r	$p \wedge q$	$(p \wedge q) \wedge r$	$q \wedge r$	$p \wedge (q \wedge r)$
T	T	T	T	T	T	T
T	T	F	T	F	F	F
T	F	T	F	F	F	F
T	F	F	F	F	F	F
F	T	T	F	F	T	F
F	T	F	F	F	F	F
F	F	T	F	F	F	F
F	F	F	F	F	F	F

Same

71.

p	q	$p \to q$	$\sim p$	$\sim q$	$(\sim q) \to (\sim p)$
T	T	T	F	F	T
T	F	F	F	T	F
F	T	T	T	F	T
F	F	T	T	T	T

Same

73. Contradiction **75.** Contradiction **77.** Tautology
79. $(\sim p) \vee p$ **81.** $(\sim p) \vee \sim(\sim q)$
83. $(p \vee (\sim p)) \wedge (p \vee q)$ **85.** Either I am not Julius Caesar or you are no fool. **87.** It is raining and I have forgotten either my umbrella or my hat. **89.** Contrapositive: "If I do not exist, then I do not think." Converse: "If I am, then I think."
91. (A); it is the contrapositive of the given statement.

93. $h \to t$
$\underline{h}$
$\therefore t$
Valid; Modus Ponens

95. $r \to u$
$\underline{\sim r}$
$\therefore \sim u$
Invalid

97. $g \to m$
$\underline{\sim m}$
$\therefore \sim g$
Valid: Modus Tollens

99. $m \vee b$
$\underline{\sim b}$
$\therefore m$
Valid; Disjunctive Syllogism

101. $s \vee a$
$\underline{a}$
$\therefore \sim s$
Invalid

103. John is green. **105.** John is not a swan. **107.** Peter is a gentleman. **109.** Their truth tables have the same truth values for corresponding values of the variables. **111.** A and B are both contradictions. **113.** Answers will vary. Example: Let p: "You have smoker's cough," and let q: "You smoke."
115. Answers will vary. Example: Let p: "It is summer in New York," and let q: "It is summer in Seattle."

Index

Index of Applications (*continued*)